MINERAL REPORT 33

CANADIAN MINERALS YEARBOOK 1983-1984
Review and Outlook

Energy, Mines and
Resources Canada

Minerals

Énergie, Mines et
Ressources Canada

Minéraux

©Minister of Supply and Services Canada 1985

Available in Canada through

Authorized Bookstore Agents
and other bookstores

or by mail from

Canadian Government Publishing Centre
Supply and Services Canada
Ottawa, Canada K1A 0S9

Catalogue No. M38-5/33E Canada: $29.50
ISBN 0-660-11816-5 Other Countries: $35.40

Price subject to change without notice

Foreword

This issue of the Canadian Minerals Yearbook looks retrospectively at the Canadian mineral industry during 1983 and 1984 and prospectively at the direction the industry is likely to take in the foreseeable future. The present edition is the latest in a series of official documents published under various titles since 1886 when the Government of Canada first reported comprehensively on the country's mineral industry.

The General Review chapter deals with the main events and trends in the Canadian economy over the two years covered in this Yearbook. This section also deals with general developments and overall patterns in the mineral industry over that time. Commodity review sections provide the same type of information as in past issues. Outlook sections, however, have been expanded to place greater emphasis on projections for the future of the industry. With this change, the Yearbook should prove to be a useful tool for a broader audience than in previous years.

The basic statistics on Canadian production, trade and consumption were collected by the Information Systems Division, Mineral Policy Sector, Energy, Mines and Resources Canada, and by Statistics Canada, unless otherwise stated. Company data were obtained directly from company officials through surveys or correspondence or from corporate annual reports. Market quotations are mainly from standard marketing reports.

Additional copies of the Yearbook can be purchased from the Canadian Government Publishing Centre. Reprints of individual chapters and Map 900A, Principal Mineral Areas of Canada, may be obtained free of charge from:

Publication Distribution Office
Mineral Policy Sector
Energy, Mines and Resources Canada
580 Booth Street
Ottawa, Ontario
K1A OE4

Previous editions have been deposited in various libraries.

Energy, Mines and Resources Canada is grateful to all those who contributed information used in compiling this report.

May 13, 1985

Editors: G.E. Thompson,
G. Cathcart
Production Editor: M. Nadeau
Graphics and Cover: N. Sabolotny

Text and tables in this yearbook were typeset on Micom 2001 equipment by the Word Processing Unit of the Mineral Policy Sector, Energy, Mines and Resources Canada and reproduced by offset lithography.

Front Cover:
Geologist studies mineral sample in open-pit mine, Pine Point, N.W.T. (George Hunter photo)

Contents

1. General Review
2. International Review
3. Regional Review
4. Canadian Reserves of Selected Mineral Commodities
5. Canadian Reserves, Development and Exploration
6. Aluminum
7. Antimony
8. Arsenic
9. Asbestos
10. Barite and Celestite
11. Bentonite (not printed)
12. Beryllium
13. Bismuth
14. Cadmium
15. Calcium (not printed)
16. Cement
17. Cesium (not printed)
18. Chromium (not printed)
19. Clays and Clay Products
20. Coal and Coke
21. Cobalt
22. Columbium (Niobium)
23. Copper
24. Crude Oil and Natural Gas
25. Diatomite (not printed)
26. Ferrous Scrap
27. Fluorspar (not printed)
28. Gold
29. Graphite (not printed)
30. Gypsum and Anhydrite
31. Indium (not printed)
32. Iron Ore
33. Iron and Steel
34. Lead
35. Lime
36. Lithium
37. Magnesium
38. Manganese
39. Mercury (not printed)
40. Mica
41. Mineral Aggregates
42. Molybdenum
43. Nepheline Syenite and Feldspar
44. Nickel
45. Phosphate
46. Platinum Metals
47. Potash
48. Rare Earths
49. Rhenium
50. Salt
51. Selenium and Tellurium
52. Silica
53. Silicon, Ferrosilicon, Silicon Carbide and Fused Alumina
54. Silver
55. Sodium Sulphate
56. Stone
57. Sulphur
58. Talc, Soapstone and Pyrophyllite
59. Tantalum
60. Tin
61. Titanium and Titanium Dioxide
62. Tungsten
63. Uranium
64. Vanadium (not printed)
65. Zinc
66. Zirconium (not printed)
67. Principal Canadian Metal Producers
68. Statistical Report

Conversion Factors
Imperial Units to Metric (SI) Units

Ounces to grams	x	28.349 523
Troy ounces to grams	x	31.103 476 8
to kilograms	x	.031 103 476
Pounds to kilograms	x	.453 592 37
Short tons to tonnes	x	.907 184 74
Gallons to litres	x	4.546 09
Barrels to cubic metres	x	.158 987 220
Cubic feet to cubic metres	x	.028 346 85

Source: Canadian Metric Practice Guide

General Review

L. LEMAY

THE ECONOMY IN 1984

1984 was a year of partial economic recovery in Canada. The expectations at the beginning of the year that stable interest rates would shape a strong, robust return to growth did not entirely materialize. The key Bank of Canada rate started the year at 9.96 per cent and forecasters predicted a strong, sector-wide recovery, but by mid-year the Bank rate had climbed considerably. It peaked in July at 13.26 per cent thus stiffling any momentum built up in the economy. Strength in business activity continued into the third quarter of 1984 due mainly to a surge in exports, specifically to the U.S. market, however, it could not be maintained. The economic slowdown in Canada in the fourth quarter mirrored a slowdown in the United States. The United States economy experienced a dramatic increase in the level of growth through most of the year but succumbed to high interest rates in the last quarter.

In Canada, real growth in GNP averaged a modest 3.2 per cent on an annual basis over the first two quarters of 1984, rose in the summer to record a surprisingly strong third quarter performance of 7.6 per cent, but fell back again late in the year. The real rate of growth of GNP for the year is estimated at 4.4 per cent compared with 3.3 per cent for 1983. Inflation, one of the favourable indicators, was under control, finishing the year at less than 5 per cent compared with 6 per cent in 1983. However, the unemployment rate stayed stubbornly above 11 per cent leaving about 1.4 million Canadians out of work.

The foreign trade sector was the brightest aspect of the Canadian economy in 1984. A record merchandise trade surplus of $20 billion was reached, up from the $18 billion of 1983, contributing to an overall surplus on the current account for the third consecutive year. This trade surplus was due to a number of factors including lower world oil prices, an improvement in auto trade with increased United States demand for larger automobiles built in Canada, the relative strength of the U.S. economy and the depreciation of the Canadian dollar against the United States dollar. The Canadian dollar slid from about 80 cents (U.S.) at the beginning of the year to an historic low of 74.86 cents (U.S.) in July. It recovered to above the 76 cent level and hovered there for the rest of the year.

Merchandise exports in Canada reached a record level of $94.5 billion in the first ten months of 1984, 76 per cent of which were destined for the United States. Canada's trade surplus with the United States grew to almost $20 billion in 1984 from less than $500 million in 1979. At the same time, the surplus with the rest of the world shrank to almost zero from about $400 million.

This growth in exports was the stimulating factor for production increases in export-oriented industries such as motor vehicles and parts, pulp and paper, sawmills and crude minerals and products including nickel, copper, iron ore, coal and some non-metallic minerals.

THE MINERAL INDUSTRY IN 1984

The Canadian mineral industry in 1984 continued to be tested as it had been throughout the 1980s by a combination of circumstances that had shaken the industry to its foundation. The road to recovery has been longer and harder than anticipated and profits in 1984 for many mining companies remained elusive. Net income for the industry (excluding energy) totalled $210 million in the first half of the year, a 96 per cent improvement over 1983 but nowhere near pre-recession levels. Despite predictions of improvement, mineral commodity prices remained generally weak on world markets and demand continued to be sluggish. The industry benefited strongly from the lower Canadian dollar, with 80 per cent of its output destined for export. One industry spokesman was quoted as saying "If

---

L. Lemay is with the Mineral Policy Sector, Energy, Mines and Resources, Canada. Telephone (613) 995-9466.

the Canadian dollar were at par with the United States dollar, the mining industry in Canada would be virtually bankrupt."

Mineral industry output showed a healthy increase in 1984 compared with 1983. Value of output reached $43.1 billion, up from 38.5 billion the previous year. All sectors of the industry, including metallics, nonmetallics, structural materials and fuels recorded significant increases with the greatest improvement coming from the metallic and nonmetallic mineral sectors.

Canadian Mineral Production
1983 and 1984

	1983	1984	% Change 1984/1983
	(millions of current dollars)		
Metals	7.4	8.5	+14.9
Nonmetals	1.9	2.3	+21.1
Structurals	1.8	1.9	+ 5.5
Fuels	27.2	30.0	+10.3
Total	38.5	43.1	+12.0

On a commodity basis, the ten leading minerals in 1984 were: crude petroleum, natural gas, natural gas byproducts, coal, iron ore, zinc, copper, gold, nickel and uranium. These ten minerals represented 87 per cent of the total value of output of the industry and all except copper, gold and silver showed increases over the previous year.

On a regional basis, Alberta represented the largest share of output in Canada, reaching 60 per cent of the total or $26 billion in 1984, up from $24.1 billion in 1983. Ontario followed with 10 per cent of the total at $4.5 billion. Output was up slightly in British Columbia, totalling $3.4 billion while, Quebec, seriously affected by the continued weak demand for iron ore and asbestos remained unchanged, at $2.0 billion. With the reopening of some mining operations in the Northwest Territories during the year, value of output there showed the largest proportionate increase moving from $595 million in 1983 to $738 million in 1984.

MINERAL INDUSTRY PRICES

Soft prices for several key commodities such as copper, molybdenum, nickel, iron ore, gold and silver kept the recovery of the industry at a slower pace than expected. An increase in the volume of copper output of 9.1 per cent in 1984 over 1983 was not matched by an increase in value. After averaging 94.8 cents (Cdn.) per pound (76.9 cents U.S.) in 1983, the price of copper dropped to an average of 86.1 cents (66.5 cents U.S.) per pound in 1984. Once again, shutdowns were prevalent in the North American industry and idle capacity of approximately 1 million t of copper existed around the world. Molybdenum showed some improvement in 1984 with value of production up 24.1 per cent but prices remained weak. At about $US 3.30 per pound the price was below the cost of production at all Canadian mines.

Demand for nickel increased with an upturn in demand in the steel sector in 1984, but overcapacity still existed, preventing any significant increase in price. Nickel is now produced in over 25 countries, up from three in 1950, and state ownership of the industry has increased from 15 per cent to 40 per cent over the same period.

In the international iron ore market, prices were at record lows during the year, well below the Quebec-Labrador average production cost. Since the recession, Quebec-Labrador has not been working at much more than 50 per cent of capacity and little improvement is seen in the next few years because of world over-capacity compounded by new projects coming on-stream in Australia, Brazil and West Africa.

The price of gold tumbled throughout 1984. It reached its lowest point since June 1982 late in December, and threatened to fall below $US 300 per ounce at year-end. It averaged $US 362.68 on the LME ($Cdn 469.65) in 1984 compared with $422.60 ($Cdn $520.79) in 1983. Losing its appeal as a hedge against inflation, the price of gold has been depressed by the strength of the U.S. dollar, reduced concerns about inflation and falling commodity prices in general.

Silver suffered from an increasing glut of supply in the western world along with stagnating, even falling demand, and the price averaged $Cdn 10.87 in 1984 compared with $14.15 in 1983. Uranium prices fell to $US 17.50 per pound in 1984 from $20 in 1983, well below the late-1970s price of $34, principally due to oversupply of the commodity.

Zinc was in a more stable supply balance in 1984 and prices improved steadily from mid-1983 through most of 1984 reaching a 10-year high of 67.5 cents (Cdn.) in June. Increased demand by automobile manufacturers kept stocks down and prices averaged 63.8 cents (Cdn.) in 1984, up from 52.1 cents in 1983 and 48.7 cents in 1982. Aluminum started the year, also on a high note. Encouraged by relatively high prices in 1983, idle capacity was brought on-stream in early-1984. However, the anticipated increase in demand failed to materialize. Supply continued to rise until mid-1984 pushing prices down and world stocks of aluminum ended the year about 25 per cent above normal. A major structural change in the world aluminum market occurred late in the year with the acknowledgement by Alcan Aluminum Ltd. of Montreal, the largest producer, that its world list price was no longer the reference point. Prices are now set mainly in major commodity markets such as the LME.

Downward pressure on prices existed for nonmetallic minerals in 1984 as well. Many major nonmetallics remained in oversupply. Shipments of asbestos for example were down to 836 000 t from 858 000 t a year earlier and a long way from the peak year of 1973 when Canada exported a record 1.7 million t. New producers such as Brazil, Colombia and Greece are entering the market making it increasingly difficult for Canada to maintain its 30 per cent share of western world markets.

MINERAL TRADE AND INVESTMENT

While mineral production increased moderately in 1984, mineral exports contributed greatly to the strong Canadian surplus in overall merchandise trade. In the first nine months, exports of crude and fabricated minerals totalled $21.6 billion, a 20 per cent increase over the same period of 1983. Crude nonferrous mineral exports were up 27.6 per cent over 1983 while crude nonmetallics showed a 35.4 per cent increase. Fabricated mineral exports totalled $9.7 billion in the first nine months of 1984 up from $7.9 billion in 1983. Crude and fabricated minerals represented 27 per cent of total domestic exports in 1984, 75 per cent of which went to the United States, 9 per cent to Japan, 3 per cent to the United Kingdom and 5 per cent to the rest of the EEC.

Mineral industry imports totalled $10.1 billion in the first nine months of 1984 up from $8.2 billion in 1983. Fifty-seven per cent of total mineral imports came from the United States and if mineral fuels were excluded that percentage would increase to 71 per cent of the total.

Investment in the mineral industry in 1984 presented a brighter picture after two years of cutbacks and cost-saving measures. New capital expenditure intentions totalled $10.2 billion in 1984 for mines, quarries and oil wells up from $9.6 billion in 1983 and $9.4 billion in 1982. When repair expenditures were included that total reached $13 billion in 1984. The greatest improvement in new capital spending came from the metal mining sector. After declining 23 per cent in 1982 from a level of $2.0 billion in 1981, spending declined a further 26 per cent in 1983 reaching only $1.1 billion. In 1984, it had recovered to a level of $1.4 billion. Nonmetal mining showed a strong increase as well from a level of $322.4 million in 1983 to $477.7 million in 1984. Coal mining, on the other hand showed a sharp decline in 1984 dropping almost 65 per cent from $1.1 billion in 1983 to $395 million in 1984. Exploration activity in Ontario was up for the third consecutive year particularly in the Hemlo area. One spokesman was quoted as saying "A trip to Hemlo these days is like a tonic to dispel the blues brought on by mining's current state of depression". Four new gold mines started up in Ontario in the first half of 1984. Gold fever filtered into Quebec as well and all the way to the Yukon where the Wheaton River Valley was providing some excitement.

The Saskatchewan Mining Development Corporation (SMDC) discovered what is expected to be the richest uranium deposit in the world at Cigar Lake, about 15 km north of Key Lake. The uranium grade is estimated to be 10 per cent, four times that of Key Lake, previously believed to be the world's richest discovered deposit.

OUTLOOK

The economic outlook for 1985 calls for more moderate growth both in Canada and the United States. Once again, interest rates will be the determining factor. If measures in the United States aimed at reducing the federal deficit are successful, and as a result, interest rates fall, Canadian rates will follow thus providing business with the incentive to invest and stimulate economic activity. The continued strength of the U.S. dollar has had a negative effect on metal prices and it may be, that with a weakening in this currency, demand for metals and their prices will rise. The Canadian dollar is expected to stay firm, strengthening from 76 cents (U.S.) to an average of 78 cents for the next few years. The basis for the improvement is the record trade surpluses and an improved inflation rate. However, in order for the mineral industry to take advantage of increased trade potential it will have to contend with the structural changes and adjustments taking place throughout the world. Changing patterns in metal consumption brought on by energy conservation policies as well as technological innovations have severely affected the growth in demand for minerals. At the same time, growing foreign competition, much of it from government-controlled producers have cut into traditional market shares once taken for granted by Canadian and United States producers. Global overcapacity has existed for most metals throughout the 1980s. The situation continues to be exacerbated by debt-ridden Third World countries forced to maintain production of minerals to earn foreign exchange. Continuous currency devaluations in Third World countries have been carried out, thus improving their competitive edge. As a result, investment, the key to economic growth has not been forthcoming in the North American mineral industry.

Due to the severity of the 1981-82 recession, cost reduction has become the main goal of most businesses. The Canadian mining industry has been largely successful in this goal making major productivity gains in 1983 and 1984, through improved mining methods, as well as labour force reductions. Tighter inventory controls and increased throughput have also brought lower costs. However, some of these reduction measures can only be maintained in the short-term. These measures along with curtailment in development at existing and new mining sites increase the lead time required to respond to a change in demand by an increasing rate. Massive capital expenditures will be necessary for most companies to reach increased production targets. The full effect of mine closures will be felt especially for copper in the next 12 to 18 months and producers will have to react to an anticipated increase in demand. In the meantime, emphasis should be placed on product development and diversification. There is room for growth in new applications for mineral products in construction, and automobile manufacturers. Increased consumption in copper, aluminum and zinc is very possible in these sectors.

With a continued improvement in cost competitiveness, the next step for Canadian mining companies is a more aggressive marketing strategy to increase market access for Canadian mineral exports. The characteristics that have made Canada the second largest mineral exporter in the world still exist. The broad range of minerals, the stable political envirinment and the well-developed economic and social infrastructure are still very much in place. The challenge will be to respond quickly and effectively to the anticipated turnaround in mineral demand.

CANADA, PRODUCTION OF LEADING MINERALS, 1983 AND 1984

	1983	1984	% change 1984/1983	1983	1984	% change 1984/1983
	(000 tonnes except where noted)			($ millions)		
Metals						
Copper	653.0	712.4	+9.1	1,364.4	1,351.4	-1.0
Gold (kg)	73.5	81.3	+10.6	1,230.9	1,227.8	-0.3
Iron ore	32 959.0	41 065.0	+24.6	1,269.9	1,470.9	+15.8
Lead	272.0	259.4	-4.6	160.5	190.8	+18.9
Molybdenum (t)	10 194.0	10 865.0	+6.6	87.7	108.9	+24.1
Nickel	125.0	174.2	+39.4	781.5	1,165.2	+49.1
Silver (t)	1 197.0	1 171.0	-2.2	544.7	409.3	-24.9
Uranium (U)	6 823.0	9 693.0	+42.1	667.7	916.3	+37.2
Zinc	987.7	1 022.1	+3.5	1,135.2	1,438.0	+26.7
Nonmetals						
Asbestos	858.0	836.0	-2.6	391.3	413.0	+5.5
Gypsum	7 507.0	8 725.0	+16.2	59.3	69.2	+16.7
Potash (K_2O)	6 294.0	6 972.0	+10.8	645.8	759.3	+17.6
Salt	8 602.0	10 294.0	+19.7	172.8	214.9	+24.4
Cement	7 871.0	8 619.0	+9.5	606.1	667.1	+10.1
Clay Products	..	..	..	132.3	140.9	+6.5
Lime	2 232.0	2 280.0	+2.2	156.7	174.5	+11.4
Fuels						
Coal	44 787.0	56 800.0	+26.8	1,303.9	1,814.0	+39.1
Natural gas (000 m^3)	72 229 000.0	73 656 000.0	2.0	7,077.2	7,514.6	+6.2
Petroleum (000 m^3)	78 751.0	82 989.0	5.4	16,091.8	17,887.8	+11.2

.. Not available.

CANADA, EXPORTS OF MINERALS, CRUDE AND FABRICATED

	Year 1973	Year 1978	Year 1983	1st 9 months 1983	1st 9 months 1984	% Changes 1st 9 months 1984 / 1st 9 months 1983
	($ millions)					
Crude						
Ferrous	497.7	854.5	1,054.3	787.5	890.9	+13.1
Nonferrous	1,501.8	1,549.2	1,845.9	1,304.9	1,664.4	+27.6
Nonmetallic	595.5	1,369.7	2,103.5	1,468.2	1,987.6	+35.4
Fuels	1,998.4	4,514.8	8,727.9	6,573.2	7,358.2	+11.9
Total	4,593.4	8,288.2	13,731.6	10,133.8	11,901.1	+17.4
Fabricated						
Ferrous	598.7	1,696.1	2,011.6	1,445.0	1,964.2	+35.9
Nonferrous	1,897.8	3,360.9	5,624.6	4,009.0	5,048.3	+25.9
Nonmetallic	166.2	377.1	424.8	315.9	400.1	+26.7
Fuels	311.6	1,022.7	2,815.6	2,109.8	2,303.8	+9.2
Total	2,974.3	6,456.8	10,876.6	7,879.7	9,716.4	+23.3
Total crude and fabricated minerals						
Ferrous	1,096.4	2,550.6	3,065.9	2,232.5	2,855.1	+27.9
Nonferrous	3,399.6	4,910.1	7,470.5	5,313.9	6,712.7	+26.3
Nonmetallic	761.7	1,746.8	2,528.3	1,784.1	2,387.7	+33.8
Fuels	2,310.0	5,537.5	11,543.5	8,683.0	9,662.0	+11.3
Total	7,567.7	14,745.0	24,608.2	18,013.5	21,617.5	+20.0
Total domestic exports all products	24,837.9	52,259.3	88,506.2	63,882.3	81,439.9	+27.5
Crude minerals as % of exports, all products	18.5	15.9	15.5	15.9	14.6	
Crude and fabricated minerals as % of exports, all products	30.5	28.2	27.8	28.2	26.5	
Crude mineral exports as % of mineral exports	60.7	56.2	55.8	56.3	55.1	

Source: Statistics Canada.

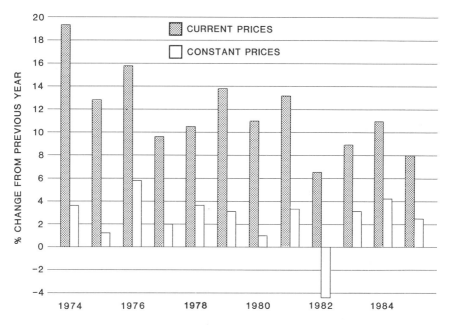

Note: Figures for 1984 and 1985 are estimates

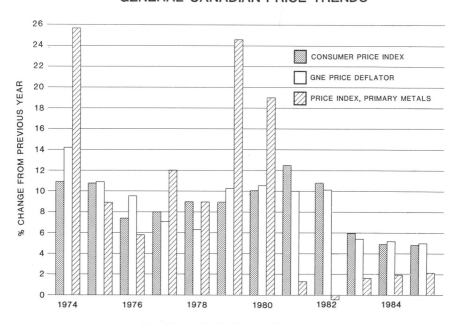

Note: Figures for 1984 and 1985 are estimates

CANADA, MINERAL PRODUCTION, 1984

% OF TOTAL BY COMMODITY

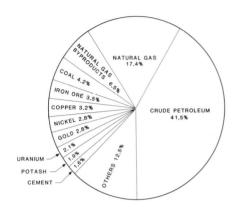

% OF TOTAL BY PROVINCE

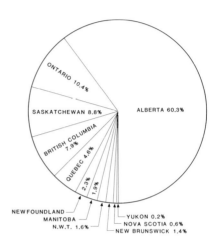

INDEXES OF GROSS DOMESTIC PRODUCT IN 1971 PRICES

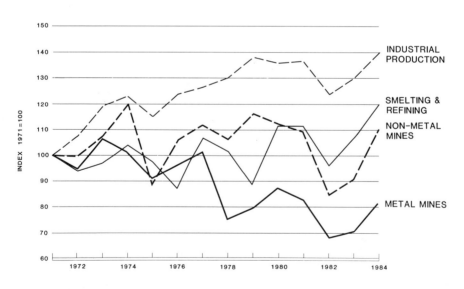

SOURCE: STATISTICS CANADA

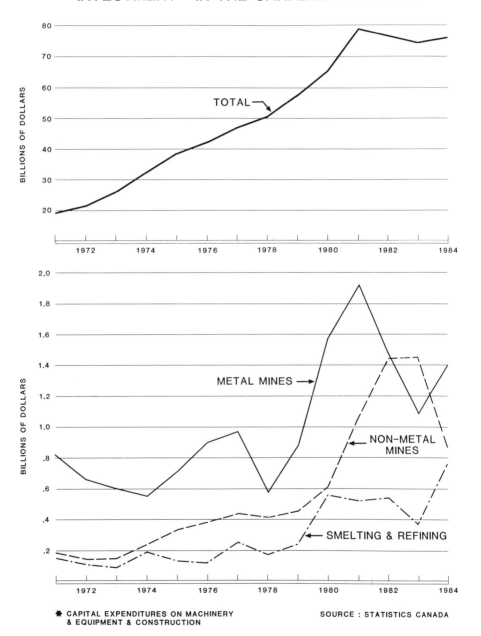

International Review

BRUCE A. McKEAN

Supply and demand elements are the touchstones of industry planning. A great deal of time and money is spent trying to predict their relative strengths so as to direct company resources to market opportunities. For minerals, there exist a number of international institutions and negotiating processes which can affect supply and demand or the expectations surrounding them. Almost all of these institutions are concerned in some way with questions of access (that is, access in its broadest sense including market transparency, tariff barriers, government procurement policies, quotas, countertrade requirements, nonconvertable currencies, etc.). Thus, they affect the environment in which future investment in mineral production and processing will occur. Not all of them make positive contributions. Nonetheless, they are important to Canada in view of our dependence on export markets and because access is more and more a vital determinant in whether a sale will be made somewhere other than on a commodity exchange.

Canadian mineral producers have to cope with established (eg. EEC, EFTA, Comecon) and emerging (Caribbean, Latin American) regional trading blocs, preferential tariffs and quotas for LDC (Less Developed Country) producers (eg. the GSP and the EEC Sysmin arrangement with the Lomé Convention countries), increasingly restrictive non-tariff barriers (including those related to the environment and health), the continuing phenomenon of tariffs escalating as value added increases, and the often very high tariff barriers of the newly industrialized countries (NIC's) and LDC's.

It is clear that government (and, increasingly, industry) must use these sometimes little known institutions and occasions to advance or defend Canadian interests. What follows is a review of the major issues and institutions, and what the Canadian mineral industry might expect from them in the near future.

CANADA/UNITED STATES FREE TRADE PROSPECTS

Periodic talks have been held with the United States since the release of Canada's Trade Policy Review in 1983. The talks have centred on the prospects for a new or revised trade regime between the two countries, i.e., covering almost three-quarters of Canada's total exports and some 50 per cent of our mineral and metal exports. In November, 1984, Canada's Minister for International Trade outlined four possible approaches:

a) a trade enhancement arrangement whereby the present trade regime is maintained but with certain shortcomings overcome;
b) an arrangement whereby the two sides liberalize and improve specific measures such as government procurement practices and emergency safeguards;
c) sectoral free trade wherein products in carefully defined sectors are traded free of all barriers; or
d) a comprehensive trade agreement which would provide for the removal of tariffs and other barriers on essentially all bilateral trade.

It is too soon to speculate on which approach, if any, may prove feasible.

Canada's mineral industry has suggested the negotiation of sectoral free trade for nonferrous metals. However, it remains unclear whether a sectoral approach might gain broad acceptance or how a "nonferrous" metal sector might be defined so as to provide for a balance of advantage for the two countries.

GENERAL AGREEMENT ON TARIFFS AND TRADE (GATT)

There is an emerging consensus in favour of a new round of multilateral trade negotiations (MTN) under GATT. This has been pushed

---

Bruce A. McKean is with the Mineral Policy Sector, Energy, Mines and Resources, Canada. Telephone (613) 995-9466.

particularly by the United States. These negotiations, which could be formally approved in 1985 and get under way in 1986 or 1987, will provide an opportunity to reduce trade barriers, tighten discipline on the use of non-tariff barriers, and facilitate industrial adjustment.

It was in anticipation of such a new MTN round that world trade ministers met in Geneva in November, 1982. There they drew up a work program for the GATT which included the creation of a working party on natural resource products, specifically in fisheries, forestry and nonferrous metals. The Canadian objective in this exercise is to ensure that all barriers impinging upon trade in these products are identified and their effects described. Of interest to the Canadian mining industry are the three studies on copper, lead and zinc completed in 1984, and the two scheduled for completion in 1985 (aluminum and nickel).

GENERAL SYSTEM OF PREFERENCES

In June 1971, members of the GATT approved a waiver of the "most favoured nation" (MFN) principle of the GATT and thus provided a legal basis for developed countries to grant preferential market access for developing countries. This has become known as the General System of Preferences (GSP). In 1984 most developed countries extended the original 10-year trial period for an additional ten years. Some countries have expanded the range of products or sizes of quotas under the GSP.

The GSP was a product of the worldwide recognition of the importance to developing countries of improved market access for their labour intensive products. Although trade statistics show OECD imports from LDC's increasing at an annual rate of 27 per cent between 1976-1980, developing countries have not been impressed by the operation of GSP schemes. In fact, experience to date in capital intensive industries such as nonferrous metal smelting and refining suggests that liberalized trade on an MFN basis would be the best objective for all raw material exporting countries. Canada supports this view.

THE UNITED NATIONS CONFERENCE ON TRADE AND DEVELOPMENT (UNCTAD)

The UNCTAD will be entering its twenty-first year of existence with a new Secretary-General, rumors of an internal reform movement, and suffering from varying degrees of disenchantment from both developed and developing countries. The depth of the problem can be seen when a comparison is made between the aspirations of the institution and its modest accomplishments over the last decade. It is not clear, and the next year will probably not tell, whether the UNCTAD will continue, halt or reverse its slide in political credibility and economic relevance. The momentum, however, remains negative.

PROCESSING, MARKETING AND DISTRIBUTION

There are a number of issues active in UNCTAD and 1985 may well be known as the year of PMD (Processing, Marketing and Distribution). The label PMD embodies a political undertaking six years ago to agree on a "framework" agreement in support of further processing goals of developing countries. At issue is the appropriateness and direction of intergovernmental action to promote such increased involvement by developing countries.

The developed countries, including Canada, have indicated a preparedness to accept a set of principles and guidelines which would allow developing countries to increase their participation in PMD over time and on economically viable terms. The latest UNCTAD documents contain many proposals which go beyond guidelines and principles. If accepted, they would, inter alia, require the "redeployment" of resource processing industries away from developed countries, the introduction of new discriminatory (in favour of developing countries) trade obligations for developed countries, the creation of a number of new international institutions, and an increase in concessional financing resources.

The initial impression is that the two sides are too far apart for anything to be achieved. The next year may confirm this. This in turn may set the stage for the later negotiation of an agreement which will allow the developing countries to use a fair and open international trading environment to realize, where it exists, the fruits of comparative economic advantage.

GLOBAL SYSTEM OF TRADE PREFERENCES

At the same time, there are other issues where either the political moment has not arrived or where the concept is still in the process of forming. An example of the latter is known as the Global System of Trade Preferences (GSTP). The LDC's are aware of the importance of international trade, of their own high tariffs, and of a feeling of solidarity over their common economic predicament. The thought is that the LDC's could increase their external trade while preserving their high tariffs against competition from developed countries if - between themselves only - they applied a formula of tariff cuts.

Preliminary economic analysis done by the OECD suggests that western trade with LDC's might drop slightly and inter-LDC trade might increase modestly. A net small increase in total international trade is forecast. This sounds promising but there remain immense political, legal and administrative hurdles ahead for the LDC's. It is possible also that, within the initial macroeconomic figures, there may be some developed countries which stand to loose more than others. Canada will, in future years, want to pay close attention to the GSTP's progress.

COMPENSATORY FINANCING

While the GSTP is an issue of the future, compensatory financing may receive considerable attention in 1985/86. The report of an Expert Group will be received by the UNCTAD early in the new year. The Group was to tackle the question of whether there is a need for an additional financial facility to compensate developing countries for shortfalls in individual commodity export earnings and, if so, what form, terms, conditions and funding should be brought to such a new facility.

The views of the developed and developing countries differ greatly. The extremes go from a continued reliance on the Compensatory Financing Facility of the IMF to assist with short-term balance of payments problems, to a new financial institution funded with $US 10 billion of new money to cover export earnings shortfalls as they occur in each commodity sector. The Expert Group (with the exception of the U.S.S.R. which will submit a minority report) is reported to have been able to achieve an internal consensus on the issue. It is likely, however, that the report will not be able to command universal support. There will probably occur a lively debate, driven by different economic and ideological interpretations of the problems faced by exporters of commodities. The prospects of an early conclusion, or of a conclusion inimical to Canadian mineral interests, is not high.

COMMON FUND FOR COMMODITIES

The Common Fund was to assist in the financing of the buffer stock operations of International Commodity Agreements. The negotiations were long (1976-1980) and arduous. The Preparatory Commission which was to set up the institution and set out how and under what conditions the Common Fund should operate, met frequently between 1980 and 1982. What emerged during this latter period were the still unresolved major areas of dispute which, in the negotiations, had been avoided through the use of generalities.

The uncertainty as to the shape and resources of the Common Fund, and the spectacular lack of success in forming new International Commodity Agreements along the lines of the International Tin Agreement, may explain the painful slowness of the ratification process, especially among LDC's. More than four years after the signing of the agreement establishing the Common Fund, neither of the two benchmarks which would bring it into operation had been achieved: 90 states ratifying and a minimum of two-thirds of the capital being represented among the ratifying states.

This year may see a modest revival in activity surrounding the Common Fund as the number of ratifications will probably exceed 100. This is no sign yet, however, that the Preparatory Commission will resume its work.

LAW OF THE SEA

As with the Common Fund, the United Nations Convention on the Law of the Sea has been negotiated. In this case, however, a Preparatory Commission is engaged in the lengthy process of developing rules for the exploration and exploitation of the deep seabed. It is also considering the question of overlapping claims among explorers for manganese nodules, specifically the state companies of France, Japan, India, and the U.S.S.R. The four existing private consortia are not involved. These have all received exploration licences under the national legislation of the United States.

The Convention closed for signature on December 9, 1984, and the United States, Federal Republic of Germany and the United

Kingdom had not signed by that date. Consideration of ratification will probably be delayed by many states (including Canada), until the results of the Preparatory Commission's work can be evaluated. The work and the evaluation will take some time.

INTERGOVERNMENTAL NICKEL DISCUSSION GROUP (INDG)

Last year saw the international emergence of the INDG proposal after five years of mainly Canadian and Australian work behind the scenes. The motivation behind the INDG is found in the profound structural changes occuring in the mineral industry and the presence of many new entrants in the nickel industry. Reliable information for companies and states upon which they can make rational investment decisions is vital if long-term stability is to be achievable. To this end, the October 1984, meeting of 31 countries with an interest in the production, consumption or trade in nickel was highly successful. This will be followed in 1985 by at least one negotiating session.

It is too early to predict success as there remain very real problems concerning the ability and willingness of a number of states to meet the minimum information gathering requirements of the INDG. The possibility of political considerations being interjected still remains, however.

COPPER STUDY GROUP

The difficult years being experienced by the copper industry - and the sometimes dramatic events such as the Section 201 action in the United States which have marked this period - have revived interest in a copper study group. There is a general acceptance that the International Lead and Zinc Study Group (ILZSG) has proven to be an efficient way to gather and share information. For reasons noted above, negotiations for a nickel discussion group along the lines of the ILZSG have been launched with reasonable prospects for success. Now, some elements of the Canadian copper industry are speaking in favour of a similar group for copper. This may lead to discussions in various international fora but at this time no formal examination of the prospects for such a group is planned.

INTEGRATED PROGRAM FOR COMMODITIES (IPC): IRON ORE AND MANGANESE

It was the IPC in the UNCTAD and the prospect of many new International Commodity Agreements with economic provisions (including buffer stocks and price undertakings) that prompted a very intensive period of international negotiations in the mid and late 1970s. Very little came of all this effort as political and economic realities clashed with too-high but strongly held expectations on the part of LDC's. The IPC remains in existance, however, and from time to time the water is tested to see if the positions of various players have changed. There was a meeting on iron ore in 1984 and there will be a meeting on manganese and perhaps iron ore in 1985. Nothing of consequence is anticipated from these meetings.

Regional Review

S. HAMILTON/G. CLEMENTS

Activity in the Canadian mineral industry during 1983 and 1984 was moderate. Mining companies took strong measures to reduce costs to survive. British Columbia, Yukon and Quebec were particularly hard hit by mine closures. Base-metal, asbestos and iron ore mines were particularly vulnerable. New mine developments have been few and have been mainly limited to coal and gold. The export market for coal is becoming increasingly competitive, forcing Canadian producers to accept contracts calling for lower prices and shipments. Precious metal prices have not improved as anticipated and marginal producers are exhibiting some signs of distress.

Federal-provincial relations in 1983 and 1984 focused on negotiating 10-year Economic and Regional Development Agreements (ERDA's), which identified priorities and strategies for cooperative development. The first ERDA was signed, with Manitoba, in November 1983, and ERDA's with all provinces were in place by the end of 1984.

Mineral development was identified as a priority in several ERDA's, and by mid-1984 Mineral Development Agreements (MDA's) had been signed with Newfoundland, Nova Scotia, New Brunswick, Manitoba and Saskatchewan Discussions with other provinces are continuing.

All MDA's encompass work to provide: geoscience data to stimulate mineral exploration; mining and processing technology to improve productivity and safety in mining operations; and, the identification of new development opportunities by means of market and economic studies. The total federal commitment for the five MDA's is $64.5 million, and the provincial commitments are $38 million.

NEWFOUNDLAND

The mineral industry is important to Newfoundland, where it accounts for about 10 per cent of the Gross Provincial Product. In 1984, the value of mineral production increased by 23.1 per cent to $993 million from 1983, of which $868 million was for iron ore, $56 million for zinc, and $23 million for asbestos.

Almost all of Newfoundland's mineral commodities are exported, and thus the industry has been severely affected by recent world recession. Weakness in iron ore demand will continue to adversely affect Newfoundland's producers as will low-cost competitors, especially those from Brazil which will have an increasing effect in the latter part of the 1980s.

The closure by Iron Ore Company of Canada (IOC) of its Schefferville, Quebec operation has adversely affected Newfoundland to the present that about 2.5 million tpy of iron ore were mined from deposits that extended into Labrador.

Base metal output has declined sharply over recent years. ASARCO Incorporated's Buchans unit was finally closed in August 1984, although a small, nearby copper deposit is to be tested before the mill is dismantled. Another property, owned by Newfoundland Zinc Mines Limited, continues to produce. Ore reserves are not extensive, but the company is actively exploring for more.

The seasonal recovery and shipment of barite from tailings at the Buchans mine continues. However, production is costly making it difficult to match overseas competition.

Transpacific Asbestos Inc. is shipping asbestos fibre from the Advocate mine at a reduced rate because of weak demand.

The former fluorspar mine at St. Lawrence, closed since 1966, reverted to the Crown in 1983. A new operator, Minworth Ltd. of Britain, has obtained $4.76 million from the federal government and $2.04

S. Hamilton and G. Clements are with the Mineral Policy Sector, Energy, Mines and Resources, Canada. Telephone (613) 995-9466.

million from the province; thus the company plans to reopen the mine in 1986.

BP Canada Inc. has discovered significant gold mineralization, on a property known as the Chetwynd deposit, in western Newfoundland. The company intends to mount a $2.5 million delineation program in 1985. This discovery indicates that the volcanic and sedimentary rocks of the Hermitage Flexure contain mineralization that had not been previously recognized.

A five-year, $22 million Canada-Newfoundland Mineral Development Agreement was signed in May 1984; 70 per cent will be contributed by the federal government and 30 per cent by the province. During the 1984 field season, geological mapping was conducted in Labrador, and on the Island, geophysical and geochemical surveys were completed. Projects concerning mining and mineral technology and economic development are being planned. This MDA continues the federal-provincial cooperation that started in the late-1960s. Since then, geological mapping and geochemical surveys, in conjunction with changes in land tenure, have been responsible for expenditures of tens of millions of dollars by the private sector on exploration and have led to the discovery of several promising mineral deposits. For example, IOC discovered the Strange Lake beryllium-yttrium rare earth deposit, north of Schefferville, while investigating an anomaly identified by an MDA-funded geochemical survey. The Chetwynd gold discovery is in an area mapped under the MDA where a provincial geologist reported discovering gold mineralization.

NOVA SCOTIA

In 1984, the value of mineral production in Nova Scotia increased by 12.6 per cent from 1983 to $293 million, of which $163 million was for coal and $44 million for gypsum.

In April 1984, an underground fire at the No. 26 colliery at Glace Bay forced the closure of the mine putting 1,200 employees out of work in an area where unemployment is already high.

The federal government announced a plan to restructure coal mining in Cape Breton in May 1984, which includes the expenditure of $325 million to expand existing mines and develop new ones. Three mines are planned to be replaced. In addition, the federal government plans to spend $2 million to study the feasibility of opening the 1B colliery which has been closed for about 30 years.

Rio Algom Limited began developing a tin mine near Yarmouth in the spring of 1984. It is expected to be in production by late 1985, employing an estimated 250 people. Development was facilitated by a change in provincial royalty legislation that will allow accelerated depreciation of new mines for the purposes of taxation.

Mineral exploration in the province was generally at low levels during 1984. Activity focused on the search for gold, tin, tungsten and barite.

Land use planning maps are being compiled to assist in the formulation of policies called for in the provincial planning act that was instituted in mid-1983.

The Canada-Nova Scotia Mineral Development Agreement was signed in June 1984, continuing and expanding work conducted under a two-year agreement which expired March 31, 1984. Under this new agreement, $16.1 million federal and $10.8 million provincial funds will be invested over five years by the parties in coordinated programs. Geoscientific, mineral technology, and economic development studies are designed to assist the industry, for example, in the exploration and development of gold and industrial mineral deposits.

NEW BRUNSWICK

In 1984, New Brunswick's mineral production increased by 16.7 per cent to $590 million, of which $347 million was for zinc, $54 million for lead and $48 million for silver.

The New Brunswick mineral industry is growing rapidly due to the discovery in the early 1970s of rich potash deposits in the Sussex region. The first mine began production in 1983, a second will follow in 1985 and a third may be completed by 1990. Other growth factors include: the expansion of the Brunswick Mining and Smelting Corporation Limited zinc-lead mining and smelting facility near Bathurst; the start of production of the Brunswick Tin Mines Limited tungsten-molybdenum mine south of Fredericton; and the rehabilitation and reopening of the antimony mine at Lake George by early-1986. The New Brunswick government estimates that the total value of mineral production could exceed $1 billion by the end of 1990, accompanied by expanding job opportunities.

Another factor that could accelerate development in the mining sector is the establishment of a pilot plant to test the sulphation roast-leach method for processing the complex base-metal ores found in the Bathurst region. This ore is fine grained and complex, resulting in relatively low recoveries of metals in the concentration stage. The project is jointly financed and managed with the province. Besides increasing recovery of metals from existing operations, this process could allow development of deposits currently considered uneconomical.

The Canada - New Brunswick Mineral Development Agreement was signed in June 1984, with spending commitments of $22.3 million by the provincial and federal governments. Its three main aspects are: geoscientific research; research and development of new technologies; and economic development studies.

QUEBEC

In 1984, the value of mineral production in Quebec remained virtually unchanged from 1983 at $2.04 billion, of which $442 million was for gold, $362 for iron ore and $301 for asbestos.

The 1980s have on the whole been a depressed period for the mining industry in Quebec. The value of mineral production has fallen sharply and has been accompanied by reductions in employment and investment. This decline is attributable partly to the continuing effects of the world-wide recession and partly to structural changes that are affecting mineral markets, particularly for iron ore and asbestos. Although the industry reached a low point in 1983, a continued relatively weak performance can be expected if the predicted economic slowdown for 1985 or 1986 takes place.

The mining industry in Quebec is diversified, with more than 15 minerals produced. On the other hand, about 60 per cent of production value derives from four minerals; gold, iron ore, asbestos and copper. Since 1980, the value of gold production has increased by 25 per cent, despite a major decline in price. However, the value of production of the other three minerals has fallen dramatically because of declining prices and volumes produced.

Mining is a major element in the economic base of northeastern and north-central Quebec, as well as the Gaspé and the Eastern Townships. Consequently, the general slowdown of the mining industry has had major regional impacts. Certain mining communities have lost part of their population and, in some cases, specifically Schefferville and Gagnon, towns have been virtually shut down.

Faced with the need to maintain some degree of economic activity in the more remote parts of the province, the Quebec government decided to invest heavily in the mining sector. Accordingly, through an assistance program designed to accelerate investment, Quebec has in the last 18 months committed nearly $120 million to assist the mining sector. This is expected to generate, mining investments of $600 million and create 4,500 temporary jobs and 2,000 permanent jobs. Some 18 projects, in the gold and base-metal sectors in particular, will benefit from this program.

An Economic and Regional Development Agreement was signed on December 7, 1984 by the federal and Quebec governments, which will provide the framework to negotiate an MDA. It will enable the two governments to examine ways to collaborate in geoscientific studies, mineral research and development and market studies.

The mining industry will continue to make an important contribution to the economy of Quebec, but it will no longer be characterized by megaprojects such as the development of the Labrador iron ore mines during the 1960s and 1970s. Rather, it will be characterized by the development of smaller, higher-grade mines. These will likely be precious metal mines or base-metal mines containing significant precious metal byproducts. Because of the smaller size, the impact of each mine will be more localized at the community level.

ONTARIO

In 1984, the value of Ontario's mineral production increased by 22.1 per cent from 1983 to $4.49 billion. Of this amount, $926 million was for nickel, $552 million for copper, $539 million for uranium, $414 million for zinc, $400 million for gold and $235 million for iron ore.

In northern Ontario some 19 communities are dependent upon the mining industry, which employs about 9 per cent of the area's experienced labour force. This work force will be under continued pressure as the industry struggles against weak commodity prices and demand.

The Griffith iron ore mine near Red Lake, owned by Stelco Inc., is scheduled to close in April 1986, affecting about 280 persons. Although various government programs will cushion the immediate effect of job loss and will aid individuals in their search for new employment, it is not likely that all will be able to find new jobs at other mining operations in northern Ontario. The impact on the Red Lake area in general, and the town of Ear Falls in particular, will be considerable.

Despite the generally gloomy outlook for mining, the exploration industry was very active in Ontario during 1983 and 1984. The main target was gold and activity was strong all across the province. No major discoveries were reported in 1984, but exploration and development took place on ground staked in the rushes of 1982 and 1983.

Three mines are nearing completion in the Hemlo area. The Teck Corporation joint venture with International Corona Resources Ltd. and the adjacent Noranda Mining Inc. properties are scheduled to start production in 1985. These two properties contain more than 30 million t of ore grading over 8 g/t. Immediately to the east, the Lac Minerals Ltd. property is to begin production in 1987, and throughput is expected to exceed 10 000 tpd by 1989. The impact of these developments is being felt in the neighbouring communities of Manitouwadge and Marathon. Within a few years, annual gold production from this region should reach 600 000 oz worth more than $250 million at current prices. A good deal of this wealth will be put into the local economy as wages and salaries.

Also near Marathon, Corporation Falconbridge Copper is continuing underground exploration work on its Winston Lake copper-zinc deposit, prior to deciding whether to bring it into production.

The Detour Lake mine, about 140 km northeast of Cochrane, has now operated for about one year. It is the fifth largest gold mine in Canada and the 500 jobs created are an important addition to the economic base of northeastern Ontario. There is no town at the mine site and services are supplied from Cochrane and Timmins.

Also at Timmins, Kidd Creek Mines Ltd. recently opened the Owl Creek gold mine and is doing underground exploration at the nearby Hoyle Pond gold deposit, which it plans to bring into production around 1987, when the Owl Creek ore will be exhausted.

In 1984, three federal-provincial regional agreements, which had provided information and assistance to the mining industry, terminated. Two agreements were specific to eastern Ontario, the upper Ottawa Valley and the Kirkland Lake area, while the more general Northern Ontario Rural Development Agreement (NORDA) affected the whole of the northern part of the province.

The mining industry in Ontario was less plagued by shutdowns and layoff in 1984 than in 1982-83. At Timmins, Pamour Porcupine Mines Limited shut down the lower-grade, higher cost portions of its gold mine salvage operations, laying off nearly 500 workers.

The Ontario mineral industry is expected to experience continued weakness in 1985 especially if the demand for mineral commodities again falls off due to the predicted slowdown of the world economies. Worldwide metal mining overcapacity and production will contribute to worsen the effects of the reduced demand.

MANITOBA

In 1984, the value of Manitoba's mineral production increased by 3.1 per cent from 1983 to $756 million, of which $239 million was for nickel, $165 million for crude petroleum, $109 million for copper and $68 million for zinc.

Base-metal mining provides the economic base for the northern communities of Flin Flon, Lynn Lake, Snow Lake, Leaf Rapids and Thompson. The low metal prices that have persisted since 1982 have caused a number of temporary and permanent mine closures and hundreds of employees have been laid off. The situation is worsened by declining ore reserves and grades. Given the poor outlook for base-metal markets, mining companies are hesitant to make the capital investments ncessary to improve output at existing mines or develop new orebodies. In particular, the existence of the town of Lynn Lake is threatened by the planned closure by 1985 of the Fox mine because of exhaustion of ore.

In June 1983, the governments of Canada and Manitoba announced a two-year $1 million agreement to conduct geoscientific studies to stimulate mineral exploration in the Lynn

Lake area. In November 1983, the two governments announced a $24.7 million five-year Mineral Development Agreement to: expand the scope of the geoscientific programs to other areas in the north; conduct research to improve health, safety and productivity at existing mines; and investigate new uses for mineral commodities that are not being fully exploited.

The only new mine that opened recently in northern Manitoba was the 1 600 tpd Trout Lake copper-zinc mine near Flin Flon. Concentrate from this mine will ease the shortage of feed to the Hudson Bay Mining and Smelting Co., Limited smelter at Flin Flon.

Sherritt Gordon Mines Limited is continuing an exploration program on the Agassiz gold deposit near Lynn Lake, supported by grants from the federal and provincial governments under the New Employment Expansion and Development (NEED) program. If the orebody proves to be economic, it is planned to coordinate development with closure of the Fox mine.

In May 1984, the Manitoba government approved a $10 million loan as part of the $30 million required to develop access to deeper ore at the Ruttan copper-zinc mine at Leaf Rapids.

The future is brighter at the Thompson nickel mining and smelting complex because of efforts to lower operating costs. This is being achieved mainly by conversion to underground bulk mining methods and the development of an open-pit mine, which will begin production in 1986. While employment levels will not be significantly affected, job security will be improved and the long-term viability of the Thompson operations will be enhanced.

Northeast of Winnipeg, near Bissett, results from an exploration program at the San Antonio gold mine, which was shut down in May 1983, are encouraging. Further work is required and gold prices must improve before the mine is reopened. In the same area, the Bernic Lake tantalum operation, shutdown since 1982, was re-activated as a pilot project to produce ceramic-grade spodumene. About 20 of the former 100 employees were rehired.

Minor amendments were made to the Mining Tax Act in Part IX of Bill 75, passed on August 18, 1983. These amendments were largely to provide greater clarification for the application of the act.

Changes to the Mining Act have been under consideration since 1980, when a commission made recommendations to improve safety conditions in Manitoba mines, many of which have already been implemented. Extensive consultation with labour and the mining industry was held to develop the proposed legislation.

The provincial government is actively pursuing the development of a potash mine and the construction of a 200 000 tpy aluminum smelter. The aluminum smelter project received a major setback in November 1984 when the potential investor, Aluminum Company of America, withdrew from a joint feasibility study. The province and Canamax Resources Inc. are proceeding with a feasibility study for the development of a potash mine near Russell. The partners are attempting to interest foreign governments to provide equity investment and a market for the potash that would be produced.

SASKATCHEWAN

In 1984, the value of Saskatchewan's mineral production increased by 33.2 per cent from 1983 to $3.78 billion. Of this amount, $2.32 billion was for crude petroleum, $378 million for uranium and $115 million for coal.

Potash sales, down since 1981, have recovered moderately. Uranium production fell sharply in 1983 due to the closure of the Eldorado Nuclear Limited operations at Uranium City, causing the loss of 830 jobs and the virtual shutdown of the community. However in 1984, uranium production expanded as the new Key Lake mine became fully operational. In the south, mining of lignite coal, mainly used by nearby thermal electric generating stations to feed the provincial power grid, has been relatively unaffected by the recession.

Expansions are under way at the Rabbit Lake uranium mining and concentrating facilities and the Belle Plaine solution mining potash operations. No further major uranium, potash or lignite coal projects are likely to commence until the late 1980s. However, encouraging results from gold exploration in the north may lead to mine developments if the outlook for future gold prices is favourable.

In May 1984, the federal and provincial governments signed a $6.38 million Mineral Development Agreement supporting geoscientific programs in the north. It will also facilitate research to improve productivity and investigate new uses for industrial mineral commodities. It is hoped that this information will lead to a number of new development opportunities.

In March 1984, the Saskatchewan government announced a number of changes to the Mineral Disposition Regulations 1961 designed to stimulate mineral exploration and development.

In the latter half of the 1980s the potash mining industry in Saskatchewan will have to compete for markets with other producers coming on-stream in New Brunswick and overseas. Uranium producers may face softening prices as deposits in Australia are brought into production. Lignite coal production for consumption in mine-mouth power stations will expand. In terms of employment and value of production, the mineral industry will likely maintain its contribution to the provincial economy, however, there is no reason to expect strong expansion.

ALBERTA

In 1984, the value of mineral production in Alberta was $25.96 billion. Of this amount, $14.93 billion was for crude petroleum, $6.98 billion for natural gas, $2.71 billion for natural gas byproducts, $557 million for sulphur and $480 million for coal.

Coal production, the principal mining activity in Alberta, is eclipsed in economic importance by the petroleum industry. However, it is important to the economy of a number of communities. The communities of Grande Cache, Edson and Hinton have been affected by layoffs at nearby mines as a result of reduced exports of thermal and coking coal, primarily to Japan. This has been partially offset by the start-up of the new Gregg River coking coal mine in April of 1983 and the Obed Mountain thermal coal mine in 1984.

Mines supplying coal for domestic electric power generation continue to produce at normal rates. In September of 1983, the Highvale mine opened at Keephills west of Edmonton. New mines are under construction at Genesee, also west of Edmonton, and at Sheerness, east of Drumheller.

Near Coleman in southeast Alberta, rising costs forced the closure of thermal coal reclaiming operations. Several proposed coal projects have been deferred until offshore markets improve.

Sulphur produced as a byproduct of sour natural gas is Alberta's other major nonfuel mineral commodity. Demand recovered during 1983-84 and prices were good. Inventories are falling as a result of increasing demand and falling production from depleting reservoir reserves.

BRITISH COLUMBIA

In 1984, the value of British Columbia's mineral production was $3.35 billion, up 15.5 per cent from 1983. Of this amount, $1.03 billion was for coal, $548 million for copper, $436 million for crude petroleum, $389 million for natural gas and $130 million for zinc. The increase in total value of production was mainly due to the start-up of shipments from the northeast coalfields.

Contractor employment in the mining industry was at an unprecedented high of 3,500 in 1983, mainly because of the construction of the northeast coal mines. Within the industry itself, employment dropped from the 1981 peak of 20,240 to 16,600 in 1983. The Mining Association of British Columbia reported that 4,060 employees, or nearly a quarter of the industry's workforce, were laid off at various times during 1983. Some of the mine closures that began in 1983 were extended into 1984 and persistent low prices for base-metals caused more mines to shut down.

During 1983, Aluminum Company of Canada, Limited (Alcan) applied to the British Columbia government for permission to proceed with the Kemano completion project, which would double Alcan's hydro generating capacity in the province and thereby permit the construction of two new aluminum smelters. By late-1984, falling aluminum prices caused Alcan to withdraw the application pending recovery of the aluminum market.

The outlook for British Columbia's large low-grade copper-molybdenum mines is not promising. Although many are being maintained on a stand-by basis, significant improvements in copper and molybdenum markets would be required to make them viable. More closures and shakeouts can be expected.

The British Columbia mining industry should see the development of a number of small-tonnage, high-grade, underground precious metals mines over the next few years. Although these mines will not have the economic impact of the large copper-molybdenum mines, they will be more labour intensive. If this type of development occurs, the availability of experienced underground miners could become a problem.

On November 23, 1984 the federal and provincial governments signed a ten-year Economic and Regional Development Agreement for coordinated initiatives. Three sectors including mineral development were identified as areas in which specific programs, should be developed.

NORTHERN CANADA

During 1983 and 1984, the Department of Indian Affairs and Northern Development (DIAND) continued to work on the development of a northern mineral policy. As part of a consultative process, eight issue papers will be released and by the end of 1984 three of these papers were available.

A new northern roads policy was approved by the federal Cabinet in 1983. Over the next five years, this policy is expected to provide an estimated $100 million for northern road construction. The level of assistance planned for northern resource roads to 1988 has been doubled.

In 1983, agreement was reached among the federal government, territorial governments and native groups on a new land use planning process for northern Canada. Organization for the implementation of land use planning proceeded through 1984.

NORTHWEST TERRITORIES

Ten mines operated in 1983 and 1984 producing gold, silver, lead, zinc, cadmium, copper and tungsten. Although depressed metal prices and labour strife resulted in intermittent closures and cutbacks at several mines, production capacity continued to increase. In general, the mines in the Northwest Territories demonstrated an ability to weather the economic downturn.

In 1984, the value of mineral production increased by 24.0 per cent from 1983 to $738 million, which included $345 million for zinc, $190 million for gold, and $65 million for lead.

The Cantung tungsten mine of Canada Tungsten Mining Corporation Limited, which was closed throughout most of 1983, reopened in December. About 120 of the 200 workers laid off were recalled. About 50 workers remained on-site during the closure.

Cominco Ltd.'s Pine Point mine was closed in January 1983. However, the mine was reopened the following June after renegotiation of contracts with: Canadian National Railways covering freight cost; the United Steel Workers covering wages, benefits and working conditions; Northern Canada Power Corporation covering power costs; and the Trail smelter covering smelter charges. Substantial improvements in costs of production were achieved. Also, a pre-production stripping program was initiated at the mine, partially funded by the federal government under the NEED program.

By 1984, employment in Northwest Territories mines returned to the 1982 levels of around 2,400 persons. Start-up of the Salmita gold mine of Giant Yellowknife Mines Limited in December 1983 created 90 new jobs.

The near-term future of mining in the Northwest Territories looks stable. There are a number of small gold and silver properties that could be brought into production with less than two years lead time. Many of these would be developed on a fly-in fly-out basis and would not require extension of infrastructure or community services.

YUKON

Mining, formerly the lead industrial activity in Yukon, is now second to tourism in terms of economic contribution. Mineral production in 1984 was valued at $60 million, a decrease of 5.4 per cent from 1983, consisting almost entirely of gold at $40 million and silver at $15 million. Because the possibilities for economic development in Yukon are mainly in natural resources, the Yukon government has established a Mining and Energy Branch.

Despite low silver prices, United Keno Hill Mines Limited kept its Elsa silver mine in production, which provided employment for about 150 workers. The company raised $13 million from an issue of flow-through shares and has reported promising results from an exploration program in the vicinity of the mine.

At the Faro mine of Cyprus Anvil Mining Corporation, which suspended production in 1982, 200 workers participated in a waste stripping program jointly financed by Dome Petroleum Ltd. and the federal government. The future of this mine remains uncertain because of low zinc, lead and silver prices. Other unresolved issues include power rates, a labour contract, and the method to transport concentrates to tidewater. The White Pass and Yukon Railway remains closed because of the loss of concentrate shipments from the Faro mine, which provided the base load.

Although several small underground gold, silver and tungsten deposits in Yukon are being evaluated, the future of the territory's mining industry in the short-term depends on the Faro mine. The output of this mine, with its large requirement for transportation, labour and power when it is operating at full capacity, comprises about 40 per cent of the Yukon economy.

The medium-term outlook of placer mining depends on the price of gold, which is currently near the economic threshold for most operators. Uncertainties involving environmental regulations for placer mining were deferred when water use permits were granted for the 1984 and 1985 seasons. Research is being carried out on various impacts of placer mining on the environment and the issue will be reviewed in 1986.

Exploration expenditures in the Yukon in 1984 were unexpectedly strong. Expenditures in 1983 were about $13 million and expenditures for 1984 may be as high as $25 million. The only property currently in the development stage is the MacTung tungsten deposit. AMAX Inc. may have this property in production in 1986.

CANADA, PROVINCES AND TERRITORIES, LEADING MINERALS, 1984P

	Value of production ($ million)	Proportion of total (per cent)	Change from 1983 (per cent)
Newfoundland			
Iron ore	867.6	87.3	21.9
Zinc	56.2	5.7	38.2
Asbestos	23.5	2.4	40.8
Total	993.5	100.0	23.1
Prince Edward Island			
Sand and gravel	0.9	100.0	22.6
Total	0.9	100.0	22.6
Nova Scotia			
Coal	162.6	55.5	11.6
Gypsum	43.7	14.9	17.8
Sand and gravel	21.6	7.4	-6.4
Cement	17.1	5.8	70.6
Total	293.0	100.0	12.6
New Brunswick			
Zinc	346.7	58.7	34.0
Lead	53.9	9.1	29.8
Silver	47.7	8.1	-47.0
Coal	30.3	5.1	2.1
Total	590.4	100.0	16.7
Quebec			
Gold	442.2	21.6	-3.4
Iron ore	362.4	17.7	-2.8
Asbestos	301.1	14.7	-6.3
Cement	146.6	7.2	14.9
Total	2,043.4	100.0	0.2
Ontario			
Nickel	925.9	20.6	68.3
Copper	552.2	12.3	20.3
Uranium	538.7	12.0	-1.4
Zinc	414.0	9.2	24.8
Total	4,493.7	100.0	22.1
Manitoba			
Nickel	239.3	31.7	3.5
Petroleum	165.0	21.8	7.9
Copper	109.3	14.5	-22.1
Zinc	68.3	9.0	21.3
Total	755.7	100.0	3.1
Saskatchewan			
Petroleum	2,323.7	61.4	30.3
Potash	x	x	x
Uranium	377.6	10.0	211.1
Total	3,785.2	100.0	33.2
Alberta			
Crude petroleum	14,927.0	57.5	8.8
Natural gas	6,981.4	26.9	6.1
Natural gas byproducts	2,717.0	10.5	3.8
Sulphur, elemental	556.9	2.1	34.4
Total	25,963.7	100.0	7.7

	Value of production ($ million)	Proportion of total (per cent)	Change from 1983 (per cent)
British Columbia			
Coal	1,026.9	30.6	78.9
Copper	547.6	16.3	-7.3
Petroleum	436.0	13.0	7.9
Natural gas	389.4	11.6	2.1
Total	3,353.7	100.0	15.5
Yukon Territory			
Gold	40.4	67.8	-19.8
Silver	15.3	25.7	122.7
Sand and gravel	1.6	2.7	7.8
Total	59.6	100.0	-5.4
Northwest Territories			
Zinc	345.2	55.8	27.9
Gold	190.0	30.7	31.4
Lead	65.0	10.5	35.7
Total	737.8	100.0	24.0
Canada			
Petroleum	17,887.8	41.5	11.2
Natural gas	7,514.6	17.4	6.2
Natural gas byproducts	2,782.9	6.4	3.8
Coal	1,814.0	4.2	39.1
Iron ore	1,470.9	3.4	15.8
Zinc	1,438.0	3.3	26.7
Copper	1,351.4	3.1	-1.0
Gold	1,227.8	2.8	-0.2
Nickel	1,165.2	2.7	49.1
Uranium	916.3	2.1	37.2
Total	43,070.7	100.0	11.8

P Preliminary; x Confidential.

Canadian Reserves of Selected Mineral Commodities

(Data available in 1984)

J. ZWARTENDYK

Any assessment of future supply of a given mineral commodity from Canadian mines requires information on current working inventories, i.e., on the amounts of ore known to be present in operating mines and on additional known tonnages in deposits that are close to being mineable profitably. The tonnages that - in 1983 and 1984 - were fairly well delineated and judged to be mineable are reported below as "reserves". The limits of what is included in reserves are further specified in each case.

	1983	1984
(A) Copper	17 021 500 t[1]	16 170 100 t[1]
Nickel	7 580 800 t	7 339 500 t
Lead	9 028 800 t	9 053 800 t
Zinc	26 077 300 t	26 449 500 t
Molybdenum	494 300 t	445 800 t
Silver	31 381 t	30 222 t
Gold	837 707 kg	1 159 773 kg

The quantities of the metals listed above are contained in ore recoverable from current mines (including those "temporarily" closed) and from deposits that had been committed for production up to January 1, 1983 and 1984 respectively.

These quantities represent proved and probable tonnages; any additional "possible" tonnages are not included.

---

[1] Metric tonne (2 204.62 pounds avoirdupois).

J. Zwartendyk is with the Mineral Policy Sector, Energy, Mines and Resources Canada. Telephone: (613) 995-9466.

(B) Iron   1 775 million t

This is the quantity of iron contained in known crude ore in producing mines[2]. Ore in undeveloped deposits is not included.

(C) Asbestos   1983   1984
                 43.5 million t 41.4 million t

This represents the fibre content (on average, a little over 5 per cent) of, respectively, 803 million t (1983) amd 760 million t (1984) of mineable ore reserves in producing mines.

(D) Potash   14 000 million t (K_2O equivalent), corresponding to 23 000 million t KCl product (standard fertilizer - exported product)

This amount would be recoverable by conventional mining (to a depth of about 1 100 m) from known potash deposits. At least an additional 42 000 million t (K_2O equivalent) would be recoverable from known deposits by solution mining at depths beyond 1 100 m; this would represent 69 000 million t of KCl product.

---

[2] Estimate updated to 1984 from "MR 170, A Summary View of Canadian Reserves and Additional Resources of Iron Ore", Energy, Mines and Resources Canada, 1977.

(E) **Uranium**

	"Reasonably Assured"	
	Proven (Measured)	Probable (Indicated)
	(t U)	

Mineable at uranium prices of:

1983[3]
$Cdn 115/kg U or less:	35 000	150 000
$115 to $170/kg U:	1 000	9 000

1984[4]
$Cdn 100/kg U or less:	30 000	162 000
$100 to $150/kg U:		41 000

The tonnages refer to uranium contained in mineable ore[3]. Unless otherwise specified, uranium "reserves" in Canada refer to the tonnages mineable at uranium prices in the low range only.

---

[3] EP 83-3, "Uranium in Canada: 1982 Assessment of Supply and Requirements", Sept. 1983, Energy, Mines and Resources Canada.
[4] Communiqué October 11, 1984, Energy, Mines and Resources Canada.

(F) **Coal**

- **Bituminous**   3 087 million t (of which 2 030 million t could be used for metallurgical purposes)

- **Sub-bituminous**   918 million t

- **Lignitic**   2 263 million t

These represent tonnages that could be profitably recovered as raw coal, given current technology and economics, from measured (proven) and indicated (probable) coal in deposits that are legally open to mining. For the purpose of making these estimates, it was assumed that coal sales would cover the costs of any required infrastructure not already in place[5].

---

[5] CANMET Report 83-2 OE, "Coal Mining in Canada: 1983", Energy, Mines and Resources Canada, 1984.

Canadian Reserves, Development and Exploration

W.H. LAUGHLIN

RESERVES

Table 1 illustrates the annually changing levels of Canadian reserves of seven major metals, in terms of the metal content of ore. These quantities were computed on the basis of information provided by mining companies. They pertain to ore tonnages that, as far as could be determined, were known at a level of assurance equivalent to "proven" (measured) and/or "probable" (indicated). Tonnages reported as "possible" (inferred) were not included. Table 2 shows a province-by-province breakdown for reserves on January 1, 1984.

While the term "reserves" is widely used to refer to that part of mineral resources that, on a given date, is well delineated and considered economically mineable, the reserves given here are confined to those in producing mines and in deposits that have been committed for production. These reserves constitute the reliable core of information. For other deposits, where concrete steps have not been taken by companies to prepare them for mining, judgments by outsiders regarding economic mineability would not form a consistent basis for reporting reserves. The purpose of the "reserves" restrictions used here is to avoid such subjective judgments.

The quantities of reserves reported cannot, by themselves, give any indication of whether or not Canada might be running out of economically mineable minerals. Future production will draw not only on the 1984 reserves but also on additional reserves yet to be developed -- from discoveries, from extensions to known orebodies and from known but currently marginal or uneconomic material.

Canada has a large number of potential supply sources that are less assured than current reserves. The most promising of these are listed in EMR's biennial mineral bulletin on Canadian reserves[1]. That publication tabulates the specific operations at which current reserves are reported. It lists also those deposits that are considered the likeliest for future development, with ratings on their relative promise. Another bulletin[2] deals with Canadian capability for metal production both from operating mines and from known deposits for which future production can be considered likely.

GOLD

At the beginning of 1984, Canadian reserves of gold were 38 per cent higher than a year earlier, largely because of new reserves in deposits committed for production during 1983.

For the first time, reserves from all three deposits in Ontario's Hemlo gold camp were counted in the national total. Reserves in other deposits newly committed for production elsewhere were also added for the first time at the beginning of 1984, notably Falconbridge's Lac Shortt (Quebec), Inco-Queenston's McBean (Ontario), Noranda's Remnor (Quebec), Flin Flon Mines' Rio (Saskatchewan), Giant Yellowknife's Salmita (N.W.T.), Société Minière Louvem's Chimo (Quebec), Westfield Minerals' Scadding (Ontario), Sigma's Sigma II (Quebec), and Lac Minerals' Lake Shore (Ontario). The McBean mine represents diversification by the giant nickel producer, Inco, into purely precious metal operations. Even though these additions were significant, at least by past standards, each was dwarfed in comparison with the individual contributions made in Ontario's Hemlo area by Noranda-Goliath-Golden Sceptre's Golden

---

[1] W.H. Laughlin, Canadian Reserves of Seven Metals as of Jan. 1, 1983, MR 201, Energy, Mines and Resources, Ottawa.

[2] Mineral Policy Sector, Canadian Mines: Perspective From 1983, MR 200, Energy, Mines and Resources, Ottawa. (1984 version in preparation).

W.H. Laughlin is with the Mineral Policy Sector, Energy, Mines and Resources, Canada. Telephone (613) 995-9466.

Giant, Teck-Corona's Corona, and Lac Minerals' Williams deposits.

Noteworthy additions to reserves of gold were also made at a number of mines where major expansion projects were under way, such as at Aiguebelle Resources' Dest-Or, Lac Minerals' Macassa, Dickenson-Sullivan's Arthur White, and Lac Minerals-SOQUEM's Doyon mines.

The large net increase in reserves at the beginning of 1984 masked sizeable reductions due to reassessments and write-offs of gold-bearing material at a number of mines and a lack of replenishment of reserves for metal produced during the previous year at several major byproduct gold operations.

ZINC

Even though, during 1983, markets for zinc fared relatively better than did those for most other metals reviewed here, reserves of zinc were barely up at the beginning of 1984 compared with a year earlier. The outstanding addition to national reserves of zinc was contributed by Cominco's Polaris mine where a large tonnage of new ore was added to the mine's inventory. Rio Algom's new development at East Kemptville (Nova Scotia) added large zinc reserves to the national total. Much new ore was also developed at Mineral Resources International's Nanisivik mine and at Teck-Amax's Newfoundland Zinc mine.

Deletion of all of the underground ore reserves at Cominco's Pine Point operations caused the single largest reduction in zinc reserves. In many mines, new ore development did not keep pace with the zinc ore produced during 1983.

LEAD

Reserves of lead in early 1984 were not much changed compared with the previous year. Because lead is commonly associated with zinc, Cominco's Polaris mine, the largest contributor to the maintenance of Canadian zinc reserves, was also the largest contributor of new lead reserves. That contribution was partly offset by Cominco's decision to cease counting as "reserves" all of the substantial underground inventory of lead ore at its Pine Point operations. Lead reserves were diminished also because several major producers had not developed as much new ore in 1983 as they had mined.

NICKEL

Reserves of nickel in early 1984 were somewhat lower than the previous year. The overall decrease at Canada's more than two dozen nickel mines owned by Inco and Falconbridge took place because the ore that was mined throughout 1983 was not matched by newly-developed reserves. Some of the cash flow that in better times might have been used to develop additional ore was directed at lowering the cost of extraction of existing reserves through the implementation of bulk mining and other innovative methods of production.

SILVER

The 1984 reserves of silver were down slightly from those of the previous year. During 1983, mining outpaced the replenishment of reserves at a dozen major silver producers across Canada. Permanent closures during 1983 and downward reassessment of reserves at a number of operations where silver is an important byproduct also diminished the Canadian total.

COPPER

Canadian copper reserves in early 1984 were about 5 per cent lower than they had been a year earlier, a reflection of the poor markets facing copper producers.

The biggest reduction in reserves resulted from large-scale reevaluation of copper-bearing material in the reserves inventories of major copper producers in British Columbia, notably at Placer Development's Gibraltar mine and at Noranda's Bell mine, as well as from some permanent mine closures. Reserves were further depressed because ore mined during 1983 at several large-scale operations had not been replenished by the beginning of 1984. The ore development activities were apparently severely restricted.

The largest contribution of new copper reserves came from Noranda's Murdochville (townsite) property, included in Canadian totals for the first time at the beginning of 1984.

MOLYBDENUM

The weak market for molybdenum over the past few years brought in its wake a decline in molybdenum reserves of more than 9 per cent from 1983 to 1984. This drop took place even though reserves reported at

Amax's Kitsault, Placer Development's Endako and Noranda's Needle Mountain, Copper Mountain and Boss Mountain mines remained unchanged during 1983 while little or no production took place at any of these mines. Reserves that had previously been allocated to some operations as a byproduct were deleted from the national total in 1984 because the molybdenum concentrates that had been expected were not produced.

Reserves were further reduced by write-offs of low-grade material at some mines in British Columbia, notably at Noranda's Brenda mine. Inadequate reserves replenishment for material mined during the course of 1983 at large-scale operations contributed also to lowering the reserves level.

OUTLOOK

Production at several major mines has been suspended indefinitely, in some cases since 1982. Weak markets for most of the major metals make the early resumption of production at these mines unlikely. For this reason, Table 1 shows estimates of 1985 reserves that exclude the "temporarily" closed mines. More mines can be expected to close if weak prices continue.

Reserves will likely continue to decline, except for gold and silver, as long as the basic market outlook does not change significantly. Should markets improve, some of the mineral material recently judged subeconomic might again be considered reserves, although some of it may have become irretrievable by that time because of revised mining procedures.

DEVELOPMENT

Energy projects dominate development statistics: of the total commitments in dollars to Canadian mineral development projects in 1984, oil and gas accounted for about one-half, coal for about one-quarter and metals (excluding uranium) for some 15 per cent. Although metal mine/mill development projects appear modest in this context, they affect more regions than do oil, gas, and coal projects. The metallic mineral industry's share of development projects is proportional to its share of the total value of Canadian mineral production (18 per cent).

Figure 1(a) illustrates annual development expenditures since 1968. These have been consistently higher than the more widely publicized expenditures on exploration; the ratio of exploration to development expenditures fluctuates in the 0.5 to 0.8 range, Figure 1(b).

During 1983 a total of $612 million was spent on mine development. Major pre-production development projects included in this total are:

Noranda's Golden Giant (Ont.)	gold
Teck's Corona (Ont.)	gold
Rio Algom's East Kemptville (N.S.)	tin
Campbell Red Lake's Detour Underground (Ont.)	gold
Westmin's H-W (B.C.)	copper, zinc
Agnico-Eagle's Dumagami (Que.)	gold, copper
Campbell Resources' S-3 Project (Que.)	copper

New commitments made in 1984 to develop additional mine/mill metal production capability in Canada amounted to $900 million, about half for new gold production and half for new or expanded base-metal production. Fifty-three per cent of the total was for plants in Quebec and 38 per cent for plants in Ontario. Table 3 shows the individual projects.

In addition to these firm commitments to develop mines and build concentrators, many development projects were under consideration with tentative plans for production. At the beginning of 1984 there were at least 30 such tentative projects with announced estimates of capital costs adding up to some $1 billion. During 1984, many plans were changed and many new ones developed; by the end of the year, there were some 40 tentative projects with capital costs of about $500 million.

The continuing commitments to new mine development in Canada through three years of low metal prices, many mine closures, and excess mineral production capacity are evidence of considerable dynamism and faith in being able to surmount serious obstacles.

EXPLORATION

Some 1,200 companies are active in exploration for metallic minerals in Canada. Fewer than 15 per cent of these are mining companies with producing mines and these companies' affiliates, but these account for well over one-half of all expenditures on such exploration in Canada.

In 1983, the continuing aura of the gold boom and expectations for higher base-metal prices encouraged predictions that the decline in exploration activity that had started in 1982 would be short-lived and that an upward trend would resume. But this did not occur, largely because metal prices in general did not rise. The price of gold dropped over 20 per cent and that of silver 35 per cent. In base-metals, only zinc was relatively sound. Nickel and molybdenum remained at the mercy of a slumping steel industry. Exploration expenditures declined further, but the expectations of economic improvement kept this decline from being more severe.

Figure 2 illustrates exploration activity in Canada expressed in terms of three yardsticks: total expenditures, new claims recorded and surface diamond drilling.[1] From 1982 to 1983, overall exploration expenditures declined 19 per cent; claim staking rose 68 per cent; and diamond drilling rose slightly. Changes differed considerably among the provinces:

Exploration Expenditures. Only the sustained activity in Quebec, Ontario and British Columbia (the gold provinces) saved the decline in exploration expenditures from being a serious retrenchment. These three provinces, accounting for 66 per cent of total exploration expenditures in Canada in 1983, made only modest gains over 1982 expenditures, but that compensated to some extent for the large declines in the prairie provinces and the territories. Preliminary information suggests no significant change in the 1984 level compared with 1983.

Claim Staking. This rose everywhere in 1983 except in Saskatchewan. The major gains were in British Columbia (a rise of 153 per cent) and Ontario (111 per cent).

For 1984, preliminary data show that staking declined on a national basis, although Newfoundland and Manitoba gained 145 and 16 per cent, respectively.

---

[1] Oil and gas are not covered by these mineral exploration statistics. In the case of new claims recorded, coal is excluded as well.

Diamond Drilling. While the overall gain in metres drilled in Canada from 1982 to 1983 was 7 per cent, this was due almost entirely to the 84 per cent gain in Ontario. Surprising were the declines in Quebec (22 per cent) and British Columbia (10 per cent), which have shared in the intensity of exploration for gold.

Preliminary data for 1984 indicate an increase of as much as 35 per cent in metres drilled for Canada, with major gains in Newfoundland, Quebec and British Columbia. These provinces saw important gold developments during the year.

In summary, exploration activities in 1983 and 1984 were sustained by interest in gold; favoured areas were those that generated most of the exploration news on gold, which were particularily in Ontario. The effect of the 1981 gold discovery at Hemlo in Ontario was profound. Exploration expenditures in Ontario in the three years following that discovery (1982-84) were 165 per cent higher than they had been in the three years preceding it (1978-80); the area of claims staked was 120 per cent larger; and diamond drilling was up 335 per cent. Hemlo brought a new dimension to geological concepts of the formation of gold deposits. Each subsequent gold discovery has been compared with the Hemlo setting.

A federal-provincial survey for 1983 showed that Canadian exploration expenditures were distributed roughly as follows:

Precious metals	45 per cent
Copper, Zinc, Lead	32 per cent
Other metals	16 per cent
Nonmetallic minerals	7 per cent

During 1984, the persistent weakness in the prices of gold and silver (a further drop of 15 per cent for each during the year) began to show its effect before year's end in curtailed and cancelled projects. The outlook for exploration in 1985 depends heavily on the behaviour of the price of gold. A price around $US 300 per ounce is not high enough to be a strong stimulant to exploration, even with low-cost bonanzas such as Hemlo to spur expectations.

TABLE 1. METAL RESERVES IN CANADA

Quantities of Metals Contained in Proven and Probable Mineable Ore[1] in Operating Mines and Deposits Committed for Production on January 1:

Metal	Units[2]	1982[3]	1983[3]	1984[3]	1985 (est)[4]
Copper	000 t	15 815	17 022	16 170	14 940
Nickel	000 t	8 013	7 581	7 339	7 305
Lead	000 t	10 244	9 029	9 054	7 730
Zinc	000 t	29 505	26 077	26 450	25 005
Molybdenum	000 t	514	494	446	95
Silver	t	32 154	31 381	30 222	30 375
Gold[5]	kg	842 215	837 707	1 159 773	1 194 080

[1] No allowance made for losses in milling, smelting and refining. [2] One t (tonne) = 1.1023113 short tons; one kg = 32.150747 troy ounces. [3] Includes reserves in mines suspended "temporarily". [4] Excludes mines suspended "temporarily" with no return to production scheduled as of January 1, 1985. [5] Excludes placer deposits.

TABLE 2. METAL RESERVES BY PROVINCE

Quantities of Metals Contained in Proven and Probable Mineable Ore[1] in Operating Mines and Deposits Committed for Production[2] on January 1, 1984

Metal	Units[3]	Nfld.	N.B.	N.S.	Que.	Ont.	Man.	Sask.	B.C.	Y.T.	N.W.T.	Canada
Copper	000 t	3.7	612.5	53.8	1075.1	6924.7	855.1	28	6617.3	0	0	16170
Nickel	000 t	0	0	0	0	5437.9	1901.5	0	0	0	0	7339
Lead	000 t	16	4105	0	2.6	192.1	20.5	1.3	1929.7	1426.2	1360.5	9054
Zinc	000 t	194.1	10226.5	90.2	422	4526.4	819.1	14.6	3563.5	2156.9	4436.4	26450
Molybdenum	000 t	0	0	0	17	0	0	0	428.8	0	0	446
Silver	t	26.4	11336.7	0	863.5	8713	862.9	19	6111.7	2051	237.9	30222
Gold[4]	kg	211.8	15146.8	0	226177.8	652342.7	38352.7	3887	136443.3	4892.4	82318.4	1159773

[1] No allowance is made for losses in milling, smelting and refining. [2] Includes reserves in mines suspended "temporarily". [3] 1 t (tonne) = 1.1023113 short tons. [4] Excludes placer deposits.

TABLE 3. MAJOR DEVELOPMENT PROJECTS ANNOUNCED DURING 1984

Operating Company	Project	Metal	Start-up year	Capital cost
				($ million)
Lac Minerals Ltd.	New Mine and mill at Hemlo, Ontario.	Gold	1986-90	250.0
	Underground development at La Mine Doyon, Quebec. Surface installations.	Gold	1985	30.0
Les Mines Selbaie	New open pit mine (A-1 orebody), Quebec.	Zinc Copper	1986	128.0*
Corporation Falconbridge Copper	5300 feet shaft and u/g development (Ansil deposit), Quebec.	Copper	1991	125.0*
	New mine development (Winston Lake, Ontario).	Zinc	1986	42.0
Noranda Inc.	Development of Murdochville townsite deposit, Quebec.	Copper	1988	84.0*
	Golden Giant mine, Ontario - increase over previous estimate	Gold	1986	42.0
Sullivan Mines Inc.	Reopening of Eldrich mine, Quebec.	Gold	1986	27.6*
Campbell Resources Inc.	Expansion of plant facilities for exploration and development, Quebec	Copper Gold	on-going	25.3*
Starrex Mining Corp.	New mine at Star Lake, Saskatchewan	Gold	?	25.0
Rio Algom Limited	Increased estimate - East Kempville mine and mill, Nova Scotia.	Tin Copper Zinc	1985	20.0
Placer Development Limited	Mill improvements for recovery of precious metals at Equity Silver mine, British Columbia	Silver Copper	1985	12.5
Abcourt Silver Mines Inc.	New mine and mill incorporating former producer (Barvue mine, Quebec).	Silver Zinc	?	12.0*
Aiguebelle Resources Inc.	Increase mine and mill capacity 600 to 1000 tpd, Quebec.	Gold	1985	12.0*
Others	Twelve additional projects, each costing less than $12 million, including four new or re-opened mines	Gold Copper Zinc Antimony	-	63.6*
			TOTAL	899.0

* Includes Quebec government grants which, for these projects, total $76.3 million.

FIGURE 1

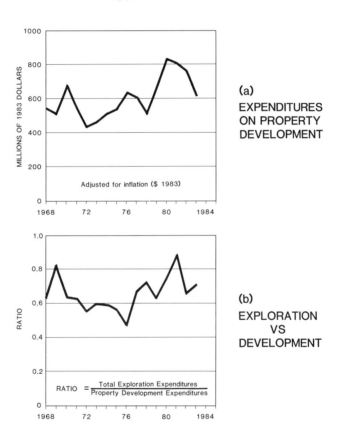

(a) EXPENDITURES ON PROPERTY DEVELOPMENT

(b) EXPLORATION VS DEVELOPMENT

$$\text{RATIO} = \frac{\text{Total Exploration Expenditures}}{\text{Property Development Expenditures}}$$

FIGURE 2
MEASURES OF EXPLORATION ACTIVITY

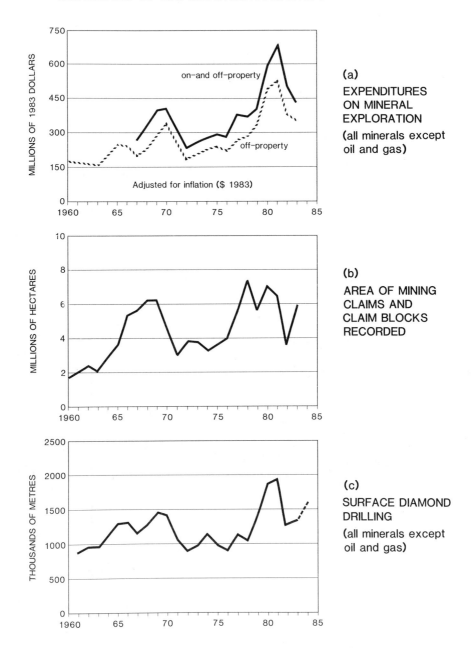

(a) EXPENDITURES ON MINERAL EXPLORATION (all minerals except oil and gas)

(b) AREA OF MINING CLAIMS AND CLAIM BLOCKS RECORDED

(c) SURFACE DIAMOND DRILLING (all minerals except oil and gas)

Aluminum

G. BOKOVAY

After a strong recovery in 1983 from the effects of the 1981-82 recession, aluminum demand began to ease somewhat during the second quarter of 1984. Unfortunately, aluminum producers continued to bring idled capacity back into production. This created a substantial build-up in aluminum inventories which began to exert considerable downward pressure on prices. Moreover, the fall of prices was exacerbated by consumer destocking in anticipation of even lower prices. In particular, the absence of Japanese buyers for long periods had a distressing effect on metal prices.

In an attempt to bolster prices, non-socialist aluminum producers began to announce production cutbacks toward the end of the second quarter of 1984. While prices did recover slightly in the fourth quarter, they had slipped back to near 1984 lows by the end of the year. Since demand is not expected to pick-up dramatically in the first quarter of 1985, improvement in the aluminum market will require additional production cutbacks.

While the majority of recent cutbacks have been deemed to be temporary in nature, some capacity may well be permanently closed in view of a major geographical restructuring which the industry has been, and is still undergoing. At a time when competition between producers has become more intense, rising energy costs particularly for fossil fuels, has resulted in substantial differences in the cost of the factors of aluminum production between regions. These cost differences have already resulted in a drastic cutback in permanent primary aluminum capacity in Japan, owing to its dependence on imported oil. In addition, widespread permanent cutbacks in primary capacity could also become a reality in the United States and western Europe.

Although Canada, with relatively large quantities of low cost hydro-electric power, will be a major beneficiary of this relocation in smelting capacity, the restructuring of the industry will introduce fundamental changes to those factors influencing long-term aluminum supply.

With changes in several of the key determinants of aluminum consumption, the outlook for aluminum is not as optimistic as in the late 1970s, although it still remains favourable.

CANADIAN DEVELOPMENTS

Two companies produce primary aluminum in Canada - Canadian Reynolds Metals Company, Limited, a subsidiary of Reynolds Metals Company of the United States and the Aluminum Company of Canada, Limited (Alcan), a subsidiary of Alcan Aluminum Limited of Montreal. Canadian Reynolds operates a 158 760 tpy capacity smelter at Baie Comeau, Quebec, while Alcan has smelters at Jonquière, Grande Baie, Isle Maligne, Shawinigan and Beauharnois in Quebec and at Kitimat in British Columbia, with a combined total capacity of 1 075 000 tpy. The most recent smelter to be constructed in Canada is located at Grande Baie. This smelter, with a capacity of 171 000 tpy, became fully operational at the end of 1983.

At the end of 1984, all Canadian smelters were operating to capacity with the exception of Alcan's Arvida works in Jonquière which was operating at about 86.5 per cent of its installed capacity of 432 000 t. This smelter had been operating at over 95 per cent of capacity until October when production was reduced by a further 37 500 tpy in response to growing metal inventories and severely depressed prices.

Alcan operates the only alumina refinery in Canada which is located at Jonquière. The plant has a capacity of 1.2 million tpy of metallurgical grade alumina and alumina derivatives. Bauxite is imported principally from related companies in Brazil and Guinea. The output of metallurgical alumina

---

G. Bokovay is with the Mineral Policy Sector, Energy, Mines and Resources Canada. Telephone (613) 995-9466.

from the Jonquière plant is consumed at Alcan's smelters in Quebec. Alcan also imports alumina for its eastern Canadian smelters from Jamaica while Alcan's Kitimat smelter is supplied with alumina principally from Australia and Japan. Alumina for the Canadian Reynolds smelter in Baie Comeau is imported from West Germany, Jamaica and the United States.

Canadian production of aluminum in 1984 is estimated at about 1 200 000 t compared to 1 091 000 t in 1983. Canadian exports of primary smelter products for the first nine months of 1984 were down to 628 000 t compared to 699 000 t recorded for the same period of 1983. However, shipments to the United States, Canada's largest export market, increased to 470 000 t from 452 000 t in 1983. The largest decline was registered in the Asian market with exports for the first three quarters of 1984 totalling 135 838 t compared to 226 805 t for the same period in 1983.

In conjunction with the successful renegotiation of its lease for the water rights on the Péribonca River in Quebec, Alcan announced a major $3 billion smelter modernization and expansion program for that province, over the next 30 years. Within this program, the most important single project will be the new Laterrière smelter in the Lac St. Jean region, which the company formally announced in April. The new $1 billion smelter will have a capacity of 248 000 tpy, and will replace about 135 000 tpy of the oldest capacity at the nearby Arvida works in Jonquière. The first of three potlines at the new smelter, each with a capacity of 82 700 tpy is expected to be in production in mid-1988.

Alcan has announced that the new smelter will employ the company's new energy efficient "275" cell technology. Operating at 275,000 amps, the new cell will use about 10 per cent less energy than those cells operating at Alcan's new Grande Baie plant.

In British Columbia, Alcan filed an application at the beginning of 1984 for permission to proceed with its Kemano Completion Project that included additional hydro-electric generating facilities and two new aluminum smelters in British Columbia. However, in late October in response to poor market conditions, Alcan asked the province of British Columbia to postpone further review of the company's application.

During 1984, Canadian Reynolds Metals Co. Ltd. continued work on an expansion of its Baie Comeau smelter begun in 1982, and which was made possible by attractive power rates offered by Quebec Hydro. The $500 million project will increase primary smelting capacity by 113 000 tpy. In addition, existing potlines will be modernized through a conversion to new Sumitomo technology. The modernization will also include the installation of new pollution control equipment, upgraded handling facilities for raw materials and finished goods, increased storage facilities and an additional casting capability. The expansion of smelting capacity at Baie Comeau, which is scheduled for completion in March 1985, is expected to create about 400 more permanent jobs.

In March 1984, state-owned Aluminum Pechiney of France signed an agreement with Quebec and Alumax Inc. of the United States for the construction of a $1.5 billion 230 000 tpy aluminum smelter at Bécancour, Quebec. With an attractive long-term power agreement with Hydro Quebec, including discounts around 50 per cent for the first few years of operation, it has been estimated that the new smelter will break even at aluminum prices around 45 cents (U.S.) per pound.

The new operating company known as l'Aluminerie de Bécancour is controlled 50.1 per cent by Aluminium Pechiney, while Alumax and the Quebec government's Société Generale du Financement (SGF) each have a 24.95 per cent interest. At the end of 1984, construction at the site was well under way, with the first 115 000 tpy potline expected to be in production in mid-1986 and the second at the end of 1987. It is expected that the smelter will employ about 845 persons.

Also in Quebec, Kaiser Aluminum and Chemical Corporation of the United States and the provincial government undertook a prefeasibility study in 1984 for a new aluminum smelter that would be built near Sept-Iles. A decision on whether to proceed is expected in the first quarter of 1985.

In Manitoba, the Aluminum Company of America (Alcoa) and the provincial government undertook a prefeasibility study in 1984 for an aluminum smelter. However, in December, Alcoa formally announced that it was abandoning its plans for the 200 000 tpy smelter in view of what it termed as the adequacy of current and planned world primary aluminum capacity to meet anticipated demand. Manitoba is discussing with other aluminum producers the possibility of constructing a smelter, with a target date of 1990.

WORLD DEVELOPMENTS

Western world inventories of aluminum rose steadily throughout 1984. The International Primary Aluminum Institute (IPAI) reported that in October, total inventories including aluminum scrap, primary and secondary ingot, metal in process and finished mill products, stood at 4.359 million t compared to 3.67 million t in December 1983. With falling metal prices, aluminum producers in the United States and western Europe instituted or announced production cutbacks totalling 737 000 tpy of capacity during the June to November period. Since new capacity continued to be brought into production elsewhere in the world and that over 300 000 t of announced cutbacks had not been implemented, average daily production in November was 33.9 thousand t. This was only marginally lower than in June when daily average production was 34.7 thousand t.

In the United States, the utilization of smelting capacity rose to about 87 per cent in the second quarter of 1984, and then fell to less than 75 per cent by year-end, with production cutbacks totalling more than 500 000 t. The Aluminum Association Inc. of the United States reported that for the first nine months of 1984, imports of aluminum in all forms were up by 39.8 per cent to 1.18 million t from the same period in 1983, due to the effects of the strong U.S. dollar.

Production cutbacks in the United States began in May with the announcement by Martin Marietta that it was cutting production at The Dalles, Oregon plant. Most major producers including Alcoa, Kaiser, Reynolds, Alumax and Consolidated Aluminium subsequently made similar announcements.

During 1984, the U.S. aluminum industry underwent significant ownership changes. In October, the U.S. Justice Department approved a modified proposal whereby Alcan would acquire a significant portion of the assets of Arco Aluminum. A proposal made by Alcan at the beginning of 1984 had raised anti-trust objections on the grounds that it would lessen competition in the U.S. market. While the accord allows Alcan to acquire Arco's 163 000 tpy smelter at Sebree, Kentucky, Arco's share of the alumina refinery at Aughinish, Ireland, it is restricted to a 40 per cent interest in Arco's new Logan County, Kentucky rolling mill. Arco later stated that it wishes to dispose of its 60 per cent share of the Logan County mill, its remaining smelter at Columbia Falls, Montana and its share of the Alpart alumina refinery in Jamaica.

A second major sale in 1984 involved the takeover of certain of the aluminum assets of Martin Marietta by Comalco Limited of Australia. Comalco will acquire the 168 000 tpy capacity Goldendale, Washington smelter, a rolling mill and recycling facility at Lewisport, Kentucky and an alumina unloading facility at Portland, Oregon. Martin Marietta was reportedly trying to sell its remaining 82 000 tpy smelter at The Dalles, Oregon and its 635 000 tpy alumina refinery in the U.S. Virgin Islands. In late November, the company announced that it was closing The Dalles smelter since no buyer had been found.

In Australia two new alumina refineries began operations in 1984. These were the 500 000 tpy Wagerup plant of Alcoa of Australia in western Australia which had been mothballed since its completion in 1982 and the 1.2 million tpy Worsley refinery of Reynolds Metals, BHP Minerals, Shell Co. of Australia and Kobe Alumina Associates.

Two new smelters also became fully operational in 1984. Comalco Limited commissioned the second potline of its Boyne Island smelter in Queensland. The smelter which has a capacity of 206 000 tpy cost $A 680 million. The second new smelter, at Tomago in New South Wales, is owned by Aluminium Pechiney, Gove Alumina, the A.M.P. Society of Australia, VAW of Australia and Hunter Douglas. This smelter which cost $A 650 million has a capacity of 230 000 tpy.

During 1984, Alcoa of Australia Ltd. announced the restart of construction to its hortland aluminum smelter. Alcoa suspended work on the project in 1982 because of depressed world markets. The $A 1.15 billion smelter will have a capacity of 300 000 tpy. The first potline is expected to be completed in late 1986 and the second in 1988. Other participants in the project include the Victorian government, Hyundai Corp. of South Korea and the Commonwealth Superannuation Fund Investment Trust.

During 1984, it was announced that Reynolds Metals, the western Australia government, the ICC-Kukje Group of Korea and the Griffin Metal Coal Mining Co. would participate in a feasibility study for a new 220 000 tpy aluminum smelter in western Australia. The smelter which would be built south of Perth, at Kemerton, will cost about $A 800 million. Power costs for the smelter will reportedly be A 1.6 cents per kWh initially and then increased to 2.2 cents by 1990.

In September, Alcan announced that it was postponing the start-up of the third potline at its Kurri Kurri smelter in New South Wales. The company said that the delay in bringing the 50 000 tpy potline into production was due to poor market conditions.

In Brazil, it was announced in August that the Alumar expansion project would proceed. The Alumar development is owned by Alcoa Aluminio and Billiton and will have the smelting capacity increased from the recently installed 100 000 tpy to 245 000 tpy, through the construction of a second potline and by technical improvements to existing facilities. The expansion, which should be completed by 1986 will involve some Brazilian investment.

During 1984, work continued on the Albras/Alunorte project owned by CVRD of Brazil and the Japanese consortium, Nippon Amazon. The first of four 80 000 tpy potlines is expected to be in production in October 1985. A 800 000 tpy alumina refinery is also being built.

In Venezuela, Aluminio Del Caroni S.A. (Alcasa) announced a modernization and expansion of its plant at Puerto Ordaz that will increase production by 80 000 tpy. The smelter currently has a capacity of 120 000 tpy. Venezuela's other producer, Venalum, which is owned by the state and Japanese companies, was also reported to be considering an expansion of its smelter from 280 000 tpy to 350 000 tpy. However, with the apparent difficulties that had arisen between Venalum and Japanese buyers on a new price agreement for aluminum exported to Japan, this expansion could be delayed.

Venezuela is proceeding with the development of a bauxite mine in Los Pijiguaos. The $US 450 million project is expected in production by mid-1986 and will have the potential to produce up to 4.5 million tpy. Reserves in the area have been estimated at between 4 and 5 billion t. Venezuela currently imports all of its bauxite requirements from Brazil, Guyana, Sierra Leone and Surinam.

In 1984, Reynolds Metals announced that it was phasing out its bauxite mining operations in Jamaica operated by Reynolds Jamaica Mines. However, the Jamaican government and aluminum producers reached an agreement during the year for a new tax regime applicable to the mining and export of bauxite. The agreement includes a production levy indexed to 6 per cent of the average realized price for primary aluminum and a royalty payment of 50 cents (U.S.) per dry tonne of bauxite. The agreement also provides incentives for increased production.

Also during 1984, Jamaica and Colombia announced that they would cooperate to build an aluminum smelter in Colombia using Colombian coal for power and Jamaican alumina, if the feasibility study is favourable. The smelter with a capacity of 140 000 tpy could come on-stream by 1990.

In India, the state-owned National Aluminum Company Ltd. (NALCO) is proceeding with a bauxite/alumina/aluminum development in the State of Orissa. The project includes an 800 000 tpy alumina refinery at Damanjodi and a 218 000 tpy smelter at Angul. The project, which is scheduled for completion in 1986 or 1987, should allow India to become self-sufficient in aluminum.

In Ghana, the government and the Volta Aluminum Co. owned by Kaiser and Reynolds reached an agreement on the price of electrical power for the Volta smelter. However, the 200 000 tpy smelter remains closed due to a shortage of electricity. Also in 1984, the U.S.S.R. and Ghana signed an agreement whereby the U.S.S.R. will aid Ghana to develop a deposit of bauxite at Kibi, in the eastern part of the country.

In Japan, the government is continuing its efforts to trim the size of its primary aluminum capacity. The most recent announced objective is to reduce primary aluminum capacity from 700 000 t to 350 000 t during the three year period from 1985 to 1987. Japanese smelting capacity in the late 1970s was approximately 1.6 million tpy.

In line with the latest goal, Nippon Light Metal Co. Ltd. announced that it would close its Tomakomai smelter on April 1, 1985, while Sumitomo Aluminum Smelting Co. Ltd. said that it would close its Toyo smelter early in the new year. The two smelters have a combined capacity of about 171 000 tpy. In order to compensate companies for the permanent cutback in capacity, the government of Japan will limit the import duty for the first 350 000 t of aluminum imported in 1985 to 1 per cent. Imports in excess of this level will be subject to the normal 9 per cent rate of duty.

In Europe, the major aluminum producers including Aluminium Pechiney of France; Vereinigte Aluminium-Werke (VAW) of West Germany; Norsk Hydro, Elkem and ASV of Norway; Schweizerische Aluminium AG (Alusuisse); and Alluminio Italia announced significant temporary cutbacks in primary aluminum production during the second half of 1984. In addition, a national plan for the aluminum industry in Italy was unveiled at the end of 1984 which includes the permanent closure of the Balzano smelter in 1985 and the Porto Marghera smelter in 1986. However, Norsk Hydro announced at the end of 1984 that it was proceeding with a 50 000 tpy capacity increase to its Karmoy smelter and was also planning to increase the capacity of the Sor Norge aluminum smelter which is jointly-owned with Alusuisse.

PRICES

Monthly average prices on the LME which had risen to a high of 73.2 cents (U.S.) per pound in September 1983 declined through the first half of 1984 to reach 45.8 cents in September. This was followed by a small improvement during which prices climbed to 55 cents by the middle of November. For the remainder of the year prices eased once again with aluminum, at the end of December, trading at 47 cents (U.S.) per pound.

The usefulness of producer list prices was dealt a severe blow in 1984 with the announcement by Alcan that it was withdrawing its international list price. The AWP price had served as a basis for various world aluminum supply contracts including past Japanese contracts with Indonesia and Venezuela but had not been relevant for the past few years.

During 1984, Pechiney began publishing its PIP index which plots the movement in aluminum prices which have been achieved on Pechiney's own sales to independent customers. According to Pechiney the index is intended to be a pricing system which avoids the volatility of day to day LME pricing and also the lack of market applicability inherent with producer list prices.

Late in 1984, the International Bauxite Association announced that it was recommending a minimum $US 35 per t cif for base-grade bauxite in 1985 and a minimum price of $US 225 per t cif for metallurgical base-grade alumina. Prices in 1984 for high quality bauxite were about $25-30 per t fob while the price of alumina was about $120 to $140 per t fob.

USES

Aluminum has various characteristics including low density, high strength and corrosion resistance, which makes it suitable for use in alloyed and unalloyed forms, in a wide variety of products. In the building and construction industry, major uses for aluminum include residential siding; window and door frames; screens; awnings and canopes; bridge, steel and highway equipment and mobile homes. In the transportation sector, aluminum is widely used in the manufacture of buses, trucks, trailers and semi-trailers and is the principal metal in aircraft. In addition, aluminum is being increasingly used in passenger cars as manufacturers move to reduce the weight of their vehicles. Since 1974 the amount of aluminum used in a typical American car has risen from 70 lbs to around 135 lbs in 1983.

In the electrical field, aluminum largely replaced copper in the wiring and power transmission in the 1960s. While aluminum has maintained the market for power transmission applications, local restrictions and consumer resistance have lessened the demand for aluminum in electrical wiring. Aluminum has however gained acceptance in various communications and computer applications.

The fastest growing market for aluminum in the 1970s was containers and packaging, including cans and foil. However, with increased recycling of aluminum cans, demand growth has slowed somewhat in recent years.

Aluminum is used to produce consumer goods and is also used in the manufacture of a wide variety of machinery and equipment, and in several important applications in the chemical industry.

OUTLOOK

Aluminum is considered to be entering the mature stage of its demand life cycle, since there appears to be no major new applications for the metal that should dramatically boost demand. Furthermore, increased competition from plastics, composite materials, high strength steels will limit aluminum demand. Consequently it is expected that aluminum demand will grow at an average annual rate of between 2 and 2.5 per cent in the next decade, compared to a 6.5 per cent rate of growth achieved during the 1965-1980 period.

Although the amount of aluminum used in beverage cans is expected to grow outside the United States, the growth of demand for primary aluminum from this sector within the large U.S. market will begin to decline. Reasons for this include greater recovery rate for used beverage containers, the production of thinner walled cans, and increased competition from steel cans and plastic cans which are expected to be in production in 1985. Efforts to develop a commercially acceptable aluminum food can have not yet proven successful.

Although, aluminum prices are expected to remain competitive with copper in the electrical field, aluminum is not expected to re-establish itself in the electrical wiring market.

While there has been much publicity concerning the potential application of newly developed aluminum lithium alloys in the aerospace industry, the success of these alloys will simply halt the erosion of market share to new nonmetallic materials.

Although faced with rigorous competition from other lightweight materials, aluminum will be increasingly used by the automobile industry. New applications for aluminum include radiators, heater cores, airconditioner units, exterior trim, engine blocks, cylinder heads, transmissions and wheels.

Aluminum prices are expected to recover in 1985 to about 55-60 cents (U.S.) as aluminum producers continue to implement production cutbacks. In 1986, the price should increase to about 65 cents (U.S.). Longer-term prices are not expected to rise significantly in view of the expected continued strength of the U.S. dollar and also due to a chronic oversupply situation which is likely to persist for the rest of the decade. This will arise because of a probable slow permanent closure rate for high cost capacity in the United States and Europe and also to new smelter construction which will continue in Canada, Australia, Brazil and in other developing nations. However, with the gradual replacement of old inefficient smelters, the average cost of producing aluminum, which has already dropped, will continue to fall.

Although the large integrated aluminum producers which once completely dominated the industry will remain important, the restructuring of the industry will result in an increased number of independent producers and an increased equity participation of governments.

Canada is currently a low, if not the lowest, cost aluminum producer in the world. Canada's production costs are estimated at 43 cents (U.S.) per pound while the United States and Japan are the most expensive at 64 cents. With large quantities of inexpensive hydro-electric power available and potential for substantially more, Canada should significantly increase its share of world aluminum production in the next decade.

PRICES

Month	U.S. Market	LME Cash
	¢ U.S./pound	
January	76.1	70.2
February	73.3	67.8
March	71.6	66.0
April	68.2	62.1
May	64.7	58.5
June	63.2	57.8
July	56.1	52.7
August	54.4	51.5
September	48.4	45.8
October	50.1	46.6
November	55.1	52.2
December	51.4	49.7
1984 Average	61.1	56.5

Source: Metals Week.

TARIFFS

Item No.		British Preferential	Most Favoured Nation	General	General Preferential
			(%)		
CANADA					
32910-1	Bauxite	free	free	free	free
35301-1	Aluminum pigs, ingots, blocks, notch bars, slabs, billets, blooms and wire bars, per pound	free	.5¢	5¢	free[1]
35302-1	Aluminum bars, rods, plates, sheets, strips, circles, squares, discs and rectangles	free	2.3	9	free
35303-1	Aluminum channels, beams, tees and other rolled, drawn or extruded sections and shapes	free	10.3	30	free
35305-1	Aluminum pipes and tubes	free	10.3	30	free
92820-1	Aluminum oxide and hydroxide; artificial corundum (this tariff includes alumina)	free	free	free	free

	1983	1984	1985	1986	1987
MFN Reductions under GATT (effective January 1 of year given)			(%)		
35301-1	.5¢	.4¢	.3¢	.1¢	free
35302-1	2.3	2.2	2.2	2.1	2.1
35303-1	10.3	9.7	9.1	8.6	8.0
35305-1	10.3	9.7	9.1	8.6	8.0
92820-1					

Item No.		1983	1984	1985	1986	1987
UNITED STATES (MFN)				(%)		
417.12	Aluminum compounds: hydroxide and oxide (alumina)	Remains free				
601.06	Bauxite	Remains free				
618.01	Unwrought aluminum in coils, uniform cross section not greater than 0.375 inch, per pound	2.9	2.8	2.8	2.7	2.6
618.02	Other unwrought aluminum, excluding alloys, per pound	0.5¢	0.3¢	0.2¢	0.1¢	free
618.04	Aluminum silicon, per pound	2.3	2.3	2.2	2.2	2.1
618.06	Other aluminum alloys, per pound	0.5¢	0.3¢	0.2¢	0.1	free
618.10	Aluminum waste and scrap, per pound	2.0	2.0	2.0	2.0	2.0

Sources: Customs Tariff, Revenue Canada; Tariff Schedules of the United States Annotated 1983, USITC Publication 1317; U.S. Federal Register Vol. 44, No. 241.

[1] Pending passage by Parliament of the Notice of Ways and Means Motion tabled on November 12, 1981.

TABLE 1. CANADA, ALUMINUM PRODUCTION AND TRADE, 1982-84

	1982		1983P		1984	
	(tonnes)	($000)	(tonnes)	($000)	(tonnes)	($000)
Production	1 064 795	..	1 091 213	..	924 213	..
Imports					(Jan.-Sept. 1984)	
Bauxite ore						
Brazil	1 316 216	49,561	1 263 507	47,225	1 071 124	42,065
Guinea	762 663	23,735	614 095	19,263	142 535	7,589
Guyana	387 973	13,617	337 482	11,574	415 156	14,111
Surinam	66 903	7,462	57 178	7,363	4 057	644
United States	20 327	3,618	24 829	4,499	33 206	5,544
Australia	17 623	1,726	17 923	1,845	44 653	5,933
People's Republic of China	3 057	409	14 803	900	23 669	2,204
Other countries	-	-	93	11	14 477	527
Total	2 574 762	100,128	2 329 910	92,680	1 748 877	78,616
Alumina						
Jamaica	391 815	112,177	423 782	93,542	404 545	88,831
Japan	194 368	48,561	261 340	57,705	210 543	49,544
Australia	257 481	60,190	256 852	54,308	217 622	50,048
West Germany	56	30	108 186	29 026	110 260	35,269
United States	95 562	32,281	12 822	6,133	30 088	11,641
Other countries	-	-	199	135	26 057	4,671
Total	939 282	253,239	1 063 181	240,849	999 115	240,004
Aluminum and aluminum alloy scrap	36 757	31,758	54 666	53,984	48 579	51,880
Aluminum paste and aluminum powder	1 675	4,725	1 625	5,873	1 492	5,768
Pigs, ingots, shot, slabs, billets, blooms and extruded wire bars	24 379	40,971	30 581	55,361	34 427	68,546
Castings	1 129	10,080	729	8,956	709	10,068
Forgings	616	10,931	456	7,187	577	9,999
Bars and rods, nes	3 453	9,617	3 250	10,046	4 959	15,337
Plates	5 930	18,926	6 010	17,906	7 234	24,410
Sheet and strip up to .025 inch thick	13 241	37,903	18 894	54,335	12 669	42,124
Sheet and strip over .025 inch up to .051 inch thick	7 629	23,777	12 356	37,693	10 731	38,769
Sheet and strip over .051 inch up to .125 inch thick	34 702	79,493	44 922	100,942	71 751	183,826
Sheet over .125 inch thick	27 957	62,796	27 618	61,141	29 669	81,212
Foil or leaf	501	1,661	666	2,248	683	2,701
Converted aluminum foil	..	10,398	..	11,169		11,495
Structural shapes	1 656	7,120	2 595	9,775	2 119	8,915
Pipe and tubing	1 160	5,175	1 430	6,267	1 179	5,791
Wire and cable, not insulated	7 295	4,921	1 459	4,414	1 019	3,725
Aluminum and aluminum alloy fabricated materials, nes	..	48,536	..	56,093		51,199
Total aluminum imports	..	408,788	..	503,390		615,765
Exports						
Pigs, ingots, shot, slabs, billets, blooms and extruded wire bars						
United States	418 662	658,945	581 123	993,205	470 084	934,452
Japan	161 160	208,737	140 351	218,739	82 638	148,839
People's Republic of China	168 017	190,223	98 989	139,105	29 996	50,740
Hong Kong	46 323	64,488	22 492	38,154	4 861	9,300
Thailand	6 264	9,267	16 992	28,799	5 899	12,172
West Germany	7 820	12,104	11 816	22,452	15 640	32,775
Malaysia	3 119	4,555	11 059	19,629	5 142	10,418
Israel	5 758	8,705	6 706	11,566	3 903	8,293
South Korea	5 269	7,861	5 808	10,644	95	208
Taiwan	11 992	14,616	4 568	6,743	1 999	3,542
Colombia	3 600	4,887	3 998	6,711	1 993	3,877
Other countries	58 381	85,387	21 501	34,676	6 606	13,691
Total	896 365	1,269,775	925 403	1,530,423	628 856	1,228,307

6.8

TABLE 1. (cont'd)

	1982		1983P		1984 (Jan.-Sept.)	
	(tonnes)	($000)	(tonnes)	($000)	(tonnes)	($000)
Exports (cont'd)						
Castings and forgings						
United States	4 870	39,055	7 140	52,173	5 865	47,862
Total	5 241	50,522	7 692	64,404	6 369	56,446
Bars, rods, plates, sheets and circles						
United States	20 901	54,968	38 671	97,816	55 024	157,165
Total	25 456	66,698	45 365	111,132	58 838	168,155
Foil						
United States	612	2,000	1 337	4,495	1 253	4,950
Total	964	3,184	1 443	4,895	1 264	5,002
Fabricated materials, nes						
United States	7 667	29,453	8 843	27,999	8 493	29,482
Total	10 540	38,185	10 606	33,371	10 072	35,365
Ores and concentrates						
United States	23 000	10,041	40 347	17,804	36 287	16,869
Total	27 746	13,142	44 993	20,427	39 461	18,889
Scrap						
United States	53 394	56,779	71 925	84,859	71 707	98,474
Total	62 610	64,959	80 911	95,572	75 337	103,209
Total aluminum exports	..	1,506,465	..	1,860,224	..	1,615,373

Sources: Statistics Canada; Energy, Mines and Resources Canada.
P Preliminary; - Nil; .. Not available; nes Not elsewhere specified.

TABLE 2. CANADA, CONSUMPTION OF ALUMINUM AT FIRST PROCESSING STAGE, 1980-82

	1980	1981	1982
		(tonnes)	
Castings			
Sand	1 788	1 397	1 241
Permanent mould	8 500	9 358	9 541
Die	20 452	18 777	19 629
Other	135	-	-
Total	30 875	29 532	30 411
Wrought products			
Extrusions, including tubing	94 129	89 057	70 116
Sheet, plate, coil and foil	112 890	138 905	99 633
Other wrought products (including rod, forgings and slugs)	83 001	71 210	67 638
Total	290 020	299 172	237 387
Other uses			
Destructive uses (deoxidizer), non-aluminum base alloys, powder and paste	8 505	8 285	5 725
Total consumed	329 400	336 989	273 523
Secondary aluminum[1]	39 723	48 453	35 938

	Metal entering plant			On hand December 31		
	1980	1981	1982	1980[2]	1981[3]	1982[3]
Primary aluminum ingot and alloys	297 515	292 100	225 156	92 659	83 088	78 191
Secondary aluminum	27 691	31 791	35 255	3 447	1 859	2 090
Scrap originating outside plant	42 166	46 305	44 271	16 037	1 596	1 483
Total	367 372	370 196	304 682	112 143	86 543	81 764

[1] Secondary metal totals not included in above consumptions. [2] Derived from opening inventory plus metal entering plant minus consumption, minus secondary aluminum. [3] Actual numbers reported by consumers.
- Nil.

TABLE 3. CANADA, ALUMINUM SMELTER CAPACITY

(as of December 31, 1984)

	Annual tonnes
Aluminum Company of Canada, Limited	
Quebec	
Grande Baie	171 000
Jonquière	432 000
Isle-Maligne	73 000
Shawinigan	84 000
Beauharnois	47 000
British Columbia	
Kitimat	268 000
Total Alcan capacity	1 075 000
Canadian Reynolds Metals Company, Limited	
Quebec	
Baie Comeau	158 760
Total Canadian capacity	1 233 760

Source: Compiled from company reports by Energy, Mines and Resources Canada.

TABLE 4. ESTIMATED WORLD PRODUCTION OF BAUXITE, 1982 AND 1983

	1982	1983P
	(million tonnes)	
Australia	23.6	24.5
Guinea	11.8	13.0
Jamaica	8.2	7.7
Brazil	4.2	5.2
Surinam	3.3	3.0
Greece	2.9	2.4
India	1.9	1.8
Guyana	1.8	1.1
France	1.7	1.7
Other market economy countries	7.4	6.6
Total market economy countries	66.8	67.0
Central economy countries[1]	15.0	15.0
World total	78.1	78.6

Source: World Bureau of Metal Statistics.
[1] Includes Yugoslavia.
P Preliminary.

TABLE 5. ESTIMATED NON-COMMUNIST WORLD PRODUCTION OF ALUMINA, 1982 AND 1983

	1982	1983	1st Qtr 1984	2nd Qtr 1984	3rd Qtr 1984
	(million tonnes)				
Europe[1]	4.46	4.35	1.24	1.34	1.35
Africa	0.58	0.56	0.13	0.13	0.14
Asia	1.81	1.89	0.50	0.53	0.53
North America	5.27	5.07	1.40	1.46	1.51
South America	3.48	4.17	1.07	1.05	1.20
Australasia	6.63	7.31	2.00	2.14	2.32
Total	22.23	23.35	6.34	6.66	7.04
of which nonmetallic uses	1.97	2.06	0.53	0.62	0.59

Source: International Primary Aluminum Institute.
[1] Excludes Yugoslavia.

TABLE 6. WORLD PRIMARY ALUMINUM PRODUCTION AND CONSUMPTION, 1982 AND 1983

	Production		Consumption	
	1982	1983P	1982	1983P
	(000 tonnes)			
United States	3 274.0	3 353.0	3 648.0	4 209.9
Europe[1]	3 309.3	3 311.3	3 501.0	3 589.5
Japan	350.7	255.9	1 636.8	1 801.2
Canada	1 064.8	1 091.2	273.5	337.5
Australia and New Zealand	548.1	695.2	242.1	221.7
Asia (excluding Japan and People's Republic of China)	675.1	717.3	774.6	775.6
Africa	501.2	433.4	187.4	170.1
America (excluding United States and Canada)	796.1	944.5	521.8	514.6
Sub-total	10 519.3	10 801.8	10 785.2	11 620.1
Central economy countries	3 437.4	3 486.1	3 501.4	3 528.3
Total	13 956.7	14 287.9	14 286.6	15 148.4

Sources: American Bureau of Metal Statistics; Energy, Mines and Resources Canada.
[1] Excludes Yugoslavia.
P Preliminary.

Antimony

J. BIGAUSKAS

INTRODUCTION

Antimony is a silver-white brittle metal generally found as the sulphide mineral stibnite Sb_2S_3, or its oxidized equivalents but may also be present in trace amounts in lead ores, or in association with gold, tungsten and silver ores. High-grade lump ores are separated by hand cobbing - a relatively labour intensive method - while more disseminated ores are selectively processed by flotation and sold as concentrates typically containing 60 per cent antimony by weight. Antimony in antimonial lead alloys is also widely recycled from lead-acid battery scrap.

Major mine producers in the non-socialist world are Bolivia and the Republic of South Africa, while major producers in the socialist world are the People's Republic of China and the U.S.S.R.

CANADIAN DEVELOPMENTS

Since 1981 primary antimony output in Canada has been mostly a byproduct of the refining of lead. Cominco Ltd., which operates a lead smelter and refinery at Trail, British Columbia, produces a primary antimonial-lead product. Trail produces most of its antimonial lead from lead concentrates obtained from Cominco's Sullivan mine at Kimberley, British Columbia. Other sources are lead-silver ores and concentrates shipped to Trail by custom shippers. The lead bullion produced from smelting of these ores and concentrates contains a small amount of antimony which subsequently collects either in anode residues from the electrolytic refining of the lead bullion or in furnace drosses. These residues and drosses are treated to yield antimonial-lead alloy, to which refined lead may be added to produce marketable products of the required grade.

Slag containing antimony is removed by oxidation from lead bullion at Brunswick Mining and Smelting Corporation's pyrometallurgical lead refinery at Belledune, New Brunswick. The slag, which also contains arsenic, is skimmed off and fed into an antimony reverberatory furnace along with some coke. From this, lead bullion is returned to the circuit while antimony - containing slag is stockpiled for sale. This product is sold without further upgrading. Brunswick's No. 12 zinc-lead-silver-copper mine near Bathurst, New Brunswick provides lead concentrate feed which contains a small amount of antimony.

Production of primary antimony in Canada totalled 510 t in 1984, some 25 t more than in 1983. Antimony contained in antimonial lead is also recycled by secondary lead smelters which process scrap lead-acid batteries.

Operations at the original orebody of Consolidated Durham Mines & Resources Limited (renamed Durham Resources Inc. in 1984) near Fredericton, New Brunswick ceased in May 1981 when antimony reserves in the Hibbard orebody were exhausted. The company produced antimony concentrates which were of premium quality and found ready markets in the United States and Europe. An extensive diamond drilling program in 1980 and 1981 to search for more ore at deeper horizons outlined an antimony-bearing zone containing an estimated 774 000 t averaging 4.15 per cent antimony. This zone appears to be the downward extension of the mined zone.

Durham completed dewatering of the mine in September and began deepening the inclined shaft from 240 m to 430 m. The rehabilitation of the mine is expected to cost $Cdn 3.7 million. Production, which is targeted for the third-quarter of 1985, will be about 360-390 tpd. A feasibility study is being conducted to evaluate the possibility of adding a roaster-smelter operation at a capital cost of $Cdn 3-4 million.

Canadian consumption of antimony in 1983 was 217 t. In 1984, consumption is expected to be slightly lower.

J. Bigauskas is with the Mineral Policy Sector, Energy, Mines and Resources Canada. Telephone (613) 995-9466.

WORLD DEVELOPMENTS

Major producers of antimony ores and concentrates are Bolivia and the Republic of South Africa, with some 40 and 25 per cent of western world production, respectively. Other producers include Mexico, Thailand, Turkey, Yugoslavia, Austria and the United States which collectively account for another 30 per cent of mine production in the western world. Large quantities of antimony metal are also exported by the People's Republic of China to consumers around the world, including other socialist nations. The Xikuangshan mine in Hunan province is the largest antimony producer. Antimony metal in three grades is supplied as well as antimony oxides and crude oxides.

In 1983, western world mine production was 26 235 t antimony content (World Metal Statistics). Dropping production in Bolivia during 1984 is expected to continue the trend toward declining western world mine production since 1980. An estimated 24 000 t was produced in 1984.

In Bolivia widespread strikes and difficulties in obtaining foreign currency were cited as reasons for a decline in mine output by Bolivia's major producer, Empresa Minera Unificado S.A. (EMUSA). Estimated production in 1984 was expected to be 30 per cent lower than in 1983. The company re-opened its Caracota mine temporarily but in June it was reported that the mine would be closed for at least three months to deepen the shaft and develop reserves.

Empresa Mineral Bernal Hermanos proceeded with a planned expansion of its antimony smelter near Tupiza. Four more rotary furnaces were to be installed in 1984 along with a new cupel furnace for fuming metal to oxide. The new equipment will upgrade crude oxide to higher quality products. The expansion of the Hermanos smelter also required the development of the 1 500 tpy Rosa de Oro mine.

The state-owned smelter of Empresa Nacional de Fundiciones (ENAF) was expected to raise overall production of antimony metal and oxides by 2 000 t in 1984.

As a result of Bolivian developments at the mining, smelting and refining stages, exports of antimony in concentrates were expected to decline to 8 500 to 9 000 t in 1984 and to 7 200 t or less in 1985. Production and sales of antimony metal and oxides are expected to increase.

A lead-silver smelter employing Kivcet technology built at Karachipampa in Potosi did not operate during 1984 due to lack of feed. The joint venture project by Comibol (Corporacion Minera de Bolivia) and ENAF was designed with a capacity of 24 000 tpy of lead, 2 000 tpy of antimony with additional amounts of zinc, silver and tin.

With the tightening of Bolivian concentrate and ore exports in 1984 the non-socialist world's second largest producer, the Republic of South Africa, found more opportunities to export its own products. Sales in the first half of 1984 depleted Consolidated Murchison Ltd.'s (CML) stocks of concentrate. Current production - about 20 000 tpy of concentrate - was expected to supply remaining demand. Generally most of CML's mine output is processed at Antimony Products (Pty.) Ltd.'s antimony trioxide plant.

A new producer, Cia Minera Norcro, commenced production of oxide ore and lump sulphides from an open-pit located in the Honduras near Santa Rita. Production began in January. Full capacity (300 tpm) is not expected to be reached until mid-1985 or 1986. The ore is processed further to antimony oxide by Anzon America Inc. in the United States.

As part of its usual production cycle, Metallurgie Hoboken-Overpelt SA/NV of Belgium resumed production of antimony metal from September until January or February 1985. Hoboken's lead smelter/antimony plant has the capacity to produce 4 000 tpy of antimony mostly from imported complex lead and lead/copper concentrates and intermediate residues and slags which contain antimony. Pennarroya S.A. produces sodium antimonate and antimony oxide as byproducts from its lead smelting operation in France. A large proportion of antimony in Europe is recycled by secondary lead smelters from lead-acid battery scrap. Battery alloys on average contain higher amounts of antimony in Europe than in North America where calcium-lead alloys now dominate the automotive battery market.

In Japan most antimony in metallic form is produced as a byproduct of lead bullion refining or recycled in lead-acid battery alloy metal. Mine production of antimony ores ceased in 1971, and as a result production of antimony metal dropped sharply. Japan relies primarily upon the People's Republic of China for metal and Bolivia for antimony concentrates. Antimony trioxide is then produced from these materials at several plants.

Although the United States is not a large mine producer of antimony, production of antimony trioxide and antimony salts from raw materials, and recycling of antimony in antimonial lead are very significant. ASARCO Incorporated produces antimony oxide at its Omaha lead refinery. The refinery is fed by two smelters which process imported and domestic complex lead concentrates and other materials. Recycling of battery scrap accounts for 50 per cent of apparent United States domestic antimony consumption and some 90 per cent of secondary antimony is recovered as antimonial lead. Antimony metal is imported mainly from Bolivia and the People's Republic of China. Ores and concentrates are mostly from Bolivia and Mexico, and oxides are purchased mainly from the Republic of South Africa.

Beginning December 6, 1984 the United States General Services Administration (GSA) commenced monthly domestic sales of a maximum of 135 t of antimony metal from the strategic stockpile. In 1983, 900 t was offered at one auction. The staggered sale of antimony in 1984 was planned to minimize effects on the market.

PRICES

The New York Dealer price for antimony metal as quoted by "Metals Week" rose from a monthly average of $US 1.26 per pound in January to $1.65 per pound in September and dropped to $1.40 per pound average in December, 1984. The annual average price in 1984 was $US 1.51 per pound, much higher than 1983s $0.91 average.

The price of clean concentrates and lump sulphides (60 per cent Sb) also as reported in "Metal Bulletin", rose from $US 16.50-17.25 per t Sb and $16.75-17.50 per t to $US 27.75-29.00 and $28.75-31.00 respectively, by the end of the year.

USES

Antimony is used principally in alloyed form and in the form of oxides. Antimony hardens and strengthens lead and inhibits chemical corrosion. These characteristics created a large use for the metal in antimonial-lead storage batteries. However, the introduction of the low antimonial-lead battery and the lead-calcium battery has reduced this application particularly in North America. Antimonial-lead alloys are also used for power transmission and communications cable sheathing, type metal, solder, ammunition, chemical pump and pipe linings, tank linings, sheets and antifriction bearings.

Antimony trioxide, pentoxide or oxychloride, and sodium antimonate are used as fire retardants in plastics, textiles and rubber. This end use is now the largest for antimony.

Antimony trioxide or sodium antimonate may also be used as refining and decolouring agents in certain types of glass. Sodium antimonate is normally used in the manufacture of television screens. Antimony trioxide is also used in the manufacturing of white pigments. Pentasulphide (Sb_2S_5) is a vulcanizing agent for the production of red rubber compounds. Burning antimony sulphide creates a dense white smoke for visual control in sea markers, in signaling and in fireworks.

High-purity metal is used in the production of indium antimonide and aluminum antimonide intermetallic materials for semiconductors.

OUTLOOK

Growth in the antimony-based pigment and fire-retardant markets is projected by manufacturers at 4 to 6 per cent annually in the next three to four years. On the other hand, demand for antimony metal in lead-acid batteries is expected to stagnate as battery alloy compositions in North America, Europe and Japan continue to shift, as a whole, to calcium-based alloys. Despite the poor outlook for the battery market, one Bolivian producer estimates that demand for antimony in general will continue to grow at 3 to 4 per cent per annum.

Nevertheless, with healthy growth in consumption and with the continuing trend of further processing by major mine producers such as Bolivia, demand for ores and concentrates and other materials may improve even more. The re-opening of Canada's sole antimony ore producer will likely depend upon favourable conditions in this segment of the market.

TARIFFS

Item No.		British Preferential	General Preferential	Most Favoured Nation	General	
			%			
CANADA						
33000-1	Antimony or regulus of, not ground, pulverized or otherwise manufactured	free	free	free	free	
33502-1	Antimony oxides	free	free	6.3	25	
MFN Reductions under GATT (effective January 1 of year given)			1983	1984	1985	1986	1987
				%			
33502-1			6.3	4.7	3.1	1.6	free
UNITES STATES (MFN)			1983	1984	1985	1986	1987
				(¢ per pound)			
601.03	Antimony ore		Remains free				
632.02	Antimony metal unwrought etc. (Duty on waste and scrap temporarily suspended)		0.5	0.4	0.3	0.1	free
EUROPEAN ECONOMIC COMMUNITY (MFN)			1983				
26.01	Antimony ore		free				
81.04	1. Antimony unwrought waste and scrap		free				
	2. Antimony, other		8				

Sources: The Customs Tariff 1983, Revenue Canada - Customs and Excise Schedules of the United States Annotated 1983, USITC Publication 1317; U.S. Federal Register Vol. 44, No. 241; Official Journal of the European Communities, Vol. 25, No. L318, 1982.

TABLE 1. CANADA, ANTIMONY PRODUCTION AND IMPORTS, 1983 AND 1984 AND CONSUMPTION, 1982 AND 1983

	1983		1984P	
	(kilograms)	($000)	(kilograms)	($000)
Production				
Ontario	..	72	..	..
New Brunswick	..	589	..	827
British Columbia	..	1,432	..	2,187
Total	..	2,093	510 000	3,014
Imports				
			(Jan. - Sept.)	
Antimony oxide				
United Kingdom	576 000	2,132	576 000	2,243
United States	263 000	979	253 000	1,115
Belgium-Luxembourg	141 000	410	93 000	387
France	21 000	48	20 000	74
Total	1 001 000	3,570	1 001 000	3,819

	1982	1983P
	(kilograms)	
Consumption[1]		
Antimony metal used for, or in the production of:		
Antimonial lead	54 754	80 561
Babbit	12 329	14 694
Type metal	5 686	6 843
Solder	4 765	5 952
Other commodities	83 500	109 302
Total	161 034	217 352
Held by consumers on December 31[2]	39 799	26 106

Sources: Statistics Canada; Energy, Mines and Resources Canada.
[1] Antimony content of primary and secondary antimonial-lead alloys. [2] Available data, as reported by consumers.
P Preliminary; .. Not available due to confidentiality.

TABLE 2. CANADA, CONSUMPTION AND CONSUMERS' STOCKS OF ANTIMONY[1], 1970, 1975, AND 1978-83

	Consumption		On hand at end of year	
	Antimony metal	Antimonial-lead alloy[2]	Antimony metal	Antimonial-lead alloy[2]
	(kilograms)			
1970	518 007	635 212	131 501	91 563
1975	454 164	723 155	116 760	170 478
1978	347 906	1 000 732	101 814	91 049
1979	463 423	931 990	39 976	87 473
1980	369 732	643 983	42 389	51 405
1981	209 829	691 180	35 105	151 400
1982	161 034	605 502r	39 799	76 979
1983	217 352	560 705	26 106	130 104

Sources: Statistics Canada; Energy, Mines and Resources, Canada.
[1] Available data, as reported by consumers. [2] Antimony content of primary and secondary antimonial-lead alloys.
r Revised.

TABLE 3. WORLD MINE PRODUCTION OF ANTIMONY, 1982-84

	1982	1983	1984 (Jan-June)
	(tonnes)		
Europe			
Austria	670	970	394
France	332	111	-
Italy	339	-	-
Spain	461	489	247
Yugoslavia	1 517	950	576
Africa			
Morocco	845	454	226
South Africa	9 134	6 302	3 152
Zimbabwe	233	143	72
Asia			
Malaysia	144	133	66
Thailand	972	1 740	870
Turkey	1 237	1 059	530
America			
Bolivia	13 612	9 951	..
Mexico	1 565	2 519	1 258
Peru	724	375	186
United States	456	760	..
Other	961	964	..
Australia			
Australia	1 146	528	264
Total	33 408	26 235	..
Other			
China, P.R.	12 000	13 000	..
Czechoslovakia	700	900	..
U.S.S.R.	6 500	6 500	..
Other Europe	500	500	..
World Total	53 108	47 135	..

Sources: World Metal Statistics
- Nil. .. Not available.

TABLE 4. INDUSTRIAL CONSUMPTION OF PRIMARY ANTIMONY IN THE UNITED STATES BY PRODUCT PRODUCED, 1982 AND 1983

	1982	1983
	(tonnes)	
Metal Products		
Ammunition	267	159
Antimonial lead	719	835
Bearing metal and bearings	130	130
Cable covering	23	28
Castings	8	8
Collapsible tubes and foil	1	W
Sheet and pipe	24	39
Solder	112	140
Type metal	10	9
Other	61	64
Non-Metal Products		
Ammunition primers	18	15
Fireworks	5	4
Flame-retardants	4 385	5 628
Ceramics & glass	1 232	1 136
Pigments	299	180
Plastics	953	1 318
Rubber products	200	64
Other	93	108
Total	7 185	9 864

Source: U.S. Bureau of Mines.
W - Withheld, confidential.

TABLE 5. ANTIMONY PRICES, N.Y. DEALER[1]

	1982	1983	1984
	($US/lb)		
January	1.18	0.91	1.26
February	1.12	0.95	1.24
March	1.11	0.99	1.52
April	1.18	0.94	1.56
May	1.17	0.95	1.62
June	1.10	0.88	1.56
July	1.07	0.80	1.52
August	1.03	0.80	1.60
September	1.00	0.79	1.65
October	0.99	0.81	1.65
November	0.96	0.95	1.56
December	0.93	1.18	1.40
	1.07	0.91	1.51

Source: Metals Week.
[1] 99.5 - 99.6% metal, cif U.S. ports, 5 ton lots, duty paid.

Arsenic

D.G. LAW-WEST

Arsenic occurs as a minor constituent of complex ores which are mined primarily for their copper, lead, zinc, silver and gold content, although globally, copper ores are the main source of arsenic. It is collected in the form of impure arsenic trioxide in dusts and residues from the roasting of these ores. Ninety-six per cent of arsenic is consumed as arsenic trioxide and other arsenic compounds. Only about 4 per cent is consumed as metallic arsenic. In the literature, arsenic trioxide is commonly referred to as arsenic.

Prior to 1983 a limited supply and increasing demand from the agricultural and wood preservative sectors created a tight market for arsenic trioxide and consumers were placed on allocation. In 1983, demand for arsenic trioxide dropped sharply due to the reduction in consumption for agricultural usage. The resulting oversupply forced prices down nearly 20 per cent to $US 0.73 per kilogram from $US 0.88 in 1982. During 1984 arsenic prices have steadied somewhat, at about $US 0.70.

CANADIAN DEVELOPMENTS

Canadian arsenic production is mainly obtained from the treatment of arsenious gold ores.

Campbell Red Lake Mines Limited in the Red Lake District of Ontario and Giant Yellowknife Mines Limited in the Northwest Territories, recover impure arsenic trioxide from dust and residues collected during the roasting of gold ores.

Both these operations use similar recovery technology, including electrostatic precipitation of dust, cooling of the arsenic-containing gases and collection of arsenic trioxide in a baghouse.

Giant Yellowknife, which had been storing arsenic containing residues underground, started shipping its residue production to a United States customer in 1981. The first year shipments amounted to 1 205 t, followed by 1 500 t in 1982 and 800 t in 1983. The decrease during 1983 reflected the drop in demand for arsenic in the agricultural sector. The company is currently considering plans to recover the arsenic residue from the underground storage areas. Campbell Red Lake also ships its arsenic residue to the United States.

Cominco Ltd. produced arsenic trioxide at its Con mine in the Northwest Territories until 1970, after which, gold ore being mined no longer contained arsenopyrite. The arsenic residue produced to that date had been stored in large ponds which were estimated to contain some 22 600 t of arsenic trioxide, as well as 400 kg of gold and 1 000 kg of silver. In 1981 Cominco began construction of a $13 million arsenic trioxide recovery plant which would use the stored residues as feed material and would permit the recovery of the contained gold and silver. The plant began start up operations in late 1983 and production was intermittent through 1984. Problems were encountered due to inconsistency of the feed material. The plant has a rated production capacity of 15 tpd of 99.5 per cent arsenic trioxide, most of which will be exported.

Cominco, through its Electronic Materials Division, produces high purity gallium arsenide at its Trail British Columbia plant. The company first produced gallium arsenide in 1981. Growth in demand for this product led to a plant expansion which was completed in 1983, and in 1984 further expansion of production facilities was started.

INTERNATIONAL DEVELOPMENTS

World production of arsenic trioxide for the period 1970 to 1983 is shown in Table 1.

Two companies, Boliden Aktiebolag in Sweden and ASARCO Incorporated in the United States, are the major producers of

---

D.G. Law-West is with the Mineral Policy Sector, Energy, Mines and Resources, Canada. Telephone (613) 995-9466.

metallic arsenic in the non-communist world. A few countries produce minor amounts of high purity metallic arsenic for use in the electronics industry.

Boliden is the non-communist world's largest producer of metallic arsenic with a present capacity of 1 600 t. Boliden is also of the world's largest producer of arsenic trioxide.

In the United States, ASARCO recovers arsenic trioxide from its copper smelter at Tacoma, Washington. The arsenic plant also produce metallic arsenic. In mid-1984 ASARCO announced that its copper smelter at Tacoma Washington would close by mid-1985, however, the company plans to continue operating the arsenic plant at a much reduced rate. About three quarters of the United States refinery production of arsenic trioxide comes from imported base-metal concentrates and impure arsenic trioxide and the remaining portion comes from domestic sources.

The United States is a major consumer of arsenic trioxide accounting for about 50 per cent of the western world's total consumption. In 1983 the U.S. imported some 10 200 t of arsenic trioxide for domestic consumption mainly from Sweden, Canada and Mexico.

Recently the People's Republic of China has become a large exporter of arsenic metal to the U.S. China increased its exports from 31 t in 1982 to 128 t in 1983. Sweden exported at 108 t to the U.S. in 1983.

The Philippines may soon become a major supplier of arsenic when the arsenic treatment plant, part of the Philippine Associated Smelting and Refinery Corp. (PASAR) $17 million roaster complex, is completed. The plant scheduled for start-up in early 1985 is expected to handle 5 000 tpm of high arsenic concentrate.

The El Indio Mine in Chile, started up its arsenic trioxide plant late in 1983 and after some start-up problems is producing some 300-400 t per month of 97 % arsenic trioxide.

USES

Arsenic trioxide is the basic raw material for the production of metallic arsenic, arsenic alloys and other compounds, both organic and inorganic. The toxicity of arsenic containing compounds ranges from low to extremely high, depending on the chemical state. All must be handled with care. Environmental issues are an important factor in the recovery and use of arsenic and its compounds, and regulations on emissions to the atmosphere and on uses are being established by many countries.

Arsenic metal is used as an alloy, mainly with lead and copper, to improve strength, corrosion resistance, and other physical and chemical properties. A growing use for high-purity arsenic metal is in arsenides of gallium, indium and other metals used in electronics.

A major use of arsenic trioxide is in the preparation of compounds for use in agricultural herbicides and dessicants. The main applications are for weed control in tropical and subtropical climates, as defoliants in cotton cultivation and as weed killers in citrus orchards and rubber plantations.

Demand for arsenic compounds as wood preservatives has increased sharply in the last few years, and is expected to continue to grow, particularly if the arsenic compounds remain relatively cheap.

There are a number of minor uses for arsenic. One is in the glass industry, where arsenic trioxide is used as a decolorizing and refining agent.

ARSENIC PRICE QUOTATIONS

	Metal 99 % As	Trioxide Mexican 99,13 % As_2O_3 fob Laredo	Trioxide 95 % As_2O_s fob Tacoma
	cents US/kg		
	-	2	4
1970	-	13	10.4
1975	-	36	no quote
1980	-	101	90
1981	606	172	88
1982	540	130	88
1983	496	99	73
1984	-	93	70

Source: USBM and Metals Week

OUTLOOK

The outlook for 1985 is somewhat uncertain. The reduced output of arsenic trioxide for the ASARCO plant could create some tightness in the market. However, this could be largely offset by increased supply from the People's Republic of China. As well, both the Philippines and Chile are expected to start up arsenic trioxide production facilities. Should the demand for arsenic trioxide from the agricultural sector again be weak, oversupply could occur and prices would be put under pressure. Demand for high purity arsenic metal is expected to continue to increase.

TABLE 1. WORLD PRODUCTION OF ARSENIC TRIOXIDE (WHITE ARSENIC) 1970-1983

	1970	1975	1980	1981	1982	1983
	(tonnes)					
U.S.S.R.	7 149	7 348	7 711	7 700	7 900	8 000
Mexico	9 140	6 121	6 532	6 500	4 800	5 000
France	10 193	8 165	5 262	5 300	5 200	5 000
Sweden	16 400	15 967	4 082	4 100	4 000	4 000
Peru	772	1 325	2 540	2 500	2 200	2 200
Namibia	4 062	6 663	1 996	2 000	1 900	2 000
Other countries[1e]	16 223	14 340	6 262	10 600	9 000	8 000
Total	63 939	59 929	34 385	38 700	45 293	39 883

Sources: U.S. Bureau of Mines, Yearbook 1970-1980; Mineral Commodity Summary 1982, 1983, 1984.
[1] Includes the United States but not the Peoples' Republic of China; United States production figures are withheld by the United States to avoid disclosing individual company data; estimates for the years 1970-1975 from Roskill Information Services Ltd.; for the years 1976-1981 estimated by Nonferrous Section, Mineral Supply Branch, Energy, Mines and Resources Canada.
[e] Estimated.

Asbestos

G.O. VAGT

Shipments of asbestos (chrysotile) in 1983-84 remained weak as a result of worldwide recessionary conditions particularly in the construction industry, foreign exchange shortages in the developing countries, uncertainties regarding future environmental regulations, and adverse publicity associated with past exposure to asbestos dust in the workplace. Total shipments in 1983 were 857 504 t valued at $391.3 million and in 1984 were 836 000 t valued at $413.0 million, according to preliminary figures.

The entire asbestos industry is undergoing the most critical period in its history. Since 1981, greatly reduced mine production, coupled with high inventories of more than 150 000 t, have resulted in shortened work periods, layoffs, prolonged shutdowns, and the closure of one operation. Employment in the mining sector of this industry has been depressed to under 4,000 from about 8,000 in 1979. Exports, generally accounting for about 95 per cent of production, amounted to 604 000 t valued at $378 million during the first 9 months of 1984, compared to 551 000 t valued at $327 million during the same period in 1983.

Canada recognizes that asbestos dust in the workplace is potentially hazardous. However, it takes the position that via enforcement of appropriate regulations to rigorously control exposure, the use of chrysotile in most activities from mining through manufacturing does not pose undue risk to workers or the public. This position is generally characteristic of the mainstream international approach except in the United States where the Environmental Protection Agency plans to propose regulations to ban key asbestos-containing products and to ultimately phase out all remaining asbestos uses.

CANADIAN DEVELOPMENTS

Companies in the industry are operating at about 55-60 per cent of capacity overall, with individual operations ranging from 35-40 per cent to 80 per cent. Quebec accounts for 85-90 per cent of total output.

Asbestos Corporation Limited (ACL) discontinued mining in May, 1983 at its seasonal operation in Ungava. During the past 10 years approximately 2 million t of concentrate (about 700 000 t of asbestos) were shipped from Asbestos Hill to Nordenham, Federal Republic of Germany, for final milling. The closure will reportedly help preserve jobs at the company's operations in Thetford Mines-Black Lake. Most of the assets in the Federal Republic of Germany were sold in mid-1984. ACL and Bell Asbestos Mines held discussions initially in mid-1984 to discuss consolidating operations. Losses at ACL amounted to $68 million from 1981 to the first six months of 1984 inclusive.

Johns-Manville Canada Inc. sold its Jeffrey operation to a group of private investors. The $117 million sale, approved in September 1983, was part of a reorganization plan submitted by the parent Manville Corporation in 1982 when it filed for protection under Chapter XI of the U.S. Bankruptcy Code. The former parent is now essentially removed from the asbestos business.

A feasibility study for Brinco Mining Limited concluded that an extension of the present Mt. McDame orebody is mineable by underground methods. A commitment to proceed with development would be required by mid-1985 if development is to be fully completed when the open-pit reserves are exhausted by about 1991.

Baie Verte Mines Inc., a new company owned by Transpacific Asbestos Inc., of Toronto, has been successfully marketing asbestos mainly in Pacific Rim countries. This operation, formerly Advocate Mines Limited, was rejuvenated following expropriation by the Government of Newfoundland and the availability of government loans and loan guarantees. Shipping and cash-flow were cited recently as problems leading

---

G.O. Vagt is with the Mineral Policy Sector, Energy, Mines and Resources, Canada. Telephone (613) 995-9466.

to a decision in late-1984 to shut down operations for one month extending into 1985.

The Société nationale de l'amiante (SNA) continued its research and development on new products and processes involving chrysotile. Substantial efforts have been directed toward modifying the surface characteristics of asbestos, or other materials, to attenuate physiological response. Present work involves treatment of the fibres with phosphorous oxychloride ($POCl_3$) to fix magnesium phosphate having low solubility and high thermal stability. The resulting product is called chrysophosphate.

The Asbestos Institute became operational in mid-1984. This industry-managed asbestos research institute, financed jointly by the Canadian asbestos industry and the Governments of Canada and Quebec, will be responsible for product and health research, market development and the dissemination of information on the safe uses of asbestos. Thus, the Institute, with operating funds up to a total of $18.75 million over the next five years, takes over and expands the functions of the Canadian Asbestos Information Centre, the Institute for Research and Development (IRDA) and the Institute for Occupational and Environmental Health (IOEH).

HEALTH AND REGULATIONS

The Royal Commission on Matters of Health and Safety Arising from the Use of Asbestos in Ontario (ORCA report) issued its comprehensive three-volume report in May 1984. Some important highlights of the report are:

1) Distinction between types of fibres for regulatory purposes, prohibition being recommended for crocidolite and amosite, and chrysotile being subjected to a level of 1.0 fibres/cm^3, which is in fact already in effect in Ontario.

2) There should be a distinction between types of processes when considering regulations for manufacturing plants.

3) There is no evidence of significant health risk to the general public from exposure to asbestos in the ambient air and in buildings. For workers and maintenance people in the immediate vicinity of loose asbestos there is a potential hazard.

4) There is no health risk from ingestion. Concern about asbestos in drinking water, beverages and food is not justified. The Commission recommends lifting the ban on the use of asbestos filters in the production of beer, wine and liquor.

It was decided in 1983 by the Ontario Ministry of Labour to postpone finalization of its proposed asbestos regulation for the construction industry until the ORCA report was reviewed. The present general occupational regulation, following the British standard, establishes different permissible exposure limits according to fibre type. These limits are, 1.0 f/cm^3 for chrysotile, 0.5 f/cm^3 for amosite and 0.2 f/cm^3 for crocidolite as measured over a 40-hour time-weighted period by the membrane filter method.

Federal emission regulations pursuant to the Clean Air Act as defined by Environment Canada require that the concentration of asbestos fibres contained in emissions to the ambient air at a mine or mill from crushing, drying or milling operations, or from dry rock storage, shall not exceed 2 f/cm^3.

WORLD DEVELOPMENTS AND INTERNATIONAL REGULATIONS

Based on an estimated 1983 world production of 4.2 million t of fibre, major producers and their approximate percentage share of production are: U.S.S.R., 54; Canada, 20; Republic of South Africa, 5; and Zimbabwe, 4.5. Canadian asbestos fibre exports account for approximately 65 per cent of total world exports. Expansions to production facilities in Russia are under way, reportedly to serve the needs of industrial and residential construction.

The Occupational Safety and Health Administration (OSHA) issued an Emergency Temporary Standard (ETS) in late 1983 without prior public hearings. This rarely used approach called for an immediate reduction in the permissable level of worker exposure from 2.0 f/cm^3 to 0.5 f/cm^3. Following a court challenge, a unanimous three-judge panel ruled that OSHA did not invoke its ETS powers properly and that the record, as a whole, does not indicate that the risk the ETS sought to eliminate was "grave", as OSHA itself has defined it. OSHA's normal rule-making route then proceeded throughout 1984.

In most countries there has been increasing acceptance for the "controlled use" principle as an approach to asbestos regulation. However, the U.S. Environmental Protection Agency (EPA) announced in late-1983 that it planned to propose a ban in 1984 on certain important asbestos product categories and a "staged production cap" on the manufacturing of the remaining products. The timing was not realized and efforts were continuing at the end of 1984.

There is also much apprehension over the removal of asbestos in American schools. Allotted funds to the end of 1984 for asbestos removal totalled $50 million, however, there is lack of agreement on cost sharing by the levels of government involved. EPA has not yet formulated acceptable procedures for removal, and further, the entire issue is controversial because several authorities have stated that there is no evidence of cancer where there is a low level of exposure in buildings.

Asbestos-related product liability lawsuits alleging past damages continued and now number about 24,000 cases. These are unique to the United States. The American Bar Association in 1983 urged Congress to find a way to handle the suits, stipulating that the problem should not be left any longer to state and federal courts applying the varied laws of 50 states.

In the United Kingdom, the governments Health and Safety Commission recommended in mid-1983 strict new control limits for asbestos in the workplace. New limits, particularly for chrysotile, came as a surprise because the 1.0 f/cm^3 figure was just established early in the year. Effective August 1 1984, limits would be; crocidolite, 0.2 f/cm^3; amosite, 0.2 f/cm^3 and chrysotile, 0.5 f/cm^3. Similarly, beginning June 1, 1984, the import and use of crocidolite and amosite, and products containing them, would formally be banned.

The Council of the European Communities (CEC) approved in 1983 directives on asbestos regulations regarding marketing and use (DG III) and protection in the workplace (DG V). The control limit for exposure to asbestos fibre (DG V), other than crocidolite, over an eight-hour sampling period will be 1.0 f/cm^3. Member states are to adopt laws necessary to comply with the workplace directive before Jan. 1, 1987. In the case of asbestos-mining activities the compliance date is Jan. 1, 1990. For crocidolite, the limit value will be 0.5 f/cm3. The marketing and use of products containing chrysotile is permitted with proper labelling, but prohibitions on the marketing and use of many crocidolite-containing products will apply to products manufactured after Jan. 1, 1986. However, member states may exclude from the prohibition certain products, including fibres for asbestos cement pipe, acid and heat resistant seals, packings and gaskets, and torque converters.

In the Federal Republic of Germany, manufacturers in the fibre-cement industry continued to reduce voluntarily the asbestos content in asbestos-cement products. Also, substitutes for many products, including friction materials, are being promoted aggressively. Sprayed-on applications have been banned since 1979.

In Denmark, the Danish Labor Inspectorate proposed new threshhold limit values that are specific to natural mineral fibres as well as to asbestos. Limits, effective in January 1985 are as follows: for asbestos fibres, excluding crocidolite which is banned, 0.5 f/cm^3, for natural mineral fibres including wollastonite, attapulgite and zeolites, 0.5 f/cm^3; and for synthetic mineral fibres including glass, mineral and slag wool, 1.0 f/cm^3.

A draft code of practice on the safe use of asbestos was approved in October 1983 at an International Labor Office (ILO) meeting. The code, drawn up by experts selected by governments, workers and employers throughout the world, advocates the continued use of asbestos in line with the application of standards and controls that minimize the risk of exposure to asbestos dust in the workplace. The subject code and the safe use of asbestos have been included in the agenda for the 1985 ILO conference and the final instruments are to be presented for ratification at the 1986 Conference.

The World Health Organization largely completed a criteria document on asbestos and man-made mineral fibres in 1984. These types of documents are highly regarded for their objectivity and are particularly relevant to many countries in the process of establishing regulations for fibrous minerals.

The Air Management Policy Group of the Organization for Economic Cooperation and Development (OECD) held meetings to prepare a policy document on asbestos in the ambient air. This document will be reviewed for release in 1985.

Adverse publicity associated with asbestos continues in Australia. According to a recent budget speech, funding has been provided to establish the extent of asbestos usage in Commonwealth buildings.

Woodsreef Mines Limited developed a method of recovering fibre from both ore and mill tailings at its Australian mine using a wet-milling process. The company planned to restart commercial operations in 1985 with a new wet mill. Production using the dry-process ceased in January, 1983.

An industry/government asbestos mission visited Saudi Arabia, Kuwait, United Arab Emirates (Dubai) and Egypt in May, 1984. A Chinese Asbestos Study Group from the Ministry of Geology and Mineral Resources, Beijing, PRC visited the mining region in Quebec during October, 1984.

PRICES AND CONSUMPTION

The average unit prices received for asbestos have not increased since 1981; these in fact have declined in real terms. The general economic recession affecting the construction and automotive sectors has been a major factor depressing demand and increasing competition for available markets. About 70 per cent of production is construction-product related (asbestos/fibre cement, papers, felts, flooring and others), with friction materials accounting for about 10 per cent.

Price competition has led to much discounting because supply has far exceeded demand. Also, in addition to the traditional competitors, Brazil, Greece, and Colombia have recently entered world markets.

Alternate Fibres and Materials

The controversial health issue and stricter regulations on use have resulted in rapid growth in the use of alternate fibres and products. These have made significant inroads even though the cost/performance ratio may be much in favor of asbestos.

In April, 1984, at a two-day symposium on asbestos replacements held in Manchester, England, it was concluded that asbestos remains the most technical- and cost-efficient material in the face of proposed replacements. However, given the sheer volume of research to develop adequate substitutes in all product areas, including brake linings, flooring materials, papers and felts, all markets in fact, are genuinely threatened.

OUTLOOK

The general economic recovery may not improve prospects in the asbestos industry given continued weakness in foreign demand, uncertain regulatory trends and the utilization of substitutes where possible. Clearly, future demand will largely depend on the industry's success in stemming the negative public perception of asbestos in the industrialized countries and the emerging concerns in developing countries.

Given the firm commitments in some countries to minimize the use of asbestos, Canadian mine production during this decade may continue at today's depressed level, or even decrease further to around 700 000 tpy. With current available production capacity at about 1.2 million tpy, the over-supply situation, along with expected continuing weak prices, indicates that additional rationalization may take place in the Canadian industry. Clearly, a major challenge will be to keep production costs down allowing companies to sell competitively, particularly in developing countries where there is growth potential. Although there are established requirements for asbestos-cement products in construction and irrigation projects in these countries, foreign exchange and debt problems will probably continue to be major obstacles hindering potential growth.

Industry considers that there is a pressing need to develop new products and applications to offset the effects of phasing out of some mature products and of inroads made by some viable alternatives. Thus, coordinated research and development efforts should strive to improve the performance, safety and reliability of existing products and technologies. An important base on which to build is the reputation of Canadian asbestos fibre world wide.

TABLE 1. CANADA, ASBESTOS PRODUCTION AND TRADE, 1982-84

	1982		1983P		1984P	
	(tonnes)	($000)	(tonnes)	($000)	(tonnes)	($000)
Production (shipments)[1]						
By type						
Crude, groups 1, 2 and other milled	-	-	-	-	..	..
Group 3, spinning	13 007	15,640	13 599	17,252	..	..
Group 4, shingle	217 840	172,550	271 374	199,019	..	..
Group 5, paper	163 707	89,659	163 980	89,584	..	..
Group 6, stucco	149 982	46,197	157 958	49,090	..	..
Group 7, refuse	289 713	40,749	250 593	36,348	..	..
Group 8, sand	-	-	-	-		
Total	834 249	364,795	857 504	391,293	836 000	412,978
By province						
Quebec	745 475	298,143	744 486	321,212	695 000	301,118
British Columbia	76 084	57,032	81 653	53,396	94 000	88,360
Newfoundland	12 690	9,620	31 365	16,686	47 000	23,500
Total	834 249	364,795	857 504	391,294	836 000	412,978
Exports					(Jan. - Sept.)	
Crude						
Japan	494	148	772	267	1 093	351
United States	61	9	96	14	144	42
United Kingdom	-	-	34	8	35	6
Singapore	-	-	18	19	-	-
Argentina	-	-	11	14	-	-
Belgium-Luxembourg	-	-	-	-	17	4
West Germany	-	-	-	-	17	5
Total	555	157	931	323	1 306	408
Milled fibre (groups 3, 4 and 5)						
West Germany	68 853	50,541	23 243	23,865	18 922	18,563
Japan	36 214	24,728	30 099	23,511	23 577	17,887
United States	34 850	30,186	33 150	30,254	39 584	37,115
France	30 497	25,415	29 525	27,588	17 761	15,236
India	28 208	24,099	27 955	23,324	18 860	17,160
United Kingdom	25 056	23,632	20 916	21,266	15 625	16,044
Mexico	20 413	16,315	12 616	11,526	13 890	13,300
Italy	16 529	16,059	14 122	14,325	13 989	14,471
Australia	14 768	14,211	8 473	8,740	8 442	9,070
Malaysia	11 302	9,935	13 847	11,649	6 218	5,812
Thailand	10 014	7,813	14 527	11,796	13 087	9,716
Spain	10 512	10,496	2 854	2,632	1 870	1,767
Belgium-Luxembourg	7 871	7,292	9 329	8,690	6 626	6,448
Austria	6 808	5,470	12 228	9,772	8 325	7,544
Other countries	132 545	111,263	131 184	121,553	115 087	104,864
Total	454 440	377,455	384 068	350,491	321 863	294,997
Shorts (groups 6, 7, 8 and 9)						
United States	191 112	38,588	149 451	33,279	108 507	24,977
Japan	70 427	19,957	59 531	18,114	47 330	14,577
United Kingdom	18 050	4,420	12 351	3,606	9 001	2,703
West Germany	17 238	4,915	17 630	5,736	11 795	3,975
France	13 467	2,614	8 546	2,129	4 361	1,159
Mexico	12 741	3,555	6 307	1,525	6 546	1,628
India	11 059	3,429	15 719	5,549	6 359	2,194
Thailand	7 610	2,603	9 096	3,474	11 938	5,425
Taiwan	7 118	2,733	11 739	4,693	13 123	5,667
South Korea	5 857	1,390	12 539	3,416	9 872	2,344
Belgium-Luxembourg	5 773	1,972	5 449	2,123	6 161	2,619
Venezuela	4 347	1,102	2 506	548	2 715	546
Argentina	4 156	1,163	4 185	1,474	5 743	1,688
Nigeria	2 595	766	6 094	1,934	1 382	391
Switzerland	779	195	997	262	120	25
Other countries	53 372	15,618	46 773	16,226	35 898	12,556
Total	425 701	105,020	368 913	104,088	280 848	82,474
Grand total crude, milled fibres and shorts	880 696	482,632	753 912	454,902	604 017	377,879

9.5

TABLE 1. (cont'd)

	1982		1983P		1984P (Jan. - Sept.)	
	(tonnes)	($000)	(tonnes)	($000)	(tonnes)	($000)
Manufactured products						
Asbestos cloth, dryer felts, sheets						
United States		1,847		1,879		1,086
United Kingdom		505		217		429
Japan		-		93		2
Other countries		501		1,085		1,621
Total	..	2,853	..	3,274		2,138
Brake linings and clutch facings						
United States		9,691		8,069		6,524
Australia		160		112		87
Hong Kong		152		108		42
West Germany		128		72		46
Ecuador		66		-		-
France		13		21		-
Other countries		160		99		85
Total	..	10,370	..	8,481		6,784
Asbestos and asbestos cement building materials						
United States		12,805		10,416		6,150
United Kingdom		816		467		363
Australia		636		204		139
Singapore		370		66		105
Venezuela		359		-		165
Egyptian A.R.		285		100		23
Indonesia		113		171		31
South Africa		81		10		43
Malaysia		64		364		48
Other countries		1,959		791		1,462
Total	..	17,488	..	12,589		8,529
Asbestos basic products, nes						
United States		6,646		3,731		2,303
West Germany		158		117		539
Australia		37		119		-
Mexico		3		18		134
Other countries		440		223		362
Total	..	7,284	..	4,208		3,338
Total exports, asbestos manufactured	..	37,995	..	28,552		
Imports						
Asbestos, unmanufactured	573	687	454	483	291	415
Asbestos, manufactured						
Cloth, dryer felts, sheets, woven or felted		1,306		898		819
Packing		2,803		2,803		1,744
Brake linings		9,740		12,020		16,181
Clutch facings		1,224		1,348		1,460
Asbestos-cement shingles and siding		56		55		77
Asbestos-cement board and sheets		439		670		398
Asbestos building materials, nes		1,856		1,025		1,120
Asbestos basic products, nes		4,846		1,590		823
Total asbestos, manufactured	..	22,270	..	20,409		22,622
Total asbestos, unmanufactured and manufactured	..	22,957	..	20,892		23,037

Sources: Statistics Canada; Energy, Mines and Resources Canada.
1 Value of containers not included.
P Preliminary; - Nil; nes Not elsewhere specified; .. Not available.

TABLE 2. CANADIAN ASBESTOS PRODUCERS, 1984

Producers	Mine Location	Mill Capacity ore/day (tonnes)	Mill Capacity fibre/year (tonnes)	Remarks
Baie Verte Mines Inc.	Baie Verte, Nfld.	6 600	80 000	Open-pit.
Carey Canada Inc.	East Broughton, Que.	6 800	210 000	Open-pit. Mainly produces groups 6 and 7.
Asbestos Corporation Limited				Purchased in 1982 by Société nationale de l'amiante (SNA) (Quebec Crown corporation).
Asbestos Hill mine	Putuniq, Que.	5 400		Mining ceased indefinitely in 1983.
British Canadian mine	Black Lake, Que.	12 000		Open-pit, two milling plants.
King-Beaver mine	Thetford Mines, Que.	7 000	210 000	Underground and open-pit.
Normandie mine	Black Lake, Que.			Reserves exhausted. Mill processes K-B open-pit ore.
Bell Asbestos Mines, Ltd.	Thetford Mines, Que.	2 700	75 000	Underground. Purchased in 1980 by SNA (Quebec Crown corporation).
Lake Asbestos of Quebec, Ltd.	Black Lake, Que.	9 000	235 000	Open-pit.
National Mines Division	Thetford Mines, Que.	4 000		Open-pit.
JM Asbestos Inc. Jeffrey mine	Asbestos, Que.	15 000	300 000	Open-pit (effective capacity reduced by one-half).
Brinco Mining Limited Cassiar mine	Cassiar, B.C.	5 000	100 000+	Open-pit.

TABLE 3. CANADA, ASBESTOS PRODUCTION AND EXPORTS, 1978-84

	Crude	Milled	Shorts	Total
	(tonnes)			
Production[1]				
1978	1	673 910	747 897	1 421 808
1979	4	725 649	767 066	1 492 719
1980	-	690 493	632 560	1 323 053
1981	10	567 288	554 547	1 121 845
1982	-	394 554	439 695	834 249
1983P	-	448 953	408 551	857 504
1984P	..	..	..	836 000
Exports				
1978	1	689 690	708 392	1 398 083
1979	20	719 075	741 947	1 461 042
1980	-	653 358	564 379	1 217 737
1981	10	519 777	542 402	1 062 189
1982	555	454 440	425 701	880 696
1983P	931	384 068	368 913	753 912
1984P(Jan.-Sept.)	1 306	321 863	280 848	604 017

Sources: Statistics Canada; Energy, Mines and Resources Canada.
[1] Producers' shipments.
P Preliminary; .. Not available; - Nil.

TABLE 4. WORLD ASBESTOS PRODUCTION, 1983

Country	tonnes[e]
U.S.S.R.[e]	2 250 000
Canada	857 504
Rep. of South Africa	220 000
Zimbabwe	190 000
Brazil	135 000
Italy	120 000
China	110 000
Greece	100 000
United States	69 906[1]
India	25 000
Australia	20 000
Cyprus	18 000
Korea	15 000
Turkey	4 000
Swaziland	31 275[1]
Mozambique	800
Yugoslavia	10 500
Japan	4 000
Taiwan	2 500
Argentina	1 350
Bulgaria	600
Egypt	325
	4 185 760

Sources: United States Bureau of Mines and Energy, Mines and Resources Canada.
[1] Reported figure.
[e] Estimated.

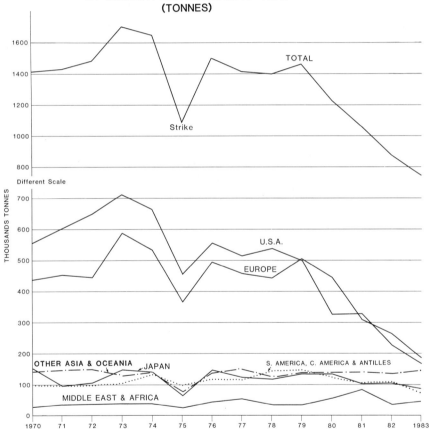

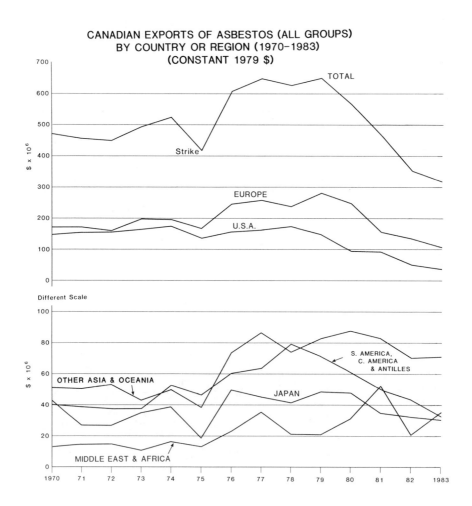

9.10

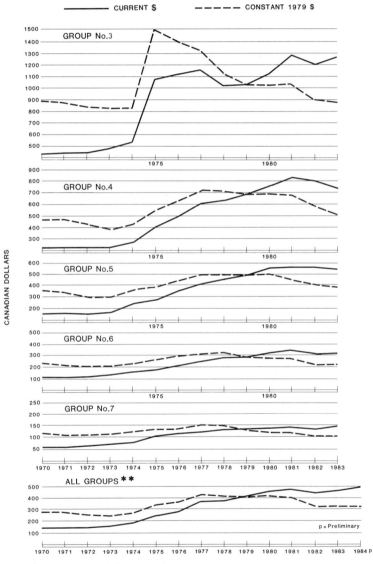

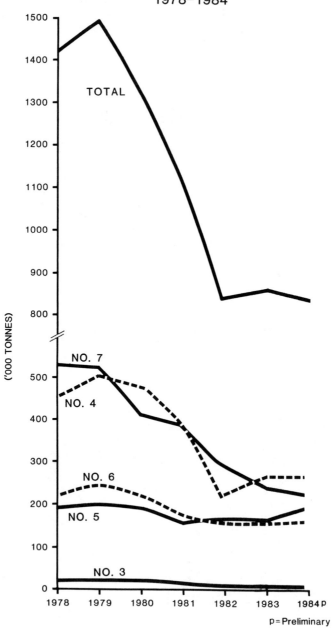

Barite and Celestite

G.O. VAGT

SUMMARY

Canadian shipments of barite in 1984 amounted to 46 884 t valued at an estimated $7.45 million. This compares to 45 465 t valued at $4.87 million shipped in 1983. The industry generally experienced moderate recovery since 1982 when shipments dropped substantially following a decline in oil- and gas-well drilling activity, particularly in western Canada. Imports of barium carbonate in 1983, one of the most important barium chemicals derived from barite, amounted to 3 699 t valued at $1,253,000. In the first nine months of 1984 imports were 2 857 t valued at $930,000.

Barite ($BaSO_4$) is a valuable industrial mineral because of its high specific gravity (4.5), low abrasiveness, chemical stability and lack of magnetic and toxic effects. Its dominant use is as a weighting agent in the oil- and gas-well drilling muds required to counteract high pressures confined by the substrata.

This mineral is found in many countries of the world and is the raw material from which nearly all other barium compounds are derived. The major world producers of barite, excluding the U.S.S.R. and most centrally planned economies are the United States, China, Morocco, India, Thailand, Mexico and Peru. Recently, China has become very important in world trade and is the leading exporter of barite to the United States.

CANADIAN DEVELOPMENTS

Barite was produced during 1983-84 from operations in British Columbia, Ontario and Nova Scotia. Production in Newfoundland was intermittent.

Mountain Minerals Co. Ltd., in eastern British Columbia, reactivated its Parson Mine in 1983 following extensive development of underground vein deposits in 1980-81. The Brisco Mine and Mineral King mine, past producers with limited reserves, did not operate. All of the crude barite from the Parson mine is shipped to the company's grinding plant at Lethbridge, Alberta. Magcobar Minerals Division of Dresser Canada Inc., started production in 1984 from the Fireside deposit near kilometre 588 of the Alaska Highway. In 1983, Baroid of Canada, Ltd. processed small quantities of old tailings at the Silver Giant property near Spillimacheen. At intervals, Baroid also processes crude barite at its grinding plant in Onoway, Alberta.

Extender Minerals of Canada Limited mines barite near Matachewan, Ontario. Production is from a vein deposit by open-pit methods and the high-quality dry-ground product is used for filler and extender pigments in paints and plastics.

ASARCO Inc. and Abitibi Price Inc., beginning in 1982, recovered barite on a seasonal basis from old tailings at the Buchans mine, in Newfoundland. There was no production in 1983 and limited output in 1984 as sales contracts were irregular. Reserves of barite in the tailings are approximately 450 000 t averaging about 35 per cent. Also in Newfoundland, at Collier Point, on the Avalon Peninsula, mining by several companies has been intermittent since 1980. In 1983 about 2 000 t were produced and in 1984 Eagle Resources produced several thousand t according to the Department of Mines and Energy.

In Nova Scotia Nystone Chemicals Ltd. mined pharmaceutical-grade barite from its deposit 2 miles northeast of Brookfield. Some drilling-grade barite was produced for the first time in 1984. Novex Mining and

---

G.O. Vagt is with the Mineral Policy Sector, Energy, Mines and Resources, Canada. Telephone (613) 995-9466.

Exploration started production at its Lake Uist mine, in Richmond County. Ore is concentrated by flotation at Enon, Cape Breton County and is used for well drilling. At Upper Bass River in Colchester County the Magcobar Division of Dresser Canada Inc. recovered barite from a small, run-of-mine operation. Ore was shipped to Walton, Hants County for final grinding.

A large, unique occurrence of barite-gold mineralization was found in 1983 to be associated with the Hemlo gold deposits 250 km east of Thunder Bay, Ontario. The barite is stratabound extending for several kilometres along strike in an Archean metavolcanic rock belt. At the Golden Giant Deposit No. 1, a Noranda Mines - Golden Giant joint venture, barite content ranges from 5 to 40 per cent. Indicated reserves of barite may be 2 million t assuming 20 million t of gold ore containing 10 per cent barite. However, no plans have been announced to recover the barite.

CONSUMPTION

In 1982 consumption of barite in Canada was 25 477 t based on estimates. About 80 per cent was used for well drilling. The balance of Canada's barite consumption was in the manufacture of paint and varnish, rubber, chemicals, brake linings and other products. Growth in demand is expected in the automotive primer paint markets and also in new plastic applications in flooring and firewall parts.

WORLD DEVELOPMENTS

World production of barite in 1983 was 5 791 million t, according to the United States Bureau of Mines. The abrupt decrease in output since 1981-82 reflected reduced well drilling as a result of the recession and a persistent oil glut.

The United States, previously by far the world's largest producer of barite, mainly from Nevada, produced an estimated 0.68 million t in 1983. Production in China was an estimated 1.0 million t in 1983 and surpassed the United States for the first time in 1983. Imports of barite to the United States during 1983 and 1984 were 1.2 million t and 1.6 million t respectively. Beginning in 1982, U.S. imports have been higher than domestic production and in 1983 net import reliance as a per cent of apparent consumption was 66 per cent. Substantial increases in shipments of crude ore from China to Gulf Coast ports mainly accounted for this shift. Lower oil- and gas-well drilling activity continued in 1984 resulting in further cutbacks in mine production and excess capacity at grinding plants. In Nevada more than 20 companies operated in 1980 whereas today only about six major operators remain. For the most part, these are fully integrated into the oil-well servicing industry.

The downward trend in drill-rig utilization in virtually all countries resulted in cutbacks of production or complete closures of numerous mining operations during 1983-84. An exception was Strontian Minerals, a member of the Minworth group of companies. It started producing drilling-grade barite from its barite, lead-zinc-silver mine and processing plant near Strontian in Scotland. Plant production capacity totals 50 000 tpy and the barite will serve the North Sea drilling market. A range of grades for non-drilling-uses is expected to be produced in the future when the operation changes from quarry to underground methods.

PRICES

Substantial overcapacity and low ocean-freight rates in the 1983-84 period resulted in some published prices falling by as much as 25 per cent for unground material priced in the range of $US 50 a t. Low-cost crude barite exported from China to the United States played an important role in depressing western world prices. Prices in the range of $US 200-400 for barite used in smaller quantities in chemical and filler/extender markets remained about the same.

USES

Principal specifications for barite used in well-drilling usually require a minimum specific gravity of about 4.2, a particle size of 90-95 per cent minus 325 mesh, and a maximum of 250 ppm soluble alkaline earths, as calcium.

Barite is used in paint as a special filler or "extender pigment". This is a vital constituent that provides bulk, improves consistency of texture, surface characteristics and application properties, and controls prime pigment settling and viscosity. Specifications for barite used in the paint industry call for 95 per cent $BaSO_4$, particle size at least minus 200 mesh, and a high degree of whiteness or light reflectance. Final "wet milled" and "floated" products result in smooth micro-crystalline surfaces that prevent agglomeration, thus allowing easy dispersal in water as well as in oil-soluble binders. When barite is used in highly pigmented distemper or latex paints, a degree of light scattering is attributed to the barite, therefore allowing it to function as a pigment.

The glass industry uses barite to increase the workability of glass, to act as a flux, assist decolouration and increase the brilliance or lustre of the product. Specifications call for a minimum of 96 to 98 per cent $BaSO_4$, a particle size range of 40 to 140 mesh and usually a magnetically separated ore is used with iron often reduced to 0.1 per cent. However, producers of fine glassware use precipitated barium carbonate to circumvent impurity problems often associated with natural barite.

The specifications for natural barite used as a filler in rubber goods vary, but the main factors are whiteness and particle size range. For general filler and extender uses most manufacturers want a product that is virtually all minus 325 mesh. Colour is important to many users.

OUTLOOK

The demand in 1985 for barite is expected to increase based on recent strength in oil- and gas-well drilling activity. In 1984 9,149 wells or 10.4 million m were drilled in Canada according to preliminary statistics. This compares with 7,064 wells drilled or 8.3 million m drilled in 1983. In the medium- to long-term, factors influencing exploration and development activity include, the expected financial returns from exports of natural gas, the timing and nature of cooperative agreements on offshore ownership and the relative level of activity in the conventional source and offshore regions.

With more delineation drilling needed to confirm reserves offshore, additional supplies of barite will be needed. There is good potential for discovery and development of barite deposits in most regions; in Nova Scotia and Newfoundland several companies are involved at various stages of property development from drilling to small-scale production. However, sources from abroad will likely continue to compete with domestic producers as long as excess capacity and low ocean-freight rates prevail.

PRICES

United States prices of barite as reported in Engineering and Mining Journal[1], of December 1984.

	$ per short ton
Unground	
Chemical and glass grade: Hand picked, 95% $BaSO_4$, not over 1% Fe	90.00
Magnetic or flotation, 96-98% $BaSO_4$, not over 0.5% Fe	106.00
Imported drilling mud grade, specific gravity 4.20 - 4.30, cif Gulf ports	32.00-50.00
Ground	
Water ground, 95% $BaSO_4$ 325 mesh, 50-lb bags	80.00-165.00
Dry ground, drilling mud grade, 83%-93% $BaSO_4$, 3-12% Fe, specific gravity 4.20-4.30	55.00-115.00
Imported Specific gravity 4.20-4.30	55.00-75.00

[1] Published by McGraw-Hill.

CELESTITE

SUMMARY

There has been no Canadian production of celestite ($SrSO_4$), the main source of strontium, since Kaiser Celestite Mining Limited, a subsidiary of Kaiser Aluminum & Chemical Canada Investment Limited, closed its mining operation at Loch Lomond, Nova Scotia and its strontium products plant at Point Edward, Nova Scotia, in 1976.

NORTH AMERICAN SCENE

North American consumers continue to depend totally on imports of strontium minerals. The strontium mining industry in the United States has been dormant since 1959 and Mexico and West Germany are the major suppliers of celestite and strontium compounds to the U.S. market.

Consumption of strontium compounds in the United States in 1984 was an estimated

24 000 t valued at $US 15.9 million. In 1984, one of the two major U.S. strontium compound producers closed its barium/strontium carbonate plant. From a 1979 base, demand for strontium in the United States is expected to increase at an annual rate of about 1.2 per cent through 1990, according to the United States Bureau of Mines.

USES

Celestite is used to produce commercial strontium compounds, principally strontium carbonate and strontium nitrate. In the sulphate form it is used for purifying electrolytic zinc. Strontium carbonate is primarily used in glass faceplces for colour television picture tubes where it improves the absorption of X-rays emitted by the high voltage tubes. Other uses include pyro-technics and signals, and ferrite ceramic permanent magnets used in small electric motors.

PRICES

United States prices of celestite according to Chemical Marketing Reporter, December 24, 1984

	$ per short ton
Strontium carbonate glass grade, bags, truckload, works	$ 655.00
	($ per 100 pounds)
Strontium nitrate, bags, carlot, works	$ 24.00

TARIFFS

Item No.		British Preferential	Most Favoured Nation	General	General Preferential
			(%)		
CANADA					
49205-1	Drilling mud and additives	free	free	free	free
68300-1	Barytes	free	10	25	free
92818-1	Barium oxide, hydroxide peroxide	9.4	7.5	25	5
92842-1	Barium carbonate	10	13.8	25	9
93207-5	Lithopone	free	11.5	25	free

MFN REDUCTIONS UNDER GATT (effective January 1 of year given)	1983	1984	1985	1986	1987
			(%)		
92818-1	7.5	5.6	3.8	1.9	free
92842-1	13.8	13.4	13.1	12.8	12.5
93207-5	11.5	11.3	11.0	10.8	10.5

UNITED STATES (MFN)

Barium carbonate:		
472.02	Natural, crude (witherite)	free
472.06	Precipitated	0.5¢ per pound
Barium sulfate:		
472.10	Natural, crude (barytes)	$1.27 per ton
472.12	Natural, ground (barytes)	$3.25 per ton
472.14	Precipitated (blanc fixe)	0.2¢ per pound
473.72	Lithopone	2.5%
473.74	Lithopone	4.6%

		1983	1984	1985	1986	1987
				(%)		
472.04	Barium carbonate, natural ground (witherite)	5.1	4.9	4.7	4.4	4.2

Sources: The Customs Tariff, 1983, Revenue Canada, Customs and Excise. Tariff Schedules of the United States Annotated 1983, USITC Publication 1317. U.S. Federal Register Vol. 44, No. 241.

TABLE 1. CANADA, BARITE PRODUCTION AND TRADE, 1982-84 AND CONSUMPTION, 1981 AND 1982

	1982		1983		1984P	
	(tonnes)	($000)	(tonnes)	($000)	(tonnes)	($000)
Production (mine shipments)	..	2 359	45 465	4,869	46 884 (Jan.-Sept. 1984)	7,450
Imports						
United States	8 558	1 185	4 602	697	5 254	501
Ireland	11 500	319	24 690	900	-	-
Netherlands	398	108	655	204	464	125
Morocco	-	-	-	-	10 593	890
Other	3 001	541	5	1	-	-
Total	23 457	2 153	29 952	1,802	16 311	1,516
Exports						
United States	470	315	795	155	634	209
United Kingdom	6	4	-	-	-	-
Japan	6	12	-	-	-	-
Total	482	331	795	155	634	209

	1981r	1982P
Consumption[1]		
Well drilling[e]	89 622	20 000
Rubber goods	1 192	946
Paint and varnish	1 598	1 737
Other[2]	1 585	2 794
Total[e]	93 997	25 477

Sources: Energy, Mines and Resources Canada; Statistics Canada.
[1] Available data reported by consumers with estimates by Energy, Mines and Resources Canada. Does not include inventory adjustments. [2] Other includes bearings and brake linings, chemicals, floor covering, adhesives, explosives, asbestos products, etc.
P Preliminary; r Revised; e Estimated; .. Not available; - Nil.

TABLE 2. CANADA, BARITE PRODUCTION TRADE AND CONSUMPTION, 1970, 1975, AND 1980-84

	Production[1] ($)	Imports	Exports (tonnes)	Consumption[e]
1970	1,388,125	6 827	90 305	50 106
1975	2,305,819	4 479	45 606	40 229
1980	4,380,000	45 157	645	138 829
1981	5,124,000	16 278	405	94 027
1982	2,359,000	23 457	482	25 477
1983	4,869,000	29 952	..	..
1984P (9 mos.)	7,450,000	16 311	..	..

Sources: Energy, Mines and Resources Canada; Statistics Canada.
[1] Mine shipments.
P Preliminary; [e] Estimated; .. Not available.

TABLE 3. WORLD PRODUCTION OF BARITE 1981-83[1]

Country[2]	1981	1982P	1983[e]
	(tonnes)		
Afghanistan[3]	1	2	2
Algeria	89	102	109
Argentina	49	36	40
Australia	41	40[e]	40
Belgium[e]	40	40	40
Bolivia[4]	2	1	1
Brazil	116	120[e]	118
Burma[5]	10[e]	20	20
Canada[e]	80	30	28
Chile	259	292	300
China[e]	798	898	1 000
Colombia	4	4[e]	4
Czechoslovakia	61[e]	61[e]	61
Egypt	2	3[e]	4
France	191	156	150
Germany, Democratic Republic[e]	35	35	35
Germany, Federal Republic	165	362	250
Greece[6]	47	47	46
Guatemala	5	2	2
India	354	326	300
Iran[e]	75	80	85
Ireland	260[e]	260[e]	218
Italy	177	180	150
Japan	56	60	70
Kenya	6[e]	-	-
Korea, North[e]	100	-	-
Malaysia	19	25[re]	20
Mexico	318	364	350
Morocco	465	538	275
Pakistan	24	26	28
Peru	409	375	163
Philippines	2	9[e]	2
Poland	85	80[e]	100
Portugal	1	1[e]	1
Romania	79[e]	78[e]	78
South Africa, Republic of	3	4	6
Spain	53	50	50
Thailand	307	331	188
Tunisia	25	31	20
Turkey	186	107	80
U.S.S.R.[e]	508	517	517
United Kingdom	63	81	85
United States[7]	2 585	1 674	684
Yugoslavia	45	45[e]	40
Total	8 200	7 493	5 760

Source: United States Bureau of Mines.
[1] Table includes data available through May 30, 1984. [2] In addition to the countries listed, Bulgaria also produces barite, but available information was inadequate to make reliable estimates of output levels. [3] Year beginning March 21 of that stated. [4] Series represents exports only. Bolivia also produces barite for domestic consumption, but available data are not adequate for formulation of estimates or levels of production to meet internal needs. [5] Data are for fiscal years beginning April 1 of that stated. [6] Barite concentrates. [7] Sold or used by producers.
[e] Estimated; P Preliminary; r Revised.

Beryllium

G. BOKOVAY

Beryllium is a gray corrosion - resistant metal with a specific gravity of 1.85, which is between that of aluminium and magnesium. The metal has a tensile strength considerably greater than either of these two metals, a good modulus/density ratio, a high melting point (1290°C), elevated temperature strength and excellent nuclear moderating and reflecting properties.

While there has been no discernible trend in beryllium consumption during the last 15 years, demand for the metal should increase in view of the expected increase in utilization of beryllium - copper alloys in the electrical and electronics industries.

Although Canada does not produce either beryllium metal or beryllium minerals, this situation could change in the next decade in light of recent discoveries of relatively high-grade beryllium mineralization. Expected growth in beryllium demand along with the apparent desire of the United States to conserve its domestic sources of supply for the metal, could encourage development of some of these resources.

PRODUCTION OF BERYLLIUM MINERALS

Beryllium is produced commercially from two minerals, beryl and bertrandite. Beryl ($3BeO \cdot Al_2O_3 \cdot 6SiO_2$), which occurs in pegmatite dykes, is hand picked and cobbed to remove gangue minerals adhering to the beryl crystals. The separation of beryl by mechanical or flotation techniques is difficult because the densities of beryl and associated minerals are similar. Since beryl production is labour intensive, beryl is principally mined in developing countries. The People's Republic of China and Brazil are considered to be the most important producers.

Beryl has been mined in Canada although there has been no production in recent years.

Bertrandite ($Be_4 Si_2 O_7 (OH)_2$), which has been mined in the United States since 1969 by Brush Wellman Inc., has become the most important source of beryllium metal in the western world. The company operates an open pit bertrandite mine in the volcanic tuff beds of the Topaz - Spor Mountain area of western Utah.

World reserves of beryllium minerals are very large in relation to current rates of extraction. At the end of 1983, Brush Wellman estimated its bertrandite ore reserves at about 4.5 million t grading approximately 0.22 per cent beryllium. At the current mining rate of about 105 000 tpy, these reserves would last about 40 years. World reserves outside China, which are estimated by Roskill Information Services at 419 000 t of contained beryllium, could satisfy current beryllium consumption for a thousand years.

CANADIAN DEVELOPMENTS

In recent years, important discoveries of beryllium minerals have been made in Canada. At Thor Lake, which is located about 65 miles southeast of Yellowknife, Northwest Territories, Highwood Resources Limited has discovered a significant deposit of the mineral phenacite ($BeSiO_4$) as well as niobium, tantalum, zirconium, columbium and rare earths. To date, drilling on the "T" zone has defined 435 000 t grading 1.4 per cent BeO while 1 200 000 t grading .66 per cent BeO has been identified in the "South" zone.

Highwood is currently undertaking a pre-feasibility study, including metallurgical testing, to assess the economic viability of an open pit beryllium mine and concentrator at the site.

At a mining rate of about 225 tpd of ore and an expected recovery rate of 85 to

---

G. Bokovay is with the Mineral Policy Sector, Energy, Mines and Resources, Canada. Telephone (613) 995-9466.

90 per cent, the company plans to produce about 3 000 tpy of concentrate containing about 10-13 per cent BeO. Since the beryllium ore also contains yttrium, a separate circuit for this metal is being considered.

It is reported that the Dene Nation in the Northwest Territories is opposed to the potential development. The probable mine-site is located on lands which are subject to a native land claims settlement that is being negotiated between the federal government and the Dene Nation.

In 1984, Bearcat Explorations undertook an exploration program at its 80 per cent owned Hellroaring Creek property near Kimberly, British Ccolumbia, where the mineral phenacite has also been identified. Exploration in the 1960s identified about 500 000 t of ore averaging about 1 per cent BeO.

At Strange Lake on the Quebec - Labrador border, the Iron Ore Company of Canada (IOC) in 1979 discovered a high grade deposit of yttrium and zirconium which also contains columbium, beryllium and rare earth elements. The deposit is reported to be amenable to open pit mining. Although IOC has undertaken market studies and metallurgical testing, no decision has apparently been made regarding future development of the deposit.

WORLD PRODUCTION AND DEVELOPMENTS

The Unites States and the U.S.S.R. are the only known producers of beryllium metal, although it is suspected that the Peoples Republic of China is also a producer. Production data for the United States is withheld to conform with confidentiality regulations, since Brush Wellman Inc. is the sole producer.

In the United States, Brush Wellman produces beryllium hydroxide at a plant at Delta, Utah, from bertrandite concentrate from the company's nearby open-pit mines and from imported beryl ore. Some of the ore is processed on a toll basis for Cabot Corporation. The output of the Delta plant, which is impure beryllium hydroxide, is sent to the company's Elmore Ohio plant for further processing.

At Delta, a new beryl furnace designed to handle lower grade beryl ore (7 per cent BeO) was brought on-stream in 1984.

However, certain technical problems arose that forced the company to restart its old beryl furnace.

In response to increased beryllium demand, Brush Wellman has in recent years been increasing the amount of imported beryl processed at its Delta plant. However, with reports that the company has reduced its purchases on the spot market, beryl consumption in 1984 will probably be equal to or lower than in 1983.

In 1984, Brush Wellman announced that a new bulk product plant would be built at Elmore, Ohio to produce Be-Cu rod and tubular products. In addition the company also plans to expand extrusion and billet capacity. Brush also announced plans for a $US 57 million expansion that will almost double it Be-Cu strip capacity. This includes a $30 million casting, rolling, pickling and annealing unit at Elmore, $10 million for additional strip finishing and improvement to existing capacity at Reading, Pa and $15 million for a finishing mill in Europe. During 1984, the Cabot Wrought Product Division of Cabot Corporation brought a new wide strip Be-Cu finishing unit at Elkhart, Ind. into production. At the time of the company's announcement, the backlog of customer orders was about 30 weeks.

In 1984, Brush Wellman Inc. announced an expansion to its beryllia ceramic production facility at Tucson, Arizona. The expansion, which was expected to be completed by the end of 1984, will increase capacity by 50 per cent.

Exposure to small concentrations of beryllium dust has been recognized as the cause of berylliosis, a serious chronic lung disorder. More restrictive exposure limits were first proposed in the United States in 1975. These proposals were strongly opposed by the U.S. industry on the grounds that they were technically impossible to meet. To date, changes to U.S. regulations have not been implemented.

USES

Although beryllium metal and its intermediates are relatively expensive, its unique properties have found application in a wide range of goods, although in small quantities. The consumption of beryllium in metal, alloy and oxide forms is estimated as

40 per cent in nuclear reactors and aerospace applications, 36 per cent for electrical equipment, 17 per cent for electronic equipment and 7 per cent for miscellaneous uses. It is estimated that 75 per cent of the beryllium is consumed in beryllium alloys principally beryllium-copper, 17 per cent as beryllium metal while 8 per cent is consumed as beryllium oxide.

Beryllium copper alloys which consume a major portion total beryllium produced, may contain anywhere from 0.25 to 2.15 per cent beryllium. In general, beryllium-copper alloys are much stronger and harder than pure copper, and still have good electrical conductivity. In addition these alloys have good corrosion resistance and low magnetic susceptibility.

A list of applications for wrought or a Be-Cu alloys include: precision mechanical and electro mechanical springs; electrical/electronic connectors, sockets, contacts, relays, and switches; diaphrams; dies; injection molds for plastics; non sparking tools; undersea cable housings; bushings; welding tips; and certain applications in oil and gas drilling housings, measurement systems and drill string components. In the United States, Be-Cu master alloys and products are supplied by Brush Wellman Inc. and Cabot Corporation.

Brush's beryllium - nickel alloys have many of the same properties as Be-Cu alloys except that they can be used in higher temperature applications for miniature electronic connector components, springs and mechanical fasteners.

In 1984, Brush introduced a new family of copper alloys containing small amounts of beryllium and cobalt. These alloys, which will be used in the electronic and electrical industries, will compete with phosphar bronze and similar copper alloys.

When beryllium is added to aluminum or magnesium, the resultant alloys are known to more easily fabricated than the host metals and have improved oxidation resistance. Although proven uses are few, (Be-Al alloys have been used in certain aerospace applications and for cooking utensils) these alloys offer some potential for increased beryllium consumption in the future.

Beryllium has also been identified as a potentially useful additive to high-tensile stainless steels although current production is thought to be limited.

For products made of beryllium metal, powder metallurgy is the preferred method of fabrication since coarse crystals develop when the metal is cast. Use of this expensive metal is justified by its superior strength and stiffness relative to density in structural aerospace applications, inertial guidance system components, space borne optics and aircraft brakes.

Beryllium metal's high moderating ratio and high neutron reflection properties have led to its use in nuclear aplications despite embrittlement after long exposure to radiation.

Beryllium oxide ceramics have excellent insulating properties, high thermal conductivity and thermal shock resistance. Applications include high density electronic circuits, radar, electronic counter-measure systems, cellular radiotelephone systems, microwave communication equipment, graphic lasers, computerized tomography scanners, automotive ignition systems and microwave ovens.

STOCKPILES

Beryllium is of strategic importance and the U.S. General Services Administration (GSA) maintains the metal in its National Defense Stockpile. During 1984, the GSA took delivery of 60 000 pounds of beryllium metal and has contracted a similar amount for delivery in 1985.

The USBM reported that at the end of 1983, the GSA held over 16 000 t of beryl ore, 208 t of beryllium metal and 6 700 t of beryllium-copper master alloy (4 per cent Be).

OUTLOOK

Demand for beryllium metal was strong in 1984 and is expected to remain strong in 1985. For the first 9 months of 1984, Brush Wellman's sales of beryllium and other specialty metals rose 40 per cent from 1983, to $US 195.5 million. The company also reported that order back-logs were at an all-time high.

The highest long-term growth for beryllium is expected to be in beryllium copper alloys in both the wrought and cast sectors. However, the movement to miniaturization in the electronics industry could be a limiting factor.

Although the outlook for beryllium metal and beryllium oxide is somewhat less optimistic than for beryllium alloys, growth is nevertheless expected. Expansion of the electronics industry will result in increased demand for beryllium ceramics while increased utilization of the metal is expected in the aerospace industry.

Overall, demand for beryllium in all forms is expected to grow at between 4 and 5 per cent annually until the mid 1990s. Since a significant portion of beryllium metal production is consumed in the form of defense related products, overall beryllium demand has fluctuated in response to defense contracts. Although these fluctuations can be expected to continue, the growing share of beryllium utilized in alloy form in non-defense applications should help to stabilize overall consumption.

TARIFFS

Item No.		British Preferential	Most Favoured Nation	General	General Preferential
			%		
CANADA					
34907-1	Copper beryllium alloys	4.6	4.5	25	free
35101-1	Beryllium metal	free	4.5	25	free

MFN Reductions under GATT (effective January 1 of year given)	1983	1984	1985	1986	1987
			%		
34907-1	4.5	4.4	4.3	4.1	4.0
35101-1	4.5	4.4	4.3	4.1	4.0

UNITED STATES (MFN)

417.90	Beryllium oxide or carbonate	3.7%			
601.09	Beryllium ore	free			
628.05	Unwrought beryllium, waste and scrap	8.5%			
628.10	Beryllium, wrought	9.0%			

		1983	1984	1985	1986	1987
				%		
612.20	Beryllium copper master alloy	8.3	7.7	7.2	6.6	6.0
417.92	Other beryllium compounds	4.4	4.2	4.0	3.9	3.7

Sources: The Customs Tariff, 1983 Revenue Canada, Customs & Excise; Tariff Schedules of the United States Annotated (1983), USITC Publication 1317; U.S. Federal Register, Vol. 44, No. 241.

TABLE 1. ESTIMATED WORLD BERYLLIUM PRODUCTION[1]

	1980	1981	1982	1983	1984e
	(tonnes)				
United States					
from Bertradite[2]	143.6	144.4	105.8	135.0	150
from Imported beryl[3]	39.8	49.9	61.9	51.2	50
Total	183.4	194.3	167.7	186.2	200.0
U.S.S.R.[3]	46.7	46.7	46.7	49.0	50.0
Total[4]	230.1	241.0	214.4	235.2	250.0

[1] Includes the beryllium metal equivalent of beryllium alloys and oxide. [2] Beryllium content of concentrate produced as reported by Brush Wellman Inc. 1983 Annual Report - 65 per cent recovery of metal assumed. [3] Reported by USBM, 1983 Beryllium preprint - 65 per cent recovery of metal from beryl assumed. [4] Excludes China.

TABLE 2. PRICES ($US)

Beryl ore; Cif Atlantic - Ports 10% to 12% per Stu. (Eff. 19/4/82) $90.00 - $130.00

Beryllium oxide
 Uox powder GCR, GCHG (Brush Wellman) (Eff. 23/1/84) $52.50

Beryllium (containing alloys; 4% BeCu)

275 C BeCu casting alloy (10-1-84)... $5.55
165 C BeCu casting alloy (10-1-84)... $4.95
245 BeCu casting alloy (10-1-84)..... $5.30
10C BeCu casting alloy (10-1-84)..... $4.95
(Containing less than 0.5 per cent Mg)
Per lb. contained beryllium $140

Beryllium 97%

Powder blend (200 grade)
 5 000 lb. lots (Eff. 14-2-84)...... $178
 Vac. cast ingots (Eff. 6-2-84)..... $213
 fob Elmore, Ohio

Beryllium aluminum (1-4-84)

 fob Reading, Pa., Detroit and Elmore, Ohio (100,000 lb)..........$230

Beryllium copper (10-1-84)

 Strip (No.25)...................... $7.10
 Rod, bar and wire (No.25)......... $7.60

Beryllium copper master alloy

fob Reading, Pa., Detroit, Michigan, Elmore, Ohio, per lb. contained beryllium for 5 lb. ingot (Eff. 10-1-84)............... $140

+ Beryllium copper alloys reflect a $1 per pound base copper price. Prices will fluctuate weekly based on a predetermined copper composite price chosen by the individual companies.

Source: American Metal Market.

Bismuth

J. BIGAUSKAS

INTRODUCTION

Bismuth (chemical symbol Bi), is generally produced as a byproduct of some lead, copper and tin mines in various parts of the world. The important bismuth-bearing minerals are: bismite, an oxide; bismuthinite or bismuth glance, sulphides; bismutite, a carbonate. The abundance of bismuth in the earth's crust is estimated to be 0.00002 per cent - about the same as silver or cadmium.

Most of bismuth in copper concentrates accompanies the copper matte during smelting. In the conversion of copper matte to blister copper, most of the bismuth is fumed and subsequently collected in the baghouse with lead, arsenic or antimony. These dusts are then processed by lead smelters.

Bismuth in lead concentrates stays with the impure lead bullion during reduction in the blast furnace. The first stages of pyrometallurgical refining produce a bismuth-containing dross and then a bismuth-lead alloy which may be refined further to remove impurities, precious metals and remaining lead and zinc. Finally, bismuth is given a final oxidation with air and caustic soda or chlorine to yield a product with final purity of 99.99+ per cent.

Alternatively, at electrolytic lead refineries, treatment of lead anodes leaves pure lead on the cathode and a cell slime. The slime is processed to yield an impure metal, and then the metal is refined. Bismuth is leached from roasted tin concentrates with hydrochloric acid, precipitated as bismuth oxychloride, reduced to metal by wet or dry methods and then refined further.

CANADIAN DEVELOPMENTS

Bismuth metal is recovered at Cominco Ltd.'s Trail, British Columbia electrolytic lead refinery and is refined to 99.99+ per cent purity. Some bismuth is refined further at Cominco Electronic Materials Division for use in electronics and research applications. The ores of the Sullivan mine, the major source of lead concentrate feed to the smelter/refinery, contain traces of bismuth.

Bismuth-lead alloys are produced at Brunswick Mining and Smelting Corporation's pyrometallurgical lead refinery at Belledune, New Brunswick. Most feed is from the Brunswick No. 12 mine in the form of a lead concentrate with small amounts of contained bismuth.

Bismuth concentrates are also produced at Terra Mines Ltd.'s Silver Bear operations in the Northwest Territories. Total production of bismuth in Canada in 1984 was approximately 220 t, 33 t less than in 1983.

Consumption of bismuth in Canada dropped in 1983 to about 7 t from 10 t in previous years. Consumption was expected to be slightly higher in 1984.

WORLD DEVELOPMENTS

Western world consumption of bismuth can only be estimated, since statistical coverage is incomplete. Preliminary estimates indicate that Japanese consumption rose from 339 t in 1983 to nearly 500 t in 1984. Consumption in the United States, the largest market, likely showed a substantial increase perhaps to 1 500 t, due partly to broader survey coverage. European consumption was estimated to be around 1 000 t while consumption in other countries likely totalled several hundred tonne in 1984.

Reported metal exports to socialist countries totalled about 400 t in 1984. These included Peru's 200 t sale of bismuth to the People's Republic of China in May and a further 200 t sale to the U.S.S.R. in August.

Major western world mine producers of bismuth-containing ores and concentrates are

---

J. Bigauskas is with the Mineral Policy Sector, Energy, Mines and Resources, Canada. Telephone (613) 995-9466.

Australia, Mexico, Peru, Canada and Japan. These countries produce bismuth from a variety of ores which may contain lead, silver, copper or zinc. Bolivia until 1980 produced from bismuth ores, but now primarily produces byproduct bismuth in other ores.

Major western world producers of byproduct bismuth at the smelting and refining stage are Mexico, Japan, Peru, the United Kingdom, Bolivia, Canada, United States, the Federal Republic of Germany, France and Belgium.

Production of refined bismuth at the refining stage can be highly variable from year to year because of the byproduct nature of production. Estimated western world production of refined bismuth in 1984 was about 3 300 t.

Technical problems at Met-Mex Penoles S.A.'s Torreon, Mexico refinery in February, 1984 made it necessary to reprocess material to meet specifications. This resulted in at least a one month delay in shipments. However, the company expected to produce its full 500 t capacity for 1984, despite this problem and despite a reported interruption in bismuth production in May.

Domestic production of bismuth metal in Japan, according to MITI estimates, was expected to have remained stable at close to 580 t in 1984 compared to 569 t in 1983. Bismuth exports, however, were expected to have fallen to 150 t in 1984, from 248 t in 1983.

Minpeco, the metal marketing company of Peru, declared a **force majeure** on bismuth needles early in the year because of flooded road and rail links. Although the **force majeure** was lifted in mid-May, shipments bound for the United States did not arrive until June.

Peko-Wallsend Ltd. of Australia cut back its mine production of bismuth substantially in 1981 because of declining prices but potentially could increase output given higher and sustained prices for its major products, copper and gold, at its Tennant Creek, Australia operation. Bismuth is also contained in small amounts in lead concentrates and is removed as crusts during refining of lead. Australia is the western world's second largest mine producer of lead.

In the United Kingdom, Mining & Chemical Products Ltd. (MCP) still processed bismuth ores and bullion to 99.99 per cent and 99.997 per cent refined metal grades for metallurgical and pharmaceutical purposes, respectively. Since production in Bolivia was curtailed in 1980 and since output from Peko-Wallsend in Australia was reduced, a drop in MCP's production to a level of a "few hundred tonnes per year" has been reported by the company.

Improving conditions in the bismuth market led Mines et Produits Chimiques de Salsigne of France to make a decision to return to bismuth production in March 1985 at an expected 100 tpy. In 1979 annual mine output was estimated at 60 tpy bismuth from a complex ore principally mined for its gold content. The company has been selling metal from stocks accumulated at its nearby smelter.

In Bolivia, the new lead-silver smelter of Corporacion Minera de Bolivia (Comibol) and Empresa Nacional de Fundiciones (ENAF) did not produce in 1984 because of a lack of feed. In addition to some 24 000 tpy of lead, the plant is expected to produce some bismuth. Since output was curtailed at Bolivia's bismuth mines in 1980, production has been limited to byproduct bismuth largely associated with tin.

Restarting of Comibol's bismuth refinery is contingent on a market assessment which is expected by February 1985. Resumption of operations at the refinery and at the Quechista mine could be as early as March. Some 200 t of bismuth could be produced in 1985 and perhaps 500 tpy thereafter.

ASARCO Incorporated is the only refined bismuth producer in the United States. Production is largely a byproduct from the processing of imported complex lead concentrates and domestic concentrates and materials which contain bismuth.

PRICES

Reported major sales of bismuth metal by Peru to the Soviet Union and to the People's Republic of China depleted stocks and set the stage for notable price increases later in 1984.

The MCP-Peko Limited published producer price in Europe and the United States remained at $2.30/lb throughout 1983 and up until March 15, 1984. Thereafter it rose to $US 6.50 per pound. The European Free Market price averaged about $US 1.60 per pound in 1983 but climbed steadily from a level of $US 1.75-1.79 at the beginning of 1984 to $US 6.40-6.55 by year-end.

USES

Bismuth is used in a wide variety of applications. Fusible alloys - bismuth alloyed with tin, lead, and cadmium, for example - melt at low temperatures. They are reusable for applications such as blocking complex shapes for machining; grinding and polishing of lenses; and die assembly. These alloys are also used as moulds for plastic extrusion, as filler for tube-bending, as solder alloys, and as thermal fuses.

Bismuth is added to iron, steel and aluminum to improve machineability. Bismuth compounds are used in pharmaceuticals, cosmetics and chemicals. For example, bismuth salts are used as a treatment for indigestion and stomach ulcers. Bismuth oxychloride, either deposited on mica or in dispersed form, provides a pearlescent lustre to lipsticks, face powders, blushes, nail color, eyeshadow and hairspray. Catalysts containing bismuth are also used for production of acrylonitrile and acroleine.

Research is on-going for possible uses in thin metallic films for optical disk information storage; in immiscible alloys for data storage; in an organometallic fungicide; in silver-bismuth oxide electric contact material and in lithium-silver bismuth chromate battery cells.

OUTLOOK

A production shortfall of at least 200 t is predicted by some experts for 1985. If true, bismuth prices can be expected to rise further in the short-term. On the other hand, substitution particularly in chemical and pharmaceutical applications is likely to be relatively rapid and this may dampen a price rise. The possibility of resumed production in Bolivia and France in 1985 also suggests that a sustained high price is unlikely.

TARIFFS

Item No.		British Preferential	Most Favoured Nation	General Preferential	General
CANADA					
33100-1	Bismuth, metallic, in its natural state	free	free	free	free
35106-1	Bismuth metal, not including alloys, in lumps, powders, ingots or blocks	free	free	free	25%

Item No.		1983	1984	1985	1986	1987
UNITED STATES (MFN)				(%)		
601.66	Bismuth ores	Remains free				
632.10	Bismuth metal unwrought and waste and scrap	Remains free				
632.64	Alloys of bismuth: containing by weight not less than 30 per cent lead	Remains free				
632.66	Other alloys of bismuth	7.3	6.8	6.4	5.9	5.5
633.00	Bismuth metal wrought	7.3	6.8	6.4	5.9	5.5

Sources: The Customs Tariff, 1983, Revenue Canada; Customs and Excise; Tariff Schedules of the United States Annotated 1982, USITC Publication 1200; U.S. Federal Register Vol. 44, No. 241.

TABLE 1. CANADA, BISMUTH PRODUCTION, 1983-84, AND CONSUMPTION, 1981-1983

	1983		1984	
	(kilograms)	($)	(kilograms)	($)
Production, all forms[1]				
New Brunswick	165 608	847,400	159 000	1,879,380
British Columbia	47 427	215,319	36 134	427,104
Ontario	8 105	41,472	-	-
Northwest Territories	31 883	163 142	24 500	289,590
Total	253 023	1,267,333	219 634	2,592,074

	1981	1982	1983
		(kilograms)	
Consumption, refined metal (available data)			
Fusible alloys	7 547	7 598	7 241
Other uses	2 547	2 476	1 019
Total	10 094	10 074	7 241

Sources: Statistics Canada; Energy, Mines and Resources Canada.
[1] Refined bismuth metal from Canadian ores, plus recoverable bismuth content of bullion and concentrates exported.
P Preliminary; - Nil.

TABLE 2. CANADA, BISMUTH PRODUCTION AND CONSUMPTION, 1970, 1975 AND 1978-84

	Production all forms[1]	Consumption[2]
	(kilograms)	
1970	267 774	11 135
1975	156 605	29 267
1978	145 104	25 665
1979	136 733	25 177
1980	149 366	10 271
1981	167 885	10 094
1982	189 000	10 074
1983	253 023	7 241
1984P	219 634	..

Sources: Statistics Canada; Energy, Mines and Resources Canada.
[1] Refined bismuth metal from Canadian ores, plus recoverable bismuth content of bullion and concentrates exported. [2] Refined bismuth metal reported by consumers.
P Preliminary; .. Not available.

TABLE 3. UNITED STATES CONSUMPTION OF BISMUTH BY PRINCIPAL USES, 1981-83

	1981	1982	1983
	(kilograms, bismuth content)		
Pharmaceuticals[1]	629 384	519 250	500 640
Metallurgical additives	139 266	56 532	237 144
Fusible alloys	297 990	259 277	282 491
Other alloys	11 772	9 700	9 072
Experimental uses	97	223	762
Other uses	6 806	6 148	6 472
Total	1 085 315	851 130	1 036 581

Source: U.S. Bureau of Mines.
[1] Includes industrial and laboratory chemicals.

TABLE 4. MONTHLY AVERAGE PRICES - BISMUTH

	1982	1983	1984
	(Major producer, $US per pound)		
January	2.30	2.30	2.30
February	2.30	2.30	2.30
March	2.30	2.30	2.49
April	2.30	2.30	2.75
May	2.30	2.30	3.59
June	2.30	2.30	4.00
July	2.30	2.30	4.00
August	2.30	2.30	4.72
September	2.30	2.30	5.00
October	2.30	2.30	5.54
November	2.30	2.30	6.50
December	2.30	2.30	6.50
Year	2.30	2.30	4.14

Source: Metals Week.

Cadmium

M.J. GAUVIN

Cadmium metal is recovered principally as a byproduct of zinc smelting and refining. In 1984 zinc metal production in the non-socialist world was at an all-time high surpassing the previous high reached in 1979. Similarly cadmium metal production in 1984 exceeded the previous high attained in 1979.

Cadmium is a relatively rare element in the earth's crust, occurring most commonly as the sulphides greenockite and hawleyite which are found associated with zinc sulphide ores, particularly sphalerite. There are no ores specifically mined for cadmium. Reserves at any time are a function of zinc reserves.

Smelter residues from which cadmium is extracted may be stockpiled in times of low demand with the result that refined cadmium production is not always directly related to production of the principal metals. During the past six years, cadmium production in Canada has varied from 2.1 to 2.6 kilograms of cadmium to each tonne (t) of zinc metal produced.

Cadmium metal is produced in varying shapes and degrees of purity for various uses. The most common forms are balls, sticks, slabs, ingots, rods and sponge.

In 1983 and 1984, Canada was the non-socialist world's third largest producer of cadmium metal, after Japan and the United States. The next three largest producers were Belgium, the Federal Republic of Germany and Australia. Production of cadmium in the non-socialist world, as reported by the World Bureau of Metal Statistics, increased in 1983 to 13 589 t from 12 733 t in 1982. While data is not yet available, 1984 non-socialist world production is estimated to increase about 7.8 per cent over that of 1983, and Canadian production is estimated at 1 500 t.

Consumption of refined cadmium in Canada, as reported by consumers to Statistics Canada, recorded a small decrease in 1982 to 33 818 kilograms from that reported in 1981. However, consumption, as measured by producers shipments to domestic consumers in 1984, is estimated at 120 000 kg during 1984 compared with 91 310 in 1983 a 7 per cent increase from that reported for 1982.

USES

Cadmium is a soft, ductile, silver-white electropositive metal with a valence of two. It is used mainly for electroplating iron and steel products to protect them against oxidation. The high ductility of cadmium is an advantage where the plated parts are to be formed. The good soldering characteristics of cadmium plate is an advantage in electrical applications. A cadmium coating, like a zinc coating, protects metals that are lower in the electromotive series by physical enclosure and by sacrificial corrosion. Cadmium is usually preferred to zinc as a coating because it is more ductile, can be applied more uniformly in recesses of intricately shaped parts, has a more aesthetic appearance and gives greater protection with the same thickness of plating than with zinc plating.

The second largest use, according to the Statistics Canada survey, is in the manufacture of pigments and chemicals. Cadmium sulphides are used for yellow to orange colours and cadmium sulphoselenides for pink, red and maroon. Cadmium-containing pigments demonstrate good reflectance, heat stability and colour intensity. Cadmium compounds are used as stabilizers in the production of plastics, and cadmium phosphors are used for picture tubes in television sets.

M.J. Gauvin is with the Mineral Policy Sector, Energy, Mines and Resources, Canada. Telephone (613) 995-9466.

Cadmium-bearing batteries such as nickel-cadmium, silver-cadmium and mercury-cadmium, have the advantage of long life, maximum current delivery with a low voltage drop, small size, excellent performance under a wide temperature range and a low rate of self-discharge. They find wide use in aircraft, satellites, missiles, calculators, and a broad assortment of portable tools and appliances.

Other uses of cadmium are for catalysts in the production of primary alcohols and esters, low-melting point alloys used in fire detection, bearing alloys, brazing alloys and solders and copper hardeners for railway catenary and trolley wires.

PRICES

North American prices, which are quoted on a delivered basis, are best represented by the "U.S. Producer" quotations published in Metals Week. European prices by the "European sticks, free market price" quoted by Metal Bulletin. All quoted prices are for cadmium having a minimum purity of 99.95 per cent.

Published U.S. producer prices were $1.00 a pound at the beginning of 1983, rising gradually to $1.25 a pound in September. This price held until March 1984 then peaked at $2.25 during the April to June period, dropped off to $1.55 a pound in August and then maintained that level for the balance of the year.

OUTLOOK

In the long-term, cadmium supply will continue to be dependent on trends established by the zinc idustry. As the level of metal production is determined by the amount of zinc metal production, periods of oversupply will develop. It is expected that greater usage in its traditional markets and possible new uses would gradually absorb the excess supply.

TARIFFS

Item No.		British Preferential	Most Favoured Nation	General	General Preferential
			(%)		
CANADA					
32900-1	Cadmium in ores and concentrates	free	free	free	free
35102-1	Cadmium metal, not including alloys, in lumps, powders, ingots, or blocks	free	free	25	free
UNITED STATES					
601.66	Cadmium in ores and concentrates		free		
632.14	Cadmium metal, unwrought, waste and scrap		free		
632.86	Cadmium alloys, unwrought containing by weight 96% or more but less than 99% of silicon		9%		

Item No.		1983	1984	1985	1986	1987
			(%)			
632.88	Cadmium alloys, unwrought, other	7.3	6.8	6.4	5.9	5.5
633.00	Cadmium metal, wrought	7.3	6.8	6.4	5.9	5.5

EUROPEAN ECONOMIC COMMUNITY (MFN)

Item No.		1983	Base Rate	Concession Rate
			(%)	
26.01	Cadmium in ores and concentrates	free	free	free
81.04	Cadmium metal, unwrought, waste and scrap	4	4	4
	Cadmium metal, other	6	6	6

JAPAN (MFN)

Item No.		1982	Base Rate	Concession Rate
			(%)	
26.01	Cadmium in ores and concentrates	free	free	free
81.04	Cadmium metal:			
	Unwrought	6.2	10	5.1
	Waste and scrap	6.0	10	4.8
	Powders and flakes	6.6	10	5.8
	Cadmium metal, other	8.6	15	6.5

Sources: The Customs Tariff and Commodities Index, January, 1983. Revenue Canada; Tariff Schedules of the United States Annotated 1983, USITC Publication 1317; U.S. Federal Register, Vol. 44, No. 241; Official Journal of the European Communities, L 318, Vol. 25; Customs Tariff Schedules of Japan, 1983.

TABLE 1. CANADIAN PRIMARY CADMIUM STATISTICS, 1981-84

	1981	1982	1983P	1984e
	(tonnes)			
Mine production[1]	834	886	1 107	1 200
Metal production	1 293	1 162	1 296	1 500
Metal capacity	1 800	1 800	1 800	1 800
Metal shipments:				
Domestic	131	85	91	120
Exports	1 182	731	1 611	1 450

Sources: Statistics Canada; Energy, Mines and Resources Canada.
P Preliminary e Estimated; [1] All forms.

TABLE 2. CANADA, CADMIUM METAL CAPACITY, 1984

Company and Location	Annual Capacity
	(tonnes)
Cominco Ltd.	
Trail, British Columbia	640
Canadian Electrolytic Zinc Limited	
Valleyfield, Quebec	550
KCML Inc.	
Hoyle, Ontario	450
Hudson Bay Mining and Smelting Co., Limited	
Flin Flon, Manitoba	160
Total Canada	1 800

Sources: Mining & Mineral Processing Operations in Canada, 1983. Energy, Mines and Resources, Canada.

TABLE 3. CANADA, CADMIUM PRODUCTION AND EXPORTS 1982, 1983, AND 1984 AND CONSUMPTION 1981 and 1982

	1982		1983P		1984P	
	(kilograms)	($)	(kilograms)	($)	(kilograms)	($)
Production						
All forms[1]						
Ontario	549 006	1,663,000	834 000	2,551,000		
British Columbia	147 656	447,000	136 000	418,000		
Quebec	135 479	410,000	90 000	275,000		
Manitoba	37 479	114,000	38 000	117,000		
Newfoundland	9 818	30,000	–	–		
Saskatchewan	6 617	20,000	9 000	27,000		
Total	886 055	2,684,000	1 107 000	3,388,000	..	..
Refined[2]	1 162 390	..	1 296 078	..	1 500 000	
Exports					(January – September)	
United States	378 645	1,161,000	776 432	2,978,000	610 415	2,663,000
United Kingdom	319 555	770,000	495 481	1,078,000	389 111	1,286,000
Netherlands	10 151	40,000	87 996	128,000	9 112	26,000
Belgium-Luxembourg	65	1,000	4 536	20,000	5 000	8,000
Other countries	61 114	154,000	666	28,000	343	38,000
Total	769 530	2,126,000	1 365 111	4,232,000	1 013 981	4,021,000

	1981	1982
	(kilograms)	
Consumption		
Cadmium metal[3]		
Plating	16 039	15 404
Solders	1 387	247
Other uses[4]	16 666	18 167
Total	34 092	33 818

Sources: Statistics Canada; Energy, Mines and Resources Canada.
[1] Production of refined cadmium from domestic ores, plus recoverable cadmium content of ores and concentrates exported. [2] Refined metal from all sources and cadmium sponge. [3] Available data reported by consumers. [4] Mainly chemicals, pigments and alloys other than solder.
P Preliminary; – Nil; .. Not available.

TABLE 4. CANADA, CADMIUM PRODUCTION, EXPORTS AND DOMESTIC SHIPMENTS, 1970, 1975 AND 1979-84

	Production		Exports Cadmium Metal	Producers' Domestic Shipments
	All Forms[1]	Refined[2]		
	(kilograms)			
1970	1 954 055	836 745	702 630	157 307
1975	1 191 674	1 142 508	637 797	98 820
1979	1 209 459	1 454 954	1 292 515	120 926
1980	1 033 000	1 302 955	1 095 825	88 232
1981	833 788	1 293 265	1 452 904	131 175
1982	886 055	1 162 390	769 530	84 910
1983	1 107 000	1 296 078	1 365 111	91 310
1984e	1 200 000	1 500 000	1 400 000	120 000

Sources: Statistics Canada; Energy, Mines and Resources Canada.
[1] Production of refined cadmium from domestic ores plus recoverable cadmium content of ores and concentrates exported. [2] Refined metal from all sources and cadmium sponge.
e Estimated.

TABLE 5. CADMIUM METAL PRICES, 1983, 1984

	Average Monthly Prices			
	Metals Week		Metal Bulletin	
	U.S. Producer	New York Dealer	European Sticks free market	Cominco
Month 1983	($US/lb)		($US/lb)	($Cdn/lb)
January	1.000	0.700	0.561-0.669	1.15
February	1.000	0.813	0.840-0.943	1.15
March	1.000	0.850	0.910-1.000	1.15
April	1.000	0.850	0.923-1.000	1.15
May	1.086	0.810	0.900-0.970	1.15
June	1.150	0.800	0.865-0.930	1.15
July	1.150	0.800	0.840-0.904	1.15
August	1.159	0.824	0.871-0.925	1.15
September	1.250	0.885	0.912-0.950	1.18
October	1.250	0.852	0.814-0.904	1.25
November	1.250	0.770	0.747-0.792	1.25
December	1.250	0.878	0.871-0.913	1.25
Average	1.129	0.819	0.840-0.900	1.10
1984				
January	1.25	0.910	0.86-0.90	1.25
February	1.25	1.056	1.05-1.13	1.25
March	1.50	1.320	1.45-1.56	1.67
April	2.25	1.705	1.68-1.74	1.75
May	2.25	1.687	1.61-1.67	2.13
June	2.25	1.543	1.51-1.60	2.25
July	1.82	1.326	1.25-1.35	2.25
August	1.55	1.285	1.27-1.33	1.78
September	1.55	1.324	1.29-1.34	1.75
October	1.55	1.204		1.75
November				
December				
Average				

Sources: Metals Week, Cominco Ltd., Metal Bulletin

TABLE 6. WESTERN WORLD CADMIUM METAL PRODUCTION, 1981-83

Continent and Country	1981	1982	1983
		(tonnes)	
Europe			
Austria	55	49	46
Belgium	1 176	1 001	1 217
Finland	621	566	616
France	664	580	447
Germany, F.R.	1 192	1 030	1 094
Italy	489	475	386
Netherlands	518	497	513
Norway	117	102	117
Spain	303	286	278
United Kingdom	278	354	340
Yugoslavia	208	174	48
Africa			
Algeria	30	30	30
Namibia	-	110	51
Zaire	230	281	308
Asia			
India	113	131	131
Japan	1 977	2 021	2 215
Republic of Korea	300	320	460
Turkey	10	10	10
America			
Canada	1 293	1 162	1 296
Mexico	633	674	847
Peru	312	425	443
United States	1 871	1 351	1 382
Other America	28	94	210
Australia	1 031	1 010	1 104
Other	-	-	-
Western World	13 449	12 733	13 589

Sources: World Metal Statistics, September 1984
Energy Mines and Resources Canada.
- Nil.

Cement

D.H. STONEHOUSE

SUMMARY 1983-1984

In 1983, for the fourth consecutive year, cement shipments from Canadian plants were down from the year previous. A slight recovery in 1984 can be credited for the most part to increased exports. Canadian consumption of portland cement bears direct relation to activity in the construction industry. With Canadian construction being anything but buoyant over the past three years, domestic cement consumption has shown no appreciable growth. The lack of mega projects, particularly in western Canada, has steadily reduced demand from that region while demand from eastern Canada has remained fairly steady, assisted in some measure by increased housing activity in 1983. Poor overall construction performance has brought about some rationalization within the Canadian cement industry. A few plants were closed for extended periods during 1983 and again through 1984, a number of kilns were shut down and in general the industry operated at about 50 per cent of rated clinker-producing capacity. One plant (Ciment Québec Inc.) brought on new capacity late in 1982 to bring Canadian total production capacity to 16.54 million tonnes (t).

Exports of Canadian cement and clinker are mainly to the United States, in particular to the states of New York and Michigan. An economic recovery, begun in late 1982 in the United States, created strong demand for many construction materials. Canadian cement production efficiencies and the strong American dollar combine to make Canadian cement and clinker competitive in bordering states, quite different than being imported as a supplement to United States production. Imports from Mexico and from off-continent sources have added to the concerns of the United States cement producers. Protectionist measures have been considered. Of particular concern to Canadian exporters is the Buy America provision within the United States Surface Transportation Assistance Act, 1982 (STAA). STAA provides substantial funding for highway and bridge projects in the United States, representing about 6 per cent total United States cement consumption. Through 1983 and early-1984, Canadian exporters were effectively excluded from supplying these projects by Buy America restrictions on foreign cement. Congress lifted these restrictions in March, 1984 and Canadian cement now enjoys full access to STAA-funded projects. The United States cement industry, however, continues to lobby Congress to restore Buy American restrictions on foreign cement. Legislation to this effect failed to win approval in the last Congress, but will likely be re-proposed during 1985.

The major exporters continued to strengthen their United States base during 1983 and 1984. With the acquisition of General Portland Inc. of Dallas, Texas in 1982, Canada Cement Lafarge Ltd. became the largest cement producer in North America with an annual capacity of 11.663 million t. Early in 1983 a corporate reorganization established Dallas-based Lafarge Corporation which wholly-owns both Canada Cement Lafarge Ltd. and General Portland Inc. The move was designed to give the corporation access to the United States money market and to maintain the overall 52 per cent control of both companies by Lafarge Coppée of Paris, France. Late in 1984 General Portland announced it would temporarily close its 600 000 tpy Miami, Florida plant and would replace production with cement imported from Mexico to supplement production from its Tampa plant and from other sources.

St. Lawrence Cement Inc., through its United States subsidiary, Independent Cement Corp., has acquired or built a cement distribution system in northeastern United States over the past three years. Early in 1984 the company purchased the Catskill, New York cement plant and termi-

---

D.H. Stonehouse is with the Mineral Policy Sector, Energy, Mines and Resources, Canada. Telephone (613) 995-9466.

nals from Lone Star Industries Inc. for a reported $US 30 million.

St. Marys Cement Company has two United States affiliates - St. Marys Wyandotte Cement Inc. and St. Marys Wisconsin Cement Inc. The former operates a 300 000 tpy grinding plant near Detroit, the latter a 150 000 tpy grinding plant in Milwaukee and distribution terminals in Green Bay, Wisconsin and Waukegan, Illinois.

CANADIAN DEVELOPMENTS

The Canadian cement industry is strongly regionalized on the basis of market availability. Capacity is concentrated near growth areas and, fortunately for some, these areas are convenient to foreign market access as well. Some plants were located to take advantage of existing United States markets and to be in a position to utilize waterborne, high-bulk transportation facilities. A typical feature of the Canadian cement industry is its diversification and vertical integration into related construction and construction materials fields. Many cement manufacturers also supply ready-mix concrete, stone, aggregates and concrete products such as slabs, bricks and prestressed concrete units.

Cement manufacture is energy-intensive. It is obvious that research should be concentrated in this area, and specifically within the pyroprocessing sector where over 80 per cent of the energy is consumed. Raw material grinding and finish grinding are being studied to determine optimum particle size for energy consumed.

Energy conservation programs adopted by the Canadian cement industry more than reached the goal of a 9 to 12 per cent reduction in energy consumption per unit of production, based on 1974 calculations. In 1983 the average plant consumption of energy of all types was 4 896 mega joules a tonne, a 21.3 per cent fuel saving over 1974.

A change in the fuel mix from 1974 to 1983 is noted. In 1974 natural gas accounted for 49.5 per cent, petroleum products 39.7 per cent and coal and coke 10.8 per cent. For 1983 natural gas usage was 36.0 per cent of the total energy requirements while petroleum products were 12.3 per cent and coal and coke rose to 51.7 per cent.

The dry process now accounts for over 70 per cent of Canadian portland cement capacity. In 1983 over 80 per cent of production was from dry process plants.

Energy conservation demonstration projects have been funded through the Conservation and Non-Petroleum Sector of Energy, Mines and Resources. The industry is represented on the Industrial Minerals Task Force on Energy Conservation and continues to play an active role in this voluntary organization.

In terms of the energy required to make concrete components and to build concrete structures, along with energy requirements to service and maintain them, they are not so energy-intensive as the nearly 5 giga joules required per t of cement would at first indicate. Through the Canada Centre for Mineral and Energy Technology, a branch of Energy, Mines and Resources and through the Building Research Division of the National Research Council a continuing program of concrete research is managed. Concrete research has generally been confined to strength determination, durability, placement and curing. Currently, great emphasis is being placed on researching the use of super-plasticizers in concrete. Super-plasticizers, a group of admixtures described chemically as naphthalene or melanine sulphonate polymers, have been found to provide greater workability over short time spans or to provide high strength by permitting lower water-cement ratios.

Portland cement used in Canada should conform to the specifications of CAN 3-A5-M83, published by the Canadian Standards Association (CSA). This standard covers the five main types of portland cement. Masonry cement produced in Canada should conform to the CAN 3-A8-M83. Blended hydraulic cements are covered by CAN 3-A362-M83. The cement types manufactured in Canada, but not covered by the CSA standards, generally meet the appropriate specifications of the American Society for Testing and Materials (ASTM).

The three plants in the **Atlantic region** constitute just over 5 per cent of total clinker producing capacity. All three obtain raw materials at or near the plant site. North Star Cement Limited purchases gypsum from Flintkote Holdings Limited, which quarries at Flat Bay about 65 km south of Corner Brook while National Gypsum (Canada) Ltd. supplies the Brookfield plant of Canada Cement Lafarge Ltd. (CCL) from its Milford, Nova Scotia quarry. CCL's New

Brunswick plant was closed during 1983 and 1984 for extensive periods. This plant has been the company's principal source of oil well cement. The region consumed 391 700 t of cement in 1983 according to Canadian Portland Cement Association data.

In the **Quebec region** the five clinker-producing plants have 25 per cent of the Canadian total in an area that has 26.1 per cent of Canadian population and which, in 1983, consumed about 1.2 million t of portland cement or 20 per cent of total consumption. At its St. Constant plant, south of Montreal, CCL experimented with the use of waste tires and rubber as an alternate fuel, as part of a program administered by the federal departments of Environment and Energy, Mines and Resources.

Miron Inc. continued a program to utilize methane gas from a garbage disposal project on its property with the goal of eventually securing as much as 40 per cent of its fuel requirements from this source. The plant's boiler room was operated in 1983 on methane gas. Input garbage has the energy potential to operate the company's two kilns. St. Lawrence Cement Inc. also continued its energy-saving programs during 1983-84 but concentrated its expenditures on downstream integrated operations in the ready mix concrete field and in acquiring distribution facilities in the United States through its subsidiary, Independent Cement Corporation headquartered in Albany, New York. Ciment Québec Inc. began full operation of its new suspension-preheater-precalciner system in 1983, adding about 735 000 tpy to capacity.

Portland cement consumption increased in the **Ontario region** where 40 per cent of the nation's clinker-producing capacity is concentrated. Canada Cement Lafarge Ltd. has brought into production about 3 million t of new cement capacity over the past seven years and currently over half of its operating kilns are less than 10 years old. The limestone for CCL's Bath, Ontario plant is quarried on-site while silica is supplied from Potsdam sandstone at Pittsburgh about 65 km east of Bath and iron oxide is purchased from Hamilton. Gypsum is from Nova Scotia. The Woodstock plant has experimented with the use of selected, processed garbage as fuel. The plant obtains limestone on site, silica from Indusmin Limited, iron oxide from Stelco Inc. and gypsum from southern Ontario mines.

At Picton, Lake Ontario Cement Limited operates one of the largest cement plants in North America. The four-kiln plant supplies cement and clinker to its United States subsidiaries - Rochester Portland Cement Corp. in New York state and Aetna Cement Corporation in Michigan - and cement to its Ontario markets.

At its Mississauga plant, St. Lawrence Cement Inc. has continued to research energy saving techniques. The company obtains limestone from Ogden Point, 160 km east of Toronto on the shore of Lake Ontario and gypsum is purchased from Nova Scotia or from southern Ontario mines.

The Bowmanville plant of St. Marys Cement Limited was expanded in 1973 with the addition of a second kiln. With the acquisition of Wyandotte Cement Inc., the company began shipments of clinker through a newly constructed lakefront loading facility at Bowmanville. The original plant at St. Marys, constructed in 1912 to serve the Toronto area, has been expanded and modernized over the years, most recently with the installation of a 680 000 tpy kiln and four-stage suspension preheater.

Federal White Cement's plant at Woodstock, can produce up to 100 000 tpy of white cement.

Two companies, Canada Cement Lafarge Ltd. and Genstar Cement Limited operate a total of five clinker producing plants in the **Prairie region** and three in the **Pacific region** along with two clinker grinding plants. This **Western region** has 30 per cent of clinker producing capacity, including the recently completed expansion at Genstar's Edmonton, Alberta plant. Consumption of portland cement in the western provinces accounted for 38 per cent of Canadian total. Recent expansions at Edmonton and at Exshaw increased capacity by about 1.3 million tpy through 1981.

Genstar Cement Limited continued to increase the productive capacity at its Cadomin limestone property which supplies the Edmonton plant through a 4 500 t unit train and materials handling system. A limestone quarry at Mafeking, Manitoba, near the Manitoba-Saskatchewan border, supplies limestone to Genstar's Regina plant, while the Winnipeg plant is supplied from Steep Rock, Manitoba.

CCL's Winnipeg plant obtains limestone from the company's quarry at Steep Rock on

Lake Manitoba, gypsum from Westroc Industries Limited at Amaranth, silica from Beausejour and clay adjacent to the plant site at Fort Whyte. Raw material for the Exshaw plant is mainly from the plant site but for gypsum from Westroc and iron oxide from Cominco Ltd. Limestone from Texada Island supplies the company's Vancouver plant at Richmond. Their Kamloops plant is supplied from resources close to the plant site.

WORLD DEVELOPMENTS

Cement markets are regional and centred in developing urban areas where construction activity is concentrated, or in areas where mining or heavy engineering construction projects are being carried out. The normal market area of a given cement-producing plant depends on the amount of transportation cost that the selling price can absorb. A potential large volume of sales could warrant a secondary distribution terminal; water transportation to a distribution system could extend a plant's market area even farther. Because raw materials for cement manufacture are generally widespread, most countries can supply their own cement requirements if the market volume warrants a plant. Few countries rely entirely on imports for their cement needs. However, some countries rely heavily on export markets for their surplus cement production in order to operate facilities economically.

Specialty cements such as white cement can be transported greater distances than ordinary grey portland cement because the transportation costs do not represent as high a proportion of the landed price, and because quantities required are generally much smaller than for portland cement.

Cement shortages in countries experiencing a buoyant surge in construction have led to exceptions to the norm and resulted in cement being shipped unusual distances.

The state of the portland cement industry in the United States, and a surprisingly large demand for cement in construction, particularly in the west and mid-west, created improved export opportunities for Canadian portland cement during the late 1970s, peaking in 1979.

Significant quantities of portland cement were imported into the United States from such countries as Australia, Columbia, Denmark, France, Korea, Mexico, Norway, Spain and Venezuela as well as from Canada.

Cembureau, The European Cement Association, has published Cement Standards of the World - Portland Cement and its Derivatives, in which standards are compared. Cembureau's World Cement Directory lists production capacities by country and by company.

USES

Portland cement is produced by burning, usually in a rotary kiln, an accurately proportioned, finely ground mixture of limestone, silica, alumina and iron oxide. The three basic types of portland cement, Normal Portland, High-Early-Strength Portland, and Sulphate-Resisting Portland, are produced by most Canadian cement manufacturers.

Cement has little use alone but, when combined with water, sand, gravel, crushed stone or other aggregates in proper proportions acts as a binder, cementing the materials together as concrete. Concrete has become a widely used and readily adaptable building material which can be poured on site in large engineering projects, or used in the form of delicate precast panels or heavy, prestressed columns and beams in building construction.

Kiln discharge, in the shape of rough spheres, is a fused, chemically complex mixture of calcium silicates and aluminates termed clinker, which is mixed with gypsum (4 to 5 per cent by weight) and ground to a fine powder to form portland cement. By close control of the raw mix, the burning conditions and of the use of additives in the clinker-grinding procedure, finished cements displaying various desirable properties can be produced.

Moderate Portland Cement and Low-Heat-of-Hydration Portland Cement, designed for use in concrete to be poured in large masses, such as in dam construction, are manufactured by several companies in Canada. Masonry cement (generic name) includes such proprietary names as Mortar Cement, Mortar Mix (unsanded), Mason's Cement, Brick Cement and Masonry Cement. The latter product produced by portland cement manufacturers, is a mixture of portland cement, finely ground high-calcium limestone (35 to 65 per cent by weight) and a plasticizer. The other products do not necessarily consist of portland cement and limestone, and may include a mixture of portland cement and hydrated lime and/or other plasticizers.

OUTLOOK

The mining industries which supply the materials of construction fared no better during 1983 and 1984 than did either the construction industry or the mining industry in general. Total plant shutdowns of unprecedented extent were not uncommon in the cement industry.

The Canadian Construction Association predicts that near-term growth in construction will be less than the average growth in the economy with non-residential building and heavy construction activity being above average but offset by no growth in residential building and engineering work. The medium- and long-term outlook is for average growth as engineering projects pick up, non-residential building activity remains steady and the residential building sector continues to show little if any growth. The capital investment intentions for 1984 of major Canadian companies surveyed during 1984 were increased from 1983 intentions in both the manufacturing and non-manufacturing sectors. Statistics Canada's review of private and public investment in Canada has been adjusted to indicate capital and repair expenditures on construction in 1983 of $56.0 billion, about the same as in 1982, while 1984 expenditures rose to $56.9 billion.

A healthy economy would permit the construction industry and that portion of the mining industry which depends on it to plan five to ten years ahead with obvious benefits in efficiency, rather than to invest with short-term survival as the main incentive.

The cement industry in Canada is capable of meeting immediate demands and is also capable of expansion to meet even greater demand from domestic and foreign markets should opportunities be presented. The pattern of consumption of portland cement established during 1983-84 will likely persist for a few years or until the development of mega projects once again alters the current demand for cement.

Conservation of energy and raw materials within the cement industry is of worldwide concern and provides a theme around which major developments in the industry have taken place. Of particular note is the emphasis on blended cements and the utilization of slag, ash and other byproducts. Even greater additions to production capacities than those witnessed during the past few years will be needed to meet demand in many developing countries.

TARIFFS

Item No.		British Preferential	Most Favoured Nation	General	General Preferential
			(cents per hundred pounds)		
CANADA					
29000-1	Portland and other hydraulic cement, nop; cement clinker	free	free	6	free
29005-1	White, nonstaining Portland cement	3.9	3.9	8	2 2/3

MFN Reductions under GATT (effective January 1 of year given)	1983	1984	1985	1986	1987
	(cents per hundred pounds)				
29005-1	3.9	3.8	3.8	3.7	3.7

UNITED STATES (MFN)

Item No.		
511.11	White, nonstaining Portland cement per 100 pounds including weight of container	1¢
511.14	Other cement and cement clinker	free
511.21	Hydraulic cement concrete	free

	1983	1984	1985	1986	1987
	(% ad valorem)				
511.25 Other concrete mixed, per cubic yard	6.2	5.9	5.6	5.2	4.9

Sources: The Customs Tariff, 1983, Revenue Canada, Customs and Excise; Tariff Schedules of the United States Annotated 1983, USITC Publication 1317; U.S. Federal Register Vol. 44, No. 241.

TABLE 1. CANADA, CEMENT PRODUCTION AND TRADE, 1982-84

	1982		1983		1984P	
	(tonnes)	($000)	(tonnes)	($000)	(tonnes)	($000)
Production[1]						
By province						
Ontario	2 800 000	215,208	2 900 565	203,243	3 100 000	222,859
Quebec	2 307 000	129,987	2 170 977	127,567	2 675 000	146,634
Alberta	1 468 000	112,830	1 189 610	126,860	1 025 000	116,450
British Columbia	776 000	70,352	76 570	66,282	810 000	84,297
Nova Scotia	..	27,670	..	10,014	..	17,084
Manitoba	275 000	21,137	289 672	31,983	365 000	42,891
Saskatchewan	206 000	15,833	184 268	21,154	165 000	17,766
New Brunswick	..	13,066	..	8,965	..	9,676
Newfoundland	..	6,443	..	10,035	..	9,453
Total	8 418 000	612,536	7 870 878	606,101	8 618 000	667,110
By type						
Portland	8 152 000	..	7 614 832	..	8 334 600	..
Masonry[2]	266 000	..	256 046	..	284 000	..
Total	8 418 000	612,526	7 870 878	606,101	8 618 600	667,110
					(Jan.-Sept. 1984)	
Exports						
Portland cement						
United States	1 464 650	66,829	1 499 751	71,574	1 476 123	73,530
Saudi Arabia	285 339	12,446	40 093	1,735	–	–
Algeria	–	–	19 076	1,112	1 510	290
Other countries	2 152	248	2 161	286	5 641	534
Total	1 752 141	79,523	1 561 081	74,707	1 483 274	74,354
Cement and concrete basic products						
United States	..	30,103	..	44,443	..	42,797
Other countries	..	1,870	..	1,935	..	1,664
Total	..	31,973	..	46,378	..	44,461
Imports						
Portland cement, standard						
United States	469 643	32,508	227 251	16,119	194 414	12,539
United Kingdom	–	–	170	14	170	14
Total	469 643	32,508	227 421	16,132	194 584	12,553
White cement						
United States	4 716	386	1 457	230	1 279	206
Japan	477	83	1 167	187	1 079	172
Other	50	6	249	31	153	19
Total	5 243	475	2 873	458	2 511	396
Aluminous cement						
United States	14 251	2,833	3 338	1,173	2 518	882
South Africa	8	11				
Total	14 259	2,844	3 338	1,173	2 518	882

TABLE 1. (cont'd)

	1982		1983		1984P	
	(tonnes)	($000)	(tonnes)	($000)	(tonnes)	($000)
					(Jan-Sept 1984)	
Cement, nes						
United States	35 291	4,797	19 069	2,776	15 070	2 213
Japan	250	44	200	33	100	17
United Kingdom	67	9	32	11	8	5
West Germany	30	4	7	1	7	1
France	14	2	-	-	-	-
Italy	6	1	13	3	9	2
Switzerland	-	65	4	1		
Other	-	-	5	9	-	-
Total	35 658	4,857	19 330	2,835		
Total cement imports	231 614	131,594	252 962	20,598	15 194	2 238
Refractory cement and mortars						
United States	..	12,308	..	13,374	..	9,747
Ireland	..	625	..	503	..	503
Austria	..	157	..	203	..	203
West Germany	..	88	..	239	..	220
United Kingdom	..	21	..	111	..	43
Other countries	..	65	..	26	..	22
Total	..	13,264	..	14,456	..	10,740
Cement and concrete basic products, nes						
United States	..	2,712	..	3,969	..	3,141
Japan	..	64	-	-	-	-
Austria	..	11	-	-	-	-
France	..	3	..	1	-	-
United Kingdom	..	2	..	1	-	-
Other countries	..	9	..	30	..	29
Total	..	2,801	..	4,001	..	3,171
Cement clinker						
United Kingdom	180	63	-	-	-	-
United States	36	29	53	2	53	2
Japan	-	-	-	-	-	-
Total	216	92	53	2	53	2

Sources: Statistics Canada; Energy, Mines and Resources Canada.
[1] Producers' shipments plus quantities used by producers. [2] Includes small amounts of other cement.
P Preliminary; .. Not available; - Nil; nes Not elsewhere specified.

TABLE 2. CEMENT PLANTS, APPROXIMATE ANNUAL GRINDING CAPACITY, END OF 1984

Company	Plant	Wet, Dry, Pre-heater	Fuel (Coal Oil Gas)	No. of Kilns	Grinding Capcity	Clinker Capacity
					(000 tpy)	
Atlantic						
Canada Cement Lafarge Ltd.	Brookfield, N.S.	D	C,O	2	485	458
	Havelock, N.B.	D	C,O	2	315	300
North Star Cement Limited	Corner Brook, Nfld.	Dx	O	1	250	120
Atlantic Region Total				5	1 050	878
Quebec						
Canada Cement Lafarge Ltd.	St. Constant	D	O,G	2	955	902
Ciment Québec Inc.	St. Basile	W,Dc	O	3	575	1 106
Miron Inc.	Montreal	D	O,G	2	1 000	840
St. Lawrence Cement Inc.	Beauport	W	C,O	2	550	598
(Independent Cement Inc.)	Joliette	D	C,O	4	1 000	976
Quebec Region Total				13	4 080	4 422
Ontario						
Canada Cement Lafarge Ltd.	Woodstock	W	C,G	2	535	505
	Bath	Dx	O,G	1	1 000	943
Federal White Cement	Woodstock	D	O	1	100	100
Lake Ontario Cement Limited	Picton	D,Dx	C,G	4	744	1 419
St. Lawrence Cement Inc.	Clarkson	W,Dc	C,O,G	3	2 400	1 700
St. Marys Cement Limited	Bowmanville	W	C	2	790	600
	St. Marys	W,Dx	O,G	3	800	990
Ontario Region Total				16	6 270	6 257
Prairies						
Canada Cement Lafarge Ltd.	Fort Whyte, Man.	W	O,G	2	565	532
	Exshaw, Alta.	D,Dc	G	3	1 230	1 184
	Edmonton, Alta.				220	
Genstar Cement Limited	Winnipeg, Man.	W	O,G	1	325	310
	Regina, Sask.	D	O,C	1	375	214
	Edmonton, Alta.	W,Dc	G	4	2 040	1 186
Prairies Region Total				11	4 755	3 426
British Columbia						
Canada Cement Lafarge Ltd.	Kamloops	D	G	1	190	180
	Richmond	W	O,G	2	555	522
Genstar Cement Limited	Tilbury Island	Dx	O,G	1	1 000	855
B.C. Region Total				4	1 745	1 557
CANADA TOTAL (9 companies)				49	17 900	16 540

Source: Market and Economic Research Department, Portland Cement Association.

TABLE 3. CANADA, CEMENT PLANTS, KILNS AND CAPACITY UTILIZATION, 1977-84

	Clinker Producing Plants	Kilns	Approximate Cement Grinding Capacity[1] (tpy)	Portland and Masonry Cement Production[2] (t)	Clinker Exports[3] (t)	Approximate Total Production[4] (t)	Capacity Utilization (%)
1977	22	49	14 885 000	9 639 679	775 145	10 414 824	72
1978	24	51	15 985 000	10 558 279	1 077 274	11 635 553	72
1979	24	51	15 985 000	11 765 248	1 530 537	13 295 785	83
1980	23	47	16 363 000	10 274 000	726 087	11 000 087	67
1981	23	48	16 771 000	10 145 000	524 006	10 669 006	64
1982	23	48	16 771 000	8 418 000	290 329	8 708 329	50
1983	23	49	17 900 000	7 870 878	404 793	8 275 671	46
1984	23	49	17 900 000	8 618 600	350 000e	8 968 600e	50

Sources: Statistics Canada, U.S. Bureau of Mines, Portland Cement Association (PCA)

[1] Includes two plants that grind only. [2] Producers' shipments and amounts used by producers. [3] Imports to United States from Canada. [4] Cement shipments plus clinker exports.
e Estimated.

TABLE 4. CANADA, HOUSE CONSTRUCTION, BY PROVINCE, 1982 AND 1983

	Starts			Completions			Under Construction		
	1982	1983	% Diff.	1982	1983	% Diff.	1982	1983	% Diff.
Newfoundland	2 793	3 281	17.4	2 331	3 176	36.2	3 373	3 494	3.5
Prince Edward Island	248	673	171.3	98	548	459.1	196	316	61.2
Nova Scotia	3 691	5 697	54.3	3 174	5 069	59.7	2 506	2 984	19.0
New Brunswick	1 680	4 742	182.2	1 427	3 487	144.3	1 122	2 346	109.0
Total (Atlantic Provinces)	8 412	14 393	71.1	7 030	12 280	74.6	7 197	9 140	26.9
Quebec	23 492	40 318	71.6	21 526	35 681	65.7	14 164	18 320	29.3
Ontario	38 508	54 939	42.6	40 437	55 287	36.7	31 009	30 243	-2.4
Manitoba	2 030	5 985	194.8	1 633	4 076	149.6	1 149	3 048	165.2
Saskatchewan	6 822	7 269	6.5	5 666	8 090	42.7	4 583	3 667	19.9
Alberta	26 789	17 134	36.0	31 364	24 693	21.2	17 663	8 336	52.8
Total (Prairie Provinces)	35 641	30 388	-14.7	38 663	36 859	-4.8	23 395	15 051	-35.6
British Columbia	19 807	22 607	14.1	26 286	22 901	-12.8	13 290	12 176	-8.3
Total Canada	125 860	162 645	29.2	133 942	163 008	21.7	89 055	84 930	-4.6

Source: Canada Mortgage and Housing Corporation.

TABLE 5. CANADA, VALUE OF CONSTRUCTION[1] BY TYPE, 1982-84

	1982	1983	1984
	($ millions)		
Building Construction			
Residential	13,581	16,683	17,240
Industrial	3,044	2,502	2,739
Commercial	7,064	6,228	5,817
Institutional	3,092	3,198	3,183
Other building	2,062	1,989	2,139
Total	28,843	30,600	31,118
Engineering Construction			
Marine	480	404	414
Highways, airport runways	4,310	4,270	4,328
Waterworks, sewage systems	2,244	2,402	2,391
Dams, irrigation	314	295	306
Electric power	4,866	4,673	3,827
Railway, telephones	2,390	2,531	2,811
Gas and oil facilities	9,706	8,115	9,141
Other engineering	2,912	2,808	2,635
Total	27,222	25,498	25,853
Total construction	56,065	56,098	56,971

Source: Statistics Canada.
[1] Actual expenditures 1982, preliminary actual 1983, intentions 1984.

TABLE 6. CANADA, VALUE OF CONSTRUCTION[1] BY PROVINCE, 1982-84

	1982			1983			1984		
	Building Construction	Engineering Construction	Total	Building Construction	Engineering Construction	Total	Building Construction	Engineering Construction	Total
	($000)								
Newfoundland	414,429	750,073	1,164,502	496,177	920,309	1,416,486	529,042	904,131	1,433,173
Nova Scotia	681,430	884,462	1,565,892	850,097	1,113,145	1,963,242	935,167	1,263,679	2,198,846
New Brunswick	619,611	462,089	1,081,700	749,843	414,249	1,164,092	866,945	503,785	1,370,730
Prince Edward Island	86,981	72,006	158,987	106,406	70,694	177,100	117,272	79,046	196,318
Quebec	5,547,556	4,672,040	10,219,596	6,693,708	4,388,346	11,082,054	7,183,496	4,352,134	11,535,630
Ontario	8,897,137	5,510,574	14,407,711	10,015,802	4,819,861	14,835,663	10,498,275	5,031,231	15,529,506
Manitoba	764,362	657,850	1,422,212	986,418	656,087	1,642,505	1,083,361	698,296	1,781,657
Saskatchewan	1,165,189	1,343,933	2,509,122	1,451,012	1,413,370	2,864,382	1,410,011	1,515,541	2,925,552
Alberta	6,053,165	8,349,406	14,402,571	4,761,621	7,044,529	11,806,150	3,920,440	7,281,719	11,202,159
British Columbia, Yukon and Northwest Territories	4,613,640	4,519,456	9,133,096	4,488,816	4,657,297	9,146,113	4,574,138	4,223,386	8,797,524
Canada	28,843,500	27,221,889	56,065,389	30,599,900	25,497,887	56,097,787	31,118,147	25,852,948	56,971,095

Source: Statistics Canada.
[1] Actual expenditures 1982, preliminary actual 1983, intentions 1984.

Clays and Clay Products

M. PRUD'HOMME

Clays are a complex group of industrial minerals, each generally characterized by different mineralogy, occurrence and uses. All are natural, earthy, fine-grained minerals of secondary origin, composed mainly of a group of hydrous aluminum phyllosilicates and may contain iron, alkalis and alkaline earths. The clay minerals, formed by the chemical weathering or alteration of aluminous minerals are generally classified into four major groups based on detailed chemistry and crystalline structure - the kaolinite group, the smectite group (montmorillonite group of some usages), the clay-mica group and the chlorite group. Clay deposits suitable for the manufacture of ceramic products may include non-clay minerals such as quartz, calcite, dolomite, feldspar, gypsum, iron-bearing minerals and organic matter. The non-clay minerals may or may not be deleterious, depending upon individual amounts present and on the particular application for which the clay is intended.

The commercial value of clays, and of shales that are similar in composition to clays, depends mainly on their physical properties - plasticity, strength, shrinkage, vitrification range and refractoriness, fired colour, porosity and absorption - as well as proximity to growth centres in which clay products will be consumed.

Brick and drain tile manufacturing included in the heavy clay products category accounts for nearly 80 per cent of the total value of output by clay products manufacturers using material from domestic sources.

USES, TYPE AND LOCATION OF CANADIAN DEPOSITS

Common Clays and Shale. Common clays and shales are the principal raw materials available from Canadian deposits for the manufacture of structural clay products. They are found in all parts of Canada, but deposits having excellent drying and firing properties are generally scarce and new deposits are continually being sought.

The clay minerals in common clays and shales are chiefly illitic or chloritic. The material is sufficiently plastic to permit molding and vitrification at low temperature. Suitable common clays and shales are utilized in the manufacture of heavy clay products such as common brick, facing brick, structural tile, partition tile, conduit tile, quarry tile and drain tile. There are no specific recognized grades of common clay and shale. Specifications are usually based upon the physical and chemical tests of manufactured products. The raw materials utilized in the heavy clay industry usually contain up to 35 per cent quartz. If the quartz, together with other nonplastic materials, exceeds this percentage, the plasticity of the clay is reduced and the quality of the ware is lowered. If calcite or dolomite is present in sufficient quantities, the clay will fire buff and the fired strength and density will be adversely affected.

Most of the surface deposits of common clays in Canada are the result of continental glaciation and subsequent stream transport. Such Pleistocene deposits are of interest to the ceramic industry and include stoneless marine and lake sediments, reworked glacial till, interglacial clays and floodplain clays.

In eastern Canada, shales are consumed in large quantities for manufacturing cement near Corner Brook in western Newfoundland, and at Havelock in King County, New Brunswick. Common clay from glacial drift is used in Ontario as a source of silica and alumina in the local manufacture of grey portland cement at Woodstock and St. Mary's. In Manitoba, shales and clays from glacial Lake Agassiz are extracted to produce lightweight aggregates. In Alberta, local glacial clays from Regina are used for manufacturing cement, lightweight aggregates

---

M. Prud'homme is with the Mineral Policy Sector, Energy, Mines and Resources, Canada. Telephone (613) 995-9466.

and mineral wood insulation. In British Columbia, altered volcanic ash is extracted at Barnhard Vale for cement, and in Quesnel for use as a natural pozzolan.

The shales provide the best source of raw material for making brick. In particular, those found in Cambrian, Ordovician and Carboniferous rocks in eastern Canada, and Jurassic, Cretaceous and Tertiary rocks in western Canada, are utilized by the ceramic industry.

China Clay (Kaolin). China clay is a white clay composed mainly of kaolinitic minerals formed from weathered igneous rocks. Some deposits occur in sedimentary rocks as tabular lenses and discontinuous beds or in rocks that have been hydrothermally altered. Commercial china clays are beneficiated to improve their whiteness when used as fillers and their whitefiring characteristics when used in ceramics. None of the crude kaolins known to exist in Canada have been developed, primarily because of beneficiation problems and the small size of some deposits.

China clay is used primarily as a filler and coating material in the paper industry, a raw material in ceramic products, and a filler in rubber and in other products. In the ceramic industry china clay is used as a refractory raw material. In prepared whiteware bodies such as wall tile, sanitaryware, dinnerware, pottery and electrical porcelain, quantities of nepheline syenite, silica, feldspar and talc are used as well.

Several occurrences of kaolin in Canada have attracted attention. In British Columbia, a deposit of clay similar to a secondary kaolin occurs along the Fraser River near Prince George. In southern Saskatchewan, there is a source of sandy kaolinised clay with off-white coloured fines which would be suitable for filler or coating material if it could be beneficiated adequately. Known deposits occur near Fir Mountain, Flintoft, Knollys and Wood Mountain from which clays in the Lower Whitemud Formation have been analyzed to produce a whiter commercial china clay. In Manitoba, various kaolinitic-rock deposits have been reported at Arborg, on Deer Island (Punk Island) and Black Island in Lake Winnipeg, and in the northwest at Cross Lake and Pine River; the Swan River Formation also has been studied as a potential source of kaolin. In Ontario, extensive deposits of kaolin-silica sand mixtures occur along the Missinaibi and the Mattagami rivers. In 1982, a feasibility study was done to evaluate the use of borehole mining in the Moose River basin, north of Kapuskasing, where silica-kaolin deposits lie under deep overburden. The distance from markets and the difficult terrain and climate still hinder development of these deposits. In Quebec, kaolin deposits have been actively mined in the past as a coproduct of a silica operation, near St-Remi-d'Amherst, in Papineau County. Occurrences near Chateau-Richer in Montmorency County and Pointe Comfort in Gatineau County have been studied as a potential source of kaolin for alumina, suitable for aluminous cement and refractories. Field investigations have been carried out to delineate known occurrences, especially near Thirty-one Mile Lake.

Ball Clay. Ball clay is defined as a fine-grained, highly plastic and mainly kaolinitic sedimentary clay. Natural colours range from white to brown, blue, grey and black, usually related to carbonaceous material. Fired colours may be white to offwhite. They are extremely refractory materials and have less alumina and more silica than kaolin. Ball clays occur in beds or lenticular units characterized by complex variation, both vertically and laterally.

Ball clays occurring in Canada are mineralogically similar to high-grade, plastic fire clay and are composed principally of fine-particle kaolinite, quartz and mica. These clays are known to occur in the Whitemud and the Ravenscrag Formations - Willowbunch Member - of southern Saskatchewan. Clay production takes place near Claybank, Eastend, Estevan, Flintoft, Readlyn, Rockglen, Willowbunch and Wood Mountain.

Fire Clay (Refractory Clay). Fire clay is a detrital clay mainly composed of kaolinite with a high content of alumina and silica. It usually occurs in sedimentary rocks as lenticular bodies. These clays may range in plasticity from essentially that of ball clay to nonplastic varieties such as flint clay. They are formed by alteration of aluminous sediments deposited in a swampy environment or following transportation and concentration of clayey material.

Fire clay is used in the manufacture of products requiring high resistance to heat such as fire brick, insulating brick and refractory mortar. The refractory suitability is determined by the pyrometric cone equivalent (PCE) test. Canadian fire clays are used principally for the manufacture of medium- and high-duty fire brick and refractory specialties. Known Canadian fire clays are not sufficiently refractory for the manufacture of superduty refractories without the addition of some very refractory material such as alumina.

Various grades of good-quality fire clay occur in the Whitemud Formation in southern Saskatchewan and on Sumas Mountain in British Columbia. Fire clay, associated with lignite as well as with kaolin-silica sand mixtures, occurs in the James Bay watershed of northern Ontario along the Missinaibi, Abitibi, Moose and Mattagami rivers. At Shubenacadie, Nova Scotia, some seams of clay are sufficiently refractory for medium-duty fire clay. Clay from Musquodoboit, Nova Scotia, has been used by some foundries in the Atlantic provinces, and the properties and extent of this clay were investigated by the Nova Scotia Department of Mines. Ontario and Quebec have no producing domestic sources of fire clay and import most of their requirements from the United States.

Stoneware Clay. Stoneware clays are intermediary between low-grade common clays and the high-grade kaolinitic clays. They are typically a mixture of kaolinitic clay minerals and micaceous clay minerals. Stoneware clays must be capable of being fully vitrified at a relatively low temperature.

Stoneware clays are used extensively in the manufacture of sewer pipe, flue liners, and facing brick. They are widely used by amateur and studio potters.

The principal source of stoneware clay in Canada is the Whitemud Formation in southern Saskatchewan and southeastern Alberta. Stoneware clays also occur near Abbotsford on Sumas Mountain, at Chimney Creek Bridge, Quesnel and Williams Lake, British Columbia; near Swan River in Manitoba; and in Nova Scotia, at Musquodoboit and at Shubenacadie where it is used principally for manufacture of buff-facing bricks.

Bentonite and Fuller's Earth. Bentonite consists primarily of montmorillonite clay, and is formed from volcanic ash, tuff or glass, other igneous rocks, or from rocks of sedimentary origin. Sodium bentonite has strong swelling properties and possesses a high dry-bonding strength. Calcium bentonite of the non-swelling type, exhibits adsorptive characteristics. Fuller's earth contains mainly smectite-group clay minerals and is very similar to non-swelling bentonite. It is formed by alteration of volcanic ash or by direct chemical precipitation of montmorillonite in shallow marine basins. Fuller's earth is characterized by absorptive properties, catalytic action, bonding power and cation-exchange capacities.

Drilling Mud and Activated Clays. Drilling mud contains about 10 per cent swelling bentonite. Synthetic bentonites may also be used for special muds. The swelling properties of a bentonite used as a drilling mud may be improved by adding soda ash in a drying process to substitute calcium cations by sodium cations. Activated clays are non-swelling bentonites that are acid leached to remove impurities and to increase reactive surface and bleaching power. They are used for decolouring mineral oils and as catalysts.

Bentonite, Fuller's earth and Activated clays are covered at intervals in a separate mineral review.

CANADIAN INDUSTRY

Clays. Production of clays is captive to its use in lightweight aggregates, cement and mineral wool insulation, which consumed mainly common clay, stoneware clay and ball clay. In Canada, there is no commercial production of china clay, and all requirements are imported mainly from the United States (97 per cent) and the United Kingdom (3 per cent) into Ontario (56 per cent), Quebec (35 per cent), Manitoba (4 per cent) and British Columbia (3 per cent). Demand for china clay depends principally on the paper industry which accounts for more than 75 per cent of its use. The traditional method of shipping kaolin is dry, but transportation as a 70 per cent solid-slurry is increasing. Consumption of china clay has risen in response to certain needs in the refractories industry. It may be used in its natural form, or when calcined, to compete with mullite in some uses. Average prices for imported kaolin have increased at an average annual rate of 6 per cent from 1982 to 1984. The current price is about $125 per t.

Fire clay, with an estimated unit value of $71.20 per t in 1984 is imported for use in refractories from the United States (98 per cent) into Ontario (74 per cent) and Quebec (21 per cent).

Clay products. Production of clay products includes structural materials - such as bricks and tiles - sewer pipes, flue linings, drain tiles, earthenwares, tablewares, sanitaryware and pottery. In 1984, nearly 35 companies were producing more than 95 per cent of total output. Production of clay products has risen from $95 million in 1982 to $141 million in 1983, mainly due to increased output in eastern Canada. Imports of structural materials account for only 5 per cent of total clay products and are shipped into Ontario (50 per cent) and British Columbia (40 per cent).

The average import price in 1984 was about $190 per thousand bricks, a small increase from $176 per thousand bricks in 1983.

Imports of ceramic tiles account for 25 per cent of the value of imported manufactured clay products. Sources are Italy (47 per cent), Spain (17 per cent) and Japan (15 per cent) into Ontario (55 per cent), Quebec (25 per cent) and British Columbia (16 per cent). Imports of tablewares account for about 45 per cent of the value of manufactured clay products. Sources are United Kingdom (36 per cent) and Japan (27 per cent).

Toronto Brick Co. closed its brick plant near Toronto in the Fall of 1984. In Quebec, La Compagnie 124984 Canada has been reactivating the former Quéabrique brick plant near Westbury to produce facing bricks. Also, La brique Citadelle, in Beauport, has carried out a modernization program since 1983 to upgrade handling outlet and improve productivity. Blue Mountain Pottery Ltd. of Collingwood, Ontario, is diversifying into the field of hi-tech ceramics relating to computers and communication equipment. Expertise of the Ontario Research Foundation is being utilized. Since 1983, the federal department of Energy, Mines and Resources has conducted a potential opportunity study for industrial minerals used in the ceramic sector, including abrasives, glass, clay products and refractories. With the cooperation of 60 major Canadian manufacturers, more than 50 industrial minerals were surveyed in terms of consumption, source of supply and specifications.

Refractories. Refractories are produced in Canada by 16 major manufacturers of basic and alumina-silica products. Special refractories such as refractory mineral wool and carbon-compound mortars are also produced. These products are imported mainly from the United States (98 per cent) and include alumina bricks into Ontario (98 per cent) and magnesite bricks into Ontario (83 per cent), Quebec (10 per cent), Nova Scotia (5 per cent) and British Columbia (5 per cent). Trade in refractories since 1982 has increased sharply. Imports increased at an average annual rate of 28 per cent and exports increased 30 per cent. Refractories account for 78 per cent of total exports of all products, mainly to American markets (96 per cent).

WORLD REVIEW

World production of kaolin in 1983 was around 20 million t, an increase of 5 per cent from 1982. Major world producers are the United States (32 per cent), the United Kingdom (15 per cent) and the U.S.S.R. (15 per cent). Ball clay production is dominated by the United States, the United Kingdom and Czechoslovakia. While plastic refractory clay is produced widely, fire flint clay is restricted to Australia, Austria, China, France, Hungary, South Africa, United States and U.S.S.R.

Extensive development of kaolin deposits occurred in 1983 in Australia, Austria, Benin, Bulgaria, Cameroon, Nigeria, Oman, Spain and Sweden.

In the United States, shipments of clays in 1983 amounted to 90.8 million short tons valued at $US 931 million, an increase of 16 per cent in tonnage and 13 per cent in value compared to 1982. China clay accounts for 18 per cent of total production and 63 per cent of output in terms of value. Shipments of clay refractories amounted to $US 595,300 in 1983. American clay producers operated between 50 to 70 per cent of capacity. From a 1982 base, demand for clays is expected to increase at an average annual rate of 2 to 4 per cent to 1990. In anticipation of growth, the productive capacity for United States kaolin increased steadily since the early-1980s. This caused an excess in supply and generated lower profit margins for water-washed kaolin producers. In comparison, a shortage of capacity occurred for calcined grades used for paper filling and coating. Opening of new calcined kaolin plant facilities in Georgia were announced in 1984 by J.M. Huber Co., Wilkinson Kaolin Associates and Thiele Kaolin Co. Engelhard Corp. announced in late-1984 that it will purchase for $US 1 million Freeport Kaolin's operation in Gordon, Georgia. The purpose will be to increase the company's markets for pigments and extenders by expanding the capacity by

450 000 tons to 1.5 million tons. Kentucky-Tennessee Clay Co. the largest American producer of ball clay, was acquired by Ranchers Exploration and Development Corp. In Alabama, Donoho Clay Co. started a $US 1.4 million expansion plan to increase production of refractory clay.

OUTLOOK

Clays and clay products are materials mainly characterized by high bulk, low unit value and sensitivity to transportation costs. Therefore, they are very sensitive to fluctuations in the general economic climate. In 1983, the increase in value of production was due to an increase in residential housing starts. This led to higher production of bricks in 1984. Construction activity rose in all provinces except Alberta where high vacancy rates persisted. From 1985 to 1990, total construction expenditures are forecast to grow at an annual rate varying between 2.5 per cent to 4.5 per cent. Expenditures in the industrial and non-residential building sectors are expected to be well above the average growth in the economy, while only modest growth is anticipated in the residential sector. A steady economic recovery would permit the construction materials sector to expand production where necessary and establish long-term plans to meet demand more efficiently. With confidence in lower interest rates, improved economic conditions and stimulative government housing programs, there could be a resurgence in housing starts.

Refractory product plants were on tight production schedules in 1984 as demand apparently shifted away from refractory clay products to alumina-based refractories for the steel, foundries, aluminum and cement industries. Meanwhile, acceptance of magnesia-carbon bricks is increasing in the United States. Consumption of refractories may rise slightly in the short-term, especially in Quebec, due to new smelter-related projects linked to the low cost of electrical energy.

The paper industry will remain the most important market for kaolin producers, both in terms of volume and value. As paper production is expected to increase steadily, consumption of coating-grade kaolin will in turn continue to increase through 1987. However, the outlook for filler grade clays is for little growth in the short term. Calcium carbonate is the major substitute for kaolin in the paper industry and consumption of super-fine ground calcite is expected to increase by 7 per cent per year. Given the present structure of the paper industry in North America it is unlikely that this substitution will happen in the short-term.

PRICE QUOTATIONS FOR BALL CLAY AND KAOLIN

United States clay prices, according to Chemical Marketing Reporter, December 24, 1984.

	$US per short ton
Ball clay, fob Tennessee	
Airfloated, bags, carload	49.00
Crushed, moisture repellent, bulk carload	24.00
Kaolin, fob Georgia	
Dry-ground, airfloated, soft	60.00
NF powdered, colloidal, 50 lb bags, 5,000 lb lots	.24/lb
Waterwashed, fully calcined, bags, carload	255.00
Waterwashed, uncalcined, delaminated paint grade, 1 micron average	182.00
Uncalcined, bulk, carload	
No. 1 coating	94.00
No. 2 coating	75.00
No. 3 coating	73.00
No. 4 coating	70.00
filler, general purpose	58.00

Industrial Minerals, November 1984 quotation (£1.00=$US 1.50-1.70)

	£ per tonne
Kaolin, refined, bulk, fob works	
Coating clays	60-110
Filler clays	40-60
Pottery clays	25-65

TARIFFS

Item No.		British Preferential	Most Favoured Nation	General	General Preferential
			(%)		
CANADA					
29500-1	Clays, including china clay, fire clay and pipe clay not further manufactured than ground	free	free	free	free
29525-1	China clay	free	free	25	free
28100-1	Firebrick containing not less than 90 per cent silica; magnesite firebrick or chrome firebrick; other firebrick valued at not less than $100 per 1,000, rectangular shaped, not to exceed 100 x 25 in.³ for use in kiln repair or other equipment of a manufacturing establishment	free	free	free	free
28105-1	Firebrick, nop, of a class or kind not made in Canada, for use in construction or repair of a furnace, kiln, etc.	free	free	15	free
28110-1	Firebrick, nop	5	8.4	22.5	5
28200-1	Building brick and paving brick	8.1	7.5	22.5	5
28205-1	Manufactures of clay or cement, nop	10.8	10.3	22.5	6.5
28210-1	Saggars, hillers, bats and plate setters, when used in the manufacture of ceramic products	free	free	free	free
28215-1	Grog, crushed or ground, but not further manufactured for use exclusively as refractory material per ton	free	free	$1.15	free
28300-1	Drain tiles, not glazed	free	13.9	20	free
28400-1	Drain pipes, sewer pipes and earthenware fittings therefore; chimney linings or vents, chimney tops and inverted blocks glazed or unglazed, nop	15	15.7	35	10
28405-1	Earthenware tiles, for roofing purposes	free	13.9	35	free
28415-1	Earthenware tiles, nop	12.5	16.3	35	10.5
28416-1	Biscuits larger than 135 cm² in dimension containing not less than 50 per cent by weight of silica, kaolin and limestone for use in the manufacture of ceramic wall tiles (expires June 30, 1984)	free	free	35	free
28500-1	Tiles or blocks of earthenware or of stone prepared for mosaic flooring	15	16.3	30	10.5
28600-1	Earthenware and stoneware, viz: demijohns, churns or crocks, nop	16.7	15.7	35	10

TARIFFS (cont'd)

Item No.		British Preferential	Most Favoured Nation	General	General Preferential
			(%)		
CANADA (cont'd)					
28700-1	All tableware of china, porcelain, semi-porcelain or white granite, excluding earthenware articles	free	15	35	free
28705-1	Articles of chinaware, for mounting by silverware manufacturers	12.5	13.9	22.5	9.0
28710-1	Undecorated tableware of china, porcelain, semi-porcelain for use in the manufacture of decorated tableware	free	8.4	35	free
28715-1	Undecorated whiteware not less than 1/8 inch in thickness for use in the manufacture of decorated heavy duty tableware (expires June 30, 1984)	free	free	35	free
28800-1	Stoneware and Rockinghamware and earthenware, nop	15.7	15.7	35	10
28805-1	Chemical stoneware	free	8.4	35	free
28810-1	Hand forms of porcelain for manufacture of rubber gloves	free	free	35	free
28900-1	Baths, bathtubs, basins, closets, closet seats and covers, closet tanks, lavatories, urinals, sinks and laundry tubs of earthenware, stone, or cement, clay or other material, nop	12.5	15	35	10
30000-1	Crucibles and covers, nop	free	8.4	15	free
44515-1	Porcelains all of one piece, over eighty-six inches in length or having an outside diameter greater than twenty-four inches, for use in the manufacture of electric instrument and power transformers (expires June 30, 1984)	free	free	37.5	free
44518-1	Electric insulators of all kinds, nop	12.8	12.1	27.5	8.0
44518-2	Ceramic insulator spark plug cores not further manufactured than burned and blazed	12.8	free	27.5	free
44518-3	Porcelain or ceramic insulators, nop and complete parts thereof	15	15	27.5	10
44519-1	Porcelain or ceramic bushings for use in the manufacture of hermetically sealed power capacitors (expires June 30, 1984)	free	free	27.5	free
49203-1	Ceramic discs for use in the manufacture of carrier assemblies for multi-orifice valves (expires June 30, 1984)	free	free	20	free
62430-1	Statues and statuettes of porcelain or earthenware	free	13.9	30	free

TARIFFS (cont'd)

Item No.		British Preferential	Most Favoured Nation	General	General Preferential
			(%)		
MFN Reductions under GATT (effective January 1 of year given)		1983	1984 1985 (%)	1986	1987
28110-1	Firebrick, nop	8.4	8.0 7.6	7.2	6.8
28200-1	Building brick and paving brick	7.5	6.9 6.3	5.6	5.0
28205-1	Manufactures of clay or cement, nop	10.3	9.7 9.1	8.6	8.0
28300-1	Drain tiles, not glazed	13.9	12.9 12.0	11.1	10.2
28400-1	Drain pipes, sewer pipes and earthenware fittings therefore; chimney linings or vents, chimney tops and inverted blocks glazed or unglazed, nop	15.7	14.6 13.5	12.4	11.3
28405-1	Earthenware tiles, for roofing purposes	13.9	12.9 12.0	11.1	10.2
28415-1	Earthenware tiles, nop	16.3	15.3 14.4	13.4	12.5
28500-1	Tiles or blocks of earthenware or of stone prepared for mosaic flooring	16.3	15.3 14.4	13.4	12.5
28600-1	Earthenware and stoneware, viz: demijohns, churns or crocks, nop	15.7	14.6 13.5	12.4	11.3
28700-1	All tableware of china, porcelain, semi-porcelain or white granite, excluding earthenware articles	15.0	14.6 13.5	12.4	11.3
28705-1	Articles of chinaware, for mounting by silverware manufacturers	13.9	12.9 12.0	11.1	10.2
28710-1	Undecorated tableware of china, porcelain, semi-porcelain for use in the manufacture of decorated tableware	8.4	8.0 7.6	7.2	6.8
28800-1	Stoneware and Rockinghamware and earthenware, nop	15.7	14.6 13.5	12.4	11.3
28805-1	Chemical stoneware	8.4	8.0 7.6	7.2	6.8
28900-1	Baths, bathtubs, basins, closets, closet seats and covers, closet tanks, lavatories, urinals, sinks and laundry tubs of earthenware, stone, or cement, clay or other material, nop	15.0	14.6 13.5	12.4	11.3

UNITED STATES (MFN)		(¢ per long ton)			
521.41	China clay or kaolin	33.0			
521.81	Other clays, not beneficiated	free			
521.84	Other clays, wholly or partly beneficiated	50.0			
		1983	1984 1985 (¢ per long ton)	1986	1987
521.71	Common blue clay and other ball clays, not beneficiated	40.0	39.5 39.0	38.5	38.0
521.74	Common blue clay and other ball clays wholly or partly beneficiated	81.0	80.0 79.0	78.0	77.0

Sources: The Customs Tariff, 1983, Revenue Canada, Customs and Excise; Tariff Schedules of the United States Annotated (1983), USITC Publication 1317, U.S. Federal Register, Vol. 44, No. 241.
nop - Not otherwise provided for.
Note: In addition to the above tariffs various duties are in existence on manufactured clay products, viz., brick pottery, artware, etc.

TABLE 1. CANADA, PRODUCTION OF CLAYS AND CLAY PRODUCTS FROM DOMESTIC SOURCES, 1982-84

	1982	1983P	1984e
	($000)		
Production from domestic sources, by provinces			
Newfoundland	860	1,381	1,600
Nova Scotia	4,500	5,900	6,700
New Brunswick	2,200	3,200	3,550
Quebec	14,047	20,667	20,430
Ontario	52,229	74,673	86,130
Manitoba	1,735	3,395	2,300
Saskatchewan	3,349	3,572	3,740
Alberta	11,220	12,207	8,775
British Columbia	5,853	7,335	7,680
Total Canada	95,993	132,329	140,905
Production[1] from domestic sources, by products			
Brick - soft and stiff mud process and dry press	71,643	98,982	105,680
Drain tile	4,651	4,764	5,635
Sewer pipe	(2)	(2)	(2)
Flue linings	2,296	3,308	4,230
Pottery glazed or unglazed (including coarse earthenware, stoneware and all pottery)	(2)	(2)	(2)
Other products	11,702	17,600	18,315
Small establishments not reporting detail	5,701	7,675	7,045
Total	95,993	132,329	140,905

Source: Statistics Canada.
[1] Producers' shipments. Distribution for 1983 estimated by Energy, Mines and Resources Canada. (2) Included in "Other products".
P Preliminary; e Estimated.

TABLE 2. CANADA, IMPORTS AND EXPORTS OF CLAYS, CLAY PRODUCTS AND REFRACTORIES, 1982

	1982		1983P		1984P	
	(tonnes)	($000)	(tonnes)	($000)	(tonnes)	($000)
					(Jan.-Sept.)	
Imports						
Clays						
China clay, ground or unground	205 952	22,254	249 829	28,534	195 042	24,445
Fire clay, ground or unground	33 574	2,782	30 065	2,315	33 107	2,373
Clays, ground or unground nes	105 856	7,803	89 117	6,932	75 476	5,839
Bentonite	238 069	12,340	187 218	9,545	234 422	11,071
Fuller's Earth	1 081	75	536	75	3 177	488
Drilling mud	11 355	3,095	44 964	7,951	3 648	2,297
Clays and earth, activated	13 369	9,714	12 203	10,304	9 040	7,104
Subtotal, clays	609 256	58,063	613 932	65,656	553 912	53,617
Clay Products	(M)		(M)		(M)	
Brick-building, glazed	1 224	190	1 991	351	1 182	224
Brick-building, nes	13 818	2,544	25 208	4,223	23 940	4,387
Building blocks and hollow tiles	..	1,541	..	761	..	693
Brick acid-proof	..	131	..	89	..	59
Clay bricks, blocks and tiles, nes	..	3,150	..	4,211	..	3,861
Ceramic tiles	(m^2)		(m^2)		(m^2)	
under 2 1/2" x 2 1/2"	705 566	5,492	544 942	3,984	453 835	3,571
over 2 1/2" x 2 1/2"	5 651 402	38,217	7 148 879	46,072	6 973 033	46,202
Subtotal, bricks, blocks, tiles	..	51,265	..	59,691	..	58,997
Tableware, ceramics	..	92,679	..	93,068	..	77,298
Sanitaryware	..	166	..	148	..	48
Artware	..	24,233	..	25,449	..	22,853
Porcelain, electric insulators	..	31,666	..	21,002	..	23,704
Chemical stoneware, exc. laboratory	..	1,166	..	1,154	..	999
Pottery settings and firing supplies	..	975	..	710	..	329
Pottery basic products, nes	..	1,705	..	1,933	..	2,449
Clay end-products, nes	..	668	..	1,451	..	1,475
Subtotal, ceramics	..	153,258	..	144,915	..	129,155
Refractories						
Fire brick and shapes						
Alumina	14 050	13,007	20 177	16,347	19 248	16,281
Chrome	190	222	533	492	733	962
Magnesite	12 461	11,174	18 928	19,093	15 463	18,598
Silica	2 984	2,406	3 027	2,671	3 018	2,772
nes	102 916	35,240	112 133	38,395	97 741	39,898
Refractory cements and mortars	..	13,264	..	14,456	..	12,918
Plastic fire brick and ramming mixture	..	1,342	..	1,933	..	929
Crude refractory materials, nes	9 457	1,831	7 148	1,213	7 059	1,562
Grog (refractory scrap)	5 339	778	4 655	476	3 764	401
Foundry facings	..	1,012	..	1,865	..	1,583
Refractories, nes	..	6,978	..	7,287	..	10,671
Subtotal, refractories	..	87,254	..	104,228	..	106,575
Total clays, clay products and refractories	..	349,840	..	374,490	..	348,344
By main countries						
United States	..	173,461	..	186,351	..	177 158
Japan	..	40,751	..	38,644	..	43,421
United Kingdom	..	48,746	..	53,126	..	37,785
Italy	..	24,726	..	26,490	..	26,832
West Germany	..	11,691	..	17,807	..	13,890
Spain	..	8,081	..	7,856	..	9,351
Taiwan	..	6,342	..	6,919	..	7,218
Greece	..	4,341	..	4,339	..	4,217
South Korea	..	3,904	..	3,780	..	4,186
France	..	6,090	..	5,713	..	3,763
People's Republic of China	..	5,215	..	3,831	..	2,950
Brazil	..	1,818	..	3,744	..	2,807
Hong Kong	..	1,579	..	2,116	..	2,220
Other	..	13,095	..	13,774	..	12,546
Total	..	349,840	..	374,490	..	348,344

TABLE 2. (cont'd)

	1982		1983P		1984P	
	(tonnes)	($000)	(tonnes)	($000)	(tonnes)	($000)
					(Jan.-Sept.)	
Exports						
Clays, ground and unground	557	40	272	66	501	123
Clay products	(M)		(M)		(M)	
Building brick, clay	2 138	467	2 352	641	1 711	428
Clay bricks, blocks, tiles, nes	..	2,085	..	1,496	..	1,605
Subtotal, bricks, blocks, tiles	..	2,552	..	2,137	..	2,156
High-tension insulators and fittings	..	4,392	..	3,447	..	3,173
Tableware, nes	..	9,718	..	8,770	..	6,140
Subtotal, porcelain, tableware	..	14,110	..	12,217	..	9,313
Refractories						
Fire brick and shapes	33 041	20,287	32 181	20,480	28 309	16,297
Crude refractory materials	40 839	150	241 127	955	422 231	1,733
Refractory nes	..	13,388	..	20,159	..	24,799
Subtotal refractories	..	33,825	..	41,594	..	42,829
Total clays, clay products and refractories	..	50,527	..	56,014	..	54,298
By main countries						
United States	..	30,606	..	41,426	..	41,946
Cuba	..	1,217	..	743	..	1,871
Dominican Republic	..	645	..	2,129	..	1,087
South Africa	..	875	..	734	..	967
Algeria	..	12	..	566	..	749
Australia	..	955	..	612	..	689
New Zealand	..	936	..	828	..	654
United Kingdom	..	723	..	747	..	381
Colombia	..	913	..	2,073	..	257
Trinidad-Tobago	..	1,491	..	696	..	156
Belgium-Luxembourg	..	364	..	370	..	109
Other countries	..	11,790	..	5,090	..	5,432
Total	..	50,527	..	56,014	..	54,829

Source: Statistics Canada.
P Preliminary; .. Not available; nes Not elsewhere specified; M = Thousands; m^2 = Square metres.

TABLE 3. CANADA, SHIPMENTS OF REFRACTORIES, 1980-82

	1980		1981		1982	
	(tonnes)	($000)	(tonnes)	($000)	(tonnes)	($000)
Monolithics	42 852	19,555	25 103	14,026	28 948	18,404
Fire brick and shapes	134 671	73,664	122 413	66,034	87 066	52,781
Cement and mortars	39 402	13,842	56 558	18,026	46 004	15,198
All other products	...	28,596	...	34,002	...	26,753
Total	...	135,657	...	132,088	...	113,136

Source: Statistics Canada.
... Figures not appropriate or not applicable.

TABLE 4. CANAD, CLAYS, CLAY PRODUCTS AND REFRACTORIES, PRODUCTION AND TRADE, 1970, 1975, 1979-83

Year	Production			Refactory Shipments[1]	Imports	Exports
	Domestic Clays	Imported Clays[2]	Total			
	($million)					
1970	51.8	33.6	85.4	42.3	81.2	15.6
1975	78.4	59.1	137.5	65.0	177.4	25.1
1979	121.5	71.4	192.9	139.7	323.1	61.2
1980	108.5	83.4	191.9	135.7	386.2	63.8
1981	119.1	85.1	204.2	132.1	432.0	65.7
1982	96.0	63.4	159.4	113.1	349.8	50.5
1983P	123.2	..	..	..	374.5	56.0

Source: Statistics Canada.
[1] Includes fire brick and shapes, refractory cements, mortars, and monolithics, plus all other products shipped. [2] Includes electrical porcelains, glazed floor and wall tile, sanitaryware, pottery, art and decorative ware plus all other products.
P Preliminary; .. Not available.

TABLE 5. CANADA, CONSUMPTION (AVAILABLE DATA) OF CLAYS, BY INDUSTRIES, 1981-83

	1981	1982	1983P
		(tonnes)	
China Clay			
Pulp and paper products[1]	85 555	96 333	97 235
Ceramic products	9 764	6 680	10 267
Paint and varnish	5 955	5 510	6 082
Rubber and linoleum	4 033	5 951	6 568
Other products[2]	21 917	74 513	21 176
Total	127 224	188 987	141 328
Ball Clay			
Ceramic products misc.	18 694	11 084	19 749
Refractories	2 743	11 969	2 578
Other[3]	127 979	78 951	45 049
Total	149 416	102 004	67 376
Fire Clay			
Foundries	11 731	8 936	8 829
Refractories	14 929	14 546	5 840
Other[4]	2 467	4 183	9 458
Total	29 127	27 665	24 127

[1] Includes paper and paper products and paper pulp. [2] Includes refractory brick mixes, cements, glass fibre and wools, adhesives, foundry, wire and cable and other miscellaneous products. [3] Includes structural clay products, adhesives, miscellaneous chemicals, petroleum refining, paint and varnish and other miscellaneous products. [4] Includes abrasives, ceramic products, concrete products, paint and varnish, petroleum refining, and rubber products.
P Preliminary.

TABLE 6. KAOLIN: WORLD PRODUCTION, 1981-83, MAJOR COUNTRIES

	1981	1982P	1983e
		(000 tonnes)	
United States	6 950	5 770	6 530
United Kingdom	3 800	3 560	3 600
U.S.S.R.e	2 540	2 630	2 630
Colombia[1]	810	810	810
Spain[3]	790	700	700
Czechoslovakia	510	530	520
West Germany	470	450	500
India[1]	390	530	500
Brazil[2]	470	490	490
Romania	410	410	410
France	330	350	350
Others	3 040	2 380	3 130
Total	20 510	19 140	20 170

Source: U.S. Bureau of Mines, 1983, clays, Preprints of Minerals Yearbook, S. Ampian.
[1] Crude saleable kaolin. [2] Processed.
[3] Included crude and washed kaolin.
e Estimated; P Preliminary.

TABLE 7. MAJOR CANADIAN MANUFACTURERS OF STRUCTURAL CLAY PRODUCTS AND REFRACTORIES, 1984, BY PROVINCE

Company	Plant Location	Products	Raw Material	Size[1] and Remarks
NEWFOUNDLAND				
Trinity Brick Products Ltd.	St. John's	brick, building	shale	(B)
NEW BRUNSWICK				
L.E. Shaw Limited	Chipman	brick, facing; tiles, drainage and partition	shale	(E)
NOVA SCOTIA				
L.E. Shaw Limited	Lantz	bricks, blocks and tiles	common clay ball clay	(E)
QUEBEC				
Brique Citadelle Ltée	Beauport	bricks, facing and building, drain tiles and flue linings	shale	(-)
Compagnie 124984 Canada Ltée	Westbury	brick, facing	common clay	(A) formerly Quéabrique, also East Angus Brick & Tiles
Didier Corp. Refractaires	Becancour	bricks and shapes monoliths and mortars	alumina-silica silica	(E)
Domtar Inc. Construction materials div.	Laprairie	bricks, building and facing	shale	(G)
Dresser Canada Inc. Canadian refractories div.	Grenville	bricks and shapes monoliths	alumina-silica and basic (MgO)	(F)
Duquesne Refractories Ltd.	Montreal	monoliths and mortar	alumina-silica, silica and carbon	(A)
La Brique de Scott Ltée.	Scott-Junction	bricks and tiles	common clay	Closed since 1982
La Briquerie St-Laurent Inc.	Laprairie	bricks, building and facing	shale	(C)
Montreal Terra-Cotta (1966) Ltée	Deschaillons	bricks, building, tiles and flue linings	shale, common clay	(B)
Quigley Canada Inc.	Lachine	bricks and shapes cements	fire clay, basic (MgO)	(A)

TABLE 7. (cont'd)

Company	Plant Location	Products	Raw Material	Size[1] and Remarks
ONTARIO				
Amos C. Martin Ltd.	Park Hill	drain tiles	shale	(-)
A.P. Green Refractories				
Acton div.	Acton	bricks and shapes	alumina-silica	(D)
Weston div.	Weston	monoliths	alumina-silica	
Babcock and Wilcox Refractories	Burlington	bricks and shapes monoliths, mineral wool	alumina-silica kaolin	(C)
Barrie Brick Co. Ltd.	Midhurst	bricks, building	-	(-)
Bimac Canada Metallurgical Inc.	Burlington	bricks and shapes	-	(B)
BMI Refractories Inc.	Smithville	monoliths and mortars	-	(-)
Brampton Brick Ltd.	Brampton	bricks, building and facing	shale	(-)
Canada Brick Co. Ltd.				
Burlington div.	Burlington	bricks, building	shale	(E)
F.B. McFarren div.	Streetsville	bricks, building	shale	
Streetsville div.	Streetsville	bricks, building	shale	
Dochart Clay Products Co. Ltd.	Arnprior	tiles	common clay	(B)
Domtar Inc. Construction materials div.				
Mississauga div.	Mississauga	bricks, building	shale	(G)
Ottawa div.	Ottawa	bricks, building	shale	
Dresden Tiles Yard (1981) Ltd.	Dresden	bricks, facing and building, tiles and flue liners	-	(A)
General Refractories Co. of Canada Ltd.	Smithville	bricks and shapes mortars	alumina-silica, basic (MgO)	(D)
Georges Coultis and Sons Ltd.	Thedford	tiles, drain tiles	-	(B)
Halton Ceramics Ltd.	Burlington	blocks and tiles	common clay and shale	(A)
Hamilton Bricks Ltd.	Hamilton	bricks, building	shale	(B)
Kaiser Refractories Co.	Oakville	monoliths and mortars	alumina-silica and basic	(C)
Meaford Tiles Ltd.	Wallenstein	-	-	(-)

TABLE 7. (cont'd)

Company	Plant Location	Products	Raw Material	Size[1] and Remarks
ONTARIO (cont'd)				
National Sewer Pipe Ltd.	Oakville	flue linings and sewer pipes	shale and fire clay	(-)
North American Refractories Ltd.	Haldimand	monoliths and mortars	alumina-silica	(B)
Norwich Brick and Tile Yard (1975) Ltd.	Norwich	-	-	(-)
Paisley Brick and Tile Co. Ltd.	Paisley	-	-	(-)
Plibrico (Canada) Ltd.	Burlington	monoliths and mortars	alumina-silica	(E)
R&I Ramtite Canada (Ltd.) C-E Refractories	Welland	monoliths and mortars; bricks	alumina-silica and basic	(C)
Toronto Brick Co.	Toronto	brick, facing and building	shales	(D), closed in 1984
MANITOBA				
I-XL Industries Ltd. Red River Brick and Tiles div.	Lockport	bricks and tiles	common clay	(E)
SASKATCHEWAN				
A.P. Green Refractories (Canada)	Claybank	bricks and shapes	alumina-silica	(B)
Estevan Brick Ltd.	Estevan	bricks, building and facing	ball clay	(C)
I-XL Industries Ltd. Western Clay Products div.	Regina	bricks facing flue lining and sewer-pipe	-	Closed in 1982
ALBERTA				
I-XL Industries Ltd.				(E)
Medicine Hat Brick and Tiles div.	Medicine Hat	bricks, blocks, tiles	common clay	
Medicine Hat Sewer Pipe div.	Medicine Hat	sewer pipes and flue lines	common clay	
Northwest Brick and Tiles div.	Edmonton	bricks, building	common clay	
Redcliff Pressed Bricks div.	Redcliff	brick, facing and fire brick	common clay fire brick	
BRITISH COLUMBIA				
Clayburn Refractories Ltd.	Abbotsford	bricks, mortars and monoliths	alumina-silica	(D)
Fairey and Company Ltd.	Surrey	bricks and shapes, monoliths, mortars	alumina-silica	(A)

Sources: Statistics Canada; Mineral Policy Sector; Energy, Mines and Resources Canada.
[1] Size keys: (A) up to 25 employees; (B) 25-49; (C) 50-99; (D) 100-199; (E) 200-499; (F) 500-999; (G) over 1000 employees.
- Information not available.

Coal and Coke

J.A. AYLSWORTH

The Canadian coal industry continued to set new records in 1983 and 1984 despite difficult market conditions brought on by the recent world recession and changes in energy supply-demand situations in many countries using coal as an energy fuel. Coal production, domestic consumption, exports and imports reached record levels in Canada in 1983 and again in 1984. Production and exports increased as a result of decisions taken, particularly by Japan, in 1980 and 1981 during a period of tight markets and supply shortfalls in the United States, Australia and Poland. New production capacity brought on-stream in Canada and Australia in 1983 combined with a resolution of delivery and port problems in the United States, Australia and Poland increased the supply of coal available for trade in international markets beginning in 1982. As this market adjustment was occurring the recession of the early 1980s resulted in reduced demand for coal in most major coal markets, putting significant downward pressures on both the volumes and prices of coal traded. The resulting supply-demand imbalance is expected to continue throughout the 1980s and will be further influenced by major new export capacity coming on in the U.S.S.R. and Columbia.

COAL PRODUCTION AND DEVELOPMENTS

Canadian coal production is forecast to approach 57 million t of clean or saleable coal in 1984, up about 27 per cent over 1983. Production of coking coal increased due to five new mines which came on-stream in British Columbia and Alberta in 1983 and completed a full year's operation in 1984. Thermal coal production was up in 1984 over 1983 due to increased domestic demand, increased exports, and the development of one new mine in Alberta.

Coal production in Nova Scotia was up by about 4 per cent in 1984 in spite of a fire in May which resulted in the closure of No. 26 colliery in Glace Bay. As a result of the fire, output from this mine was only 200 000 t in 1984, down from nearly 700 000 t in 1983. Output from all Cape Breton Development Corporation's (CBDC) mines in 1984 was up marginally over 1983 due to significant increases in production from the Lingan mine. Exports by CBDC fell by about 50 per cent to 500 000 t as a result of the closure of No. 26 colliery. In late May the federal government unveiled a set of initiatives that will provide over $300 million of new capital investment to increase coal production and improve CBDS's commercial performance. The initiatives include: development of the Lingan-Phalen Colliery and expansion of the Victoria Junction preparation plant; completion of an exploratory tunnel to permit a detailed evaluation of the Donkin-Morien coal project; and construction of a wash plant and railway loading dock for the Prince mine. The new productive capacity created by these investments will ensure that sufficient coal is available to meet Nova Scotia's growing thermal coal requirements throughout the 1980s.

Coal production in New Brunswick grew by 3 per cent in 1984 to 575 000 t. Most of this coal is sold to the provincial utility for the generation of electricity. Coal production is forecast to remain at about this level throughout the remainder of this decade.

The five active coal mines in Saskatchewan are forecast to have produced a record 9.7 million t of coal in 1984, up 25 per cent over the 1983 level. The majority of this coal is marketed to mine mouth power stations, but a significant amount is shipped to power stations in western Ontario.

Alberta remains Canada's leading coal producing province with a 1984 estimated output of 22.8 million t, up 5 per cent over 1983. The production of sub-bituminous coal, which is used to generate electricity in mine mouth plants, accounted for 67 per cent of this total. The remaining output is bituminous coal which is sold domestically in Alberta and Ontario and internationally in Asian, European and other markets.

J.A. Aylsworth is with the Energy Sector, Energy, Mines and Resources Canada. Telephone (613) 995-1118.

Two new mines came on-stream in Alberta in 1983-84. A new coking coal mine operated by Gregg River Resources Ltd. (owned by Manalta Coal Ltd. and Japanese interests) produced about 700 000 t of coal in the latter part of 1983 and 2.0 million t of coal during its first full year of operation in 1984. All of this coal is marketed to steel mills in Japan. In the last quarter of 1984, the Obed Marsh Thermal Coal Project, (the Union Oil Company of Canada Limited) began production of thermal coal. Output in 1984 was 250 000 t, down somewhat from the planned level because of start-up problems. Output in 1985 is forecast to exceed one million t, destined primarily for European and Asian markets, and output could eventually reach 3 million t. During 1984 some Obed Marsh coal was shipped to Ontario Hydro power stations.

British Columbia's eight mines are forecast to have produced a record 20.6 million t of coal in 1984, up 75 per cent from the 1983 level of 11.7 million t. The main factor underlying this major increase was the first full year's production from four new mines which came on-stream in 1983. The Bullmoose (Teck Corporation) and the Quintette mines (Denison Mines Limited) in northeastern British Columbia produced 1.9 and 3.6 million t of coal respectively in 1984, compared with 200 000 and 100 000 t in 1983. Production from these mines should increase in 1985 as the Quintette operation overcomes start-up problems which constrained 1984s output. New mines in southeastern British Columbia, including the Greenhills mine (Westar Ltd.) and a new Line Creek mine (Crow's Nest Resources Limited), also contributed to the record British Columbia output, producing respectively 2.3 and 2.6 million t of coal in 1984.

CANADIAN COAL UTILIZATION

Domestic coal utilization is expected to reach 50 million t in 1984, up 13 per cent from 1983. The utility sector remains the major domestic market in Canada, consuming 41 million t of coal in 1984, an increase of 13 per cent over 1983. This sector accounts for approximately 80 per cent of Canadian coal consumption and is forecasted to grow in relative and absolute terms through the remainder of this decade.

Thermal coal consumption grew by 50 per cent in Nova Scotia in 1984 and reached a record level of 2 million t. This increase reflected the first full year's operation of the MW Lingan #3 generating unit by the Nova Scotia Power Corporation and nearly half a year's operation of Lingan #4. Coal consumption will increase again in 1985 and in the latter years of the decade. Approximately 60 per cent of the province's electricity is now generated from coal compared with 19 per cent in 1979.

Most of the coal consumed in New Brunswick is used by the provincial utility, New Brunswick Electric Power Commission, to generate electricity at its Dalhousie and Grand Lake stations. Consumption in 1984 is forecast to be a record 600 000 t, up 6 per cent from 1983. Coal demand is forecast to remain at about this level for the foreseeable future.

Thermal coal consumption in Ontario approached a record level of 14 million t in 1984, up 8 per cent over 1983. About 3.6 million t, or 25 per cent of this total came from Canadian mines in British Columbia, Alberta and Saskatchewan with the other 75 per cent imported from the United States. The unexpected increase in 1984 occurred because of the shutdown of Units I and II of the Pickering nuclear generating station and because of higher than forecast electricity demands. Coal requirements may increase in 1985 as the new 200 MW Atikokan coal-fired station in western Ontario beginsits first full year of commercial operation, but the trend in the last half of this decade is for coal requirements to fall as more nuclear capacity comes on-stream. Forecasts indicate that by 1990 the amount of electricity generated from coal in Ontario will only be about one-half of 1984s level. However, thermal coal demand in Ontario may grow again in the 1990s.

Coal utilization in Manitoba was estimated at 160 000 t in 1984, up from 109 000 t in 1983. Although it is forecast that demand may exceed 250 000 t in 1985, coal consumption is unlikely to increase to significant levels in the future.

Thermal coal consumption grew by 14 per cent in Saskatchewan in 1984 reflecting increased electricity demand. Consumption is estimated at 7.5 million t, divided primarily between the Boundary Dam and Poplar River generating stations. A small amount of coal was also consumed at the Estevan generating station. Forecasts for 1985 indicate that coal demand will remain at about 1984's level, but may increase later in the decade. Late in 1984, Saskatchewan Power Corporation sold its Poplar River coal mine near Coronach to Manalta Coal Ltd. of Calgary. The Corporation will sign a 30 year contract to purchase coal from Manalta to fuel the nearby Poplar River generating station.

Thermal coal consumption in Alberta is forecast to be 16.7 million t in 1984, up 15 per cent from 1983. Most of this increase reflects the first full year's operation of TransAlta Utilities Corporation's new 800 MW Keephills generating station, which consumed 3.3 million t of coal in 1984. Coal demand will continue to increase for the remainder of the 1980s, although at a slower pace than previously forecast. Two new 750 MW generating stations are scheduled to be brought on-stream by 1990. The first 375 MW unit of the Sheerness Generation Station is to begin commercial operation by January 1986. The Genesee station near Edmonton is scheduled to come on-stream later in the decade.

Consumption of coking coal increased by 15 per cent to 6.8 million t in 1984 as Canada's steel industry recorded increased sales in both domestic and export markets. The majority of the coal used in Canada to produce coke is imported from the United States. In 1984 Canada's three major Ontario-based steel companies (Dofasco Inc. and Stelco Inc. in Hamilton, and The Algoma Steel Corporation, Limited in Sault Ste. Marie) imported 6.6 million t of coking coal from the United States, up 20 per cent from 1983. These imports are expected to increase slowly throughout the remainder of the decade. Sydney Steel Corporation's ovens in Sydney, Nova Scotia, were shut-down in December 1983. They remain in a hot underfire mode, not producing coke but heated with propane, awaiting start-up. While these ovens produced no coke in 1984 plans call for a resumption of production in late-1985. A small amount of coal was used to produce coke in western Canada during 1984.

EXPORTS AND IMPORTS

The international coal trade scene remains unbalanced with an excess of production capacity competing for slowly growing markets. Statistics for 1983 show that in spite of depressed international demands, Canada remained a net coal exporter, with exports reaching 17 million t and imports 14.6 million t. Exports of coking coal to Japan, which account for about two-thirds of all Canada's exports, remained at 10.9 million t in 1983, unchanged from 1982. This was a major achievement given that Japanese coking coal imports declined by 8 per cent in 1983 due to declining domestic and international steel demand. Nevertheless, in spite of the encouraging statistics, 1983 was a difficult year, due to the price cuts forced on Canada's exporters by the oversupplied market. Coking coal prices fell by approximately 15 per cent and virtually all existing mines were forced to accept volume cutbacks.

Preliminary statistics for 1984 indicate that Canadian coal exports will register an unprecedented increase of 48 per cent in 1984 over 1983. This increase is primarily the result of output from five new mines that began production in late-1983 and achieved nearly full production levels in 1984. In addition one new mine began producing thermal coal, primarily for the export market, late in 1984.

The six new mines include: the Quintette mine (Denison Mines Limited) and the Bullmoose mine (Teck Corporation) in northeastern British Columbia; the Line Creek coking coal mine (Crow's Nest Resources Limited) and Greenhills mine (Westar Mining Ltd.) in southeastern British Columbia; the Gregg River mine (Manalta Coal Ltd.) and the Obed Marsh mine (Union Oil Company of Canada Limited) in south central Alberta. Combined production from these mines in 1984 totalled about 12.7 million t compared with 3.4 million t in 1983. Virtually all of this output was destined for Japanese markets. Canadian coal will account for about 25 per cent of Japanese steel industry imports in 1984, compared with about 17 per cent in 1983.

Total exports in 1984 are forecast to be 25.2 million t compared with 17 million t in 1983. All of this increased tonnage will be exported through the three existing coal terminals in the Port of Vancouver's jurisdiction and through the new Ridley Terminal Inc. port near Prince Rupert. Exports through the three Vancouver terminals in 1984 totalled 19.5 million t compared with 16 million t in 1983. In its first year of operation, Ridley Terminals Inc. exported 5.2 million t of coal. Another 500 000 t of coal was exported through Sydney, Nova Scotia, to European, Latin American and other markets.

Although the volume of exports increased in the calendar year 1984, coal exporters in Canada and elsewhere are still experiencing difficult times. Average price fob, the port of export for existing Canadian producers in 1984, was down about 1 per cent over 1983. This normal price reduction on top of the major price cuts in 1983 resulted in renewed efforts by companies to cut costs and increase productivity. Most of the new mines, which had not been subject to the major price cutbacks of 1983 because they were not yet producing coal, agreed to contract adjustments late in 1984 which resulted in price decreases that took account of the current market conditions.

By the end of the year it became clear that 1984 crude steel production by the Japanese steel industry would surpass the 1983 level of 1 billion t by at least 5 per cent. This gave rise to the hope that the volume and price cuts of the last few years might be over and that producers might soon begin to realize some real price increases and recoup some volumes lost during the previous three years.

However, the large oversupply of coal still available in both coking and thermal coal markets and the uncertainty in world steel and energy markets forced first the United States, then Canadian, and finely Australian producers to unusually early contract settlements for fiscal year 1985-86. Several of the established producers in these three countries settled on new contracts with Japanese steel mills in October and November, months ahead of normal contract negotiations. The contracts basically left the 1984 prices unchanged but, in the case of some of the existing Canadian producers, provided for small increases in volumes of coal to be exported. This, coupled with the increases in volumes to be shipped from some of the new Canadian producers, should result in another record year for exports in 1985. Further increases in exports are anticipated for the last half of the 1980s, although the rate of year-to-year increases will moderate.

Two export oriented coal mines in British Columbia are under active development and hope to be producing in the next few years. The Quinsam Coal project on Vancouver Island (Brinco Limited) is actively looking for markets for its thermal coal. Trial shipments of anthracite coal from the Mount Klappan project (Gulf Canada Resources Inc.) in northwestern British Columbia were completed by year-end and more are scheduled for 1985. Other export coal projects under consideration for later in the 1980s include the McLeod River and Mercoal projects (Manalta Coal Ltd.) in Alberta and the Telkwa coal project (Crow's Nest Resources Limited) in north central British Columbia. The expansion of port capacity at Westshore Terminals Ltd. in 1984 and the opening of Ridley Terminals Inc. near Prince Rupert provided 22 million t of new export capacity for the Canadian coal industry, thereby ensuring adequate port capacity well into the 1990s.

Canadian coal imports are estimated to have reached 18.6 million t in 1984, up 27 per cent from the 1983 level of 14.6 million t. Most of this increase was accounted for by Ontario Hydro's imports which totalled 11.4 million t in 1984, up about 45 per cent from 1983. The three Canadian steel companies also increased their imports of coking coal in 1984 over 1983. Forecasts suggest that total imports will decline in the remainder of this decade due to Ontario Hydro's declining requirements for coal. Coking coal imports are forecast to increase slowly over the next few years.

TABLE 1. SUMMARY OF COAL SUPPLY BY TYPE AND VALUES, 1980-84

	1980		1981		1982		1983		1984[3]	
	(000 t)	($000)	(000 t)	($000)	(000 t)	($000)	(000 t)	($000)	(000 t)	($000)
DOMESTIC[1]										
Bituminous										
Nova Scotia	2 726	132,750	2 539	133,226	3 052	174,474	2 986	144,000	3 110	162,600
New Brunswick	439	17,269	524	23,308	499	24,450	558	29,000	575	30,300
Alberta	6 830	246,771	6 895	272,238	6 978	337,742	7 315	371,000	7 630	384,200
British Columbia	10 156	457,959	11 781	590,935	11 768	654,130	11 697	588,000	20 600	1,026,900
Total	20 151	854,749	21 739	1,019,707	22 396	1,190,796	22 556	1,132,000	31 915	1,604,000
Sub-bituminous										
Alberta	10 542	55,402	11 551	42,559	13 021	88,022	14 464	112,000	15 170	131,300
Lignite										
Saskatchewan	5 971	32,381	6 798	55,305	7 494	73,520	7 760	95,000	9 715	114,700
Total	36 664	942,532	40 088	1,117,571	42 811	1,352,398	44 780	1,339,000	56 800	1,850,000
IMPORTED[2]										
Bituminous & Anthracite Briquettes	15 860	953,998	14 836	991,994	15 773	1,132,000	14 667	1,031,000	18 600	..
Total Coal Supply	52 524	1,896,530	54 924	2,109,565	58 584	2,484,338	59 447	2,370,000	75 400	

Sources: Statistics Canada; Energy, Mines and Resources Canada.
[1] fob mines; [2] Value at United States ports of exit. [3] Preliminary figures or estimates.
.. Not available.

TABLE 2. PRODUCER'S DISPOSITION OF CANADIAN COAL[1], 1983

Destination	Nova Scotia		New Brunswick		Saskatchewan		Alberta		British Columbia		Canada	
	Cok	Th	Cok	Th	Cok	Th	Cok	Th	Cok	Th	Cok	Th
	(kilotonnes)											
Newfoundland	–	1	–	–	–	–	–	–	–	–	–	1
Prince Edward Island	–	12	–	–	–	–	–	–	–	–	–	12
Nova Scotia	91	1 835	–	–	–	–	–	–	–	–	91	1 835
New Brunswick	–	4	–	558	–	–	–	–	–	–	–	562
Quebec	–	28	–	–	–	–	–	–	–	–	–	28
Ontario	–	–	–	–	–	962	–	1 418	–	531	–	2 911
Manitoba	–	–	–	–	–	232	–	1	–	21	–	254
Saskatchewan	–	–	–	–	–	6 566	–	87	–	50	–	6 703
Alberta	–	–	–	–	–	–	–	14 485	–	1	–	14 486
British Columbia	–	–	–	–	–	–	–	1	25	34	25	35
Total Canada	91	1 880	–	558	–	7 760	–	15 992	25	637	116	26 827
Japan	50	–	–	–	–	–	3 743	421	6 355	276	10 148	697
Other	959	100	–	–	–	–	506	446	2 934	1 221	4 399	1 767
Total shipments	1 100	1 980	–	558	–	7 760	4 249	16 859	9 314	2 134	14 663	29 291
Grand total	3 080		558		7 760		21 108		11 448		43 954	

Sources: Statistics Canada; Energy, Mines and Resources Canada.
[1] Saleable coal (raw coal, clean coal and middling sales).
Cok coking coal; Th thermal; – Nil.

TABLE 3. SUMMARY OF COAL SUPPLY-DEMAND, 1974-84

Year	CANADA PRODUCTION				IMPORTS			Domestic Consumption	Exports
	Bituminous	Sub-Bituminous	Lignite	Total	Anthracite	Bituminous	Total Available		
	(million tonnes)								
1974	12.5	5.1	3.5	21.1	0.4	12.0	33.5	24.9	10.5
1975	15.8	6.0	3.5	25.3	0.4	15.4	41.1	25.5	11.4
1976	14.4	6.4	4.7	25.5	0.3	14.3	40.1	28.2	11.9
1977	15.3	7.9	5.5	28.7	0.4	15.0	44.1	30.8	12.4
1978	17.1	8.3	5.1	30.5	0.3	13.8	44.6	31.7	14.0
1979	18.4	9.6	5.0	33.0	0.2	17.3	50.5	34.8	13.7
1980	20.2	10.5	6.0	36.7	0.3	15.5	52.5	37.3	15.3
1981	21.7	11.6	6.8	40.1	0.4	14.4	54.9	38.4	15.7
1982	22.3	13.0	9.5	42.8	0.3	15.5	58.6	41.5	16.0
1983	22.5	14.5	7.8	44.8	0.3	14.4	59.5	43.6	17.0
1984[1]	31.9	15.2	9.7	56.8	..	..	..	49.7	25.2

Sources: Statistics Canada; Energy, Mines and Resources Canada.
[1] Preliminary figures or estimates.
.. Not available.

TABLE 4. CANADA, COAL PRODUCTION, IMPORTS, EXPORTS AND CONSUMPTION, 1979-84

	Production	Imports	Exports	Domestic Consumption
	(000 tonnes)			
1979	33 013	17 524	13 698	34 764
1980	36 664	15 829	15 269	37 333
1981	40 088	14 836	15 705	38 367
1982	42 811	15 773	16 004	41 478
1983	44 780	14 667	17 011	43 649
1984[1]	56 800	18 600	25 200	49 700

Sources: Statistics Canada; Energy, Mines and Resources Canada.
[1] Preliminary figures or estimates.

TABLE 6. EXPORT DEMAND FOR CANADIAN COAL, 1983

Country	1983	
	(000 t)	($000)[1]
Belgium-Luxembourg	26	1,981
Denmark	296	14,992
West Germany	779	51,099
Greece	29	1,144
Italy	68	4,741
Netherlands	165	10,736
Sweden	218	14,999
Algeria	51	3,480
Hong Kong	210	10,906
Pakistan	139	10,828
People's Republic of China	17	1,302
Japan	10 996	829,863
South Korea	2 356	160,619
Philippines	35	1,361
Taiwan	552	39,233
Argentina	19	1,353
Brazil	716	54,028
Chile	120	9,158
United States	183	10,540
Total	16 978	1,232,000

Source: Statistics Canada.
[1] fob port of export Canadian dollars.

TABLE 5. COAL USED BY THERMAL POWER STATIONS IN CANADA, BY PROVINCES, 1966-84

	Nova Scotia	New Brunswick	Ontario	Manitoba	Saskatchewan	Alberta	Total Canada
				(000 tonnes)			
1966	799	294	3 500	79	1 116	1 360	7 148
1967	758	275	4 435	38	1 334	1 427	8 267
1968	646	240	5 523	179	1 354	2 128	10 070
1969	676	150	6 424	51	1 123	2 378	10 802
1970	548	113	7 696	503	1 969	2 951	13 780
1971	689	271	8 560	446	1 996	3 653	15 615
1972	663	281	7 599	410	2 145	4 113	15 211
1973	585	193	6 615	386	2 806	4 474	15 059
1974	606	292	6 721	132	2 902	4 771	15 424
1975	571	248	6 834	323	3 251	5 345	16 572
1976	730	207	7 612	979	3 521	5 996	19 045
1977	572	198	8 795	1 113	4 304	7 461	22 443
1978	771	151	9 097	341	4 585	8 029	22 914
1979	644	198	9 901	73	4 956	9 181	24 956
1980	1 052	315	10 779	240	4 972	10 424	27 782
1981	1 126	515	11 460	332	4 935	11 445	29 813
1982	1 300	548	12 484	184	5 897	13 242	33 656
1983	1 400	564	13 025	109	6 625	14 492	36 216
1984[1]	2 000	600	14 000	160	7 500	16 700	40 960

Sources: Statistics Canada; Energy, Mines and Resources Canada.
[1] Preliminary figures or estimates.

TABLE 7. SUMMARY OF COAL DEMAND, 1979-83

	1979	1980	1981	1982	1983
			(000 tonnes)		
DEMAND					
Thermal Electric					
Canadian Coal	16 104	19 314	20 998	24 033	26 748
Imported Coal	8 857	8 468	8 815	9 623	9 468
Total	24 961	27 782	29 813	33 656	36 216
Metallurgical					
Canadian Coal	1 272	961	784	229	102
Imported Coal	6 593	6 279	5 593	5 347	5 481
Total	7 865	7 240	6 377	5 576	5 583
General Industry					
Canadian Coal	963	1 190	962	1 075	667
Imported Coal	751	955	1 044	986	1 003
Total	1 714	2 145	2 006	2 061	1 670
Space Heating					
Canadian Coal	200	166	171	185	180
Imported Coal	24	-	-	-	-
Total	224	166	171	185	180
Exports					
Canadian Coal	13 698	15 269	15 705	16 004	17 011
Total					
Canadian Coal	32 237	36 900	38 620	41 526	44 708
Imported Coal	16 225	15 702	15 452	15 956	15 952
Total Coal Demand	48 462	52 602	54 072	57 482	60 660

Sources: Statistics Canada; Energy, Mines and Resources Canada.
- Nil.

TABLE 8. CANADA, COKE PRODUCTION AND TRADE, 1973-83

	Production		Imports		Exports	
	Coal Coke	Petroleum Coke	Coal Coke	Petroleum Coke	Coal Coke	Petroleum Coke
			(tonnes)			
1973	5 369 861	286 530	357 815	637 664	367 916	1 960
1974	5 443 427	274 412	509 058	746 033	260 892	24 940
1975	5 277 837	270 685	546 456	572 557	96 081	161 576
1976	5 289 185	678 432	287 249	591 859	169 895	136 970
1977	4 845 066	921 363	382 827	986 678	198 727	157 191
1978	4 967 664	1 014 076	553 349	973 985	217 595	134 762
1979	5 775 141	1 105 433	520 534	980 657	228 601	125 416
1980	5 249 744	1 156 444	626 923	908 322	319 554	150 200
1981	4 659 007	1 098 397	653 645	935 929	190 879	200 149
1982	3 999 117	1 083 129	453 915	650 810	129 793	104 897
1983	4 120 002	986 730	576 649	759 954	45 606	65 323

Cobalt

R.G. TELEWIAK

Consumption of cobalt in the western world was close to 17 500 t in 1984 compared to 14 800 t in 1983. The United States is the largest market for cobalt, and strong economic activity in this country in 1984 was a major factor in the 18 per cent increase in world consumption. The other major markets of western Europe and Japan, showed a much lower increase.

Demand from the superalloy sector, which accounts for about one-third of the total, was particularly strong. This was led by the manufacture of new commercial and military jet engines, as well as normal replacement of jet engine parts, primarily turbine blades.

CANADIAN DEVELOPMENTS

The two mine producers of cobalt, Inco Limited and Falconbridge Limited, recover cobalt as a byproduct of nickel-copper production. Due to an improvement in the nickel market, production of nickel increased by close to 30 per cent in 1984 over 1983. This resulted in a similar increase in cobalt production.

Inco Limited started production at its cobalt refinery at Port Colborne, Ontario in May 1983 and by year-end was producing at over two-thirds capacity. For most of 1984, the plant produced near its capacity of 900 tpy of electrolytic cobalt rounds.

Falconbridge Limited closed its Sudbury operations for two weeks during the summer of 1983 but in 1984 decided against taking a temporary closure. A series of rockbursts stopped production at the Falconbridge and East mines on June 20. In early July the company announced that after 35 years of operation, the Falconbridge mine would remain closed but the East mine would reopen. Production was increased at the mine to make up for this closure. At Kristiansand, Norway, Falconbridge continued to treat Sudbury matte and to refine some other feedstock from some other sources.

Output at the Fort Saskatchewan refinery of Sherritt Gordon Mines Limited was 539 t in the first nine months of 1984, a decline of 9 per cent from the comparable period in 1983. Sherritt Gordon toll refines feed for several producers and not as much material was available in 1984.

A joint venture consisting of Falconbridge and Geddes Resources Limited, carried out a drilling program in 1983 at the Windy Craggy copper-cobalt deposit in northwestern British Columbia. In late-1983 Falconbridge reduced its interest in the project to 22.5 per cent of net proceeds of production. In 1984, an underground exploration project was started. The deposit was reported to contain 318 million t of ore grading 1.5 per cent Cu and 0.08 per cent Co.

WORLD DEVELOPMENTS

Zaire, which holds about one-half of the cobalt production capacity in the western world, regained much of its lost market share in 1983, after abandoning its producer pricing policy the previous year. In 1984, cobalt supply and demand came into close balance early in the year and Zaire was able to reinstitute its producer price in March.

In July 1984, Zaire announced that the state-owned metals marketing organization, Société Zairose de Commercialisation des Minerais (Sozacom) would be dissolved. La Générale des Carrières et des Mines du Zaire (Gecamines), the state-owned mining company, also assumed responsibility for both production and marketing. The change was designed to streamline these operations. Both the World Bank and the International Monetary Fund were reported to be in favour of giving more marketing control to Gecamines.

R.G. Telewiak is with the Mineral Policy Sector, Energy, Mines and Resources Canada. Telephone (613) 995-9466.

Zambia Consolidated Copper Mines Ltd. closed its Khana cobalt plant in 1983 and started producing from a new leach-electrowin facility. In 1984, Zambia continued to operate below capacity to keep inventories under control.

The General Services Administration (GSA) in the U.S. awarded contracts to Zaire and Zambia in 1983 for delivery of 2 948 t of cobalt to the national defence stockpile, at a price of $US 5.50/lb. In 1984, the United States awarded a contract to Inco for 227 t at $US 11.70/lb. These additional purchases will raise the stockpile to 23 914 t, or 62 per cent of the U.S. government's goal.

AMAX Inc. started in 1983 to ship its cobalt slurry to Fort Saskatchewan, Alberta for toll refining by Sherritt Gordon. Due to a shortage of nickel-copper matte, AMAX closed its Port Nickel refinery in Louisiana for two months in the summer of 1983 and for five weeks in 1984. AMAX obtains feed from BCL Limited in Botswana and Agnew Mining Co. Pty. Ltd. in Australia but these two sources are insufficient to keep the AMAX refinery operating at capacity for the full year.

In Japan, Sumitomo Metal Mining Co., Ltd. suspended cobalt production in early April 1984. Feedstock for the refinery had been obtained from Marinduque Mining and Industrial Corporation (in August renamed Nonoc Mining and Industrial Corporation) in the Philippines, but this operation was closed for most of 1984. The contract between Sumitomo and Nonoc Mining was cancelled late in the year and other refineries were being examined for treatment of Nonoc's concentrate. Sumitomo will continue to produce cobalt compounds from the small quantities of cobalt recovered in its nickel refinery.

PRICES

In 1983, cobalt prices started from a low of $US 4.50-4.75/lb in early January and then peaked in April at $US 6.30-6.50. Speculative buying was a factor in the increase and when consumption did not respond as some anticipated, prices slipped to $US 5.35-5.45. By year-end they recovered to $US 5.95-6.10.

In 1984, prices advanced to $US 11.70/lb by the first week in March and then weakened slightly during the summer, before recovering to this same level by year-end. A major factor in the price rise and the relative stability of the price, was the action of the Zairean producer, Gecamines. The corporation, which is the largest producer in the world, pursued a marketing strategy to keep the price at these levels.

USES

One of the major uses for cobalt is in superalloys where it improves the strength, wear and corrosion resistance characteristics of the alloys at elevated temperatures. The major use of cobalt-base superalloys is in turbine blades for aircraft jet engines and gas turbines for gas pipelines. Cobalt-base superalloys normally contain 45 per cent or more cobalt, while nickel- and iron-based superalloys contain 8 to 20 per cent cobalt.

Although the demand for cobalt in the production of magnets has been declining in recent years, this is still an important use for cobalt. Consumption of cobalt in this sector is almost one-half of what it was in 1970.

Cobalt-base alloys are used in applications where difficult cutting is involved and high abrasion resistance qualities are required. The most important group of cobalt-base alloys is the stellite group, containing cobalt, tungsten, chromium, and molybdenum as principal constituents. Hardfacing or coating of tools with cobalt alloys provides greater resistance to abrasion, heat, impact and corrosion.

Cobalt metal powder is used as a binder in making cemented tungsten carbides for heavy-duty and high-speed cutting tools.

As a chemical product, cobalt oxide is an important additive in paint, glass, and ceramics. Cobalt is also used to promote the adherence of enamel to steel for applications such as appliances, and steel to rubber for the construction of steel-belted tires. A cobalt-molybdenum-alumina compound is used as a catalyst in hydrogenation and in petroleum desulphurization.

OUTLOOK

Over the long-term, cobalt consumption is expected to increase at an annual rate of 14.2 per cent. The price volatility over the past few years, particularly in the late-1970s when the prices peaked at over

$US 40/lb, has resulted in considerable substitution away from cobalt in certain uses and is a major factor in the forecast of a relatively modest increase.

Major consuming countries have expended considerable resources to find substitutes for cobalt in key applications. As an example of the results of these programs, the United States Air Force has developed two new cobalt-free superalloys which have excellent resistance to corrosion and oxidation. Although the precise composition have not been revealed, they are basically nickel aluminides. These alloys are being tested for application as gas turbine components for a new series of jet fighter engines.

Due primarily to the threat of substitution, the price of cobalt is not considered likely to advance significantly in real terms from its average level for 1984 of just over $US 11.50/lb. A substantially higher price would encourage more substitution and it would be in the long term interest of producers to not precipitate that type of action.

Zaire and Zambia are the two largest producers in the world, accounting for about two-thirds of cobalt capacity. The strategies which these two producers adopt, along with possible other events in these countries, will have a major impact upon supply and resultantly on prices and consumption.

PRICES

	Dec. 1983	Dec. 1984
	($)	
Cobalt metal, per lb. fob New York		
Shot, 99.5%, 250-kg drum	12.50[1]	11.70
Powder, 99%+		
300 and 400 mesh, 50-kg drums	6.91	13.24
extra fine, 125-kg drums	10.01	16.53

Source: Metals Week.
fob Free on board.
[1] Official price but transactions were taking place at considerably lower prices.

TABLE 1. CANADA, PRODUCTION TRADE 1982-84 AND CONSUMPTION 1981 AND 1982

	1982		1983P		1984P	
	(kilograms)	($)	(kilograms)	($)	(kilograms)	($)
Production¹ (all forms)					(Jan.-Sept.)	
Ontario	971 821	28,444,909	1 324 000	44,941,000	1 890 000	54,304,000
Manitoba	302 663	10,295,778	260 000	8,819,000	435 000	12,501,000
Total	1 274 484	38,740,687	1 584 000	53,760,000	2 325 000	66,805,000
Exports						
Cobalt metal						
United States	526 673	14,206,000	654 191	11,585,000	846 961	18,509
United Kingdom	1 696	17,000	107 974	3,805,000	136 716	1,923
South Africa	8 321	606,000	21 559	539,000	–	–
Belgium-Luxembourg	–	–	67 995	379,000	119 995	669
Australia	2 927	86,000	14 856	208,000	3 869	115
Other countries	45 388	984,000	18 707	330,000	4 442	156
Total	585 005	15,899,000	885 282	16,846,000	1 111 983	21,372,000
Cobalt oxides and hydroxides²						
United Kingdom	230 000	8,521,000	184 000	6,061,000	248 000	3,975,000
United States	–	–	8 000	112,000	17 000	72,000
Belgium-Luxembourg	–	–	–	–	36 000	573,000
Total	230 000	8,521,000	192 000	6,173,000	301 000	4,620,000

	1981		1982	
Consumption³				
Cobalt contained in:				
Cobalt metal	87 583		63 863	
Cobalt oxide	6 979		10 070	
Cobalt salts	6 772	..	12 456	..
Total	101 334		86 389	

Sources: Energy, Mines and Resources Canada; Statistics Canada. ³ Available data reported by consumers.
¹ Production (cobalt content) from domestic ores. ² Gross weight.
P Preliminary; – Nil; .. Not available.

TABLE 2. CANADA, COBALT PRODUCTION, TRADE AND CONSUMPTION, 1970, 1975 AND 1979-84

	Production[1]	Exports		Imports		Consumption[4]
		Cobalt metal	Cobalt oxides and hydroxides	Cobalt ores[2]	Cobalt oxides and hydroxides[3]	
		(tonnes)				
1970	2 069	381	837	..	..	148
1975	1 354	431	561	..	..	123
1979	1 640	296	445	190	46	115
1980	2 118	325	1 091	2	26	105
1981	2 080	677	601	24	20	101
1982	1 274	585	212	2	30	86
1983P	1 584	885	192	45	30	101
1984[5]	..	1 112	302	13	17	..

Sources: Energy, Mines and Resources Canada; Statistics Canada.
[1] Production from domestic ores, cobalt content including cobalt content of Inco Limited and of Falconbridge Limited shipments to overseas refineries. [2] Cobalt content. [3] Gross weight.
[4] Consumption of cobalt in metal, oxides and salts. [5] First 9 months.
P Preliminary; .. Not available.

TABLE 3. MAJOR CANADIAN PRODUCER'S SHIPMENTS OF COBALT

	1981	1982	1983
	(tonnes)		
Inco	1 642	1 148	812
Falconbridge	622	377	659
Sherritt Gordon	379	342	346

Source: Company Annual Report.

TABLE 4. WORLD PRODUCTION OF COBALT

	1981	1982	1983e
	(tonnes)		
Zaire	15 420	11 300	11 300
Zambia	3 420	3 250	3 200
Canada	2 300	1 400	1 750
Australia	1 660	1 800	1 815
Finland	1 035	930	910
Cuba	1 715	1 500	1 650
U.S.S.R.	2 180	2 270	2 360
Other	3 288	1 904	1 142
Total	30 018	24 354	24 127

Source: United States Bureau of Mines.

Columbium (Niobium)

D.G. FONG

CANADIAN DEVELOPMENTS

Niobec Inc., Canada's sole columbium producer at St. Honoré, Quebec, is jointly owned by Société québécoise d'exploration minière (SOQUEM) and Teck Corporation. Production in 1983 was 1 815 t of columbium pentoxide (Cb_2O_5) contained in pyrochlore concentrate, a 42 per cent decline from 1982. This low production result was directly related to a shutdown at the mining operations from April 1 to August 22, 1983 because of high inventories.

Niobec's mine production was estimated at 2 770 t contained Cb_2O_5 in 1984. Apart from a strike which closed the mining operation from October 19 to November 29, production was maintained at near capacity level.

Niobec deferred its plans to build a high-purity-oxide plant, due to a limited market and an underutilization of existing world capacity. The company had been considering a 907 tpy plant at an estimated cost of $3.2 million.

Camchib Mines Inc. and New Venture Equities Ltd., which took over the Martison Lake phosphate-columbium property from Shell Canada Resources Limited in 1982, continued drilling and other evaluating work in 1983 and 1984. The property is located north of Hearst in northern Ontario. Drilling has outlined 140 million t of mineralization grading 0.35 per cent Cb_2O_5 and 20 per cent phosphate (P_2O_5).

WORLD DEVELOPMENTS

SUPPLY

Ferrocolumbium production (FeCb) in 1984 by Companhia Brasileira de Metalurgia e Mineração S.A. (CBMM) of Brazil, estimated at 15 000 t Cb_2O_5, more than doubled its output of 7 320 t in 1983. Exports in 1984 were estimated at 14 000 t of FeCb and 400 t of columbium oxide (Cb_2O_5), compared with 6 680 t of FeCb and 277 t of Cb_2O_5 in 1983. The weakness in sales in 1983 was attributed to economic recession and a draw-down of consumer inventories.

CBMM is owned 55 per cent by Metropolitana de Comércio e Partiçipacões, a private Brazilian interest, and 45 per cent by Molycorp, Inc. of the United States. The company operates the world's largest columbium mine at Araxá in the State of Minas Gervais, and accounts for 90 per cent of Brazilian production. The remaining 10 per cent is produced by Mineração Catalão de Goiás S.A. in the Ouvidor-Catalão region of Goias state. The main international markets for CBMM's columbium products are Europe (42 per cent), Canada and the United States (27 per cent), Japan (17 per cent) and the U.S.S.R. (6 per cent).

In recent years, CBMM has placed a high priority on further processing. The company continued to work on its columbium metal project and has made small shipments on a trial basis. The 99.99 per cent pure metal is used as a coating for medical instruments, nuclear-magnetic resonance (NMR) spectrometers, and corrosion-resistant applications. However, the market at present is fairly small. In 1983, CBMM began the production of optical and crystal grade oxides for the optical and electronic industries. In addition, its 2 400 tpy high purity oxide plant was restarted in September 1983 after a shutdown of almost two years.

Productos Metalúrgicos S.A. (PROMETAL) and Metais Goiás S.A. (METAGO) formed a Brazilian joint venture in 1983 called Goiás Nióbio S.A. to bring a new mining and ferrocolumbium plant into operation in the state of Goiás. The deposit was reported to contain 9 million t with an average grade of 0.95 per cent Cb_2O_5. Goiásnnióbio was planning to bring the project into production at a rate of 1 100 tpy of ferrocolumbium in the latter part of 1984.

D.G. Fong is with the Mineral Policy Sector, Energy, Mines and Resources, Canada. Telephone (613) 995-9466.

A large columbium deposit was discovered in 1984 in São Gabriel da Cachoeira, Amazonas state by Companhia de Pesquisa de Recursos Minerais (CPRM), a Brazilian state mineral research organization. The deposit contains 2.9 billion t grading an average 2.81 per cent pyrochlore. This new discovery raised Brazilian columbium resources to 7.4 billion t, which is equivalent to about 92 per cent of the world's reserves.

Teledyne Wah Chang in the United States, produced high-purity columbium oxide at its Oregon plant using standard grade ferrocolumbium as feed material. The company also produced reactor-grade metal and powders. Operations during the last two years, however, have been at a reduced rate due to weak demand from the aerospace and nuclear industries. NRC Inc. continued to produce columbium oxide and metal at its Newton, Mass. plant. The company installed a new electron-beam furnace and a new rolling mill in 1982.

DEMAND

Rising steel and automobile production have helped to raise columbium consumption during the last two years. Carbon steels showed the largest increase in 1983 while high-strength, low-alloy (HSLA) steels exhibited a strong recovery in the first half of 1984. Stainless steels and superalloys, also showed a significant increase, especially in 1984.

The increase in consumption of the HSLA steels came largely from the recovery in the automotive industry although consumption has also expanded in recent years due to new applications in automotive body parts and forged components such as crank shafts and engine mounts. The strong trend toward front-wheel-drive vehicles has created an additional demand for HSLA steel because of its use in trailing axle assemblies.

Other important end-uses, such as pipeline and structural steels began to show evidence of recovery in 1984. Higher construction activity helped to augment the demand for structural steels. Although large diameter linepipe has remained depressed during the last two years the consumption for smaller diameter gathering, transmission and distribution linepipe returned to the 1981 level.

The fastest growing market for stainless steels was in vehicle exhaust systems. In the United States, the consumption of stainless steels doubled during the first five months of 1984 compared with the corresponding period in 1983.

Superalloys were also in stronger demand, largely because there was an expansion in the production of military jet engines and spare parts for commercial jet engines.

GOVERNMENT STOCKPILES

In August 1984, the General Services Administration (GSA) announced plans to purchase columbite concentrates containing 259 527 kg of Cb_2O_5 for the National Defense stockpile. Bids were invited to supply eight lots of concentrates with a minimum content of 45 per cent Cb_2O_5, a maximum of 0.10 per cent phosphate (P_2O_5), and a combined content of at least 60 per cent of Cb_2O_5 and tantalum pentoxide (Ta_2O_5). None of the bids were successful because the prices quoted in all of them were considered to be too high.

USES

The steel industry is the largest consumer of columbium, which is used in the form of ferrocolumbium as an additive agent in HSLA steels, carbon steels, and stainless steels. Although the quantity of columbium may be as low as 0.02 per cent, the yield strength and mechanical properties of the resulting steel are significantly improved. These characteristics are particularly important in applications such as large-diameter pipeline, automotive components, structural applications and drilling platforms.

High-purity columbium pentoxide is used mainly in superalloys for turbine and jet engines, which have traditionally been the second largest use after steels. A columbium addition to the cobalt-and nickel-based superalloys improves the high-temperature characteristics of these alloys. In the manufacture of high-alloy and stainless steels, columbium is used to impart resistance to corrosion at elevated temperatures, a property of particular importance in petroleum processing plants, heat exchangers for severe chemical environments and acid pressure vessels.

One of the important properties of columbium is its superior conductivity compared with other pure metals.

Superconductivity is the loss of all resistance to direct electrical current at temperatures near absolute zero. This special property of columbium allows the construction of powerful electrical generators, which are much more efficient than conventional generators with copper wire windings. Also, because of the powerful magnetic field created by the superconductors, it is used extensively in the construction of nuclear magnetic resonance (NMR) spectrometers. In addition, many potential applications in electrical devices are being developed, including new types of motors, ship engines, electric generators and switch elements for computers.

Special high-purity columbium pentoxide is produced for optical applications. Additions of columbium pentoxide to optical glass give a high refractive index and thereby allow production of thin lenses for eyeglasses. This characteristic, along with others such as lightweight and durability, enable such lenses to be competitive with plastic lenses.

PRICES

Prices for Canadian pyrochlore concentrates remained unchanged throughout 1983 and 1984. Niobec, the world's sole major supplier of concentrate since the Brazilian government banned exports of pyrochlore in 1981, quoted its price at $US 3.25 a lb. Cb_2O_5 contained in concentrate.

The quoted price for standard grade ferrocolumbium was unchanged at $US 6 a lb. contained columbium in 1983, but was lowered in the United States to $US 5.60 in February 1984, despite evidence of stronger demand. Niobium Products Company Ltd. (NPC) of Pittsburgh, a subsidiary of CBMM, initiated the price cut and stated that a reduction was necessary in order to stimulate demand and bring the U.S. price into line with European prices.

The price of high purity ferrocolumbium has been moving downward since CBMM introduced its high purity product into the U.S. market in 1982. The price fell from $US 21 a lb. in early 1983 to $US 17.70 in October 1984. NRC of Newton, Mass. raised its price for high purity columbium pentoxide in October 1983 by 10 per cent to $US 6.60 a lb. in response to an upturn in demand and low inventories. The higher price, matched later by CBMM, remained in effect for the balance of 1983 and in 1984.

OUTLOOK

Improved steelmaking technology and lower steel requirements will continue to have a negative impact on the consumption of a number of ferroalloy products. The demand for ferrocolumbium, however, is expected to fare better than most other ferroalloys because of the advantages held by columbium steels, which result in substantial material savings for many steel applications. Also, columbium steels are not as yet widely produced in Asia. As Asia is increasingly becoming a steel manufacturing centre, demand for higher grade steels will likely rise and consumption of columbium will increase. In the long run, the demand for columbium in the western world is expected to grow between 4 and 5 per cent a year.

Microalloyed columbium steels are forecast to enjoy greater popularity in many applications, particularly in the automotive industry where weight reduction continues to offer scope for achieving higher fuel efficiency. The consumption of micro-alloyed steels in each 1983 car averaged 250 lbs as compared with none in 1973. This application is expected to increase to 375 lbs per car by 1990.

The Canadian steel industry has forecasted a demand increase from the pipe and tube industry of 5 per cent in 1985, based largely on an expectation that there will be increased activity in hydrocarbon developments in both Canada and the United States.

The content of columbium per t of stainless steel has been declining. With the advent of the argon-oxygen-decarburization (AOD) steelmaking process and the ability to decarburize to low levels, there is less need for columbium in stainless steels. However, this development is being offset to some degree by the growing consumption of ferritic stainless steels, used in higher temperature applications to resist corrosion.

The demand for superalloys is forecast to improve as commercial air carriers start to buy new airplanes with more fuel-efficient engines. Consumption is also forecast to increase for replacement parts in the air carrier industry. However, it will be at least the late 1980s before usage in this sector will reach its record level of 1980. The new fuel-efficient engines operate at higher temperatures and, accordingly, require materials such as high-columbium superalloys which can withstand this condition.

On the supply side, there will be adequate production capacity to meet the forecast demand increase. CBMM is in a position to double its 25 000 tpy pyrochlore plant whenever markets warrant. The recent discovery of a large pyrochlore deposit in the Amazon region could add another major producer by next decade. This major discovery reafirms Brazil's predominance as a columbium source.

Canada also has large resources, which occur in a number of undeveloped deposits across the country. In view of the strong growth potential for columbium, some of these deposits are likely to be mined in the next decade. The development of these deposits will enhance Canada's position as an important world columbium supplier. China and Zaire have reported discoveries which could result in new producers for these countries in the near-future.

Adequate supplies to satisfy growing demand and stability of price have been key factors in the success of columbium. These factors are expected to prevail in the foreseeable future, with the result that columbium should remain cost effective as a steel additive, particulary in relation to its closest alternative, vanadium.

PRICES

Prices as quoted in Metals Week in December 1983 and 1984, U.S. currency.

	1983	1984
	($)	
Columbium ore		
Columbite, per kg of pentoxide, cif U.S. ports[1]	11.02-15.43	7.72-11.02
Canadian pyrochlore, per kg, fob mine	7.17	7.17
Ferrocolumbium, per kg Cb, fob shipping point		
Low alloy	13.23	12.48
High purity alloy	43.54	39.02
Columbium metal, per kg 99.5-99.8%, fas shipping point		
Reactor ingot	72.75-88.18	72.75-88.18
Reactor powder	79.37-105.82	79.37-105.82

[1] The range reflects variations in the ratio of columbium pentoxide (Cb_2O_5) to tantalum pentoxide (Ta_2O_5).
cif - Cost, insurance and freight; fob - Free on board; fas - Free alongside ship.

TARIFFS

Item No.		British Preferential	Most Favoured Nation	General	General Preferential
			(%)		
CANADA					
32900-1	Columbium and tantalum ores and concentrates	free	free	free	free
35120-1	Columbium (niobium) and tantalum metal and alloys in powder, pellets, scrap, ingots, sheets, plates, strips, bars, rods, tubing or wire for use in Canadian manufactures (expires June 30, 1984)	free	free	25	free
37506-1	Ferrocolumbium, ferrotantalum, ferro-tantalum-columbium	free	4.7	5	free

TARIFFS (cont'd)

MFN Reductions under GATT (effective January 1 of year given)	1983	1984	1985 (%)	1986	1987
37506-1	4.7	4.5	4.3	4.2	4.0

UNITED STATES

601.21	Columbium ore	free				

		1983	1984	1985 (%)	1986	1987
628.15	Columbium metal, unwrought, and waste and scrap (duty on waste and scrap suspended to June 30, 1982)	4.4	4.2	4.0	3.9	3.7
628.17	Columbium, unwrought alloys	6.2	5.9	5.6	5.2	4.9
628.20	Columbium metal, wrought	7.3	6.8	6.4	5.9	5.5

Sources: The Customs Tariff, 1983, Revenue Canada, Customs and Excise; Tariff Schedules of the United States Annotated 1983, USITC Publication 1317; U.S. Federal Register, Vol. 44, No. 241.

TABLE 1. CANADA, COLUMBIUM (NIOBIUM) PRODUCTION, TRADE AND CONSUMPTION, 1970, 1975 AND 1979-84

	Production[1] Cb_2O_5 Content	Imports[3] Primary forms and fabricated metals		Exports[2] Columbium Ores and Concentrates to United States	Consumption Ferrocolumbium and Ferrotantalumcolumbium, Cb and Ta-Cb Content
		Columbium	Columbium Alloys		
		(kilograms)			
1970	2 129 271	..	..	576 227	132 449
1975	1 661 567	..	..	9 682	215 910
1979	2 512 667	855	W	509 953	360 152
1980	2 462 798	877	156	655 721	486 251
1981	2 740 736	913	303	419 865	455 500r
1982	3 086 000	805	59	291 193	356 000
1983	1 744 722	967	396	543 599	352 000
1984P	2 500 000	684	104	..	..

Sources: Energy, Mines and Resources Canada; Statistics Canada; U.S. Department of Commerce.
[1] Producers' shipments of columbium ores and concentrates and primary products, Cb_2O_5 content. [2] From U.S. Department of Commerce, Imports of Merchandise for Consumption, Report FT 135. Quantities in gross weight of material. [3] 1984 imports based on nine month statistics.
P Preliminary; .. Not available; W Withheld to avoid disclosing confidential company data; r Revised.

Copper

W.J. McCUTCHEON

Canadian copper producers reacted to low copper prices by increasing productivity to reduce costs or in some cases by shutting down. Concentrate producers shipping to Pacific Rim smelters benefitted from relatively low smelting treatment and refining charges. However, some Canadian smelters in central and eastern Canada faced concentrate shortages. Byproduct and coproduct metal prices remained relatively low and these contributed to the unfavorable financial performance of many producers.

Canadian copper shipments in 1983 were 653 040 tonnes (t) of recoverable (or if exported, paid for) copper and these increased to an estimated 712 000 t in 1984, reflecting more normal copper production by Inco Limited. Approximately 150 000 tpy of copper production capacity in Canada was closed at the end of 1984, either on a temporary or permanent basis. Canadian copper consumption (defined as shipments to domestic consumers) increased from 170 000 t in 1983 to an estimated 190 000 t in 1984.

Space does not permit a detailed review of all major events in 1983 and 1984 which appear in commercial publications and in monthly reviews by Energy, Mines and Resources Canada. Some of the major events in the regions of Canada and in the world are reviewed below.

CANADIAN DEVELOPMENTS

ASARCO Incorporated mined out its Buchans mine in Newfoundland. At Noranda Inc.'s, Division Mines Gaspé at Murdochville, Quebec, the open-pit remained closed through 1983-84. Development of the E zone orebody below Murdochville commenced in 1984, and the recall of 250 workers began in August 1984 to reopen the 4 000 tpd underground mine. Corporation Falconbridge Copper resumed operations at its Lake Dufault Division in Rouyn-Noranda in April 1983. Exploration development at the 2 million t, 7 per cent Cu, 1.6 km deep Ansil orebody, began late 1984. Operation ceased at Northgate Patino Mines Inc.'s, Lemoine mine in March 1983, after producing 30 160 t of copper in seven years. BP Resources Canada started a $125 million development of the A1 zinc-copper orebody at its mine near Joutel. The 5 000 tpd pit should begin operations in 1986. The B zone underground production at 1 500 tpd should continue until 1990.

In the Sudbury district, Falconbridge Limited resumed production at its Sudbury operations in January, 1983 after a shutdown of six months. Inco ended its 10 month shutdown of its Sudbury operations in April 1983. Inco was the largest copper producer to shutdown in the western world in the 1982 recession.

Hudson Bay Mining and Smelting Co., Limited (HBMS) became a subsidiary of Inspiration Resources Corporation of New York. The Centennial and Westarm mines require new development which cannot be justified at low metal prices and were temporarily closed. The White Lake mine ore reserves were exhausted while Trout Lake mine was brought up to its rated production level of 1 600 tpd. The Rod mine produced at its rated capacity in 1984. Exploration for new reserves is under way to maintain present levels of production.

Sherritt Gordon Mines Limited announced in 1983 that its Fox copper-zinc mine would be mined out by early 1986. At its Ruttan mine, Sherritt mined higher-grade ore while it attempted to find an investor for a portion of the $25-30 million required to develop deeper higher-grade ore. When no agreement was reached, Sherritt announced that the mine would be closed in June 1984 but the Manitoba government agreed to contribute $10 million and the closure decision was rescinded. Mining of the deeper higher-grade ore planned for late 1985 should result in significant cost reductions.

Cominco Ltd. commenced production at the Lake Zone in the Highland Valley in British Columbia in January 1983, after acquiring ownership of the entire orebody. Staged expansions of the Lake Zone has been studied but low copper prices are preventing expansion. Highmont mine which ceased operations in October 1984 had remained in operation because of support prices paid by the purchasers of the concentrates. Afton Operating Corp. near Kamloops, resumed production in May 1983 but closed the smelter in August 1983 as concentrates sale became more economic. Esso Resources Canada Limited wrote down the book value of its Granduc mine by $40 million in January 1983 and closed the mine in April 1984 due to low copper prices. At the Island Copper mine at Port Hardy on Vancouver Island, installation of an in-pit moveable crusher and conveying system to reduce operating costs is scheduled for completion in early 1985. Westmin Resources Limited continued development of its HW project on Vancouver Island. The $225 million project had been delayed by a 4½ month strike and is now expected to be completed in early 1985. Brenda Mines Ltd. shut down at the end of September 1983 due to low copper and molybdenum prices, then operated from late May to mid-December 1984. Mine management at Brenda was quoted as saying that a price of between 75-80¢ (U.S.)/lb of copper and $US 4/lb for molybdenum would be required for the mine to reopen. Falconbridge mined out its Wesfrob iron-copper mine at Tasu on the Queen Charlotte Islands.

Gibraltar Mines Limited completed milling its low-grade ore stockpile in 1983. Mining recommenced following a replanning of the operations which permitted operation with a reduced waste to ore ratio. Noranda's Goldstream mine shut down in April 1984.

Smelting and Refining

As the Division Mines Gaspé mines were shut in 1983, the smelter had difficulty in obtaining sufficient concentrates. When the smelter at KCML Inc. (formerly Kidd Creek Mines Ltd.) was rebricked in 1983, concentrates were directed to Gaspé for treatment. These, plus imported concentrates permitted the Gaspé smelter to operate continuously throughout the second half of 1983 to May 1984. In December 1984, the Gaspé smelter reopened with a concentrates inventory sufficient to permit operation until May 1985. In 1984, the Quebec government announced intended legislation to reduce sulphur dioxide emissions from Noranda's Horne smelter by the end of 1988: Noranda stated that this could force the closure of the Horne smelter. Noranda's CCR Division's refinery which produced 176 900 t of refined copper in 1983 cut its operating rate by 25 per cent due to reduced blister supplies. KCML's Mitsubishi continuous reactor smelter operated at its rated capacity of 59 000 tpy of copper in 1983. In July 1984, KCML announced that it would spend $54 million to increase the capacity of the copper smelter and refinery to 90 000 tpy of copper by 1988.

The "custom" copper smelters in Canada continued to face difficult market circumstances. The Gaspé, Horne and Flin Flon smelters were originally built to treat concentrates from adjacent large mines. However, now that these orebodies are either exhausted or closed or unable to supply sufficient feed, the smelters must attempt to obtain concentrates from more distance sources. Well situated for their original purpose, these smelters are now poorly located to obtain concentrates at economic rates due to transportation costs. The world-wide shortage of concentrates has further exacerbated their problems by driving down smelting and refining charges as too much smelter and refinery capacity competes for concentrates. The need to control sulphur dioxide emissions will further erode the viability of these smelters.

Government Initiatives

Energy, Mines and Resources with assistance from industry, the provinces, labor and other departments completed a special report entitled "Canada's Nonferrous Metals Industry: Nickel and Copper" which was released in May 1984. The study's objective was "to identify the elements of a Canadian strategy which would foster an internationally viable and growing nonferrous metals industry, consistent with long standing economic development and environmental goals". The study examined the implications associated with reductions in sulphur dioxide emissions, at a time when metals markets are weak.

The Quebec government provided financial grants and loans forgiveable under certain circumstances to a wide range of the mineral industry. Copper deposits which will benefit from the program include Ansil, the E zone at Murdochville, and BP's A-1 zone.

WORLD DEVELOPMENTS

Copper producers in the western world reacted to the sustained low prices in 1983 and 1984 by reducing costs through changed work practices, mining and processing techniques, engineering, operational control, and labour productivity. While demand grew from 1983 to 1984 and much of the overhanging inventory was drawn down, copper prices remained very low as consumers perceived future price reductions and no supply problems.

In the western world, mine production of copper was 6.18 million t in 1983 and was estimated at 6.35 million t in 1984, refined production was 7.32 million t and was estimated at 7.26 million t in 1984, and consumption was 6.77 million t in 1983 and was estimated at 7.20 million t in 1984.

Chile became the leading producer in the western world. The People's Republic of China imported 485 000 t in 1983 over four times the average annual imports for 1979-82. Chinese imports in 1984 are expected to remain very high.

United States producers mine low grade deposits by world standards and thus they were hard hit by the low metal prices in 1983 and 1984. United States mine production of copper was 1.038 million t in 1983 and was estimated at 1.1 million t in 1984, compared to 1.538 million t in 1981. U.S. refined copper production was estimated at 1.5 million t in 1984 compared to 1.58 million t in 1983. As the economic recovery commenced in the United States, refined copper consumption increased from 1.664 million t in 1982 to 1.767 million t in 1983 to 2.1 million t in 1984. Domestic U.S. producers' concern with respect to increased imports, primarily from Chile, resulted in a petition for relief under section 201 of the U.S. Trade Act. The intent was to limit imports to increase U.S. prices so that U.S. producers could reopen. The petition was opposed by producers from Canada and from other major producing companies and governments, and by various U.S. copper fabricators who feared that their competitive positions would be threatened despite a tariff on wrought copper. Two commissioners' recommendations were for a 5¢ (U.S.)/lb tariff, two were for quotas and the fifth did not recommend relief. In September, President Reagan chose not to impose quotas or tariffs. About 500 000 tpy of copper mine capacity was shut down in the United States at the end of 1984.

Anamax Mining Company suspended mining operations at its Twin Butte mine at the end of January 1983. The mill will continue to process oxides at a rate of 13 600 tpy of contained copper in 1985. Atlantic Richfield Co. intends to sell its Anaconda Minerals Company copper assets. ASARCO Incorporated completed the rebuilding of its 160 000 tpy Hayden smelter in mid-1984, and in June 1984 announced that its Tacoma smelter would shut by mid-1985 due to large projected expenses to comply with environmental regulations and also due to concentrate shortages. The 105 000 tpy smelter was one of the few in the world which could treat concentrates with high levels of impurities. The company also shut down its Silver Bell Copper Mine in August 1984 where costs of 72¢ (U.S.)/lb for the first half of 1984, were reported. Pennzoil Corporation announced in November 1984 that it would sell the copper operations of its subsidiary, Duval Corporation. At the time, Duval's only operating copper property was the Sierita mine. Inspiration Resources Corporation spent $US 101 million to modernize its 100 000 tpy smelter at Miami, Arizona to meet environmental regulations. It was estimated that total environmental expenditures could reach $US 158 million by the end of 1986. Kennecott Corporation cut production at its 180 000 tpy Utah division to 53 500 tpy in July 1984. Decision about a $US 1 billion program to modernize the Utah division is expected by early 1985. The 100 000 tpy Inco flash furnace was completed in October 1984 at Chino Copper Mines. Phelps Dodge Corporation continued operations in 1983-84 although the union called a strike, in July 1983. Prior to the strike the average miner at Phelps Dodge had been receiving wages of $US 26,000/year and $US 10,500/year in benefits.

Chile became the world's largest copper producer in 1983. Production was 1.038 million t in 1983 and was estimated at 1.07 million t in 1984. Chile is the world's largest exporter of refined copper and third largest exporter of copper in concentrates. Due to lower than forecast copper prices, the government devalued the peso from 93 per U.S. dollar to 115 per U.S. dollar in September 1984 (Chilean inflation was 8.7

per cent for the first eight months of 1984). Chile earned about half of its foreign exchange by copper exports in 1983. The Corporacion Nacional del Cobre de Chile (Codelco), a state-owned corporation, produced 1.012 million t of copper in 1983 and estimated production for 1984 was 1.046 million t. Income before taxes in 1983 was $US 528 million. Present plans call for Codelco production of 1.35 million tpy by the late 1980s. Official investment plans for 1984-86 total $US 1,418 million, with at least a further $US 673 million required after 1986. Part of the investment ($US 268 million) will be financed by an Inter-American Development Bank Loan. Expansions are intended to offset grade declines, normal at most mining operations. Codelco intends to meet new demand for copper by expansions of Chilean production. Possible development of the La Escondida copper deposit in northern Chile was delayed by low copper prices, and by the sale of 50 per cent of the ownership (when BHP acquired Utah from General Electric Company). The owner of the other 50 per cent, Texaco Inc. also wishes to sell its mineral subsidiary. Development will likely be delayed until the new owners study their decisions to commit about $US 750 million each for the 380 million t deposit averaging 1.92 per cent copper. Thus, production may not proceed before 1990 or 1992.

Zaire's two copper producers, Gécamines and Sodimiza, produced 466 600 t and 40 000 t of copper respectively in 1983. Production in 1984 was expected to be equivalent to 1983 levels. Copper plus cobalt accounted for 47 per cent of Zairean exports in 1982. Gécamines plans to obtain loans for a third of the $US 750 million required for a five year program to reduce costs and increase productivity. A partially completed refinery may be finished which would decrease the present 240 000 tpy of blister exported (200 000 t of which is refined in Belgium). Management of Sodimiza, originally financed and managed by Japanese interests, was contracted out by Zaire in 1984 to Phillips Barat Kaiser Engineering Ltd. of Vancouver. Sodimiza concentrates which had been shipped to Japan are now sold to Motors Trading Corporation and smelted in Zambia.

Zambia's state-owned Zambia Consolidated Copper Corporation (ZCCM) produced 551 000 t of copper in the financial year 1983/84, down 245 000 t from a year earlier. Zambia arranged financing for ZCCM in excess of $US 500 million from the World Bank, the Africa Development Bank and the European Economic Community (EEC).

In **Portugal**, Rio Tinto Zinc Corporation Limited made an offer to acquire French interests in the Neves-Corvo deposit. The state-owned company which holds 51 per cent of the interest has first right of refusal on the 49 per cent interest being sold. The first orebody is 27 million t of 8.66 per cent copper. By 1987, 1 million tpy of ore will be mined producing over 70 000 tpy copper in concentrates.

In **Peru**, the Tintaya project is expected to start up in early 1985 producing 55 000 tpy of Cu in concentrates. Southern Peru Copper Corporation decided not to expand the Cuajone mine.

Australia's largest copper producer is Mount Isa Mines. Due to low copper prices, expenditures were cut and development of deeper ore will be delayed a year. At the Olympic Dam project, the feasibility of mining 1.25 million tpy in the late 1980s then 5 million tpy in the mid-1990s was studied. A bulk sample was shipped to Finland for flash smelting tests.

In the **Philippines**, the $US 380 million Pasar copper smelter and refinery complex was opened in July 1983. Due to low smelting and refining charges outside of the Philippines, the government used export control legislation to force some mines to supply feed to Pasar. Copper smelting charges were reduced by 5.9¢ (U.S.)/lb at the end of 1984. Expansion beyond 138 000 tpy is unlikely to be economic in the medium-term.

Japan's custom smelters and refineries reduced production due to shortages of concentrates. Forecasts for fiscal 1984/85 (and actual for fiscal 1983/84) are: refined production 0.932 million t (1.05 million t), domestic demand 1.48 million t (1.41 million t), refined imports 0.414 million t (0.267 million t) and refined exports 0.020 million t (0.132 million t). A significant proportion of Japanese output is believed to be exported to China as semi-manufactured items.

China purchased 486 000 t of copper in all forms in 1983. Purchases were estimated at 375 000 t for 1984. A 90 000 tpy smelter completed in 1983 in Guixi province remained idle due to low copper prices.

In **Papua, New Guinea,** development of the copper portion of the Ok Tedi mine, already delayed to 1987, may be put back to 1989.

STOCKS

London Metal Exchange (LME) total copper stocks rose from the start of 1983 almost uninterruptedly by over 70 per cent to just over 435 000 t at the end of 1983. Then total stocks decreased rapidly to 126 375 t at the end of 1984 while higher grade stocks went from 118 575 t to 49 400 t. Comex stocks rose from 249 000 t at the start of 1983 to 371 000 t at the start of 1984 and decreased to 277 009 t at the end of 1984. Comex plus LME stocks decreased from 806 900 t to 403 384 t during 1984. Total non-socialist world stocks at the end of October 1984 were reported at 1.65 million t compared to 2.06 million t at the end of 1983.

The LME was asked to establish a separate contract for copper wirebars, as higher-grade cathodes are increasingly favoured by purchasers. It was argued that allowing delivery of wirebars (when the purchaser probably wishes cathodes) acts to depress the price of the higher-grade cathode contract. Thus a separate contract for wirebars would likely result in lower prices for wirebars compared to higher-grade cathodes. Opposition was expressed by wirebar producers who feared a loss in revenue and no changes were made in 1984. In October, the LME fire refined contract was eliminated.

The basic copper grade for COMEX was changed to Grade 2 Electrolytic for December 1985 contracts and thereafter. Wirebars, which used to command a premium, will trade at par with the basic contract. Grade 1 electrolytic copper will command a ½¢ (U.S.)/lb premium (£ 9/t at the exchange rate prevailing when announced).

PRICES

The variations in copper prices depended upon the currency used. Fourth quarter LME prices in sterling recovered nearly to mid-1983 levels of over £ 1 100/t whereas prices in U.S. currency in the range of 57-61¢ (U.S.)/lb in December 1984 were far below the 77-79¢ (U.S.)/lb range of mid-1983. Average price for LME copper higher grade cash was 72.1¢ (U.S.)/lb in 1983 and 62.4¢ (U.S.)/lb in 1984. Historical and deflated copper prices from 1970 to 1984 are shown graphically.

OUTLOOK

Consumption is expected to increase at an average of 1.2 per cent per year in the 1980s from a 1981 base and at an average of 1.6 per cent per year in the 1990s. While no major new uses for copper have been identified, substitution away from copper in the future is expected to occur at lower rates than those in the past. Current applications of copper will continue to be the major end use. As funding is inadequate to support significant market research and development to promote new uses of copper, substitution away from copper will continue. In the short-term, western world consumption is forecast at 7.35 to 7.40 million t in 1985. In the long-term, western world copper refined consumption is forecast at 7.9 million t in 1990 and 9.2 million t in 2000. China is expected to continue to import about 375 000 tpy of copper in the short-term.

Existing mine, smelter and refinery capacity will be able to meet the projected short-term demand without requiring any new greenfields projects. Producers have increased productivity to reduce costs in the face of low prices and have expanded production where possible to lower unit costs. Mine production is expected to increase in Chile and Peru in the short-term and decrease to 900 000 t in the United States.

In the medium-term, incremental expansions are expected to meet all of the increases in demand and to compensate for permanent closures. The only greenfield project expected to be producing before 1990 is Neves-Corro in Portugal. However by the early 1990s, the producers who have reduced stripping ratios and restricted development in the 1980s in order to minimize financial loses are not expected to be able to maintain output. Thus in the early 1990s, new greenfields projects are likely to be required. The first projects to be completed would likely include the full scale Olympic Dam project in Australia, La Escondida in Chile, and Valley Copper in Canada.

Most copper producers, with the exception of Codelco, continued to lose money in 1984 or to generate very low profits. The copper production capacity completed in the 1970s when real rates of interest were low was built to fulfil a demand growth of 4 per cent per year which never materialized. The majority of the price-sensitive, high cost producers have closed, but many have been kept in a state allowing for rapid reopening if prices were to increase. As long as these mines can reopen, there is a barrier to sustained high prices. In the short-term, a price over 75¢ (U.S.)/lb would allow significant production to be restarted. Hence an average price of 68¢ (U.S.)/lb is predicted for 1985. Localized supply problems and general inventory drawdowns are likely to increase prices in the first half of 1985 above the forecast average price for the year. Premiums for higher-grade materials are likely to rise at that time. A price of 75¢ (U.S.)/lb in 1985 U.S. dollars is forecast for 1986. By 1990 as closures have become permanent and opportunities for incremental expansions have been exhausted the price in 1985 U.S. dollars should trend through a value of 88-92¢ (U.S.)/lb. Thereafter, a price rise in the 1990s above a level of 88-92¢ (U.S.)/lb in 1985 U.S. dollars is not expected to be sustainable due to technological advances in the mineral industry and increased substitution due to price. Historical average prices of approximately $US 1.25/lb (1985 dollars) are not expected to be sustainable.

Due to the oversupply of smelting and refining capacity, smelting and refining charges in 1985 and 1986 are forecast at 15-17¢ (U.S.)/lb for medium-term contracts. Spot prices will remain very low. Possible new smelting and refinery capacity in Chile, Brazil, Zaire, Australia and Portugal are expected to keep charges low.

In **Canada,** the industry joined the rest of the price-sensitive market economy producers by reducing operating costs in response to low metal prices. For example, at the two major nickel producers (who produced 15 per cent of Canada's copper in 1983 despite a shutdown by Inco) some of the mines have reduced production costs by 25 per cent in one year. In order to coordinate research and development directed at lowering costs and to best employ Canadian expertise, Inco, Noranda, KCML and Falconbridge formed a company to jointly fund research and development to improve productivity. Another company owned by Inco develops and manufactures hard rock mining equipment.

Between 1981 and 1983, the international competitive position of the Canadian copper industry weakened, partly due to very significant currency devaluations by major competitors. While some of those devaluations may have been overdue based on balance of payments considerations, these devaluations were responsible for much of the deterioration in the Canadian competitive position. However, currency devaluations are followed by inflation which later negate the competitive advantages afforded by devaluation. Devaluations also impose social strains on some economies which can lead to political resistance to further devaluations. Thus Canadian producers who have kept in operation, while unable to avail themselves of relief from low prices through large currency devaluations and who have been able to make significant cost reductions, should move back towards their more traditional competitive positions as inflation increases the costs of their competitors.

TARIFFS

Item No.		British Preferential	Most Favoured Nation	General	General Preferential
			(%)		
CANADA					
32900-1	Copper in ores and conc.	free	free	free	free
33503-1	Copper oxides	free	13.8	25.0	free
34800-1	Copper scrap, matte and blister and copper in pigs, blocks or ingots; cathode plates of electrolytic copper for melting, per lb	free	free	1.5¢	free
34820-1	Copper in bars or rods, when imported by manufacturers of trolley, telegraph and telephone wires, and electric wires and electric cables, for use only in the manufacture of such articles in their own factories	free	4.5	10.0	free
34835-1	Electrolytic copper powder electrolytic iron powder, for use in Canadian manufactures (expires June 30, 1983)	free	free	10.0	free
34845-1	Electrolytic copper wire bars, for use in Canadian manufactures per lb (expires June 30, 1983)	free	free	1.5¢	free
35800-1	Anodes of copper	free	free	10.0	free

MFN Reductions under GATT (effective January 1 year given)	1983	1984	1985	1986	1987
			(%)		
33503-1	13.8	13.4	13.1	12.8	12.5
34820-1	4.5	4.4	4.3	4.1	4.0

UNITED STATES (MFN)

602.30	Copper, ores etc.		Remains free		
612.02	Unwrought copper, etc.		- no change -		1.7
612.08	Copper waste and scrap	4.2	3.8	3.3	2.9	2.4

EUROPEAN ECONOMIC COMMUNITY (MFN)		1983	Base Rate	Concession Rate
26.01	Copper, ores and conc.	free	free	free
74.01	Copper in matte, unwrought copper, waste and scrap	free	free	free

JAPAN (MFN)

Item	Description			
26.01	Copper, ores and conc.	free	free	free
74.01	(1) Copper in matte etc.	free	free	free
	(2) Copper, unwrought			
	(a) containing not more than 99.8% by weight of copper etc.	7.9%	8.5%	7.3%
	(b) Other			
	(i) Containing by weight, not less than 25% of zinc and not less than 1% of lead	19.50yen/kg	24yen/kg	15yen/kg
	(ii) Containing more than 95% by weight of copper			
	- blister copper in bar	7.9%	8.5%	7.3%
	- other	22.50yen/kg	24yen/kg	21yen/kg
	(iii) Containing not more than 95% by weight of copper	22.50yen/kg	24yen/kg	21yen/kg
	(3) Waste and scrap			
	(a) Unalloyed	1.3%	2.5%	free
	(b) Other: containing more than 10% by weight of nickel	11.3%	22.5%	free
	(c) Other	1.3%	2.5%	free

Sources: The Customs Tariff, 1983, Revenue Canada, Customs and Excise; Tariff Schedules of the United States Annotated (1983), USITC Publication 1317; U.S. Federal Register Vol. 44, No. 241; Official Journal of the European Communities, Vol. 25, No. L318, 1982; Customs Tariff Schedules of Japan, 1983; GATT Documents, 1979.

TABLE 1. CANADA, COPPER PRODUCTION, TRADE AND CONSUMPTION, 1982-84

	1982		1983r		1984e	
	(tonnes)	($000)	(tonnes)	($000)	(tonnes)	($000)
Shipments[1]						
British Columbia	280 969	548,256	282 754	597,710	288 646	547,561
Ontario	158 220	308,735	219 803	412,140	291 107	552,231
Quebec	94 977	185,329	63 740	132,352	62 713	118,967
Manitoba	48 810	95,243	67 163	132,008	57 642	109,347
New Brunswick	13 125	25,610	11 369	19,037	7 185	13,631
Saskatchewan	4 898	9,557	6 204	13,676	4 252	8,066
Newfoundland	3 731	7,280	-	384	749	1,421
Yukon	7 510	14,654	1 904	-	-	-
Northwest Territories	215	419	102	-	79	149
Total	612 455	1,195,083	653 040	1,307,307	712 374	1,351,373
Refinery output	337 780	..	464 333	..	515 000	
Exports						(9 months export data)
Copper in ores, concentrates and matte						
Japan	182 916	236,366	212 094	270,931	165 095	210,331
South Korea	14 882	17,427	38 716	39,898	25 479	28,857
Taiwan	5 672	6,658	23 761	35,151	1 405	1,826
Norway	15 018	21,523	18 216	27,818	20 229	30,140
United States	19 508	21,602	12 255	17,608	14 956	19,203
People's Republic of China	-	-	2 516	4,064	18 837	26,285
Mexico	-	-	1 713	2,572	-	-
United Kingdom	747	567	1 775	2,538	611	1,172
Belgium-Luxembourg	1 984	2,691	1 900	2,141	246	145
Other countries	17 203	23,844	850	882	2 396	3,075
Total	257 930	330,678	313 796	403,603	249 252	321,034
Copper in slag, skimmings and sludge						
United States	1 105	215	1 708	753	2 488	1,051
Spain	247	228	-	-	-	-
United Kingdom	4	13	-	-	-	-
Total	1 356	456	1 708	753	2 488	1,051
Copper scrap (gross weight)						
United States	21 613	30,089	31 939	42,123	21 351	32,669
Japan	4 788	5,705	2 385	3,242	*	*
South Korea	1 400	1,996	510	797	*	*
Sweden	-	-	279	507	-	-
Taiwan	49	65	370	443	*	*
Belgium-Luxembourg	649	703	239	368	*	*
Spain	1 675	2,203	181	259	*	*
Other countries	1 013	1,236	203	298	2 565	3,839
Total	31 187	41,997	36 106	48,037	23 916	36,508
Brass and bronze scrap (gross weight)						
United States	6 948	8,370	11 870	16,557	8 947	12,224
Belgium-Luxembourg	2 951	3,710	694	881	*	*
Taiwan	214	255	727	829	*	*
West Germany	949	1,335	332	534	*	*
India	2 224	2,645	292	335	*	*
Japan	452	582	241	316	*	*
South Korea	596	741	155	244	*	*
Netherlands	88	93	88	137	*	*
Other countries	958	1,124	139	155	2 849	3,642
Total	15 380	18,855	14 538	19,988	11 796	15,866

TABLE 1. (cont'd.)

	1982		1983P		1984	
	(tonnes)	($000)	(tonnes)	($000)	(tonnes)	($000)
					(9 months export data)	
Copper alloy scrap, nes (gross weight)						
United States	3 037	3,031	2 094	2,473	3 178	3,800
Belgium-Luxembourg	1 079	1,348	222	268	*	*
South Korea	375	445	149	188	*	*
West Germany	88	138	40	97	*	*
Taiwan	207	84	93	46	*	*
Japan	19	22	18	29	*	*
Other countries	68	55	21	53	456	611
Total	4 873	5,123	2 637	3,154	3 643	4,411
Copper refinery shapes						
United States	93 219	170,781	93 138	190,264	147 642	277,311
People's Republic of China	-	-	67 137	135,242	12 556	22,014
United Kingdom	65 881	132,659	46 444	91,697	29 799	55,724
Netherlands	9 040	17,302	35 075	71,784	15 268	27,798
West Germany	22 194	42,291	24 496	49,494	18 652	33,755
France	10 741	20,573	13 204	25,842	10 182	18,403
Sweden	9 578	18,391	6 261	12,229	8 188	15 031
Belgium-Luxembourg	14 594	32,907	4 429	9,362	18	38
Italy	4 129	7,990	3 299	6,353	3 104	5,634
Japan	3	6	2 946	3,317	2 004	3,664
South Korea	293	631	-	-	3 996	7,282
Taiwan	83	50	-	-	3 007	5,177
Other countries	1 394	2,661	181	363	811	1,729
Total	232 621	448,992	298 528	599,698	255 297	473,560
Copper bars, rods and shapes, nes						
United States	8 084	16,571	5 531	14,396	7 814	19,394
Venezuela	1 451	3,356	1 574	3,556	1 975	4,181
Dominican Republic	416	881	826	1,986	*	*
Saudi Arabia	-	-	-	-	2 057	3,927
Ireland	-	-	397	860	*	*
Bangladesh	567	1,188	285	588	2 196	4,145
India	-	-	-	-	1 800	3,260
Other countries	1 678	3,808	194	430	1 847	3,868
Total	12 327	26,084	9 435	23,144	17 689	38,775
Copper plates, sheet and flat products						
United States	3 707	11,612	3 473	11,253	4 497	14,392
United Kingdom	125	404	162	602	*	*
Israel	-	-	89	236	*	*
West Germany	-	-	44	145	*	*
Venezuela	50	202	29	115	*	*
Other countries	27	94	30	101	149	515
Total	3 909	12,312	3 827	12,452	4 646	14,907
Copper pipe and tubing						
United States	2 327	7,288	3 685	10,832	3 633	10,781
Israel	826	2,238	1 073	2,905	897	2,381
West Germany	1 058	2,600	553	1,520	*	*
United Kingdom	536	1,634	245	992	*	*
Netherland Antilles	5	18	98	291	*	*
South Africa	1	5	26	85	*	*
Mexico	-	-	18	74	-	-
Saudi Arabia	38	146	31	48	-	-
Netherlands	5	19	3	12	-	-
Spain	-	-	2	11	-	-
Other countries	46	167	38	139	529	1,538
Total	4 842	14,115	5 772	16,909	5 059	14,700
Copper wire and cable (not insulated)						
United States	100	350	539	1,129	455	933
Saudi Arabia	38	125	122	376	-	-
Puerto Rico	-	-	128	344	-	-
New Zealand	5	29	54	113	-	-
Other countries	39	102	47	181	142	391
Total	182	606	890	2,143	597	1,324

TABLE 1. (cont'd.)

	1982		1983P		1984	
	(tonnes)	($000)	(tonnes)	($000)	(tonnes)	($000)
					(9 months export data)	
Copper alloy shapes and sections						
United States	7 873	24,264	12 376	35,858	14 102	40,650
United Kingdom	71	230	36	126	-	-
Venezuela	24	81	19	71	-	-
West Germany	--	1	15	42	-	-
Other countries	117	252	9	56	70	225
Total	8 085	24,828	12 455	36,153	14 172	40,875
Copper alloy pipe and tubing						
United States	1 616	6,038	4 630	12,564	4 047	11,580
West Germany	-	-	56	166	*	*
Bangladesh	-	-	18	87	-	-
United Kingdom	--	6	16	63	-	-
Other countries	42	170	12	41	40	112
Total	1 658	6,214	4 732	12,921	4 087	11,692
Copper alloy wire and cable, not insulated						
United States	102	529	104	292	286	660
South Africa	18	156	43	222	*	*
New Zealand	17	113	17	103	*	*
Chile	10	70	11	68	-	-
Other countries	5	37	17	54	29	195
Total	152	905	192	739	315	4,855
Copper and alloy fabricated materials, nes						
United States	658	3,183	1 337	4,855	1 518	4,890
United Kingdom	53	186	132	319	*	*
Taiwan	29	211	39	221	*	*
Mexico	-	-	16	86	-	-
Other countries	66	511	40	224	488	1,094
Total	806	4,091	1 564	5,705	1 966	5,984
Insulated wire and cable[2]						
United States	14 315	54,717	24 324	83,689	24 434	84,792
Saudi Arabia	7 805	24,736	3 196	10,397	*	*
Trinidad-Tobago	324	1,063	1 805	6,098	1 700	5,423
Puerto Rico	157	528	847	2,520	1 845	5,718
Panama	206	574	194	1,493	*	*
Singapore	660	2,408	366	1,300	*	*
Libya	31	135	306	1,138	*	*
Egyptian A.R.	1 201	3,789	317	1,100	*	*
Indonesia	108	450	261	907	*	*
Pakistan	-	-	-	-	2 990	9,082
Other countries	3 224	14,571	1 534	6,888	3 492	16,774
Total	28 121	103,418	33 302	116,338	34 461	121,789
Total exports of copper and products	..	1,038,674	..	1,301,737		
					(9 month impact data)	
Imports						
Copper in ores and concentrates	12 362	13,742	24 535	39,984	30 024	28,592
Copper scrap	33 230	34,553	64 363	68,206	48 966	55,444
Copper refinery shapes	28 028	52,760	24 559	56,126	21 453	40,979
Copper bars, rods and shapes, nes	6 061	12,406	9 626	21,508	3 829	8,436
Copper plates, sheet strip and flat products	977	3,533	1 370	4,615	3 587	10,403
Copper pipe and tubing	2 519	9,170	3 479	12,022	2 162	7,844
Copper wire and cable, not insulated	1 952	5,702	1 822	7,023	2 263	8,585
Copper alloy scrap (gross weight)	7 883	8,266	7 164	7,521	7 722	8,341
Copper powder	540	1,245	827	1,846	1 005	2,316
Copper alloy refinery shapes, bars and sections	6 732	16,449	9 835	23,695	9 410	23,829
Brass plates, sheet and flat products	2 767	8,663	3 542	11,060	3 348	10,221

TABLE 1. (cont'd.)

	1982		1983P		1984	
	(tonnes)	($000)	(tonnes)	($000)	(tonnes)	($000)
					(9 months export data)	
Imports (cont'd)						
Copper alloy plates, sheets, strip and flat products	773	4,397	1 673	7,733	1 336	6,114
Copper alloy pipe and tubing	1 884	8,978	2 851	12,859	3 121	14,654
Copper alloy wire and cable, not insulated	774	2,837	1 194	3,969	951	3,313
Copper and alloy fabricated material, nes	2 386	11,813	1 849	10,277	1 576	8,661
Insulated wire and cable	..	133,634	..	61,873	..	114,765
Copper oxides and hydroxides	288	767	201	543	183	467
Copper sulphate	4 536	2,751	873	638	2 029	1,354
Copper alloy castings	228	1,395	503	3,416	593	3,758
Total imports of copper and products	..	333,061	..	454,914	–	358,076
					(estimated consumption)	
Consumption[3]						
Refined	130 559	..	170 443	..	190 000	..

Sources: Energy, Mines and Resources Canada; Statistics Canada.
[1] Blister copper plus recoverable copper in matte and concentrate exported; totals may not add due to independent rounding.   [2] Includes small quantities of non-copper wire and cable, insulated.   [3] Producers' domestic shipments, refined copper.
- Nil;   .. Not available or not applicable;   nes Not elsewhere specified;   [r] Revised;   [e] Estimated;   -- Amount too small to be expressed.
* Minor amounts, included in other countries total: for complete details, refer to Statistics Canada Monthly Catalogue 65-004.

TABLE 2. CANADA, COPPER PRODUCTION, TRADE AND CONSUMPTION, 1970, 1975 AND 1980-84

	Production		Exports			Imports	Consumption[2]
	Shipments[1]	Refinery Output	Ore and Matte	Refined	Total	Refined	Refined
	(tonnes)						
1970	610 279	493 261	161 377	265 264	426 641	13 192	215 834
1975	733 826	529 197	314 518	320 705	635 223	10 908	185 198
1980	716 363	505 238	286 076	335 022	621 098	13 466	195 124
1981	691 328	476 655	276 810	262 642	539 452	24 778	216 759
1982	612 455	337 780	257 930	232 621	490 551	28 028	130 559
1983[r]	653 040	464 333	313 796	298 528	612 324	24 559	170 443
1984[e]	712 374	515 000	249 252*	255 297*	504 549*	21 453*	190 000

Sources: Energy, Mines and Resources Canada; Statistics Canada.
[1] Blister copper plus recoverable copper in matte and concentrate exported.   [2] Producers' domestic shipments of refined copper.
[r] Revised; [e] Estimated; * Jan. to Sept. trade data for 1984.

TABLE 3. CANADIAN COPPER AND COPPER-NICKEL SMELTERS, 1983

Company and Location	Product	Rated Annual Capacity (tonnes of ores and concentrates)	Ore and Concentrates Treated (tonnes)	Blister or Anode Copper Produced (tonnes)	Remarks
Afton Operating Corporation Kamloops, B.C.	Blister copper	22 500 t of blister copper	..	3 100	The smelter commenced commercial operation on May 1, 1978. The uniquely low-sulphur concentrate, consisting chiefly of native copper, was smelted in a top-blown rotary converter. SO_2 produced was neutralized with limestone. Smelter was closed permanently, August 1983.
Falconbridge Limited Falconbridge, Ont.	Copper-nickel matte	570 000	..	27 500	A smelter modernization program begun in 1975 was completed in 1978 at a cost of $79 million. Fluid bed roasters and electric furnaces replaced older smelting equipment. A 1 800 tpd sulphuric acid plant treats roaster gases. Matte from the smelter is refined in Norway.
Inco Limited Sudbury, Ontario	Molten "blister" copper, nickel, sulphide and nickel sinter for the company's refineries; nickel oxide sinter for market, soluble nickel oxide for market	3 630 000 1	..	66 700 2	Oxygen flash-smelting of copper concentrate; converters for production of blister copper. Roasters, reverberatory furnaces for smelting of nickel-copper concentrate, converters for production of nickel-copper Bessemer matte. Production of matte followed by matte treatment, flotation, separation of copper and nickel sulphides, then by sintering to make sintered-nickel products for refining and marketing. Electric furnace melting of copper sulphide and conversion to blister copper.
KCML Inc Timmins, Ontario	Molten "blister" copper	59 000 t of copper	118 600	53 000	Mitsubishi-type smelting, separation and converting furnaces treat continuous copper concentrate feed stream to yield molten 99 per cent pure copper which is transported by ladles and overhead cranes to two 350 t anode furnaces. Furnaces rebricked in fall of 1983. Expansion plans to 90 000 tpy

23.12

Noranda Mines Limited, Horne smelter, Noranda, Que.	Copper anodes	838 000	176 900	Three reverberatory furnaces, one of which is now considered to be permanently shut down; 5 converters; 1 continuous reactor; an 85 tpd oxygen plant to supply oxygen-enriched blast. Continuous reactor modified to produce matte instead of metal. A $35 million project to overhaul and modify the smelter, with electricity to become the plant's major energy source was completed in 1982. The new 450 tpd oxygen plant will decrease unit fuel requirements and increase capacity of the continuous reactor, and reduce fuel requirements for a reverberatory furnace.
Noranda Mines Limited, Gaspé smelter, Murdochville, Que.	Copper anodes	325 000	30 800	Equipped with one fluid bed roaster, one reverberatory furnace and two converters plus an acid plant. Treats Gaspé and custom concentrates (mine at Gaspé shut throughout 1983).
Hudson Bay Mining and Smelting Co., Limited, Flin Flon, Man.	Copper anodes	400 000	66 700	Five roasting furnaces, one reverberatory furnace and three converters. Company treats its own copper concentrates from mines at Flin Flon, Snow Lake and Whitehorse, as well as custom copper concentrates, zinc plant residues and stockpiled zinc-plant residues fed to reverberatory furnace.

1 Includes copper and nickel-copper concentrates. This capacity cannot all be fully utilized owing to Ontario government sulphur dioxide emission regulations. 2 A small portion of this copper was from Inco's Manitoba ores; operations at Sudbury restarted in April 1983 after a 10 month shutdown.
.. Not available.

TABLE 4. COPPER REFINERIES IN CANADA, 1983

Company and Location	Rated Annual Capacity (tonnes)	Output in 1983	Remarks
Noranda Mines Limited, Division CCR, Montreal East, Quebec	435 000	335 700	Refines anodes from Noranda's Horne and Gaspé smelters and from the Flin Flon smelter; also purchased scrap. Copper sulphate and nickel sulphate recovered by vacuum evaporation. Precious metals, selenium and tellurium recovered from slimes. Produces C.C.R. brand electrolytic copper wirebars, ingot bars, ingots, cathodes, cakes and billets.
Inco Limited Copper Refining Division Copper Cliff, Ont.	180 000	66 700	Casts and refines anodes from molten converter copper from the Copper Cliff smelter; also refines purchased scrap. Gold, silver, selenium and tellurium recovered from anode slimes, along with platinum metals concentrates. Recovers and electrowins copper from Copper Cliff nickel refinery residue. Produces ORC brand electrolytic copper cathodes, and wirebars.
Kidd Creek Mines Ltd. Timmins, Ontario	59 000	53 000	Molten copper from two 350 t anode furnaces is cast in a Hazelett continuous casting machine into continuous copper strip, then formed to 145 kg anodes in a blanking press. Spent and scrap anodes are remelted in a 40 t ASARCO shaft furnace. Cathodes formed in jumbo sized electrolytic tanks in a highly automated tankhouse. A decopperized precious metal slime is also marketed. Expansion plans to 90 000 tpy by 1988 announced in 1984.

TABLE 5. WESTERN WORLD MINE PRODUCTION OF COPPER, 1983 AND 1984

	1983	1984e
	(000 tonnes)	
Chile	1 257	1 290
United States	1 038	1 100
Canada[1]	654r	712
Zambia[2]	515	455
Zaire	502	520
Peru	322	370
Philippines	271	230
Mexico	206	200
South Africa	212	200
Other[3]	1 205	1 273
Total Western World[4]	6 182	6 350

Sources: World Bureau of Metal Statistics, U.S. Bureau of Mines, and Energy, Mines and Resources Canada.
[1] Shipments. [2] Excludes electrowon copper. [3] Includes estimated 65 000 t increase in Iranian production from 1983 to 1984. [4] Includes Yugoslavia.
e Estimated; r Revised.

TABLE 6. WESTERN WORLD REFINED PRODUCTION[1] OF COPPER, 1983 AND 1984

	1983	1984e
	(000 tonnes)	
United States	1 580	1 500
Japan	1 090	930
Chile	833	850
Zambia	574	530
Canada	464	515
Zaire	227	220
Peru	190	245
Australia	200	200
Other	2 160	2 275
Total Western World	7 318	7 265

Sources: World Bureau of Metal Statistics, U.S. Bureau of Mines, and Energy, Mines and Resources Canada.
[1] Primary, secondary and electrowon copper.
e Estimated.

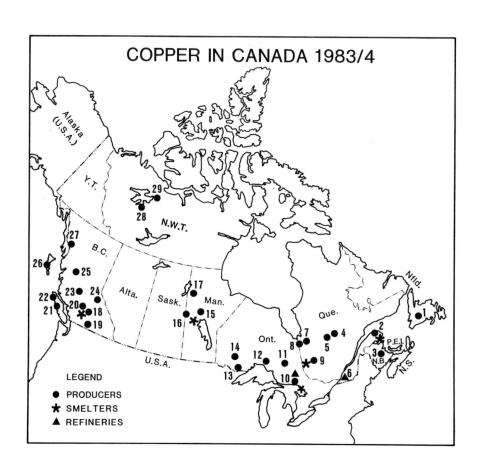

PRODUCERS IN 1983 or 1984

numbers correspond to those in map

1. ASARCO Incorporated (Buchans Unit)
2. Noranda Inc., Division Mines Gaspé (Needle Mountain)
3. Brunswick Mining and Smelting Corporation Limited (Nos. 6 and 12 mines)
   Heath Steele Mines Limited
4. Camchib Resources Inc. (Cedar Bay, Henderson and Merrill mines)
   Northgate Patino Mines Inc. (Copper Rand, Lemoine and Portage mines)
5. Corporation Falconbridge Copper, Opemiska Division (Perry, Springer and Cooke mines)
7. Noranda Mines Limited, Mattagami Division (Mattagami, Orchan, Norita mines)
8. B.P. Canada Inc., Les Mines Selbaie
9. Corporation Falconbridge Copper, Lake Dufault Division (Millenbach and Corbet mines)
   Louvem Mining Company Inc.
10. Falconbridge Limited (East, Falconbridge, Fraser, Lockerby, North and Strathcona mines)
    Inco Limited (Clarabelle, Coleman, Copper Cliff North, Copper Cliff South, Creighton, Frood, Garson, Levack, Little Stobie, Stobie, McCreedy West)
11. KCML Inc.
    Pamour Porcupine Mines, Limited (Schumacher, Ross mines)
12. Noranda Mines Limited, Geco Division
13. Inco Limited (Shebandowan mine)
14. Mattabi Mines Limited
    Noranda Mines Limited, Lyon Lake and F group mines
15. Inco Limited (Pipe No. 2 and Thompson mines)
16. Hudson Bay Mining and Smelting Co., Limited (Anderson, Centennial, Chisel, Flin Flon, Ghost, Osborne, Stall, Trout Lake Westarm and White Lake mines), Spruce Point
17. Sherritt Gordon Mines Limited Fox and Ruttan mines
18. Brenda Mines Ltd.
19. Newmont Mines Limited (Ingerbelle and Copper Mountain mines)
20. Cominco Limited (Valley Copper Lornex Mining Corporation Ltd. Afton Operating Corporation Highmont Operating Corporation
21. Westmin Resources Limited (Lynx, Myrna, Price and HW mines)
22. Utah Mines Ltd. (Island Copper mine)
23. Gibraltar Mines Limited
24. Noranda Mines Limited (Goldstream mine)
25. Equity Silver Mines Limited
26. Falconbridge Limited (Wesfrob mine)
27. Canada Wide Mines Ltd. (Granduc mine)
28. Terra Mines Ltd.
29. Echo Bay Mines Ltd.

SMELTERS

2. Noranda Mines Limited Division Mines Gaspé
9. Noranda Mines Limited
10. Falconbridge Limited
    Inco Limited
11. KCML Inc.
16. Hudson Bay Mining and Smelting Co., Limited
20. Afton Mines Ltd.

REFINERIES

6. Noranda Mines Limited, Division CCR
10. Inco Limited
11. KCML Inc.

An inventory of undeveloped Canadian copper deposits is available in the publication Canadian Mineral Deposits Not Being Mined in 1983, Energy, Mines and Resources Canada, Report MRI 198, ISBN 0-660-11580-8

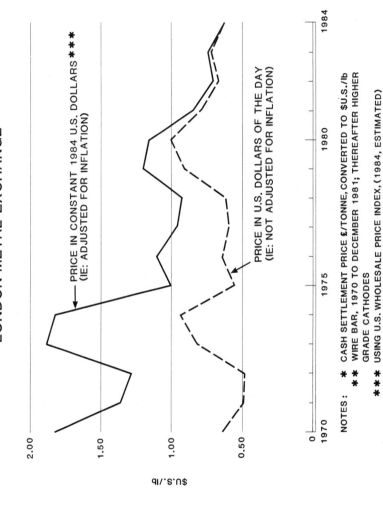

Crude Oil and Natural Gas

R. THOMAS

Over much of the past century, Canada's petroleum resource base has been developed in the prairie provinces' portion of the Western Sedimentary Basin with continuing production from the smaller pools of southwestern Ontario that were discovered in the 1860s. It has been estimated that the initial volume of crude oil in-place for the provinces is some 9.5×10^9 cubic metres (m^3), (60 billion barrels) of which $2.5 \times 10^9 m^3$ (16 billion barrels) have been established. Currently, the volume of remaining established reserves is at some $0.9 \times 10^9 m^3$ (5.6 billion barrels) indicating that over $1.65 \times 10^9 m^3$ (10 billion barrels) have been produced to date.

In the provinces, the initial volume in-place of natural gas has been estimated to be $5\,418 \times 10^9 m^3$, (191 trillion ft^3) with some $3\,979 \times 10^9 m^3$ (140 TCF) in established reserves. The current estimate of remaining reserves has been placed at some $2\,000 \times 10^9 m^3$ (71 TCF).

The recent report published by the Geological Survey of Canada (Paper 83-31) indicates the potential remaining to be discovered for crude oil in western Canada could range between 593 million m^3 at average expectation and 1.210 billion m^3 as a speculative estimate (3.736 - 7.623 billion barrels). Comparatively, the discovery potential for natural gas may be in the range of $2\,504 \times 10^9 m^3$ to $4\,930 \times 10^9 m^3$ (89 - 175 TCF).

Another resource being developed in Canada is from Alberta's vast oil sands deposits, believed to contain approximately 159 billion m^3 (1×10^{12} barrels) of crude bitumen in-place, of which some 56 billion m^3 (350 billion barrels) may be recoverable. There are currently two major commercial projects, Suncor and Syncrude, that are operating in the Athabasca field from a licenced reserve base of some 3 860 million m^3 (25 billion barrels). The combined current output from the plants averages 20 000 m^3/d (126,000 b/d) from open-pit mines. Studies are being undertaken to consider the possibility of producing from sunken shafts incorporating horizontal wells. Esso Resources Canada Limited has undertaken the recovery of bitumen in the Cold Lake deposits by in-situ recovery methods. The project at Leming, has been classified as "experimental" to date with the daily bitumen production averaging 2 600 m^3. The expansion of the project from experimental stage to the commercial stage will be constructed in six phases. Each phase will increase the daily production by 1 500 m^3 and will cost approximately $150 million. The first two phases are to begin production in 1985 and the subsequent phases will be added to maintain the production over the 25-year project life. The capital cost of the entire project is expected to be in the order of $1.5 billion. Other major heavy ore and bitumen projects are underway at Lindbergh (Dome, Sulpetro, CNG), Peace River (Shell), Wolf Lake (BP Canada, Petro-Canada) and Fort Kent (Suncor, Canadian Worldwide Energy).

In addition to the oil and gas resources previously mentioned, there remains the technique of further extracting crude oil from older existing pools through "enhanced oil recovery", which recover crude that has been left in the ground after primary and secondary (waterflood) production. Techniques used include: chemical, steam stimulation, carbon dioxide, and wet/dry thermal flooding. The Geological Survey of Canada estimates the enhanced oil recovery's potential may range from 160 to 950 million m^3 (1 to 6 billion barrels).

During 1984, the expected production of petroleum from Canada's provinces may average 280 700 m^3/d (1.77 million barrels) which includes conventional crude, synthetic crude, pentanes plus/condensate and natural gas liquids. The bulk of this output comes from western Canada with Alberta being the largest producer of oil and natural gas. It is anticipated that Alberta will account for some 243 160 m^3/d of liquid petroleum or about 87 per cent.

R. Thomas is with the Energy Sector, Energy, Mines and Resources Canada. Telephone (613) 995-1118.

The production of natural gas is expected to average 180 million m^3/d (6.4 billion ft^3/d) in 1984 of which some 53 million m^3 was exported to the United States, and the balance was consumed domestically. Again, Alberta accounted for 89 per cent of total Canada output. Ontario is the greatest consuming province of gas where much of the throughput is allocated for the industrial sector. From total domestic sales, some 54 per cent is absorbed by major industries, 26 per cent is sold to residences and the remaining 21 per cent is consumed by the commerical sector.

The last record year for drilling in the provinces was in 1980, in which some 9,200 wells were completed while a low of 5,800 completions occurred in 1982. In 1984, it is anticipated that drilling will nearly equal the amount recorded in 1980. Statistics compiled to the end of the third quarter show more than 4,200 oil wells, almost 1,300 gas wells and 1,300 dry holes for a total exceeding 6,800 completions. The current trend in drilling has shifted toward oil where markets are more firm because of domestic and export demand. Well completions for the year could total 9,100 wells and aggregate metrage attained may be 10.2 million m.

Over the past year, additional discoveries of oil in established regions were made in western Canada in areas such as: Pembina, Redwater, Swan Hills, Mitsue, and Willesden Green. Significant discoveries have been made in new oil plays as shown on the accompanying map of western Canada. In Ontario, oil and natural gas continue to be from old pools. Most of the old oil wells are on pumping and many wells are producing about 1 m^3/d. Much of the natural gas in the province is from Lake Erie, where operations are conducted on barges or jack-up rigs. There was very little oil and gas activity on land in Quebec and the Maritimes.

Canada's refining industry has seen many changes during the recent past. Throughout the country, the demand for oil has been decreasing due to slack markets, conservation and oil-substitution programs by natural gas. Several refineries closed in eastern Canada and others have rationalized their operations to fully utilize capacity. During the past two years, natural gas from western Canada has been increasingly shipped into Ontario and Quebec through the TransCanada PipeLines and the TransQuebec and Maritimes Line. The TQM pipeline transports some 13.4 million m^3/d of gas into Quebec.

In June, 1984, an agreement was reached among federal-provincial governments (Alberta and Saskatchewan) and Husky Oil Ltd. in which a heavy oil upgrader would be constructed in order to convert heavy oil and crude bitumen into a synthetic oil suitable for refining in Canada. This plant, costing around $2.0 billion, will produce 6 667 m^3/d (42,000 b/d) of synthetic crude oil upon its completion, expected in 1989. Located in Lloydminster, the feedstock would be drawn from Cold Lake (Alberta) and the Lloydminster area, bordering Alberta and Saskatchewan.

The pricing of crude oil in Canada is calculated on the basis of its quality (density and sulphur content) and year produced. The two types of crude are: New Oil Reference Price (NORP - post 1974) and Conventional Oil Oil Price (COOP - pre-1974), with the reference price for each based on a quality of 38°API and a sulphur content of 0.7 per cent. The wellhead price for a NORP crude in this category would be $252.35/m^3 ($Cdn 40.00/bbl) and a corresponding COOP price would be $185.85/m^3 ($Cdn 29.50/bbl).

The price of domestic natural gas is based on the gigajoule, where 1GJ = 0.95 million BTU. The current field gate price of gas is $Cdn 2.099/GJ and its export price is $Cdn 5.407/GJ;, or respectively $1.99/MCF and $4.14/MCF. In a late year amendment to export pricing, an agreement was reached between the federal government and six companies that these exporters could negotiate a price with American buyers that would make Canadian gas more competitive in the market place. This change in pricing policy was necessary because the volumes exported to the United States were considerably less than the amount under contract. The average negotiated prices will be in the range of $US 3.25 - $3.35/million BTU, comparable to the Canadian eastern market import price of $US 3.15/million BTU. This newly negotiated price is anticipated to account for some $US 2.4 billion in revenue out of a total expected $US 3,100 million from the sales of natural gas to the United States in the coming year.

As 1984 has seen a marked improvement in overall industry activity levels, it is expected to continue through 1985. The federal and provincial governments, through changes in policies and prices, have encouraged the petroleum sector to promote new and additional projects necessary to ensure Canada's supply of oil and natural gas.

Additional oil sands projects, enhanced recovery schemes, heavy oil upgraders, and further development of natural gas are expected as the country's resource potential is more fully realized.

SUMMARY

Heavy oils and bitumen are composed of high carbon to hydrogen ratio, hydrocarbons having a small distillable fraction, high sulphur and traces of heavy metals. They are defined: a) by viscosity being respectively below and above 10^5 mPa.s, and b) by having a density of 934-1 000 kg/m^3 (20-10° API) and above 1 000 kg/m^3 (10°API).

Deposits of oil sands (bitumen) are found in Peace River, Wabasca, Athabasca and Cold Lake, changing to heavy oils in Lloydminster and south along the Alberta-Saskatchewan border through Wainwright, Colville, Suffield and Taber.

Total bitumen in-place in the oil sands and underlying "carbonate triangle" are about one trillion barrels and over 10 billion barrels for the heavy oils. Current production of bitumen is about 33,500 b/d and could rise to 140,000 b/d from existing projects by the year 1990. Undiluted heavy oil production from Alberta and Saskatchewan is about 176,000 b/d of which over 25 per cent is from Lloydminster.

There are 86 approved heavy oil and bitmen extraction operations, including Syncrude and Suncor, of which 48 are in oil sands experimental projects, 22 heavy oil projects in Alberta and 14 projects in Saskatchewan. Technologies used are surface mining and cyclic steam injection (huff and puff) for bitumen and steam and fire-flooding for heavy oils. The use of mine shafts for bitumen recovery instead of surface mining is being tested to avoid overburden removal and sand handling.

The greatest potential for large and commercially viable production is at Cold Lake as indicated by the Esso Leming Lake, BP Wolf Lake, Amoco Elk Point and Suncor Ft. Kent projects. New surface mining projects are constrained by the large capital investment and operational costs, oil price risks, and environmental problems. A combination of mine shaft and steam thermal recovery may be the way to reduce costs.

Fifty billion barrels of heavy oils and bitumen are recoverable by existing technology. This provides a major opportunity for exports. Large amounts of capital are required and foreign investment should be encouraged allowing for exports of both domestic and crude, upgraded and synthetic oils surplus to our needs. Tax and royalty incentives should be continued to maintain competitiveness and to encourage new technologies.

CANADA'S PETROLEUM REGIONS

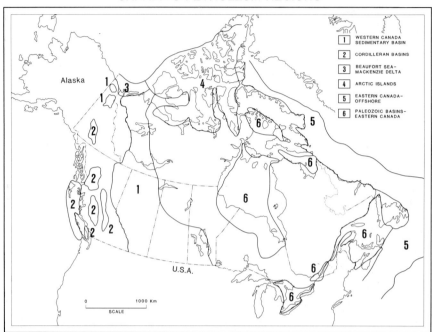

SIGNIFICANT NEW OIL PLAYS

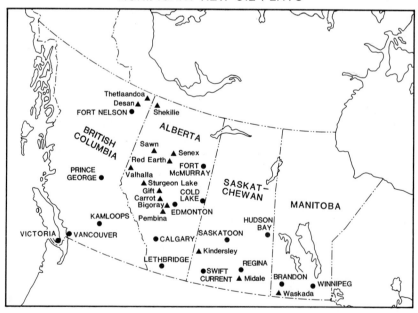

Ferrous Scrap

R. McINNIS

INTRODUCTION

1983 marked the beginning of a slight improvement in consumption of purchased scrap with volumes of 3.09 million t compared to 2.5 million t in 1982. The improvement continued in 1984 with a total estimated to be 3.8 million t.

Canadian scrap consumption including own generated scrap improved from 6.8 million t in 1982 to 7.5 million t in 1983 with further gains to a level estimated at 8.9 million t for 1984.

Demand for increased quality of product, especially in terms of the chemical analysis by scrap users, will continue as the world steel industry continues to increase the quality and strength of the steel they produce.

CANADIAN INDUSTRY STRUCTURE

The Canadian ferrous scrap industry totals approximately 600 firms. The companies collect, store and process the ferrous scrap purchased by the user industries, generally steel companies. Most of these firms are small and are involved only in the collection of scrap; dealers who are involved in the sorting and storage of scrap are fewer in number, while the capital intensive processors are the smallest in number. A processor requires such equipment as mechanical shredders, shears, presses and bundlers. They are the most important segment of the scrap industry as they produce the products needed by the user industries.

Scrap is such an important input that Canadian steel producers often have equity ownership in scrap processing companies so as to minimize supply problems.

In terms of capital equipment autobody shredders represent a significant capital investment. However the 15 machines presently installed has the capacity to process about 1.3 million cars per year.

CANADIAN DEVELOPMENTS

The Canadian ferrous scrap industry was severely depressed in 1982 and early-1983. However, during the second half of 1983 demand increased and prices improved rapidly.

In an integrated steel mill the ratio of purchased to own generated scrap varies from year to year. In 1981 this ratio was 1.0 it decreased to .89 in 1982 and fell to .89 in 1983. This ratio is not simply a function of price although price is important. Prices in 1982 were very low and relatively less purchased scrap was consumed. Steelmakers tend to maximize the use of scrap in normal markets because scrap is virtually always cheaper than the cost of producing molten pig iron and the more scrap used the lower the cost of producing steel. When demand for steel is very high scrap use is maximized to increase the amount of steel produced in the steel mill. The alternative can also be true with mills not maximizing their use of scrap in order to operate their blast furnace at a reasonable rate or to utilize contracted supplies of iron ore or coal. This latter situation may explain the low scrap to steel ratio of 1982, that is quantity of scrap consumed per tonne of steel produced.

In the case of the electric furnace steel industry the price demand relationship is much more direct as ferrous scap is virtually the entire source of raw materials. In periods of low demand and low scrap prices, electric furnace mills can produce steel for considerably less than integrated mills allowing them to capture market share and the increasing spread between the price of scrap and the price of steel helps them to remain profitable in periods of low demand.

Increasing use of continuous casting and improved basic oxygen furnace (BOF), such as the LBE equipment recently installed and in planning steel mills will reduce the levels of own generated scrap and increase the

---

R. McInnis is with the Mineral Policy Sector, Energy, Mines and Resources, Canada. Telephone (613) 995-9466.

relative demand for purchased scrap, as will the entrance of QIT Fer et Titane into the production of steel. QIT produces pig iron as a coproduct with titanium oxide in its electric furnace ilmenite smelting facilities at Sorel, Quebec. The company's decision to further process this high-purity molten iron to billets was based on the following factors; a severe downturn in the company's historic market for pig iron in Europe and the United States, the qualification of the project for government grants, and a risk sharing contract with Hydro Quebec.

This new steelmaking facility will impact on the ferrous scrap industry as a potential customer because with their capacity to produce 440 000 t of billet they could use up to 132 000 t of scrap. Further, the reduction in pig iron production could also increase demand for scrap.

Canada is more than self-sufficient in scrap but there are regional differences in supply and consumption that influence the bilateral trade in scrap that exists between Canada and the United States. Scrap surplus to eastern Canadian needs is often exported to markets in the northeastern United States while the western Canadian market which is generally deficient in local scrap generally imports it from the American northwest and central regions where there is little steelmaking capacity.

The recycling industries of both Canada and the United States share, what to a great extent is, a single market as there are few restraints to the movement of scrap across the border and the prices set in the United States have a major impact on the Canadian price of the same product. Most Canadian exports of scrap go to the United States, over 90 per cent in the last three years while virtually all of Canadian imports originate in the United States.

As the recycling industry in Canada has grown and become more mechanized and efficient an increasing surplus has been available for export. The international market for scrap is very competitive and a good market this year may disappear next year. Countries which have a history of buying significant volumes of scrap include South Korea, Spain, Italy and Japan.

USES

Most ferrous scrap is used in the production of steel in both electric furnace and integrated mills. The foundry industry is the second largest market for scrap while production of iron powder is a minor market.

Scrap is classified by grade and specification, and also by source or type. There are three major types of scrap. Home scrap, which is produced in the manufacture of steel mill products; Prompt industrial scrap, which is generated by the secondary manufacturing industry; and Obsolete scrap, which comes from worn out discarded machinery, equipment and structures.

Prompt and obsolete scrap is generally processed by the recycling industry and is available in a number of grades and to various specifications that concern chemical analysis and the shapes and section of the product. Number 1 heavy melting steel, and number 1 shredded scrap are examples of some of the products available.

The electric furnace steel industry is the largest consumer of scrap in Canada operating on virtually 100 per cent scrap feed. The scrap to finished steel ratio for the industry averages 1.1 to 1.2 depending on the grade of scrap charged that is it takes 1100 to 1200 kg of scrap to produce 1000 kg of steel. The grade also influences the time required to melt each heat and the cost of energy per tonne and thus the productivity of the furnace. The quality of the steel produced is influenced by the amount of tramp elements present in the scrap.

In the integrated steel industry, scrap is charged to both the open-hearth and BOF at about 50 per cent and 30 per cent respectively. Because scrap is a much cheaper source of iron than the blast furnace, integrated steel mills tend to maximize the scrap charged to their steel furnace.

There is considerable research and development work presently under way and completed, the purpose of which is to increase the percentage of scrap that can be used in the integrated steel industry.

OUTLOOK

The largest consumer of iron and steel scrap is the steel industry; thus consumption of scrap is a direct function of the level of steel production. In the case of the electric furnace industry scrap consumption is a direct function of steel production. However, the integrated steel industry has considerable flexibility in the percentage of scrap they feed to their steel furnaces. In a BOF a small amount of scrap is necessary to absorb the energy released when the carbon in the molten iron is removed by oxidization, this same energy can be used to melt up to about 30 per cent scrap. The scrap to iron ratio used is a function of such variables as the price of scrap, the operating rate of the steel mill and the existence of contracts for imput materials such as coal and iron ore. In a period of high demand when a mill is operating near capacity, scrap use will be maximized to increase the amount of steel produced even if the price is high. The alternate situation also applies. When demand is low the minimum output from a blast furnace is such that scrap use even if very low priced, is reduced to avoid over production. In the longer-term, both integrated and electric furnace mills are experiencing rapid technological change which will have an impact on the scrap market. In periods of normal market demand the industry tends to maximize its use of scrap because the cost of a tonne of scrap is generally much less than the cost of producing a tonne of molten pig iron. Recent research and development efforts have been focused on increasing the amount of scrap that can be used in the oxygen steel converter or BOF. Developments include systems where fuel and oxygen are blown into the converter to preheat the scrap charge and Lance Bubbling Equilibrium (LBE) equipment which blows inert gases through the bottom of a BOF type vessel. The resultant mixing improves the yield of the amount of scrap that can be charged and the quality of the steel produced. LBE equipment is being installed in a number of Canadian BOF's. The market for scrap and the amount of scrap purchased by integrated mills also varies with the amount of home scrap produced in the steel plant, and again a new process in continuous casting is effecting the ratio of home to purchased scrap. Yields from molten crude to finished steel can increase by almost 20 per cent when continuous casting is used instead of the ingot roughing mill route. At least 3 new C.C.machines will be installed in Canadian mills in the next few years.

In 1985 scrap usage is expected to increase by approximately 3 per cent per year. In the medium-term to 1990 usage should increase 4 to 5 per cent per year. In the longer-term to 1995 growth should average approximately 2 per cent per year.

Demand for higher quality scrap, especially in terms of low levels of tramp elements will likely require the installation of more equipment such as x-ray spectrometers to analyze scrap, mechanical separators, high pressure bailers and briquetters that produce a higher density product, and improved shredders that increase the separation of nonferrous and nonmetallic components, of the materials in the feed, i.e. automotive hulk.

PRICES

The composite prices in U.S. dollar per long ton delivered, for No. 1 heavy melting steel scrap, as quoted by the American Metal Market increased from $59.2 in January 1983 to $86.99 in December. Prices continued to increase in 1984 with the highest price of $96.25 occuring in February when it fell to a low of $79.17 at year-end.

TABLE 1. CANADA, EXPORTS OF STEEL SCRAP, BY PROVINCE OF LADING, 1981-83

		1981 World	1981 U.S.	1982 World	1982 U.S.	1983P World	1983P U.S.
Newfoundland	tonnes	-	-	-	-	1 910	-
	$000	-	-	-	-	170	-
Nova Scotia	tonnes	29	29	-	-	38	38
	$000	2	2	-	-	60	60
New Brunswick	tonnes	340	200	485	425	475	475
	$000	71	14	55	27	37	37
Quebec	tonnes	114 663	12 896	156 651	21 326	105 496	3 415
	$000	14,672	2,005	15,659	2,288	10,437	416
Ontario	tonnes	235 487	233 326	220 134	162 618	549 008	438 215
	$000	28,461	28,134	20,811	15,880	42,398	31,095
Manitoba	tonnes	1 472	1 472	1 410	1 410	836	836
	$000	281	281	194	194	87	87
Saskatchewan	tonnes	2 195	2 195	3	3	161	161
	$000	381	381	1	1	30	30
Alberta	tonnes	1 288	1 266	1 377	1 377	607	587
	$000	197	192	125	125	106	100
British Columbia	tonnes	90 769	87 068	85 687	84 263	130 178	128 471
	$000	9,889	9,272	7,568	7,136	11,529	11,209
Yukon	tonnes	72	72	-	-	-	-
	$000	4	4	-	-	-	-
Canada total	tonnes	446 315	338 524	465 747	271 422	788 709	572 198
	$000	53,958	40,285	44,413	25,651	64,854	43,034

Source: Statistics Canada.
P Preliminary; - Nil.

PRICE INDEX (1971 = 100)

Product	1981	1982	1983	1984
Ferrous Scrap (Average)	254.3	216.5	222.6	277.4
No. 1 Heavy Melting	243.3	207.3	208.6	268.8
Nos. 1 & 2 Bunders	245.1	216.5	222.1	240.2
Steel Foundry Scrap	294.5	246.8	254.8	319.9

Source: Statistics Canada.

TABLE 2. CANADA, IMPORTS OF STEEL SCRAP, BY PROVINCE OF ENTRY, 1981-83

		1981 World	1981 U.S.	1982 World	1982 U.S.	1983P World	1983P U.S.
Nova Scotia	tonnes	-	-	---	---	86	86
	$000	-	-	---	---	6	6
New Brunswick	tonnes	1 131	1 131	62	---	19	19
	$000	89	89	16	---	2	2
Quebec	tonnes	60 701	60 659	28 605	26 785	26 998	26 952
	$000	5,486	5,405	2,812	2,741	3,479	3,446
Ontario	tonnes	311 917	311 840	194 335	194 291	262 360	262 281
	$000	30,648	30,592	15,376	15,350	20,783	20,726
Manitoba	tonnes	55 781	55 781	8 233	8 233	25 815	25 815
	$000	4,390	4,390	514	514	1,852	1,852
Saskatchewan	tonnes	127 733	127 733	68 005	68 005	135 008	135 008
	$000	13,419	13,419	5,337	5,337	10,511	10,511
Alberta	tonnes	24 600	24 600	3 291	3 291	14 798	14 798
	$000	2,423	2,423	315	315	1,108	1,108
British Columbia	tonnes	2 005	1 956	926	926	1 489	1 483
	$000	270	265	109	109	537	536
Canada total	tonnes	583 869	583 700	303 458	301 533	466 573	466 442
	$000	56,724	56,583	24,479	24,366	38,278	38,187

Source: Statistics Canada.
P Preliminary; - Nil; --- Amount too small to be expressed.

TABLE 3.   CANADA, EXPORTS OF STAINLESS STEEL SCRAP, BY PROVINCE OF LADING, 1981-83

		1981		1982		1983P	
		World	U.S.	World	U.S.	World	U.S.
Newfoundland	tonnes	14	14	-	-	-	-
	$000	3	3	-	-	-	-
Nova Scotia	tonnes	140	122	133	13	46	5
	$000	116	102	84	11	42	12
New Brunswick	tonnes	350	281	273	10	83	-
	$000	263	221	197	6	68	-
Quebec	tonnes	2 136	1 519	4 403	1 496	2 108	1 172
	$000	1,942	1,398	3,065	894	1,696	876
Ontario	tonnes	12 011	11 377	15 982	9 890	14 905	11 328
	$000	6,953	6,277	9,138	4,366	9,310	6,718
Manitoba	tonnes	163	163	283	283	177	177
	$000	75	75	144	144	121	121
Saskatchewan	tonnes	-	-	-	-	-	-
	$000	-	-	-	-	-	-
Alberta	tonnes	39	39	223	223	137	137
	$000	26	26	168	168	74	74
British Columbia	tonnes	1 589	868	2 608	1 530	1 460	543
	$000	1,031	522	1,032	339	944	196
Canada total	tonnes	16 442	14 383	23 905	13 445	18 916	13 362
	$000	10,409	8,624	13,828	5,928	12,255	7,997

Source: Statistics Canada.
P Preliminary;   - Nil.

TABLE 4. AUTOMOBILE SHREDDERS IN CANADA

Company	Location	Capacity (tonnes/month)
Intermetco	Hamilton, Ontario	8 000
United Steel and Metal	Hamilton, Ontario	5 000
Bakermet	Ottawa, Ontario	8 000
Industrial Metals Co. of Canada	Toronto, Ontario	10 000
Zalev Brothers Limited	Windsor, Ontario	8 000
Sidbec-Feruni Inc.	Contrecoeur, Quebec	8 300
Fers et Metaux Recyclés Ltée	Longueuil, Quebec	4 000
	Laprairie, Quebec	4 000
Associated Steel Industries Ltd.	Montreal, Quebec	8 000
Native Auto Shredders	Regina, Saskatchewan	6 000
Cyclomet	Moncton, New Brunswick	4 000
Navajo Metals Ltd.	Calgary, Alberta	3 000
Stelco Inc.	Edmonton, Alberta	8 000
Richmond Steel Recycling Ltd.	Richmond, British Columbia	5 800
General Scrap & Car Shredder Ltd.	Winnipeg, Manitoba	3 000
Total		85 100

TABLE 5. CANADIAN CONSUMPTION OF IRON AND STEEL SCRAP

	1975	1976	1977	1978	1979	1980	1981	1982	1983P	1984P (Jan.-Sept.)
					(000 tonnes)					
Used in steel furnaces	5 997	5 658	5 708	7 076	7 250	7 501	6 845	5 492[2]	6 000	5 020
Used in iron foundries	544	550	524	518	604	470	500	448	500	360
Other[1]	846	824	938	865	868	770	926	837	1 000	620
Total	7 387	7 032	7 170	8 459	8 722	8 741	8 271	6 777	7 500	6 000

Sources: 1982 Annual Census of Manufactures. 1983 and 1984 Catalogue 41-001 Primary Iron and Steel.
[1] Includes mainly steel pipe mills, motor vehicle parts industries, and railway rolling stock industries. [2] The number from Catalogue 41-001 was 4,619 or within 2.3 per cent.
P Preliminary.

TABLE 5. CANADIAN CONSUMPTION OF IRON AND STEEL SCRAP

	1975	1976	1977	1978	1979	1980	1981	1982	1983P	1984P
					(000 tonnes)					(Jan.-Sept.)
Used in steel furnaces	5 997	5 658	5 708	7 076	7 250	7 501	6 845	5 492[2]	6 000	5 020
Used in iron foundries	544	550	524	518	604	470	500	448	500	360
Other[1]	846	824	938	865	868	770	926	837	1 000	620
Total	7 387	7 032	7 170	8 459	8 722	8 741	8 271	6 777	7 500	6 000

Sources: 1982 Annual Census of Manufactures. 1983 and 1984 Catalogue 41-001 Primary Iron and Steel.
[1] Includes mainly steel pipe mills, motor vehicle parts industries, and railway rolling stock industries. [2] The number from Catalogue 41-001 was 4,619 or within 2.3 per cent.
P Preliminary.

Gold

D. LAW-WEST

After eighteen months of downward pressure, the price of gold at the end of 1984 had fallen to a two year low of slightly above $US 300 per oz. This caused considerable concern in the Canadian gold mining industry as some higher cost producers became unprofitable and other producers experienced reduced cash flows. Despite weakening prices over this period, several new mines came on stream and only a few of the very high cost producers were forced to shut down.

The price of gold on the London market was in the $US 370 to $US 380 range in early 1984, fell to about $US 340, where it remained until December when the price averaged $US 320.

Despite the lower prices, gold production appears to be increasing in several countries including Canada, Australia, Brazil and the United States.

The future price of gold will depend on the strength of the major OECD economies, interest rates, inflation rates, gold sales from Comecon countries, and the ability of OPEC countries to maintain high oil prices. The short-term outlook for gold prices indicates continued weakness.

CANADIAN DEVELOPMENTS

Canada's primary production of gold in 1984 was estimated at 81 300 kg which was 9.5 per cent higher than the 73 513 kg produced during 1983. The largest increases in production occurred in the Northwest Territories, Quebec and Ontario.

In Atlantic Canada, an increase in gold production of about 150 kg in 1984, was primarily attributable to the successful efforts of New Brunswick Mining and Smelting Corporation to increase gold recovery from reprocessed slimes. In addition two small gold producing operations opened up, Anaconda Canada Exploration Ltd. and the Heath Steele Mines Ltd. in New Brunswick.

Selco a division of BP Canada announced a significant gold discovery near Port aux Basques, Newfoundland. The company expects to complete drilling and be ready to conduct a feasibility study, by mid-1985.

Gold exploration in Nova Scotia picked up significantly during 1984. Several companies secured financing for more detailed work and feasibility studies which, if positive, could re-establish the province as a gold producer.

In Quebec, which retained its position as the country's largest gold producer, production increased by some 7 per cent to 29 282 kg. There were several significant new mines which came on-stream in 1984.

Corporation Falconbridge Copper opened its Lac Shortt gold mine in northern Quebec. The mine is expected to produce 1 555 kg of gold per year from its 750 tpd mill. There is a possibility that the mill rate could be doubled in the future. The mine, with reserves estimated at 2.4 million t grading 4.8 g/t has an expected life of 8 to 9 years.

Agnico Eagle Mines continued with the development and mining at its Tebel Mine which adjoins the Eagle mine near Joutel, Quebec. The company completed the sinking of a 1 200 m production shaft and is mining on the 775 m level. Ore from the Tebel mine accounts for 50 per cent of Agnico's mill feed.

Kiena Gold Mines opened its new 1 250 tpd mill in September. Gold production from the mill is expected to be 2 080 kg per year. Prior to the completion of the mill, Kiena had been trucking its ore to the Lamaque mill under the terms of a contract that runs until January 1985. Operating costs for Kiena are expected to drop from about $US 250/oz to about $US 200/oz.

Aiguebelle Resources opened a new mill facility at the D'est-Or mine near Rouyn.

D. Law-West is with the Mineral Policy Sector, Energy, Mines and Resources Canada. Telephone (613) 995-9466.

The 1 000 tpd mill will eliminate the company's need to truck its ore 100 miles to Louvem's Manitou mill. The company expects to produce some 1 244 kg of gold in 1984 up from the 622 kg the previous year.

Société Miniere Louvem reopened the Chimo mine, after deepening the shaft from 175 m to 335 m and opening up four new mining levels. In addition the company completely rebuilt the 900 tpd Manitou mill. Initial output is expected to be 500 kg per year from a mill rate of 300 tpd. Eventual output is expected to reach 1 860 kg per year. Proven ore reserves at the mine are some 950 000 t grading 7.78 g/t. However, mining and milling operations were expected to be suspended early in 1985, to enable development of the new ore zone. When the mine reopens production will be from both existing stopes and the new zone at an initial rate of 400 tpd. This will increase to 700 tpd by the end of 1985.

Northgate Exploration increased gold production at its Copper Rand and Portage mine near Chibougamau from 1 910 kg in 1983 to some 2 050 kg in 1984. This increase is largely the result of a $12.5 million development and expansion program.

In Ontario, Dome Mines continued with an expansion program which included completely rebuilding the mill and sinking the new No. 8 shaft. Production is expected to increase by about 50 per cent by early-1985. The mill capacity was increased to 2 500 tpd, since the new shaft will allow more direct ore haulage.

Pamour Porcupine Mines announced the closure of three of its six mining operations, during the fourth quarter of 1984. The suspension will entail a reduction in workforce of 480 employees, however, the mines will be maintained on standby to permit reopening, should gold prices recover.

At the Detour Lake gold mine, a joint venture between Campbell Red Lake Gold Mines and Amoco Canada Petroleum, construction of 2,200 ft shaft was scheduled to commence by the end of 1984. The 2 750 tpd mill has been operating on ore from the open-pit mine. Production for the first half of 1984 was 920 kg of gold. Development plans call for eventual mill expansion to 4 400 tpd. This would make Detour Lake mine one of the country's largest producers.

Campbell Resources is in the process of expanding output at its Renabie mine near Missinabi, Ontario. The company expects to spend some $17 million to expand production to 1 000 tpd and to recover some 1 860 kg of gold per year by 1986. Underground drilling has outlined large reserves grading about 8.1 g/t at depth.

Westfield Mines began operation of its 200 tpd mill at its gold mine in Scadding Township near Sudbury. The company expects to produce some 1 090 kg of gold during the four year life of the operations.

The McBean mine near Kirkland Lake poured its first gold bar in mid-1984. The operation, a joint venture between Inco Limited (65 per cent) and Queenston Gold Mines (35 per cent), represents the culmination of over 50 years of exploration and development in the area. Operating at 500 tpd, the operation is expected to produce some 620 kg of gold per year. The open-pit is expected to reach its maximum depth of 80 m within 3 to 4 years. The ore is processed in the refurbished Upper Canada mill which had operated between 1939-72.

The Hemlo area of northern Ontario remained the focus for development and exploration in the province. In a joint venture, Teck Corp. (55 per cent) and International Corona Resources (45 per cent), plan to spend some $157 million to construct a 1 000 tpd mine/mill complex on their Hemlo property. Ore production is expected to begin early in 1985 and the mill is expected to begin operating in April 1985. Proven ore reserves are estimated at 8.4 million t grading 11.19 g/t of gold.

Noranda Corp. is also expected to begin production from its Hemlo property early in 1985. The company will spend about $250 million on a 915 m production shaft and a mill with initial capacity of 1 000 tpd, but which could be expanded in stages to 4 000 tpd. The proven reserves are estimated at 24 million t grading 8.71 g/t. Noranda with 50 per cent interest has optioned this property from Golden Sceptre Resources Ltd. and Goliath Gold Mines Ltd. which each hold 25 per cent interest.

Lac Minerals Ltd., has reached the 107 m level via a 1 500 m decline and plans to start sinking a 7 m diameter shaft by the year-end on its Hemlo deposit. Initial plans call for a 3 000 tpd mill with options for expansion to 6 000 tpd. Lac Minerals has proven reserves of about 42 million t grading 7.15 g/t.

The Hemlo find has stimulated exploration activities in several previously explored areas. At Cameron Lake potentially mineable reserves have been blocked out by Nuinsco Resources Ltd. - Lockwood Petroleum Inc. joint venture. Lytton Minerals has been actively drilling just west of Hemlo and has uncovered what appears to be a significant gold discovery.

The Matheson area south of Lake Abitibi has also become a recent focus of exploration where, at the end of 1984, some 16 drills were active. The most advanced project was that of Barrick Resources with six drills active.

In Saskatchewan, Flin Flon Mines officially opened its Rio mine and milling complex. Initial feed for the 125 tpd mill will come from the Rio deposit which has reserves of 163 000 t grading 9.02 g/t. Two other gold deposits are part of the producing property; the Maloney deposit with reserves of 15 000 t at 14.0 g/t and the Newcor deposit of 42 600 t grading 9.60 g gold, 27.4 g silver, 4.10 per cent zinc, and 0.4 per cent copper per t. The mill which could be expanded to 250 tpd, uses the "carbon in leach process" to recover precious metals.

Erickson Gold Mines Ltd. is in a development stage at its gold mining operations near Cassiar, British Columbia. The company is developing additional underground ore to supply the existing mill capacity of 300 tpd.

Most gold production in British Columbia is as a byproduct of base-metal operations. Some producing companies include Noranda, Placer Development, Utah Mines Ltd. and Falconbridge.

The Cariboo-Quesnel gold belt has become an area of active exploration. Dome Mines has uncovered a 1 million t at 6.2 g/t deposit. Mt. Calvery Resources has acquired several gold properties with initial showings of up to 28 g/t gold, although reserves have not yet been established.

Queenstake Resources, a major placer gold operator in the Yukon Territory, produced some 6,300 oz of gold in 1984. The company expects to increase production in 1985.

The Mount Skukum gold deposit owned by AGIP Canada may be the next gold mine in the Yukon. To date some 165 000 t of reserves grading 22.7 g/t of gold and 19.6 g/t of silver have been outlined.

Echo Bay Mines carried out an expansion program aimed at increasing gold production by 20 per cent to about 5 560 kg per year at its Lupin Mine in the Northwest Territories. In addition to increasing output the company has reduced operating costs to about $US 207 per oz. The Lupin mine was the second largest gold producer in Canada at the end of 1984.

Giant Yellowknife Mines brought its small but high-grade Salmita gold mine into production, in mid-1983. Reserves are 190 000 t grading 21.12 g/t. However, production was less than expected due to the hardness of the ore, which created milling problems and reduced recovery.

WORLD DEVELOPMENTS

In South Africa, mine production for 1984 was estimated 680 t from 677.3 t in 1983. This slight increase was accounted for by an increase in the tonnes of ore milled as opposed to mining of higher grade ore.

The South African government decided not to discontinue its State Assistance program for marginal South African gold producers. The decision was based on the continued weakness of the gold price. However, the gold mining industry continued to be adversely affected by the 20 per cent mining surcharge tax, which the government had raised from 5 per cent in 1983.

Canadian mining companies have made significant inroads into the gold mining industry in the western United States. Lacana Mining Corporation, through its 75 per cent owned Lacana Gold Inc., is currently involved in several gold mining ventures in Nevada. These are a 100 per cent interest in Relief Canyon Mine, 26.25 per cent in Pinson Mine and 29.3 per cent in Dee Gold Mine, all of which are heap leach operations.

Pegasus Gold Ltd. and Wharf Resources Ltd. are both involved in heap leaching operations in the United States. This type of mining represents about 15 per cent of the country's gold production.

Asamera Minerals (51 per cent) and Breakwater Resources (49 per cent) expect to have the Cannon Gold Mine outside of Wenatchee, Washington in production by 1985. A 250 m shaft is being sunk to obtain access to ore reserves of about 6 million t grading about 7.78 g/t. The company estimates a 20 year life at a milling rate of 2 000 tpd.

Exploration activity in Australia has been similar to that in Canada in that known gold producing areas are being intensively examined. Since 1981 five major mines in western Australia alone have been brought back into production including: Paringa-Kalgoorlie by Gold Resources Pty. in 1983 with 1.31 million t of ore grading 6.3 g/t, Lancefield Mine by Western Mining Corp. in 1982 with 460 000 t of ore grading 6.2 g/t; Horseshoe Lights by a consortium, with 1.3 million t grading 4 g/t; Griffins Find by Otter Explorations Ltd. based on 600 000 t at 3.7 g/t; and Mt. Magnet reopened by Hill 50 NL based on 1.04 million t grading 8.3 g/t.

Esso Exploration and Production Australia has received the go ahead for its Harbour Lights joint venture. Full production is expected by mid-1985 at a rate of 2 000 kg of gold per year, based on 5.5 million t grading 4 g/t.

The Kidson Gold Mine is expected to begin operation in early-1985 at an annual rate of 6 100 kg of gold and 4 850 kg of silver. Production costs have been estimated at $A 226 per oz of gold.

Gold mining at the Ok Tedi gold-copper mine in Papua-New Guinea, which started up in mid-1984, was hampered by several short closures due to untreated tailings spillages and an accident which involved a spill of sodium cyanide. Production has resumed at 17 000 tpd of ore with a planned increase to 22 500 tpd. Bearing further shutdowns first year production is expected to be about 18 000 to 20 000 kg of gold. Ok Tedi is a consortium project involving BHP Minerals (30 per cent), Amoco Minerals (PNG) (30 per cent), Papua-New Guinea government (20 per cent) and a West German consortium (20 per cent).

PRICES

Gold averaged $US 424.18 per oz during 1983 on the LME. The price started the year in a strong position of $US 481 in January strengthened to $US 491 in February then subsequently fell over $US 100 per oz and finished the year at $US 383 per oz.

The price fared even worse in 1984 with an average of $US 360 per oz. The price remained relatively strong until mid-year when the average monthly price fell from $US 378 to $US 347 between June and July. The monthly average stayed above $US 340 per oz until December when the price fell to $US 319.

CONSUMPTION AND USES

The use of new gold for jewellery, coin and industrial purposes declined in 1983 in the western world to 1 002 t from 1 073 t in 1982. Poor economic conditions in many parts of the world, combined with the effects of a strengthening U.S. dollar, were mainly responsible for the consumption decrease. The combination of these factors reduced jewellery fabrication in all regions except North America, Japan and the Middle East.

After gaining in 1981 and 1982, demand for new gold for jewellery fell 16 per cent in 1983 to 599 t. Jewellery consumption may be divided into demand by industrialized nations where it is a luxury item and by developing countries where it is often a form of savings and investment. In industrialized countries fabrication declined 9 per cent to 383 t and in developing countries the decline was 27 per cent to 215 t.

The consumption of gold in electronic components increased by 14 per cent to 96.4 t in 1983. The largest improvements occurred in Japan and the United States. While the electronics industry has cut its gold consumption per electronic unit, stronger consumer sales of equipment such as videos and home computers offset unit reductions. In addition, development of new electronic products has resulted in the expansion of the applications for gold.

Worldwide demand for gold for use in dental alloys diminished from 60 t in 1982 to 53 t in 1983. The gold content of dental alloys has been reduced over the years, primarily due to the relatively high price for gold, despite the fact that the new alloys are harder to work with and have a shorter useful life.

The use of gold in various miscellaneous applications such as decorative plating, liquid gold for decoration of glass and ceramics, rolled gold and various industrial chemicals declined by about 7 per cent from 620 t in 1982 to 58 t in 1983. The main factors contributing to the decline are the continued substitution of rolled gold by plating on such articles as pens and the replacement of gold brazing alloys by nickel alloys in jet engines. A relatively recent new application for gold which has gained some acceptance is the use of gold in windows, as protection against light and heat.

Production of gold medals, medallions and other coins dropped to 34 t in 1983 from 56 t in 1982. Sales of the American Arts medallions being minted in the United States have been disappointing since the public appears to view the medallions as collectable items and not as investments such as the Canadian Maple Leaf or the Krugerrand.

The sales of Canadian Gold Maple Leaf coins rose by 8 per cent in 1983 and a steady market appears to have been established in the United States, Japan and Europe. This coin may pick up market share as some countries ban or impede the sale of South African Krugerrands.

Investment demand includes jewellery purchases (in developing countries), hoarding of bars and bullion coins, and investor demand which include future trading as well as metal purchases on account. In 1983 the sale of bars fell by 73 per cent to 81.4 t from 303.2 t in 1982. Various factors have exerted an adverse influence on attitudes towards gold as a potential investment including the strong U.S. dollar which has kept the gold price at a high level in terms of other currencies, and attractiveness of investing in U.S. dollars.

OUTLOOK

The short-term outlook for gold is for continued weakness and a rather lacklustre performance through 1985. Based on recent past performance it seems that the continued strength of the U.S. dollar along with higher real interest rates and low inflation expectations will divert much of the investment interest away from gold. As well, the decline in the price of oil has both diminished the purchases of gold by OPEC nations and has also precipitated sales of gold which were bought in the 1970s.

Despite the current low price for gold, many countries will be increasing their gold output in the near future, namely Canada, Australia, Brazil and the United States. Newly applied technologies such as heap leaching have made numerous large low-grade gold deposits economically viable.

In Canada gold production is expected to increase substantially as the Hemlo gold producers come on-stream. In addition the current high level of gold exploration will likely result in additional gold mines being developed. Table 6 indicates a forecast of future growth in Canadian gold production.

TABLE 1. CANADA, GOLD PRODUCTION AND TRADE, 1982-84

	1982	1983	1984
		(grams)	
Production			
Newfoundland			
Base-metal mines	140 978	10 000	186 600
New Brunswick			
Base-metal mines	204 938	142 000	501 500
Quebec			
Auriferous quartz mines			
Bourlamaque-Louvicourt	4 558 818	4 157 000	
Malartic, Matagami and Chibougamau[1]	14 895 307	16 110 000	
Total	19 454 125	20 267 000	
Base-metal mines	6 376 846	5 716 000	
Total Quebec	25 830 971	25 983 000	29 282 000
Ontario			
Auriferous quartz mines			
Larder Lake[2]	3 447 365	3 610 000	
Porcupine[3]	7 389 293	8 497 000	
Red Lake and Patricia	8 110 798	8 086 000	
Total	18 947 456	20 193 000	
Base-metal mines	1 119 960	1 427 000	
Total Ontario	20 067 416	21 620 000	26 472 000
Manitoba-Saskatchewan			
Auriferous quartz mines	343 700	157 000	
Base-metal mines	1 656 399	2 256 000	
Total Manitoba-Saskatchewan	2 000 099	2 413 000	2 060 000
Alberta			
Placer operations	10 836	21 000	2 800
British Columbia			
Auriferous quartz mines	3 007 042	3 671 000	
Base-metal mines	4 527 670	4 733 000	
Placer operations	175 607	17 000	
Total British Columbia	7 710 319	8 421 000	7 550 000
Yukon			
Base-metal mines	366 313	514 000	
Placer operations	2 290 025	2 494 000	
Total Yukon	2 656 338	3 008 000	2 700 000
Northwest Territories			
Auriferous quartz mines	6 113 335	9 128 000	12 600 000
Canada			
Auriferous quartz mines	47 865 658	53 416 000	
Base-metal mines	14 393 104	14 798 000	
Placer operations	2 476 468	2 532 000	
Total	64 735 230	70 746 000	81 350 900
Total value ($)	968,012,000	1,186,411,000	1,227,846,688
Average value per oz[4] ($)	465.10	521.60	469.66

TABLE 1. (cont'd)

	1982		1983		1984	
	(kilograms)	($000)	(kilograms)	($000)	(kilograms)	($000)
Imports					(Jan.-Sept.)	
Gold in ores and concentrates						
United States	382	4,579	773	9,984	325	4,739
Peru	19	201	143	2,177	66	895
South Korea	24	380	2	31	-	-
Chile	-	-	-	-	112	1,490
Other countries	20	248	-	-	36	574
Total	445	5,408	918	12,192	539	7,648
Gold						
United States	21 141	337,004	38 788	668,938	36 405	573,924
Switzerland	3 881	57,331	1 740	30,470	711	10,709
United Kingdom	20	346	641	10,497	-	-
U.S.S.R.	-	-	200	3,320	-	-
West Germany	79	1,223	139	2,393	201	3,095
Other countries	145	2,222	93	1,560	39	710
Total	25 266	398,126	41 601	717,178	37 356	588,440
Gold alloys						
United States	20 055	254,402	13 086	218,734	7 271	102,977
Peru	329	5,076	2 444	38,641	1 900	29,760
Costa Rica	-	-	3 538	24,477	-	-
Nicaragua	4 806	20,762	3 220	20,719	3 199	15,429
Uruguay	-	-	1 637	19,770	-	-
Other countries	1 233	18,535	497	5,505	175	1,739
Total	26 423	298,775	24 422	327,846	12 545	149,400
Exports						
Gold in ores and concentrates						
Japan	3 059	31,086	3 552	45,346	2 595	30,632
United States	736	9,843	1 401	19,848	870	12,414
Taiwan	196	1,708	376	4,683	13	152
Switzerland	-	-	337	4,352	112	1,283
Other countries	1 049	10,810	607	7,448	627	8,041
Total	5 040	53,447	6 273	81,677	4 217	52,322
Gold						
United States	81 264	1,213,652	7 415	1,244,130	77 876	1,213,417
Japan	570	7,621	2 288	28,276	8 354	127,654
Hong Kong	57	1,007	546	7,908	825	12,530
Singapore	501	8,098	444	7,490	250	7,058
West Germany	4	42	449	7,254	445	7,580
Other countries	4 341	65,904	491	8,577	797	6,915
Total	86 737	1,296,324	78 368	1,303,635	88 247	1,372,157

Sources: Energy, Mines and Resources Canada; Statistics Canada.
[1] Includes Mines D'Or Lac Bachelor. [2] Includes Thunderbay area. [3] Includes Sudbury area. [4] Average of London Gold Market afternoon fixings in Canadian funds.
- Nil.

TABLE 2. CANADA, GOLD PRODUCTION BY SOURCE, 1970, 1975 AND 1979-83

	Auriferous Quartz Mines		Placer Operations		Base-Metal Ores		Total	
	(grams)	(%)	(grams)	(%)	(grams)	(%)	(grams)	(%)
1970	58 591 610	78.2	228 890	0.3	16 094 525	21.5	74 915 025	100.0
1975	37 529 456	73.0	335 077	0.6	13 568 581	26.4	51 433 114	100.0
1979	33 794 332	66.1	899 202	1.7	16 448 825	32.2	51 142 359	100.0
1980	31 928 594	63.1	2 059 727	4.0	16 631 942	32.9	50 620 263	100.0
1981	35 876 992	69.0	1 632 720	3.1	14 524 569	27.9	52 034 281	100.0
1982	47 865 658	74.0	2 476 468	3.8	14 393 104	22.2	64 735 230	100.0
1983P	53 416 000	75.5	2 532 000	3.6	14 798 000	20.9	70 746 000	100.0

Sources: Statistics Canada; Energy, Mines and Resources Canada.
P Preliminary.

TABLE 3. CANADA, GOLD PRODUCTION, AVERAGE VALUE PER GRAM AND RELATIONSHIP TO TOTAL VALUE OF ALL MINERAL PRODUCTION[1], 1970, 1975 AND 1979-83

	Total Production (grams)	Total Value ($ Cdn)	Average Value per Gram[1] ($ Cdn)	Gold as per cent of Total Value of Mineral Production (%)
1970	74 915 025	88,057,464	1.18	1.5
1975	51 433 114	270,830,389	5.27	2.0
1979	51 142 359	590,766,328	11.55	2.3
1980	50 620 263	1,165,416,873	23.02	3.7
1981	52 034 281	922,089,087	17.72	2.9
1982	64 735 230	968,012,000	14.95	2.9
1983P	70 746 000	1,186,411,000	16.77	3.3

Sources: Statistics Canada; Energy, Mines and Resources Canada.
[1] Value not necessarily based on average annual gold price.
P Preliminary.

TABLE 4. GOLD MINE PRODUCTION IN THE NON-COMMUNIST WORLD

	1973	1974	1975	1976	1977	1978	1979	1980	1981	1982	1983
							(tonnes)				
South Africa	855.2	758.6	713.4	713.4	699.9	706.4	705.4	675.1	657.6	664.3	679.7
Canada	60.0	52.2	51.4	52.4	54.0	54.0	51.1	50.6	52.0	62.5	70.7
United States	36.2	35.1	32.4	32.2	32.0	31.1	29.8	30.2	42.5	43.5	50.4
Other Africa:											
Ghana	25.0	19.1	16.3	16.6	16.9	14.2	11.5	10.8	11.6	12.0	11.8
Zimbabwe	10.5	10.4	11.0	12.0	12.5	12.4	12.0	11.4	11.6	13.4	14.1
Other	1.7	1.5	1.5	1.5	1.5	2.0	2.5	8.0	12.0	15.0	15.0
Zaire	2.5	4.4	3.6	4.0	3.0	1.0	2.3	3.0	3.2	4.2	6.0
Total Other Africa	39.7	35.4	32.4	34.1	33.9	29.6	28.3	33.2	38.4	44.6	46.9
Latin America:											
Brazil	11.0	13.8	12.5	13.6	15.9	22.0	25.0	35.0	35.0	34.8	51.0
Colombia	6.7	8.2	10.8	10.3	9.2	9.0	10.0	17.0	17.7	15.5	17.9
Dominican Republic	-	-	3.0	12.7	10.7	10.8	11.0	11.5	12.8	11.8	10.8
Chile	3.2	3.7	4.1	3.0	3.0	3.3	4.3	6.5	12.2	18.9	19.8
Other	4.7	2.2	1.9	5.0	5.0	5.2	3.7	5.9	8.1	9.0	16.5
Peru	2.6	2.7	2.9	3.0	3.4	3.9	4.7	5.0	7.2	7.2	9.9
Mexico	4.2	3.9	4.7	5.4	6.7	6.2	5.5	5.9	5.0	5.2	6.3
Nicaragua	2.8	2.4	1.9	2.0	2.0	2.3	1.9	1.5	1.6	2.9	1.7
Total Latin America	35.2	36.9	41.8	55.0	55.9	62.7	66.1	88.3	99.6	105.3	133.9
Asia:											
Philippines	18.1	17.3	16.1	16.3	19.4	20.2	19.1	22.0	24.9	26.0	33.3
Japan	6.2	4.5	4.7	4.6	4.8	4.9	4.4	4.2	3.5	3.8	3.6
India	3.3	3.2	3.0	3.3	2.9	2.8	2.7	2.6	2.6	2.2	2.2
Other	2.7	2.7	2.7	3.0	3.0	3.0	3.0	3.0	3.8	3.6	5.3
Total Asia	30.3	27.7	26.5	27.2	30.1	30.9	29.2	31.8	34.8	35.6	44.4
Europe	14.3	11.6	11.0	11.4	13.2	12.5	10.0	8.6	8.5	10.6	10.0
Oceania:											
Papau/New Guinea	20.3	20.5	17.9	20.5	22.3	23.4	19.7	14.3	17.2	17.8	18.4
Australia	17.2	16.2	16.3	15.4	19.2	20.1	18.6	17.0	18.4	27.4	32.2
Other	2.8	2.2	2.2	2.3	1.8	1.1	1.0	1.0	1.1	1.2	1.6
Total Oceania	40.3	38.9	36.4	38.2	43.3	44.6	39.3	32.3	36.7	46.4	52.2
TOTAL	1 111.2	996.4	945.3	963.9	962.3	971.8	959.2	950.1	970.1	1 012.8	1 088.2

Source: Consolidated Gold Fields PLC, Gold 1983, p. 12.
- Nil.

TABLE 5. AVERAGE ANNUAL PRICE OF GOLD, 1970, 1975 AND 1979-83

	London Gold Market[1]	
	$US	equiv. $Cdn (per troy ounce)
1970	35.97	37.55
1975	161.018	163.781
1979	306.686	359.289
1980	612.562	716.087
1981	459.715	551.178
1982	376.877	465.102
1983	422.600	520.792

[1] Annual average of London Gold Market afternoon fixing price, as reported by Sharpes Pixley Ltd.

TABLE 6. ANNUAL GOLD PRODUCTION FORECAST

	Non-communist World[1,2]	Canadian
	(tonnes)	
1979	1 153.9	51.1
1980	1 029.5	50.6
1981	1 234.8	52.5
1982	1 231.7	64.7
1983	1 265.9	73.5
1984	1 280.0p	81.3p
1985	1 310.0f	82.0f
1986	1 315.0	84.0
1987	1 315.0	87.0
1988	1 310.0	85.0
1989	1 310.0	85.0
1990	1 310.0	87.0
1991	1 290.0	90.0
1992	1 290.0	90.0
1993	1 270.0	85.0
1994	1 270.0	85.0
1995	1 270.0	85.0
1996	1 250.0	80.0
1997	1 250.0	80.0
1998	1 250.0	75.0
1999	1 250.0	75.0
2000	1 250.0	75.0

[1] Mine production; does not include recycled material. [2] Market economy country production plus sales from East Bloc Countries.
p Preliminary; f Forecast.

Gypsum and Anhydrite

D.H. STONEHOUSE

SUMMARY 1983-1984

During the last quarter of 1982, demand for gypsum wallboard by the building construction industry in the United States continued to increase in evidence of that country's recovery from the recessionary conditions of the previous two years. To meet that demand, wallboard producers required greater than usual amounts of crude gypsum from their Canadian subsidiary operations. The trend continued throughout 1983 and 1984 with the result that Canadian exports of gypsum to the United States were up by 9 per cent in 1983 and are expected to increase a further 25 per cent during 1984. Canadian wallboard producers also benefitted from the United States demand for board and increased their output by over 30 per cent to 200 million square metres in 1983 while exports to the United States rose by 84 per cent to 36.5 million square metres. Wallboard exports, mainly from Ontario and Quebec plants, are expected to increase by over 200 per cent during 1984.

CANADIAN DEVELOPMENTS

Gypsum production in Canada is in direct response to demand from the wallboard industries in Canada and the United States, which in turn satisfy demand from the building construction sector for residential, institutional and commercial construction projects. The fire retardant qualities of gypsum wallboard have encouraged its greater application in the non-residential area in recent years. This, together with increasing amounts used in renovation of older buildings, has made housing starts a less-than-accurate indicator of wallboard demand.

The portland cement industry uses as much as 5 per cent by weight of gypsum intimately ground with cement clinker to act as a set inhibitor. This could amount to nearly 0.5 million tpy in Canada.

Canadian production of crude gypsum is mainly from Atlantic Canada where major deposits, principally in Nova Scotia and Newfoundland, have been worked for many years by Canadian subsidiaries of U.S. gypsum products producers. The region accounts for over 75 per cent of Canadian gypsum production and for the major portion of exported gypsum which usually is about 70 per cent of total production; however nearly 80 per cent of production will be exported in 1984. Shipments are made from quarries in the Atlantic region to wallboard plants and portland cement plants in Quebec and Ontario. New Brunswick production is used locally by a cement producer. Ontario production is used on-site except that from the Westroc Industries Limited mine at Drumbo which is shipped to its Mississauga wallboard plant. Manitoba production, and output from Windermere and Falkland in British Columbia, supply the prairie markets and most of the British Columbia markets. Imports from Mexico and the United States are used by both wallboard and cement producers in British Columbia.

Because gypsum is a relatively low-cost, high-bulk mineral commodity it is generally produced from deposits situated as conveniently as possible to areas in which markets for gypsum products exist. Exceptions occur if deposits of unusually high quality are available, even at some distance from markets, if comparatively easy and inexpensive mining methods are applicable, or if low-cost, high-bulk shipping facilities are accessible. Nova Scotia and Newfoundland deposits meet all three of these criteria and have been operated for many years by, and for, United States companies in preference to some known but unexploited United States deposits. During 1984, Little Narrows Gypsum Company Limited and Georgia Pacific Corporation each began development of new quarry areas.

In Canada occurrences besides those currently being exploited are known in the southwest lowlands, west of the Long Range Mountains in Newfoundland; throughout the central and northern mainland of Nova Scotia as well as on Cape Breton Island; in the southeastern counties of New Brunswick; on

---

D.H. Stonehouse is with the Mineral Policy Sector, Energy, Mines and Resources, Canada. Telephone (613) 995-9466.

the Magdalen Islands of Quebec; in the Moose River, James Bay and southwestern regions of Ontario; in Wood Buffalo National Park, in Jasper National Park, along the Peace River between Peace Point and Little Rapids, and north of Fort Fitzgerald in Alberta; on Featherstonhaugh Creek, near Mayook, at Canal Flats, and Loos in British Columbia; on the shores of Great Slave Lake, the Mackenzie, Great Bear and Slave rivers in the Northwest Territories; and on several Arctic islands.

There was no change in the number of producing mines or plants in 1983. Eleven open-pit mines and two underground mines produced about 7.5 million t in total. Gypsum products were produced at 17 plants. During the year, CGC Inc. and Westroc Industries Limited jointly announced that a merger of the two companies had been arranged through the respective parent companies - United States Gypsum Company (U.S.G.) and British Plaster Board Industries Limited. U.S.G. was to own 70 per cent of the new company. Initial inquiries were made at the Foreign Investment Review Agency in May 1983 and when no decision was made by October both companies withdrew their request for a ruling on a merger. United States Gypsum Company sold the roofing business of its subsidiary, CGC Inc., to Canroof Corp. in 1983. Plants are in Montreal, Toronto and Winnipeg.

In 1981, Domtar Inc. purchased the inactive mine and wallboard plant of Grand Rapids Gypsum Company in Michigan. The increased demand for wallboard prompted Domtar to reopen the operation and to install new rock grinding facilities. The plant has a capacity of over 11 million square metres of board a year.

The demand for wallboard provided CGC Inc. with incentive to reopen its St. Jerome, Quebec plant in early 1984 after a 20-month closure because of poor domestic markets. Domtar Inc. did not reactivate its Montreal East plant as the capacity of the new plant at the Caledonia, Ontario minesite is adequate to meet foreseeable requirements. The Montreal East plant serves as a distribution terminal for all eastern Canadian shipments.

WORLD DEVELOPMENTS AND TRADE

Gypsum occurs in abundance throughout the world but, because its use is dependent on the building construction industry, developments are generally limited to the industrialized countries. Reserves are extremely large and are conservatively estimated at over 2 billion t. After the United States, Canada is the world's second largest producer of natural-gypsum. Together they produce about 24 per cent of world output.

During 1984 shipments of crude gypsum were made from Spain to eastern and western United States ports.

Gypsum products, particularly wallboard, have limited market range because of high unit weight, friability, high transportation costs and relatively low unit values. These factors generally dictate that markets are supplied from the closest producer. There are exceptions, however, and gypsum wallboard has been shipped not only between the United States and Canada over rather surprisingly great distances but shiploads of wallboard have been received at United States southeast ports from European producers. The Canada-United States trade is usually in truckload lots of 20 to 25 t for delivery to warehousing or to job sites.

Imports of crude gypsum, mainly into British Columbia from Mexico and the United States were between 100 000 t and 125 000 t during 1983 and 1984. Imports of wallboard and gypsum products increased to close to 500 000 square metres (m^2) in 1983 from about 350 000 m^2 in 1982, but declined to just over 200 000 m^2 in 1984.

USES

Gypsum is a hydrous calcium sulphate ($CaSO_4 \cdot 2H_2O$) which, when calcined at temperatures ranging from 120° to 205°C, releases three-quarters of its chemically combined water. The resulting hemihydrate of calcium sulphate, commonly referred to as plaster of paris, when mixed with water, can be moulded, shaped or spread and subsequently dried, or set, to form a hard plaster product. Gypsum is the main mineral constituent in gypsum wallboard, lath and tile. Anhydrite, an anhydrous calcium sulphate ($CaSO_4$), is commonly associated geologically with gypsum.

Crude gypsum is crushed, pulverized and calcined to form stucco, which is mixed with water and aggregate (sand, vermiculite or expanded perlite) and applied over wood, metal or gypsum lath to form interior wall finishes. Gypsum board, lath and sheathing are formed by introducing a slurry of stucco, water, foam, pulp and starch between two unwinding rolls of absorbent paper, the result is a continuous "sandwich" of wet board. As the stucco hardens, the board is cut to predetermined lengths, dried, bundled and stacked for shipment.

Grinding, calcining and drying are the main energy-using steps in the manufacture of gypsum wallboard. In the interests of energy conservation and process cost reduction in general, significant savings have been achieved by recycling heat from calcining kettles for use in preheating and in board drying. One-step grinding and calcining as an alternative to either the batch kettle or the continuous kettle has been adopted by one producer. There is also a trend towards using less calcined gypsum in board while using greater amounts of foam and more effective dispersing agents to obtain a lighter-weight unit with equal or greater strength.

Keene's cement is made by converting crushed gypsum to insoluble anhydrite by calcining at temperatures as high as 700°C, usually in rotary kilns. The ground calcine, mixed with a set accelerator, produces a harder and stronger plaster product than ordinary gypsum plaster.

Crude gypsum is also used in the manufacture of portland cement where it acts as a retarder to control set. It is used as a filler in paint and paper manufacture, as a substitute for salt cake in glass manufacture and as a soil conditioner.

Byproduct gypsum, produced from the acidulation of phosphate rock in phosphate fertilizer manufacture, has not been utilized in Canada despite available technology from European countries and from Japan. In these countries, byproduct gypsum is used in the manufacture of gypsum products, by cement manufacturing plants, and also for soil stabilization. Recent experiments in France have produced paper with a 20 per cent phosphogypsum content as filler. Studies have indicated that a potential radiation hazard exists in the use of phosphogypsum produced from sedimentary phosphate rock which can contain significant quantities of uranium and radium. Fluorogypsum is a byproduct of the manufacture of hydrofluoric acid. Cooperative research programs have been conducted to determine the suitability of using waste fluorogypsum from Allied Chemical Canada Ltd.'s, Amherstburg, Ontario plant at St. Lawrence Cement Inc.'s Clarkson, Ontario cement plant.

The use of lime or limestone to desulphurize stack gases from utility or industrial plants burning high-sulphur fuel will also result in production of large amounts of waste gypsum in the form of a sludge which will present disposal problems if profitable uses are not developed.

Canadian Standards Association (CSA) standards A 82.20 and A 82.35 relate to gypsum and gypsum products.

OUTLOOK

Total construction in Canada in 1983 was valued at $56 billion, about the same as in 1982. This is expected to increase to $56.9 billion for 1984. Building construction is expected to account for a traditional 54.5 per cent of total value as recovery in the residential building sector offsets declines in the industrial and commercial sectors. Housing starts increased to 162,645 units in 1983 from a low 125,860 in 1982 but will be in the 133,000 range for 1984. Construction of homes, apartments, schools and offices will continue in the building construction sector and the need for gypsum-based building products will rise steadily. Although new construction materials are being introduced, gypsum wallboard will remain popular because of its low price, ease of installation and well-recognized insulating and fire-retarding properties. The present structure of the gypsum industry in Canada is unlikely to change greatly in the near future. Building materials plants have sufficient capacities to meet the short-term, regional demand for products and to supply at least some of the unusually high demand from the United States.

The Canadian Construction Association (CCA) predicts above average growth in the non-residential sector in the near term followed by average growth in engineering and non-residential projects in the medium term. CCA predicts no growth in the residential sector in the medium to long term.

ANHYDRITE

Production and trade statistics for anhydrite are included with gypsum statistics. Anhydrite is produced by Fundy Gypsum Company Limited at Wentworth, Nova Scotia, and by Little Narrows Gypsum Company Limited at Little Narrows, Nova Scotia. According to the **Nova Scotia Annual Report on Mines 1983,** production of anhydrite in that year was 108 807 t. Most of this was shipped to the United States for use in portland cement manufacture and as a peanut crop fertilizer. Cement plants in Quebec and Ontario also used some Nova Scotia anhydrite.

TARIFFS

Item No.		British Preferential	Most Favoured Nation	General	General Preferential
CANADA					
29200-1	Gypsum, crude	free	free	free	free
29300-1	Plaster of paris, or gypsum, calcined, and prepared wall plaster, weight of package to be included in weight for duty; per hundred pounds	free	5.0¢	12.5¢	free
29400-1	Gypsum, ground, not calcined	free	free	15%	free
28410-1	Gypsum tile	12.8%	12.1%	25%	8.0%

MFN Reductions under GATT (effective January 1 of year given)	1983	1984	1985	1986	1987
29300-1	5.0¢	4.8¢	4.5¢	4.3¢	4.0¢
28410-1	12.1%	11.4%	10.7%	9.9%	9.2%

UNITED STATES (MFN)

512.21	Gypsum crude	free				
		1983	1984	1985	1986	1987
512.24	Gypsum, ground calcined, per ton	50¢	48¢	46¢	44¢	42¢
245.70	Gypsum or plastic building boards and lath, ad valorem	4.2%	3.8%	3.3%	2.9%	2.5%

Sources: The Customs Tariff, 1983, Revenue Canada, Custom and Excise; Tariff Schedules of the United States Annotated (1983), USITC Publication 1317; U.S. Federal Register, Vol. 44, No. 241.

TABLE 1. CANADA, GYPSUM PRODUCTION AND TRADE, 1982-84

	1982		1983		1984P	
	(tonnes)	($000)	(tonnes)	($000)	(tonnes)	($000)
Production (shipments)						
Crude gypsum						
Nova Scotia	4 480 000	30,500	5 397 000	37,064	6 461 000	43,677
Ontario	574 000	5,350	907 000	11,354	1 104 000	13,408
British Columbia	415 000	5,468	460 000	4,917	504 000	4,859
Newfoundland	409 000	3,284	553 000	3,731	430 000	3,610
Manitoba	109 000	2,006	190 000	2,231	226 000	3,600
Total	5 987 000	46,608	7 507 000	59,297	8 725 000	69,154
					(Jan.-Sept. 1984)	
Imports						
Crude gypsum						
Spain	-	-	-	-	83 914	2,876
Mexico	83 102	2,806	97 444	2,949	34 343	1,090
United States	10 742	264	3 479	128	3 510	81
Hong Kong	-	-	16	1	57	2
Total	93 844	3,069	100 939	3,078	121 824	4,049
Plaster of paris and wall plaster						
United States	18 627	3,654	24 717	4,630	16 480	3,385
France	175	34	-	-	12	1
United Kingdom	15	3	-	-	20	4
Italy	16	3	-	-	6	2
Other countries	93	30	11	3	3	1
Total	18 926	3,724	24 728	4,633	16 521	3,393
	(square metres)		(square metres)		(square metres)	
Gypsum lath, wallboard and basic products						
United States	349 862	643	485 614	722	227 391	522
Other countries	-	-	5 942	8	-	-
Total	349 862	643	491 556	730	227 391	522
Total imports gypsum and gypsum products		7,436		8,441		7,964
	(tonnes)		(tonnes)		(tonnes)	
Exports						
Crude gypsum						
United States	4 775 780	28,716	5 186 529	33,331	4 839 315	37,494
Other	-	-	503	6	1 022	8
Total	4 775 780	28,716	5 187 032	33,337	4 840 337	37,502
	(square metres)		(square metres)		(square metres)	
Gypsum lath, wallboard and basic products						
United States	13 808 620	12,898	25 836 909	28,435	55 025 773	78,075
Saudi Arabia	224 507	576	154 418	485	60 853	189
Algeria	31 639	46	195 192	189	121 991	236
Bermuda	111 219	139	70 344	114	60 306	76
Other countries	209 016	261	223 072	296	224 937	416
Total	14 385 001	13,920	26 479 935	29,519	55 513 860	78,992
Total exports of gypsum and gypsum products		42,636		62,856		116,494

Sources: Energy, Mines and Resources Canada; Statistics Canada.
P Preliminary; - Nil.
N.B. Totals may not add due to rounding.

TABLE 2. CANADA, GYPSUM MINING AND GYPSUM PRODUCTS MANUFACTURING OPERATIONS, 1983-84

Company	Location	Operation
Newfoundland		
Flintkote Holdings Limited	Flat Bay	Open-pit mining of gypsum
Atlantic Gypsum Limited	Corner Brook	Wallboard manufactors
Nova Scotia		
Domtar Inc.	McKay Settlement	Open-pit mining of gypsum by contract
	Windsor	Plaster and "Gypcrete" manufacture
Fundy Gypsum Company Ltd.	Wentworth and Miller Creek	Open-pit mining of gypsum and anhydrite
Georgia-Pacific Corporation	River Denys	Open-pit mining of gypsum
Little Narrows Gypsum Company Limited	Little Narrows	Open-pit mining of gypsum and anhydrite
National Gypsum (Canada) Ltd.	Milford	Open-pit mining of gypsum
New Brunswick		
Canada Cement Lafarge Ltd.	Havelock	Open-pit mining of gypsum for cement manufacture
Quebec		
CGC Inc.	Montreal	Wallboard manufacture
	St-Jerome	Wallboard manufacture - closed mid-1982, reopened early 1984
Domtar Inc.	Montreal	Wallboard plant now used only as distribution terminal
Westroc Industries Ltd.	Ste. Catherine d'Alexandrie	Wallboard manufacture
Ontario		
CGC Inc.	Hagersville	Underground mining and wallboard manufacture
Domtar Inc.	Caledonia	Underground mining and wallboard manufacture
Westroc Industries Ltd.	Drumbo	Underground mining
	Clarkson	Wallboard manufacture
Manitoba		
Domtar Inc.	Gypsumville	Open-pit mining
	Winnipeg	Wallboard manufacture
Westroc Industries Ltd.	Amaranth	Open-pit mining
	Winnipeg	Wallboard manufacture
Saskatchewan		
Genstar Corporation	Saskatoon	Wallboard manufacture
Alberta		
Domtar Inc.	Calgary	Wallboard and "Gypcrete" manufacture
Genstar Corporation	Edmonton	Wallboard manufacture
Westroc Industries Ltd.	Calgary	Wallboard manufacture
British Columbia		
Domtar Inc.	Vancouver	Gypsum products manufacture
Genstar Corporation	Vancouver	Gypsum products manufacture
Westroc Industries Ltd.	Windermere	Open-pit mining
	Vancouver	Gypsum products manufacture

TABLE 3. WORLD PRODUCTION OF GYPSUM, 1982 AND 1983

	1982	1983e
	(000 tonnes)	
United States	9 560	11 068
Canada	5 987	7 507
France	6 169	6 169
U.S.S.R.	5 443	5 443
Spain	5 262	5 352
Iran	4 990	4 717
United Kingdom	2 722	2 903
West Germany	2 268	2 268
People's Republic of China	3 538	3 538
Mexico	1 542	1 724
Italy	1 633	1 724
Other market economy countries	19 732	17 236
Other central economy countries	4 549	4 536
World total	73 395	74 215

Sources: Energy, Mines and Resources Canada; United States Bureau of Mines Mineral Commodity Summaries, January 1984.
e Estimated.

TABLE 4. CANADA, GYPSUM PRODUCTION, TRADE AND CONSUMPTION, 1970, 1975, 1979-83

	Production[1]	Imports[2]	Exports[2]	Apparent Consumption[3]
	(tonnes)			
1970	5 732 068	35 271	4 402 843	1 364 496
1975	5 719 451	55 338	3 691 676	2 083 113
1979	8 098 166	152 953	5 474 765	2 776 354
1980	7 336 000	154 717	4 960 240	2 530 477
1981	7 025 000	143 500	5 094 873	2 073 627
1982	5 987 000	93 844	4 775 780	1 305 064
1983	7 507 000	100 939	5 187 032	2 420 907

Sources: Energy, Mines and Resources Canada; Statistics Canada.
[1] Producers' shipments, crude gypsum.
[2] Includes crude and ground, but not calcined.
[3] Production, plus imports, minus exports.

TABLE 5. CANADA, HOUSE CONSTRUCTION, BY PROVINCE, 1982 AND 1983

	Starts			Completions			Under Construction		
	1982	1983	% Diff.	1982	1983	% Diff.	1982	1983	% Diff.
Newfoundland	2 793	3 281	17.4	2 331	3 176	36.2	3 373	3 494	3.5
Prince Edward Island	248	673	171.3	98	548	459.1	196	316	61.2
Nova Scotia	3 691	5 697	54.3	3 174	5 069	59.7	2 506	2 984	19.0
New Brunswick	1 680	4 742	182.2	1 427	3 487	144.3	1 122	2 346	109.0
Total (Atlantic Provinces)	8 412	14 393	71.1	7 030	12 280	74.6	7 197	9 140	26.9
Quebec	23 492	40 318	71.6	21 526	35 681	65.7	14 164	18 320	29.3
Ontario	38 508	54 939	42.6	40 437	55 287	36.7	31 009	30 243	-2.4
Manitoba	2 030	5 985	194.8	1 633	4 076	149.6	1 149	3 048	165.2
Saskatchewan	6 822	7 269	6.5	5 666	8 090	42.7	4 583	3 667	-19.9
Alberta	26 789	17 134	-36.0	31 364	24 693	-21.2	17 663	8 336	-52.8
Total (Prairie Provinces)	35 641	30 388	-14.7	38 663	36 859	-4.8	23 395	15 051	-35.6
British Columbia	19 807	22 607	14.1	26 286	22 901	-12.8	13 290	12 176	-8.3
Total Canada	125 860	162 645	29.2	133 942	163 008	21.7	89 055	84 930	-4.6

Source: Canada Mortgage and Housing Corporation.

TABLE 6. CANADA, VALUE OF CONSTRUCTION[1] BY TYPE, 1982-84

	1982	1983	1984
		($ millions)	
Building Construction			
Residential	13,581	16,683	17,240
Industrial	3,044	2,502	2,739
Commercial	7,064	6,228	5,817
Institutional	3,092	3,198	3,183
Other building	2,062	1,989	2,139
Total	28,843	30,600	31,118
Engineering Construction			
Marine	480	404	414
Highways, airport runways	4,310	4,270	4,328
Waterworks, sewage systems	2,244	2,402	2,391
Dams, irrigation	314	295	306
Electric power	4,866	4,673	3,827
Railway, telephones	2,390	2,531	2,811
Gas and oil facilities	9,706	8,115	9,141
Other engineering	2,912	2,808	2,635
Total	27,222	25,498	25,853
Total construction	56,065	56,098	56,971

Source: Statistics Canada.
[1] Actual expenditures 1982, preliminary actual 1983, intentions 1984.

Iron Ore

B.W. BOYD

Shipments of Canadian iron ore dropped in 1983 to the lowest point since 1964 as the industry, already reduced to only nine mines, operated at less than two-thirds capacity. Partial recovery, in 1984, saw shipments rise by 25 per cent, but that was still far below the average shipments of the past 20 years.

On October 5th, 1983 Falconbridge Limited permanently closed the Wesfrob mine at Tasu, British Columbia. On October 12, 1984 Sidbec-Normines Inc. announced that the Fire Lake mine and Lac Jeannine concentrator in Quebec would close by year-end and, on December 24th, 1984 Stelco Inc. announced that the Griffith mine near Red Lake Ontario would close as of April 1986. These closures will reduce the Canadian iron ore production capacity to 49.3 million t of product at six mines. By comparison, production capacity was growing rapidly in 1964 and, with 34.5 million t shipped, the 17 mines were operating at 90 per cent of capacity.

Restructuring of the Canadian industry has, in addition to mine closures, involved a drive for higher productivity, reduction of energy costs, and research and development of new iron ore products. Total employment in the industry will have declined from a peak 16,000 established in the mid-1970s to approximatly 6,900 in April 1986. The use of coke breeze at pellet plants reduced energy costs by over 10 per cent at the largest plants in 1984 and research is continuing on the use of coal-water slurries, plasma burners and self-fluxed pellets which could enhance the competitive position of the Canadian industry.

CANADIAN DEVELOPMENTS

Quebec-Labrador

The Quebec-Labrador mines, operating at 51 per cent capacity in 1983 and 68 per cent in 1984, fared better than most other North American iron ore mines.

Although the Iron Ore Company of Canada (IOC) did not close for extented periods in 1983 or 1984, it did reduce the work force progressively to meet lower production requirements. From a total workforce of 3,710 at the end of 1982, the number of employees contracted to about 2,500 by the end of 1984. The largest number of terminations were due to closure of the Schefferville mines and cut-backs at the port of Sept-Îles in 1983.

After the closure of Schefferville, a stock of over 3 million t of direct shipping ore remained at Sept-Îles. Since then, about 1 million tpy of this product has been shipped from Sept-Îles, with the result that shipments from IOC have exceeded production for the past two years.

The addition of coke breeze in the pellet mix has allowed significant savings on energy at IOC's Carol Lake plant. Burner tests using coal-water slurry as a fuel in the pellet indurating line proceeded in 1984; further tests and plasma burner trials are planned for 1985.

The ownership of IOC changed in 1983 with the trade of Dofasco Inc.'s 16.0 per cent share of the Eveleth Mine in Minnesota for Armco Inc.'s 6.07 per cent share in IOC.

Wabush Mines closed for 50 days in 1983 and, although about 50 workers were permanently laid-off, the actual workforce was reduced by five times that number. In 1984, there was no planned shutdown and only 4 days were lost due to a strike prior to agreement on a new 3-year labour contract.

The addition of coke breeze to concentrate in the pellet plant is now a standard practice at Wabush's Pte. Noire plant, and significant energy savings are also expected from the use of coal-water

---

B.W. Boyd is with the Mineral Policy Sector, Energy, Mines and Resources, Canada. Telephone (613) 995-9466.

slurry as a burner fuel, which was tested in the autumn of 1984. A major research and pilot plant proposal on the removal of manganese from the Wabush concentrate was prepared in 1984. The process, if brought to commercial scale, would remove a limiting factor in marketing Wabush iron ore and provide a valuable manganese byproduct.

Quebec Cartier Mining Company (QCM) was closed for 72 days in the summer of 1983 and its unionized employees were laid off temporarily for the last 2 months of the year. In 1984, the summer shutdown lasted 58 days. A new labour contract was signed in June, which differed from the agreement at the IOC and Wabush mines in that the cost of living allowance was replaced by a productivity bonus.

The closure of the Sidbec-Normines mine at Fire Lake and the concentrator at Gagnon, Quebec followed major financial losses to its owners since production began in 1977. The mine was closed for 80 days in 1983 and 70 days in 1984, while the pellet plant was operated at little more than one-half capacity for the 2 years. At the time of the announcement of the permanent closure, there were 610 company employees at Gagnon and Fire Lake, but with some early retirements and a transfer of some workers to QCM's Mt. Wright mine where an increase in production levels is planned, about 300 permanent lay-offs are expected to result. QCM has agreed to operate the Sidbec-Normines pellet plant at Port Cartier under a lease arrangement. The 295 pellet plant workers will transfer to QCM but the 35 office staff at Port Cartier will be laid off.

Ontario

The four Ontario mines, considered as a group, operated at 60 per cent capacity in both 1983 and 1984.

The Adams and Sherman Mines, owned by Dofasco Inc., closed for 3 months in 1983 and 5 weeks in 1984. The Algoma Steel Corporation, Limited's mine at Wawa, Ontario was closed for one month in 1983 and five weeks in 1984.

The closure of the Griffith mine and pellet plant, scheduled for April 1986, will affect about 280 employees in the Red Lake area. Shipments from the mine will be replaced mainly by increased purchases from the Quebec-Labrador mines.

Small shipments from Inco Limited's stockpile of iron ore pellets at Copper Cliff continued in 1983 and 1984.

British Columbia

On October 5, 1983, Falconbridge Limited permanently closed the Wesfrob mine located at Tasu, Queen Charlotte Islands, due to depletion of economic reserves. The planned closure, which was announced in April 1983, resulted in 135 employees being laid off. All employees were given the opportunity to move at company expense. Shipments from stockpiles continued into 1984.

Other Developments

Although 1984 exports of Canadian iron ore recovered significantly from the depressed shipments of 1982 and 1983, they did not reach the levels achieved during the 1970s.

Dofasco tested self-fluxing pellets, which were produced by Sidbec-Normines, in its Hamilton, Ontario blast furnaces in 1984 and planned to do further tests in the future. These pellets are low in silica and have dolomite added. Self-fluxed pellets could be supplied by the Port Cartier, Quebec plant, which will be operated by QCM, if the test results demonstrate that the use of these pellets would be cost-effective.

Borealis Exploration Limited continued work on its magnetite ore property on Melville Peninsula in the Northwest Territories. During 1983, a campsite was established, and a road and an airstrip were built. Work done at the Ontario Research Foundation indicated that a super-concentrate of over 71 per cent iron could be achieved using wet magnetic separation without flotation. The company reported that it is continuing discussions with several resource developers.

On December 9, 1983 the Federal Minister of Finance announced that the special tax remission for housing and travel benefits received by employees working in northern Canada and isolated posts will be continued. This extension applies to employees covered by benefit plans that were arranged before November 13, 1981. The Minister in making the annoucement, said "The northern economy and, in particular, the mining industry continue to be in a weak state due to the slow recovery of world markets and mineral prices. This tax relief will provide an impetus to recovery and allow additional time for adjustment".

WORLD DEVELOPMENTS

Hamersley Holdings Limited of Australia announced in mid-1984 an agreement with the China Metallurgical Import and Export Corporation (CME's) to conduct a feasibility study for a joint venture iron ore mine in the Channar Mining area located about 20 km east of Hamersley's Paraburdoo operation. The announcement stated that the project would commence at a production of 5 million tpy and expand to 10 million tpy as demand warranted.

The Brazilian company, Cia Vale do Rio Doce (CVRD) plans to begin shipping ore from the Carajas Iron Ore Project in February 1985. Over 1 million t of iron ore, and 150 000 t of concentrate grading 40 per cent manganese are scheduled to be shipped during the year. The 890 km railroad linking the mine with new port facilities in Sao Luis should be completed in February 1985. Development of the mine is planned to proceed in stages so that full production at 35 million tpy can be attained in 1988.

In mid-1984, the Periquito mine at Itabira, Brazil began operations at 3 million tpy. Its rate of production was planned to increase in 1985 to 7 million tpy. The Timbopeba operation, also in Brazil, began production in April 1984 and was scheduled to produce 7.5 million tpy beginning in 1985.

Exports from Brazil were estimated at a record 90 million t in 1984, an increase of 30 per cent over its exports in 1983. This performance put Brazil in the lead as the world's largest exporter of iron ore, ahead of Australia.

Exports of iron ore from India recovered in 1984 to about 24 million t after a 15 per cent drop in 1983. The 3 million tpy pellet plant at the Kudremukh Iron Ore Co. Ltd. mine is due to be commissioned in mid-1985. The plant, part of a 7.5 million tpy iron ore complex, contributes to an Indian objective to increase exports of low-cost iron ore products.

In the United States, several permanent mine closures have resulted from the depressed iron ore market. Inland Steel Co. permanently closed the Black River Falls mine in Wisconsin almost two years after operations were halted in April 1982. The Hanna Mining Company announced that the Whitney mine, Minnesota would close permanently August 10, 1984; the mine had been inactive since 1977. CF&I Steel Corp.'s Sunrise mine, Wyoming, shut down since July 1980, was closed permanently in mid-1984 when the mine and plant equipment were auctioned. The United States Steel Corporation's mines near Cedar City, Utah were permanently closed in 1983.

In addition, several closures for indefinite periods have been in effect for many months and, in some cases, years. By mid-1984, however, the number of operating mines had recovered from the low point in 1982 and there was hope that future permanent closures could be offset by reopenings or recoveries at other mines.

A merger of Republic Steel Corp. and LTV Corp. combined the holdings of Republic and Jones & Laughlin Steel Corporation in the United States and Canadian iron mining ventures. The merged corporation will hold 50 per cent of Reserve Mining Co., 35 per cent of Erie Mining Co. and 16 per cent of Hibbing Taconite Co., in Minnesota; 35 per cent of the Empire Iron Mining partnership and 12 per cent of the Empire Tilden Mining Co., in Michigan; 15.6 per cent of Wabush Mines and 12.6 per cent of Iron Ore Company of Canada, in the Labrador Trough area.

DIRECT REDUCTION AND PLASMA TECHNOLOGY

Ivaco Inc., at L'Orignal, Ontario had planned to install an Inred hot metal direct reduction DR plant based on Boliden Aktiebolog technology. However, the availability of steel billets in the near future from QIT-Fer et Titane Inc. of Sorel, Quebec, eliminated the need for the DR plant.

Sidbec-Dosco Inc. at Contrecoeur, Quebec produced somewhat less direct-reduced iron in 1984 than in 1983. The operation of the Midrex Corp. designed plant remained at close to one-half its 1.2 million tpy capacity.

Midrex Corp. received and order to conduct a $250,000 feasibility study on converting one of Sidbec-Dosco Inc.'s natural gas-based Midrex direct reduction plants to electric reforming. The study will evaluate the possibility of partially replacing natural gas reforming with either plasma electric reforming or resistance electric reforming. Sidbec expects that operating costs should be substantially reduced by replacing some of the natural gas fuel with hydro-electric power. A coal-based plasma reforming alternative for completely replacing natural gas will also be considered in the study.

PRICES

The Lake Erie base price for Mesabi non-Bessemer ore recovered to the 1981 level in 1983 but dropped back in 1984. Pelletprices increased very little in 1983 and remained at that level in 1984.

World prices for iron ore, especially pellets, came under heavy pressure in 1983 and 1984, and prices declined in both years. The gap between the pellet and fines prices decreased from between 15 and 20 cents an iron unit to about 10 cents a unit. At the same time, the gap between the Lake Eric price for pellets and the price in Europe and Japan increased from about 30 cents a unit in 1981 to 40 cents a unit in 1984. With the Lake Eric price double the world price for pellets, steel companies in North America were considering alternative ways to reduce the cost of their ore requirements.

OUTLOOK

World iron ore production capacity is currently estimated at 1 800 million t. Meanwhile, production was only 774 million t in 1983, indicating capacity utilization of about 43 per cent. The growth rate for steel production in the medium-term has been forecast at 1.5 per cent and, with more use of continuous casting and higher levels of scrap recovery, iron ore consumption is likely to grow at an even slower rate. Therefore, world production capacity is projected to exceed iron ore consumption for the rest of this century.

In spite of the chronic oversupply situation, some 75 million t of annual production capacity is under construction. Brazil alone accounts for about 50 million t, and eight other countries account for the remainder.

The financial arrangement for the Carajas project in Brazil required almost $US 1.5 billion foreign debt and $US 1.2 billion domestic borrowing, which together could place a severe burden, in terms of interest payments, on Brazil as a whole.

On the other hand, operating costs at the Carajas project are relatively low because of the high-grade of ore, requiring little benefication, the ease of mining with relatively little blasting necessary, and labour costs which are a fraction of North American and European wages, in U.S. dollar terms. These two factors, high debt load and low operating costs create a situation where the highest return on investment will likely result from operation and sale at the limit of production capacity, even if the price is well below the established international level. Because of this, many analysts expect downward pressure on price to be maintained well into the 1990s as the expansions in Brazil carve out a share of the international iron ore market.

In Australia, mining costs are relatively low because of the large scale of the open pit mines and the high grade ore (average above 60 per cent iron). In the interests of maximizing profit, rather than total foreign earnings, the Australian producers will likely concentrate their efforts on market share and price, through the pursuit of opportunities for long-term contracts and joint venture development with consumers.

India also has low operating costs at a number of mines that were developed for the foreign market. The National Mineral Development Corporation was in the process of merging with the Minerals and Metals Trading Corporation (MMTC), which will create a near monopoly for the export of Indian iron ore. As a result, the aggressive marketing of iron ore, long established by Indian exporters, will likely be fused with demands for higher prices in future negotiations.

The Swedish iron ore producers, largely dependent on underground mining, face higher than average operating costs. The publicly stated policy of the Swedish exporters is to maintain a stable level of exports and to concentrate on gaining a premium price by producing high quality products, such as self-fluxing olivine pellets.

Other major iron ore exporters, such as Liberia, Mauritania, Chile, Peru, South Africa and Venezuela are expected to favour volume of exports and market share as priorities, and to place less importance on higher prices.

For Canadian exporters of iron ore, the medium-term will remain difficult. The grade of ore at operating mines ranges from 18 to 39 per cent iron, and all of the ore must be concentrated before shipping. Accordingly, Canadian operating costs are higher than in Brazil and Australia, and depressed ocean freight rates are likely to continue to reduce the shipping advantage normally enjoyed by Canadian suppliers.

The performance at Canadian mines, therefore, will be very dependent on price negotiations. Canadian concentrates and pellets are recognized as having high quality and reliable standards, but this has had little effect on price in the past. However, the potential to improve quality and offer premium iron ore products, as practised in Sweden, could be exploited further in Canada because of the sound technological base already in place. Rationalisation to reduce costs, initiatives to preserve market share in the key markets, and capitalizing on short-term imbalances in the more distant markets appears to be the most likely strategy of Canadian iron ore exporters, and one that could produce a reasonable return over the long-term.

Within the North American market, steel producers have increased their iron ore consumption from the low levels of 1982 and 1983. A further increase in consumption is expected in 1985, but growth thereafter will likely be very slow. For Canadian and United States iron ore mines, there will likely be some loss of market to imports from Brazil, and the price will remain under pressure because of the large surplus production capacity and low prices outside North America.

Producers
(numbers refer to numbers on map above)

1. Iron Ore Company of Canada, Knob Lake Division (Schefferville)
2. Iron Ore Company of Canada, Carol Division (Labrador City)
2. Scully Mine of Wabush Mines (Wabush)
3. Quebec Cartier Mining Company (Mount Wright)
4. Sidbec-Normines Inc. (Gagnon, Fire Lake)
5. Iron Ore Company of Canada, Sept-Iles Division (Sept-Iles)
5. Wabush Mines, Pointe Noire Division (Pointe Noire)
5. Quebec Cartier Mining Company and Sidbec-Normines Inc. (Port Cartier)
6. Sherman Mine of Dofasco Inc. (Temagami)
7. Adams Mine of Dofasco Inc. (Kirkland Lake)
8. Algoma Ore division of The Algoma Steel Corporation, Limited (Wawa)
9. The Griffith Mine (Bruce Lake)
10. Wesfrob Mines Limited (Moresby Is.)

TABLE 1. CANADA, IRON ORE PRODUCTION AND TRADE, 1982-84

	1982		1983		1984P	
	(tonnes)[1]	($000)	(tonnes)[1]	($000)	(tonnes)[1]	($000)
Production (mine shipments)						
Newfoundland	15 806 000	572,386	18 404 585	711,727	21 669 849	867,622
Quebec	12 984 000	428,335	10 246 761	372,880	14 745 000	362,448
Ontario	3 633 000	180,905	3 810 509	172,239	4 478 480	235,065
British Columbia	602 000	14,307	496 823	13,078	172 000	5,775
Total[2]	33 198 000	1,201,256	32 958 678	1,269,924	41 065 329	1,470,910
					(Jan.-Sept. 1984)	
Imports						
Iron ore						
United States	3 359 303	192,294	3 977 869	231,976	3 377 629	201,732
Brazil	-	-	35 232	1,267	104 607	2,915
Netherlands	-	-	2	2	-	-
Italy	-	-	6	1	14	1
Total	3 359 303	192,294	4 013 109	233,246	3 482 250	204,648
Exports						
Iron ore, direct shipping						
United States	1 231 718	28,380	824 886	17,589	1 124 958	22,200
Italy	373 614	7,411	344 469	6,364	168 175	3,168
Belgium and Luxembourg	87 462	2,186	61 778	1,236	69 007	1,252
United Kingdom	-	-	57 030	1,084	53 671	1,020
Total	1 692 794	37,977	1 288 163	26,273	1 415 811	27,640
Iron ore, concentrates						
Japan	2 871 436	69,286	2 986 123	70,089	2 350 891	50,071
West Germany	1 818 107	44,463	2 015 388	53,069	1 601 726	36,491
United Kingdom	2 022 900	50,488	1 757 046	41,289	791 509	17,255
Netherlands	3 440 354	83,292	1 255 670	30,556	2 335 860	52,942
United States	1 772 368	64,579	1 188 091	28,538	1 350 572	31,167
France	1 058 685	28,771	925 331	22,898	783 747	15,954
Yugoslavia	127 243	3,489	319 995	10,842	408 558	13,786
Italy	812 854	19,825	442 408	10,516	241 558	5,162
Philippines	288 926	7,223	307 588	7,222	208 081	4,525
Austria	105 465	2,795	213 388	6,035	87 691	1,697
Belgium and Luxembourg	424 342	11,567	203 893	4,893	154 349	3,536
Portugal	49 365	1,721	110 036	2,723	90 399	2,512
Pakistan	125 707	3,007	99 745	2,192	131 508	2,870
Spain	252 822	7,046	61 471	1,977	-	-
Other countries	50 132	1,749	51 936	1,429	234 442	5,892
Total	15 220 706	399,301	11 938 109	294,268	10 770 891	243,858
Iron ore, agglomerated						
United States	5 950 756	335,958	6 852 094	376,612	6 839 108	395,168
United Kingdom	1 771 653	109,957	3 040 908	180,045	1 847 642	79,962
Netherlands	1 010 763	61,600	606 964	31,602	533 103	22,672
Italy	348 246	16,489	470 470	22,458	477 257	22,633
West Germany	819 771	47,749	144 648	8,704	662 837	26,327
France	-	-	141 274	3,851	-	-
Japan	81 390	4,965	148 645	3,383	-	-
Other countries	305 146	16,736	147 577	6,004	158 349	6,396
Total	10 287 725	593,454	11 552 580	632,659	10 518 296	553,158

TABLE 1. (cont'd.)

	1982		1983		1984P	
	(tonnes)[1]	($000)	(tonnes)[1]	($000)	(tonnes)[1]	($000)
Iron ore, nes						
Netherlands	-	-	304 853	7,148	-	-
United Kingdom	-	-	186 253	4,191	-	-
Yugoslavia	-	-	99 999	5,387	-	-
United States	80 147	2,801	59 080	1,596	25 729	708
Other countries	23	1	98 923	2,442	-	-
Total	80 170	2,802	749 108	18,764	25 729	708
Total exports, all classes						
United States	9 034 989	431,718	8 924 151	424,335	9 340 367	449,243
United Kingdom	3 794 553	160,445	5 041 237	226,609	2 692 822	98,237
Netherlands	4 451 117	144,892	2 167 487	69,306	2 868 963	75,614
West Germany	2 637 878	92,212	2 160 036	61,773	2 264 563	62,818
Japan	2 952 826	69,286	3 134 768	73,472	2 350 891	50,071
Italy	1 534 714	43,725	1 257 347	39,338	886 990	30,963
Belgium and Luxembourg	511 804	28,871	265 671	6,129	256 796	6,895
France	1 058 685	28,771	1 066 605	26,749	783 747	15,954
Philippines	288 926	7,223	307 588	7,222	208 081	4,525
Yugoslavia	127 243	3,489	419 994	14,229	408 558	13,786
Other countries	888 660	22,902	783 076	22,802	668 949	17,260
Total	27 281 395	1,033,534	25 528 070	971,964	22 730 727	825,364
Consumption of iron ore at Canadian iron and steel plants	11 999 449	..	13 102 908	..	14 620 016	..

Sources: Energy, Mines and Resources Canada; Statistics Canada; American Iron Ore Association.
[1] Dry tonnes for production (shipments) by province; wet tonnes for imports and exports. [2] Total iron ore shipments include shipments of byproduct iron ore.
P Preliminary; - Nil; .. Not available; nes Not elsewhere specified.

TABLE 2. CANADA, IRON ORE PRODUCTION (SHIPMENTS), 1981-84

Company and Location	Ore Mined	Product Shipped	1981	1982	1983	1984P
Adams Mine, Kirkland Lake, Ont.	Magnetite	Pellets	1 231	964	865	1 134
Algoma Ore division of The Algoma Steel Corp. Ltd., Wawa, Ont.	Siderite	Sinter	1 485	871	1 247	1 250
Caland Ore Company, Limited Atikokan, Ont.	Hematite and goethite	Pellets Concentrate	- 142	- -	- -	- -
Griffith Mine, Bruce Lake, Ont.	Magnetite	Pellets	1 538	910	790	970
Iron Ore Company of Canada Schefferville, Que.	Hematite, goethite and limonite	Direct shipping	2 833	1 675	1 366	1 527
Carol Lake, Lab.	Specular hematite and magnetite	Concentrate Pellets	7 091 10 057	5 609 5 830	5 618 6 590	5 751 8 190
Sept Iles, Que.	Schefferville "treat ore"	Pellets	1 348	129[1]	235	157
Quebec Cartier Mining Company, Mount Wright, Que.	Specular hematite	Concentrate	13 139	9 048	6 683	9 637
Sidbec-Normines Inc. Fire Lake and Lac Jeannine, and Port Cartier, Que.	Specular hematite	Concentrate Pellets (standard) Pellets (low silica)[2]	50 3 500 1 344	47 3 122 681	- 3 211 495	- 4 951 -
Sherman Mine, Temagami, Ont.	Magnetite	Pellets	1 142	850	760	1 124
Wabush Mines, Wabush, Labrador and Pointe Noire, Que.	Specular hematite and magnetite	Pellets	5 291	3 048	5 180	6 202
Wesfrob Mines Limited, Queen Charlotte Islands, B.C.	Magnetite	Pellet feed Fine magnetite	537 39	726 37	492 -	172 -
Byproduct producer						
Inco Limited, Sudbury, Ont.	Pyrrhotite	Pellets Magnetite concentrate	54 126	- -	- -	- -
Total			50 947	33 547	33 532	41 065

[1] Stockpile ore. [2] Included with standard pellets in 1984.
- Nil; P Preliminary.

TABLE 3. RECEIPTS AND CONSUMPTION OF IRON ORE AT CANADIAN IRON AND STEEL PLANTS, AND INVENTORIES, 1983 AND 1984

	1983	1984e
	(tonnes)	
Receipts imported	4 230 377	5 231 331
Receipts from domestic sources	8 558 519	10 476 690
Total receipts at iron and steel plants	12 788 890	15 708 021
Consumption of iron ore	13 102 894	14 620 691
Inventory at docks, plants, mines and furnace yards, December 31	12 491 962	10 123 416
Inventory change	-4 341 623	-2 368 546

Source: American Iron Ore Association.
e - based on 11 month data.

TABLE 4. WORLD IRON ORE PRODUCTION, 1981-83

	1981	1982	1983e
	(000 tonnes)		
U.S.S.R.	242 023	243 953	244 868
Brazil	99 979	110 038	99 573
Australia	85 958	87 787	83 316
People's Republic of Chinae	70 107	70 006	71 123
India	41 150	40 947	40 642
United States	74 375	35 968	38 610
Canada (mine shipments)	49 551	33 549	33 532
Republic of South Africa	28 348	24 588	22 353
France	21 642	19 407	18 289
Liberia	19 711	18 187	15 241
Sweden	23 267	16 155	14 225
Venezuela	15 546	11 685	10 160
Other countries	88 092	82 706	79 252
Total	859 749	794 625	770 034

Sources: U.S. Bureau of Mines Mineral Commodity Summaries, 1983, 1984; Energy, Mines and Resources Canada.
e Estimated.

TABLE 5. CANADIAN CONSUMPTION OF IRON-BEARING MATERIALS BY INTEGRATED[1] IRON AND STEEL PRODUCERS, 1983

Material Consumed	Sinter Plants at Steel Mill	Direct Reduction Plants	Consumed In		
			Iron and Steel Furnaces		
			Production of Pig Iron	Steel Furnaces	Total in Furnaces
			(tonnes)		
Iron Ore					
Crude and concentrate	40 882	112 037	42 238		42 238
Pellets	6 455	687 465	10 901 151	56 588	10 957 738
Sinter	16 243	-	1 226 546	-	1 226 546
Sinter produced at steel plant	-	-	169 714	-	169 714
Direct reduced iron	-	-	-	516 097	516 097
Other iron-bearing materials					
Flue dust	9 947	-	-	-	-
Mill scale, sinder, slag	85 691	-	334 398	4 112	338 511
Total					13 250 844

Source: Company data.
[1] Dofasco Inc.; Sidbec-Dosco Inc.; Sydney Steel Corporation; The Algoma Steel Corporation, Limited; Stelco Inc.
- Nil

TABLE 6. LAKE ERIE BASE PRICE OF SELECTED ORES AT YEAR-END, 1970, 1975 AND 1980-84

	1970	1975	1980	1981	1982	1983	1984
				($US)			
Mesabi Non-Bessemer[1]	10.63	18.21	28.05	32.02	31.73-32.01	32.25-32.53	30.03-31.53
Old Range Non-Bessemer[1]	10.87	18.45	28.30	32.26	32.26	32.78	32.78
Pellets (per natural iron unit)[2]	0.262	0.464	0.725	0.792	0.792-0.855	0.805-0.869	0.805-0.869
Direct Reduced Pellets[3]						115-135	115-135

Sources: Skillings Mining Review; Iron Age.
[1] $US per gross ton, 51.5 per cent of iron natural, at rail of vessel, lower lake ports. [2] $US per gross ton natural iron unit. One iron unit equals 1 per cent of a ton; an ore containing 60 per cent iron, therefore, has 60 iron units. [3] $US per metric tonne.

TABLE 7. SELECTED PRICES OF IRON ORE BOUND FOR JAPAN AND EUROPE 1979-84
(U.S. cents per Fe Unit DMT, FOB)

Ore	Market	Source	%Fe	1979	1980	1981	1982	1983	1984
Fines (including concentrate)	Europe	Rio Doce	(64)	23.5	28.1	28.1	32.5	29.0	26.15
		Iscor	(65)	22.4	26.9	26.9	31.4	27.9	20.60
		Kiruna	(66)	26.6	34.5	33.0	34.7	30.1	27.7
		Carol Lake		23.7	29.3	29.3	33.0	29.3	26.8
		Mt. Wright	(66)	24.0	29.75	29.75	33.0	29.3	26.8
	Japan	Rio Doce		21.6	25.4	26.9	30.5	27.5	24.3
		Iscor		21.6	25.0	26.9	30.5	27.0	23.9
		Hamersley		22.7	27.6	29.7	34.2	30.5	26.7
		Carol Lake	(65)	21.4	25.1	27.0	29.8	26.7	23.4
Lump	Europe	Rio Doce		26.6	31.2	31.2	-	-	-
		Iscor	(65)	25.5	31.9	31.9	35.9	31.3	24.0
	Japan	Rio Doce		21.6	25.4	26.9	30.5	27.9	24.6
		Iscor	(65)	24.7	28.6	30.9	35.0	30.6	27.2
		Hamersley		25.7	31.2	34.2	40.0	34.9	30.9
Pellets	Europe	Rio Doce		40.2	47.1	43.1	47.5	39.0	36.0
		Kiruna		42.2	49.9	48.5	50.2	41.0	38.6
	Japan	Rio Doce (Nibrasco)		46.0	50.3	55.2	53.6	42.9	37.3
		Savage River		37.9	46.2	48.9	53.4	-	38.3

Sources: The Tex Report, Metal Bulletin and Japan Commerce Daily.
- Not available; DMT dry metric tonne; FOB free on board.

TABLE 8. CAPACITY AND PRODUCTION OF DIRECT REDUCED IRON (DRI), 1983

Country	Capacity (million tpy)	Production (million t)
Argentina	.930	.949
Brazil	.315	.255
Burma	.020	.010[e]
Canada	1.625	.538
India	.180	.042
Indonesia	2.300	.500[e]
Iran	.330	.000
Iraq	.485	.000
Mexico	2.025	1.498
New Zealand	.150	.155
Nigeria	1.020	.162
Peru	.100	.026
Quatar	.400	.383
Saudi Arabia	.800	.351
South Africa	.225	.076
Sweden	.070	.020[e]
Trinidad	.840	.283
U.S.S.R.	.417	.015[e]
United States	1.090	.000
Venezuela	4.452	2.468
West Germany	1.280	.070[e]
	19.054	7.801

Source: Midrex Corp., North Carolina, United States.
[e] Estimated.

Iron and Steel

R. McINNIS

SUMMARY

Nineteen eighty three marked the beginning in North America of a consumer-led recovery that gained momentum throughout the year and continued in 1984. Capital expenditure began to increase by mid-1984. In other parts of the western world the recovery began later than in North America, and was much weaker in some regions such as western Europe.

International trade was characterized by national efforts to gain protection from imported steel. Investigations into steel trade resulted in a variety of restrictions to trade such as quotas and anti-dumping duties, especially in North America.

Operating rates of Canadian mills increased from 45 per cent at the beginning of January 1983 to 69 per cent at the end of December 1983. The industry reached 75 per cent of its production capacity in March 1984. However, the average rate of capacity utilization was 68 per cent in 1984.

Canadian crude steel production in 1983 increased 7.2 per cent to 12.1 million t, and increased a further 14.2 per cent in 1984 to 14.5 million t.

Rolled steel shipments from domestic mills, including ingot and semis, increased by 6.9 per cent in 1983 to 10.0 million t, with a further increase of 18.7 per cent to an estimated 11.8 million t in 1984.

Canadian trade figures also reflected the upturn in the global economy. Exports increased by 13.0 per cent to 3.2 million t in 1983 and 12.1 per cent to 3.6 million t in 1984. Imports increased even faster at 60 and 47 per cent in the last two years to 1.5 and 2.2 (estimate) million t respectively.

Employment levels decreased from an average of 49,470 in 1982 to 46,666 in 1983, but rebounded to 49,868 in September 1984. Employment statistics before April 1983 are not fully comparable with later statistics because of changes in Statistics Canada survey methods.

A world oversupply of steel will persist in the medium-term in spite of planned plant closures. The installation of modern equipment based on new cost-saving technologies at existing plants will increase productivity in developed nations. The developing nations will continue to add to their capacity, increasing world output while growth in steel consumption will be modest.

CANADIAN DEVELOPMENTS

Capital expenditures in the Canadian steel industry totalled $198 million in 1983, a considerable drop from the $416 million of 1982 which was also lower than previous years. The demand for steel strengthened in 1984 and capital expenditure intentions improved to $ 226.5 million. Many new projects have been announced and capital expenditures should continue to increase in 1985.

Although the consumption of steel increased throughout 1983, total production was only slightly better than in 1982. The recovery gathered strength during 1984 as demand for consumer durables, especially automobiles, increased significantly. Highlights by company are as follows:

Stelco Inc. The 80-inch hot strip mill at Stelco's Lake Erie Works began operating in May 1983. This mill completes the first phase of this new integrated works.

The company completed a reline of D blast furnace at Hilton Works, with start-up coinciding with the shutdown of F furnace for relining. The #1 bloom and billet mill was permanently closed. Increased quality and productivity were highlighted in 1983 capital expenditures. Improvements were made on the process for desulphurization of the hot metal fed to the Basic Oxygen Furnace (BOF), a microprocess control system was installed on the 5-strand cold reduction mill, and Lance Bubbling

---

R. McInnis is with the Mineral Policy Sector, Energy, Mines and Resources Canada. Telephone (613) 995-9466.

Equilibrium Process (LBE) was installed on a second (#5) BOF. In October 1984, the company announced a major 5-year upgrading program that will cost approximately $400 million at the Hilton Works. This will involve improvements to the basic oxygen steelmaking facilities, installation of a continuous slab caster and a continuous slab/bloom caster, and the modernization of the company's #1 bar mill. Completion is scheduled for 1987.

Lake Erie, Montreal, and Edmonton Works operated at capacity while the older Hilton Works operated at somewhat less than 50 per cent of capacity in 1983 and at 50 per cent in 1984, an improvement over the 40 per cent rate of 1982. Increasing sales of steel and a rundown of inventories were deciding factors in the company's decision to recall 1,500 employees. The Edmonton Steel Works operated at capacity because low scrap prices and reduced freight rates made it economical to ship cast billets to Hamilton.

An internal reorganization in 1983 grouped several existing branches into an autonomous business unit called Stelco Fastner and Forging Company, with responsibilities for manufacturing and marketing fastners and forgings. Effective November 1, 1984, Stelco Pipe and Tube Co. was organized as another profit centre and made responsible for the manufacturing and marketing of pipe and tubular products.

Dofasco Inc. Increasing orders for steel, particularly for flat products, enabled Dofasco to recall by the third quarter of 1983 all 2,100 employees that had been laid off in 1982.

Capital expenditures in 1983 were slightly more than $52 million, most of which was required for the completion of the company's $91 million No. 2 hot strip rolling mill. Expenditures for 1984 were concentrated on the #4 continuous pickle line, which was completed in March 1984, and on the conversion of the #1 galvanizing line to the production of Galvalume. In November 1984, the company announced a $600 million expansion and modernization program that will include the installation of continuous casting equipment. The program is scheduled for completion in 1987.

The company's specilization in the production of high-value sheet steel products that are used in the production of automobiles and other consumer durables was largely responsible for profitability in both 1983 and 1984. Shipments in 1983 were 6.6 per cent higher than in 1982 and up 22.5 per cent in the first nine months of 1984 compared to the same period in 1983.

The Algoma Steel Corporation, Limited. Algoma continued to delay construction of its new seamless tube mill, although the delivery of machinery continued throughout much of 1983. A 72-year old merchant mill was closed on October 28 and a grinding ball mill on September 17, 1983. Both of these non-competitive mills are located at Sault Ste. Marie. The production of merchant pig iron was discontinued.

Capital expenditures were reduced to $32 million in 1983 compared to $ 184.5 million in 1982. Capital expenditures in 1984 were to be limited to $50 million and used for improvements in the quality of continuous cast billets, for improvements in steelmaking and for a product rationalization program. Future capital expenditures will stress cost competitiveness and product quality rather than higher production capacities.

The company's specialization in heavy structurals, plate and line pipe used in capital equipment construction and the oil industry contributed to the difficulties and losses the company faced in 1983 and 1984. A revival of capital investment, especially for new tar sands projects and the twinning of railway lines, should improve the demand for Algoma's products in 1985.

IPSCO Inc. Low demand for tubular products was responsible for reduced operating rates in 1983 and resulted in temporary employee layoffs. Sales improved in fiscal 1984 to $280 million from $193 million in 1983, and earnings doubled to $14.6 million, a level still far below average earnings in 1980-82. The increased oil and gas drilling activity in Canada and the overall improvement in the North American economy were important factors in improved markets. Company sales were also improved by the broader product mix and the higher percentage of flat rolled products now produced.

IPSCO was actively engaged in the purchase, sale and expansion of business interests. The sale of Wescan Pipe Protectors Ltd., a 51 per cent owned subsiduary, was announced in September 1983. An offer to purchase $17.7 million for the assets of a bankrupt pipe company, Steel Corp. Ltd. of Red Dear, Alberta was in November.

Higher quality and productivity improvements have been the goals of recent capital expenditures. A $ 10 million expansion of the company's Regina Steel Works was announced in September, 1983. Funds totalling $28.5 million were raised by a rights issue and these will be used for a modernization and expansion program that will include the installation of a continuous caster, a slab reheat furnace and related rolling mill modifications at a projected cost of $63 million. Completion is scheduled for late-1986.

IPSCO Inc. became the company's formal name on April 21, 1984 when it was changed from Interprovincial Steel and Pipe Corporation Ltd.

Atlas Steels division of Rio Algom Limited. Poor demand and low prices for specialty and stainless steels characterized 1983. However, improved markets for automobiles and consumer durables increased orders for some of the company's products late in the year. The company's 1983 operating loss was significantly lower than that in 1982. Atlas' products are widely used in the production of capital goods and, although the consumer-led recovery continued in 1984, there was no significant increase in capital expenditures until the last quarter of the year. Profitable operating rates were reached in 1984.

Capital expenditures were minimized in 1983-84 and efforts were directed to cost reduction, product development and quality improvement.

Sydney Steel Corporation (Sysco). On December 15, 1983, an explosion in Sysco's only operating blast furnace caused the death of three workers and resulted in a temporary layoff of plant workers while the furnace was being repaired.

Stage I of the company's $96.2 million modernization program was scheduled for completion by March 31, 1985. The largest component of this phase, the rebuilding and modernization of a blast furnace, was completed. The furnace was lit in September 1984 and the inital production target was attained. The company was satisfied with the quality of the iron being produced. Stage 2 of the modernization program was being planned in 1984.

The company operated seven shifts a week and employed 1,200 people during 1983. At year-end 1984, the employment level had increased to 1,475, including 140 working on construction. The operating level in 1984 varied from 7 to 9 shifts per week, depending on shipping requirements.

Sysco shipped rails to Mexico and Mozambique, in addition to domestic sales to Canadian National Railways. The elimination of the Crow rate for grain shipments and the upgrading of rail lines, including double tracking, strengthened the domestic market for rail.

Slater Steels Corporation. This is the new name for the corporation formerly called Slater Steel Industries Limited. The company's divisions have also been renamed to reflect its specialization in steel production. Burlington Steel became Hamilton Specialty Bar Division, Joslyn Stainless Steels became Fort Wayne Specialty Alloys Division and Crucan became Sorel Forge Division. Salcan, the division that specilizes in the manufacture of hardware for telephone and electric lines, was not renamed. Efforts have been made to sell this division as it is outside the company's main area of specilization.

Fiscal 1983 was a low point in the company's financial history, with sales 26 per cent below the level of the previous year. Sales of the company's products, including forgings from the company's recently purchased Sorel Forge Division, increased by over 25 per cent in 1984. This excellent improvement was however, not enough to return the corporation to its pre-recession levels. The company was building on its strength and skill in steelmaking and processing, with 1984 capital expenditures of $5.57 million going primarily for upgrading of equipment and processes. The corporation is well positioned to benefit from any further improvements in the economy.

Sales in the first three months of the latest fiscal year ending June 30, 1984 improved 43 per cent over the same period in the previous year.

Slater's 20.2 per cent interest in IPSCO Inc. was sold in November 1984.

Lake Ontario Steel Company Limited (LASCO). Plant and process modernization continued in 1983-84 and included the following: installation of a new speed control system on the rolling mill that allows both higher rolling speeds and greater uniformity of section dimensions; addition of another strand to the existing continuous casting machine; installation of oxygen fuel burners to the arc furnace to increase the energy input to the furnace, and thereby reducing melt times and improving the productivity of the furnace; development of a ladel injection practice that improves the chemistry and the quality of the steel produced.

QIT-Fer et Titane Inc. (QIT) announced plans to diversify into the production of high-quality steel billet, a decision which will require an investment program totalling $154 million. The company produces pig iron as a coproduct with titanium slag in its illmenite smelting facilities at Sorel, Quebec. This high purity iron will allow the company to produce steel with a relatively small capital investment. The new facilities will include continuous casting machinery for the production of billets.

Important considerations in reaching this investment decision included a severe downturn in the company's historic market for pig iron in Europe and the United States; the qualification of the project for a $25 million grant from the Quebec Government; and a risk sharing contract with Hydro Québec that guarantees a 20-year period of electric power at an attractive rate. If the new product mix proves to be profitable for QIT, part of the rebate on electricity costs will be paid back. However, if another recessionary period develops and the plant is unprofitable, Hydro Québec will maintain the special low rates.

Ivaco Inc. Expansion remained a priority with Ivaco and this goal was pursued by a combination of capital expenditures at existing plants and acquisitions. In November 1983, Ivaco acquired 51 per cent of the shares of Laclede Steel Company of St. Louis, Missouri. This new subsidiary added 800 000 t of steel capacity and the facilities to manufacture bar, flat bar, plate, strip, wire and continuously welded pipe.

A recent purchase of Dofasco shares has moved Ivaco to the position of the largest single shareholder of Dofasco, with 12 per cent of issued shares.

Capital investment in 1983 at $16.6 million was down relative to previous years. However, in the first 9 months of 1984, expenditures were up sharply to $29.5 million. Capital expenditures associated with steelmaking were mainly on ladel metallurgy technology for the refinement and alloying of steel outside of the electric furnace. Such facilities help to optimize the flow of metal to the continuous casting machine.

An arrangement with Boliden AB of Sweden regarding the proposal construction of an INRED smelting plant at l'Orignal, Ontario was cancelled. An agreement to purchase 225 000 tpy of high quality billets for a period of three years from QIT Fer et Titane's new plant in Quebec has reduced Ivaco's need to produce high-purity iron.

WORLD DEVELOPMENTS

A rapid change within the steel industries of the western world occurred in 1983-84. A worldwide surplus of steel persisted even though some economies began to recover from the recession of the previous year. Strong demand for consumer durables led the recovery in North America in early 1983. However, an increase in capital spending did not begin until the last half of 1984. The recovery began later in the rest of the western world.

Steel industries in developed nations reacted to the depressed market and oversupply by closing obsolete plants and rationalizing the rest of their facilities. In the United States, steel production capacity was reduced from the 1980 peak of 146 million t to 123 million t in 1984. A further 10 million t is likely to be closed in the next few years. In the European Community (EC), the industry rationalization and plant closures had been formalized under the seven year old d'Avignon plan wherein member countries agreed to stop subsidizing their steelmakers and to reduce capacity by 26.7 million t to 141.9 million t by 1985.

The rationalization that occurred in the developed nations has reduced the cost of producing steel, partially by a reduction of manning rates but also by investment in new steelmaking technologies such as continuous casting. Many companies have specialized in higher-quality, higher-value-added products, and more capital investment will be made in the near future. For many companies the breakeven point has been lowered considerably so they can now operate profitably at much lower levels of capacity utilization.

Developing nations have responded to the oversupply of steel by reducing their rate of investment. Many planned steelmaking complexes have been put on hold.

In summary, the western world's steel capacity was approximately 600 million t at year end 1984, still considerably higher than consumption of about 440 million t for 1984. This latter level of consumption follows two years of recovery from the low of 1982, a 2.9 per cent increase in 1983 and over 12 per cent in 1984. Such considerations suggest that fierce competition in steel markets will persist in the near future, especially if the rate of economic recovery falters and a recessionary period develops in 1985-86 as forecast by many forecasters.

Trade patterns in steel have also undergone considerable change in recent years with the entry of developing nations to international markets. This development had a disproportionate impact on the market because it occurred at a time of global excess capacity, especially in Europe.

The U.S. market was particularly susceptible to imported steel due to the increase in value of the American dollar and the relatively high cost of producing steel in the United States. Imports captured 21 per cent of the market in 1983 and had in excess of 25 per cent in the first six months of 1984. In response to low and unprofitable operating rates in combination with high levels of import penetration, companies in North America petitioned their governments for protection from low-priced imported steel. The United States won concessions from its trading partners on stainless and alloy steels in 1983, and an accord to limit exports of carbon steel to the United States was reached with the EC and Japan. In spite of these arrangements, large tonnages of imported steel, often from developing nations, continued to enter the U.S. market. Consequently, the U.S. industry increased its lobby to have imports controlled, with the result that the United States International Trade Commission (ITC) recommended that the President impose a system of tariffs and quotas. However, the President instead instructed his officals to negotiate voluntary restraint agreements with the main steel producing countries so that imports would be limited to 20.5 per cent of the U.S. market for the next five years. At year-end 1984, voluntary restaint agreements had been negociated with seven countries: Japan, South Korea, Brazil, Mexico, Spain, Austrialia and South Africa.

Anti-dumping duties were imposed on imported steel in Canada and a number of European countries.

OUTLOOK

The consumer-led recovery that began in Canada in early 1983 resulted in a significant increase in demand for steel, especially for the sheet steel used in the production of consumer durables such as automobiles and appliances. This recovery showed signs of pausing in the last quarter of 1984. However, a slight increase in the capital investment market became evident at that time. Projects in Canada that will require significant tonnages of steel include the expansion of rail line capacity by twinning sections of railroads, the expansion of capacity for extracting oil from the tar sands, as well as a resurgence in oil exploration. These projects should increase the consumption of plate, structural, rails and tube mill products.

Canadian demand for sheet steel should also be stimulated by recent decisions to build new automobile manufacturing capacity, in the form of both new plants and expansions at existing plants. These recently announced expansions should increase steel demand even in a slower market for cars because the vehicles and parts produced will displace imports, and provide export sales under the auto pact.

Canada is expected to remain a net exporter of steel, with the bulk of its exports going to the United States. There remains considerable potential to increase the sale of semifinished steel for rolling and further processing by U.S. firms.

Total Canadian consumption is expected to remain flat or decline slightly until the end of 1985. Growth rates to 1995 are forecast at 2 per cent per annum, reaching a 1995 total of 17.8 million t. However, a significant increase in interest rates during this period could result in a slowing of the economy and a decline in the demand for steel.

In the medium-term to 1990, an average annual increase in domestic steel production of 1.8 per cent is forecast. At this rate of growth, production should reach the level of the peak year 1979 by about 1991. The longer term rate of growth in production to 1995 is expected to be 1.7 per cent per annum.

A worldwide oversupply of steel will likely persist for the next 10 years as developing nations increase their capacity and developed nations continue to have capacity in excess of domestic consumption. The availability of low-priced imported steel will likely continue to depress the price of domestic steel in Canada.

The relative steel intensity (tonnes of steel consumption per million 1975 U.S. dollars of GNP) which dropped dramatically from the early 1970s to the early 1980s will continue to decline, but at a considerably lower rate. The downsizing of the automobile has reached a level where future size reductions will be minimal. However, materials substitution for steel will continue.

RELATIVE STEEL INTENSITY

Country	1970-73	1980-83	1990-95
(tonnes of steel consumption per million 1975 U.S. Dollars of GNP)			
United States	92.8	56.2	46.5
Canada	88.5	61.1	56.2
Japan	161.5	98.9	78.1
United Kingdom	116.4	62.4	50.6
West Germany	98.8	59.7	53.4
Brazil	89.5	72.0	65.7
South Korea	115.5	208.2	210.1

However, world supply and demand should be in better balance by the end of the decade as consumption in developing nations, which are actively building infrastructure, will consume a much higher percentages of their domestic production and the rationalization and modernization of the steel industries in developed nations will be well advanced.

Present trends that are likely to continue include: the importing of more semifabricated steel for further processing in the United States, continuing growth in the electric furnace industry because technical developments will allow such mills to make higher quality steel in a greater variety of shapes and sizes, and ongoing closures of excess capacity in industrialized nations. The remaining plants will be much more productive and efficient because of capital investment in new steelmaking technology. As the developing nations invest in the infrastructure necessary to become developed nations, more of their steel production will be consumed domestically and their need to export will be reduced. However, developing nations will continue to produce an increasing percentage of the world's steel.

PRICES

Economic recovery and increasing demand for steel allowed the Canadian steel industry to reduce the discounting that was so prevalent during 1982. By year-end 1984, some minor price increases in flat products were in place. These increases were quite low because of the availability of low-priced imported steel, which continued to have an impact on domestic prices.

Price changes are indicated by the steel Industry Price Index (Statistics Canada, Catalogue 62-011 Iron and Steel Mills D527101 (1971=100)). For 1982, the index average was 314.7. It increased to an average of 319.2 in 1983 and was 328.4 in August of 1984.

Premium medium volatile bituminous coal, imported from the United States on a long-term contract basis, was $Cdn 80-85 a t cif Ontario steel mills in 1983, compared with $84-90 a t at year-end 1982. By year-end 1984, it was $78-80 a t.

TABLE 1. CANADA, GENERAL STATISTICS OF THE DOMESTIC PRIMARY IRON AND STEEL INDUSTRY, 1982-84

		1982	1983P	1984 (9 Months)
Production				
Volume indexes				
Total industrial production	1971=100	122.8	129.7	140.0
Iron and steel mills[1]	1971=100	106.2	104.2	120.0
		($ million)	($ million)	($ million)
Value of shipments, iron and steel mills[1]		6,095.9	6,294.7	5,696,230
Value of unfilled orders, year-end (Sept. in 1984), iron and steel mills		494.4	712.3	890,603
Value of inventory owned, year-end (Sept. in 1984), iron and steel mills		1,741.3	1,858.6	1,942,728
		(number)	(number)	(number)
Employment, iron and steel mills[1]				
Administrative		12,871	12,454[3]	11,927
Hourly rated		36,599	34,212	37,921
Total		49,470	46,660	49,868
Employment index, all employees 1961=100		142.9	..	..
Average hours per week, hourly rated		38.1	40.0	..
		($)	($)	($)
Average earnings per week, hourly rated		501.77	568.81	591.86
Average salaries and wages per week, all employees		528.89	592.75	616.53
		($ million)	($ million)	($ million)
Expenditures, iron and steel mills[1] (investment intentions in 1984)				
Capital: on construction		63.1	17.8	10.4
on machinery		381.6	180.5	216.1
Total		444.7	198.3	226.5
Repair: on construction		39.3	29.6	33.7
on machinery		624.7	525.5	543.9
Total		664.0	551.1	577.6
Total capital and repair		1,108.7	753.4	804.1
		($ million)	($ million)	($ million)
Trade, primary iron and steel[2]				
Exports		1,831.4	1,492.9	1,638.2
Imports		1,128.7	1,049.9	1,218.0

Sources: Statistics Canada; Energy, Mines and Resources Canada.
[1] S.I.C. Class 291 - Iron and Steel Mills: covers the production of pig iron, steel ingots, steel castings, and primary rolled products, sheet, strip, plate, etc. [2] Includes pig iron, steel ingots, steel castings, semis, hot and cold-rolled products, pipe, wire and forgings. Excludes sponge iron, iron castings. [3] A new survey on labour statistics was started in March 1982 - statistics in later years are not fully comparable.
P Preliminary; r Revised; .. Not available.

TABLE 2. CANADA, PIG IRON PRODUCTION, SHIPMENTS, TRADE AND CONSUMPTION, 1981-83

	1981	1982	1983P
	(tonnes)		
Furnace capacity January 1[1]			
Blast	11 272 000	12 432 000	9 907 000
Electric	525 000	600 000	600 000
Total	11 797 000	13 032 000	10 507 000
Production			
Basic iron	9 007 942	7 463 457	..
Foundry iron[2]	735 557	536 692	..
Total	9 743 499	8 000 149	8 566 621
Shipments	738 698	559 529	530 669
Imports			
Tonnes	6 964	2 262	4 855
Value ($000)	1,200	540	951
Exports			
Tonnes	466 358	485 621	348 281
Value ($000)	101,785	96,420	69,973
Consumption of pig iron			
Steel furnaces	9 589 451	7 926 396	8 544 591
Consumption of iron and steel scrap			
Steel furnaces	7 378 826	5 618 834	6 222 820

Sources: Statistics Canada: Primary Iron and Steel (monthly).
[1] The capacity figures as of January 1 in each year take into account both new capacity and obsolete capacity anticipated for the year. [2] Includes malleable iron.
P Preliminary; .. Withheld to avoid disclosing company proprietory data.

TABLE 3. CANADA, CRUDE STEEL PRODUCTION, SHIPMENTS, TRADE AND CONSUMPTION, 1981-83

	1981	1982r	1983P
	(tonnes)		
Furnace capacity, January 1[1]			
Steel ingot			
Basic open-hearth	3 742 250	3 812 250	3 622 250
Basic oxygen converter	11 746 200	12 010 340	12 285 640
Electric	4 526 000	5 367 410	5 387 135
Total	20 014 450	21 190 000	21 295 025
Steel castings	392 990	536 197	471 444
Total furnace capacity	20 407 440	21 726 197	21 766 469
Production			
Steel ingot			
Basic open-hearth	1 999 248	1 645 891	920 771
Basic oxygen	8 679 354	7 248 158	8 495 536
Electric	3 958 669	2 868 247	3 312 068
Total	14 637 271	11 762 296	12 728 375
Continuously cast, included in total above	4 770 276	3 894 604	4 801 761
Steel castings[2]	173 952	109 078	104 102
Total steel production	14 811 223	11 871 374	12 832 477
Alloy steel in total	1 659 287	1 032 265	928 306
Shipments from plants			
Steel castings	159 691	104 721	93 721
Rolled steel products	11 999 291	9 349 217	9 997 656
Total	12 158 982	9 453 938	10 091 377
Steel ingots included with rolled steel products above	583 705	816 938	949 655
	(000 tonnes)		
Exports, equivalent steel ingots	3 545.2r	3 592.9	2 796.8
Imports, equivalent steel ingots	3 412.3r	1 214.6	1 346.2
Indicated consumption, equivalent steel ingots	14 678r	9 493	10 437

Source: Statistics Canada.
[1] The capacity figures as of January 1 in each year take into account both new capacity and obsolete capacity anticipated for the year. [2] Produced mainly from electric furnaces.
P Preliminary; r Revised.

TABLE 4. PRODUCER SHIPMENTS[1] OF ROLLED STEEL[2], 1982 AND 1983

	1982	1983	Growth
	(000 tonnes)		(per cent)
Ingots and semis	525.0	747.7	42.4
Rails	412.1	458.6	11.3
Wire rods	898.0	1 024.9	14.1
Structural shapes	331.6	387.9	17.0
Concrete reinforcing bar	542.7	526.4	-3.0
Other hot-rolled bars	753.1	850.9	13.0
Track material	57.2	50.0	-12.6
Plate	1 122.6	1 014.6	-9.6
Hot-rolled sheet and strip	1 998.9	2 112.7	5.7
Cold finished bars	68.3	90.0	31.8
Cold reduced sheet, strip other and coated	1 709.5	1 784.4	4.4
Galvanized sheet	930.2	949.6	2.1
Total	9 349.2	9 997.7	6.9
Alloy steel in total shipments	700.8	481.3	-31.3

Source: Statistics Canada: Primary Iron and Steel (monthly).
[1] Includes producer exports. [2] Includes ingots and semis, but not steel castings; comprises both carbon and alloy steels.

TABLE 5. DISPOSITION OF ROLLED STEEL PRODUCTS[1], 1982 AND 1983

	1982	1983	Growth
	(tonnes)		(per cent)
Wholesalers, warehouses and steel service centres	1 230 570	1 551 091	26.0
Automotive vehicles and parts	1 082 718	1 640 179	51.5
Agricultural equipment	93 065	89 748	-3.6
Contractors products	345 534	400 053	15.8
Metal building systems	38 351	38 572	0.6
Structural steel fabricators	666 223	717 438	7.7
Containers	404 365	412 345	2.0
Machinery and tools	343 977	367 626	6.9
Wire, wire products and fasteners	596 678	775 804	30.0
Natural resources and extractive industries	177 869	154 767	-13.0
Appliances and utensils	92 886	119 643	28.8
Stamping, pressing and coating	336 595	390 767	16.1
Railway operating	245 800	286 803	16.7
Railroad cars and locomotives	53 126	50 193	-5.5
Shipbuilding	25 634	18 732	-26.9
Pipes and tubes	1 095 312	1 039 508	-5.1
Miscellaneous	47 671	39 875	-16.4
Total domestic shipments	6 876 374	8 093 144	17.7
Producer exports[2]	2 472 843	1 904 512	-23.0
Total producer shipments	9 349 217	9 997 656	6.9

Sources: Statistics Canada; Primary Iron and Steel (monthly).
[1] Includes ingots and semis, but excludes steel castings, pipe and wire. [2] Total rolled steel exports amounted to 2 819.7 and 2 283.6 million t in 1982 and 1983, respectively.

TABLE 6. CANADA, VALUE[1] OF TRADE IN STEEL CASTINGS, INGOTS, ROLLED AND FABRICATED PRODUCTS, 1981-83

	Imports			Exports		
	1981r	1982r	1983p	1981r	1982r	1983p
			($000)			
Steel castings	41,009	26,178	24,295	16,092	13,144	7,657
Steel forgings	30,017	23,569	24,554	80,743	71,223	72,575
Steel ingots	25,379	3,028	1,523	53,499	20,837	31,456
Rolled products						
Semis	33,687	8,795	12,062	209,695	51,296	133,746
Other	1,406,682	642,086	669,783	1,035,416	1,182,381	876,329
Fabricated						
Pipe and tube	465,424	365,823	246,543	524,521	298,889	179,499
Wire	76,300	58,792	70,166	100,093	94,925	121,667
Total steel	2,078,498	1,128,271	1,048,926	2,020,059	1,732,695	1,422,929

Source: Statistics Canada.
[1] The values in this table correspond with the tonnages shown in Table 7.
p Preliminary; r Revised.

TABLE 8. CANADA, TRADE IN STEEL[1] BY COUNTRY, 1981-83

	Imports			Exports		
	1981r	1982r	1983p	1981r	1982r	1983p
			(000 tonnes)			
United States	1 138.3	462.8	586.8	2 880.1	1 711.9	2 343.9
ECSC[2] countries	1 041.3	320.7	370.1	101.8	364.9	55.6
Japan	419.3	230.6	188.3	1.0	7.9	1.3
Other	509.1	232.9	191.2	598.4	1 148.7	296.1
Total	3 108.0	1 247.0	1 336.4	3 581.3	3 233.4	2 696.9

Source: Statistics Canada.
[1] Comprised of steel castings, ingots, semis, finished steel, forgings, pipe and wire.
[2] European Coal and Steel Community includes the European Economic Community members (Belgium, Denmark, France, Ireland, Italy, Luxembourg, Netherlands, United Kingdom, West Germany and effective 1981, Greece).
p Preliminary; r Revised.

TABLE 7. CANADA, TRADE IN STEEL BY PRODUCT[1], 1981-83

	Imports			Exports		
	1981	1982r	1983p	1981r	1982r	1983p
	(000 tonnes)					
1. Steel castings (including grinding balls)	23.7	13.6	15.2	13.7	8.1	4.1
2. Ingots	72.0	4.3	1.7	220.5	81.5	122.7
3. Semi-finished steel blooms, billets, slabs	95.0	12.2	35.4	674.1	176.7	456.3
4. Total (1+2+3)	190.7	30.1	52.3	908.3	266.3	583.1
5. Finished steel						
A) Hot-rolled						
Rails	35.0	25.7	16.1	174.1	94.6	25.2
Wire rods	195.0	112.9	137.2	323.9	342.7	276.5
Structurals	364.4	120.4	162.2	264.5	201.4	226.8
Bars	127.2	95.4	127.5	265.3	204.5	275.6
Track material	6.8	6.0	4.0	18.2	12.3	13.0
Plate	662.9	213.1	144.2	287.4	238.3	139.6
Sheet and strip	654.4	100.9	135.9	246.1	630.8	251.5
Total hot-rolled	2 045.7	674.4	727.1	1 579.5	1 724.6	1 208.2
B) Cold-rolled						
Bars	18.6	11.3	13.4	18.4	19.3	26.6
Sheet and strip	153.2	59.3	70.6	99.7	308.5	76.9
Galvanized	110.9	62.3	56.5	156.2	345.3	209.0
Other[1]	152.2	104.8	128.7	176.6	163.8	183.9
Total cold-rolled	434.9	237.7	269.2	450.9	836.9	496.4
6. Total finished steel (A+B)	2 480.6	912.1	996.3	2 030.4	2 561.5	1 704.6
7. Total rolled steel (2+3+6)	2 647.6	928.6	1 033.4	2 925.0	2 819.7	2 283.6
8. Total steel (4+6)	2 671.3	942.2	1 048.6	2 938.7	2 827.8	2 287.7
9. Total steel (raw steel equivalent)[2]	3 412.3	1 214.6	1 346.2	3 545.2	3 592.9	2 796.8
10. Fabricated steel products						
Steel forgings	6.3	5.8	7.1	41.3	32.2	34.1
Pipe	365.0	249.7	217.4	497.1	277.1	241.6
Wire	65.4	49.3	63.3	104.2	96.3	133.5
11. Total fabricated	436.7	304.8	287.8	642.6	405.6	409.2
12. Total castings, rolled steel and fabricated (8+11)	3 108.0	1 247.0	1 336.4	3 581.3	3 233.4	2 696.9

Source: Statistics Canada.
[1] Includes steel for porcelain enameling, terneplate, tinplate and silicon steel sheet and strip.
[2] Calculation: finished steel (row 6) divided by 0.77, plus steel castings, ingots and semis (row 4).
p Preliminary; r Revised.

TABLE 9. WORLD RAW STEEL PRODUCTION, 1982 AND 1983

	1982	1983P		1982	1983P
	(million tonnes)			(million tonnes)	
U.S.S.R.	147.2	152.0	Mexico	7.1	7.0
Japan	99.5	97.2	North Korea	5.8	5.9
United States	67.6	76.6	Australia	6.4	5.6
People's Rep. of China	37.1	39.9	Taiwan	4.2	5.0
West Germany	35.9	35.7	Netherlands	4.4	4.5
Italy	24.0	21.7	Austria	4.3	4.4
France	18.4	17.6	Sweden	3.9	4.2
Poland	14.8	16.4	Yugoslavia	3.8	4.1
Czechoslovakia	15.0	15.1	Turkey	2.8	3.8
United Kingdom	13.7	15.0	Hungary	3.7	3.8
Brazil	13.0	14.7	Luxembourg	3.5	3.3
Romania	13.1	13.5	Argentina	2.9	2.9
Spain	13.1	12.7	Bulgaria	2.6	2.8
South Korea	11.8	11.9	Finland	2.4	2.4
Canada	11.9	12.8	Venezuela	2.3	2.3
India	11.0	10.3	Iran	1.2	1.2
Belgium	10.0	10.2	Others	11.2	12.7
East Germany	7.2	7.5			
South Africa	8.2	7.1	Total	644.8	663.9

Source: International Iron and Steel Institute.
P Preliminary.
Note: Totals may not add due to rounding.

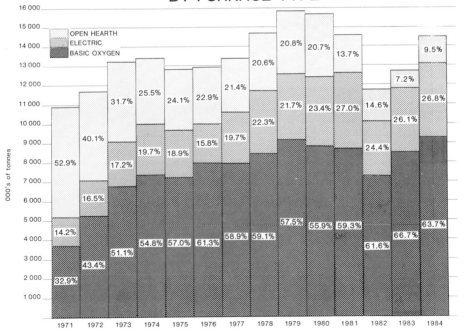

CANADA PRODUCTION OF STEEL BY FURNACE TYPE

TABLE 10. CANADA, ROLLED STEEL SUPPLY AND DEMAND, 1979-83

	Producer or mill shipments[1]	Exports[2]	Imports[3]	Apparent rolled steel consumption[4]	Raw steel production[5]
	(000 tonnes)				
1979	12 230	2 132	1 811	11 909	16 078
1980	12 097	3 019	1 116	10 194	15 901
1981	11 999	2 925r	2 648r	11 722r	14 811
1982	9 349	2 820r	929r	7 458r	11 871
1983P	9 998	2 284	1 033	8 747	12 832
% Change 1983/1982	6.9	-19.0	11.2	17.3	8.1

Source: Statistics Canada.
[1] Comprises domestic shipments plus producer exports. A portion of domestic shipments to warehouses and steel service centres is also exported. Excludes steel castings amounting to 200 000 t in 1979, 198 000 t in 1980, 160 000 t in 1981, 105 000 t in 1982 and 94 000 t in 1983. [2] Total exports includes producer exports plus exports from warehouses and steel service centres. Excludes exports of pipe, wire, forgings and steel castings. [3] Excludes imports of pipe, wire forgings and steel castings. [4] Excludes apparent consumption of steel castings. [5] Includes production of steel castings amounting to 223 353 t in 1979, 217 266 t in 1980, 173 952 t in 1981, 109 078 t in 1982, and 104 102 in 1983.
P Preliminary; r Revised.

TABLE 11. PRICES FOR RAW MATERIALS AND SELECTED STEEL PRODUCTS, 1982 AND 1983[1]

	Currency	1982	1983
Raw Materials			
Iron ore pellets, Lake Superior base price, per metric iron unit[2]	$US	0.792-0.855	0.792-0.855
Coal, metallurgical, imported for Ontario steel mills, per tonne	$Cdn	91.00	80.50
Scrap, Number 1 heavy melting, per tonne	$US	54.95	77.92
Direct reduced iron, per tonne	$US	115.00	115.00
Basic pig iron, per tonne	$US	234.79	234.79
Steel Price Index 1971=100	1981	1982	1983
Structural steel shapes, unfabricated, heavy and intermediate	290.9	302.9	303.3
Steel and strip, hot rolled carbon	281.6	306.5	316.2
Sheet and strip, cold reduced, carbon, alloy and silicon	280.1	308.2	317.2
Plate, carbon and alloy	321.4	351.3	351.9

Sources: Statistics Canada; Skillings Mining Review; Iron Age; Energy, Mines and Resources Canada.
[1] Prices in effect at end of December of each year. [2] One iron unit equals one per cent of a tonne. Hence, iron ore pellets with a grade of 65 per cent iron would contain 65 iron units per tonne.

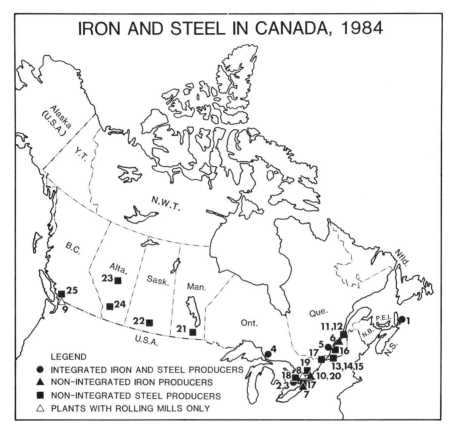

Integrated iron and steel producers
(numbers refer to locations on map above)

1. Sydney Steel Corporation (Sydney)
2. Dofasco Inc. (Hamilton)
3. Stelco Inc. (Hamilton and Nanticoke)
4. The Algoma Steel Corporation, Limited (Sault Ste. Marie)
5. Sidbec-Dosco Incorporated (Contrecoeur)

Non-integrated iron producers

6. QIT-Fer et Titane Inc. (Sorel)
7. Canadian Furnace Division of Algoma (Port Colborne)

Plants with rolling mills only

8. Stanley Strip Steel Division of Stanley Canada Inc. (Hamilton)
9. Pacific Continuous Steel Limited (Delta)

Non-integrated steel producers

10. Courtice Steel Limited
11. Stelco Inc. (Contrecoeur)
12. Atlas Steels division of Rio Algom Limited (Tracy)
13. Sorel Forge Division of Slater Steels Corporation
14. Canadian Steel Foundries Division of Hawker Siddeley Canada Inc. (Montreal)
15. Canadian Steel Wheel Limited (Montreal)
16. Sidbec-Dosco Inc. (Montreal and Longueuil)
17. Ivaco Inc. (L'Orignal)
18. Atlas Steels Division of Rio Algom Limited (Welland)
19. Hamilton Specialty Bar Division of Slater Steels Corporation (Hamilton)
20. Lake Ontario Steel Company Limited (LASCO) (Whitby)
21. Manitoba Rolling Mills division of AMACA International Limited (Selkirk)
22. IPSCO Inc. (Regina)
23. Stelco Inc. (Edmonton)
24. Western Canada Steel Limited (Calgary)
25. Western Canada Steel Limited (Vancouver)

Lead

J. BIGAUSKAS

The non-socialist world industry in 1984 was marked by work stoppages at major mines in the United States and Australia, which in turn affected smelter operations. At the same time consumption of refined lead rose to an estimated 3.9 million t compared to 3.8 million t in 1983. The most notable increases occurred in Japan and Europe. As a result, the U.S. producer price of lead rose to an average 25.5 cents per pound - a marked improvement over 1983s 22 cent per pound average.

CANADIAN DEVELOPMENTS

Lead is principally mined in New Brunswick, British Columbia and the Northwest Territories, but smaller amounts are also produced as a byproduct of base-metal and precious metal mines in Ontario, Manitoba and the Yukon Territory. The former major Yukon lead producer - Cyprus Anvil Mining Corporation - remained shut-down, and therefore production from this Territory in 1984 was limited. Although production from Newfoundland was reported in 1984, the closure of ASARCO Incorpo-rated's Buchans mine will likely mean that Newfoundland will no longer be a producer of lead concentrate.

Primary metallurgical works are located in Belledune, New Brunswick and Trail, British Columbia. Capacities of these plants are 72 000 tpy and 136 000 tpy of refined lead, respectively. Secondary lead plants with a combined capacity of 123 000 tpy are located in Quebec, Ontario, Manitoba, Alberta and British Columbia.

In 1984, Canadian mines produced 290 000 t of lead in concentrates, some 40 000 t more than in 1983. Production of refined lead from primary lead plants totalled 173 000 t which was 5 000 t less than in 1983. Secondary lead production rose to about 79 000 t from 58 000 t.

At the Buchans, Newfoundland copper-lead-zinc mine of Abitibi-Price Inc. (51 per cent) and ASARCO Incorporated (49 per cent), production ceased permanently in September because of exhaustion of reserves. Production had been discontinued at the 56 year-old mine in December 1981, but it was reopened in July 1983 to salvage the remaining reserves in its deepest levels.

At Brunswick Mining and Smelting Corporation Limited's No. 12 mine, planned major modifications were under way at the concentrator. The five-year program, started in 1984, will see major changes to all three concentrator lines. Substantial savings in energy costs, simplification of circuit control and reagent control are the major benefits of the project, although improved metallurgical recovery may also be an added benefit. At the mine itself, development of the 1 000 m level continued although production tonnages are mainly from 575, 725 and 850 m levels.

The Little River Joint Venture mine of Heath Steele Mines Limited and ASARCO Incorporated was shutdown in April 1983. Only limited output was obtained during 1984. Some production may come from a new open-pit in 1985.

Small quantities of lead in concentrate are produced at base-metal mines in Ontario, mainly as a byproduct of zinc and copper mining. These operations are Kidd Creek Mines Ltd.'s mines (now KCML Inc.); Mattabi Mines Limited's operation and Noranda Inc.'s Lyon Lake and "F" Group mines near Sturgeon Lake; and Noranda Inc.'s Geco Division near Manitouwadge. Similarly, in Manitoba, Hudson Bay Mining and Smelting Co. Limited recovers lead concentrates from its zinc and copper mines.

British Columbia's Sullivan Mine continues to produce substantial amounts of lead for the smelting operation at Trail, British Columbia. Pillar extraction from the mechanized sections of the mine have displaced conventional ore removal methods above the 3900 level since a program to modernize the mine began in 1978.

J. Bigauskas is with the Mineral Policy Sector, Energy, Mines and Resources, Canada. Telephone (613) 995-9466.

Currently production by mechanized methods contributes over 60 per cent of output.

Lead-silver concentrates are produced at a number of smaller mines in British Columbia. The more notable lead producers are Teck Corporation's Beaverdell mine, Dickenson Mines Limited's Silmonac mine and Westmin Resources Limited's Lynx and Myra mines.

A promising silver-zinc-lead discovery on the Yukon-British Columbia border contains an estimated 5 million t of mineralization which averages 394 g silver and 18 per cent combined zinc and lead per t. The Midway property is owned 51 per cent by Regional Resources Ltd. and 49 per cent jointly by Canamax Resources Inc. and Procan Exploration Company. An exploration decline was begun late in 1984.

With the shutdown of Cyprus Anvil Mining Corporation's (CAMC) Faro, Yukon Territory lead-zinc mine, the Yukon Territory's production of lead was limited to lead-silver concentrates from the Elsa mine of United Keno Hill Mines Limited. CAMC began a two-year, over-burden-stripping program in June 1983 with financial assistance from the federal and the Yukon territorial governments.

In the Northwest Territories, Cominco Ltd. announced the shutdown of its Polaris zinc-lead mine for one month starting in mid-December 1984 because of low prices for its products. The Polaris mine, located on Little Cornwallis Island, has a capacity of 100 000 tpy of zinc and 30 000 tpy of lead in concentrates.

Normal operations continued at Pine Point Mines Limited's, Northwest Territories zinc-lead mine while union and company officials negotiated a new labour contract. The old agreement expired in April 1984 and a new three-year contract was settled in early-July. The company's N-81 orebody was being prepared for full production in 1986.

Cominco Ltd. has proposed the modernization and expansion of its lead smelter in Trail, British Columbia. The conventional sinter-blast furnace with a capacity of 120 000 tpy would be replaced in two stages with state-of-the-art Kivcet furnaces with total capacity of 160 000 tpy. The smelter, built originally in 1899, suffers from high energy and maintenance costs. The new Kivcet direct-smelting process would signifi-cantly improve the competitiveness of the operation.

Canadian consumption of refined lead as measured by producers' shipments estimated at 120 000 t in 1984, higher than the 96 000 t lead in 1983.

WORLD DEVELOPMENTS

Non-socialist world consumption of refined lead from all sources rose to an estimated 3.9 million t in 1984, or 0.1 million t more than in 1983. Production of refined lead from primary and secondary lead plants in the non-socialist world also was about 3.9 million t - not significantly different from the level in 1983. Mine production of lead showed a year-over-year drop of 5 per cent to an estimated 2.3 million t of lead content.

In the United States - lengthy strikes had a major effect on operations in 1984. Mine workers at St. Joe Lead Co.'s Viburnum Division in Missouri went on strike on April 11. Subsequently, lack of feed caused the company to close its 204 000 tpy lead smelter at Herculaneum on May 18. The strike was resolved mid-December and workers were then recalled.

At Amax Lead Company of Missouri's Buick mine and Boss, Missouri lead smelter (AMAX - Homestake Lead Tollers), management personnel operated the smelter with stockpiled concentrate after workers went on strike June 1. The company was able to fulfil 50-75 per cent of its contractual requirements until October, but by November shipments were reported to be one-third of pre-strike output. On December 31,1984, the strike ended after workers accepted a three-year proposal by the company.

As a result of the indefinite shutdown of the Ozark Lead Co. mine in March 1983, feed shortages continued to affect operations at Asarco Incorporated's 100 000 tpy Glover, Missouri lead smelter. Workers went on strike at the plant at the beginning of October. Supervisory personnel operated the plant with stocks on hand and shipments were maintained at reduced levels. Terms for a new labour contract were agreed to on November 27, and the smelter was reopened the following week.

Offsetting the difficulties of these operations was the opening of St. Joe Lead Co.'s new Viburnum No. 35 lead-zinc-copper mine. About 41 000 tpy of lead in concentrate are expected to be produced from the operation.

The Mexican government has said that it intends to spend $US 1.7 billion in the mining sector as part of a major five-year plan to increase the sector's contribution to the country's GNP from 1.3 per cent to 6 per cent. The investment will mainly be made in state mining companies. Mine projects with a total capacity of some 22 000 tpy lead content are expected to start-up in 1985 and 1987. Underground development work is in progress at three other potential lead producers.

Early in March, Peru's Empresa Minera del Centro del Peru S.A. (Centromin, Peru) declared **force majeure** on refined lead shipments from its 90 000 tpy La Oroya plant because of severe rainstorms and landslides which blocked rail and road links. Shipments were resumed early in April. In July, the company once again declared **force majeure** due to a two-week strike.

Bolivia's Comporacion Minera de Bolivia (Comibol) and Empresa Nacional de Fundiciones (ENAF) completed construction of their new 22 000 tpy Kivcet lead smelter at Karachipampa in March. Start-up was delayed for the entire year because of lack of suitable feed and lack of further financing for the project.

Two mine projects were completed in Yugoslavia during 1984 - one an expansion by 4 000 tpy of lead-in-concentrate and in the other a new mine with the same capacity. Yugoslavia is currently the largest European mine producer and with continuing investments in the industry is likely to continue this role for many years.

In Italy, SAMIM S.p.A.'s planned new lead smelter with capacity of 84 000 tpy and SAMETON S.p.A.'s new electrolytic refinery are expected to start up by 1986.

In Greece, the Laurium, Attica primary lead smelter formerly owned by Pechiney Ugine Kuhlmann of France was recommissioned in February by Emmel SA. Output of 20 000 tpy of lead will be consumed domestically. The plant had been closed since 1982.

Supplies of 99.99 per cent refined lead to Europe were disrupted by a dispute over concentrate prices for Société des Fonderies de Plomb de Zellidja's 65 000 tpy lead plant in Morocco. The plant remained idle from January 1 to late-February when the government intervened to resolve the dispute.

Industrial disputes in late-March 1984 forced suspension of mine production at Broken Hill, New South Wales, Australia. This caused a shutdown of The Broken Hill Associated Smelters Pty. Ltd.'s (BHAS) 250 000 tpy lead smelter - the largest in the world - in early-May. Shipments of lead bullion were, however, maintained. Workers voted to return to work on May 18, and operations at both the mine and smelter returned to normal in mid-June.

Asian mine production, although much less significant than total metal production, rose in 1984 to some 140 000 t of lead-in-concentrate. Metal production is also expected to rise to about 475 000 t of which 210 000 t is recycled scrap. Most Asian mine and metal production presently takes place in Japan.

PRICES

The average U.S. producer price of refined lead, as quoted by Metals Week, was 25 cents (U.S.) per pound in January 1984 and remained around this level until June when it rose to 28 cents per pound. The average monthly price peaked in July at 30.5 cents. Thereafter the price declined to a 22 cent average in October, rose briefly in November to 25 cents and fell again to 22 cents in December. The average U.S. producer price for 1984 was 22.5 cents per pound - better than the U.S. 22 cent average in 1983.

USES

Lead's malleability allows it to be rolled to thicknesses of 5 cm to 0.01 mm and in varying widths and shapes for use in gaskets, washers, impact extrusion blanks, soundproofing radiation protection and architectural applications. Lead can also be extruded in the form of pipe, rod, wire or other cross-sections and can also be extruded around power cables. Flux-cored, tin-lead solders and cable sheathing are typical extrusions. The low melting point of lead allows simple casting for massive counterweights, sailboat keels and minute die castings for instruments. Type metal is noted for its ability to reproduce fine detail. Storage battery grids may be cast or rolled and together with battery posts and battery oxides represent the largest use for lead. Calcium, antimony, tin or arsenic are generally added to impart castability, strength or hardness to the alloys.

Lead shot may be used in ammunition or for mass or sound/radiation shielding where accessibility is a problem. Lead and lead alloy powder particles and flakes are added to grease and pipe joint compounds, powder metallurgy products - such as bearings, brake linings, clutch facings - solder pastes and may be incorporated into rubber and plastics for soundproofing curtains.

When added to steel, brass or bronze, lead improves machineability. Alloyed with tin, lead is used as a hot-dip coating alloy known as terne-coated steel.

Lead oxides and other compounds are also used in paints, pigments, glazes and a wide variety of chemicals. Tetraethyl lead - a gasoline additive - continues to decline in importance but still represents a significant market particularly for primary refined lead.

OUTLOOK

Consumption of lead in the non-socialist world is expected to grow by an average 1.4 per cent per annum to the year 2000. Lead-acid batteries are expected to become an increasingly important end use as other uses, particularly tetraethyl lead, decline in relative significance. In Canada, regulations were passed limiting the amount of lead in leaded gasoline to 0.29 g/litre by January 1, 1987. The U.S. Environmental Protection Agency (EPA) has proposed a limit equivalent to 0.02 g/l by 1986. Cable sheathing will likely maintain a market niche in spite of inroads made by substitute materials. Architectural applications of sheet lead have generally been more important in European countries, although markets may develop elsewhere. Chemicals which contain lead are likely to remain fairly steady as an end-use while alloys are likely to see some decline over the next few years. Because of the increasing importance of the starting-lighting-ignition (SLI) battery as an end-use, markets are expected to become increasingly cyclical - a reflection of its product life and servicing requirements of vehicle populations.

Overall consumption growth is expected to be strongest (4.5 per cent per annum) in the Asian region particularly in the newly industrialized countries. Growth in the Japanese market is expected to be better than in the rest of the industrialized world - about 2 to 3 per cent per annum - but certainly less than the historical trend. Growth propects in Central and South America (1.8 per cent) in Africa and the Middle East (3.5 per cent) are less promising than before but will be better than in the traditional markets of Europe and the United States. Demand is expected to grow at less than 1 per cent per annum in both of these markets.

Production of refined lead in the non-socialist world is expected to rise relatively slowly to just above 4 million t in 1985, 4.4 million t in 1990 and 4.7 million t by 1995. In the year 2000 production could approach 5 million t. Projects in Italy and Peru and those known to be under consideration in Morocco, India, Iran, Taiwan and Canada may add some 250 000 tpy to metal capacity in the long term. Recycling is not expected to grow as rapidly as in the past mainly because of the expected decline in lead content of lead-acid batteries with technological improvements. Thus, mine production is expected to grow to about 2.5 million t lead content in 1985 and 2.8 million t in 1990. In the year 2000 non-socialist world mine production may exceed 3 million t of lead in concentrate.

Because of existing overcapacity at the mining and refining stages and because of relatively unoptimistic prospects for demand growth, the price of lead is not expected to return to its historical average level of 42 cents per pound (constant 1982 U.S. dollars). Nevertheless, a gradual increase to 26-27 cents per pound (constant 1982 U.S. dollars) is expected in the short-term. Increasing cost of production in the secondary sector may put upward pressure on the price of lead in the long-term to perhaps 32 cents and 34 cents per pound by 1990 and 2000, respectively.

Lead's varied markets, and producing regions, its recycleability, its joint mine production with other metals - particularly zinc and silver - are all important facets which must be considered for determining its outlook. These perhaps are reasons why the lead market is one of the most difficult to analyze. Factors which will likely be important in the future will be the increasing dependence upon the lead-acid battery as the dominant end use, slowing consumption growth in the industrialized countries in contrast to newly industrialized and developing countries, slowing expansion of the secondary industry in relation to the 1960s and 1970s, and the more optimistic outlook for zinc - a major coproduct, which comes almost entirely from mine sources.

Given all of these developments research and development on new applications for lead are clearly a priority. Recent improvements in lead-acid battery technology will ensure that it will remain the most competitive cell in the starting-lighting-ignition market. Therefore it will remain an important end-use for lead in spite of the tendency to use less lead per cell. While prospects for lead-acid traction batteries for electric vehicles are becoming more elusive, other battery applications such as utility load-levelling may still offer long-term markets. Experiments on the use of an organic lead compound as an asphalt stabilizer may lead to significant markets particularly in North America.

In the face of overcapacity and comparatively slow growth in the lead market, the key for existing producers will be to improve competitiveness of operations. Canadian mines which produce lead, on average, enjoy relatively favourable unit costs. Mining methods and equipment employed at most operations are comparatively modern. These factors in combination with Canada's well-trained labour force, make the industry one of the most competitive in the world.

Nevertheless the most important factor underlying Canada's competitive position in world lead markets is the quality of lead-bearing orebodies. Over 90 per cent of Canada's lead is produced from mixed ores in contrast to the rest of the world where only about two-thirds is produced from this source. Of course, those mines which are more reliant on lead as a source of revenue are more vulnerable to cyclical swings in lead prices.

At the smelting and refining stage, Canada is less competitive in terms of operating costs, but this is offset to some extent by recovery of byproduct silver. To improve competitiveness, modernization of smelters is being considered by the operators of both primary plants.

TARIFFS

Item No.		British Preferential	Most Favoured Nation	General	General Preferential
CANADA					
32900-1	Ores of lead	free	free	free	free
33700-1	Lead, old scrap, pig and block	free	free	1¢/lb	free
33800-1	Lead in bars and in sheets	4.6%	4.5%	25%	3%
33900-1	Manufacturers of lead not otherwise provided for	14.8%	13.9%	30%	free[1]

	1983	1984	1985	1986	1987
MFN Reductions under GATT (effective January 1 of year given)			(%)		
33800-1	4.5	4.4	4.3	4.1	4.0
33900-1	13.9	12.9	12.0	11.1	10.2

UNITED STATES (MFN)

602.10	Lead bearing ores per lb of lead content	0.75¢
624.02	Lead bullion	3.5%
624.03	Other	3.5%

	1983	1984	1985	1986	1987
			(%)		
624.04 Lead waste etc.	3.0	2.8	2.7	2.5	2.3

EUROPEAN ECONOMIC COMMUNITY: (MFN)		1983	Base Rate (%) (unless otherwise specified)	Concession Rate
26.01	Lead ores & concentrates	free	free	free
78.01	Lead unwrought	3.5	3.5	3.5
	Lead waste & scrap	free	free	free
JAPAN (MFN)				
26.01	Lead ores & concentrates	free	free	free
78.01	Lead unwrought			
	Unalloyed	6.8	7.5	6.0
	Alloyed	7.7	12.0	6.5
	Other	5.9	7.0	4.7
	Lead waste & scrap	3.5	5.0	3.2

Sources: The Customs Tariff, 1983, Revenue Canada, Customs and Excise; Tariff Schedules of the United States Annotated (1983), USITC Publication 1317; U.S. Federal Register Vol. 44, No. 241; Official Journal of the European Communities, Vol. 25, No. L 318, 1982; Customs Tariff Schedules of Japan, 1983.

[1] Pending passage by Parliament of the notice of Ways and Means motion tabled on November 12, 1981 entries entitled to the General Preferential Tariff will be accepted "Subject to Amendment".

TABLE 1. CANADA, LEAD PRODUCTION AND TRADE, 1982-84, AND CONSUMPTION 1982 AND 1983

	1982		1983		1984P	
	(tonnes)	($000)	(tonnes)	($000)	(tonnes)	($000)
Production						
All forms[1]						
British Columbia	83 657	60,651	112 942	66,659	84 186	61,936
New Brunswick	81 475	59,069	70 346	41,518	73 250	53,890
Northwest Territories	63 955	46,367	81 161	47,901	88 355	65,003
Ontario	5 697	4,130	6 473	3,820	8 223	6,050
Newfoundland	1 180	855	-	-	3 527	2,594
Yukon	35 493	25,733	520	307	1 139	838
Manitoba	730	530	519	307	722	531
Total	272 187	197,335	271 961	160,512	259 402	190,842
Mine output[2]	341 212	..	251 383	..	290 000	..
Refined production[3]	174 310	..	178 043	..	173 000	..
					(Jan. - Sept. 1984)	
Exports						
Lead contained in ores and concentrates						
Belgium-Luxembourg	22 387	8,310	53 545	9,518	33 303	7,183
West Germany	16 573	5,238	15 049	3,021	14 235	3,312
United States	11 401	4,072	6 440	2,416	7 194	2,165
United Kingdom	3 051	430	4 914	1,009	940	234
Italy	2 690	781	3 702	684	2 626	672
Japan	36 928	10,643	-	-	-	-
Other countries	13 714	5,424	1 810	418	1 438	360
Total	106 744	34,898	85 460	17,066	59 736	13,926
Lead pigs, blocks and shot						
United States	53 105	34,329	63 661	35,144	59 300	41,064
United Kingdom	37 042	23,325	28 781	12,699	22 080	10,595
Belgium and Luxembourg	17 527	11,976	13 008	6,981	5 818	3,687
U.S.S.R.	10 999	5,797	12 498	6,337	1 000	3,729
Italy	7 179	4,907	4 736	2,643	1 963	1,329
West Germany	4 061	2,539	5 551	2,497	1 461	759
Other countries	16 213	11,119	19 030	10,006	4 817	2,566
Total	146 126	93,992	147 265	76,307	96 439	63,731
Lead and alloy scrap (gross weight)						
United States	6 253	2,254	4 960	1,925	3 269	1,542
South Korea	731	167	756	165	136	66
Spain	36	11	758	123	-	-
United Kingdom	98	71	332	119	138	56
West Germany	2 242	758	363	96	394	149
Taiwan	550	174	125	25	473	82
Other countries	5 980	1,554	236	98	307	184
Total	15 890	4,989	7 530	2,551	4 717	2,081
Lead fabricated materials nes						
United States	5 977	4,624	10 696	6,727	11 981	9,138
Taiwan	199	114	299	163	-	-
Italy	101	67	251	144	-	-
Japan	100	68	59	61	187	190
U.S.S.R.	-	-	-	-	500	300
Other countries	264	159	81	48	296	233
Total	6 641	5,032	11 386	7,143	12 964	9,851

TABLE 1. (cont'd)

	1982		1983		1984P	
	(tonnes)	($000)	(tonnes)	($000)	(tonnes)	($000)
					(Jan.-Sept. 1984)	
Imports						
Lead pigs, blocks and shot	5 661	3,894	2 550	1,642	4 226	2,961
Lead oxide, dioxide and tetroxide	839	938	1 409	1,419	857	980
Lead fabricated materials nes	1 753	2,304	1 298	1,526	565	820
Lead in concentrates	34 389	16,384	18 515	6,528	20 504	10,688
Lead in crude ores	22	5	34	5	-	-
Lead in dross, skimmings and sludge	81	23	271	47	46	10
Lead and lead alloy scrap (gross weight)	54 527	14,697	58 072	8,748	33 023	5,223

	1982			1983		
	Primary	Secondary[5]	Total	Primary	Secondary[5]	Total
	(tonnes)					
Consumption[4]						
Lead used for, or in the production of:						
Antimonial lead	1 471r	x	x	1 499	x	x
Battery and battery oxides	25 855r	6 708r	32 563r	27 792	5 555	33 347
Cable covering	x	x	x	x	x	x
Chemical uses: white lead, red lead, litharge, tetraethyl lead, etc.	16 623	4 643	21 266	14 834	5 543	20 377
Copper alloys; brass, bronze, etc.	110	24	134	197	89	286
Lead alloys:						
solders	1 752	7 495	9 247	1 812	7 633	9 445
others (including babbitt, type metals, etc.)	64	14 204	14 268			
Semi-finished products: pipe, sheet, traps, bends, blocks for caulking, ammunition etc.	4 217	x	x	4 799	x	x
Other lead products	4 944	x	x	2 977	x	x
Total, all categories	55 036r	48 020r	103 056r	54 068	41 944	96 012

Sources: Energy, Mines and Resources Canada; Statistics Canada.
1 Lead content of base bullion produced from domestic primary materials (concentrates, slags, residues, etc.) plus estimated recoverable lead in domestic ores and concentrates exported. 2 Lead content of domestic ores and concentrates produced. 3 Primary refined lead from all sources. 4 Available data, as reported by consumers. 5 Includes all remelt scrap lead used to make antimonial lead.
P Preliminary; - Nil; .. Not available; x Confidential, but included in "other"; r Revised; nes Not elsewhere specified.

TABLE 2. CANADA, LEAD PRODUCTION, TRADE AND CONSUMPTION, 1970, 1975, 1979-84

	Production		Exports			Imports	
	All forms[1]	Refined[2]	In ores and concentrates	Refined	Total	Refined[3]	Consumption[4]
			(tonnes)				
1970	353 063	185 637	186 219	138 637	324 856	1 995	84 765
1975	349 133	171 516	211 909	110 882	322 791	1 962	89 193
1979	310 745	183 769	151 485	117 992	269 477	2 133	98 018
1980	251 627	162 463	147 008	126 539	273 547	2 602	106 836
1981	268 556	168 450	146 090	119 815	265 905	9 220	110 931
1982	272 187	174 310	106 744	146 126	252 870	5 661	103 056[r]
1983	271 961	178 043	85 460	147 265	232 725	2 550	96 012
1984P	259 402	173 000	59 736[5]	96 439[5]	156 175[5]	4 226[5]	..

Sources: Energy, Mines and Resources Canada; Statistics Canada.
[1] Lead content of base bullion produced from domestic primary materials (concentrates, slags, residues, etc.) plus the estimated recoverable lead in domestic ores and concentrates exported.
[2] Primary refined lead from all sources. [3] Lead in pigs and blocks. [4] Consumption of lead, primary and secondary in origin. [5] January to September 1984.
P Preliminary; .. Not available; [r] Revised.

TABLE 3. CANADA, PRIMARY LEAD METAL CAPACITY, 1984

Company and Location	Annual Rated Capacity (tonnes of refined lead)
Brunswick Mining and Smelting Corporation Limited Belledune, New Brunswick	72 000
Cominco Ltd. Trail, British Columbia	136 000
Canada total	209 000

TABLE 4. LEAD-PRINCIPAL USES, NON-SOCIALIST WORLD, 1983

	Europe[e]	United States	Japan
		(per cent)	
Batteries	40	70	58
Cable Sheathing	5	1	6
Pipe and Sheet	20	2	5
Chemicals[1]	25	15	19
Alloys	5	5	4
Other	5	7	8

[1] Including tetraethyl lead.
[e] Estimated.

TABLE 5. MONTHLY AVERAGE LEAD PRICES

	United States Producer (U.S. ¢ per lb)	Canadian Producer (Cdn. ¢ per lb)	LME Settlement (£ per tonne)
1983			
January	22.0	27.8	302
February	21.1	26.3	297
March	20.7	25.8	297
April	21.2	25.8	298
May	20.2	24.8	276
June	19.4	24.0	264
July	19.3	24.5	264
August	19.4	23.9	265
September	21.7	26.4	270
October	25.4	30.8	280
November	25.2	30.8	274
December	24.4	30.5	280
Year Average	21.7	26.8	280
1982			
January	25.1	31.2	282
February	24.1	30.0	280
March	25.0	31.1	316
April	26.4	33.0	339
May	25.4	33.0	326
June	28.2	34.8	352
July	30.5	41.5	374
August	28.2	38.0	356
September	24.2	33.4	320
October	22.3	31.5	339
November	25.2	33.4	356
December	21.9	31.2	350
Year Average	25.5	33.5	332

Source: Metals Week, Northern Miner.

TABLE 6. CANADA, LEAD-BEARING DEPOSITS CONSIDERED MOST PROMISING FOR FUTURE PRODUCTION

Company and Location	Deposit Name	Indicated Tonnage (000 tonnes)	Per Cent Lead	Lead Content (000 tonnes)
New Brunswick				
Billiton Canada Ltd. and Gowganda Resources Inc.	Restigouche	2 200	5.13	112
Caribou-Chaleur Bay Mines Ltd.	Caribou	44 600	1.70	757
Cominco Ltd.	Stratmat 61	2 040	2.44	50
Key Anacon Mines Limited	Middle Landing	1 690	3.03	51
KCML Inc. and Bay Copper Mines Limited	Halfmile Lake	14 000	2.52	340
		64 530	2.03	1 310
British Columbia				
Regional Resources Ltd./Canamax Resources Inc./Procan Exploration Company	Midway Project	3 900	6.0e	230
Cyprus Anvil Mining Corporation	Cirque	36 000	2.2	800
		39 900	2.6	1 030
Yukon Territory				
Cyprus Anvil Mining Corporation	DY Zone	21 000	5.6	1 296
	Swim Lake	4 500	4.0	180
Hudson Bay Mining and Smelting Co., Limited	Tom	8 000	8.6	690
Aberford Resources Ltd. and Ogilvie Joint Venture	Jason	14 100	7.09	997
Placer Development Limited and United States Steel Corporation	Howard's Pass	120 000	2.1	2 500
Sulpetro Minerals Limited and Sovereign Metals Corporation	MEL	5 000	2.0	100
		172 600	3.3	5 763
Northwest Territories				
Cadillac Explorations Limited	Prairie Creek	1 800	17.16	311
Cominco Ltd. and Bathurst Norsemines Ltd.	Seven deposits	19 000	0.75	140
KCML Inc.	Izok Lake	11 000	1.4	150
Westmin Resources Limited, Du Pont Canada Inc. and Phillip Brothers (Canada) Ltd.	X-25	3 400	3.3	110
	R-190	1 300	6.2	79
		36 500	2.2	790
Canada		310 000	2.9	8 900

Canadian Reserves of Copper, Nickel, Lead, Zinc, Molybdenum, Silver and Gold, as of January 1, 1983, MR 201. Energy, Mines and Resources Canada, 1984.
e Estimated.

TABLE 7. WESTERN WORLD LEAD STATISTICS, 1981-84

	1981	1982	1983	1984
		(tonnes)		
Mine production (Pb content)	2 463	2 550	2 446	2 300
Metal production				
- primary	2 236	2 269	2 342	2 300
- secondary[1]	1 798	1 649	1 565	1 600
Metal consumption[2]	3 899	3 796	3 805	3 900

Source: International Lead and Zinc Study Group, Energy, Mines and Resources Canada estimates.
[1] Excluding recovery of secondary materials by remelting without undergoing futher treatment before re-use. [2] Total consumption of refined pig lead including the lead content of antimonial lead. Pig lead and lead alloys recovered from secondary materials by remelting alone without undergoing further treatment are excluded.
[e] Estimated.

TABLE 8. NON-SOCIALIST WORLD LEAD INDUSTRY, 1984[e]

	Production			Consumption[2]
	Mine	Primary Metal	Secondary[1] Metal	
		(000 tonnes)		
Europe	420	860	700	1 650
North America	64	640	560	1 230
Central America	220	140	30	100
South America	250	110	70	140
Africa	250	90	30	100
Asia	140	270	210	650
Oceania	400	190	30	70
Non-socialist World	2 320	2 300	1 630	3 940

[1] Excluding recovery of secondary meterials by remelting without undergoing further treatment before re-use. [2] Total consumption of refined pig lead including the lead content of antimonial lead. Pig lead and lead alloys recovered from secondary materials by remelting alone without undergoing further treatment are excluded.
[e] Estimates.

Princial mine producers
(numbers refer to locations on map above)

1. ASARCO Incorporated (Buchans Unit)

2. Brunswick Mining and Smelting Corporation Limited

3. Heath Steele Mines Limited (Little River Joint Venture) KCML Inc.

4. Noranda Inc. (Geco Division)

5. Mattabi Mines Limited
   Noranda Inc. (Lyon Lake, "F" Group)

6. Hudson Bay Mining and Smelting Co., Limited

7. Cominco Ltd. (Sullivan mine)
   Teck Corporation (Beaverdell mine)

8. Dickenson Mines Limited (Silmonac mine)

9. Westmin Resources Limited (Lynx and Myra)

10. United Keno Hill Mines Limited (Elsa)

11. Pine Point Mines Limited

12. Nanisivik Mines Ltd.

13. Cominco Ltd. (Polaris mine)

Metallurgical Plants

A. Brunswick Mining and Smelting Corporation Limited, Smelting Division, Belledune

B. Cominco Ltd., Trail

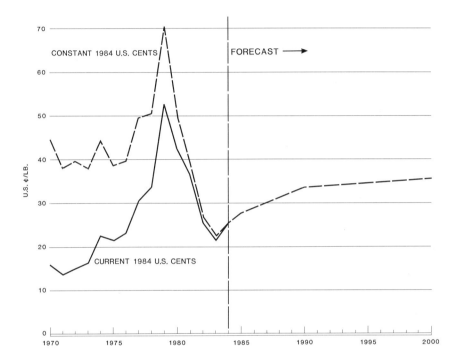

Lime

D.H. STONEHOUSE

SUMMARY 1983-84

The principal markets for lime produced in Canada are in the steel, pulp and paper and mining industries. Demand from these industries was no greater in 1983 than in 1982, but was increased marginally during 1984. Significant markets for lime have yet to develop in the environmental control field in Canada, although the possibility of use in water and sewage treatment and in the removal of sulphur dioxide from smelter gases and from thermal power plant emissions is becoming more likely.

During 1983, Domtar Inc. initiated major changes in its Lime Division. In March, Domtar sold its limestone quarrying operation on Texada Island in British Columbia to Oregon Portland Cement Company which company was subsequently taken over by Ash Grove Cement Co., now Ash Grove Cement West, Inc. The quarry will continue to supply stone to cement, pulp and paper and lime industries in both Canada and the United States. The Bellefonte, Pa. limestone and lime plant, which Domtar closed in June 1982 because of reduced demand from the steel industry, was reopened under new management (Con Lime, Inc.) in May 1983 after having been purchased in April 1983 by a local investor. In November 1983 Domtar sold its Joliette, Quebec lime plant to Jolichaux Inc., a subsidiary of Graymont Limited of Vancouver. Graymont also controls Domlim Inc. which operates plants at Lime Ridge and at St. Adolphe de Dudswell in Quebec. To complete its withdrawal from the lime business Domtar sold its lime plant and limestone quarry at Beachville, Ontario to BeachviLime Limited, effective October 1, 1984.

In late 1983, Dickenson Mines Limited announced an arrangement to acquire the assets of the Havelock Lime group of companies in Havelock, New Brunswick. Sixty per cent of the assets were to be transferred by the end of January 1984.

For the next three years, Dickenson and Havelock will operate the lime works as a joint venture, after which Dickenson will take over the remaining 40 per cent.

Steetley Industries Limited expanded into the United States with the purchase of Ohio Lime Co. and of National Gypsum Company's dolomite lime plant at Gibsonburg, Ohio in 1979. In 1983, Ohio Lime Co. acquired the Millersville, Ohio plant of the J.E. Baker Co., extending Steetley's U.S. influence.

Canada's lime producing capacity remained in the order of 12 000 tpd, sufficient in quantity and well located to meet foreseeable demand.

Dolomitic limestone and magnesite deposits have been investigated as sources of magnesia. The most recent development in this area is that of Baymag Mines Co. Limited which has quarried a high-grade magnesite at Eon Mountain in British Columbia since 1982. The ore is calcined in a refurbished kiln at Canada Cement Lafarge Ltd.'s Exshaw, Alberta plant to produce caustic magnesia and refractory grade MgO. The Canadian Refractories Division of Dresser Canada Inc. has produced refractory products for many years from a magnesitic dolomite at Kilmar, Quebec.

CANADIAN DEVELOPMENTS

Lime is a high-bulk, comparatively low-cost commodity and it is uncommon to ship it long distances when the raw material for its manufacture is available in so many localities. The preferred location for a lime plant is obviously near the principal lime markets, adjacent to a source of high-quality raw material and close to a supply of energy. The more heavily populated and industrialized provinces of Ontario and Quebec together produce over 80 per cent of Canada's total lime output, with Ontario contributing about two-thirds of Canada's

D.H. Stonehouse is with the Mineral Policy Sector, Energy, Mines and Resources, Canada. Telephone (613) 995-9466.

total. Production figures do not include some captive production such as that from pulp and paper plants that burn sludge to recover lime for reuse in the causticization process.

Exports of lime from Canada have continued to decline since 1979 when over 490 000 t were exported, principally to the United States. In 1983, about 216 000 t, mainly from Ontario producers, were exported

Freight costs can represent a large part of the consumer's cost. Production costs have increased significantly as a result of higher energy costs. The industry, on average, uses about 6.4 gigajoules per t of production. New plants have incorporated preheater systems, and the need to replace some of the older less-efficient production capacity with fuel-conserving equipment is well recognized. A new-design, short-rotary kiln (65 metres) and preheater system can reduce energy consumption to about 5.1 gigajoules per t of product. The manufacturer of the new lime kiln for Domlim Inc. at St. Adolphe de Dudswell, Quebec states that with an on-line computerized process control system the "multi-column, parallel-flow, regenerative, vertical kiln" at 360 tpd rated capacity would consume less than 4.2 gigajoules per t.

Average Canadian prices for high calcium quicklime and for high calcium hydrated lime, fob plant, in bulk, were $66.14 a t and $69.06 a t respectively at the end of 1983. At mid-1984 these priceshad risen to $70.11 and $73.19 a t.

USES

Carbonate rocks are basic to industry. They form about 15 per cent of the earth's crust and fortunately are widely distributed and easily exploitable. The principal carbonate rocks utilized by industry are limestones - sedimentary rocks composed mainly of the mineral calcite ($CaCO_3$) - and dolomites - sedimentary rocks composed mainly of the mineral dolomite ($CaCO_3 \cdot MgCO_3$). Commonly termed limestones, they can be classified according to their content of calcite and dolomite. Their importance to the construction industry is not only as building stone and aggregate but as the primary material in the manufacture of portland cement and lime. Limestones are also used as flux material, in glass manufacture, as refractories, fillers, abrasives, soil conditioners and in the manufacture of a host of chemicals.

Quicklime (CaO or CaO·MgO) is formed by the process of calcination, in which limestones are heated to the dissociation temperature of the carbonates (as low as 402°C for $MgCO_3$ and as high as 898°C for $CaCO_3$) and held at that temperature over sufficient time to release carbon dioxide. Although the word "lime" is used generally, and wrongly, to refer to pulverized limestone as well as to forms of burned lime, it should refer only to calcined limestone (quicklime) and its secondary products, slaked lime and hydrated lime. Slaked lime is the product of mixing quicklime and water, hydrated lime is slaked lime dried and, possibly, reground.

Calcining is done in kilns of various types, but essentially those of vertical or rotary design are used. Of comparatively recent design are the rotary hearth, travelling grate, fluo-solid and inclined vibratory types. The high cost of energy has made it imperative to include preheating facilities in any new plant design, and environmental regulations have necessitated the incorporation of dust collection equipment.

The metallurgical industry provides the largest single market for lime. With increased application of the basic oxygen furnace (BOF) in the steel industry, lime consumption increased greatly in certain areas of the United States and Canada. An increase in the demand for steel will result in the need for more fluxing lime and will encourage the development of captive sources by steel producers. The pulp and paper industry is currently the second-largest consumer of lime, most of which is used in the preparation of digesting liquor and in pulp bleaching. Any reduction of activity in either of these two industry sectors, brought on by strikes or lack of product demand, can have an immediate and serious effect on the lime industry, at least regionally. Developments in mechanical fiberizing in the pulp industry could reduce the current lime requirements of this industry significantly.

The uranium industry uses lime to control hydrogen-ion concentrations during uranium extraction, to recover sodium carbonate and to neutralize waste sludge. In the production of beet sugar, lime is used to precipitate impurities from the sucrate. It is used also in the manufacture of many materials such as calcium carbide, calcium cynamide, calcium chloride, fertilizers, insecticides, fungicides, pigments, glue, acetylene, precipitated calcium carbonate, calcium hydroxide, calcium sulphate, magnesia and magnesium metal.

The rapidly-growing concern for the safeguarding and treatment of water supplies and the appeal for enforced anti-pollution measures should result in greater use of lime for water and sewage treatment. The removal of sulphur dioxide (SO_2) from hydrocarbon fuels, either during the burning procedure, or from stack gases by either wet or dry scrubbing, could necessitate the use of lime. This may become a major market for this commodity as SO_2 emission regulations are developed. Lime is effective for this purpose, inexpensive, and can be regenerated in systems where the economics would so dictate. The creation of large amounts of gypsum waste sludge during SO_2 removal will present a disposal problem. Paradoxically, the lime industry is itself caught up in the clean-up campaigns sponsored by various levels of government, particularly efforts directed at dust removal.

Soil stabilization, especially for high-ways, offers a potential market for lime. However, not all soils have the physical and chemical characteristics to react properly with lime to provide a dry, impervious, cemented and stable roadbed. Hydrated lime added to asphalt hot-mix prevents the asphalt from stripping from the aggregate. This could become more important as new technologies relating to asphalt maintenance and repair are adopted and as the sources of good clean aggregate become scarce.

The use of lime-silica bricks, blocks and slabs has not been as popular in Canada as in European countries, although lightweight, cellular, insulating masonry forms have many features attractive to the building construction industry.

OUTLOOK

The short-term outlook for the lime industry in Canada is directly related to economic recovery in general which will benefit the steel, pulp and paper and mining industries, the principal consumers of lime. In the longer term environmental legislation to control acid rain and other pollutants could have a revitalizing influence on lime production.

TARIFFS

CANADA

Item No.		British Preferential	General Preferential	Most Favoured Nation	General
29010-1	Lime	free	free	free	25%

UNITED STATES (MFN)

512.11	Lime hydrated			free	
512.14	Lime other			free	

Sources: The Customs Tariff, 1983, Revenue Canada, Customs and Excise; Tariff Schedules of the United States Annotated (1983), USITC Publication 1317.

TABLE 1. CANADA, LIME PRODUCTION AND TRADE, 1982-84

	1982		1983P		1984	
	(tonnes)	($000)	(tonnes)	($000)	(tonnes)	($000)
Production[1]						
By type						
Quicklime	2 017 000	128,332	1 961 000	..		
Hydrated Lime	180 000	13,749	165 000	..		
Total	2 197 000	142,081	2 126 000	139,638	2 299 162e	172,451e
By province						
Ontario	1 466 000	89,887	1 418 000	89,546		
Quebec	352 000	26,691	297 000	21,673		
Alberta	151 000	9,888	159 000	10,717		
British Columbia	104 000	6,816	129 000	8,726		
Manitoba	..	4,838	..	5,436		
New Brunswick	..	3,961	..	3,540		
Total	2 197 000	142,081	2 126 000	139,638	2 299 162e	172,451e
					(Jan.-Sept. 1984)	
Imports						
Quick and hydrated						
United States	15 875	1,500	22 822	2,232		
France	88	43	22	41		
Total	15 963	1,543	22 844	2,273	15 988	1,796
Exports						
Quick and hydrated						
United States	280 760	17,850	215 521	14,279	138 811	10,129
Other countries	487	103	421	87	510	114
Total	281 247	17,953	215 942	14,366	139 321	10,243

Sources: Energy, Mines and Resources Canada; Statistics Canada.
[1] Producers' shipments and quantities used by producers.
P Preliminary; .. Not available. e Estimated.

TABLE 2. CANADA, LIME PRODUCTION, TRADE AND APPARENT CONSUMPTION, 1970, 1975, 1978-83

	Production[1]			Imports	Exports	Apparent Consumption[2]
	Quick	Hydrated	Total			
			(tonnes)			
1970	1 296 590	224 026	1 520 616	30 649	181 994	1 369 271
1975	1 533 944	199 195	1 733 139	30 099	234 034	1 529 204
1978	1 857 580	176 631	2 034 211	31 130	478 552	1 586 789
1979	1 662 405	196 920	1 859 325	41 480	490 863	1 409 942
1980	2 364 000	190 000	2 554 000	40 901	403 166	2 191 735
1981	2 359 000	196 000	2 555 000	23 144	432 845	2 145 299
1982	2 017 000	180 000	2 197 000	15 963	281 247	1 931 716
1983P	1 961 000	165 000	2 126 000	22 844	215 942	1 932 902

Sources: Energy, Mines and Resources Canada; Statistics Canada.
[1] Producers' shipments and quantities used by producers. [2] Production, plus imports, less exports.
P Preliminary.

TABLE 3. CANADIAN LIME INDUSTRY, 1983-1984

Company	Plant Location	Type of Quicklime
New Brunswick		
Havelock Processing Ltd.	Havelock	High-calcium[2]
Quebec		
Domlim Inc.	Lime Ridge	High-calcium[2]
	St. Adolphe de Dudswell	High-calcium
Jolichaux Inc.	Joliette	High-calcium[2]
Quebec Sugar Refinery[1]	St.-Hilaire	High-calcium
Ontario		
The Algoma Steel Corporation, Limited[1]	Sault Ste. Marie	High-calcium and dolomitic
Allied Chemical Canada, Ltd.[1]	Amherstburg	High-calcium
BeachviLime Limited	Beachville	High-calcium
Guelph DoLime Limited	Guelph	Dolomitic[2]
Timminco Limited[1]	Haley	Dolomitic
Domtar Inc.[3]	Beachville	High-calcium[2]
Reiss Lime Company of Canada, Limited	Spragge	High-calcium
Stelco Inc.	Ingersoll	High-calcium
Steetley Industries Limited	Dundas	Dolomitic
Manitoba		
Alberta Sugar Company[1]	Fort Garry	High-calcium
Steel Brothers Canada Ltd.	Faulkner	High-calcium
Alberta		
Canadian Sugar Factories Limited[1]	Taber	High-calcium
	Picture Butte	High-calcium
Steel Brothers Canada Ltd.	Kananaskis	High-calcium[2]
Summit Lime Works Limited	Hazell	High-calcium and dolomitic[2]
British Columbia		
Steel Brothers Canada Ltd.	Kamloops	High-calcium
BP Resources Canada Limited	Fort Langley	High-calcium

[1] Production for captive use. [2] Hydrated lime produced also. [3] Effective October 1, 1984 Domtar Inc. sold the Beachville plant to BeachviLime Limited.

TABLE 4. CANADA, CONSUMPTION OF LIME, QUICK AND HYDRATED, 1981 AND 1982
(PRODUCERS' SHIPMENTS AND QUANTITIES USED BY PRODUCERS, BY USE)

	1981		1982P	
	(tonnes)	($000)	(tonnes)	($000)
Chemical and metallurgical				
Iron and steel plants	1 237 519[2]	74,529	940 204[2]	60,803
Pulp and paper mills	271 945	16,378	248 298	16,058
Water and sewage treatment	22 760[3]	1,371	85 313	5,517
Nonferrous smelters	117 632[2]	7,085	126 597[2]	8,187
Cyanide and flotation mills	(4)	(4)	41 412[2]	2,678
Sugar refineries	25 841	1,556	34 729	2,246
Other industrial[1]	690 040	41,557	617 300	39,921
Agricultural	17 370	1,046	20 752[3]	1,342
Road stabilization	9 338[3]	565	(4)	(4)
Other uses	162 555	9,787	82 395	5,329
Total	2 555 000	153,874	2 197 000	142,081

Sources: Statistics Canada; Energy, Mines and Resources Canada.
[1] Includes glassworks, fertilizer plants, tanneries, uranium plants and other miscellaneous industrial uses. [2] Figures represent quicklime only. Figures for hydrated lime are included in "other industrial" to avoid disclosing confidential company information. [3] Figures represent hydrated lime only. Figures for quicklime are included in "other uses". [4] Confidential figures are included in "other industrial works".
P Preliminary.

TABLE 5. WORLD PRODUCTION OF QUICKLIME AND HYDRATED LIME INCLUDING DEAD-BURNED DOLOMITE SOLD AND USED, 1982 AND 1983

	1982P	1983e
	(000 tonnes)	
U.S.S.R.	25 038	25 401
United States	12 802	13 063
West Germany	7 983	8 165
Japan	7 983	8 618
Poland	7 502	7 530
Brazil	4 990	5 171
Mexico	3 992	..
Romania	3 538	..
France	3 402	3 538
East Germany	3 538	..
Belgium	2 722	2 812
United Kingdom	3 003	..
Czechoslovakia	3 084	..
Yugoslavia	2 707	..
Canada	2 197	2 126
Italy	2 300	2 449
Other countries	15 175	36 107
Total	111 956	114 980

Sources: Energy, Mines and Resources Canada; Statistics Canada; U.S. Bureau of Mines Minerals Yearbook Preprint 1982; U.S. Bureau of Mines, Mineral Commodity Summaries, 1984.
P Preliminary; e Estimated; .. Included in other countries.

Magnesium

G. BOKOVAY

After several years of weak demand and substantial cutbacks on the part of magnesium producers in the non-socialist world, market conditions for magnesium, which began to improve in the second half of 1983, continued to strengthen into 1984. With the reduction of magnesium inventory levels to more manageable levels, western world producers were estimated to be operating at about 80 per cent of installed capacity at the end of 1984 compared to an average operating rate of 63 per cent in 1983.

Magnesium consumption in 1984 is expected to be significantly higher than that recorded for 1983 and much better than the depressed conditions experienced in 1982. Although conditions have improved, increased re-cycling of aluminum alloys containing magnesium, particularly beverage cans, is one of the principal reasons for the apparent decline in the demand for magnesium since the late 1970s.

While the recovery rate for magnesium within aluminum alloy scrap is expected to increase with greater recycling, magnesium demand in other sectors will more than compensate for this declining market in the next decade. In particular, magnesium will find greater application in diecast products for the automotive industry and in desulphuriation applications in the steel industry.

As a structural material, the future for magnesium is bright as consumers re-examine the metal's desirable qualities including extremely lightweight, high strength and stiffness and excellent heat dissipation, particularly in light of recent improvements in the corrosion resistance of magnesium alloys.

Existing western world primary magnesium capacity is considered to be adequate in meeting projected demand until the end of the decade. However, if new low cost production technology being considered for introduction into Canada is successfully developed, then this could encourage the installation of additional capacity.

Whether or not the new technology becomes a reality, the magnesium industry will remain a relatively energy intensive industry and one which will increasingly view the availability of low cost energy as a major determinant for a production location decision. With a large hydroelectric potential, Canada is expected to become a much larger force in world magnesium production before the end of the century.

CANADIAN DEVELOPMENTS

The Chromasco Division of Timminco Limited is Canada's only producer of primary magnesium. The company operates a plant at Haley, Ontario which is located about 110 kilometers west of Ottawa.

Chromasco uses the Pigeon magnesium process in which calcined dolomite is reduced by ferrosilicon in a vacuum retort. The ferrosilicon used in the process is produced by Chromasco at a plant located at Beauharnois, Quebec, while the dolomite is mined at the plant site.

Although the capacity of the individual vacuum retorts is quite low and the cost of their maintenance is reported to be quite high, the process is relatively energy efficient and the output is of extremely high purity.

Chromasco currently markets four grades of primary magnesium ranging from 99.8 to 99.98 per cent purity. As well, the corporation markets a wide range of magnesium alloys, including its AZ91X high purity diecasting alloy which contains a maximum of 0.004 per cent iron, 0.001 per cent nickel, 0.0001 copper and 0.01 per cent silicon.

In addition to magnesium, calcium and strontium are also produced at Chromasco's

---

G. Bokovay is with the Mineral Policy Sector, Energy, Mines and Resources Canada. Telephone (613) 995-9466.

Haley plant. Magnesium capacity at the plant is estimated to be about 10 000 tpy. After operating at 60 per cent of capacity in 1983, production was gradually increased in 1984 and full production was reached in September.

In Quebec, Metamag-SNA Inc., which is 90 per cent owned by Société nationale de l'amiante, continued work in 1984 on a process to extract magnesium metal from asbestos tailings. With favourable initial test results, a new and larger pilot plant is being considered.

During 1984, MPLC (Mineral Processing Licencing Corp.) Holdings S.A. of Switzerland was reported to be holding discussions with the Province of Quebec about the possibility of constructing a new magnesium production facility. MPLC, which holds the patent to a new single stage production technology that utilizes magnesite as a feedstock for the production of magnesium chloride, claims that process is very energy efficient and will significantly lower production costs. The company plans to eventually build three 100 000 tpy capacity magnesium plants, one each to serve the U.S., European and Japanese markets. According to the company, each plant would be comprised of 10 000 tpy capacity modules each costing about $38 million. Development work on the new process is currently being carried out by MPLC at a pilot plant located near London in the United Kingdom.

The latest negotiations involving MPLC follow earlier reports that the company and the Alberta government were involved in similar negotiations. However, these discussions are reported to have been terminated.

In 1982, MPLC purchased the assets of American Magnesium Co. which had closed its 6 000 tpy capacity plant at Snyder, Texas in 1980. To date, the company has neither resumed production nor moved the plants equipment to another location. In late 1983, it was reported that MPLC was holding discussions with Interstat Resources about the possibility of using the former's technology for Interstat's Pine Flat Mountain, California orebody which contains among other commodities, 24.5 per cent magnesium oxide.

WORLD DEVELOPMENTS

On the basis of data collected for the first half of 1984, the International Magnesium Association (IMA) expected that non-socialist world production of magnesium could reach 212 000 t for the entire year. This would represent a 24 per cent increase over the 171 000 t reported by the IMA for 1983. Current non-socialist world magnesium capacity is estimated at 262 000 tpy.

According to the IMA, magnesium inventories were drawn down by about 30 000 t in 1983 and were estimated at 24 200 t on June 30, 1984. However, inventories on September 30, 1984 were reported to have increased to 32 900 t.

The United States, which is the largest magnesium producer in the world, has three primary magnesium producers. Dow Chemical Co., the largest U.S. producer, operates a 115 000 tpy primary magnesium plant at Freeport, Texas. The plant utilizes an electrolytic process to produce magnesium from magnesium chloride from seawater.

The plant, which had been operating at around 50 per cent of capacity since the second quarter of 1982, increased production to around 65 per cent of capacity in October 1983 and to 75 per cent of capacity in July of 1984. The company had been reported to be considering incremental increases in production to year-end. In 1984, Dow completed the first phase of a modernization program started in 1980 and which included the installation of more energy efficient electrolytic cells. The modernization is reported to have resulted in some increase in capacity.

The second largest U.S. producer, Amax Magnesium Corp., operates a primary magnesium plant at Rowley, Utah. An electrolytic process is used to extract magnesium from magnesium chloride derived from the natural brines of the Great Salt Lake.

With the resurgence of magnesium demand, the Amax facility has been operating at capacity since mid-1984. The plant had operated at about 50 per cent of capacity for part of 1982 and most of 1983.

Since purchasing the plant in 1980, Amax has increased capacity to 35 000 tpy through engineering improvements and the partial startup of a fourth cell room. The company plans to increase this to over 40 000 tpy by 1986 should demand continue to improve.

Although the salinity of the Great Salt Lake has been declining because of greater runoff, the company maintains that this has not affected production.

The third U.S. producer, Northwest Alloys, a subsidiary of the Aluminum Company of America, operates a plant at Addy, Washington that uses the Magnatherm process in which magnesium is produced by reducing dolomite with ferrosilicon. The capacity of the plant is about 22 500 tpy although this will be increased to 25 000 tpy in 1985 with the addition of a tenth furnace. The plant operated at full capacity for most of 1984. It had operated at about two thirds of capacity from November 1982 to September 1983 when the first of three idled furnaces was restarted.

Norsk Hydro, the second largest non-socialist magnesium producer operates a primary magnesium plant at Porsgrunn, Norway. The plant produces magnesium by the electrolysis of magnesium chloride from seawater and magnesium brines imported from West Germany.

The Porsgrunn plant operated at about 60 per cent of capacity during 1983 at which time a major modernization program was undertaken to install new energy efficient cells. The program, which has increased capacity from 50 000 to 60 000 tpy was completed in the first half of 1984. At the end of 1984, Norsk Hydro was reportedly operating at a level of around 50 000 tpy.

The U.S.S.R. is the second largest magnesium producer in the world, with 1983 output estimated by the U.S.B.M. at 82 500 t. The known production facilities are at Zaporozh'ye in the Ukraine, at Bereznike in the Urals and at Usf-Kamenogorsk in Kazukhstan. The U.S.S.R. is reported to have developed a new type of diaphramless magnesium electrolytic cell that significantly reduces energy requirements.

PRICES

The U.S. price for magnesium ingot of 99.8 per cent purity, in 10 000 lb lots, as reported by American Metal Market, has risen steadily during the last several years, despite considerable variations in the level of demand and inventory levels. After remaining at $US 1.34/lb through 1982, prices increased to $1.38/lb in May 1983. In January 1984, prices increased to $1.43/lb and are currently being reported at $1.48/lb.

The price of the important diecasting alloy AZ81B at the end of 1984 was reported to be between $1.26 and $1.30/lb. With the price of secondary aluminum diecasting alloy reported to be around 60 cents US per lb, magnesium is more expensive for diecasting applications. Since on a volume basis magnesium is only two thirds the weight of aluminum, magnesium remains competitive as long as its price does not exceed 150 per cent of the price of aluminum.

At the end of 1984, Chromasco quoted a price of about $Cdn 1.84/lb for 99.8 per cent purity ingot. The price of 99.9 per cent purity ingot was $1.85 while 99.98 per cent purity ingot was $2.40. (Prices in Canada are quoted on a per kilogram basis).

USES

According to the IMA, the largest single application for magnesium, representing over 57 per cent of non-socialist magnesium consumption in 1983, is as an alloying agent with aluminum. The addition of magnesium to aluminum imparts greater tensile strength, increased hardness, better welding properties and better corrosion resistance. One of the most important applications for aluminum alloys containing magnesium has been in beverage cans, with each can containing about 1.9 per cent magnesium. While the popularity of the aluminum can has not diminished recycling of can scrap has risen dramatically thereby reducing the demand for magnesium.

One of the potential new uses for aluminum-magnesium alloys is in the aluminum foil industry. The addition of magnesium increases the strength of the foil and thereby permits a thinner product.

The second largest use for magnesium is for structural applications of which pressure diecast products, constitute the most important component. After increasing from 21 000 t in 1982 to 28 000 t in 1983, magnesium consumption in the form of pressure diecastings is expected to reach 37 000 t in 1984 and 64 000 t in 1988.

As automobile manufacturers attempt to improve the fuel efficiency of their products, the use of lightweight parts including those of diecast magnesium is growing. Among some new automotive applications for magnesium are brake and clutch pedal support brackets, clutch housings, gear shift levers, head lamp assemblies, grill covers, air cleaner covers and valve covers. A new experimental car, manufactured by Volvo, utilizes 110 pounds of magnesium in various applications.

While the greater use of magnesium by the automobile industry is no doubt the result of the U.S. Environmental Protection Agency's fuel economy requirements, high purity magnesium alloys can be used in applications that were once considered too corrosive for the metal. In response to concern about corrosion, both Dow Magnesium and Amax in the U.S. have announced the development of new higher purity alloys and Chromasco also intends to place more emphasis on its high purity products.

Aside from automotive applications, magnesium finds application in the manufacture of portable tools, luggage and sports gear. Magnesium usage in electronic equipment and particularly computers has grown substantially and can be expected to continue.

Magnesium is also used as a deoxidizing and desulphurizing agent in the ferrous industry, and is used to produce ductile or nodular iron. Magnesium is also used as a reducing agent in the production of titanium, zirconium and other reactive metals. Pure magnesium metal is used frequently for cathodic corrosion protection of steel structures, especially underground pipes and tanks. There are many uses for magnesium in the chemical industry including the making of Grignard reagents used in the production of tetraethyl lead for gasoline, although this use has declined in recent years as government move to cut the use of these additives. Magnesium is also used in the fuel cladding material in Magnox-type nuclear reactors.

Potential new applications for magnesium that are currently being investigated include fiber-reinforced magnesium castings, hydrogen storage systems utilizing magnesium hydride, and a magnesium battery.

To support research designed to improve magnesium processing and application technology, the International Magnesium Development Corp. has been established. In particular, the corporation will be looking at new magnesium alloys and corrosion prevention.

OUTLOOK

While there are good prospects for significant growth for magnesium usage in diecasting and desulphurizing applications, the anticipated decline in some other applications, such as aluminum alloying, alloying will moderate potential growth in demand.

As described earlier, magnesium's price is currently at a level that makes it much more expensive for diecasting applications than aluminum. Although aluminum prices are expected to recover somewhat in the next year, it is expected that aluminum markets will remain under some pressure in the medium term. Since it is likely that the magnesium prices will remain at current levels in real terms, it seems unlikely that diecasters will switch to magnesium on the basis of price alone. However, with the excellent casting characteristics of magnesium, the higher cost of the metal may be compensated by the production of thinner-walled products.

The IMA expects that by 1988, almost all North American steel plants and many in Europe will utilize magnesium desulphurization techniques. This will mean a 16 per cent increase in annual magnesium demand from this sector until then with future growth expected but a slower rate.

Overall, magnesium demand is expected to grow at an average annual rate of about 4.0 per cent until 1990 and 3.5 per cent in the 1990s.

TARIFFS

Item No.		British Preferential	Most Favoured Nation	General	General Preferential
		(%)			
CANADA					
35105-1	Magnesium metal, not including alloys, in lumps, powders, ingots or blocks	4.8	4.7	25	3
34910-1	Alloys of magnesium; ingots, pigs, sheets, plates, strips, bars, rods and tubes	4.6	4.5	25	free
34911-1	Magnesium alloy ingots, for use in the production of magnesium castings (expires 30/6/83)	free	free	25	free
34912-1	Hardener alloys for use in the manufacture of magnesium castings (expires 30/6/83)	free	free	25	free
34915-1	Magnesium scrap	free	free	free	free
34920-1	Sheet or plate, of magnesium or alloys of magnesium, plain, corrugated, pebbled, or with a raised surface pattern, for use in Canadian manufactures (expires 30/6/83)	free	free	25	free
34925-1	Extruded tubing, of magnesium or alloys of magnesium, having an outside diameter of five inches or more, for use in Canadian manufactures (expires 30/6/83)	free	free	25	free

MFN Reductions under GATT (effective January 1 of year given)		1983	1984	1985	1986	1987
				(%)		
35105-1	Magnesium metal, not including alloys, in lumps, powders, ingots or blocks	4.7	4.5	4.3	4.2	4.0
34910-1	Alloys of magnesium; ingots, pigs, sheets, plates, strips, bars, rods and tubes	4.5	4.4	4.3	4.1	4.0
UNITED STATES		1983	1984	1985	1986	1987
				(%)		
628.55	Magnesium, unwrought, other than alloys and waste and scrap	15	13.5	12	10	8
628.57	Magnesium, unwrought, alloys, per pound on magnesium content	7	6.8	6.7	6.6	6.5
		¢ per lb of magnesium content + % ad valorem				
628.59	Magnesium metal, wrought, per pound on magnesium content	5.5¢ / 3.0%	5.2¢ / 2.9%	5.0¢ / 2.8%	4.7¢ / 2.6%	4.5¢ / 2.5%

Sources: Customs Tariff, 1983, Revenue Canada, Customs and Excise; Tariff Schedules of the United States Annotated (1983), USITC Publication 1317; U.S. Federal Register Vol. 44, No. 241.

TABLE 1. CANADA, CONSUMPTION OF MAGNESIUM, 1977-82

	1977	1978	1979	1980	1981	1982P
	(tonnes)					
Castings and wrought products[1]	879	952	1 447	1 412	619	574
Aluminum alloys and other uses[2]	5 343	3 001	3 003	4 000	5 768	4 431
Total	6 222	3 953	4 450	5 412	6 387	5 005

[1] Die, permanent mould and sand castings, structural shapes, tubing, forgings, sheet and plate. [2] Cathodic protection, reducing agents, deoxidizers and other alloys.
P Preliminary.

TABLE 2. CANADIAN IMPORTS/EXPORTS OF MAGNESIUM METAL

	Imports (t)	Exports (t)
1980	3 419	5 316
1981	3 249	6 221
1982	1 972	4 501
1983	3 714	2 500
1984 (first nine months)	2 960	2 653

Source: Statistics Canada.

TABLE 3. WORLD PRIMARY MAGNESIUM PRODUCTION, 1981, 1982 and 1983

	1981	1982	1983
	(000 tonnes)		
Canada	8.8	7.9	7.8
United States	129.9	89.9	104.7
U.S.S.R.e	76.0	77.0	80.0
Norway	47.6	35.9	29.9
France	7.3	9.6	9.7
Italy	10.8	9.7	9.8
China, P.R.	7.0	7.5	8.5
Japan	5.7	5.6	6.0
Yugoslavia	3.9	4.2	4.7
Poland	0.5	0.5	-
Brazil	-	0.3	0.5
India	0.1	0.1	0.1
Total	297.6	248.4	261.7

Source: Metallgesellschaft AG.
e Estimated; - Nil.

TABLE 4. PRIMARY MAGNESIUM PRODUCTION BY WORLD ZONE

Period	Area 1 United States & Canada	Area 2 Latin America	Area 3 Western Europe	Area 4 Africa & Middle East	Area 5 Asia & Oceania	Total
			(tonnes)			
1978	143 900	-	54 700	-	11 300	209 200
1979	156 400	-	58 700	-	11 400	226 500
1980	163 000	-	64 400	-	9 200	236 600
1981	138 400	-	64 400	-	5 700	208 500
1982	97 800	-	52 800	-	5 800	156 400
1983	109 000	-	51 000	-	6 000	166 000
1984 (first nine months)	113 300	700	53 200	-	4 800	172 000

Source: International Magnesium Association.
- Nil.

TABLE 5. PRIMARY PRODUCERS SHIPMENT BY WORLD ZONE

Period	Area 1 United States & Canada	Area 2 Latin America	Area 3 Western Europe	Area 4 Africa & Middle East	Area 5 Asia & Oceania	Total
			(tonnes)			
1982	85 761	8 347	60 591	1 278	17 731	173 708
1983	98 600	9 600	60 400	2 400	33 400	204 400
1984 (first nine months)	84 500	5 700	49 400	1 100	21 600	162 300

Source: International Magnesium Association.

TABLE 6. WESTERN WORLD PRIMARY MAGNESIUM CONSUMPTION PATTERN, 1983

Use	North America	Latin America	Western Europe	Africa/ M. East	Asia/ Oceania	Total 1983
			(000 t)			
Aluminium alloying	53	2	34	2	26	117
Nodular iron	4	-	4	-	1	9
Desulphurization	10	-	2	-	1	13
Chemical/reduction	19	1	3	-	3	25
Pressure diecasting	5	7	15	-	1	28
Other structural	7	-	1	-	-	8
Other	2	-	1	-	1	4
Total	99	10	60	2	33	204

Source: International Magnesium Association.
- Nil.

Manganese

D.R. PHILLIPS

SUMMARY

The consumption of ferromanganese is directly related to the consumption of steel and ferrous castings. Canadian ferroalloy companies have remained competitive due to low energy costs, modern plants and technological know-how. The investment by Elkem A/S in the Canadian ferrosilicon industry, through its acquisition of Union Carbide Canada Ltd., could strengthen the latter organization's international marketing capability and make accessible the specialized ferroalloy technology necessary to remain competitive.

Manganese is essential in the production of nearly all types of steel, and approximately 95 per cent of all manganese produced is consumed by the iron and steel industry. Accordingly, the demand for manganese ores is essentially determined by the world production of iron and steel. Manganese is considered to be a strategic commodity because of its critical role in iron and steel making, for which there are no acceptable substitutes.

CANADIAN DEVELOPMENTS

Canada has no domestic producers of manganese ore although several low-grade deposits have been identified in Nova Scotia, New Brunswick and British Columbia. The largest of these deposits, located near Woodstock, New Brunswick is reported to contain about 45 million t of mineralization grading 11 per cent manganese and 14 per cent iron. Although processes have been developed to utilize such low-grade deposits, the commercial viability of putting them in production has not been demonstrated.

The two ferromanganese producers in Canada, Elkem Metal Canada Inc. (Elkem), previously Union Carbide Canada Limited, and Timminco Limited, previously Chromasco Limited, use imported metallurgical-grade manganese ore as feed material. These companies have plants at Beauharnois, Quebec and both sell their production mainly to domestic steel producers.

Following the 1981 acquisition of the U.S. ferroalloy facilities of Union Carbide Corp., Elkem A/S exercised its legal option and acquired the assets of Union Carbide Canada Limited in 1984. Elkem planned to have its Canadian marketing office based in Toronto. The owners also negotiated a new power contract with Hydro-Québec. This acquisition by Elkem A/S could strengthen the formerly held Union Carbide organization because of Elkem A/S' worldwide operations and its strong position in international markets.

Elkem's main component of the Beauharnois plant is a 30 megawatt (MW) electric arc furnace, the largest in the western world. This furnace has a nominal capacity of 120 000 tpy of standard grade ferromanganese. However, actual production of ferromanganese is considerably less as the furnace is also used to produce silicomanganese. In 1983, the furnace was operated well below capacity but, as a result of the increased demand for steel, production was expanded in 1984 to nearly full capacity.

Timminco Limited, previously Chromasco Limited, operated its plant in Beauharnois at approximately 80 per cent of capacity in 1983 and near capacity in 1984. The plant had shut down three of its four furnaces in 1982.

Canada also imports manganese metal, an important additive in specialty steels and aluminum alloys. The main consumers of manganese metal are Atlas Steels division of Rio Algom Limited, Aluminum Company of Canada, Limited (Alcan) and Reynolds Aluminum Company of Canada Ltd.

High-purity manganese dioxide and battery-grade manganese ores are imported into Canada by various companies including

---

D.R. Phillips is with the Mineral Policy Sector, Energy, Mines and Resources Canada. Telephone (613) 995-9466.

Duracell Inc., Gould Manufacturing of Canada, Ltd. (Industrial Battery Division), Cominco Ltd. and Canadian Electrolytic Zinc Limited (CEZ).

WORLD DEVELOPMENTS

World manganese ore production decreased about 3 per cent to an estimated 22.4 million t in 1983 from 1982, mainly due to the decreased demand from the steel industry. Ore production in 1984 was estimated to exceed that of 1983 production.

In 1983, the U.S.S.R. continued to be the world's largest producer of manganese ore, accounting for about 47 per cent of world production, followed by the Republic of South Africa, which produced approximately 13 per cent. Brazil accounted for 9 per cent of total production while the People's Republic of China, Gabon, India, Mexico and Australia collectively accounted for approximately 27 per cent of production.

South Africa remained the largest supplier of manganese to the western world by producing an estimated 2.89 million t of ore in 1983, approximately 45 per cent less than that produced in 1982. The production for 1984 was estimated to be approximately the same as 1982, which was close to capacity.

The recent acquisitions by Elkem A/S of ferroalloy facilities in the United States and Canada make it the world's largest independent ferroalloy producer.

Australian production of manganese ore in 1983 was estimated at 1.4 million t which was an increase of approximately 27 per cent over 1982. Australia is a major exporter of manganese ore and ranks in sixth place as a producer. Due to the effect of the last recession, the Australian ferromanganese industry operated at approximately 60 per cent of capacity in 1983 and 1984. Mine capacity is equivalent to approximately 2.5 million tpy of saleable ore.

Australia exported approximately 75 per cent of its manganese ore production to Europe, the U.S.S.R., Japan and the Republic of Korea. These countries normally account for approximately 90 per cent of Australian exports. The sale of ferromanganese ore to the U.S.S.R. was made by the Broken Hill Proprietary (BHP) of Australia.

The development of the Carajas manganese deposit in Brazil was reported on schedule. Trial shipments were started in late-1984 and a pier for handling the ore at Itaqui was under construction. Ore reserves at Carajas are reported to be some 60 million t grading more than 40 per cent manganese. It was estimated that the Igarape Azul Mine, also in the Carajas district, was producing sufficient manganese ore to supply the Brazilian battery industry. This mine has a production capacity of approximately one million t of manganese ore.

USES

The excellence of manganese as a desulphurizer has made this metal an irreplaceable input in the steel industry. Sulphur in steel tends to migrate to the grain boundaries and causes steel to crack and tear during hot rolling and forming. Manganese combines with the sulphur to produce manganese sulphide inclusions, which do not migrate to the grain boundaries. The metal also acts as a deoxidizer during the steelmaking process.

Manganese is usually added to steel in the form of a ferroalloy such as ferromanganese or silicomanganese. Steel mills in Canada use about 5.8 kilograms (kg) of manganese per t of crude steel produced.

Specialty steels frequently contain manganese to increase strength and hardness. Manganese metal is normally used in preference to ferromanganese for making these specialty steels because it provides better control of the manganese and impurities content.

Hadfield steels, a type of specialty steel, contains between 10 and 14 per cent manganese. These steels are extremely hard and tough, and are particularly suited for applications such as rock crusher parts and teeth in earth-moving machinery.

Iron used for castings is desulphurized with manganese. Otherwise, the sulphur causes surface imperfections and makes precision casting difficult.

Also, manganese is used to form alloys with nonferrous metals: aluminum-manganese alloys are noted for their strength, hardness and stiffness; manganese-magnesium alloys are hard, stiff and corrosion resistant; and manganese bronzes have properties desirable in specific applications such as ship propellers.

Manganese has many nonmetallurgical applications including its use in dry-cell batteries. In this role, manganese dioxide provides oxygen to combine with hydrogen, which permits the battery to operate at maximum efficiency. Manganese ores used for batteries must grade above 85 per cent manganese dioxide and have a low iron content. Very few natural manganese dioxide ores can meet these specifications and, as a result, most batteries contain a blend of natural ore and synthetic manganese dioxide.

A common classification of manganese ores gives rise to the following ore types: (1) Manganese ores containing more than 35 per cent manganese: these are used in the manufacture of both low- and high-grade ferromanganese. Although battery-grade ores are included in this class, these ores must contain no less than 85 per cent manganese dioxide. (2) Ferruginous manganese ores containing 10 to 35 per cent manganese and used in the manufacture of spiegeleisen. (3) Manganiferous iron ores containing 5 to 10 per cent manganese and used to produce manganiferous pig iron.

All types of manganese ores can be employed in the production of manganese chemicals such as: potassium permanganate, a powerful oxidant used in the purification of public water supplies; manganese oxide, an important addition to welding rods and fluxes; and an organometallic form of manganese, which inhibits smoke formation and improves the combustion of fuel oil. Various manganese chemicals are employed to produce colour effects in face bricks and, to a lesser extent, to colour or decolour glass and ceramics.

PRICES

Annual price negotiations for metallurgical grade manganese ore are normally concluded from April to June of each year. Prices are mainly determined on the basis of manganese content. However, several other factors including quantity, delivery schedules, tariffs and other related supply aspects of the ore are taken into account, including the physical character of the ore.

Australia's manganese producer started price negotiations in March 1983 and settled in April at $US 55.20 fob per t of high-grade manganese ore (48 per cent Mn). In the latter month the Gabonese producer concluded a price of $US 67.68 cif with Japan, and the South African producers settled for $US 67.50 cif.

Contract negotiations again progressed slowly in 1984. However, by the end of May 1984, agreement was reached at a price of $US 67.68 cif.

OUTLOOK

The outlook for world production of manganese ore in 1985 indicates that there will be a slight increase over 1984 in response to an anticipated increased demand for steel. It is forecasted that by the year 2000, developing countries, which presently export all or most of their ore, will become increasingly involved in secondary processing.

In the long-term it is likely that the U.S.S.R. and the Republic of South Africa will continue to be the major suppliers of manganese ore. Brazil, India, Gabon, Australia and Mexico are expected to increase their share of the market. Those LDC's which have manganese reserves and low cost energy are expected to increase their ore production to meet their future domestic demand for the processing of secondary manganese products, especially ferromanganese.

World ferromanganese capacity is expected to decline within the next decade. This decline will result from a reduction of output from obsolete plants in industrialized nations, partially offset by the establishment of new efficient facilities in those countries where ore and low cost energy are available. Australia, Brazil and Mexico can be expected to substantially increase their ferroalloy capacity in the next decade.

Since manganese has no economical substitute in its main use, overall manganese ferroalloy production is projected to increase, at the rate of 1.2 per cent per year up to 1990, in parallel with growth in steel production.

Given the world reserves of manganese, the possibility of future mining of manganese modules and the decreased demand for manganese in steel and iron due to technological developments, supply shortages of manganese ore and ferromanganese are likely to develop in the foreseeable future.

PRICES

United States prices in U.S. currency, as published by **Metals Week**,

	December 1982	December 1983	December 1984
	$		
Manganese ore, per long ton unit (22.4 lb) cif U.S. ports, Mn content			
Min. 48% Mn (low impurities)	1.58-1.68	1.44-1.47	1.44-1.47
Ferromanganese, fob shipping point, carload lots, lump, bulk			
Standard 78% Mn, per long ton	490.00	490.00	- LPS -
	(cents)		
Medium-carbon, 80-85% Mn, per lb Mn	46.00	41.00-46.00	41.00
Silicomanganese, per lb of alloy, fob shipping point, 65-68% Mn, 16-18.5% Si, 0.2% P, 2% C	24.50	21.00	23.50
Manganese metal, per lb of product, fob shipping point			
Regular, minimum 99.5% Mn	70.00	70.00	80.00
6% N, minimum 93.7% Mn	80.00	80.00	86.00

fob Free on board; cif Cost, insurance and freight; LPS - List Price Suspended.

TARIFFS

Item No.		British Preferential	Most Favoured Nation	General	General Preferential
CANADA					
32900-1	Manganese ore	free	free	free	free
33504-1	Manganese oxide	free	free	free	free
35104-1	Electrolytic manganese metal	free	free	20%	free
37501-1	Ferromanganese, spiegeleisen and other alloys of manganese and iron, not more than 1% Si, on the Mn content, per lb.	free	0.5¢	1.25¢	free
37502-1	Silicomanganese, silico-spiegel and other alloys of manganese and iron, more than 1% Si, on the Mn content, per lb.	free	0.73¢	1.75¢	free

MFN Reductions under GATT (effective January 1 of year given)	1983	1984	1985	1986	1987
	(cents)				
37501-1	0.5	0.4	0.4	0.4	0.4
37502-1	0.73	0.73	0.72	0.71	0.70

UNITED STATES (MFN)

Item No.		MFN
601.27	Manganese ore, including ferruginous manganese ore and manganiferous iron ore, all the foregoing containing over 10% Mn	free
632.30	Manganese metal, unwrought	14.0%

Item No.		1983	1984	1985	1986	1987
		(%)				
606.26	Ferromanganese, not containing over 1% C, per lb Mn content	2.6	2.5	2.4	2.4	2.3
606.28	Ferromanganese containing 1 to 4% C, per lb Mn content	1.4	1.4	1.4	1.4	1.4
606.30	Ferromanganese containing over 4% C, per lb Mn content	1.6	1.6	1.5	1.5	1.5
632.28	Manganese metal waste and scrap	9.8	8.8	7.7	6.7	5.6

Sources: The Customs Tariff, 1983, Revenue Canada, Customs and Excise; Tariff Schedules of the United States Annotated 1983 USITC Publication 1317; U.S. Federal Register Vol. 44, No. 241.

TABLE 1. CANADA, MANGANESE, TRADE AND CONSUMPTION, 1982-84

	1982		1983		1984P (Jan.-Sept. 1984)	
	(tonnes)	($000)	(tonnes)	($000)	(tonnes)	($000)
Imports						
Manganese in ores and concentrates[1]						
South Africa	10 746	2,011	8 037	1,112	20 227	2,539
Brazil	19 935	4,023	9 976	1,853	-	-
United States	3 158	1,209	3 316	1,166	2 596	965
Gabon	37 816	7,508	20 931	2,848	12 199	1,653
Total	71 655	14,751	42 260	6,978	35 022	5,157
Manganese metal						
South Africa	430	769	2 051	2,961	1 260	1,829
People's Republic of China	150	204	300	380	225	292
United States	201	341	265	374	225	306
Other countries	-	-	36	68	36	68
Total	781	1,314	2 652	3,782	1 746	2,495
Ferromanganese, including spiegeleisen[2]						
United States	11 319	11,243	8 498	8,229	9 099	6,334
West Germany	-	-	2 300	836	7 100	4,582
South Africa	11 335	5,985	2 031	1,223	3 676	7,735
Mexico	541	433	3 640	2,364	2 195	1,467
France	1 693	675	1 301	1,462	1 785	1,985
Norway	200	120	489	227	400	222
Total	25 088	18,456	18 259	14,342	24 255	16,325
Silicomanganese, including silicospiegeleisen[2]						
South Africa	960	482	-	-	6 077	2,636
Brazil	-	-	7	3	2	3
Norway	1 537	866	-	-	-	-
United States	380	372	453	329	4	251
Total	2 877	1,720	460	332	6 083	2,890
Exports						
Ferromanganese[2]						
United States	11 440	4,549	2 631	902	1 185	342
Puerto Rico	157	81	-	-	-	-
United Kingdom	141	17	-	-	-	-
Total	11 738	4,647	2 631	902	1 185	342
Consumption						
Manganese ore						
Metallurgical grade	127 450	..	..	..	..	..
Battery and chemical grade	3 376	..	..	..	..	..
Total	130 826	..	..	..	..	..

Sources: Energy, Mines and Resources Canada; Statistics Canada.
[1] Mn content; [2] Gross weight.
P Preliminary; - Nil; .. Not available.

TABLE 2. CANADA, MANGANESE IMPORTS, EXPORTS AND CONSUMPTION, 1970, 1975, 1979-83

	Imports			Exports	Consumption	
	Manganese Ore[1]	Ferro-Manganese	Silico-Manganese	Ferro-Manganese	Ore	Ferromanganese and Silicomanganese
	(gross weight, tonnes)					
1970	115 052	17 891	975	510	153 846	97 952
1975	69 773	35 701	5 732	1 168	160 976	95 869
1979	45 150	83 700	21 876	12 043	64 699	89 429
1980	95 161	26 704	20 901	11 278	157 680	95 796
1981	119 746	36 656	12 669	57 040	288 908	83 958
1982	71 655	25 088	2 877	11 738	130 826	69 166
1983P	42 260	18 259	460	2 631	..	..

Sources: Energy, Mines and Resources Canada; Statistics Canada.
[1] Mn content.
P Preliminary; .. Not available.

TABLE 3. WORLD PRODUCTION OF MANGANESE ORES, 1981-83

	Mn (%)	1981	1982r	1983e
		(000 tonnes)		
U.S.S.R.	30-33	9 153	9 824	10 432
Republic of South Africa	30-48+	5 039	5 216	2 886
Brazil	38-50	2 042	1 300	2 087
Gabon	50-53	1 488	1 512	1 857
People's Republic of China[e]	20+	1 597	1 597	1 597
Australia	37-53	1 449	1 132	1 353
India	10-54	1 526	1 448	1 320
Mexico	27+	578	509	350
Ghana	30-50	225	132	191
Morocco	50-53	110	94	74
Hungary	30-33	71	93	85
Japan	24-28	87	82	77
Bulgaria	30-	45	50	45
Yugoslavia	30+	31	31	30
Other countries[1]	..	102	61	44
Total	..	23 543	23 081	22 428

Source: U.S. Bureau of Mines, "Mineral Yearbook", 1982.
[1] Includes 19 countries, each producing less than 24 000 tpy.
e Estimated; r Revised; .. Not available.

Mica

M. PRUD'HOMME

SUMMARY

Canadian mica shipments are estimated at 12 000 t for 1984, an increase of approximately 9 per cent over 1983. Almost 50 per cent of these shipments are exported, primarily to the United States, Japan and Europe. Since 1982, the unit value of ground mica imports rose at an average annual rate of 4 per cent, to $272.7 per t in 1984. Although fabricated mica imports have dropped by approximately 4.5 per cent per year since 1981, they should rise in 1984.

Canada is the world's leading producer of ground and flake phlogopite. Marietta Resources International Ltd., in a joint-venture with la Société Minéralurgique Laviolette Inc., operate the only active mine in Quebec's Suzor township and, in Boucherville, near Montreal, produce varieties of mica used in gypsum caulking products, in plastics and paints.

Mica production and consumption should increase in the near future. The most promising growth potential is in the plastics industry, especially for automotive applications.

THE MICA GROUP

The micas comprise a series of phyllosilicates with a variable chemical composition, but distinct physical properties, such as basal cleavage. The term "mica" primarily refers to muscovite $KAl_2 (AlSi_3O_{10}) (OH)_2$, biotite $K (Mg, Fe)_3 (Al Si_3O_{10}) (OH)_2$ and phlogopite $K Mg_3 (Al Si_3 O_{10}) (OH)_2$. The micas are complex hydrated aluminous silicates that crystallize in the monoclinic system.

Color varies from black to virtually colorless. Hardness is approximately 2 to 3 on Mohs' Scale, and density ranges from 2.7 to 3.1.

Essentially, only the muscovite and phlogopite varieties are of economic importance. Muscovite is a common constituent of acid igneous rock, such as granites, pegmatites and aplites. Phlogopite is particularly common in ferromagnesian basic rock, such as the pyroxenites, meta-sedimentary crystalline limestones, peridotites and dunites.

The micas are marketed in various forms, ranging from blocks and splittings to scrap, flake, ground and micronized mica.

Sheet mica is extracted from enormous crystals and worked by hand to obtain blocks, sheets and splittings. These grades are classified according to the size, thickness and color of the sheets. Mica sheets are valued by the electrical and electronics industries for their dielectric, optical and mechanical properties.

Scrap mica is obtained from sheet mica waste. It is generally reduced to a powder or flakes for the manufacture of mica paper and micanite or filler. These micas are classified according to particle size and are wet-ground or dry-ground. Flake mica is extracted as a coproduct of feldspath, kaolin and lithium; certain types are found in schist deposits with a high mica content.

SITUATION IN CANADA

Production and deposits

Canada has been producing mica since 1886; production was continuous until 1966, when the last phlogopite shipment was taken from the Blackburn mine in Cantley, Quebec. The Lacey mine near Sideham, Ontario was a major phlogopite-producing site until 1948. From 1941 to 1953, Canada was one of the largest producers of sheet muscovite, which was extracted from the Purdy mine near Mattawa, Ontario. In 1977, mica production resumed in Canada following the development of a large phlogopite deposit in Quebec's Suzor township. Since then, Canadian production has relied on a single active producer. However, since 1982, several exploration and development projects have been undertaken in Ontario and Quebec.

M. Prud'homme is with the Mineral Policy Sector, Energy, Mines and Resources Canada. Telephone: (613) 995-9466.

The only active mine is near Parent in the township of Suzor, Laviolette county. Since 1974, Marietta Resources International Ltd., in a joint-venture with la Société Minéralurgiqu Laviolette Inc., has been exploiting a schist containing approximately 80-90 per cent phlogopite, 4-8 per cent pyroxene, 2-6 per cent perthite and traces of apatite, calcite and chlorite. Phlogopite reserves are estimated at more than 27 million t of homogenous ore to a depth of 60 m.

The ore is open-pit mined and crushed by a Kennedy Jaw crusher to smaller than 20 cm, then shipped by rail twice each year to the Boucherville processing plant near Montreal. The ore is then dry-ground, concentrated in a pneumatic separator and classified according to four sizes: -10 +20 mesh, -20 +40 mesh, -40 +100 mesh, and over 100 mesh.

Production capacity varies from 25 000 to 35 000 tpy according to the final quality. For surface-treated mica, the capacity is approximately 6 000 tpy. Flake mica, registered under the name of Suzorite, is used as a reinforcing agent in plastics and composites. The ground phlogopite varieties act as fillers in asphalt products, gypsum caulking products and drill muds.

Traces of mica have been discovered at several sites in Canada. Muscovite is particularly common in pegmatite intrusions. Interesting occurrences were located in the following Ontario townships: Addington, Calvin, Canney, Chapman, Chisholm, Christie, Clarendon, Davis, Deacon, Hungerford, Kaladar, Lennox, Mattawa, McKonkey, Olrig and Sheffield; in Quebec, muscovite is found in the counties of Abitibi-Temiscamingue, Charlevoix, Dubuc and Saguenay; in British Columbia, near Yellowhead Pass, the Big Ben district of the Columbia River, and in the Fort Grahame district.

In Canada, phlogopite is virtually confined to the northeastern belt of the Grenville series. The major occurrences of phlogopite are found in Quebec in the counties of Argenteuil, Gatineau, Hull, Labelle, Laviolette, Montcalm and Papineau, and in Ontario in the counties of Frontenac and Lanark.

Exploration and development

Since 1982, commercial interest in mica has intensified and led to several projects, primarily in Ontario and Quebec. In Ontario's Kaladar township, a quartzite schist deposit rich in muscovite was the focus of an exploration exploration campaign by Koizumi (Canada) Ltd. The deposit, named Kaladar Aimko Property, contains almost 60 per cent muscovite associated with quartz with traces of biotite and hematite. Approximately 5,200 short tons of bulk samples have already been shipped to Japan, where they were processed in a pilot plant to produce mica for the plastics industry. In Clarendon township, Ontario, another deposit has attracted special attention since 1980. The Ontario Ministry of Natural Resources carried out exploration work in 1982 and 1983. Pelitic schists contain approximately 40 to 50 per cent muscovite with varying amounts of biotite, feldspath, garnet, staurolite, kyanite, sillimanite and magnetite.

In Chasseur Township, Quebec, SOQUEM carried out improvement work on a deposit belonging to Provinces & Explorations Inc. Drilling estimated the reserves at approximately 380 000 t per vertical m with a 50 per cent phlogopite content. In Lamy township, exploration was undertaken; reserves have already been estimated at more than 10.2 million t of ore containing 60 to 75 per cent phlogopite. In Suzor township, on the property of Marietta Resources International Ltd., EM-6 electromagnetic surveys were carried out in 1982 and 1983 in order to determine the reserves. In Dandurand township, Frédéric Exploration collected mica samples in 1982. Mineralogical assays at the Centre de Recherche Minérale in Quebec City set the mean flake diameter at approximately 1.1 mm.

In British Columbia, a micaceous schist deposit was evaluated by Brinco under a Tournigan Mines option in 1982.

Research and expansion

From 1979 to 1982, the Centre de Recherche Minérale in Quebec City carried out laboratory and pilot plant research on mica processing. Various grinding and delamination techniques were tested to obtain a fine-grain material with a high diameter/thickness ratio. Since 1982, the Industrial Materials Research Institute (IMRI) of the National Research Council completed applied studies on the use of mica in thermoplastics and on resistance to fatigue of engineering plastics. In 1983, Marietta Resources International Ltd. invested approximately $1 million in an expansion program that will allow it to increase its production capacity of fine mica to approximately 15 000 tpy; this grade S mica is used to reinforce plastics in the automobile industry.

USES AND SPECIFICATIONS

Sheet mica is used mainly in the electrical and electronics industries, and in small quantities for thermal insulation. Sheet muscovite is used to manufacture micanite, mica paper and fabricated products, such as capacitors and communicator segments. Since the dielectric properties of musconite are better than those of phlogopite, transparent mica is the most common variety used in those sectors. The specifications for sheet mica comply with the standards of the American Society for Testing and Materials (ASTM). Designation ASTM-D351-62 specifies quality according to stain, inclusions and imperfections. Designation ASTM-D2131-65 specifies the required characteristics for mica product manufacture. Designation ASTM-D748-59 defines the requirements for the electrical, physical and visual properties of the mica sheets used in capacitors.

Sheet mica is graded according to thickness: blocks must be thicker than 0.007 in. (0.18 mm), films between 0.0008 and 0.004 in., and splittings approximately 0.0011 in.

Ground and micronized micas are used as reinforcing agent or filler in drill muds. The major user industries include gypsum caulking products, asphalt roofing products, paints, rubber products and plastics.

Drywall products and joint cement compounds constitute the major utilization range for mica. Mica prevents cracking and provided good workability due to its structural quality. Product size should be less than 150 microns and be free of abrasive grits. Muscovite is occasionally preferred over phlogopite as it is virtually colorless. The principal substitutes for mica are talc, clay and asbestos.

Mica is used in asphalt roofing products as dusting agent. It is also used as a filler in asphalt mixtures to improve their resistance to weather. Dry-ground mica varies in size from 850 to 75 microns (20 to 200 mesh sieve).

Paints require fillers to improve surface qualities. Mica reduces shrinkage, prevents cracking and improves resistance to weather. It is used in exterior paints, anticorrosive emulsions and oil-based metal primers. Wet-ground or micronized mica should be transparent. Required particle sizes are of the order of 100, 160 and 325 mesh sieves, and more than 30 microns.

Producers of rubber materials use mica as a dusting and releasing agent. It is also used as a filler to reduce gas penetration and shrinkage during moulding. The mica is generally in 850 to 150 micron flakes.

Plastics are a recent application for micas. Flakes are used as a reinforcing agent along with fibrous substances, such as fiberglass, wollastonite and asbestos. Micas with a high diameter/thickness ratio (High Aspect Ratio, HAR) are used in polypropylenes, polyethylenes and phenolic plastics. The resulting plastics have high flexural and tensile strength, and low permeability and good resistance to weathering. Delaminated micas are treated with coupling agents to improve their cohesion with resins. Typical loading range from 20 to 50 per cent in plastic mixtures. These ground micas have particle size varying from 425 to 45 microns.

Other principal uses of the mica varieties include: wet-ground mica; wallpaper and lubricants, dry-ground mica; drill muds, insulating panels, welding electrodes, acoustical products, adhesives, extinguisher powders and composite cement materials.

CONSUMPTION AND TRADE IN CANADA

In Canada, mica is primarily consumed by the construction industry. More than 90 per cent of the mica is used in gypsum caulking products and paints. The rubber, plastics and drill muds industries share the remaining 10 per cent.

Canada imports ground muscovite from the United States for asphalt roofing products and gypsum products, especially for manufacturers in the western provinces.

Canadian imports have been declining since 1981. In 1983, imports of ground and micronized mica was valued at $687,000 for a volume of 2 632 t, which account for only 19 per cent of the total import value. These imports are overshadowed by the fabricated mica products coming in equal volumes from the United States and France. More than 80 per cent of these imports were traditionally from the United States. Canadian exports to the United States have been climbing since

1982, reaching almost 4 700 t in 1983. The principal sector served is the plastics industry. Japan, Europe and South America are the other export markets for Canadian mica.

WORLD PRODUCTION AND REVIEW

World production of mica can be broken down according to mica type. India is the most important source of sheet muscovite, followed by Brazil, Argentina and Madagascar. The United States is the largest producer - and consumer - of ground and flake mica. It produces muscovite by wet and dry methods, generally as a coproduct of kaolin, lithium or feldspath. Canada dominates the world market in the production of flake, ground and micronized phlogopite; Argentina produces a small volume of sheet phlogopite.

In 1983, the world production of mica rose by approximately 13 per cent to almost 237 000 t as a result of an increase in U.S. production, which accounts for approximately 54 per cent of the world total.

United States: Principal uses are gypsum caulking products (47 per cent) and paints (15 per cent). The market for gypsum products has been captive, especially since the purchase of the Diamond Mica Co. of North Carolina by U.S. Gypsum, in 1979. Harris Mining Co. and J.M. Huber Corp. will increase their production capacity in 1985 to serve the plastics industry. In 1982, Mineral Industrial Commodities of America Inc. (MICA) of New Mexico began construction of a new plant to increase its total production of ground mica.

Finland: Kemira Oy announced the construction of a plant that, will begin producing mica for the plastics industry in autumn 1985. This plant, which generates an investment of approximately 14.9 million Markkaa, will process the phlogopite that is currently dumped from the apatite mine in Siilinjaervi.

India: In 1983, the Mica Trading Corporation of India Ltd. (MITCO) started work on several development projects: the production of crushed mica of the order of 1 micron, the manufacture of mica paper using Japanese technology, and the production of insulation with a mica paper base with the participation of a West German firm.

OUTLOOK FOR THE MICA INDUSTRY

The sheet mica producers' strategy is to further develop the production of fabricated mica. These high-value-added products still require extensive manual labour, and a major change should happen in the future.

In North America, the production capacity of ground mica exceeds demand; however, optimistic forecasts of increased demand in the 1980s have resulted in work being started on several expansion projects. The economic performances of the ground and flake mica industry are linked to those of the construction, plastics, drill mud and insulation industries.

In the construction sector, manufacturers of paints, asphalt roofing products and gypsum caulking products consume more than 90 per cent of all mica used in Canada.

Capital investment and repair expenditures in the construction sector have been rising since 1982; they reached $56,370 million in 1984, an increase of 1.6 per cent over 1983. According to the Canadian Construction Association, investment in the residential sector represents 30 per cent of investment in the construction industry. In the 1984-1990 period, slow growth is anticipated, of the order of 0.2 to 0.6 per cent, i.e. a smaller increase than the average annual growth rate for the Canadian economy, estimated at 3 per cent. In the non-residential sector, which represents 25 per cent of total investment in the construction industry, average growth (estimated at 4.8 to 6.5 per cent for that period) far exceeds the average annual growth of the economy.

In Canada, the consumption of mica in construction-related industries should equal the growth rate. In the United States, demand for mica in the gypsum products and paint industries will record average increases of 3.8 per cent and 3.1 per cent, respectively, for the 1983-88 period.

The plastics industry appears to be the favored sector for any major growth in mica consumption. Mica is used in the plastics intended for the automobile sector, for which demand has been increasing for the last few years. Indeed, the quantity of plastics consumed in a standard American car rose from 162.5 pounds in 1976 to 200 pounds in 1983, and should increase to 225 pounds in 1990 and 250 pounds in 1992. The proportion of plastics in a car will rise at an average annual rate of 9.2 per cent over the 1987-92 period. Moreover, according to Chase Econometrics, the North American automobile industry should record average annual growth of 1.8 to 2.6 per cent for the 1985-90 period. In view of these increases in the demand for plastics, mica could benefit both in the short- and in the long-term from a changing market, particularly as a reinforcing agent in certain polymers (especially polypropylenes, polyurethanes and polyesters), the use of which should increase considerably in this sector. Some estimates set the consumption of mica for plastics in 1988 at four times the 1983 figure.

The oilwell drilling industry, a traditional market for mica, is relying more on mica substitutes. In the short-term, little growth is expected from this market. The insulation industry should maintain its mica consumption despite the pressure exerted by the producers of substitute materials and minerals.

PRICES, 1983

Average price[1] for wet-ground and dry-ground mica in the United States.

	$US per short ton
Wet-ground mica:	397
Dry-ground mica:	118
By uses:	
drill muds	105
paints	164
joint cement	146

Prices for mica in the United States, according to the Chemical Marketing Reporter[2].

	$US per lb
Wet-ground mica: Paints, per carload, 325 mesh sieve, fob, works	.6¾
Dry-ground mica: Joint cement, plastics, 50-pound bags, carload, works	.07½
Dry-ground mica: Roofing products, 20 to 80 mesh sieves, point of shipping	.07
By uses; carload, fob, works	
rubber products	.16½
wallpaper	.22

	$Cdn per short ton
Prices for phlogopite[3], fob, carload	
miconized mica	220-380
Surface-treated mica	590-720
Flake or ground mica	225-370

---

[1] U.S. Bureau of Mines, 1983, mica.   [2] CRM, May 30, 1983.   [3] Marietta Resources International Ltd., April 15, 1983.
fob Free on board.

TARIFFS

Item No.		British Preferential	Most Favoured Nation (MFN)	General	General Preferential
			(per cent)		
CANADA					
29600-1	Mica schist	free	free	free	free
29650-1	Mica, phlogopite and muscovite, unmanufactured, in blocks, sheets, splittings, films, waste and scrap	free	free	25	free
44550-1	Raw low loss mica, sheets and punchings of low loss mica	free	free	25	free
UNITED STATES (MFN)					
516.11	Untrimmed phlogopite		free		
516.31	Split block mica		free		
516.41	Other		free		
516.51	Mica splittings		free		
516.61	Mica, not over 0.006" in thickness, not cut or stamped to dimensions, shape or form		free		

Item No.		1983	1984	1985	1986	1987
				(per cent)		
516.21	Phlogopite, waste and scrap	5.1	4.9	4.7	4.4	4.2
516.24	Other mica, waste and scrap	2.4	2.4	2.4	2.4	2.4
516.81	Ground or pulverized mica	5.2	3.8	3.3	2.9	2.4

Sources: The Customs Tariff, 1983, Revenue Canada, Customs and Excise; Tariff Schedules of the United States Annotated 1983, USITC Publication 1317; U.S. Federal Register, Vol. 44, No. 241.

Note: Various other tariffs are in effect on cut and stamped mica and on mica manfuacturing.

TABLE 1. CANADA, MICA IMPORTS, 1981-84

	1981		1982		1983		1984P	
	(tonnes)	($000)	(tonnes)	($000)	(tonnes)	($000)	(tonnes)	($000)
							(Jan.-Sept.)	
Unmanufactured, scrap and schist mica								
United States	..	14	..	24	..	52	..	39
India	..	-	..	134	..	-	..	-
Subtotal	..	14	..	158	..	52	..	39
Ground mica								
United States	2 994	735	2 378	590	2 632	680	1 786	487
Subtotal	2 994	735	2 378	590	2 632	680	1 786	487
Mica in blocks, films and sheets								
United States	138	270	481	250	157	191	78	106
India	1	1	1	2	1	6	-	-
Subtotal	139	271	482	252	158	197	78	106
Total sheet and ground mica	3 133	1,007	2 860	842	2 790	877	1 864	593
Fabricated mica nes								
United states	..	2,598	..	2,230	..	1,385	..	1,688
France	..	287	..	420	..	1,118	..	550
Great Britain	..	34	..	88	..	115	..	145
India	..	8	..	18	..	54	..	28
West Germany	..	-	..	-	..	-	..	2
Switzerland	..	5	..	7	..	2	..	-
Hong Kong	..	-	..	3	..	-	..	-
Subtotal	..	2,932	..	2,766	..	2,674	..	2,413
Total unmanufactured, sheet, ground and fabricated mica	..	3,953	..	3,766	..	3,603	..	3,045

Sources: Energy, Mines and Resources Canada; Statistics Canada.
Note: As figures have been rounded off, sums may not correspond to the totals indicated.
P Preliminary; nes - Not elsewhere specified; .. Not available; - Nil.

TABLE 2. MICA CONSUMPTION IN CANADA, 1980-83

	1980	1981	1982	1983
	(tonnes)			
Gypsum products	790	545	1 204	1 722
Paint and varnish	1 678	1 483	1 402	948
Rubber	24	54	30	52
Other products[1]	84	177	109	280
Total	2 576	2 259	2 745	3 002

[1] Includes electrical apparatus, foundries, paper and paper products, floor coverings, plastics and other miscellaneous products.

TABLE 3. WORLD PRODUCTION OF MICA, ALL VARIETIES, 1981-83

	1981	1982	1983	Remarks
	(tonnes)			
United States[1]	120 630	96 140	126 980	Muscovite, flake and scrap
U.S.S.R.[e]	47 160	48 070	48 980	All varieties
India[e]	29 010	21 540	19 050	Muscovite, exports and local consumption
Korean Republic	9 980	20 350	14 400	Scrap, coproduct of kaolin and feldspath
Canada[2]	12 000	11 000	12 000	Phlogopite, flake and ground
France[e]	6 800	6 480	5 990	Muscovite, coproduct of kaolin
Spain	3 520	3 430	3 400	Muscovite, coproduct of kaolin
South Africa	2 390	1 760	2 670	Muscovite, flake
Brazil	1 950	1 080	610	Muscovite, sheet
Argentina	500	280	330	Muscovite, sheet and scrap
Madagascar	380	280	330	Phlogopite, sheet and scrap
Other countries	5 300	2 430	2 540	
World total[3]	240 620	212 910	237 270	

[1] Excluding the production of sericite, estimated at 35 370 tonnes in 1983. [2] Shipments estimate. [3] In addition to these countries, Morocco, Taiwan and Zimbabwe produced almost 3 280 tonnes in 1981; Romania, the People's Republic of China and Pakistan are other mica producers.
[e] Estimate.

TABLE 4. MICA[1] TRADE AND CONSUMPTION IN CANADA 1970, 1975 AND 1979-83

	Imports	Consumption
	(tonnes)	
1970	3 422	2 611
1975	5 111	3 718
1979	3 131	4 498
1980	2 597	2 576
1981	3 133	2 259
1982	2 860	2 745
1983P	2 790	3 002

Sources: Energy, Mines and Resources, Canada; Statistics Canada.
[1] Sheet and ground mica.
P Preliminary.

Mineral Aggregates

D.H. STONEHOUSE

SUMMARY 1983-84

Demand for mineral aggregates is created by activity in the construction industry. During 1983 and 1984 construction expenditures in Canada were about the same as in 1982 in current dollars, which means put in place construction was less than in 1982. Increased building construction in 1983, in particular in the residential sector where housing starts rose to 162,645 units from a 1982 low of only 125,860, were offset by decreased expenditures in engineering categories. Housing starts were greatly reduced through 1984 while expenditures in heavy construction rose slightly. Production of sand, gravel and crushed stone in Canada during both 1983 and 1984 was about 300 million tpy. Although the amount of mineral aggregates imported and exported is quite small, Canada is consistently a net importer of these materials.

Until recently, none of the principal lightweight aggregates (vermiculite, pumice and perlite) was mined in Canada. Imports, mainly from the United States, supplied the requirements for use in both lightweight concrete and gypsum products, for loose insulation applications and for horticulture uses. During 1983, Aurun Mines Ltd. developed a perlite property in the Empire Valley area of British Columbia and in 1984 processed approximately 1 000 t for market trials. Total imports of crude lightweight aggregate materials were greater in 1983 than in 1982 but actual production of the lightweight, expanded or exfoliated material was severely reduced as demand for these commodities in construction tapered off.

The constraints to development of aggregate properties have not lessened. Property owners do not want quarries or gravel pits nearby nor would they like to see the prices increase to compensate for greater hauling distances. An awareness of the importance of mineral aggregates to the construction industry has been heightened by an appreciation of the extent and rate of urban expansion and the realization that already large deposits of aggregate material have been made inaccessible by the growth of towns and cities or by legislation. Surveys to determine the quality and quantity of construction aggregate deposits within easy reach of many rapidly-expanding, major communities in Canada are either planned, in progress or completed. The industry has been hesitant to invest in new plant sites or major new equipment in face of the uncertain economic conditions and the uncertain impact of some new and pending legislation.

Of particular note is the development of the Ontario Government's new Planning Act and the importance given to mineral aggregates policy within the Act. Municipal plans will be required not only to protect existing pits and quarries but to identify and preserve presently untapped aggregate reserves for future development. Concern has been registered from agricultural and environmental interests that aggregates have been given greater priority than they deserve and that touted future shortages are both exaggerated and unfounded.

The fact remains that sand, gravel and stone are non-renewable resources which continue to be vital to the economy. For these reasons mineral aggregate market and resource studies have been proposed as part of mineral development programs under the Economic and Regional Development Agreements (ERDA) arranged between federal and provincial governments.

CANADIAN DEVELOPMENTS

Sand and Gravel

During 1983-84 production of sand and gravel averaged 225 million tpy increasing per capita consumption to over 9 tpy. Average unit values were slightly increased as well to about $2.68 a t.

Sand and gravel deposits are widespread throughout Canada, and large

---

D.H. Stonehouse is with the Mineral Policy Sector, Energy, Mines and Resources, Canada. Telephone (613) 995-9466.

producers have established "permanent" plants as close to major consuming centres as possible. In addition to large aggregate operations usually associated with some other phase of the construction industry such as a ready-mix plant or an asphalt plant, there are many small producers serving localized markets. These are often operated on a seasonal or part-time basis. Many larger operations are short-term, intermittently serving as a supply arm of a heavy construction company, and provide material for a given project. Provincial departments of highways operate regional or divisional quarries to supply roadbed material for new and repair work. Exploitation by such a large number of widely diversified groups not only makes control difficult, it also provides great obstacles to the collection of accurate data concerning both production and consumption of sand, gravel and stone.

Estimates have indicated that available sand and gravel supplies in some regions will be depleted by the 1990s. This could make outlying deposits not only attractive but necessary to the continued operation of the Canadian construction industry in certain areas. Transportation charges represent from 35 to 58 per cent of consumer costs for over 75 per cent of sand and gravel consumption in southern Ontario, where 90 per cent is moved by truck, according to the Ontario Ministry of Natural Resources. Predicted shortages could encourage exploitation of underwater deposits and could make underground mining of crushed stone attractive.

Crushed Stone

The large number of stone-producing operations in Canada precludes describing within this review individual plants or facilities. Many are part-time or seasonal operations, many are operated subsidiary to construction or manufacturing activities by establishments not classified to the stone industry, and some are operated directly by municipal or provincial government departments producing stone for their own direct use. Quarries removing solid rock by drilling, blasting and crushing are not likely to be operated for small, local needs as are gravel pits and are, therefore, usually operated by large companies associated with the construction industry. Depending on costs and availability, crushed stone competes with gravel and crushed gravel as an aggregate in concrete and asphalt, and as railway ballast and road metal. In these applications it is subject to the same physical and chemical testing procedures as the gravel and sand aggregates.

Detailed information about the aggregates industries can be obtained through the individual provincial departments of mines or equivalent. Most provinces have accumulated data relative to occurrences of stone of all types and in many cases have published such studies. The federal government, through the Geological Survey of Canada, has also gathered and published a great number of geological papers pertaining to stone occurrences.

Lightweight Aggregates

Four categories generally used to classify the lightweight aggregates combine elements of source, processing methods and end-use. Natural lightweight aggregates include materials such as pumice, scoria, volcanic cinders and tuff. Manufactured lightweights are bloated or expanded products obtained by heating certain clays, shales and slates. Ultra-lightweights are made from natural mineral ores, such as perlite and vermiculite, which are expanded or exfoliated by the application of heat and used mainly as plaster aggregate or as loose insulation. Fly ash, which is obtained from the combustion of coal and coke and slag, which is obtained from metallurgical processes, are classed as byproduct aggregates.

Perlite: Perlite is a variety of obsidian or glassy volcanic rock that contains 2 to 6 per cent of chemically combined water. When the crushed rock is heated rapidly to a suitable temperature (760°C to 980°C) it expands to between 4 and 20 times its original volume. Expanded material can be manufactured to weigh as little as 30 to 60 kg/m^3, with attention being given to preblending of feed to the kiln and retention time in the kiln.

In Canada, imported perlite is expanded and used mainly by gypsum products manufacturers in plaster products such as wallboard or drywall, and in fibre-perlite roof insulation board, where its value as a lightweight material is augmented by its fire-resistant qualities. It is also used as a loose insulation and as an insulating medium in concrete products. Perlite, vermiculite, and expanded shale and clay are becoming more widely used in agriculture as soil conditioners and fertilizer carriers.

Imports of crude perlite for consumption in Canada are from New Mexico and Colorado deposits, worked by such companies as

Manville Corporation, United States Gypsum Company, United Perlite Corp. and Grefco, Inc.

Aurun Mines Ltd. has begun to produce perlite from a deposit near Empire Valley in British Columbia. During 1984 the company constructed a processing plant near Vancouver. Export markets are being investigated.

Pumice: Pumice is a cellular, glassy lava, the product of explosive volcanism, usually found near geologically-recent or active volcanoes. It is normally found as a loosely compacted mass composed of pieces ranging in size from large lumps to small particles. It is not the lightest of the lightweight aggregates, but when utilized as a concrete aggregate, particularly for the manufacture of concrete blocks, it exhibits strength, density and insulating values that have made it a preferred material.

In Canada, a number of concrete products manufacturers use pumice imported from Greece or from the northwestern United States, mainly in the manufacture of concrete blocks. A major use for pumice, as yet unexplored in Canada, has been in highway construction, where lightweight aggregate surfaces have been shown to have exceptional skid resistance.

Pumicite, distinguished from pumice by its finer size range (usually minus 100 mesh), is used in concretes mainly for its pozzolanic qualities. (A pozzolan is a siliceous material possessing no cementitious qualities until finely ground, in which form it will react with calcium hydroxide in the presence of moisture to form insoluble calcium silicates.)

Extensive beds of pumicite have been noted in Saskatchewan and British Columbia.

Vermiculite: The term vermiculite refers to a group of micaceous minerals, hydrous magnesium-aluminum silicates, that exhibit a characteristic lamellar structure and expand or exfoliate greatly upon being heated rapidly. Mining is normally by open-pit methods, and beneficiation techniques include the use of hammer mills, rod mills, classifiers, screens, dryers and cyclones. Exfoliating is done in oil- or gas-fired, vertical or inclined furnaces, usually close to the consuming facility to obviate the higher costs associated with shipping the much-bulkier expanded product. Required temperatures can vary from 1 100°C to 1 650°C depending on the type of furnace in use. A controlled time and temperature relation is critical in order to produce a product of minimum bulk density and good quality.

The expansion process has been improved technologically to enable production of various grades of expanded vermiculite as required. The uses to which the product is put depend on its low thermal conductivity, its fire-resistance and, more recently, on its lightweight qualities.

Canadian consumption is mainly as loose insulating material, with smaller amounts being used as aggregate in the manufacture of insulating plaster and concrete. The energy situation will undoubtedly result in continued increases in domestic fuel costs, and greater use of insulation in both new construction and older buildings will continue to tax the production capability of manufacturers for some time.

The major producer of vermiculite is the United States. The principal company supplying Canada's imports is W.R. Grace and Company, from operations at Libby, Montana and from the Enoree region of South Carolina. Canada also imports crude vermiculite from the Republic of South Africa, where Palabora Mining Co. Ltd. is the major producer. Minor amounts of vermiculite are produced in Argentina, Brazil, India, Kenya and Tanzania.

Vermiculite occurrences have been reported in British Columbia, and deposits near both Perth and Peterborough in Ontario have been investigated but, as yet, no commercial deposits have been developed in Canada.

Clay, shale and slag: Common clays and shale are used throughout Canada as raw material for the manufacture of lightweight aggregates. Although the Canadian industry began in the 1920s in Ontario, it did not evolve significantly until the 1950s when it grew in support of demands from the construction industry. The raw materials are usually quarried adjacent to the plant sites at which they are expanded. Clays receive little beneficiation other than drying before being introduced to the kiln in which they are heated. Shales are crushed and screened before burning.

In steelmaking, iron ore, coke and limestone flux are melted in a furnace. When the metallurgical process is completed, lime has combined with the silicates and aluminates of the ore and coke and formed a nonmetallic product (slag) which can be subjected to controlled cooling from the molten state to yield a porous, glassymaterial. Slag has many applications in the construction industry. The statistics relative to expanded slag production are included in those of clay and shale.

Although Canada does not produce large amounts of fly ash, the technology of fly ash processing and utilization is well advanced. The largest single use for fly ash is as a cementitious material, in which application its pozzolanic qualities are utilized. Use of fly ash as a lightweight aggregate could become increasingly important. Ontario Hydro produces over 400 000 tpy of fly ash from three coal-fired stations. Experimentation continues toward successful utilization of this material.

PRICES

There is no standard price for sand, gravel and crushed stone. In addition to supply-demand factors, prices are determined regionally, or even locally, by production and transportation costs, by the degree of processing required for a given end use and by the quantity of material required for a particular project. Increased land values, reduction of reserves and added rehabilitation expenditures should result in higher prices.

Prices for graded, washed and crushed sand, gravel and crushed stone will show a slow but steady increase, based on greater property costs, more sophisticated operating techniques and equipment, pollution and environmental considerations, and higher labour and transportation costs.

USES

The principal uses for sand and gravel are in highway construction and as concrete aggregate. Individual home construction triggers the need for about 300 t of aggregate per unit while apartment construction requires only about 50 t per unit, according to an Ontario Ministry of Natural Resources study.

The construction industry utilizes 95 per cent of total stone output as crushed stone mainly as an aggregate in concrete and asphalt, in highway and railway construction and as heavy riprap for facing wharves and breakwaters. Specifications vary greatly, depending on the intended use, and many tests are required to determine the acceptability of aggregates for certain applications. Particle size distribution of aggregates, as assessed by grading tests or sieve analysis, affects the uniformity and workability of a concrete mix as well as the strength of the concrete, the density and strength of an asphalt mix, and the durability, strength and stability of the compacted mass when aggregates are used as fill or base-course material. Of importance also are tests to determine the presence of organic impurities or other deleterious material, the resistance of the aggregate to abrasion and to freeze-thaw cycles, the effects of thermal expansion, absorption, porosity, reactivity with associated materials and surface texture.

The use of sand and gravel as backfill in mines continues, along with increasing use of cement and mill tailings for this purpose. Abrasive sands, glass sand, foundry sands and filter sands are also produced.

The use of lightweight concrete in commercial and institutional projects has facilitated the construction of taller building and the use of longer clear spans in bridges and buildings. Additional advantages from the use of lightweight aggregates lie in the fact that they supply thermal and acoustical insulation, fire resistance, good freeze-thaw resistance, low water absorption and a degree of toughness to the concrete product. Disadvantages stem from the fact that in production of both manufactured and ultra-lightweight aggregates heat processing is required. As the cost of fuel increases, the competitiveness of these types will be reduced unless the insulation values more than offset the heat units consumed in processing.

All types of lightweight aggregates are used in Canada, but only expanded clays, shale and slag are produced from materials of domestic origin. Vermiculite is imported mainly from Montana, although a small amount is brought in from the Republic of South Africa. Perlite is imported mainly from New Mexico and Colorado, and pumice is imported from Oregon and Greece. Most processed lightweight aggregate is utilized in the construction industry, either as loose insulating material or as aggregate in the manufacture of lightweight concrete units. The scope of such applications has not yet been fully investigated.

Any lightweight material with acceptable physical and chemical characteristics could substitute for the mineral commodities generally used. The most significant substitute for vermiculite, for instance, is styrofoam or polyurethane, which offers insulating value and comparable strength. However, these materials are petroleum-based and higher fuel prices could limit their use. Mineral wool is a competitive insulation material but its manufacture requires a pyro-processing stage, as does the production of perlite and vermiculite. Transportation costs for high-bulk, lightweight materials are high; those materials, such as perlite and vermiculite, that can be transported to a consuming centre prior to expansion, have obvious advantages.

There are as yet no Canadian Standards Association (CSA) specifications for the lightweight aggregates. Production and application are based on the American Society for Testing and Materials (ASTM) designations as follows: ASTM Designations C 332-66 - Lightweight Aggregates for Insulating Concrete; C 330-75a - Lightweight Aggregates for Structural Concrete; and C331-69 - Lightweight Aggregates for Concrete Masonry Units.

OUTLOOK

Urban expansion has greatly increased demand for sand and gravel in support of major construction. Paradoxically, urban spread has not only tended to overrun operating pits and quarries, but has extended at times to areas containing mineral deposits, thereby precluding the use of these resources. Further complications have arisen in recent years as society has become increasingly aware of environmental problems and the need for planned land utilization. Municipal and regional zoning must be designed to determine and regulate the optimum utilization of land, but must not be designed to provide less than optimum resources utilization. Industry must locate its plants so as to minimize any adverse effects on the environment from their operations. Also, provision must be made for rehabilitation of pit and quarry sites in order to ensure the best sequential land use. The frequency with which small quarries and pits materialize to supply short-lived, local demands, leaving unsightly properties, has prompted action by municipal and provincial governments to control or to prohibit such activity.

Ideally, the exploitation of sand, gravel and stone deposits should be done as part of the total land-use planning package, such that excavations are designed to conform with a master plan of development and even to create new land forms. Inventories indicating the potential available reserves of sand, gravel and stone should be prerequisite to legislation regulating land use. Surveys to locate such resources are being carried out in many provinces in order to optimize their use and to choose the best possible distribution routes to consuming centres. It should be observed that controls and zoning can reduce reserves of these resources significantly.

On average, total aggregate consumption will rise in line with population increases, housing requirements and construction in general. Sand and gravel consumption will continue in competition with crushed stone and, in some applications, with lightweight aggregates. New reserves must be located, assessed and made part of any community development planning or regional zoning, with optimum land and resource utilization in mind.

TABLE 1. CANADA, TOTAL PRODUCTION OF STONE, 1982-84

	1982		1983		1984P	
	(000 t)	($ 000)	(000 t)	($ 000)	(000 t)	($ 000)
By province						
Newfoundland	357	1,763	279	1,431	415	1,608
Nova Scotia	679	4,638	1 296	7,784	1 510	9,400
New Brunswick	2 261	11,556	2 087	11,310	2 005	10,940
Quebec	25 060	106,989	27 303	121,154	28 237	124,581
Ontario	23 582	100,278	27 843	122,272	29 500	131,335
Manitoba	2 345	11,670	1 137	5,452	1 675	9,300
Alberta	264	3,161	286	3,457	300	3,275
British Columbia	4 310	21,926	4 915	27,084	4 885	27,500
Northwest Terrtories	323	1,268	2 409	14,601	2 420	15,750
Canada	59 181	263,249	67 555	314,545	71 047	333,689
By use[1]						
Building stone						
Rough	230	4,828	..	..	..	..
Monumental and ornamental stone	38	4,002	..	..	..	..
Other (flagstone, curbstone, paving blocks, etc.)	26	1,027	..	..	..	..
Chemical and metallurgical						
Cement plants, foreign	598	1,461	..	..	..	..
Lining, open-hearth furnaces	38	141	..	..	..	..
Flux in iron and steel furnaces	742	2,861	..	..	..	..
Flux in nonferrous smelters	114	1,126	..	..	..	..
Glass factories	169	2,271	..	..	..	..
Lime kilns, foreign	512	1,903	..	..	..	..
Pulp and paper mills	295	2,706	..	..	..	..
Sugar refineries	108	586	..	..	..	..
Other chemical uses	137	2,840	..	..	..	..
Pulverized stone						
Whiting (substitute)	71	2,863	..	..	..	..
Asphalt filler	41	238	..	..	..	..
Dusting, coal mines	7	171	..	..	..	..
Agricultural purposes and fertilizer plants	1 037	10,562	..	..	..	..
Other uses	687	2,153	..	..	..	..
Crushed stone for						
Manufacture of artificial stone	7	154	..	..	..	..
Roofing granules	253	16,776	..	..	..	..
Poultry grit	28	721	..	..	..	..
Stucco dash	15	993	..	..	..	..
Terrazzo chips	3	184	..	..	..	..
Rock wool	-	-	..	..	..	..
Rubble and riprap	1 730	6,421	..	..	..	..
Concrete aggregate	4 671	17,571	..	..	..	..
Asphalt aggregate	4 540	17,766	..	..	..	..
Road metal	17 997	62,795	..	..	..	..
Railroad ballast	2 626	12,823	..	..	..	..
Other uses	22 461	80,318	..	..	..	..
Total	59 181	258,261	..	..	..	..

P Preliminary; .. Not available; - Nil.
[1] The 1982 value of production includes companies transportation costs not applicable in the by use category.

TABLE 2. CANADA, PRODUCTION OF SAND AND GRAVEL BY PROVINCE, 1982-84

	1982		1983		1984P	
	(000 t)	($000)	(000 t)	($000)	(000 t)	($000)
Newfoundland	2 839	9,317	4 057	18,389	3 715	16,150
Prince Edward Island	1 136	1,774	1 173	726	1 156	890
Nova Scotia	5 309	17,302	8 136	23,076	7 600	21,600
New Brunswick	6 206	8,359	5 668	10,830	5 410	10,275
Quebec	41 932	66,060	37 006	71,167	30 518	59,510
Ontario	62 256	156,525	68 316	174,933	65 300	166,000
Manitoba	10 284	28,054	9 909	26,537	10 950	27,500
Saskatchewan	8 512	21,001	7 999	21,014	9 500	25,500
Alberta	46 092	129,664	43 789	126,354	44 000	127,600
British Columbia	24 618	74,520	40 769	112,456	36 000	100,450
Yukon and Northwest Territories	7 088	..	6 385	33,917	6 500	35,050
Canada	216 274	554,608	233 408	619,400	220 649	590,525

P Preliminary; .. Not available.

TABLE 3. AVAILABLE DATA ON CONSUMPTION OF SAND AND GRAVEL, BY PROVINCE, 1981 AND 1982

		Atlantic Provinces	Quebec	Ontario	Western Provinces[1]	Canada
				(000 tonnes)		
Roads	1981	12 631	34 944	38 110	47 599	133 284
	1982	11 525	26 430	36 292	62 441	136 688
Concrete aggregate	1981	1 018	4 268	11 688	10 538	27 512
	1982	1 029	3 037	9 265	9 106	22 437
Asphalt aggregate	1981	1 446	3 020	3 653	7 492	15 611
	1982	1 479	3 462	4 016	6 560	15 517
Railroad ballast	1981	199	348	82	2 299	2 928
	1982	168	152	777	1 699	2 796
Mortar sand	1981	18	409	1 332	426	2 185
	1982	37	307	865	354	1 563
Backfill for mines	1981	19	204	1 404	152	1 779
	1982	1	601	557	23	1 182
Other fill	1981	828	24 884	8 160	7 468	41 340
	1982	931	7 719	7 289	13 388	29 327
Other uses	1981	167	6 652	1 232	1 477	9 528
	1982	294	224	1 312	993	2 823
Total sand and gravel	1981	16 326	74 729	65 661	77 451	234 167
	1982	15 464	41 932	60 373	94 564	212 333

[1] As of 1982 the western provinces include the Yukon and Northwest Territories.

TABLE 4. CANADA, EXPORTS AND IMPORTS OF SAND AND GRAVEL AND CRUSHED STONE, 1981-84

	1981r		1982r		1983		Jan.-Sept. 1984	
	(tonnes)	($)	(tonnes)	($)	(tonnes)	($)	(tonnes)	($)
Exports								
Sand and gravel								
United States	239 641	649,000	168 179	624,000	83 931	328,000	82 475	418,000
Bermuda	78 888	262,000	16	2,000	11 497	80,000	-	-
Indonesia	5	25,000	-	-	-	-	-	-
St. Pierre and Miquelon	37	11,000	-	-	19	2,000	19	2,000
France	49	4,000	335	34,000	49	4,000	575	10,000
Other countries	13	2,000	162	25,000	137	18,000	274	47,000
Total	318 633	953,000	168 692	685,000	95 633	432,000	83 343	477,000
Crushed limestone								
United States	1 758 290	6,007,000	1 516 889	8,475,000	1 390 795	8,375,000	912 248	5,137,000
Other countries	-	-	602	8,000	-	-	16	2,000
Total	1 758 290	6,007,000	1 517 491	8,483,000	1 390 795	8,375,000	912 264	5,139,000
Imports								
Sand and gravel, nes								
United States	1 439 686	6,068,000	1 172 707	5,248,000	878 545	4,362,000	970 784	4,536,000
West Germany	7 178	16,000	2 219	5,000	36	6,000	715	3,000
Sweden	-	-	4 341	10,000	-	-	-	-
Other countries	-	-	18	3,000	33	4,000	13	2,000
Total	1 446 864	6,084,000	1 179 285	5,266,000	878 614	4,372,000	971 512	4,541,000
Crushed limestone								
United States	2 526 469	14,769,000	1 485 428	9,003,000	1 799 861	8,447,000	1 393 239	6,453,000
Other countries	394	12,000	-	-	-	-	-	-
Total	2 526 863	14,781,000	1 485 428	9,003,000	1 799 861	8,447,000	1 393 239	6,453,000
Crushed stone, nes								
United States	33 108	1,266,000	71 313	1,239,000	43 889	1,092,000	37 896	1,034,000
Sweden	342	66,000	-	-	-	-	-	-
Other countries	676	49,000	67	5,000	97	13,000	285	28,000
Total	34 126	1,381,000	71 380	1,244,000	43 986	1,105,000	38 181	1,062,000

Source: Statistics Canada.
r Revised; - Nil; nes Not elsewhere specified.

TABLE 5. LIGHTWEIGHT AGGREGATE PLANTS IN CANADA 1983

Company	Location	Commodity	Remarks
Atlantic Provinces			
Annapolis Valley Peat Moss Co. Ltd.	Berwick, N.S.	Perlite, Vermiculite	Processed mainly for use in horticulture
Avon Aggregates Ltd.	Minto, N.B.	Expanded Shale	Processed for concrete products industry.
Quebec			
Masonite Canada Inc.	Gatineau	Perlite	Processed for use in ceiling tile manufacture.
Domtar Inc.	Montreal	Perlite, Vermiculite	Processed material purchased for use in gypsum plaster and wallboard at all company plants.
F. Hyde & Company, Limited	Montreal	Vermiculite	Processed for use in horticulture and as loose insulation.
Miron Inc.	Montreal	Pumice	Purchased for concrete block manufacture.
Perlite Industries Inc.	Ville St. Pierre	Perlite	Processed for use in horticulture and as industrial filler.
V.I.L. Vermiculite Inc.	Lachine	Vermiculite	Processed for use in horticulture and as loose insulation.
Ontario			
CGC Inc.	Hagersville	Perlite	Processed for use in gypsum plaster.
National Slag Limited	Hamilton	Slag	Used in concrete blocks and as slag cement.
V.I.L. Vermiculite Inc.	Rexdale	Vermiculite	Processed for use in horticulture and as loose insulation.
W.R. Grace & Co. of Canada Ltd.	St. Thomas	Vermiculite	Vermiculite processed for use in horticulture and as loose insulation.
	Ajax	Vermiculite, Perlite	Perlite processed for use in gypsum plaster and in horticulture.
Prairie Provinces			
Apex Aggregate	Saskatoon, Sask.	Expanded clay	Processed for concrete block manufacture.
Cindercrete Products Limited	Regina, Sask.	Expanded clay	Processed for concrete products industry.
Consolidated Concrete Limited	Calgary, Alta.	Expanded shale	Processed for concrete products industry.
	Edmonton, Alta.	Expanded clay	Processed for concrete products industry.
Genstar Corporation, Edcon Block Division	Edmonton, Alta.	Expanded clay	Processed for concrete block manufacture.
Kildonan Concrete Products Ltd.	Winnipeg, Man.	Expanded clay	Processed for concrete products industry.
	Winnipeg, Man.	Vermiculite, Perlite	Perlite processed for use in gypsum plaster and in horticulture.
W.R. Grace & Co. of Canada, Ltd.	Edmonton, Alta.	Vermiculite, Perlite	Vermiculite processed for use in horticulture and as loose insulation.
British Columbia			
Ocean Construction Supplies Ltd.	Vancouver	Pumice	Purchased for concrete block manufacture.

TABLE 6. CANADA, IMPORTED RAW MATERIALS PURCHASED, 1982 AND 1983

	1982		1983	
	(tonnes)	($)	(tonnes)	($)
Pumice, perlite and vermiculite[1]	40 617	5,961,961r	47 160	5,267,013

Source: Company data.
[1] Combined to avoid disclosing confidential company data.
r Revised.

TABLE 8. CANADA, CONSUMPTION OF SLAG, PERCENTAGE BY USE, 1981-83

Use	1981	1982	1983
Concrete block manufacture	46.0	38.0	27.0
Ready-mix concrete	2.0	4.0	2.0
Loose insulation	1.0	1.0	1.0
Slag cement	51.0	57.0	70.0

Source: Company data.

TABLE 10. CANADA, CONSUMPTION OF EXPANDED PERLITE, PERCENTAGE BY USE 1981-83

Use	1981	1982	1983
Insulation			
in gypsum products	11.3	20.6	21.9
in other construction materials	46.9	34.9	28.0
Horticulture and agriculture	23.9	33.7	34.6
Loose insulation and miscellaneous uses	17.9	10.8	15.5

Source: Company data.

TABLE 9. CANADA, CONSUMPTION OF EXPANDED CLAY AND SHALE, PERCENTAGE BY USE, 1981-83

Use	1981	1982	1983
Concrete block manufacture	76.7	78.7	80.6
Precast concrete manufacture	6.5	11.5	7.8
Ready-mix concrete	14.6	4.3	6.5
Horticulture and miscellaneous uses	2.2	5.5	5.1

Source: Company data.

TABLE 11. CANADA, CONSUMPTION OF EXFOLIATED VERMICULITE, PERCENTAGE BY USE 1981-83

Use	1981	1982	1983
Insulation			
loose	55.2	45.8	30.2
in concrete and concrete products	8.8	0.5	0.4
in gypsum products	3.0	1.7	0.5
Horticulture	23.3	48.2	46.3
Miscellaneous uses	9.7	3.8	22.6

Source: Company data.

TABLE 7. CANADA, PRODUCTION OF LIGHTWEIGHT AGGREGATES, 1982 AND 1983

	1982		1983	
	(m^3)	($)	(m^3)	($)
From domestic raw materials				
Expanded clay, shale and slag	260 247	5,832,343	204 264	5,049,810
From imported crude materials				
Expanded perlite and exfoliated vermiculite[1]	395 540	12,991,301	216 266	10,796,688
Total	655 787	18,823,644	420 530	15,846,498

Source: Company data.
[1] Combined to avoid disclosing confidential company data.

TABLE 12. CANADA ROCK-, MINERAL- AND GLASS-WOOL PRODUCERS, 1984

Company	Location	Remarks
Atlantic Provinces		
Fiberglas Canada Inc.	Moncton, N.B.	New in 1975. Capacity 15 000 tpy. Closed March 31, 1984.
Quebec		
Fiberglas Canada Inc.	Candiac	Expanded in 1977.
Manville Canada Inc.	Brossard	15 000 tpy capacity.
Ontario		
Fiberglas Canada Inc.	Sarnia	Expanded in 1978. New electric furnace is largest of kind.
	Toronto	New plant in 1979.
CGC Inc.	Mount Dennis (Toronto)	Using slag from Hamilton.
Holmes Insulations Inc.	Sarnia	Slag - Detroit.
Bishop Building Materials Ltd.	Toronto	Slag - Hamilton.
Graham Fiber Glass Limited	Erin	New by 1979. Capacity 10 000 tpy.
Roxul Company	Milton	A division of Standard Industries Ltd.
Ottawa Fibre Industries Ltd.	Ottawa	
Prairie Provinces		
Fiberglas Canada Inc.	Clover Bar, Alta. (Edmonton)	Expanded in 1977.
Manville Canada Inc.	Innisfail, Alta.	New in 1978. Capacity 6 000 tpm. New energy-efficient mechanical fiberizing technology in use.
Alberta Rockwool Corporation	Calgary, Alta.	
British Columbia		
Fiberglas Canada Inc.	Mission	New in 1980. Capacity 45 000 tpy.
Pacific Enercon Inc.	Grand Forks	

TABLE 13. CANADA, VALUE OF CONSTRUCTION[1] BY PROVINCE, 1982-84

	1982			1983			1984		
	Building Construction	Engineering Construction	Total	Building Construction	Engineering Construction	Total	Building Construction	Engineering Construction	Total
				($000)					
Newfoundland	414,429	750,073	1,164,502	496,177	920,309	1,416,486	529,042	904,131	1,433,173
Nova Scotia	681,430	884,462	1,565,892	850,097	1,113,145	1,963,242	935,167	1,263,679	2,198,846
New Brunswick	619,611	462,089	1,081,700	749,843	414,249	1,164,092	866,945	503,785	1,370,730
Prince Edward Island	86,981	72,006	158,987	106,406	70,694	177,100	117,272	79,046	196,318
Quebec	5,547,556	4,672,040	10,219,596	6,693,708	4,388,346	11,082,054	7,183,496	4,352,134	11,535,630
Ontario	8,897,137	5,510,574	14,407,711	10,015,802	4,819,861	14,835,663	10,498,275	5,031,231	15,529,506
Manitoba	764,362	657,850	1,422,212	986,418	656,087	1,642,505	1,083,361	698,296	1,781,657
Saskatchewan	1,165,189	1,343,933	2,509,122	1,451,012	1,413,370	2,864,382	1,410,011	1,515,541	2,925,552
Alberta	6,053,165	8,349,406	14,402,571	4,761,621	7,044,529	11,806,150	3,920,440	7,281,719	11,202,159
British Columbia, Yukon and Northwest Territories	4,613,640	4,519,456	9,133,096	4,488,816	4,657,297	9,146,113	4,574,138	4,223,386	8,797,524
Canada	28,843,500	27,221,889	56,065,389	30,599,900	25,497,887	56,097,787	31,118,147	25,852,948	56,971,095

Source: Statistics Canada.
[1] Actual expenditures 1982, preliminary actual 1983, intentions 1984.

TABLE 14. CANADA, VALUE OF CONSTRUCTION[1] BY TYPE, 1982-84

	1982	1983	1984
	($ millions)		
Building Construction			
Residential	13,581	16,683	17,240
Industrial	3,044	2,502	2,739
Commercial	7,064	6,228	5,817
Institutional	3,092	3,198	3,183
Other building	2,062	1,989	2,139
Total	28,843	30,600	31,118
Engineering Construction			
Marine	480	404	414
Highways, airport runways	4,310	4,270	4,328
Waterworks, sewage systems	2,244	2,402	2,391
Dams, irrigation	314	295	306
Electric power	4,866	4,673	3,827
Railway, telephones	2,390	2,531	2,811
Gas and oil facilities	9,706	8,115	9,141
Other engineering	2,912	2,808	2,635
Total	27,222	25,498	25,853
Total construction	56,065	56,098	56,971

Source: Statistics Canada.
[1] Actual expenditures 1982, preliminary actual 1983, intentions 1984.

Molybdenum

D.G. FONG

SUMMARY

Western world molybdenum production in 1984 increased by 67 per cent to 78 000 t while consumption advanced 14 per cent to 62 000 t. The supply increase was due to the resumption of operations at the major primary producers and the start-up at two new mines all located in the United States. The increase in consumption was especially significant in western Europe and the United States. However, this increase was more than offset by the expanded output, resulting in a large increase in inventories. The molybdenum market showed signs of stability following a price recovery during the early part of 1983. However, prices fell drastically during the last quarter of 1984 as a result of intense competitive selling activity by major producers.

Canada's 1984 molybdenum production, estimated at 9 070 t, remained unchanged while shipments were up eight per cent to about 10 965 t. With all three primary producers remaining on prolonged shutdown and two byproduct producers on intermittent operation, Canada's molybdenum output in 1984 represented only 38 per cent of capacity utilization.

CANADIAN DEVELOPMENTS

Noranda Mines Inc. closed its Boss Mountain mine in British Columbia, on February 15, 1983. The mine, which had undergone a mine- and mill-expansion and was operating at 50 per cent capacity prior to the shutdown, was maintained on a standby basis. The company reopened its Gaspé, Quebec mine in September 1984 after an extended period of shutdown. The latter had been closed since June 20, 1982.

Brenda Mines Ltd., a Noranda subsidiary, reopened its Peachland, British Columbia mining operation at the end of May 1984 after an eight-month shutdown. Following a brief operation, the mine was again closed in December because of a sharp decline in metal prices and a buildup of molybdenum inventories. During the summer months of 1984, Brenda constructed additional tailing dams, channels for the diversion of runoff water, and an evaporation system to secure the area from excessive water runoff. At the end of 1983, ore reserves at Brenda were 99 million t averaging 0.148 per cent copper (Cu) and 0.032 per cent molybdenum (Mo).

Teck Corporation closed indefinitely its Highmont copper-molybdenum mine in British Columbia on October 19, 1984. Highmont's production was sold on long-term contracts, which provided for customer price support payments. Its floor price for molybdenum contained in concentrate was $US 7.50 per lb. Payments were also received as nonrecurse project loans that will be repayable out of future profits from the Highmont mine.

The Highmont mine is owned 50 per cent by Teck Corporation, 30 per cent by Redclay Holdings Limited and 20 per cent by Metallgesellschaft Canada Limited. Early in 1983, a change in mining from the West Pit to the larger East Pit resulted in an increase in copper output and a decline in molybdenum. The production of molybdenum during 1983 was 1 542 t, a reduction of 20 per cent from the previous year.

Teck also has a 21.7 per cent interest in the neighbouring Lornex Mining Corporation Ltd. However, the majority share (68.1 per cent) is held by Rio Algom Limited. Lornex operates Canada's largest porphyry copper concentrator. Production of molybdenum during 1984 estimated at 3 400 t, remained at a high level, as in 1983. The record levels of output during the past two years were due to higher tonnage throughputs and better mill head grades.

Placer Development Limited kept its Endako Mine in British Columbia closed throughout 1983 and 1984. The mine has

---

D.G. Fong is with the Mineral Policy Sector, Energy, Mines and Resources Canada. Telephone (613) 995-9466.

been maintained on standby since its closure in June 1982. Placer was considering re-opening the mine because of declining inventories. However, no decision had been reached by the end of 1984. The company has been selling molybdenum from inventories and byproduct molybdenum production at Gibraltar mine, another of its subsidiaries in British Columbia. Placer was planning to begin production of high purity molybdenum oxide in 1985 at its Equity Silver mines near Houston, British Columbia, in addition to operating its own lubricant plant at the Endako mine site.

Amax of Canada Limited, a subsidiary of AMAX Inc. in the United States, placed the Kitsault Mine in British Columbia on an extended care and maintenance basis in July 1983 as a result of continuing weakness in the molybdenum market. Operations at Kitsault were temporarily suspended in November 1982 and it has been closed since that date. At capacity, the mine could produce 4 080 to 4 536 tpy of molybdenum.

In October 1984, Billiton Canada Ltd. reduced production by approximately 50 per cent at its Mount Pleasant tungsten-molybdenum mine. This New Brunswick mine, owned jointly by Brunswick Tin Mines Limited and Billiton, commenced commercial production at the end of 1983 and was to produce 1 600 tpy of tungsten and 350 tpy of molybdenum, both in concentrates. However, the mine had not reported any molybdenum concentrate production by the end of 1984.

WORLD DEVELOPMENTS

Consumption in the western world declined 3 per cent in 1983 to 54 400 t of molybdenum whereas production dropped 42 per cent to 46 705 t. The sharp decline in output was due to extended mine shutdowns and production cutbacks at both the primary and byproduct producers in Canada and the United States. In 1984, however, western world production was up to an estimated 78 000 t while consumption increased to 62 000 t. Shipments to eastern bloc countries, estimated at 10 000 t, remained at about the same level as in 1983.

In the United States, the consumption of molybdenum rose 20 per cent in 1984 and demand was firm in all major market areas, especially the stainless steels. A recovery in the stainless steel industry began to improve in 1983. Also, the demand in western Europe improved significantly during 1984, especially in Sweden, West Germany and the United Kingdom. In the Japanese market, however, increases in demand for molybdenum were much more moderate.

In spite of increased consumption in 1984, the large inventory in the western world continued to increase, a setback from 1983. This inventory increase was due to the resumption of production in 1984 at the major primary producers, start-up at two new mines, and increased output at some of the byproduct producers.

AMAX Inc. of the United States restarted production at the Henderson mine in January and the Climax mine in April. Both of the AMAX mines, each with a rated capacity of 22 680 tpy of molybdenum had been closed since October 2, 1982. During 1984, both mines were operating below capacity; the Henderson mine operated two production lines and the Climax mine only one, with ores derived from underground. The open-pit at Climax remained closed due to high production costs.

Mining at Molycorp, Inc.'s Goat Hill mine began in October 1983. This new underground mine, brought on-stream at a cost of $US 250 million, is located on the same property as the Questa open-pit which was closed in 1982 due to ore exhaustion. The Goat Hill mine has a capacity to produce about 9 070 tpy of molybdenum contained in concentrate, which is roasted at the company's Washington, Pennsylvania, plant.

Amoco Minerals Company began commercial production at its Thompson Creek mine, Idaho, in early-1984. The new 25 000 tpd open-pit mine, at a cost of $US 375 million, has an annual production capacity in the range of 8 160 to 9 070 t of molybdenum, but it was operating at half capacity during the first year of operation. Thompson Creek has a reserve of 193 million t at a grade of 0.187 per cent molybdenite (MoS_2).

Duval Corporation operated its Sierrita, Arizona mine in 1983 at a low production rate and kept the Mineral Park and Esperanza mines idled. In late-1984, however, the company, in an effort to reduce unit production costs, raised its mining rate at Sierrita to over 100 000 tpy of ore from 85 000 tpd, and restarted the crushing plant at the Esperanza property. The increased ore output from Sierrita was hauled to Esperanza to be crushed and then returned to Sierrita for concentration.

Kennecott Minerals Company cut back production by 13 per cent at its Utah Copper mine in Bingham during the first half of 1984. In July the production rate was further reduced to one-third of capacity as a result of continued deterioration of the copper market. Molybdenum production in 1984 was reduced to about 2 087 tpy as compared with a normal level of output of 5 000 tpy.

At year end 1984, Kennecott and Anaconda Minerals Corporation signed a letter of intent to jointly run the Utah operations. Under the agreement, Kennecott would get 96 per cent of the output from its Bingham mine and the adjacent Carr Fork mine, which is owned by Anaconda. However, the arrangement must be approved by the U.S. Justice Department.

In 1983, United States Borax & Chemical Corporation continued development work at the Quartz Hill deposit near Ketchikan, Alaska. Work was completed on metallurgical tests and a 16 km access road from the mine to tidewater. The company was planning to bring the deposit into production at a rate of 54 500 tpd of ore, or about 18 000 tpy of molybdenum. However, excessive capacity world-wide prompted U.S. Borax to announce an indefinite delay in the project.

Molybdenum output from Corporation Nacional del Cobre de Chile (CODELCO) in 1984 was up 8 per cent to about 16 300 t. Production in 1983-84 was significantly lower than the record 20 000 t level CODELCO achieved in 1982. Much of CODELCO's molybdenum was derived from the Chuquicamata mine, the largest copper mine in the world. At Chuquicamata, the start-up of a molybdenum roaster in 1982 resulted in less concentrate available for toll roasting elsewhere in the western world.

USES

Molybdenum is used in a wide range of products as an alloy additive, a chemical compound, a pure metal and as lubricants. Approximately 90 per cent of all molybdenum consumed in the western world is used in metallurgical applications including steel, cast iron, and special alloys. The remaining 10 per cent is used in non-metallurgical applications such as chemicals, catalysts and lubricants.

As a refractory additive in steel, molybdenum imparts hardenability, strength, toughness and resistance to corrosion and abrasion. Tool steels, stainless steels, high-strength steels, heat resisting steels and a wide range of alloy steels are important consumers of molybdenum. Depending on type and specification, molybdenum is added in amounts ranging from less than 0.1 per cent to nearly 10 per cent. Molybdenum can be added as a sole agent but, more often, it is used in combination with other additive metals.

Molybdenum is an important alloying element in most types of tool steels. Among the tool steel additives, molybdenum and tungsten both promote red hardness and increase wear resistance in high speed steels. The performance of these steels is proportional to the percentage of the elements. However, molybdenum produces more carbide than tungsten per unit weight added, and thus can replace tungsten at a rate of almost one to two. For some hot work tool steels and high speed steels, the molybdenum content can be as high as 9 to 10 per cent.

Additions of molybdenum to austenitic and ferritic stainless steels enhance the resistance to corrosive acids and seawater. These steels are finding increasing use in heat exchangers for severe chemical environments, seawater condenser tubings, caustic evaporators, and heat resisting steels operating under stress and high temperatures.

In high-strength, low-alloy steels, the addition of molybdenum increases the yield and tensile strength, and improves toughness and weldability. Steels with these properties are especially useful in structural applications and in Arctic-grade large-diameter pipelines. The consumption intensity of molybdenum in pipeline steels has declined, especially in Japan and western Europe where pipeline manufacturers have switched to non-molybdenum steels, even for the Arctic-grade pipelines. This increase in substitution to other ferroalloy additives was brought about mainly by the high prices and short supply of molybdenum in the late 1970s.

Molybdenum is an important constituent of many high performance alloys that are extremely resistant to heat, corrosion and wear. These alloys are used extensively in aerospace engineering, chemical processing plants, and high temperature furnace and foundry parts.

Molybdenum compounds are used as catalysts in petroleum refining and chemical processing industries. Molybdenum orange, an important molybdenum pigment, is used in printing inks, dyes and corrosion resistant primer. Pure molybdenum disulphide is an excellent dry lubricant and is used as an oil additive. The lamellar structure of the molybdenum disulphide helps to reduce friction and thus prolongs engine life. In recent years, non-metallurgical applications have been experiencing a faster growth rate than other uses.

PRICES

Molybdenum prices on the spot market improved from $US 5.40 a kg of oxide at the begining of 1983 to a high of $US 7.94-9.81 at the end of the first quarter. The sharp price increase was mainly due to a scarcity of molybdenum concentrates in the spot market. Meanwhile, North American producers raised their prices on oxide to $US 11.02 a kg. Subsequently, prices began to weaken and producer prices for oxide at the end of the year were being quoted at $US 9.26-9.81 while the spot price declined to $US 8.27-8.93 a kg.

By year-end 1983, major molybdenum producers including AMAX and Placer Development discontinued publication of producer list prices. Accordingly, the dealer price became the only published reference price in the molybdenum market.

The molybdenum market showed signs of stability during the first three quarters of 1984. Dealer prices for oxide hovered around the $US 8.05-8.93 level. However, prices dropped abruptly during the last quarter of 1984, apparently sparked by rumours of distress sales of molybdenum by the major producers. At year-end, the price of oxide in the dealer market was quoted at $US 6.17-6.72 a kg.

OUTLOOK

The demand for molybdenum will continue to rely on steel production intensity, especially in the alloy, stainless and tool steels sector which together account for about 75 per cent of total molybdenum consumption. Although consumption improved substantially in 1984 as a result of the economic recovery, it is unlikely that demand will return to the record level of the 1970s before the second half of the 1990s.

The consumption of molybdenum is forecast to increase by 7 or 8 per cent in 1985 and slow to 2 to 3 per cent in 1986. However, a significant increase in interest rates would have a dampening impact on the economic recovery, and hence, the growth rate of molybdenum demand.

Among the end uses, growth is projected for the transportation sector where fuel efficiency and weight reduction considerations will result in a rising consumption of high strength steels. The oil and gas industries, which are expanding their use of stainless steels will also provide a major growth market for molybdenum.

The superalloy and electronics sectors have fully recovered from the economic downturn and are showing signs of strong growth, especially in military, aerospace and high technology applications. Also, the capital goods market is recovering steadily after a very poor performance during the recessionary years.

At the 1984 rate of production, western world producer inventories are expected to increase by 9 070 t in 1985 to about 79 000 t. The excess supply, mainly due to the large increase in output in the United States, will result in a continuing market weakness and could lead to production cuts in 1986. In the long run, production restraint will be an ongoing industry challenge if market stability is to be achieved.

Despite the weak molybdenum market in recent years, the average growth rate for consumption during the next two decade is expected to be comparable with that of the past decade, due in large part to the increasing use of molybdenum in low-alloy steels. A stable and dependable supply of molybdenum at relatively low prices will favor the expanding use of molydenum as compared with many of its substitutes.

Despite the rapid growth of steel production in the developing countries and a continuing decline in steel output in the advanced industrial nations, a major shift of molybdenum consumption to the developing countries is not expected to take place in the next two decades. This is because the specialized technology for making alloy steel is concentrated in the advanced industrial countries; most of the steel production in developing countries will continue to be common carbon steels.

World production capacity for molybdenum is more than adequate to satisfy the projected demand increase well into the next decade. Also, new copper developments that will give rise to byproduct molybdenum and mine expansions in Chile and Peru will result in additional supplies of molybdenum in years to come. Moreover, there are a number of well defined molybdenum deposits, especially in North America, which could be developed when the world's supply and demand approach a reasonable balance.

PRICES

Prices in $US per kilogram of contained molybdenum, fob shipping point unless indicated otherwise, December 31.

	1983	1984
	($)	
Byproduct concentrates (MoS_2)	7.28-7.94	5.95-6.39
Export oxide (MoO_3) in cans	8.65-13.23	8.65-13.23
Dealer oxide (MoO_3) in cans; min. 57% Mo	8.27-8.93	6.06-6.61
Ferromolybdenum[1] Dealer export (fas port)	9.81-10.03	7.83-8.05

Source: Metals Week.
[1] Price based on molybdenum content.
fob Free on board, fas Free alongside ship.

TARIFFS

Item No.		British Preferential	Most Favoured Nation	General	General Preferential
			(%)		
CANADA					
32900-1	Molybdenum ores and concentrates	free	free	free	free
33505-1	Molybdenum oxides	10.0	13.8	25.0	9.0
37506-1	Ferromolybdenum	free	4.7	5.0	free
35120-1	Molybdenum metal in powder, pellets, scrap, ingots, sheets, strips, plates, bars, rods, tubing or wire, for use in Canadian manufactures	free	free	25.0	free
92847-1	Molybdates	10.0	12.1	25.0	8.0
	Temporary reduction, June 3, 1980 to June 30, 1987	free			free
92856-1	Molybdenum carbides	9.4	7.5	25.0	5.0
	Temporary reduction, June 3, 1980 to December 31, 1986	free			free

MFN Reductions under GATT (effective January 1 of year given)	1983	1984	1985	1986	1987
			(%)		
33505-1	13.8	13.4	13.1	12.8	12.5
37506-1	4.7	4.5	4.3	4.2	4.0
92847-1	12.1	11.4	10.7	9.9	9.2
92856-1	7.5	5.6	3.8	1.9	free

TARIFFS (cont'd)

UNITED STATES

601.33	Molybdenum ore (per lb on Mo content)	10.5¢	10.1¢	9.8¢	9.4¢	9.0¢
419.60	Molybdenum compounds	3.7	3.5	3.4	3.3	3.2
606.31	Ferromolybdenum	5.9	5.6	5.2	4.9	4.5
628.70	Molybdenum metal, waste and scrap	8.3	7.7	7.1	6.6	6.0
628.72	Molybdenum metal, unwrought	8.1¢/ lb on Mo content +2.5	7.6¢/ lb on Mo content +2.3	7.2¢/ lb on Mo content +2.2	6.7¢/ lb on Mo content +2.0	6.3¢/ lb on Mo content +1.9
628.74	Molybdenum metal, wrought	9.6	8.8	8.1	7.3	6.6
417.28	Ammonium molybdate	5.3	5.0	4.8	4.5	4.3
418.26	Calcium molybdate	4.8	4.8	4.8	4.7	4.7
421.10	Sodium molybdate	4.4	4.2	4.1	3.9	3.7
423.88	Molybdenum carbide	3.2	3.1	3.0	2.9	2.8

EUROPEAN ECONOMIC COMMUNITY (MFN)

		1983	Base Rate (%)	Concession Rate
26.01	Molybdenum ores and conc	free		
28.28	Molybdenum oxides and hydroxides	6.7	8.0	5.3
73.02	Ferromolybdenum	6.3	7.0	4.9
81.02	Molybdenum metal			
	A. Unwrought: powder	6		
	other	5		
	B. Wrought: bars, angles, plates, sheets, strip, wire	8		
	C. Other	10		
28.47	Molybdates	8.9	11.2	6.6
28.56	Molybdenum carbides	8.6	9.6	8.0

JAPAN (MFN)

26.01	Molybdenum ores and conc			
	A. Quota	free		
	B. Other	3.8	7.5	free
28.28	Molybdenum trioxide	3.8	5.0	3.7
73.02	Ferromolybdenum	5.3	7.5	4.9
81.02	Molybdenum metal			
	A. Unwrought, powders and flakes	3.8	5.0	3.7
	B. Waste and scrap	3.8	5.0	3.7
	C. Other	5.3	7.5	4.9
28.47	Molybdates	5.3	7.5	4.9
28.56	Molybdenum carbides	3.8	5.0	3.7

Sources: The Customs Tariff, 1983, Revenue Canada, Customs and Excise; Tariff Schedules of the United States Annotated 1983, USITC publication 1317; U.S. Federal Register, Vol. 44, No. 241; Official Journal of the European Communities, Vol. 25, No. L318, 1982; Customs Tariff Schedules of Japan, 1983.

TABLE 1. CANADA, MOLYBDENUM PRODUCTION AND TRADE, 1982-84, AND CONSUMPTION, 1982-83

	1982		1983		1984P	
	(tonnes)	($000)	(tonnes)	($000)	(tonnes)	($000)
Production (shipments)[1]					(Jan.-Sept.)	
British Columbia	13 584	155,112	10 179	87,564	10 865	108,916
Quebec	377	4,030	15	146	100	1,060
Total	13 961	159,142	10 194	87,710	10 965	109,976
Exports						
Molybdenum in ores, concentrates and scrap[2]						
United Kingdom	1 574	21,215	2 452	24,374	719	6,783
Belgium-Luxembourg	3 000	48,578	1 969	22,216	834	7,465
Netherlands	2 344	43,104	2 097	19,511	969	9,173
West Germany	2 314	23,207	2 006	16,404	1 173	10,273
Japan	3 185	53,492	1 274	14,940	1 947	22,954
United States	2 249	31,341	437	3,584	244	2,241
Austria	393	1,036	404	2,900	..	..
Chile	468	3,928	517	2,422	396	3,651
Other countries	916	12,217	128	1,426	94	853
Total	17 443	238,118	11 284	107,777	6 349	63,393
Imports						
Molybdic oxide (containing less than 1 per cent impurities)	193	2,740	141	1,486	209	2,165
Molybdenum in ores and concentrates (Mo content)	3 027	40,119	233	1,833	..	..
Ferromolybdenum alloys	77	1,017	34	323	162	1,873

	1982		1983	
	(kilograms)	($)	(kilograms)	($)
Consumption (Mo content)				
Addition agents	512 553	..	387 874	..
Electrical and electronics	4 960	..	2 009	..
Other Uses[3]	153 855	..	100 234	..
Total	671 368	..	490 117	..

Sources: Energy, Mines and Resources Canada; Statistics Canada.
[1] Producers' shipments (Mo content of molybdenum concentrates, molybdic oxide and ferromolybdenum). [2] Includes molybdenite and molybdic oxide in ores and concentrates. [3] Alloy, pigment and ceramics.
P Preliminary; .. Not available.

TABLE 2. CANADA, MOLYBDENUM PRODUCTION, TRADE AND CONSUMPTION, 1970, 1975 AND 1979-84

	Production[1]	Exports[2]	Imports Molybdic oxide[3]	Ferro-molybdenum[4]	Consumption[5]
			(kilograms)		
1970	15 318 593	13 763 800	33 500	29 619	1 036 940
1975	13 323 144	15 710 300	56 400	269 281	1 436 883
1977	16 567 555	15 326 100	192 100	74 330	1 149 736
1978	13 943 405	13 421 000	329 500	55 294	1 268 640
1979	11 174 586	11 481 900	335 900	153 945	1 249 944
1980	11 889 000	14 584 500	361 700	53 618	1 055 107r
1981	12 850 000	13 664 000	423 000	36 069	1 311 863
1982	13 961 000	17 444 000	193 000	6 840	671 368
1983	10 194 000	11 284 000	141 000	34 000	490 117
1984P	10 965 000	8 465 000	313 000	243 000	..

Sources: Energy, Mines and Resources Canada; Statistics Canada; except where noted.
[1] Producers' shipments (Mo content of molybdenum concentrates, oxide and ferromolybdenum).
[2] Mo content, ores and concentrates. [3] Gross weight. [4] United States exports to Canada, reported by the U.S. Bureau of Commerce, Exports of Domestic and Foreign Merchandise (Report 410), over 50 per cent molybdenum. [5] Mo content of molybdenum products reported by consumers.
P Preliminary for production and estimate for trade data; .. Not available; r Revised.

TABLE 3. CANADA, MINE PRODUCTION, 1983

Company and Mine Name	Location	Type of Producer	Mill Capacity (tpd)	Ore Milled		Concentrates Produced		Contained Mo (tonnes)
				Tonnes	Grade (% Mo)	Tonnes	Grade (% Mo)	
Amax of Canada Limited Kitsault Mine	Alice Arm B.C.	Primary	10 886	-	-	-	-	-
Brenda Mines Ltd.	Peachland, B.C.	Coproduct	27 200	8 185 403	0.032	3 629	55.52	2 015
Gibraltar Mines Limited	McLeese Lake, B.C.	Byproduct	37 195	13 437 210	0.010	797	54.57	435
Highmont Mining Corporation	Highland Valley, B.C.	Coproduct	22 680	8 799 692	0.024	2 910	53.75	1 674
Lornex Mining Corporation Ltd.,	Highland Valley, B.C.	Byproduct	72 575	28 766 769	0.016	6 351	53.23	3 381
Noranda Mines Limited, Boss Mountain Division	Williams Lake, B.C.	Primary	2 631	29 772	0.187	93	54.98	51
Mines Gaspé Division Needle Mountain and Copper Mountain	Holland Twp. Gaspé, Que.	Byproduct	32 800	-	-	-	-	-
Placer Development Limited, Endako Mine	Endako, B.C.	Primary	29 937	-	-	-	-	-
Utah Mines Ltd., Island Copper mine	Port Hardy, B.C.	Byproduct	38 100	16 330 081	0.017	3 426	46.65	1 599
Total								9 155

Sources: Energy, Mines and Resources Canada; Company annual reports.
- Nil.

TABLE 4. WORLD PRODUCTION OF MOLYBDENUM IN ORES AND CONCENTRATES, 1981-83

Country	1981	1982P	1983e
	(tonnes Mo content)		
United States	63 458	37 671	13 608
Canada	12 850	13 961	10 523
Chile	15 360	20 000	14 515
U.S.S.R.e	10 705	11 022	..
People's Republic of Chinae	1 996	1 996	..
Peru	2 488	2 565	2 722
Republic of Korea	465	400	..
Bulgariae	150	150	..
Japane	74	98	..
Philippines	80	60	..
Mexico	451	450	..
Mongolia	..	..	14 515
Total	108 077	88 373	55 883

Sources: Energy, Mines and Resources Canada; U.S. Bureau of Mines, Minerals Yearbook, Preprint, 1982; U.S. Bureau of Mines, Mineral Commodity Summaries, 1984.
P Preliminary; e Estimated; .. Not available.

TABLE 5. PRINCIPAL MOLYBDENUM PRODUCERS IN THE WESTERN WORLD, 1984

Company	Country	Installed Capacity (000 tpy Mo)
AMAX Inc.	United States	45
Corporacion Nacional del Cobre de Chile (CODELCO-Chile)	Chile	20
Duval Corporation	United States	10
Amoco Minerals Company	United States	9
Molycorp, Inc.	United States	9
Placer Development Limited	Canada	7.7
Anaconda Minerals Corporation	United States	6.8
Mexicana de Cobre, S.A.	Mexico	5.4
Kennecott Minerals Company	United States	5
Noranda Inc.	Canada	4.5
Southern Peru Copper Corporation	Peru	4.5
Amax of Canada Limited	Canada	4
Lornex Mining Corporation Ltd.	Canada	3.6
Newmont Mining Corporation	United States	2.2
Teck Corporation	Canada	2.0
Utah Mines Ltd.	Canada	1.6
Others		4.5
Total		144.8

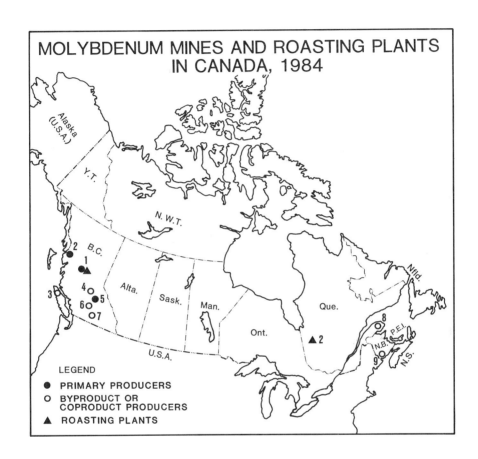

Mines

1. Placer Development Limited (Endako mine)
2. Amax of Canada Limited (Kitsault mines)
3. Utah Mines Ltd. (Island Copper mine)
4. Gibraltar Mines Limited
5. Noranda Inc. (Boss Mountain Division)
6. Lornex Mining Corporation Ltd.
   Highmont Mining Corporation
7. Brenda Mines Ltd.
8. Noranda Inc. (Gaspé Division)
9. Mount Pleasant Mines Limited

Roasting Plants

1. Placer Development Limited (Endako mine)
2. Eldorado Gold Mines Inc. (Duparquet)

Nepheline Syenite and Feldspar

MICHEL A. BOUCHER

SUMMARY

Nepheline syenite is produced commercially as an industrial raw material for the manufacture of glass and ceramics, mainly by Canada and Norway, and to a smaller extent by Brazil and the U.S.S.R. However, only Canada and Norway export nepheline syenite in large quantities. The producers in these countries have very defined markets with Canada exporting mainly to the United States and Norway exporting mainly to western Europe.

In 1983 and 1984, Canadian production and exports continued to decrease from the high level reached in 1979. The decline in production was due mainly to lower demand for container glass, which must compete with plastics and aluminum, and to an increase in glass waste recycling. Lower exports were the result of the closing of glass plants in the United States in recent years, the acceptance of lower quality feldspathic material by United States manufacturers, and recent deregulation of transport in the United States that makes producers of aplite and feldspar (which are substitutes for nepheline syenite in glass production) in the United States more competitive with Canadian nepheline syenite.

Prices remained stable in 1983 and 1984.

CANADIAN DEVELOPMENTS

Production

Nepheline syenite is produced in Ontario by two companies - Indusmin Limited, a subsidiary of Falconbridge Nickel Mines Limited and IMC Industry Group (Canada) Ltd. a subsidiary of International Minerals & Chemical Corporation (IMC). The two companies have an estimated combined production capacity of 800 000 tonnes (t) of finished product a year, with Indusmin being the largest producer. Nepheline syenite is mined by these companies from two adjacent deposits located on Blue Mountain in Methuen Township, Peterborough County, 175 km northeast of Toronto. Operating capacity at the two plants was between 60 and 65 per cent in 1983 and 1984, and no major capital expenditures were reported at either operation during those years. The nepheline syenite is upgraded to low-iron and high-iron glass grades, and to ceramic grades by primary and secondary crushing, drying, screening, high-intensity magnetic separation, pebble milling, and for ultra-fine grades (for use as filler in paints, plastics, etc.) using fluid energy mills and air classification.

Feldspar was not produced in Canada in 1983 nor in 1984.

CONSUMPTION

The glass and glass fibre industries account for some 70 per cent of nepheline syenite consumption in Canada. As stronger growth is expected in the future in filler and extender pigment applications (plastics, paints, etc.) than in the glass industry further diversification by the producers into these markets is expected. It may take years, however, before substantial tonnages in these applications can be attained. Producers will have to attempt to demonstrate the special properties of nepheline syenite over readily available and excellent materials such as kaolin, calcium carbonate, etc.

Consumption of nepheline syenite in Canada by glass producers continued to be influenced by the use of plastics and aluminum, by increased recycling of glass waste, and by the development of thinner glass containers.

The major consumers of nepheline syenite in Canada are Dominion Glass Company Limited, Consumers Glass Company Limited, Fiberglass Canada Limited, American Standard Incorporated, and Crane Canada Inc.

Michel A. Boucher is with the Mineral Policy Sector, Energy, Mines and Resources, Canada. Telephone (613) 995-9466.

TRADE

Canada has a large trade surplus in nepheline syenite and feldspar with the United States. While exports of nepheline syenite to the United States average about 350 000 tpy, imports of feldspar from the United States average about 4 000 tpy. Nearly all feldspar imported into Canada comes from North Carolina where both Indusmin and IMC have mining operations.

PRICES

The value of production per t increased by 10 per cent in 1983 and 5 per cent in 1984. However, the value of exports per t increased by only 7.7 per cent in 1983 and 3.3 per cent in 1984 (based on 9 month figures), barely offsetting inflation.

Prices for nepheline syenite in 1984 ranged from a low of $22 per t for sand, to a high of $93 per t for filler and extender grade. Listed prices for nepheline syenite products have remained the same since 1982 while feldspar prices were only increased in 1984.

USES

Nepheline syenite is preferred to feldspar as a source of alumina and alkalis for glass manufacture. Its use results in more rapid melting of the batch at lower temperatures than with feldspar thus reducing fuel consumption, lengthening the life of furnace refractories, and improving the yield and quality of glass. Other industrial uses for nepheline syenite include ceramic glazes, enamels, fiberglass and fillers in paints, papers, plastics and foam rubber.

Feldspar is the name of a group of minerals consisting of aluminum silicates of potassium, sodium and calcium. It is used in glassmaking as a source of alumina and alkalis, in ceramic bodies and glazes, in cleaning compounds as a moderate abrasive and as a flux coating on welding rods. High calcium feldspars, such as labradorite, and feldspar-rich rocks, such as anorthosite, find limited use as building stones and for other decorative purposes. Potash feldspar is an essential ingredient in the manufacture of high voltage porcelain insulators. Dental spar, which is used in the manufacture of artificial teeth, is a pure white potash feldspar, free of iron and mica.

OUTLOOK

Only a modest increase in sales of nepheline syenite can be expected in 1985. This is mainly due to improvements in the ceramic and fiberglass markets.

In the medium-term, if the North American economy does not improve, some small producers of feldspar in eastern United States can be expected to close.

In the longer-term glass container producers will continue to compete with plastics and aluminum producers. Fiberglass is bulky and consequently expensive to transport and may over the years loose markets to more compact material, consequently slow growth can be expected in this area. The use of nepheline syenite in ceramics and as a filler is expected to grow substantially, but consumption tonnages will remain small for several years compared with tonnages used in glass and fiberglass.

There are opportunities for new feldspar or nepheline syenite development in western Canada as all the production currently is concentrated in eastern Canada and eastern United States.

PRICES OF FELDSPAR AND NEPHELINE SYENITE IN U.S. CURRENCY

	1983	1984
	($/tonne)	
FELDSPAR		
Ceramic grade, bulk		
FOB Spruce Pine, NC, 170-250 mesh	45.46	48.50
FOB Monticello, Ga, 200 mesh, high potash	76.31	81.00
FOB Middleton, Con, -200 mesh	55.65	58.68
Glass grade, bulk		
FOB Spruce Pine, NC, 97.8% + 200 mesh	30.30	32.34
FOB Middleton, Con, 96% + 200 mesh	41.05	42.98
FOB Monticello, Ga, 200 mesh, high potash	56.20	59.50
NEPHELINE SYENITE		
Canadian, CL-car lots TL-truck lots		
Glass gr., 30 mesh, bulk CL/TL, low iron	28.65-31.40	28.65-31.40
Glass gr., 30 mesh, bulk CL/TL, high iron	22.04-25.62	22.04-25.62
Ceramic gr., 200 mesh, bagged 10-ton lots	60.60-62.81	60.60-62.81
Filler/extender grade	73.83-93.67	73.83-93.67

Source: Industrial Minerals, December 1983 and December 1984.

TARIFFS

Item No.		British Preferential (%)	Most Favoured Nation (%)	General (%)	General Preferential (%)
CANADA					
29600-1	Feldspar, crude	free	free	free	free
29625-1	Feldspar, ground but not further manufactured	free	6.5	30	free
29640-1	Ground feldspar for use in Canadian manufactures	free	free	30	free

	1983	1984	1985	1986	1987
MFN Reductions under GATT (effective January 1 of year given)			(%)		
29625-1	6.5	6.3	6.0	5.8	5.5

UNITED STATES (MFN)

522.31	Crude feldspar	free

	1983	1984	1985	1986	1987
			(%)		
522.41 Feldspar, crushed, ground or pulverized	3.2	3.1	3.0	2.9	2.8

Sources: The Customs Tariff, 1983 Revenue Canada Customs and Excise; Tariff Schedules of the United States, Annotated 1983, USITC Publication 1317. U.S. Federal Register Vol. 44, No. 241.

TABLE 1. CANADA, NEPHELINE SYENITE PRODUCTION, EXPORTS AND CONSUMPTION, 1982-84

	1982		1983		1984P	
	(tonnes)	($)	(tonnes)	($)	(tonnes)	($)
					(Jan.-Sept. 1984)	
Production (shipments)	550 480	17,323,776	523 249	18,130,692	485 012	17,671,351
Exports						
United States	373 932	13,557,000	345 245	13,469,000	259 637	10,514,000
Netherlands	24 490	1,014,000	20 995	1,019,000	14 094	630,000
Italy	6 834	495,000	8 614	658,000	4 766	378,000
United Kingdom	4 751	256,000	8 926	472,000	295	34,000
Australia	1 537	121,000	8 943	294,000	1 241	95,000
Spain	269	18,000	1 927	105,000	679	49,000
Other countries	2 975	304,000	3 649	293,000	3 161	308,000
Total	414 788	15,765,000	398 299	16,310,000	283 873	12,008,000
Consumption[1]						
Glass and glass fibre	75 852					
Insulation	10 898					
Ceramic products	10 465					
Paints	3 669					
Others[2]	1 725					
Total	102 609	..	..	..		

Sources: Statistics Canada; Energy, Mines and Resources, Canada.
[1] Available data, as reported by consumers. [2] Includes frits and enamel, foundry, plastics, rubber products, electrical apparatus and other minor uses.
P Preliminary; .. Not available.

TABLE 2. CANADA, NEPHELINE SYENITE PRODUCTION AND EXPORTS, 1970, 1975 AND 1979-83

	Production[1]	Exports
	(tonnes)	
1970	454 110	351 940
1975	468 427	356 629
1979	605 699	471 056
1980	600 000	448 468
1981	588 000	476 281
1982	550 480	414 788
1983	523 249	398 299

Sources: Energy, Mines and Resources, Canada; Statistics Canada.
[1] Producers' shipments.

TABLE 3. CANADA, ESTIMATED FELDSPAR CONSUMPTION, 1981 AND 1982

	1981	1982
	(tonnes)	
Consumption		
Whiteware	4 410	2 585
Other products[1]	196	205
Total	4 606	2 790

[1] Includes porcelain enamel, artificial abrasives and other minor uses.

TABLE 4. CANADA, IMPORTS AND CONSUMPTION OF CRUDE OR GROUND FELDSPAR, 1975 AND 1979-83

	Imports	Consumption
	($)	(tonnes)
1975	..	5 630
1979	501,000	4 588
1980	385,000	4 051
1981	642,000	4 606
1982	251,000	2 790
1983	309,000	..

Sources: Statistics Canada; Energy, Mines and Resources Canada.
.. Not available.

TABLE 5. WORLD PRODUCTION OF FELDSPAR, 1982 AND 1983

	1982	1983e
	(tonnes)	
United States	557 919	625 957
Italy	399 161	399 161
West Germany	340 194	344 730
France	181 437	190 509
Mexico	117 934	127 006
Spain	99 790	108 862
Brazil	95 254	99 790
Other countries	1 307 254	1 324 491
Total	3 098 943	3 220 506

Source: United States Bureau of Mines Mineral Commodity Summaries, 1984.
e Estimated.

Nickel

R.G. TELEWIAK

Strong growth in the United States economy was the principal factor in nickel consumption in the western world increasing by an estimated 12 per cent in 1984, compared to 1983. This was the second consecutive year of growth in consumption and brought the level to an estimated 560 000 t, which was slightly below the record high reached in 1979.

Growth in the consumer goods sector was the leading factor in nickel demand trends in 1983 but, with economic recovery shifting into a more mature phase, demand from the capital sector was more important in 1984. Unlike other economic recoveries, however, nickel demand from the capital sector tended to reflect modernization of existing plants, as opposed to major new plant construction and expansion.

Producers around the world carried out extensive cost-saving programs in 1983 and 1984. The nickel price on the London Metal Exchange (LME) averaged $US 2.12 in 1983 and $2.16 in 1984, compared to $2.70 in 1981; and this low price and the expectation that the price will continue to be under pressure for several years, due to global overcapacity, forced producers to take dramatic measures to reduce costs. While some of the measures, such as selective mining, are short-term, many others will result in long-term cost-savings due to improved mining and processing techniques and equipment.

CANADIAN DEVELOPMENTS

Inco Limited and Falconbridge Limited both substantially reduced production costs and by late-1984, Inco announced that the break-even price for its nickel operations was $US 2.20 and Falconbridge announced that its costs were marginally below $2.00. By comparison, Falconbridge stated its costs were $3.40 in 1981; no estimate is available for Inco. Costs in all segments of the operations were reduced, as evidenced by 3.8 per cent decrease in mining costs from 1981 to 1984 to $Cdn 28.00 per t announced by Inco.

In June 1984, a series of rockbursts occurred at the Falconbridge and East mines of Falconbridge at Sudbury. Four fatalities resulted and the Falconbridge mine, after 55 years of operation, was permanently closed. Production subsequently resumed at the East mine and the adjacent Falconbridge shaft and certain haulage areas, were made safe for the hoisting of ore from this mine.

Inco resumed production at Sudbury, Port Colborne and Shebandowan in April 1983 after a nine month shutdown brought on by low prices and high inventories. Four-week summer closures were taken in 1983 and 1984. Faced with excess refining capacity, Inco decided to indefinitely suspend electrolytic production at Port Colborne. Utility nickel will still be produced.

Development work commenced in September 1983 on Inco's Thompson open-pit in Manitoba, with production scheduled to begin early in 1986. The Pipe open-pit was mined out in 1984 but with some ore stockpiled from this pit, along with some ore from the Thompson underground mine, there will adequate feed to the mill through 1985, although production is likely to be somewhat lower. Grade at the Thompson open-pit is 2.7 per cent, which is more than three times that of the Pipe open-pit and production from the Thompson pit will certainly enhance the already cost-competitive nature of these operations.

Sherritt Gordon Mines Limited increased annual capacity in 1983 of its Fort Saskatchewan, Alberta refinery to 21 000 t of contained nickel from 17 500 t. In 1984, the plant operated near capacity for most of the year, although production was adversely affected in October due to technical difficulties. Inco Thompson continued to be the major source of feed.

R.G. Telewiak is with the Mineral Policy Sector, Energy, Mines and Resources Canada. Telephone (613) 995-9466.

WORLD DEVELOPMENTS

Producers continued to operate well below capacity in response to depressed market conditions. Production increased in 1984 over 1983 but producer inventories still decreased due to higher consumption.

In Australia, Queensland Nickel Pty. Ltd. increased production in the first half of 1984 at its Greenvale mine and refinery, to 70 per cent of capacity from 50 per cent. The complex is capable of producing 21 800 tpy of contained nickel in oxide sinter. Agnew Mining Co. Pty Limited re-evaluated its plans to expand production at its nickel mine in western Australia, given the low prevailing nickel price.

Société Métallurgique Le Nickel (SLN) in New Caledonia was expected to have produced close to 35 000 t of nickel in 1984, compared to 26 000 t a year earlier. Late in the year, production was disrupted for two weeks at the Thio mine by Melanesian militants, who have been compaigning for independence. Production continued at SLN's other mine at Kouaoua and some intermittent shipments from this mine, along with stockpiled ore, kept smelter production at normal rates. The third Demag furnace had been brought back on-stream on August 1, 1984.

Marinduque Mining & Industrial Corporation at Surigao in the Philippines completed its oil to coal conversion plan in June 1983, and in December closed due to lack of working capital. Some spare parts and other supplies were lacking. In August 1984, the Development Bank of the Philippines and the Philippine National Bank, which together had held 87 per cent equity in the company as well as outstanding loans of $1.4 billion, decided to foreclose. A new company, Nonoc Mining & Industrial Corporation, wholly-owned by the banks, took 100 per cent equity in the operation and eliminated the loans. Production subsequently resumed but was then disrupted from early-September until November as the result of typhoon damage.

Ni-Cal Developments Ltd. reported in October, 1984 that it had received a letter of intent from the Development Bank of the Philippines regarding a partial retrofit of the Nonoc refinery. A feasibility study was scheduled to be completed by year-end. If the study was favourable, then Ni-Cal would arrange financing for, construct and operate a single module 770 tpd ore acid leach unit, at Nonoc. Cost would be about $50 million. A full retrofit would involve a facility 10 times this size. The Ni-Cal acid leach technology has not yet been applied commercially.

Start-up problems continued to affect production at the Cerro Matoso SA ferronickel complex in Colombia. Difficulties arose due to the acidic nature of the ore which eroded the furnace lining at the matte-slag interface. Cerro Matoso reduced its electrical power to the furnace, blended the ore more thoroughly, altered the manner in which the ore was loaded into the furnace and took other measures to reduce the problem. Cerro Matoso expects to have a long-term solution in place in 1985, when an Outokumpu Oy cooling system should be operational.

AMAX Inc. obtains matte for its Port Nickel, Louisiana refinery from BCL Ltd. in Botswana and Agnew Mining in Australia, but these sources are insufficient to keep the refinery operating at capacity. AMAX closed the complex for two months in the summer of 1983 and for five weeks in 1984.

Also in the United States, The Hanna Mining Company reopened its ferronickel operations in Riddle, Oregon in November 1983 and continued to operate through 1984 at about one-half capacity. Hanna installed a casting machine in the summer of 1984 and was testing two grades of nickel nuggets. One type contains 50 per cent nickel for use in the stainless steel sector and the other 60 per cent nickel for application in the plating industry.

Effective December 22, 1983, the United States banned all unfabricated nickel and nickel-bearing materials imported directly or indirectly from the U.S.S.R. because it was believed that the Soviet refined product contained some Cuban nickel. The Soviet Union annually imports about 19 000 t of nickel in oxide from Cuba for refining at Monchegorsk in the Kola Peninsula. It was some of this material that was considered to have been shipped to the United States. Unlike the ban on Cuban nickel, the ban on Soviet nickel does not apply to nickel-bearing materials which have been combined with other elements in a third country to form different metals, such as nickel alloys or stainless steel, and then exported to the United States.

In 1984, Comecon exports to the west declined to about 25 000 t from 35 000 t in 1983. The U.S. ban, along with concern by some consumers that it could be expanded to cover alloys containing nickel, likely affected exports but there were other factors as well. It was rumoured that one reason for the decline was that the U.S.S.R. had switched some of the Norilsk production from nickel to copper, in order to meet internal needs for copper.

Cuba continued development of the Punta Gorda complex and full production from the three lines - which have a capacity of 30 000 tpy of contained nickel - is planned for the end of 1985. Preliminary work commenced on a twin plant at Las Camariocas, 20 km away, with production scheduled for the late-1980s. If the planned timetable is met, Cuba will have a capacity of 100 000 t of contained nickel by 1990.

In 1984, Yugoslavia commenced production from its new Kosovo ferronickel plant, and also closed its Feni Kavadarci operation. Kosovo, with a capacity of 12 000 tpy of contained nickel, started producing on May 23. Feni, which had been open for a little over a year, was plagued by high operating costs and intermittent energy supplies. With low nickel prices prevailing, the facility was closed in July. The 2,000 workers were to be relocated to other industries in the region but the plant was being maintained for possible production in the future, if the nickel price increases substantially.

China announced that it would be modernizing and expanding its nickel plant at Jinchuan in the Gansa province. Outokumpu of Finland was granted a licensing agreement for a new 350 000 tpy flash smelter and Western Mining Corporation Ltd. obtained a consultancy agreement to advise on the design and construction of the smelter. The original plant had been brought into production in the mid-1960s.

In South Africa, Western Platinum Limited started construction of a nickel-copper-cobalt-precious metals refinery at Rustenburg. Production is expected in 1986. The plant will utilize Sherritt Gordon's sulphuric acid leach process to make 1 800 tpy of nickel in nickel sulphate and 1 100 tpy of copper cathode. Falconbridge, which owns 25 per cent of Western Platinum, will continue to refine the matte in Norway until the new refinery is operational.

NICKEL DEVELOPMENT INSTITUTE

In recent years, only limited resources have been expended on promoting the use of nickel and in researching new uses for nickel. Inco formerly led this effort but, with weak market conditions, Inco and other nickel producing companies were forced to reduce their expenditures in this area.

On May 31-June 1, 1984, a meeting of western world nickel producers was held to discuss the need for an international organization to conduct this work. An organization, the Nickel Development Institute (NiDI) was subsequently formed with headquarters in Toronto. While the members are all nickel producers, it is planned that after an initial phase, consumers will also be invited to become members.

The primary emphasis of the NiDI will be to promote the use of nickel in the major markets of the United States, Japan and western Europe. Research will also be conducted into new uses for nickel.

INTERGOVERNMENTAL NICKEL DISCUSSION GROUP

Thirty-one countries attended an exploratory meeting in October 1984 in Geneva, to discuss the need for an Intergovernmental Nickel Discussion Group (INDG) which would publish statistics and conduct special studies. These countries represented over 95 per cent of world production and 90 per cent of consumption. Canada and Australia hosted the session. As well, there were observers from the Commission of the European Communities, the GATT, the UNCTAD and the International Lead and Zinc Study Group.

There was broad recognition expressed that there are serious information gaps with respect to the world nickel economy, in terms of the quality, timeliness and international comparability of available statistics. Most delegations indicated a willingness and ability to improve the quality of nickel data they provide for publication.

Many countries expressed a preference for an organization similar to the International Lead and Zinc Study Group (ILZSG), an autonomous intergovernmental organization which has proven highly successful over its 25-year history. Some other delegations felt that further reflection would be required on the final character of the organization.

It was agreed that another meeting would be held in the spring of 1985, with UNCTAD to provide the meeting facilities and to arrange for distribution of documentation. Australia and Canada were requested to continue their roles in the preparation of documentation.

PRICES

Nickel prices advanced on the London Metal Exchange (LME) early in 1983, from an average $US 1.72 in January to $2.19 by March and remained near that level through the rest of 1983 and through 1984. In December 1984, the average LME quote was $2.18.

This price stability occurred despite some wide fluctuations in LME stock levels. At the end of February 1983 nickel stocks were 9 800 t, they then increased steadily to peak at 32 600 t in February 1984 and then declined to 7 400 t by December 29.

Overall stocks, producer, consumer, trade and commodity exchanges declined slightly through 1983 and 1984 but not enough to affect prices. Producers increased output at a rate which was fairly close to the higher consumption levels.

USES

Resistance to corrosion, high strength over a wide temperature range, pleasing appearance and suitability as an alloying agent are characteristics of nickel which make it useful in a wide range of applications. The major use is in stainless steels, which account for close to 50 per cent of consumption, followed by nickel-base alloys, electroplating, alloy steels, foundry and copper-based alloys. Nickel is extensively used as an alloying agent and is a component in some 3,000 different alloys.

Close to two-thirds of nickel consumption is in capital goods with the remainder used in consumer products. Nickel is used in chemical and food processing, nuclear power plants, aerospace equipment, motor vehicles, oil and gas pipelines, electrical equipment, machinery, batteries, catalysts, and in many other applications.

Relatively new end-use markets that will contribute to nickel's consumption growth in the future are pollution abatement equipment, cryogenic containers, barnacle-resisting copper-nickel alloy plating for boathulls, and nickel-cadmium batteries for standby power applications. The use of nickel-zinc batteries in electric cars was earlier considered to be an important nickel market which would develop in the late 1980s, but the large scale production of electric cars has been deferred. The fledgling solar energy industry could provide a market for increasing amounts of nickel alloys where there is a need for durability and corrosion resistance.

OUTLOOK

The intensity of nickel use, expressed in terms of consumption per unit of gross national product, has been declining in some major market areas in recent years. While this trend could continue in several of the major industrial economies, the intensity of use is also likely to increase in some growing markets such as India, Brazil, China and the Republic of Korea.

Nickel is expected to lose markets in certain applications to other metals, as well as to ceramics, plastics, composites and some other materials. For example, there is expected to be continued penetration of certain nickel-containing stainless steel markets by non-nickel stainless steel. However, nickel is also expected to substitute for other materials in some applications, particularly since the price of nickel is not expected to increase markedly in real terms for several years. Nickel is far cheaper to use as alloying agent than it was three or four years ago.

Superalloys containing nickel provide an attractive area for growth in certain applications. Continued research in the composition of these alloys and in the techniques of producing them should increase the market penetration of these alloys. As an example of the effects of the research, the U.S. Air Force has developed two nickel aluminide superalloys which are being tested for use in gas turbines of military jet engines. A relatively new class of materials, low-expansion nickel superalloys containing nickel-iron-cobalt, have already proven to offer good potential for use in gas turbines.

Overall, consumption of nickel in the western world is expected to grow at an annual rate of about 1.7 per cent to 1990, from a 1981 base, and then at 1.6 per cent to the year 2000. While these rates are substantially lower than the average 6 per cent annual rate which prevailed for many years prior to the early 1970s, they are still better than the flat rate of the past decade.

In response to the higher levels of demand, production in Canada is expected to increase slowly to the year 2000. There are various constraints to the rate of increases in production. Environmental limits on SO_2 emissions is one of the key ones, particularly at Inco, Sudbury. A significant sustained rise in the price of nickel will be needed to permit Inco to regain the financial capability to build a new smelter, which could permit higher output, depending upon potential new government regulations. At Falconbridge, Sudbury considerable expensive mine development work needs to be done before production can be increased to near capacity.

Nickel prices are expected to increase in real terms over the long-term, as production overcapacity is reduced. Other than for a planned expansion of capacity in Cuba, there is not expected to be much new capacity brought on-stream by 1990. With consumption increasing this will bring a better balance between available supply and demand. In constant 1984 $US, the realized price for producers could be close to $3.00 per pound by 1990 and remain at that level through the 1990s.

In the short-term we believe primary nickel consumption will increase less quickly than it has in the past two years. With the strong stainless steel sector in 1984, there will be more scrap available in 1985 and this will have a negative impact upon demand for primary nickel. According to a recent OECD report, GNP growth is expected to be 3 per cent in the United States in 1985, compared to 6.75 per cent in 1984 and 5 per cent in Japan in 1985, compared to 5.75 in 1984. Growth in other major OECD countries is also forecast to be relatively strong. On this basis, primary nickel consumption is forecast to increase between 2 and 3 per cent in 1985.

Nickel prices are expected to firm in 1985 as market fundamentals improve, and nickel on the LME could average in the $US 2.30-2.45 range.

TARIFFS

Item No.		General Preferential	British Preferential	Most Favoured Nation	General
CANADA					
32900-1	Nickel ores	free	free	free	free
33506-1	Nickelous oxide	9%	10%	13.8%	25%
35500-1	Nickel and alloys containing 60% or more nickel by weight, not otherwise provided for, viz: ingots, blocks and shot; shapes or sections, billets, bars and rods, rolled, extruded or drawn (not including nickel processed for use as anodes); strip; sheet and plate (polished or not); seamless tube	free	free	free	free
35505-1	Rods containing 90% or more nickel, when imported by manufacturers of nickel electrode wire for spark plugs, for use exclusively in manufacture of such wire for spark plugs in their own factories	free	free	free	10%
35510-1	Metal alloy strip or tubing, not being steel strip or tubing, containing not less than 30% by weight of nickel and 12% by weight of chromium, for use in Canadian manufactures	free	free	free	20%
35515-1	Nickel and alloys containing 60% by weight or more of nickel, in powder form	free	free	free	free
35520-1	Nickel or nickel alloys, namely: matte, sludges, spent catalysts and scrap and concentrates other than ores	free	free	free	free
35800-1	Anodes of nickel	free	free	free	10%
37506-1	Ferronickel	free	free	4.7%	5%
44643-1	Articles of nickel or of which nickel is the component material of chief value, of a class or kind not made in Canada, when imported by manufacturers of electric storage batteries for use exclusively in manufacture of such storage batteries in own factories.	5.5%	8.8%	8.4%	20%

MFN Reductions under GATT (effective January 1 of year given)	1983	1984	1985	1986	1987
	(%)				
33506-1	13.8	13.4	13.1	12.8	12.5
37506-1	4.7	4.5	4.3	4.2	4.0
44643-1	8.4	8.0	7.6	7.2	6.8

TARIFFS (cont'd)

Item No.		General Preferential	British Preferential	Most Favoured Nation	General
UNITED STATES					
419.72	Nickel oxide		free		
423.90	Mixtures of two or more inorganic compounds in chief value of nickel oxide		free		
601.36	Nickel ore		free		
603.60	Nickel matte		free		
606.20	Ferronickel		free		
620.03	Unwrought nickel		free		
620.04	Nickel waste and scrap		free		
620.32	Nickel powders		free		
620.47	Pipe and tube fittings if Canadian article and original motor vehicle equipment		free		

Item No.		1983	1984	1985	1986	1987
				(%)		
419.70	Nickel chloride	4.4	4.2	4.0	3.9	3.7
419.74	Nickel sulfate	4.1	3.9	3.7	3.4	3.2
419.76	Other nickel compounds	4.4	4.2	4.0	3.9	3.7
426.58	Nickel salts: acetate	4.4	4.2	4.0	3.9	3.7
426.62	Nickel salts: formate	4.4	4.2	4.0	3.9	3.7
426.64	Nickel salts: other	4.4	4.2	4.0	3.9	3.7
620.08	Nickel plates and sheets, clad	9.0	8.3	7.5	6.8	6.0
620.10	Other wrought nickel, not cold worked	4.3	4.1	3.9	3.7	3.5
620.12	Other wrought nickel, cold worked	5.9	5.6	5.3	5.0	4.7
620.16	Nickel, cut, pressed or stamped to nonrectangular shapes	7.3	6.8	6.4	5.9	5.5
620.20	Nickel rods and wire, not cold worked	4.4	4.2	4.0	3.9	3.7
620.22	Nickel rods and wire, cold worked	5.9	5.6	5.3	5.0	4.7
620.26	Nickel angles, shapes and sections	7.3	6.8	6.4	5.9	5.5
620.30	Nickel flakes, per pound	2.5¢	1.9¢	1.2¢	0.6¢	free
620.40	Pipes, tubes and blanks, not cold worked	2.8	2.7	2.6	2.6	2.5
620.42	Pipes, tubes and blanks, cold worked	3.5	3.4	3.3	3.1	3.0
620.46	Pipe and tube fittings	6.3	5.6	5.0	4.3	3.6
620.50	Electroplating anodes, wrought or cast, of nickel	4.4	4.2	4.0	3.9	3.7
642.06	Nickel wire strand	5.9	5.6	5.3	5.0	4.7
657.50	Articles of nickel, not coated or plated with precious metal	7.3	6.8	6.4	5.9	5.5

Sources: The Customs Tariff and Commodities Index, January 1983, Revenue Canada; Tariff Schedules of the United States Annotated 1983, USITC Publication 1317; U.S. Federal Register, Vol. 44, No. 241.

TABLE 1. CANADA, NICKEL PRODUCTION, TRADE AND CONSUMPTION, 1982-84

	1982		1983		1984P	
	(tonnes)	($000)	(tonnes)	($000)	(tonnes)	($000)
Production[1]						
All forms						
Ontario	62 564	429,271	94 621	595,165	138 417	925,871
Manitoba	26 017	171,665	27 215	171,186	35 778	239,320
Total	88 581	600,936	121 836	766,351	174 195	1,165,191
Exports					(Jan. - Sept. 1984)	
Nickel in ores, concentrates and matte[2]						
Norway	19 736	136,888	22 812	116,908	24 646	151,815
United Kingdom	7 299	50,925	17 271	116,654	20 356	140,889
Japan	2	5	4	33	-	-
United States	-	-	-	-	119	341
Total	27 037	187,818	40 087	233,595	45 121	293,045
Nickel in oxides						
United States	4 733	36,363	5 501	44,631	5 895	..
EEC	5 285	40,599	2 237	18,149	1 590	..
Other countries	3 109	23,888	3 429	27,821	5 966	..
Total	13 127	100,850	11 167	90,601	13 451	98,249
Nickel and nickel alloy scrap						
United States	2 123	7,141	2 524	10,176	2 990	13,214
Netherlands	622	775	329	1,526	3 328	17,579
Austria	-	-	61	410	-	-
South Korea	92	630	19	79	131	804
Other countries	433	1,430	54	145	1 394	6,877
Total	3 270	9,976	2 987	12,336	7 843	38,474
Nickel anodes, cathodes, ingots, rods						
United States	36 937	251,851	37 370	232,424	32 754	..
EEC	12 974	88,462	17 364	107,996	11 032	..
Other countries	12 403	84,569	12 215	75,971	14 654	..
Total	62 314	424,882	66 949	416,391	58 440	351,004
Nickel and nickel alloy fabricated material, nes						
United States	8 385	65,751	7 745	62,465	6 439	49,925
South Africa	244	2,334	676	5,332	10	96
Belgium-Luxembourg	256	1,427	603	3,367	405	3,076
Hong Kong	-	5	540	3,015	27	183
United Kingdom	259	2,133	221	1,862	237	1,629
Japan	460	1,769	134	1,028	260	1,760
Other countries	804	7,852	447	3,165	425	3,245
Total	10 408	81,271	10 366	80,234	7 803	59,914
Imports						
Nickel in ores, concentrates and scrap						
Australia	4 496	20,867	6 601	23,492	2 718	11,483
United States	9 324	12,568	12 316	15,011	6 745	10,548
United Kingdom	434	556	2 106	3,676	5 272	8,476
Belgium-Luxembourg	5 744	5,733	3 650	3,173	1 937	1,855
Norway	-	-	2 731	2,245	-	-
Other countries	2 357	2,908	1 877	1,672	391	605
Total	22 355	42,632	29 281	49,269	17 073	21,967

TABLE 1. (cont'd.)

	1982		1983		1984P	
	(tonnes)	($000)	(tonnes)	($000)	(tonnes) (Jan.-Sept. 1984)	($000)
Nickel anodes, cathodes, ingots, rods						
Norway	1 603	11,107	1 045	5,808	2 054	15,717
United States	908	5,454	654	4,185	741	5,022
United Kingdom	37	314	444	3,850	39	266
Netherlands	18	78	122	815	-	-
Other countries	22	168	92	472	22	141
Total	2 588	17,121	2 357	15,130	2 856	21,146
Nickel alloy ingots, blocks, rods and wire bars						
United States	969	6,891	607	6,487	449	5,270
Dominican Republic	-	-	347	692	-	-
West Germany	1	6	42	269	26	163
Belgium-Luxembourg	-	-	2	13	-	-
Total	970	6,897	998	7,461	475	5,435
Nickel and alloy plates, sheet, strip						
United States	934	8,411	424	5,626	382	5,249
West Germany	388	2,802	508	3,078	372	2,561
Sweden	-	-	26	161	5	29
Other countries	2	40	1	10	6	49
Total	1 324	11,253	959	8,875	765	7,888
Nickel and nickel alloy pipe and tubing						
Sweden	600	6,881	325	4,518	246	2,519
United States	314	5,329	106	2,041	100	1,677
West Germany	108	1,752	70	958	45	695
Other countries	48	466	24	389	28	435
Total	1 070	14,428	525	7,906	419	5,326
Nickel and alloy fabricated material, nes						
United States	582	14,172	516	11,147	350	8,540
United Kingdom	212	2,133	125	1,050	26	473
West Germany	34	381	66	498	49	447
Japan	3	8	4	16	1	4
Other countries	7	77	-	15	10	89
Total	838	16,771	711	12,726	436	9,553
Consumption[3]	6 637	..	..	..	..	..

Sources: Energy, Mines and Resources Canada; Statistics Canada.
[1] Refined nickel and nickel in oxides and salts produced, plus recoverable nickel in matte and concentrates exported.
[2] For refining and re-export. [3] Consumption of nickel, all forms (refined metal and in oxides and salts) as reported by consumers.
P Preliminary; - Nil; .. Not available; nes Not elsewhere specified.

TABLE 2. CANADA, NICKEL PRODUCTION, TRADE AND CONSUMPTION, 1970, 1975, 1979-83

	Production[1]	Exports				Imports[2]	Consumption[3]
		In Matte etc.	In Oxide Sinter	Refined Metal	Total		
				(tonnes)			
1970	277 490	88 805	39 821	138 983	267 609	10 728	10 699
1975	242 180	84 391	38 527	91 164	214 082	12 847	11 308
1979	126 482	42 735	17 190	84 809	144 734	2 516	8 336
1980	184 802	42 647	16 989	88 125	147 761	4 344	9 676
1981	160 247	53 841	14 390	79 935	148 166	2 335	8 603r
1982	88 581	27 037	13 127	62 314	102 478	2 588	6 637
1983P	121 836	40 087	11 167	66 949	118 203	2 357	5 015

Sources: Energy, Mines and Resources Canada; Statistics Canada.
[1] Refined metal and nickel in oxide and salts produced, plus recoverable nickel in matte and concentrates exported;   [2] Refined nickel, comprising anodes, cathodes, ingots, rods and shot;   [3] Consumption of nickel, all forms (refined metal, and in oxides and salts), as reported by consumers.
P Preliminary;   r Revised.

TABLE 3. CANADIAN PROCESSING CAPACITY, 1984

	Inco			Falconbridge	Sherritt Gordon
	Port Colborne	Sudbury	Thompson	Sudbury	Fort Saskatchewan
	(tpy of contained nickel)				
Smelter	n.a.	127 000[1]	81 600	45 000	n.a.
Refinery	65 000[2]	56 700	55 000	n.a.	17 500

[1] Reduced from 154 200 t due to a government regulation on SO_2 emissions imposed in 1980.
[2] Electrolytic nickel portion of refinery closed in 1984, only utility being produced at year end.
n.a. Not applicable.

TABLE 4. WORLD MINE PRODUCTION OF NICKEL, 1982 AND 1983

	1982	1983
	(tonnes)	
U.S.S.R.	170 000	175 000
Canada[1]	88 600	121 800
Australia	88 600	78 700
New Caledonia	60 100	45 000
Cuba	37 600	39 000
Indonesia	48 500	38 400
South Africa	20 500	20 500
Dominican Republic	6 000	20 200
Botswana	17 800	18 200
People's Republic of China	12 000	15 000
Other	33 900	78 100
Total	625 100	649 900

Sources: Energy, Mines and Resources Canada; World Bureau of Metal Statistics.
[1] Refined nickel and nickel in oxide and salts produced, plus recoverable nickel in matte and concentrates produced.

TABLE 5. WORLD CONSUMPTION OF NICKEL, 1982 AND 1983

	1982	1983
	(tonnes)	
U.S.S.R.	138 000	140 000
United States	94 300	127 800
Japan	106 700	114 800
Germany, F.R.	57 700	63 000
France	31 800	32 500
Italy	24 000	22 500
United Kingdom	22 500	21 800
People's Republic of China	19 000	19 000
Sweden	15 000	16 400
India	11 000	13 000
Other	60 500	101 000
Total	624 500	671 300

Sources: World Bureau of Metal Statistics; Energy, Mines and Resources Canada; U.S. Department of the Interior.

TABLE 6. FORECAST OF CANADIAN MINE PRODUCTION

Year	Inco Sudbury[1]	Inco Thompson	Falconbridge Sudbury	Total
		(tonnes)		
1985	106 000	35 000	33 000	174 000
1986	107 000	42 000	33 500	182 500
1987	108 000	45 000	33 500	186 500
1988	108 000	49 000	34 000	191 000
1989	109 000	50 000	35 000	194 000
1990	110 000	50 000	35 500	195 500
1991	110 000	50 000	33 000	193 000
1992	110 000	50 000	33 000	193 000
1993	111 000	50 000	34 000	195 000
1994	111 000	50 000	34 000	195 000
1995	111 000	50 000	35 000	196 000
1996	112 000	50 000	35 000	197 000
1997	113 000	50 000	35 000	198 000
1998	114 000	50 000	36 000	200 000
1999	115 000	50 000	36 000	201 000
2000	116 000	50 000	37 000	203 000

[1] Includes Shebandowan.

TABLE 7. NICKEL PRICES IN UNITED STATES DOLLARS PER POUND, 1983

	1st Quarter Average	2nd Quarter Average	3rd Quarter Average	4th Quarter Average	Annual Average
Cathodes					
- New York dealer[1]	1.98	2.31	2.27	2.17	2.18
- major producer	3.20	3.20	3.20	3.20	3.20
LME CASH	1.94	2.22	2.20	2.10	2.12
Briquettes/Western	3.20	3.20	3.20	3.20	3.20
Falconbridge, ferronickel[1]	3.18	3.18	3.18	3.18	3.18
Hanna, ferronickel[1]	3.16	3.16	3.16	3.16	3.16

Source: Metals Week.
[1] Per pound of contained nickel.

NICKEL IN CANADA, 1984

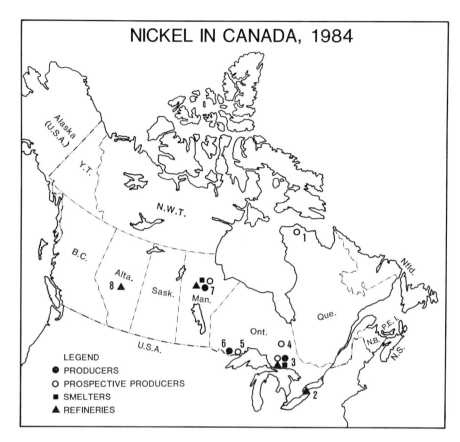

Producers, prospective producers, smelters and refineries
(numbers refer to locations on map above)

Producers

3. Falconbridge Limited
   (East, Falconbridge, Fraser, Lockerby, North, Strathcona
   Inco Limited
   Clarabelle, Copper Cliff North, Copper Cliff South, Creighton, Frood, Garson, Levack, Little Stobie, McCreedy West and Stobie)
6. Inco Limited (Shebandowan mine)
7. Inco Limited (Pipe open pit and Thompson)

Prospective Producers

1. New Quebec Raglan Mines Limited
3. Falconbridge Limited
   (Craig, Lindsley, Onaping, Onex, Thayer mines)
   Inco Limited (Coleman, Crean Hill, Murray, Totten)
4. Teck Corporation (Moncalm Township)
5. Great Lakes Nickel Limited (Pardee Township)
7. Inco Limited (Thompson open pit, Soab, North, Soab, South, Birchtree, Pipe No. 1)

Smelters

3. Falconbridge Limited (Falconbridge)
3. Inco Limited (Sudbury)
7. Inco Limited (Thompson)

Refineries

2. Inco Limited (Port Colborne)
3. Inco Limited (Sudbury)
7. Inco Limited (Thompson)
8. Sherritt Gordon Mines Limited (Fort Saskatchewan)

44.13

Phosphate

G.S. BARRY

Naturally occurring rock deposits are the most common source of phosphorus; other sources are bones, guano, and some types of iron ores that yield byproduct basic slag containing sufficient phosphorus to warrant grinding and marketing.

Phosphate rock, contains one or more suitable phosphate minerals, usually calcium phosphate, in sufficient quantity for use, either directly or after beneficiation, in the manufacture of phosphate products. Sedimentary phosphate rock, or phosphorite, is the most widely used phosphate raw material. Apatite, which is second in importance, occurs in many igneous and metamorphic rocks.

Phosphate rock is graded either on the basis of its P_2O_5 equivalent (phosphorus pentoxide) or its $Ca_3(PO_4)_2$ content (tricalcium phosphate of lime or bone phosphate of lime - TPL or BPL). For comparative purposes, 0.458 unit P_2O_5 equals 1.0 unit BPL, and 1 unit of P_2O_5 contains 43.6 per cent phosphorus.

Approximately 80 per cent of world phosphorus production goes into fertilizers; other products which require the use of phosphorus include organic and inorganic chemicals, soaps and detergents, pesticides, insecticides, alloys, animal-food supplements, motor lubricants, ceramics, beverages, catalysts, photographic materials, and dental and silicate cements.

A severe world contraction in the demand for phosphatic fertilizers began in 1981 and continued until mid-1983. Since that time there is a clear steady improvement in phosphate markets which as measured by deliveries of phosphate rock rose by 11.7 per cent from 123.3 million t in 1982 to 137.8 million t in 1983. There was further improvement in the first nine months of 1984 with deliveries by major western world producers advancing by 13.5 per cent (68.9 million t in 1984 compared to 60.7 million t in 1983). It is estimated that world phosphate rock production in 1984 will exceed the record of 139.4 million t achieved in 1981 by two to three million t. A further positive element is a decrease in stocks held by producers, particularly in the USA. Stocks held by major western world producers were 34.2 million t at their peak on July 1, 1982 and declined to 22.2 million t on October 1, 1984. Among the traditional large producers and exporters, Brazil, Israel, Jordan, Morocco, Senegal, Togo, Tunisia and the USA recorded increased production while only South Africa and Syria experienced cutbacks in 1983, whereas in 1984 only South Africa and Tunisia experienced a significant decline. Iraq brought new production on stream. The United States which experienced an unprecedented drop in production of 30 per cent between 1981 and 1982, recorded in 1983 a 8.5 per cent increase in production and a 21 per cent increase in deliveries. A further production increase of 18 per cent is estimated for 1984.

World exports of phosphate rock increased by 5.9 per cent from 43.8 million t in 1982 to 46.4 million t in 1983. Exports in 1984 are expected to be over 48 million t.

OCCURRENCES IN CANADA

Known Canadian deposits are limited and fall into three main categories: apatite deposits within Precambrian metamorphic rocks in eastern Ontario and southwestern Quebec; apatite deposits in some carbonate-alkaline complexes (carbonatites) in Ontario and Quebec; and Late Paleozoic-Early Mesozoic sedimentary phosphate rock deposits in the southern Rocky Mountains. Phosphatic mineralization was also reported in the layered rocks of the Athabasca series.

The deposit of greatest economic significance is the Kapuskasing (Cargill) phosphate deposit, where early studies indicated the presence of about 60 million t of ore grading 20.2 per cent P_2O_5. The property was optioned by Sherritt Gordon Mines Limited in 1979 from International Minerals & Chemical Corporation (Canada)

G.S. Barry is with the Mineral Policy Sector, Energy, Mines and Resources, Canada. Telephone (613) 995-9466.

Limited (IMCC). The option was exercised in December 1983. Preliminary designs for an open pit at Cargill, based on IMCC's grades and tonnages, allowed for the production of 450 000 tpy of 39 per cent P_2O_5 concentrates for a minimum of 17 years. Additional drilling, test pits and bulk sample pilot plant testing confirmed the technical viability of this deposit, and a preliminary feasibility study on the mining of higher grade material produced encouraging economics despite the low phosphate prices used in the study.

Another important carbonatite deposit was discovered by Shell Canada Resources Limited near Martison Lake north of Hearst, Ontario. In December 1982 the deposit was purchased by New Venture Equities Ltd. which formed a 50-50 joint venture with Camchib Mines Inc. for further exploration and development. Camchib Mines Inc. is wholly owned by Campbell Resources Inc. The Joint Venture continued with detailed, fill-in drilling on the property and announced in August 1983 that higher grade zones of the deposit contain 57 million t grading 23 per cent P_2O_5. A $1.2 million additional drilling program was completed in 1984.

In July 1984, Sherritt Gordon Mines, Campbell Resources and New Venture Equities combined forces on a 50/50 joint venture on the two phosphate properties at Cargill and Martison Lake. The company is confident that when North American market conditions improve the development at one or both properties could be economic. Depending on these conditions a decision on a full scale feasibility study will be made in 1985.

Additional details on the Canadian phosphate deposits and industry were provided in the publication MR 193, "Phosphate Rock, an Imported Mineral Commodity" (December 1981).

CANADIAN PHOSPHATE INDUSTRY

Phosphate Rock. In 1983, Canada imported only 2.66 million t of phosphate rock; for the first nine months of 1984 imports increased by 25 per cent. The general economic recession was responsible for the low import levels. Imports averaged 3 058 463 t from 1975 to 1981. Approximately 75 per cent of the phosphate rock is used in fertilizer production, 18 per cent in elemental phosphorus production and 6 per cent in calcium phosphate production.

About 70 per cent of Canada's imports of phosphate rock from the United States has been from Florida since the late 1970s. The remainder was from western states. Purchase practices, which include commercial factors as well as the characteristics of rock used by the fertilizer plants, point to the continuation of this pattern of supply for at least several years.

Currently, eastern Canada is supplied from Florida. From 700 000 t to 750 000 t are transported by sea, with three quarters of this total being used for elemental phosphorus production and the remainder for fertilizer production in New Brunswick.

Approximately 450 000 t to 500 000 t of phosphate rock is shipped annually by rail from Florida mines to Ontario fertilizer plants because generally for this part of Canada direct unit train rail service is more advantageous than ocean shipping combined with short overland hauls. The fact that shipments in Florida do not have to be routed via the congested port of Tampa is another positive factor. Another advantage is that railroad shipments can be maintained at a schedule that allows for very low inventories. However, recession related, ocean shipping costs were so low during 1983 and most of 1984, that they were more than competitive, creating a situation where finished fertilizers could be shipped at less than the cost of shipping phosphate rock by rail.

Florida is the source of phosphate rock for about 55 to 60 per cent of the five western Canadian fertilizer plants and western U.S. states for some 40 to 45 per cent. Rock shipped from Florida via the Panama Canal to Vancouver is mainly transported as back-haul to Canadian lumber (to United States) and potash exports (to South America). The inland rail haul from Vancouver to the Edmonton area is a back-haul to exports of potash. Total shipping costs are competitive with rail haul from mines in the western United States.

Belledune Fertilizer (a Division of Noranda Inc.) produced 140 000 t of DAP in 1983 (152 000 t in 1982) at its New Brunswick fertilizer plant. Shipments were higher by some 20 000 t resulting in a substantial fall in inventories. The 1984 DAP production is estimated at over 170 000 t.

International Minerals & Chemical Corporation operated its Port Maitland fertilizer plant at just below 70 per cent capacity. IMCC produced commercial phosphoric acid, single super phosphate (SSP), triple super phosphate (TSP), mono-ammonium phosphate (MAP), and a calcium phosphate. The plant had considerable profitability problems and was closed at the end of June 1984 for an indefinite period (currently the plant is "mothballed" for a duration of about two years).

C-I-L Inc. operated its phosphate plant intermittently, one to two weeks per month averaging less than 50 per cent capacity utilization, and producing about 80 000 t of ammonium phosphates. The company uses smelter and waste sulphuric acid. C-I-L is now expanding its nitrogen capacity from 350 000 t to 750 000 t of ammonia for completion in mid-1985. The company will continue to produce phosphatic fertilizers at the current low production levels.

Cominco Ltd. shut down its Kimberley fertilizer plant for eight weeks in mid-1983 for annual maintenance and inventory control. Production of ammonium fertilizers at the Trail and Kimberley plants combined was 273 000 t in 1983 and 309 000 t in 1984.

Esso Chemical Canada completed its $400 million fertilizer plant expansion during 1983 on target and on budget. Costs for the expansion of the phosphate plant from 204 000 tpy to 370 000 tpy capacity were about $50 million. The phosphate part of the fertilizer plant was actually completed in 1982 but was never put on stream until September 1983. The associated sulphuric acid plant, also completed has not yet been commissioned since the company finds it more economic at present to buy sulphuric acid from Sherritt Gordon and other commercial sources. Since September 1983 production of ammonium phosphates was carried at almost full capacity, with part of the output being tolled for Sherritt Gordon Mines.

Western Co-operative Fertilizers Limited operated its Calgary plant throughout the year except for a two-week maintenance shutdown in mid-summer. The plant produced over 200 000 t of ammonium phosphate fertilizers in 1983 and over 250 000 t in 1984, mainly MAP. The Medicine Hat plant remained closed throughout the year. The company imports phosphate rock from Idaho.

Sherritt Gordon Mines closed its phosphate production facilities at the Fort Saskatchewan plant in September 1983 and will remain shut down for two years. During this period requirements for phosphate fertilizer will be provided by Esso Chemical under a tolling agreement. Sherritt will convert Esso's ammonia into a tonnage of urea equivalent to the tonnage of phosphates provided by Esso.

Elemental phosphorus. ERCO Industries Limited operates two thermal reduction plants in Canada where elemental phosphorus is produced by the smelting of a mixture of phosphate rock, coke and silica. One tonne of phosphorus requires the input of about 10 t of phosphate rock (60 to 67 per cent BPL), 2 t of coke and 3 t of silica.

ERCO has plants at Varennes, Quebec with a 22 500 t annual capacity (P_4) and at Long Harbour, Newfoundland with an effective capacity of about 60 000 tpy. Until recently the elemental phosphorus production from Long Harbour was almost exclusively reserved for Albright & Wilson, Inc. derivative plants in Europe, but in 1982, 1983 and 1984 a proportion was sent to Buckingham, Quebec and Port Maitland, Ontario to supplement supplies from Varennes, Quebec. In late 1984, ERCO restarted the second main furnace idle since 1982. The company also put on stream the second of two smaller furnaces to recover up to 2 000 t annually of phosphorus from the "mud" produced as a byproduct of furnace operation and "mud" stored on site in the past.

The Long Harbour, Newfoundland plant operated at full capacity during 1983. In total, the ERCO plants use from 600 000 to 650 000 tpy of Florida phosphate rock. Since the low-grade phosphate rock acceptable for thermal reduction cannot be used by the fertilizer industry, it can be purchased at relatively lower prices (per P_2O_5 unit value).

Production from Varennes, Quebec is 90 per cent or more oriented toward Canadian markets. The elemental phosphorus (P_4) produced at Varennes is shipped to two ERCO plants, one at Buckingham, Quebec and the other at Port Maitland, Ontario. At Buckingham about 9 000 tpy of P_4 is used to produce technical and food grade phosphoric acid (95 per cent H_3PO_4) and 1 000 t to produce amorphous red phosphorus and phosphorus sesquisulphide.

ERCO's Port Maitland plant operates on phosphorus from Varennes and Long Harbour, using about 12 000 tpy. It is all converted to technical grade phosphoric acid.

Coproducts of elemental phosphorus are ferrophosphorus, carbon monoxide and calcium silicate slag. Ferrophosphorus contains 20 to 25 per cent phosphorus and is used by the steel industry as a direct source of the phosphorus needed in some types of steel.

Phosphate fertilizers. Nine Canadian plants (Table 3) produce wet phosphoric acid by the dihydrate process in which 28 to 30 per cent P_2O_5 acid is the principal product and gypsum is the waste product. Three of the nine plants are now idle. At present, there is no use for the gypsum and it accumulates in large settling ponds near all the plants except one in New Brunswick where it is disposed of in the sea.

Canadian phosphoric acid plants are designed to operate on phosphate rock which grades between 69 and 72 per cent BPL (31.1 to 33.0 per cent P_2O_5). The first stage of acid production, which is digestion and filtration, produces "filter acid" grading 28 to 30 per cent P_2O_5. This product is then upgraded by evaporation to about 40 to 44 per cent acid for most in-plant use, or to 52 to 54 per cent P_2O_5 for commercial sales or specialized uses. The evaporation step is energy intensive, and the provenance of sulphuric acid has a bearing on energy consumption. Plants using elemental sulphur as the source of in-plant sulphuric acid production have their evaporation energy requirements met by heat generated in the sulphuric acid plants since the process is exothermic, (i.e., 1 t of sulphur has a BTU content equivalent to about 2 barrels of oil). Plants using commercial sulphuric acid, (e.g., produced from SO_2 smelter gases) have to generate vapour requirements with natural gas or coal-fired boilers. To balance energy requirements, an efficient dihydrate WPA plant could theoretically operate using elemental sulphur for 70 to 75 per cent of its requirements and purchased sulphuric acid for the remainder.

Most phosphate rock contains uranium. It is in small enough quantities not to present any problems for fertilizer production. In Canada, Earth Sciences Inc. (ESI) started a uranium recovery plant in Calgary in 1980. It treats phosphoric acid from the adjoining plant of Western Co-operative Fertilizers Limited, and returns the acid to the owner. The plant was placed on standby in November 1981. The plant underwent major modifications during 1982 and 1983 and was re-opened in May 1983. The recovered yellow cake is shipped to British Nuclear Fuels Ltd. in the United Kingdom for refining and then returned to the United States. In 1982 Urangesellschaft Canada Limited acquired 49 per cent interest in the Calgary ESI plant.

Capacity of Canadian phosphoric acid plants is expressed in 100 per cent P_2O_5 equivalent and the total installed annual capacity is currently estimated at 1 146 000 t. Some of this capacity is now idle and mothballed. Efficient plants can consistently operate at 90 to 95 per cent of nameplate capacity. Most Canadian plants, gauge their annual production levels to corporate marketing strategies and fertilizer demand forecasts. At times when agricultural demand is low Canadian production capacities are seriously underutilized. The recovery of P_2O_5 from phosphate rock i.e. the efficiency of conversion varies from 88 to 94 per cent.

All of the nine phosphoric acid plants in Canada are integrated to produce phosphatic fertilizers, mainly ammonium phosphates. Ammonium phosphates are produced by a neutralization reaction of phosphoric acid with ammonia and, depending on the proportions of the original constituents, either diammonium phosphate (DAP) (18-46-0) or mono-ammonium phosphate (MAP) (range from 11-48-0 to 11-55-0) are produced. Another common grade particularly in the West is 16-20-0.

Canadian fertilizer plants produce annually between 800 000 t and 950 000 t of mono-ammonium phosphates (MAP), between 250 000 t and 300 000 t of diammonium phosphate (DAP) and about 250 000 t of other ammonium and ammonium-nitrate phosphates. Up to 100 000 t of normal (SSP) and triple (TSP) superphosphate is also produced but this production is on a decline.

Calcium Phosphate. Two fertilizer plants in Canada use phosphoric acid for the production of calcium phosphates that are used mainly for supplementing the calcium and phosphorus content of animal and poultry feedstocks. The two products are: monocalcium phosphate (21 per cent phosphorus) or dicalcium phosphate (18.5 per cent phosphorus).

The phosphoric acid used for calcium phosphate production in eastern Canada was all produced by IMCC in Port Maitland, Ontario. The company used more than half for its own requirements and sold the remainder to Cyanamid Canada Inc. which has a nearby plant at Welland. After the mid-1984 closure of the IMCC plant the company supplied markets for these products through imports.

WORLD DEVELOPMENTS

World phosphate rock production in 1983 was estimated at 130.8 million t, an increase of 8.1 per cent from 1982. Western world production was 91.4 million t, an increase of 8.4 per cent from the year before. In 1984 world phosphate rock production was estimated to exceed 143 million t.

The uptrend in both production and sales was strong in 1983 and 1984, particularly since the second half of 1983 following two and a half years of recession. Major producers like the United States and Morocco increased production by 28.3 and 19.2 per cent respectively between 1982 and 1984. The U.S. producers sales exceeded production allowing for a decline in producers inventories to 14.3 million t at the end of 1983, and to 13.0 million t at the end of September 1984, approaching the low level of 1980. There was an increase in production in the USSR and China. Among lesser but significant world producers Brazil, Israel, Jordan, Norway, Senegal and Tunisia recorded substantial production increases; only South Africa recorded a decline.

The world phosphate rock industry however, still suffered from low prices since there remains a large excess of capacity. In 1983 the approximate utilization of capacity in the USA was only 71 per cent and in Africa 67 per cent. Despite this, a number of large new mines were recently put on stream or are in advanced stages of completion. Table 6 lists 43 important active and potential phosphate mine projects. The timely development of even half of these will ensure ample supply of phosphate rock well into the next century. There is however a continuous decline in the average grade and purity of phosphate rock concentrates and this trend is expected to continue making the development of deposits that will yield exceptionally high grade concentrates such as the Canadian Precambrian deposits progressively more attractive.

PRICES

Most phosphate rock is purchased under producer-consumer negotiated prices which depart from listed prices in consideration of volume, transportation conditions and local competitive conditions. Phosrock Ltd., a Florida-based marketing organization which represents about two thirds of producers for export markets lists prices as shown in Table 7. International prices are also quoted by Office Cherifien des Phosphates (OCP) fob ports of Safi or Casablanca. are usually $2 to $4 above Tampa prices, the difference reflecting competitive conditions, for "landed" prices to most European destinations.

The average unit price of phosphate rock sold or used in the USA was $US 23.98 per tonne fob plant in 1983 compared to $26.63 in 1981 and $25.50 in 1982. The average value of phosphate rock exported decreased from $US 29.83 per tonne fob mine in 1982 to $27.26 in 1983. For fertilizer year 1983-84 the average unit price was $US 23.94/t.

OUTLOOK

The outlook for 1985 is for a continuation of higher demand under ample supply conditions. Prices for phosphoric acid and phosphatic fertilizers, although higher than in 1983 and 1984 will remain low, probaby not yet exceeding those of 1982. Under the current marketing conditions prices are 25 to 40 per cent lower than normal renumerative levels. A continuous steady increase in demand after 1985, however will be mirrored by substantially better price movements.

Most experts forecast a consumption growth for phosphate fluctuating between 3.5 and 4.5 per cent annually for the next few years. The USA demand should average 2 per cent growth after the considerable improvement in 1983 and 1984.

TARIFFS

Item No.		British Preferential	Most Favoured Nation	General	General Preferential
			(%)		
CANADA					
93100-2	Phosphate rock	free	free	free	free
66345-1	Defluorinated calcium phosphates for use in the manufacture of animal or poultry feeds	free	free	free	free
93103-1	Calcium phosphate dibasic	free	free	free	free
93103-2	Calcium phosphate, disintegrated, calcined, thermophosphates, fused phosphates; superphosphates	free	free	free	free
92840-1	Phosphites, phosphorus, hypophosphites and phosphates	10	13.8	25	9
—	Sodium phosphate disbasic, and monobasic, pharmacopoeial tribasic, commercial grade; sodium pyrophosphate; sodium tripolyphosphate (temporary rate reduction 3/06/80 to 30/06/87)	free	13.8	25	free
92840-2	Di-calcium phosphate (temporary rate reduction 3/06/80 to 31/12/86)	9.4 free	7.5 7.5	25 25	5.0 free
93100-1	Fertilizers; goods for use as fertilizers	free	free	free	free
93105-1	Ammonium phosphates	free	free	free	free

MFN Reductions under GATT (effective January 1 of year given)	1983	1984	1985	1986	1987
			(%)		
92840-1	13.8	13.4	13.1	12.8	12.5
92840-2	7.5	5.6	3.8	1.9	free

UNITED STATES, Customs Tariffs (MFN)

		1983	1984	1985	1986	1987
				(%)		
420.92	Sodium phosphate containing over 45% water	2.8	2.7	2.7	2.6	2.5
421.22	Pyrophosphates	4.4	4.2	4.0	3.9	3.7
606.33	Ferrophosphorus	4.2	3.8	3.3	2.9	2.4

Sources: The Customs Tariff, 1983, Revenue Canada; Customs and Excise. Tariff Schedules of the United States Annotated (1983), USITC Publication 1317; U.S. Federal Register Vol. 44, No. 241.

TABLE 1. CANADA, PHOSPHATE ROCK IMPORTS, 1982-84, AND CONSUMPTION, 1981-84

	1982		1983		1984	
	(tonnes)	($)	(tonnes)	($)	(tonnes)	($)
					(Jan.-Sept. 1984)	
Imports						
United States	2 482 583	101,704,000	2 662 725	102,194,000	2 593 556	99,097,000
Other countries	29 141	1,503,000	-	-	-	-
	2 511 724	103,207,000	2 662 725	102,194,000	2 593 566	99,097,000

	1981	1982	1983e	1984e
	(tonnes)	(tonnes)	(tonnes)	(tonnes)
Consumption[1]				
Eastern Canada	1 364 839	1 222 520	1 132 000	1 222 000
Western Canada	2 217 847	1 159 151	1 698 000	1 807 000
Total	3 582 686	2 381 671	2 830 000	3 029 000

Sources: Statistics Canada; Energy, Mines and Resources Canada.
[1] Breakdown by Energy, Mines and Resources Canada.
e Estimated; - Nil.

TABLE 2. CANADA, PHOSPHATE FERTILIZER SHIPMENTS, 1978-84[1]

	1978/79	1979/80	1980/81	1981/82r	1982/83	1983/84
	(tonnes P_2O_5 equivalent)					
Domestic markets:						
Atlantic provinces	18 867	19 441	24 481	26 261	29 443	24 965
Quebec	23 540	20 992	28 610	34 915	43 308	37 835
Ontario	63 379	54 602	82 496	71 033	71 959	79 166
Manitoba	89 576	110 382	97 529	75 239	81 907	90 529
Saskatchewan	131 636	131 500	135 534	144 998	153 784	195 170
Alberta	140 880	131 413	149 116	152 906	157 010	161 185
British Columbia	12 440	14 204	13 308	8 998	10 970	11 311
Total Canada	480 318	482 533	531 074	514 350	548 381	600 161
Export markets:						
United States	144 670	146 813	194 565	141 411	82 478	65 790
Offshore	46 814	44 999	77 328	20 305	715	4 652
Total exports	191 484	191 812	271 893	161 716	83 193	70 442
Total shipments	671 803	674 344	802 968	676 066	631 574	670 603

Source: Canadian Fertilizer Institute.
[1] Fertilizer year: July 1 to June 30; not 100% industry coverage.
r Revised.
Note: Totals may not add due to rounding.

TABLE 3. CANADA, PHOSPHATE FERTILIZER PLANTS

Company	Plant Location	Annual Capacity (tonnes) (P_2O_5 eq.)	Principal End Products	Source of Phosphate Rock	Basis for H_2SO_4 Supply for Fertilizer Plants
Eastern Canada					
Belledune Fertilizer Div. of Noranda Inc.	Belledune, N.B.	150 000	am ph	Florida	SO_2 smelter gas
C-I-L Inc.	Courtright, Ont.	90 000	am ph	Florida	SO_2 smelter gas, pyrrhotite roast and waste acid
International Minerals & Chemical Corporation (Canada) Limited (IMCC)	Port Maitland Ont.	118 000*	H_3PO_4, ss ts, ca ph	Florida	Sulphur and SO_2 smelter gas
		358 000			
Western Canada					
Cominco Ltd.	Kimberley, B.C.	86 700	am ph	Montana and Utah	SO_2 pyrite roast
	Trail, B.C.	77 300	am ph	Utah	SO_2 smelter gas
Esso Chemical Canada	Redwater, Alta.	370 000**	am ph	Florida	Sulphur
Sherritt Gordon Mines Limited	Fort Saskatchewan, Alta.	50 000***	am ph	Florida	Sulphur
Western Co-operative Fertilizers Limited	Calgary, Alta.	140 000	am ph	Idaho	Sulphur
	Medicine Hat, Alta.	65 000****		Idaho	
		788 000			
Canada: installed capacity		1 146 000			
effective capacity:					
end of 1982		915 000			
end of 1983		1 031 000			
end of 1984		913 000			

P_2O_5 eq. Phosphorus pentoxide equivalent; am ph Ammonium phosphates; ss Single superphosphate; ts Triple superphosphate; ca ph Food supplement calcium phosphate; H_3PO_4 phosphoric acid for commercial sales.
* Shutdown and mothballed for an indefinite period - July 1984. ** Expansion from 204 000 t to 370 000 t completed in 1982 and put on stream, Sept. 1983. *** Shutdown temporarily for two years - Sept. 1983 to Sept. 1985. **** Shutdown for an indefinite period - May 1982.

TABLE 4. CANADA, TRADE IN SELECTED PHOSPHATE PRODUCTS, 1982-84

	1982		1983P		1984 9 months	
	(tonnes)	($)	(tonnes)	($)	(tonnes)	($)
Imports						
Calcium phosphate						
United States	18 216	9,917,000	34 446	16,413,000	29 901	13,620,000
Other countries	52	37,000	733	280,000	946	351,000
Total	18 268	9,954,000	33 179	16,692,000	30 847	13,971,000
Fertilizers:						
Normal superphosphate, 22 per cent or less P_2O_5						
United States	188	56,000	1 368	254,000	18	3,000
Triple superphosphate, over 22 per cent P_2O_5						
United States	31 947	7,143,000	54 755	11,562,000	30 804	6,945,000
Phosphatic fertilizers, nes						
United States	216 588	61,344,000	303 022	82,682,000	197 410	58,903,000
Belgium-Luxembourg	901	547,000	673	408,000	843	484,000
Israel	183	149,000	299	213,000	182	111,000
United Kingdom	1	--	-	-	-	-
Singapore	4	10,000	-	-	-	-
Netherlands	16	8,000	101	39,000	58	21,000
Other countries	-	-	86	50,000	58	24,000
Total	217 693	62,057,000	304 181	83,392,000	198 551	59,541,000
Chemicals:						
Potassium phosphates						
United States	1 243	1,322,000	1 327	1,782,000	1 194	1,725,000
France	110	118,000	139	150,000	136	149,000
Israel	131	115,000	216	195,000	138	152,000
Netherlands	8	10,000	23	26,000	35	42,000
West Germany	-	-	24	37,000	-	-
Total	1 492	1,566,000	1 729	2,190,000	1 503	2,068,000
Sodium phosphate, tribasic						
United States	408	281,000	521	476,000	267	223,000
France	177	65,000	249	98,000	166	63,000
Belgium-Luxembourg	-	-	40	61,000	-	-
Netherlands	51	21,000	84	35,000	55	21,000
People's Republic of China	-	-	119	49,000	174	75,000
Total	636	367,000	1 013	719,000	662	382,000
Exports						
Nitrogen phosphate fertilizers, nes						
United States	272 086	62,198,000	193 724	43,052,000	128 962	28,388,000
Costa Rica	-	-	5 190	1,716,000	2 929	887,000
Jamaica	-	-	3 564	1,035,000	7 183	2,123,000
Other countries	-	-	36	15,000	28 985	7,598,000
Total	272 086	62,198,000	202 514	45,818,000	168 059	38,996,000

Source: Statistics Canada.
P Preliminary; - Nil; nes Not elsewhere specified; -- Too small to be expressed.

TABLE 5. WORLD PHOSPHATE ROCK PRODUCTION, 1981-84

	1981	1982	1983	1984e
	\multicolumn{4}{c}{(000 tonnes product)}			
WORLD TOTAL	139 348	122 913	134 621	143 000
West Europe	372	389	538	600
Finland	201	232	381	
Sweden	128	131	107	
Turkey	43	26	50	
East Europe	26 200	27 200	27 700	28 000
U.S.S.R.	26 200	27 200	27 700	
North America	53 624	37 414	42 187	48 000
United States	53 624	37 414	42 187	
Central America	252	379	389	400
Mexico	252	379	389	
South America	2 791	2 779	3 229	3 500
Brazil	2 764	2 732	3 208	
Colombia	15	18	18	
Peru	12	29	3	
Africa	33 200	29 907	33 796	34 400
Algeria	858	946	893	
Egypt	720	708	647	
Morocco/Sahara	19 696	17 754	20 106	
Senegal	1 927	975	1 250	
South Africa	3 034	3 173	2 742	
Togo	2 244	2 035	2 081	
Tanzania	-	-	20	
Tunisia	4 596	4 196	5 924	
Zimbabwe	125	120	133	
Asia	21 408	23 251	25 077	27 000
China	10 862	11 728	12 500	
Christmas Island	1 423	1 328	1 095	
India	562	560	600	
Iraq	-	363	1 199	
Israel	2 373	2 711	2 969	
Jordan	4 244	4 431	4 749	
North Korea	500	500	500	
Syria	1 321	1 455	1 229	
Vietnam	110	160	220	
Sri Lanka	13	15	16	
Oceania	1 501	1 594	1 705	1 500
Australia	21	235	21	
Nauru	1 480	1 359	1 684	

Sources: Phosphate Rock Statistics, 1983, ISMA Ltd.;   United States Bureau of Mines (USBM), Mineral Commodity Summaries 1984.
e Estimated.
Totals may not add due to rounding.

TABLE 6. PHOSPHATE, MAJOR NEW PROJECTS AND EXPANSION PROGRAMMES

Company	Location	Design Production (ore, unless otherwise stated)	Completion Date	Capital ($ million)	Type of Operation	Remarks
CANADA						
Sherritt Gordon	Kapuskasing, Ont.	450 000 tpy phosphate			P/Cn	Evaluating Cargill property under option from IMCC. Partners being sought to fund acquisition of property and further development work. Feasibility of mining higher grade material at reduced rate under study.
New Ventures – Camchib Mines	Martison Lake, Ontario	Phosphate	late-1980s – early-1990s		P/Cn	A high grade deposit with 57 million t grading 23% P2O5 plus niobium. Additional drilling and feasibility study in 1984.
U.S.A.						
Amax Phosphate	De Soto/Manatee Co., Fla.	3.6 million tpy phosphate	1980s	250	P/Cn	Pine Level property development deferred.
C.F. Mining	Hardee Co., Fla.	2 million tpy phosphate	1985-86		P/Cn	Developments due on stream end of 1985.
Chevron	Vernal, Utah	1.8 million tpy phosphate or less	1986	250	P/Cn	Expansion from 680 000 tpy to feed Rock Springs plant in Wyoming. Project downsized from $400 million.
Estech/Royster	Duette, Fla.	2.7 million tpy phosphate	1986	135	P/Cn	Problems in obtaining Manatee County.
Grace/IMC	Hillsboro Co., Fla.	5 million tpy phosphate	1984-85	615	P/Cn	Four Corners mine development in progress. Start in fiscal 1985.
Mobil Chemical	Ft. Meade, Fla.	3.1 million tpy phosphate	late-1980s		P/Cn	Dames & Moore study for new mine at South Ford Meade.

TABLE 6. (cont'd)

Company	Location	Design Production (ore, unless otherwise stated)	Completion Date	Capital ($ million)	Type of Operation	Remarks
NC Phosphate Corp.	Aurora, North Carolina	3.6 million tpy phosphate	1987	333	P/Cn	Canvas Creek development by Agrico led j.v.
J.R. Simplot	Afton, Idaho	2 million tpy phosphate	1984		P/Cn	New mine development at Smoky Canyon as replacement for the Conda mine.
BRAZIL						
Industria de Fosfatados Catarinense	Anitapolis, Santa Catarina	216 000 tpy P_2O_5	1985	170	P/Cn	To supply ILM fertilizer plants.
Norfértil	Olinda, Pernambuco	214 000 tpy phosphate (72% TCP)	1985	53	P/Cn	Feasibility studies completed. Run of mine capacity of 600 000 tpy. Reserves 20 million t.
COLOMBIA						
IAN (Govt.)	Berlin, Caldas	U. Phosphate				Underground development work leading to prefeasibility study in collaboration with UNDP.
MEXICO						
Rofomex	Bahia de Magdalena	1.5 million tpy phosphate concs.			P/Cn	Production scheduled to start in the near future.
PERU						
Probayovar	Bayovar, Piura	3 million tpy phosphate	1988-89	700	P/Cn	Feasibility study completed by Jacobs. Investor interest now being sought.

PACIFIC

Company	Location	Product	Year	Status	Comments
Raro Moana	Tumoto Is., French Polynesia	Phosphate			Sofremines feasibility studies on Maraiva deposit.

BURUNDI

Company	Location	Product	Year	Status	Comments
Govt.	Matonga-Bandaga	Phosphate			British Sulphur Corp. prefeasibility study for World Bank completed end-1983.

EGYPT

Company	Location	Product	Year	Status	Comments
El Nars Phosphate Co.	Sebaiya West	770 000 tpy ore 440 000 tpy Phosphate concs.	1984	P/Cn	Mine expansion and 4 000 tpd beneficiation plant. Seltrust Engineering acting as purchasing agents and project managers for Abu Zaabel Fertilizer & Chemical Co.
Govt.	Abu Tartur, El Kharga	2 million tpy phosphate concs. (28% P_2O_5)	1988	U	Experimental mine in operation, 2nd state expansion to 3.5 million tpy unlikely before 1985.
Misr Phosphate Co.	Hamrawein	600 000 tpy phosphate concs. (33% P_2O_5)	1986-87		To increase production from present 180 000 tpy ore to 1.2 million tpy from ore (22% P_2O_5).
Red Sea Phosphate Co.	Abu Sheigela	400 000 tpy phosphate concs.	1985	U/P	Replacing Quesir/Safaga mines.

MALI

Company	Location	Product	Year	Status	Comments
Govt./BRGM	Tilemsi Valley	240 000 tpy phosphate	1980s	P	Feasibility study involving initial production of 20 000 t from Tamaguilelt section prepared.

MAURITANIA

Company	Location	Product	Year	Status	Comments
SNIM/BRGM	Bofal	Phosphate		P	Prefeasibility studies in progress.

MOROCCO

Company	Location	Product	Year	Status	Comments
OCP	Ben Guerir	6 million tpy phosphate	1989	P	Doubling of capacity planned.

45.13

TABLE 6. (cont'd)

Company	Location	Design Production (ore, unless otherwise stated)	Completion Date	Capital ($ million)	Type of Operation	Remarks
OCP	Meskala, Essaouria	Phosphate	1990		P/Cn	USSR assisted feasibility studies underway.
OCP	Recette IV, Mera El Ahrech	3 million tpy phosphate	1984		P	Part of overall expansion of Khouribga mines from 18 million t to 24 million tpy by 1985-86.
OCP	sidi Hajaj	3 million tpy phosphate	1985		P/Cn	Beneficiation tests in progress. Ore will require flotation. To supply Jorf Lasfar complex.
OCP	Youssoufia	8.4 million tpy phosphate	1985		U/P/Cn	Underground mine expansion and calcination plant for "black phosphate".
NIGER						
Govt.	Tapoa, Say	Phosphate				Feasibility studies continuing on possible j.v., with Nigeria.
SENEGAL						
Cie. Senegalaise de Phosphates de Taiba (CSPT)	Taiba	2.1 million tpy phosphate	1984		P	Expansion of capacity of Keur Mor Fall pit from 1.6 million tpy.
CSPT	Tobène	Phosphate	1990s		P	To be developed as replacement capacity for Keur Mor Fall.
ISRAEL						
Negev Phosphates	Arad	0.5 million tpy phosphate	1984	35		Expansion of capacity as Makhtesh mine nears depletion.
JORDAN						
Jordan Phosphate Mines Co. (JPMC)	Ruseifa, Al-Hasa, Wadi	Phosphate	1980s		P	Plans to increase production at existing operations by 50%.
JPMC	Shidiyeh, Aqaba	10 million tpy phosphate	1990		P	French consortium led by Sofremines studying project. Reserves 1 billion

PAKISTAN

Company	Location	Capacity	Year	Type	Notes
Sarhad Dev. Authority (SDA)	Hazara district, N.W. frontier province	200 000 tpy phosphate	1986	19	P/Cn	British Mining Consultants have completed feasibility studies.

PHILIPPINES

Philphos	Isabel, Leyte	100 000 tpy phosphate	1984		P	Planning development of small phosphate mine to feed fertilizer complex.

SRI LANKA

State Mining Corp./ Agrico Chemical Co.	Eppawala	585 000 tpy phosphate products	1987	40	P/Cn	Expansion from 150 000 tpy to feed TSP plant at Trincomalee (10%). Most of the rock will be exported.

SYRIA

General Co. for Phosphates & Mines	Gadir al Hamal	5 million tpy phosphate	1985		P/Cn	Expansion from present output level of 1.5 million tpy.

TUNISIA

Cie. des Phosphates de Gafsa (CPG)	Gafsa	10 million tpy phosphate concs.	1990		P/Cn	Tenders invited for feasibility studies on development of reserves in the Oum el Kheceb and South Sehib areas. Includes 3 new mines -- Oum El Khjer, Kef Eddour and Jellabia-M'Zinda.
Soc. Phosphates de Sra Ouertane	Sra-Ouertane, Kef	700 000 tpy phosphate concs.	1987		P/Cn	Jacobs Engineering conducting feasibility studies on new development. Ultimately capacity to rise to 10 million tpy phosphate concs. by 2000.

TABLE 6. (cont'd)

Company	Location	Design Production (ore, unless otherwise stated)	Completion Date	Capital ($ million)	Type of Operation	Remarks
TURKEY						
Etibank	Maziday	250 000 tpy phosphate	1985		P/Cn	Developing Semikan reserves.
UGANDA						
Tororo Industrial Chemicals & Fertilizer Co. (Ticaf)	Sukulu, Sulukwe Hills	Phosphate				Detailed engineering study being undertaken by Beardon Potter/SEMA (France) on development of phosphate project.
UPPER VOLTA						
Buvogmi/CDF Engenierei	Abobo Djouna/ Kodjari	Phosphate				Studies continuing with technical and financial support from the F.R.G.
ZAMBIA						
Zimco	Kaluwe	Phosphate				Feasibility studies and pilot plant testing by Serrana (Brazil) and Kemira Oy (Finland). Project not feasible economically but kept pending.

Source: Mining Magazine, January 1984 with update by the Industrial Minerals Division, Energy Mines and Resources Canada.
P - Placer or opencast; U - Underground; Cn - Concentrator

TABLE 7. LISTED EXPORT PRICES[1] FOR FLORIDA PHOSPHATE ROCK, 1982-84

Grade	January 1982	Early[2] 1983	Late 1984
	($US per tonne fob Tampa or Jacksonville)		
75% BPL	57	35	39
72% BPL	53	31	34
70% BPL	50	29	31
68% BPL	48	28	30
64% BPL	46	26	25

Source: Phosphate Rock Export Association, Tampa, U.S.A.
[1] These prices do not include the charge for severance tax in Florida (i.e. $US 2.38 in 1984). [2] List prices for 1982 and early 1983 were not posted but indicative prices for the two periods are available.

Platinum Metals

G. BOKOVAY

For the lesser known platinum group metals (PGMs), osmium, rhodium, ruthenium and iridium, 1984 was a spectacular year as industrial and speculative demand pushed prices to new or near all-time highs. On the other hand, platinum prices declined throughout the year.

With the recovery of the U.S. economy, the strong U.S. dollar and the easing of inflation, platinum like gold, lost much of its attractiveness as an investment and speculative medium. Although industrial demand for platinum has been increasing, continuing surpluses of the metal have so far prevented a price recovery from occurring.

Palladium prices did appreciably better than those of platinum during 1983 and 1984, owing to a significant number of new applications and its increasing substitution for higher priced platinum and gold in other areas. However, with the significant erosion of palladium prices toward the end of 1984, this metal may have finally succumbed to the influence of weakening investor confidence that plagued some of the other precious metals.

Although the PGMs may have assumed some of the investment and speculative attributes of gold and silver, primary demand for these metals has and will remain dependant upon their industrial uses. Since the utilization of PGMs in existing industrial applications will likely increase substantially and that new uses will continue to be developed, the outlook for the platinum group is very positive. The prospects for platinum are somewhat less optimistic in view of the potentially fewer new applications for this metal and continuing substitution by palladium.

CANADIAN DEVELOPMENTS

Platinum group metals are produced in Canada by Inco Limited and Falconbridge Limited as byproducts from the mining of nickel-copper ores. Although the bulk of the PGMs are recovered from operations in the Sudbury basin, small amounts of these metals are also produced by Inco at Thompson, Manitoba.

The residue from the refining of nickel-copper matte, which contains platinum group metals, is shipped by Inco to its refinery at Acton in the U.K. for the extraction and refining of PGMs. Falconbridge ships a nickel-copper matte containing PGMs to its refinery at Kristiansand, Norway.

Canadian production of PGMs in 1984 was estimated at 10 830 kg, of which 80 per cent was produced by Inco.

Canadian reserves of PGMs are estimated at about 280 million g of which 90 per cent are platinum and palladium in almost equal proportions.

Besides reserves associated with the nickel-copper ores of the Sudbury basin or Thompson, Boston Bay Mines has a PGM property in the Lac Des Iles area, near Thunder Bay, Ontario. The company has outlined two zones with an estimated average grade of 0.185 oz/st of PGMs plus recoverable values of copper, gold and nickel.

WORLD DEVELOPMENTS

The major world producers of platinum group metals are the U.S.S.R., the Republic of South Africa and Canada. Minor producers include Japan - from imported nickel ores - Colombia, Finland, United States, Yugoslavia and Zimbabwe. Estimated 1984 world primary production of PGMs was about 213 million g which was about 6 per cent higher than in 1983.

The U.S.S.R., the largest producer of PGMs in the world, derives these metals principally as a byproduct of nickel-copper production. About half of Soviet output is exported to the west.

It is reported that about 85 to 90 per cent of Soviet production is produced from six mines in the Noril'sk region of Northern

---

G. Bokovay is with the Mineral Policy Sector, Energy, Mines and Resources, Canada. Telephone (613) 995-9466.

Siberia. The U.S. Bureau of Mines estimates that the PGM content of Noril'sk ores is as follows: 25 per cent platinum, 67 per cent palladium and the remaining 7 per cent composed of iridium, rhodium, ruthenium and osmium.

The other principal source of PGMs in the U.S.S.R., accounting for about 10 per cent of Soviet product, are found in the nickel-copper ores of mines in the Kola Pennisula in the northwestern part of the country. PGMs are also recovered from placer deposits in the southern Urals, once the major source of U.S.S.R. output.

In the Republic of South Africa, the world's second largest producer, there are three producers: Rustenburg Platinum Holdings Ltd., Impala Platinum Holdings Limited and Western Platinum Mines Limited. Unlike production in Canada or in the U.S.S.R., South African PGMs are derived from ores that are mined primarily for their platinum metal content. In addition, South African ores differ significantly from Soviet mineralization to the extent that they have a much higher ratio of platinum to palladium. The bulk of South Africa ores, which come from the Merensky Reef of the Bushveld Igneous Complex in the Transvaal, are thought to contain precious metals in the following proportion: 60 per cent platinum, 25 per cent palladium, 8 per cent other PGMs and 3 per cent gold. In addition, these ores also contain appreciable nickel and copper.

Rustenburg Platinum, the largest South African producer, has four active mining operations in the Merensky reef. The processing and refining of PGM concentrates and refinery residues is carried out at the Wadeville refinery at Gemistown in South Africa and another at Royston in the U.K. Both refineries are operated by the Matthey Rustenburg Refiners Group which is jointly owned by Rustenburg Platinum and the Johnson Matthey Group. A major fire at the Royston refinery on April 17, 1984 closed the plant until the end of June. However, with increased production at the Wadeville refinery in South Africa, overall production was reported to have been only minimally affected.

Matthey Rustenburg is considering the construction of a new PGM refinery in South Africa using a Solvex process that could significantly reduce refining expenses. An experimental Solvex refinery, which was opened at the Royston plant in 1983, was reported to be operating satisfactorily.

Rustenburg Platinum has the capacity to produce about 1 300 000 oz of platinum and 545 000 oz of palladium annually plus smaller quantities of the other PGMs. The company which was operating at about 65 per cent of capacity in 1983, was thought to be producing at significantly higher levels in 1984.

Impala Platinum, the second largest producer, has four active mining operations on the Merensky Reef. The company operates a base metal refinery at Springs, near Johannesburg, which processes matte containing nickel, copper and small quantities of PGMs. The PGM residue is also refined at Springs.

Impala's annual capacity is estimated at about 1 050 000 oz of platinum, 542 000 oz of palladium and 147 000 oz of ruthenium. At the beginning of 1984, the company was operating at about 70 to 75 per cent of capacity.

Western Platinum, the smallest producer in South Africa, operates one mine on the western limb of the Bushweld Complex. The company, which is owned by Lonrho S.A. Ltd., Falconbridge and Superior Oil Company, mines ore from both the Merensky and UG2 reefs in a ratio of about 2:1. Production by Western Platinum, which began in 1973, has been steadily increasing. For the fiscal year ending on Sept. 30/84, production of PGM in matte was 263 182 oz compared to 213 989 oz in fiscal year 1983.

Western Platinum ships matte to Falconbridge's refinery in Norway where nickel, copper and cobalt are extracted. The platinous sludge from this process is then returned to South Africa where platinum metals are extracted and refined at Lonrho's refinery at Brakpan. This facility is currently being expanded to handle rhodium, ruthenium and iridium. In addition, Western Platinum is proceeding with the construction of its own matte treatment plant in South Africa.

In the United States, test mining at PGM rich Stillwater Complex in Montana yielded positive results. Participants in the project, Chevron USA Inc., Anaconda Minerals Co. and Manville Corporation, were reported to be undertaking a feasibility study for full scale development. The project which could be in operation by late-1987, would have an annual output of 150 000 oz of palladium and 50 000 oz of platinum. A decision on whether or not to proceed is expected in the spring of 1985.

Although, European Economic Community members have so far been unable to agree on the timing of automobile emission regulations that would require PGM catalytic converters on vehicles, several related developments in Europe will undoubtedly boost demand for platinum metals used to manufacture automobile catalysts. Late in 1984, European Community ministers agreed to make lead free gasoline, which is used in automobiles with catalytic converters, available in all member countries by 1989. Earlier in the year, the Federal Republic of Germany announced that lead free gasoline would be available by 1986 and that road tax incentives will be in operation at the same time for cars fitted with pollution control catalysts. To stimulate the purchase of cars with catalysts, the price of non leaded fuel will be sold at a discount to other fuels.

Recycling

It is estimated that about 10 per cent of the western world supply the platinum and 20 per cent of the palladium supply is derived from the recycling of industrial PGM scrap. Although the potential recovery of platinum metals from spent automotive catalysts could yield as much as 300 000 oz/y, the high cost of collecting and transporting catalysts or other scrap such as that generated by the electronics industry, to a processing facility have discouraged the necessary investment. In addition, falling prices for platinum, in particular, have discouraged secondary recovery.

PRICES

Platinum

Platinum prices like those of gold fell throughout 1983 and 1984 as investor interest decreased in the face of a strengthening U.S. dollar and diminishing rates of inflation. Although industrial demand has been increasing, increased consumption has failed to remove platinum surpluses from the market.

New York dealer prices for platinum which averaged $US 461 per oz in January 1983 fell to $374.50 in January 1984 and to $303.00 in December 1984. At the end of December, platinum was trading at $291.00 to $293.00.

Palladium

Palladium prices rose throughout 1983 and into 1984 on the strength of growing industrial demand for the metal itself and as a substitute in many applications for higher priced precious metals. However, palladium prices which had risen from an average price of $US 125.10 per oz in January 1983 to an average price of $164.10 in December of 1983 fell substantially in the middle of 1984 and again at the end of the year. The average dealer price for December was $134.00.

Other PGMs

Prices for rhodium, ruthenium, iridium and osmium rocketed upwards during 1984 due to strong speculative buying. This was fueled by good prospects for industrial demand in new and expanding applications, possible widening of U.S. stockpiling objectives and concerns over the supply adequacy of lesser known platinum group metals.

Unlike platinum and palladium which are at least principal products in the case of South African production, the other PGMs are all byproducts and as such, their supply is essentially inelastic.

Even in applications where overall PGM production might be stimulated, such as the three-way platinum/palladium/rhodium automotive catalysts being proposed for the European Community, the proportion of rhodium to platinum required for these products is considerably higher than that typically found in PGM ores.

During 1984, concern over the supply of the lesser PGMs was heightened by the labour unrest which was experienced by Impala Platinum early in the year and the major fire in April at Matthey Rustenburg's Royston refinery which was known to have affected the production of rhodium, iridium and ruthenium. In October reports of another fire at Englehard Corporation's refinery at Cinderford in the U.K. may have added to the speculation into the adequacy of PGM stocks.

Leading the price increases for the lesser PGMs was osmium, the rarest metal of the group, which increased from about $US 130 per oz in January 1984 to a trading range of $900-$1,000 per oz at the end of the year. Meanwhile, ruthenium increased from $US 25-27 in early January 1984 to $160-175 by year-end while rhodium increased from $US 265-270 to $890-910 during the same period. In addition, iridium increased from $US 285-305 to $460-475.

USES

Platinum group metals are used in a wide variety of applications in pure form and in a host of alloys which require both the combination of different PGMs together or with other metals. The diversity of uses for these metals reflects their varied and unique attributes which include: chemical inertness and corrosion resistance, special magnetic properties, stable catalytic and thermoelectric properties, excellent reflectivity, stable electrical contact resistance and good high temperature oxidation resistance. The major uses of PGMs are in the automotive, jewelry, chemical, electrical, petroleum and glass industries.

One of the largest uses for PGMs, particularly platinum, is in the production of automobile catalysts. Although platinum is the principal PGM used in these catalysts, its importance in this application has been somewhat reduced owing to the substitution of lower priced palladium for at least some of the total PGM requirement. In addition to platinum and palladium, auto catalysts designed to control nitrous oxides as well as hydrocarbons contain rhodium. Depending on engine size, anywhere from 1.5 to 4.0 g of PGMs are contained in a single catalytic converter.

In addition to their use in controlling automobile exhausts, the production of lead free gasoline, which is required to avoid poisoning of auto catalysts, also uses PGM catalytic agents. Also in the refining industry PGM catalysts are used in hydrocracking and isomerization applications.

While the consumption of PGMs for automobile catalysts is largely in the United States, their use in jewelry, which constitutes the second largest use for platinum, is particularly large in Japan and has also been growing in the Federal Republic of Germany. Iridium and ruthenium are also widely used in jewelry.

In the chemical industry, PGMs are widely used as catalysts with the most important being platinum, ruthenium and palladium. Important specific applications include the production of nitric acid and hydrogen cyanide. It was reported that British Petroleum was developing a new ammonia production catalyst that uses ruthenium. The process is supposedly more efficient than the traditional one which uses an iron catalyst. PGMs are also used in the manufacture of equipment that is exposed to highly corrosive environments including anodes used in electrolytic manufacturing processes for such products as chlorine and caustic soda.

One of its major markets for palladium, is in the electronics industry where is is used in the manufacture of printed circuits, electrical contacts and electrical furnaces. In addition, PGMs and particularly palladium, are extensively used in the dental and medical field. The most important applications include dental alloys, orthodontic and prothodontic devices, hypodermic needles, electrodes, casings for pacemakers and as essential ingredients for certain chemotherapeutic agents used to treat certain cancers.

Other important applications for PGMs include: themocouples used for high temperature measurement; the manufacture of glass, glass fibre and synthetic fibres; fuel cells; permanent magnets; and catalytic applications in the pharmaceutical and food processing industries.

In addition to uses by industry or in the manufacture of jewelry, Rustenburg and Impala began in 1983 to market an array of platinum coins, wafers and small bars to stimulate investment demand.

OUTLOOK

Within the next 12 to 18 months, it is expected that increasing industrial demand for platinum and palladium will cause prices for platinum and palladium to rise. Increased PGM production, particularly from South Africa, will likely result in some downward pressure on prices for other PGMs.

Consumption of the platinum group metals in the next decade is expected to increase at about 3 per cent per annum in view of an expected increase in the number of new applications and more widespread use of these metals in existing areas. Furthermore, demand should also be buoyed by continuing stockpiling activities by the U.S. government.

Assuming that the development of a clean-burn type of engine technology does not make PGM automotive catalysts obsolete, demand in this area is expected to remain strong in view of expected more widespread and stringent emission standards. These include the adoption of emission regulations in Australia in 1986, diesel engine standards which will take effect in North America in 1987 and probable European legislation which will take effect in the early-1990s. In the case of the latter, it is estimated that the European automobile catalyst market could require 500 000 to 700 000 oz/y of platinum.

Demand for palladium should continue to grow at a faster rate than platinum in view of the multitude of new applications particularly in the electronics industry for this metal, and also because of its increasing substitution for platinum and gold. As a result the price ratio between platinum and palladium which is presently about 2.4:1 could fall to about 2.0:1, even though South African production of palladium will increase as the more palladium rich UG2 reef begins to account for a larger share of production.

This price ratio is supported by the fact that in the electronics industry, palladium is roughly half as efficient to use as platinum.

While the U.S.S.R. will remain an important supplier of PGMs, particularly palladium, to the west, its sales are expected to flatten out somewhat because of the probable increase in PGM consumption by Comecon countries and also due to an expected slower rate of growth in nickel production.

Canadian production of platinum metals is expected to increase somewhat in the next several years in response to higher nickel production levels.

Since it is considered that with existing recycling technology, it is uneconomic to recover PGMs when prices for platinum and palladium are below $US 300 and $150 respectively, little expansion of PGM recycling is expected in the short-term.

TARIFFS

Item No.		British Preferential	Most Favoured Nation	General	General Preferential
CANADA					
36300-1	Platinum wire and platinum bars, strips, sheets or plates; platinum, palladium, iridium, osmium, ruthenium and rhodium, in lumps, ingots, powder, sponge or scrap	free	free	free	free
48900-1	Crucibles of platinum, rhodium and iridium and covers therefore	free	free	15%	free
UNITED STATES (MFN)					
601.39	Precious metals ores		free		
605.02	Platinum metals, unwrought, not less than 90% platinum		free		

MFN Reductions under GATT (effective January 1 of year given)		1983	1984	1985	1986	1987
				(per cent)		
605.03	Other platinum metals, unwrought	14.1	12.6	11.2	9.7	8.2
605.05	Alloys of platinum, semi-manufactured, gold-plated	17.5	15.6	13.8	11.9	10.0
605.06	Alloys of platinum, semi-manufactured, silver-plated	9.3	8.6	7.9	7.2	6.5
605.08	Other platinum metals, semi-manufactured, including alloys of platinum	14.1	12.6	11.2	9.7	8.2
644.60	Platinum leaf	14.1	12.6	11.2	9.7	8.2

Sources: The Customs Tariff and Commodities Index, January 1983. Revenue Canada; Tariff Schedules of the United States Annotated 1983, USITC Publication 1317; U.S. Federal Register Vol. 44, No. 241.

TABLE 1. PLATINUM METALS, PRODUCTION AND TRADE, 1982-84

	1982		1983		1984P	
	(grams)	($)	(grams)	($)	(grams)	($)
Production[1]						
Platinum, palladium, rhodium, ruthenium, iridium	7 104 814	82,252,861	6 965 000	67,885,000	10 831 000	..
Exports					(Jan.-Sept. 1984)	
Platinum metals in ores and concentrates						
United Kingdom	7 057 534	52,621,000	5 675 451	55,775,000	5 779 000	59,555,000
United States	104 352	882,000	81 243	834,000	31 000	515,000
Total	7 161 886	53,503,000	5 756 694	56,609,000	5 810 000	60,070,000
Platinum metals, refined						
United States	519 273	4,591,000	2 471 140	31,479,000	3 040 000	27,399,000
United Kingdom	220 182	821,000	352 931	2,426,000	24 000	191,000
Japan	139 966	161,000	30 046	575,000	-	-
Brazil	1 182	22,000	17 107	292,000	-	-
Other countries	64 974	176,000	4 106	45,000	1 000	1,000
Total	945 577	5,771,000	2 875 330	34,817,000	3 065 000	27,591,000
Platinum metals in scrap						
United States	1 148 000	14,925,000	906 169	11,728,000	1 478 000	24,717,000
United Kingdom	376 197	3,266,000	220 399	2,826,000	2 212 000	16,505,000
West Germany	61 772	200,000	-	-	420 000	5,478,000
Total	25 750 000	18,391,000	1 126 568	14,554,000	4 110 000	46,700,000
Re-export[2]						
Platinum metals, refined and semiprocessed	8 242	170,000	276 000	4,384,000	..	..
Imports						
Platinum lumps, ingots, powder and sponge						
United States	139 966	2,010,000	24 416	418,000	118 000	1,903,000
Switzerland	-	-	17 511	265,000	-	-
United Kingdom	98 474	1,595,000	10 047	162,000	213 000	3,398,000
Total	238 440	3,605,000	51 974	845,000	331 000	5,301,000
Other platinum group metals						
United States	183 106	602,000	347 768	1,902,000	186 000	1,546,000
United Kingdom	15 552	40,000	169 483	792,000	39 000	599,000
West Germany	24 883	76,000	-	-	-	-
Total	223 541	718,000	517 251	2,694,000	225 000	2,145,000
Total platinum and platinum group metals						
United States	323 072	2,612,000	372 184	2,320,000	304 000	3,449,000
United Kingdom	114 026	1,635,000	179 530	954,000	252 000	3,997,000
Switzerland	-	-	17 511	265,000	-	-
West Germany	24 883	76,000	-	-	-	-
Total	461 981	4,323,000	569 225	3,539,000	556 000	7,446,000
Platinum crucibles[3]						
United States	447 921	6,615,000	483 783	8,415,000	547 000	9,613,000
West Germany	-	-	218	4,000	-	-
Total	447 931	6,615,000	484 001	8,419,000	547 000	9,613,000
Platinum metals, fabricated materials, not elsewhere specified						
United Kingdom	259 216	4,307,000	406 833	5,168,000	116 000	1,889,000
United States	521 045	3,518,000	724 182	2,976,000	471 000	2,119,000
West Germany	7 807	24,000	6 874	118,000	2 000	19,000
Belgium-Luxembourg	43 452	4,083,000	-	-	-	-
Switzerland	995	15,000	-	-	-	-
Total	832 515	11,947,000	1 137 889	8,262,000	589 000	4,027,000

Sources: Energy, Mines and Resources Canada; Statistics Canada.
[1] Platinum metal, content of concentrates, residues and matte shipped for export. [2] Platinum metals, refined and semiprocessed, imported and re-exported in the same form as when imported. [3] Includes spinners and bushings.
P Preliminary; - Nil; .. Not available.

TABLE 2. CANADA, PLATINUM METALS, PRODUCTION AND TRADE, 1970, 1975 AND 1979-83

	Production[1]		Exports				Imports[4]	
			Domestic[2]		Re-exports[3]			
	(grams)	($)	(grams)	($)	(grams)	($)	(grams)	($)
1970	15 005 188	43,556,597	15 327 731	44,174,000	634 480	2,365,735	1 889 381	3,123,000
1975	12 417 099	56,493,077	15 530 930	50,244,000	538 899	2,928,000	1 896 410	6,061,000
1979	6 156 716	56,333,561	6 641 432	54,686,000	43 172	359,000	826 886	6,546,000
1980	12 776 000	159,088,000	13 524 725	191,569,000	9 176	68,000	1 064 578	14,347,000
1981	11 902 283	136,186,021	11 094 424	110,838,000	498	10,000	687 604	8,573,000
1982	7 104 814	82,252,861	8 107 463	59,274,000	8 242	170,000	461 981	4,323,000
1983P	5 195 000	67,885,000	8 632 024	91,426,000	.276 000	4,384,000	569 225	3,539,000

Sources: Energy, Mines and Resources Canada; Statistics Canada.
[1] Platinum metals, content of concentrates, residues and matte shipped for export. [2] Platinum metals in ores and concentrates and platinum metals, refined. [3] Platinum metals, refined and semiprocessed, imported and re-exported. [4] Imports, mainly from United States and United Kingdom, of refined and semiprocessed platinum metals, derived from Canadian concentrates and residues, a large part of which is re-exported.
P Preliminary.

TABLE 3. WORLD MINE PRODUCTION OF PLATINUM METALS, 1981-84

	1981	1982	1983e	1984e
	(grams)			
U.S.S.R.[e]	104 196 647	108 862 169	111 973 000	111 973 000
Republic of South Africa[e]	96 731 813	80 869 040	80 869 000	87 090 000
Canada	11 902 283	9 104 814	6 965 000	10 831 000
Japan	1 128 092	1 345 941	..	..
Colombia	460 363	373 242	..	..
Australia	470 347	436 382	..	..
United States	227 615	249 854	249 000	..
Other countries	227 895	224 101	2 177 000	2 756 000
Total	215 345 055	199 465 543	202 233 000	212 650 000

Sources: U.S. Bureau of Mines, and Energy, Mines and Resources, Canada.
[e] Estimated; .. Not available but included in "other".

Potash

G.S. BARRY

Production and shipments of potash to all markets in 1983 and 1984 were much higher than in the depressed year of 1982. Production was up 13.6 and 26.5 per cent for 1983 and 1984 respectively and shipments were up 28.5 and 10.8 per cent for 1983 and 1984. The volume of shipments was particularly strong in the second half of 1983; was steady but much below expectations in the first half of 1984; was strong again in the third quarter of 1984 but failed to follow through in the last quarter of 1984. The surge in offshore demand however was steady whereas North American sales showed wide fluctuations.

Producers started 1983 with high stocks (1 486 200 t K_2O) and were carrying this high inventory until July, since the universally expected pick-up in spring sales did not materialize. Those initial conditions necessitated some cutbacks in production during January and February and the summer months. An unexpected boost in sales for the last five months of 1983 resulted in a rapid reduction of inventories to 861 500 t K_2O. In 1984 inventories increased again to end the year at about 1 570 000 t.

The reported average price received for potash in 1983 by Canadian producers was $Cdn. 102.60 per t K_2O compared to $Cdn. 118.78 in 1982 and $Cdn. 151.25 in 1981. The price received in 1984 was estimated at $Cdn. 108.90 per t K_2O.

Employment in the Saskatchewan potash mining industry was 3,697 in 1983 compared to 4,075 in 1982. There was not much change in 1984. In addition up to 300 were employed on contract at an expansion project. In New Brunswick employment was about 260 at the operating mine and rose from 170 in 1983 to just over 600 in 1984, mainly contracted, at the mine under construction.

DOMESTIC DEVELOPMENTS

At the end of 1983, Canadian installed potash production capacity was 9 160 000 t in Saskatchewan and a nominal 100 000 t in New Brunswick. The largest share of capacity, 41.7 per cent, is held by Potash Corporation of Saskatchewan (PCS), a provincial Crown corporation, followed by 20.2 per cent for International Minerals & Chemical Corporation (IMC) the largest private producer in the western world.

PCS curtailed all new expansions but continued with its major expansion program at the Lanigan mine which was 63 per cent completed by the end of 1983 and 80 per cent completed by the end of 1984 with a target for total completion in 1986. To date over $350 million has been expended against a total estimate of $475 million. Capacity of the Lanigan mine will be 1 740 000 t K_2O. PCS introduced at Lanigan the world's largest underground continuous mining machine. It is known as the "Orebiter". It combines the high productivity of a four-rotor borer (lower part) with the flexibility of a boom-type miner with two cutting heads mounted on top. The Orebiter cuts an opening averaging 5.2 by 8.0 m at one pass; weighs 364 t and has a rated capacity of 1 000 t per hour. During 1983, PCS instituted in all of its mines, except one, a working schedule of 10 days for every 14-day period for underground mining, thus reducing by 1 shift the traditional 4 shifts per week working schedule. The surface plants remained on a continuous schedule. The Lanigan, Cory, Rocanville and Allan mines were not producing in January and February and for about half a month each either in July or August 1983. In 1984 PCS mines were closed for short maintenance periods during July or August and for Christmas week. An inflow of water occurred in November 1984 at the Rocanville mine. At the end of December the situation was under control but still serious. A primary plug was being installed at the affected entry and drilling was in progress to permit supplementary grouting from the surface. The mine was closed for production a few days before Christmas and is expected to reopen as soon as the plug installation is completed. PCS announced that it will construct a "demonstration plant" to

G.S. Barry is with the Mineral Policy Sector, Energy, Mines and Resources, Canada. Telephone (613) 995-9466.

produce 30 000 tpy of potassium sulphate at its Cory division. The plant is expected to be in production in 1985 and the capital expenditures will be in excess of $6 million. The potassium sulphate will be produced through a chemical reaction of sodium sulphate and potassium chloride (Glaserite process). Natural sodium sulphate is also abundant in Saskatchewan.

The International Minerals & Chemical Corporation reduced production for two months early in 1983 and for two short intervals in mid-year at its K1 and K2 mines, but for the last four months the company operated at full capacity. In 1984 the company produced near normal levels, shutting the mine for a two week maintenance period in June-July. The company announced that a potash production expansion remains on hold until there is clear evidence of a long-term improvement in market.

Cominco Ltd. produced 620 000 t K_2O (1 018 000 t KCl) during 1983, but had to close its mine at mid-year for nine weeks, six for inventory control and three for maintenance. In 1984, the company produced 1 235 000 t (KCl). It closed the plant in July for three weeks for normal maintenance. Central Canada Potash, a subsidiary of Noranda Inc., closed its mine for three months from July to September 1983. During 1984 the mine was closed from July 14 to September 6 and for the Christmas week. The company produced 970 000 t (KCl) in 1983 and 1 074 000 t (KCl) in 1984. The company is currently re-examining its options for expanding capacity in the late-1980s. Potash Corporation of Saskatchewan, a subsidiary of Ideal Basic Industries, Inc. produced 682 000 t of muriate of potash (KCl) in 1983. The company closed its mine for the month of September. Kalium Chemicals, a division of PPG Canada Inc., continued to operate at below recently expanded capacity levels for the first nine months of 1983, thereafter operating at full capacity. Kalium began work on expanding its capacity at the Belle Plaine mine from 1 055 000 t to 1 320 000 t K_2O at a cost of $100 million. Completion is scheduled for the end of 1986. During expansion, improvements to the process will be carried out principally raising energy efficiency, returning a larger proportion of waste salt to underground cavities and the installation of compaction machines. The company established a 41 700 t storage depot in Toledo, Ohio, and is building another facility in Madison, Wisconsin.

For the period in effect from July 1st, 1979 to June 30, 1984 the Saskatchewan potash producer paid resource revenues to the province under separate but similar agreements known as the "Potash Resource Payment Agreement" (PRPA). The payment system was described in detail in the 1980 potash review. The private potash companies felt that the provisions of these agreements require some improvements and hoped to negotiate a better five-year term with the new provincial government. However, following extensive negociations during 1984, the government decided to extend the previous agreement for six months to the end of 1984, at which time a further extension could be considered if necessary. During the past five-year PRPA the province of Saskatchewan collected approximately $700 million from the potash industry.

In Manitoba, two concession areas are potential sites for future potash mines. IMC suspended further planning and consultations with the provincial government on their site until market conditions improve. Meanwhile on the second site, Canamax Resources Inc., a subsidiary of AMAX Inc., and the Manitoba Government commissioned a feasibility study to be completed by the end of June 1985 for a 1.8 million tpy (KCl) potash mine. Reserves are approximately 440 million t grading better than 25 per cent K_2O. If a positive decision is made a mine with a capacity of about 1.2 million tpy K_2O could be in production in the early-1990s. Costs of construction in 1984 dollars would be in excess of $500 million.

In New Brunswick, the Potash Company of America (PCA) commenced production and plant tune-up at its new mine near Sussex in August 1983. Later in that year, product was railed to the company's port storage facilities at Saint John and the first shipment to Europe was made in January 1984. The company experienced some start-up difficulties which were partially remedied during 1984. Evaporator capacity was increased, and modifications were carried out on the continuous miners and the conveyor system. Full production capability is targeted for mid-1985. Shipments of byproduct salt to International Salt Co. (N.Y.) which took place intermittently during the development stage were resumed. PCA intends to mine approximately 440 000 tpy of commercial common salt. Capital expenditures for the PCA mine were approximately $225 million. Port facilities at Saint John were $40 million of which the company paid half. The PCA mine was designed for a capacity of

545 000 tpy K_2O but will be operated at a reduced nominal base of 380 000 tpy K_2O for at least the first two full years of production. The upgrading of capacity in response to market conditions can be accomplished rapidly.

The Denison-Potacan Potash Company announced in February 1983 that it will proceed to bring into production its Clover Leaf mine near Salt Springs by late-1985. Reserves at this location are 254 million t of high-grade ore of which 45 million t grading 28.5 per cent K_2O was outlined through underground exploration. The nameplate capacity of the mine will be 780 000 tpy. The first shaft at the mine was completed in 1982 and the second production shaft was started in March 1983 and will be completed in mid-1985. All work at the development site is slightly ahead of schedule. Capital expenditures for the project are estimated at above $300 million.

The British Petroleum Company Limited (BP) completed a drilling program on its Millstream concession near Sussex in 1983. The company encountered beds of potash at depths between 950 and 1 050 m. Tonnage and grade indications on two mineable beds are very encouraging. In 1984, a pilot hole for a shaft was completed without encountering any problems. In early-1985 BP is expected to apply for a mining licence which will permit a decision on sinking an exploration shaft. There is a possibility that production on this site, the third potash basin in New Brunswick, will start sometime in the 1990s.

Potash bearing intersection were also reported in Nova Scotia where two companies did some limited drilling in the Bras d'Or lake area. There are also some indications of potash presence in the saline formations along the west coast of Newfoundland and on the Madeleine Islands of Quebec.

INTERNATIONAL DEVELOPMENT

For 1983 the United States Bureau of Mines estimates world potash production at 26.7 million t K_2O which includes a sales level for Canada of 6.2 million t as a proxy for production. The International Fertilizer Association (IFA) estimates world production at 26.2 million t which includes a high estimate of 9.3 million t for the U.S.S.R. Energy Mines and Resources Canada estimates world 1983 production at only 25.3 million t allowing for 5.9 million t in Canada and only 8.4 million t in the U.S.S.R.

World production in 1984 is estimated at 28 500 000 t, a 13 per cent increase over 1983. Production in the U.S.S.R. is estimated at 9 million t. Canada and the U.S.S.R. contributed most to this increase.

Brazil - PETROBRAS Mineracao S.A. awarded a contract for the potash handling system to a firm in the Federal Republic of Germany. The mine has a projected 500 000 tpy KCl capacity and will achieve commercial production in 1985 or 1986. Petromisa holds another interesting potash deposit near Fazendinha in the Amazonian basin. During 1984 the company awarded a $700,000 feasibility study to a joint venture consisting of Paulo Abib Engenharia, Mines de Potasse d'Alsace and Patrick Harrison and Company Ltd., a Canadian firm. Petromisa estimated reserves of 525 million t at an average depth of 1 050 m.

Chile - Corporacion de Fomento de la Production asked for bids for the development of potash in the Salar de Atacama. Preliminary studies indicated the potential for a production of 500 000 tpy of KCl and 150 000 tpy of sulphate of potash with the aid of solar evaporation. A provisional award was made to Amax Chemical Corporation. Its final acceptance involves the exploitation of lithium. It is estimated that the exploration phase of the potash project will take three years.

China - Jacobs International of Ireland completed a feasibility study for a 200 000 tpy KCl potash plant at the dry salt Chaerhan Lake in western China. Another project for a 1 million tpy plant 200 km to the west of the first location is also under consideration. The region requires considerable infrastructure and costs of production will be high. The development is not likely to occur before the end of this decade.

Ethiopia - A two-year feasibility study on potash production in the Danakil Depression was started by Entreprise Minière et Chimique for the Ethiopian-Libyan Mining Co. The region produced some carnallite before the Second World War. In 1965 and 1966 a shaft was sunk at Musley into sylvinite beds by the Ralph M. Parsons Company of the United States but the 500 000 tpy project was abandoned in 1968.

France - In October 1983 the French Assembly approved the Rhine Treaty which limits the disposal of waste salt tailings into the Rhine, currently amounting to between 6 and 7 million tpy. Measures to this effect will be implemented within 18 months. Well tests for deep subsurface brine disposal were conducted near Reiningue but commercial injection is being delayed by protests from local residents. The Theodore mine of Mines de Potasse d'Alsace (MPDA) will be closed in 1985 or 1986. The Amelie surface plant will be converted from the dissolution-recrystallization process to flotation at a cost of $Cdn. 50 million. An additional $75 million will be spent on other MPDA's modernization programs.

Germany, Federal Republic of - The Friedrichshall Potash Works mine, purchased from Kali-Chemie A.G. by Kali & Salz A.G. in 1981 was joined up underground to the Bergmannssegen-Hugo Potash Works mine. Surface operations at Friedrichshall were then closed and Bergmannssegen-Hugo facilities will serve the output of both mines. The Siefried-Giesen Potash Works mine ceased production of muriate of potash but will continue the output of kainite ore. Production at Kali und Salz in 1984 was back to normal levels of 2.6 to 2.7 million t K_2O.

Israel - In early 1984 the Dead Sea Works Ltd. completed its expansion program at Sodom and now has a capacity of 2.1 million tpy of muriate of potash equivalent to 1.28 million tpy of K_2O. The DSW units comprise 250 000 tpy of flotation product, 1 million tpy from the hot leach plant and two cold leach units of 450 000 tpy each. Production at the very high cost flotation plant will be reduced once the cold leach plant is in full operation. DSW transports most of its product by trucks and train to the port of Ashdod and some through the port of Eilat on the Red Sea. Between 1984 and 1987 the company will construct a conveyor belt from Sodom to Nahal Zin that will link to existing rail and replace costly truck haulage on steep grades.

Jordan - After the completion of the Arab Potash Co. Ltd. (APC) 1.2 million tpy KCl plant on the Dead Sea in 1982 output rose more slowly than originally anticipated. The current estimates are: 260 000 t in 1983, 500 000 t in 1984, 700 000 t in 1985, and full production expected by 1987. However, APC has to make a major additional investment on the addition of a fourth carnallite pan (C4) before the envisaged capacity is reached. Without C4 maximum capacity would be 1.1 million tpy.

Mexico - Fertilazantes Mexicanos SA has reactivated its potash project at Cerro Prieto near Mexicali on the California border. The extraction will be from brines, using as energy geothermal steam available nearby. The projected capacity is 80 000 tpy KCl to be completed by late-1986 or early-1987. A byproduct capacity of 125 000 tpy of common salt and 81 000 tpy of calcium chloride is also envisaged.

Spain - The Spanish potash industry is undergoing major reorganization. Instituto Nacional de Industria (INI) announced that it will close its subsidiary Potassas de Navarra SA in December 1985. This company currently operates a mine near Pamplona which has a theoretical capacity of 325 000 tpy K_2O but recently operated at only half this level. The deposit is low-grade (12 per cent K_2O) and structurally complex. INI acquired a 51 per cent interest in Minas de Potasas de Suria S.A. from Solvay & Cie in 1982 with the option to buy the remaining 49 per cent share. The company plans to expand capacity by 50 000 tpy K_2O to 220 000 tpy K_2O at the Suria mine by 1986. As the mining will be concentrated in the Catalonia district, there is a possibility of opening a new mine between Suria and Llobregat in the more distant future. Explosivos Rio Tinto (ERT) also plans to substantially increase investments in its Cordona and Llobregat mines so that by the end of this decade a substantial part of the lost capacity will be restored.

Thailand - Thailand has two potash bearing saline basins, the Khorat and the Sakhon-Nakhon basins. The Department of Mineral Resources (DMR) started a pilot project in 1982 to demonstrate the feasibility of carnallite exploitation near Chaiyaphum in the Khorat Basin. An inclined access drift was sunk but had to be abandoned because of high water inflow in 1983 after the expenditure of $5 million. The total project budget was $11.1 million of which $8.9 is a World Bank loan. DMR is considering the option of starting at a new site, 10 km away by sinking a vertical shaft to reach the potash horizon at about 150 m below surface, but a positive decision is not likely. Negotiations were meanwhile concluded for the exploration on a 3 500 square km potash concession in the Sakhon-Nakhon basin, in the province of Khon Kaen, between DMR and joint venture partners, Duval Corporation of USA (Pennzoil Company), CRA Exploration Pty Ltd. of Australia and Siam Cement of Thailand. The operating company will be Thai Potash Co. Ltd. Success could lead to development of a mine at a cost in excess of $350 million. In October 1984, the Thai government signed another agreement with the Thai Agrico Potash Co. Ltd. for a 2 333 square km concession in the Udon Thani province, 450 km northeast of Bangkok. The company is committed to spend a minimum of $US 3 million over five years on potash exploration. The Thai deposits present a geological challenge since the disposition of sylvinite ore in predominantly carnallitic potash is generally discontinuous.

Tunisia - The World Bank is making a $13.6 million loan to Tunisia for studies in the phosphate industry and a study on the feasibility of extracting potash from underground brines in the Sabkha el Melah depression near the new port of Zarzis. French companies will carry out the survey.

United States - Production in 1983 declined by 20 per cent to 1.4 million t K_2O, the lowest level since 1951, but rebounded to an estimated level of 1.6 million t in 1984. About 85 per cent of production comes from six underground mines in New Mexico where the average content of ore mined is 12.9 per cent K_2O compared to 24.8 per cent K_2O in Canada. The remaining production is from a solution mine in Utah and brines in Utah and California. On January 19, 1983, Mississippi Chemical Corporation shut down its potash mine near Carlsbad, N.M. for an indefinite period which may in fact become a permanent closure. This brings down New Mexico mines number to five. During the year, some New Mexico mines were shut during the July to September period to reduce excessive inventories.

The two brine potash producers in Utah experienced production difficulties related to severe weather conditions and flooding. Kaiser Aluminum & Chemical Corporation, Chemicals division, had to shut down its plant for one year starting September 30, 1983. Great Salt Lake Minerals & Chemicals Corp. lost its brine evaporation ponds and will take up to three years to complete the recovery of these facilities.

PPG Canada Inc. completed a drilling program on deep potash-bearing salt formations in Mecosta County, central Michigan. The company is currently constructing a test facility employing similar solution mining techniques as at Belle Plaine, Saskatchewan. The Michigan deposits are at a depth of some 3 km or less compared to 2 km or less in Saskatchewan.

In California, Kerr-McGee Chemical Corporation continued to produce muriate and sulphate of potash after only a short interruption of about three months when the company announced and then rescinded a decision to permanently suspend the production of its main product - muriate of potash.

On March 30, 1984, Amax Chemical Inc. and the Kerr-McGee Chemical Corporation filed a petition with the International Trade Commission (ITC) alleging that subsidized producers in Israel, the German Democratic Republic, Spain and the U.S.S.R. are dumping potash in the United States. These allegations were two-pronged; countervailing duties (CVD) were sought on account of subsidies and ad valorem duties (A/D) were sought for dumping. In May, ITC issued a preliminary ruling in favour of the plaintiffs. However, the CVD action against the U.S.S.R. and GDR was dismissed by the Department of Commerce on grounds that bounties or grants within the meaning of Section 303 of the tariff act of 1930 cannot be found in non-market economies. A net subsidy of 7.54 per cent for Spain and 8.71 per cent for Israel were preliminary determinants in June; these CVD were subsequently reduced and finally dropped when on October 22, 1984, ITC ruled no injury. On the A/D action the Department of Commerce found dumping margins of 187.03 per cent for the U.S.S.R.; 112.12 per cent for GDR; 43.65 per cent for Spain; and minimal for Israel. The final determinations for dumping on potash from U.S.S.R., GDR and Spain were scheduled for the first week of January 1985, but the U.S.S.R. and GDR were granted 60-day extensions.

U.S.S.R. - During 1983 U.S.S.R. potash exports declined by 18 per cent to 2.3 million t K_2O. There was a substantial decline of 33 per cent in exports to East European countries and a 15 per cent increase to the rest of the world. There are indications that there was a substantial increase in domestic demand, perhaps approaching 5.6 million t which may include some increase in inventories. Nevertheless, U.S.S.R.'s production in 1983 could not have been much higher than 8.4 million t K_2O, much below the 9.3 million t estimated by IFA. This suggested 8.4 million estimate of 1983 production will also allow for a "loss" of about 0.5 million t or 6.3 per cent of production. Such a "loss" is compatible with the average "loss" of 5.7 per cent computed for the U.S.S.R. for 1970-82 period.

During 1983, production started at two units of the Novosolikamsk mine in the Urals. This mine was originally scheduled to start in 1980 but ran into many difficulties, in particular with the operation of oversize crystallizers. The originally targeted capacity for Novosolikamsk was 2.9 million t K_2O but it had to be substantially downsized, apparently to about 1.6 million t K_2O with three units of 900 000 tpy KCl each (540 000 tpy K_2O). The third unit will be completed under the 1985-90 five-year development plan. Another source reports that each unit at the Novosolikamsk plant has a capacity of only 375 000 tpy K_2O rather than 540 000 t.

In September 1983, an earth dam burst near the Stebnik potash mine in the Ukraine, releasing brines to the Dniester river and causing widespread ecological damage. The Byeloruskali complex will add a 50 000 tpy potassium sulphate plant to be completed in 1987 or 1988. A plant in the same location based on the glaserite process is apparently not in operation. The U.S.S.R. reported the successful pilot plant testing of solution mining at Karlyuk, Turkmen S.S.R. and intends to construct a full scale 700 000 tpy KCl solution mine.

The U.S.S.R. and GDR together currently export two-thirds of their excess potash to COMECON countries and the remainder to the rest of the world. By 1988, these two countries are expected to consume internally 7.5 million t K_2O; export to other COMECON countries 4.0 million t and have available for export to the rest of the world about 2.6 to 2.7 million t. This level of 'outside' exports is most likely going to be maintained for a few years.

United Kingdom - Cleveland Potash Ltd. continued to improve productivity at its Boulby mine operating at an effective capacity of 300 000 tpy K_2O. A new tunnel and conveying system between the working panels and the shaft was installed in massive salt below the potash zone. The company is also doubling the compaction facility from 120 000 tpy to 240 000 tpy at a cost of $1.5 million. Another $800,000 will be invested to recover potash from brines at an initial pilot stage of 9 000 tpy, which could be eventually expanded to 27 000 tpy. The new prototype plant will involve a refrigeration technique whereby potash crystals will be separated from the frozen brine.

PRICES

Typical contract prices for Canadian potash (standard grade) moving out of Vancouver were $75 to $77/t KCl at the beginning of 1983; $73 to $75/t in mid-year and edged upward to $78-$81/t at the end of the year. Prices negotiated for the first half of 1984 were in the low eighties. Despite sales on good volume the prices weakened by mid-1983. Expected strong domestic sales for the fall of 1984 did not materialize and the prices in North America were still very weak at the end of the year when standard grade was quoted at $50/t fob mine. An increase is expected by March 1985.

OUTLOOK

Good recovery in the second half of 1983 allowed for a substantial reduction of inventories but was not accompanied by a sustained price rise. Slower than expected 1984 spring sales in North America put a damper on the expansion, but further gains in sales are expected in 1985 with continuing price recovery.

For 1986 to 1987 adequate supply of potash should be the rule rather than the exception and prices will remain firm. Thereafter new sources of supply should be fairly well in step with rising demand and chances of another period of expansion to overcapacity levels remain remote until the end of the 1980s. In the early 1990s producers and prospective producers in all provinces should aim at coordinated, timely developments since the penalty for a misjudgement in world capacity requirements could be very onerous.

TABLE 1. CANADA, POTASH PRODUCTION, SHIPMENTS AND TRADE, 1982-84

	1982		1983		1984P	
	(tonnes)	($)	(tonnes)	($)	(tonnes)	($)
Production, potassium chloride						
Gross weight	8 750 493	..	9 703 531	..	12 275 000	..
K_2O equivalent	5 351 786	..	5 929 567	..	7 500 000	..
Shipments						
K_2O equivalent	5 308 532	630,561,741	6 293 747	645,767,272	6 972 407	759,270,149
					(Jan.-Sept. 1984)	
Imports, fertilizer potash						
Potassium chloride						
United States	1 878	682,000	2 270	1,205,000	1 605	767,000
United Kingdom	3	2,000	7	2,000	4	1,000
Total	1 881	684,000	2 277	1,207,000	1 609	768,000
Potassium sulphate						
United States	20 045	3,524,000	15 411	3,297,000	2 201	628,000
Potassic fertilizer, nes						
United States	57 650	6,258,000	47 367	6,257,000	7 650	1,344,000
Potash chemicals						
Potassium carbonate	1 113	728,000	1 555	904,000	1 529	908,000
Potassium hydroxide	3 407	1,776,000	2 936	1,990,000	1 811	1,339,000
Potassium nitrate	2 444	1,096,000	4 343	1,869,000	2 653	1,295,000
Potassium phosphate	1 492	1,566,000	1 729	2,190,000	1 503	2,068,000
Potassium silicates	686	617,000	813	621,000	535	458,000
Total potash chemicals	9 142	5,783,000	11 376	7,574,000	8 031	6,068,000
Exports, fertilizer potash						
Potassium chloride, muriate						
United States	4 741 203	452,572,000	5 656 827	481,183,000	4 835 000	459,880,000
Japan	536 244	55,591,000	869 316	96,032,000	533 606	64,625,000
India	447 700	44,152,000	428 242	47,525,000	500 270	60,917,000
Brazil	211 808	21,789,000	287 419	34,249,000	423 916	52,325,000
People's Republic of China	66 660	7,818,000	536 539	59,420,000	403 510	49,444,000
South Korea	309 031	30,467,000	323 236	35,897,000	308 729	37,708,000
Australia	204 912	21,419,000	197 214	21,894,000	201 198	24,626,000
Singapore	228 291	22,602,000	118 260	13,128,000	144 433	17,549,000
France	-	-	48 790	5,118,000	131 123	11,815,000
Malaysia	12 185	1,186,000	37 352	4,131,000	96 730	12,067,000
Indonesia	128 883	12,788,000	16 480	1,853,000	94 151	11,277,000
Taiwan	64 040	6,347,000	86 227	9,538,000	80 851	9,723,000
Other countries	270 536	28,044,000	357 932	40,084,000	461 222	77,995,000
Total	7 221 493	704,775,000	8 963 834	850,052,000	8 214 829	854,792,000

Sources: Statistics Canada; Energy, Mines and Resources Canada.
P Preliminary; - Nil; .. Not available; nes Not elsewhere specified.

TABLE 2. CANADA, POTASH PRODUCTION AND SALES BY GRADE[1] AND DESTINATION, 1982 AND 1983

	1983						1982
	Standard[2]	Coarse	Granular	Soluble	Chemical[3]	Total	Total
			(tonnes K_2O equivalent)				
Production	1 509 928	2 239 121	1 498 848	624 619	55 885	5 928 401	5 207 878
Sales							
Canada	14 238	273 420	87 045	10 201	..	384 904	272 799
United States	286 356	2 197 484	1 138 910	523 100	..	4 145 850	3 202 377
Offshore							
Argentina	–	–	–	68	..	68	–
Australia	3 650	83 552	41 637	–	..	128 839	120 598
Bangladesh	49 898	–	–	–	..	49 898	23 145
Belgium	29 846	–	6 584	–	..	36 430	12 721
Brazil	52 227	25 320	137 075	–	..	214 622	128 945
Chile	24 434	–	–	6 847	..	31 281	19 646
China	446 297	–	–	–	..	446 297	140 855
Costa Rica	6 665	1 822	–	–	..	8 487	–
France	15 090	–	–	–	..	15 090	–
India	277 085	–	–	–	..	277 085	223 816
Indonesia	6 231	–	–	–	..	6 231	59 148
Ireland	–	–	15 323	–	..	15 323	–
Italy	–	–	–	8 911	..	8 911	–
Jamaica	–	2 585	–	–	..	2 585	–
Japan	131 260	72 970	37 608	106 402	..	348 240	346 107
Korea	182 023	–	–	8 711	..	190 734	217 335
Malaysia	86 338	–	–	–	..	86 338	83 258
Mexico	–	–	–	–	..	–	12 780
New Zealand	10 179	–	–	–	..	10 179	27 167
Philippines	42 356	–	–	–	..	42 356	39 864
Romania	32	–	–	–	..	32	23
South Africa	12 410	9 979	–	–	..	22 389	23 671
Sri Lanka	31 699	–	–	–	..	31 699	18 240
Swaziland	–	–	17 445	–	..	17 445	18 595
Taiwan	34 627	–	–	–	..	34 627	56 746
Thailand	–	–	–	–	..	–	3 025
United Kingdom	807	–	–	–	..	807	688
Offshore total	1 443 154	196 228	255 672	130 939	..	2 025 993	1 576 373
Total sales	1 743 748	2 667 132	1 481 627	664 240	..	6 556 747	5 051 549

Source: Potash and Phosphate Institute.
[1] Common specifications are: standard –28 to +65 mesh, special standard –35 to +200 mesh, coarse –8 to +28 mesh, granular –6 to +20 mesh, each grading a minimum of 60 per cent K_2O equivalent, soluble and chemical grade a minimum of 62 per cent K_2O equivalent. [2] Standard includes Special Standard, sales of which were 125 449 t K_2O equivalent in 1982, and 140 406 in 1983. [3] Chemical sales are included in standard grade sales and totalled 60 008 t in 1983.
– Nil; .. Not available.

TABLE 3. CANADA, POTASH PRODUCTION AND TRADE, YEARS-ENDED JUNE 30, 1966, 1971, 1976-84

	Production[2]	Imports[1,2]	Exports[2]
	(tonnes K_2O equivalent)		
1966	1 748 910	31 318	1 520 599
1971	3 104 782	26 317	3 011 113
1976	4 833 296	16 445	4 314 150
1977	4 803 015	24 289	4 175 473
1978	6 206 542	26 095	5 828 548
1979	6 386 617	21 819	6 256 216
1980	7 062 996	20 620	6 432 124
1981	7 336 973	35 135	6 933 162
1982	6 042 623	25 437	5 400 662
1983	5 378 842	21 846	4 864 219
1984	7 155 599	17 934	6 730 733

Source: Potash and Phosphate Institute, Canadian Fertilizer Institute.
[1] Includes potassium chloride, potassium sulphate, except that contained in mixed fertilizers. [2] Change of data source. Prior to 1978 figures were obtained from Statistics Canada.

TABLE 4. CANADA, POTASH PRODUCTION AND SALES BY QUARTERS, 1984 AND 1983

	Total (1983)	1st quarter	2nd quarter	3rd quarter	4th quarter
		(000 tonnes)			
Production	5 928.4	2 087.6	2 048.6	1 552.3	2 060.1
Sales					
North America	4 530.8	1 020.3	1 012.8	1 383.8	1 104.6
Offshore	2 026.0	678.5	630.1	711.7	524.4
Ending Inventory	861.5	1 020.3	1 656.1	1 112.8	1 543.0

Source: Potash and Phosphate Institute.

TABLE 5. CANADA, POTASH SALES BY PRODUCT AND AREA, 1982 AND 1983

		Agricultural					Industrial			Total
		Standard	Coarse	Granular	Soluble	Total	Standard	Soluble	Total	Sales
				(tonnes K$_2$O equivalent)						
Alberta	1982	708	1 546	15 822	1 289	19 365	3 042	110	3 152	22 517
	1983	590	4 746	19 934	1 680	26 950	1 961	469	2 430	29 380
British Columbia	1982	15	1 544	3 920	3 310	8 789	–	–	–	8 789
	1983	–	217	5 687	32	5 936	–	–	–	5 936
Manitoba	1982	132	7 040	13 949	760	21 881	–	–	–	21 881
	1983	22	4 618	10 551	634	15 825	–	–	–	15 825
New Brunswick	1982	–	6 184	–	–	6 184	–	–	–	6 184
	1983	–	11 325	2 493	–	13 818	–	–	–	13 818
Nova Scotia	1982	–	6 025	–	–	6 025	–	–	–	6 025
	1983	276	4 668	–	–	4 944	–	–	–	4 944
Ontario	1982	366	86 108	32 723	354	119 551	1 710	4 072	5 782	125 333
	1983	902	189 111	6 253	1 407	197 673	2 953	2 824	5 777	203 450
Prince Edward Island	1982	–	10 460	–	–	10 460	–	–	–	10 460
	1983	401	4 783	5 192	–	10 376	–	–	–	10 376
Quebec	1982	–	44 982	13 208	–	58 190	306	–	306	58 496
	1983	500	51 941	31 705	–	84 146	318	–	318	84 464
Saskatchewan	1982	1 446	800	3 513	1 446	7 205	4 188	1 721	5 909	13 114
	1983	12	2 011	5 230	794	8 047	5 728	2 361	8 089	16 136
Newfoundland	1982	–	–	–	–	–	–	–	–	–
	1983	409	–	–	–	409	166	–	166	575
Totals	1982	2 667	164 689	83 135	7 159	257 650	9 246	5 903	15 149	272 799
	1983	3 112	273 420	87 045	4 547	368 124	11 126	5 654	16 780	384 904

Source: Potash and Phosphate Institute.
– Nil.

TABLE 6.   CANADA, POTASH INVENTORY, PRODUCTION, DOMESTIC SHIPMENTS AND EXPORTS, 1983

	Beginning Inventory	Production	Domestic Shipments		Exports			Total Shipments
			Agri-cultural	Non-agri-cultural	United States		Offshore	
					Agri-cultural	Non-agri-cultural		
	(000 tonnes K$_2$O)							
January	1 486.2	357.8	22.5	0.8	267.5	14.2	154.2	459.2
February	1 417.0	263.5	11.4	1.5	215.8	10.7	119.5	358.9
March	1 272.2	626.5	41.3	1.0	276.4	18.0	143.1	479.8
April	1 386.0	599.1	31.7	0.7	252.3	14.6	174.2	473.5
May	1 515.3	521.5	62.1	0.9	370.8	13.6	177.1	624.5
June	1 409.2	540.6	23.1	1.6	318.6	16.9	96.6	456.8
Sub-total		2 909.0	192.1	6.5	1 701.4	88.0	864.7	2 852.7
July	1 494.0	313.0	3.3	1.1	239.4	10.9	125.9	380.6
August	1 432.9	351.4	16.8	2.6	437.4	12.6	278.8	748.2
September	1 060.7	484.6	48.4	1.8	475.6	15.0	189.2	730.0
October	879.7	612.9	21.5	1.5	422.3	21.5	191.7	658.5
November	765.4	629.1	39.4	2.1	424.5	14.3	171.6	651.9
December*	759.7	628.3	46.6	1.2	264.5	18.3	204.2	534.8
Sub-total		3 019.3	176.0	10.3	2 263.7	92.6	1 161.4	3 704.0
Total 1983		5 928.3	368.1	16.8	3 965.1	180.6	2 026.1	6 556.7
1982		5 216.4	267.9	14.9	3 065.8	175.7	1 576.8	5 101.1
% change 1983/82		+13.6	+37.4	+12.8	+29.3	+2.8	+28.5	+28.5

Source: Potash and Phosphate Institute of North America.
* Inventory at the end of December 1983 is estimated at 861 500 tonnes.

TABLE 7.   CANADA, POTASH INVENTORY, PRODUCTION, DOMESTIC SHIPMENTS AND EXPORTS, 1984

	Beginning Inventory	Production	Domestic Shipments		Exports			Total Shipments
			Agri-cultural	Non-agri-cultural	United States		Offshore	
					Agri-cultural	Non-agri-cultural		
	(000 tonnes K$_2$O)							
January	861.5	669.2	31.9	2.1	376.4	13.6	227.4	651.4
February	941.1	678.6	28.1	1.6	301.7	11.6	219.4	562.4
March	1 026.8	739.8	30.3	2.1	214.6	14.6	231.7	493.3
April	1 250.3	707.9	28.5	1.1	298.8	17.2	229.6	575.2
May	1 440.9	698.6	58.7	1.8	389.8	18.0	168.3	636.6
June	1 460.7	642.2	15.9	2.1	231.5	18.7	232.5	500.7
Sub-total		4 136.3	193.4	10.8	1 812.8	93.7	1 308.9	3 419.6
July	1 656.1	401.3	9.8	1.7	205.5	9.5	260.0	486.6
August	1 558.6	485.7	43.6	1.9	604.3	15.6	210.1	875.5
September	1 153.0	665.2	33.0	1.8	426.0	16.9	241.6	719.3
October	1 112.8	734.0	24.4	2.0	263.1	17.6	181.9	489.0
November	1 372.6	720.6	23.0	1.5	217.0	26.4	180.6	448.5
December*	1 555.4	605.5	25.0e	2.0e	412.7e	18.0e	160.4	618.1e
Sub-total		3 612.3	158.8e	10.9e	2 128.6e	104.0e	1 234.6e	3 637.0e
Total 1984		7 748.6	352.2e	21.7e	3 941.4e	197.7e	2 543.5e	7 056.6e
1983		5 928.3	368.1	16.8	3 965.1	180.6	2 026.1	6 556.7
% change 1984/83		+30.7	-3.8	+29.2	-0.6	+9.5	+25.5	+7.6

Source: Potash and Phosphate Institute of North America.
* Inventory at the end of December 1984 is estimated at 1 543 000 tonnes.
e Estimated by Mineral Policy Sector, Energy, Mines and Resources Canada.

TABLE 8. WORLD POTASH PRODUCTION

	1980	1981	1982	1983P	1984e
			(000 tonnes K_2O)		
Canada	7 303	7 147	5 352	5 930	7 700
Chile	25	21	22	22	22
China	12	20	26	25	25
France	1 894	1 828	1 706	1 539	1 700
Germany Dem. Rep.	3 405	3 497	3 200	3 341	3 400
Germany, Fed. Rep.	2 737	2 591	2 057	2 419	2 650
Israel	797	832	946	942	1 000
Italy	102	125	115	133	140
Jordan	-	-	9	168	280
Spain	658	728	694	659	700
U.S.S.R.	8 064	8 449	8 079	8 400e	9 000
United Kingdom	306	284	240	303	330
United States	2 239	2 156	1 784	1 429	1 600
	27 542	27 678	24 515	25 310	28 547

Sources: International Fertilizer Industry Association Ltd.; U.S. Bureau of Mines and Energy, Mines and Resources Canada.
P - Preliminary; e - Estimated; - Nil.

TABLE 9. CANADA, POTASH MINES - CAPACITY PROJECTIONS

	1980	1981	1982	1983	1984	1985	1986	1987	1988	1989	1990
					(000 tonnes K$_2$O equivalent)						
PCS											
- Allen (60%)	570	570	570	570	570	570	570	570	570	570	570
- Cory	830	830	830	830	830	830	830	830	830	830	830
- Esterhazy (25% of IMC)	580	580	580	580	580	580	580	580	580	580	580
- Lanigan	545	690	690	690	690	690	1 240	1 740	1 740	1 740	1 740
- Rocanville	725	750	830	1 160	1 160	1 160	1 160	1 160	1 160	1 160	1 160
Sub-total	3 250	3 420	3 500	3 830	3 830	3 830	4 380	4 880	4 880	4 880	4 880
CCP	815	815	815	815	815	815	815	815	815	815	815
Cominco	545	545	600	655	655	655	655	655	655	655	655
IMC	1 750	1 750	1 750	1 750	1 750	1 750	1 750	1 750	1 750	1 750	1 750
PPG (Kalium)	845	845	1 055	1 055	1 055	1 055	1 055	1 320	1 320	1 320	1 320
PCA	420	420	420	630	630	630	630	630	630	630	630
Kidd Creek (Allen 40%)	380	380	380	380	380	380	380	380	380	380	380
Sub-total	4 755	4 755	5 020	5 285	5 285	5 285	5 285	5 550	5 550	5 550	5 550
Total Saskatchewan	8 005	8 175	8 520	9 115	9 115	9 115	9 665	10 430	10 430	10 430	10 430
Denison, N.B.	-	-	-	-	-	200	450	650	780	780	780
PCA, N.B.	-	-	-	50	200	300	380	380	380	545	545
Total New Brunswick	-	-	-	50	200	500	830	1 030	1 160	1 325	1 325
Canada (firm)	8 005	8 175	8 520	9 165	9 315	9 615	10 495	11 460	11 590	11 755	11 755
(unspecified)	-	-	-	-	-	-	-	-	-	100	400
TOTAL	8 005	8 175	8 520	9 165	9 325	9 615	10 495	11 460	11 590	11 855	12 155

Note: Capacity means "rated" capacity; under normal conditions Canadian mines operate at about 90 per cent of rated capacity.
 - Nil.

TABLE 10. WORLD POTASH CAPACITY 1982-92

	1982	1983	1984	1985	1986	1987	1988	1989	1990	1991	1992
	(000 tonnes K₂O equivalent)										
North America											
Canada	8 520	9 165	9 315	9 615	10 495	11 460	11 590	11 855	12 155	12 700	13 200
United States	2 090	1 880	1 770	1 770	1 770	1 610	1 560	1 450	1 450	1 450	1 450
Total	10 610	11 045	11 085	11 385	12 265	13 070	13 150	13 305	13 605	14 070	14 570
Western Europe											
France	2 000	2 000	2 000	1 800	1 700	1 700	1 700	1 700	1 700	1 600	1 500
Germany, Fed. Rep.											
Italy	2 900	2 700	2 700	2 800	2 800	2 800	2 800	2 800	2 800	2 800	2 800
Spain	200	200	200	200	200	200	200	200	200	200	200
United Kingdom	800	800	800	800	800	800	800	800	800	800	800
	360	360	360	360	360	360	360	360	360	360	360
Total	6 260	6 060	6 060	5 960	5 860	5 860	5 860	5 860	5 860	5 760	5 660
Eastern Europe											
Germany, Dem. Rep.	3 600	3 600	3 600	3 600	3 600	3 600	3 600	3 600	3 600	3 600	3 600
U.S.S.R.	10 900	11 200	11 500	11 700	11 900	12 200	12 700	13 100	13 400	13 800	14 000
Total	14 500	14 800	15 100	15 300	15 500	15 800	16 300	16 700	17 000	17 400	17 600
Asia											
Israel	1 000	1 000	1 000	1 260	1 260	1 260	1 260	1 260	1 260	1 260	1 260
Jordan	–	200	300	450	600	720	720	720	720	720	720
China, People's Rep.	30	30	30	30	30	30	30	30	30	30	30
Total	1 030	1 230	1 330	1 740	1 890	2 010	2 010	2 010	2 010	2 010	2 010
Latin America											
Brazil	–	–	–	–	100	150	200	250	300	300	300
Chile	30	30	30	30	30	30	30	30	30	30	30
Mexico	–	–	–	–	10	30	30	50	50	50	50
Total	30	30	30	30	140	210	260	330	380	380	380
Other	–	–	–	–	50	50	100	100	100	100	150
World Total	32 430	33 165	33 605	34 415	35 705	37 000	37 680	38 305	38 955	39 720	40 370

Note: Under "other" is the probable small production from brines in Australia, Peru or Tunisia.
– Nil.

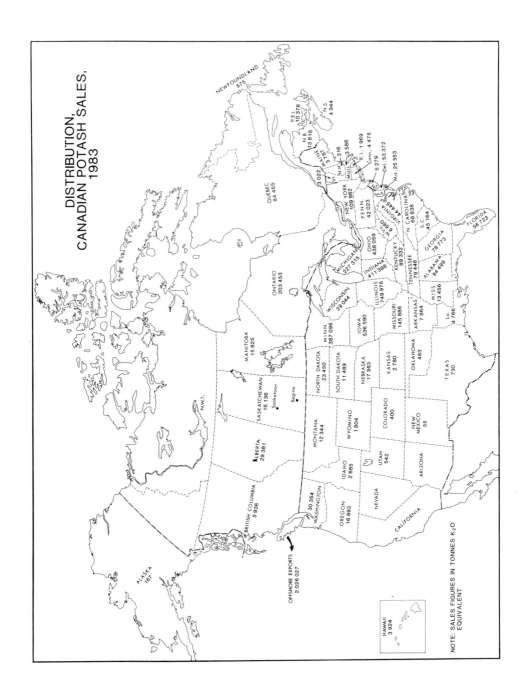

Rare Earths

D.E.C. KING

Rare earth metals or lanthanides encompass fifteen chemically similar metals, but two other metals, scandium and yttrium, closely resemble this group and are usually classed with the lanthanides. The minerals containing these metals are actually not rare, but relatively rare metals such as beryllium, zirconium, columbium, tantalum, thorium and uranium are often found associated with them in nature. The metals were originally classified "rare" because they are seldom concentrated in nature like most other elements and their widespread occurrence in the earth's crust was recognized only in recent times. The term "earth" is a retention of early terminology for insoluble natural oxides.

Lanthanide-bearing minerals contain all members of the rare earth group of elements, but some are classified as the "light" group and the remainder the "heavy" group. Cerium is generally the most abundant element in minerals containing the "light" rare earths, which are also associated with scandium. The "heavy" rare earths are usually found in minerals which often contain substantial proportions of yttrium. The rare earth metals are typically associated with alkaline igneous rocks and also occur as secondary concentrations in placer, beach sand and phosphatic sedimentary deposits. Commercial production has been derived from carbonatite occurrences, placer and beach sand deposits, uranium ores, and phosphatic rocks.

Monazite is a rare earth phosphate that contains nearly 70 per cent rare earth oxides (REO) and about 1.5 per cent yttrium oxide. Placer deposits of heavy mineral sands are the major source of monazite which is usually recovered as a byproduct of rutile, ilmenite and zircon mining operations. Only in a few cases have deposits been exploited primarily for monazite; such a deposit in South Africa was the world's major source of monazite from 1953 to 1963. Bastnaesite is a fluorcarbonate of the cerium subgroup which contains about 75 per cent REO and 0.05 per cent yttrium. It has been found in economic quantities in vein deposits, contact metamorphic zones, pegmatites and other igneous rocks. Xenotime, the yttrium phosphate isomorph of monazite is the main source of yttrium and the "heavy" rare earth elements.

World reserves of rare earth metals have been estimated at about 50 million t of contained rare earth oxides (excluding China which has large reserves of bastnaesite) while annual world consumption is only of the order of 20 000 to 25 000 t of contained REO. The total world demand for yttrium oxide is of the order of about 350 tpy. Total samarium oxide demand is of the order of 350-400 tpy.

The rare earth elements are used as metals, oxides, halides and other compounds. Because of the difficulty of separating the individual rare earth elements, most industrial applications still employ mixtures of the elements. Originally these mixtures were in the proportions occurring in the ore sources but there is a growing demand for individual rare earth elements or mixtures enriched by one specific element. However, the commercial demand for individual elements hardly ever matches the proportions occurring in ore sources so some elements are invariably in oversupply. This tends to spur research into uses for the less popular and more abundant elements. Cerium, neodymium, yttrium and lanthanum are considerably more abundant than thulium, lutetium, terbium, holmium and europium (by anywhere from 20 to over 200 times) and the remaining rare earth elements are of intermediate abundance.

New markets for specific members of the rare earth group, have resulted in increased production of all of the rare earth metals from which they must be separated because of their natural association in ores. Simultaneously, production costs for some rare earth members, produced as byproducts of the refining process, have diminished.

D.E.C. King is with the Mineral Policy Sector, Energy, Mines and Resources, Canada. Telephone (613) 995-9466.

Availability and declining costs have been important factors in the development of new uses. There is growing optimism that the rare earth metals industry will expand at a steady rate now that industrial uses are becoming more diverse.

CANADIAN INDUSTRY

While substantial resources of rare earths have been identified in Canada, raw material production in the past has been limited to relatively small outputs of byproduct residues from uranium mining and hydrometallurgical operations at Elliot Lake, Ontario. These concentrates were produced from 1966 through 1970 and from 1974 through 1977. Elliot Lake uranium ores are rich in yttrium and the "heavy" rare earth elements and, in fact, were the world's major source of yttrium concentrate during the first of these two periods. The Elliot Lake ores contain about 0.11 per cent uranium oxide (U_3O_8), 0.028 per cent thorium oxide and 0.057 per cent rare earth oxides.

Denison Mines Limited has been the largest Canadian producer of rare earth concentrates. Denison ceased production of yttrium concentrate in 1978 because its recovery had become uneconomic owing to increased costs of chemical reagents. Marketing of the yttrium concentrate had been carried out through Molycorp, Inc. and Michigan Chemical Corporation. However, shipments to the latter company had been terminated in mid-1970 when Michigan Chemical experienced difficulty in marketing the product. Denison is reviewing the possibility of resuming production.

Rio Algom Mines Limited recovered thorium and rare earth concentrate at its Nordic mill, but discontinued this activity upon transfer of uranium milling to the Quirke mill where no thorium and rare earth facilities were installed.

In addition to the large resource in Elliot Lake uranium ores, rare earths are also associated with uranium deposits at Agnew Lake, 65 kilometres (km) east of Elliot Lake (where the REO content is about twice that of Elliot Lake ores), and in the Bancroft area of Ontario.

The Highwood Resources Ltd. property at Thor Lake, located about 65 miles southeast of Yellowknife, contains several zones, each composed of somewhat different mineralization. The Lake zone is enriched in tantalum, columbium, rare earths and zirconium, the R zone contains rare earths and beryllium, the S zone is enriched in uranium, thorium and columbium and the T zone contains columbium, uranium and beryllium. Geological research on the Thor Lake area is currently being conducted at the University of Alberta with federal government support.

Drilling on the R and T zones has identified a beryllium enrichment of about 1.8 million t grading 0.75 per cent beryllium oxide (BeO). With existing resources and the expectation of further discoveries, Highwood is currently undertaking a pre-feasibility study, including metallurgical testing, to assess the economics and viability of an open pit beryllium mine and concentrator.

The Strange Lake deposit, owned by Iron Ore Company of Canada (IOC), is located near Lac Brisson on the Quebec-Labrador boundary and approximately 155 miles northeast of Schefferville. It is reported to contain large tonnages of yttrium and zirconium as well as large quantities of columbium, beryllium and rare earth elements. The deposit, discovered by IOC in 1979 in a follow-up to government surveys, is located at shallow depth and is amenable to open pit mining. The deposit is being evaluated by IOC with the objective of defining economic reserves.

Significant quantities of rare earths are found in a number of Canadian pyrochlore-bearing carbonatite deposits such as the Niobec Inc. mine near St. Honoré, Quebec and in deposits on the Manitou Islands, Lake Nipissing, Ontario.

Phosphorite formations in western Canada contain small quantities of rare earths, as do Florida phosphates imported into Canada for the production of phosphoric acid. Other potential sources include apatite-rich carbonatites in Ontario and Quebec. Rare earth elements, primarily the light element group, are associated with apatite in the Nemegos No. 6 magnetite deposit, which is located in the Chapleau area of Ontario.

Sherritt Gordon Mines Limited has been producing samarium-cobalt alloy powder at Fort Saskatchewan, Alberta since 1980 as a means of marketing cobalt with added value. Through continued research and development, Sherritt has improved the processing technology originally purchased from Canadian General Electric Ltd. in 1979, to the point that in 1984 their $SmCo_5$ (1-5) type alloy powder was of fully competitive quality. Production of type 1-5 alloy powder is in tonnage quantities, dependent to a certain extent on samarium supply, and all of this material is exported to a number of countries. Sherritt has also carried out research and development on the production of another class of samarium-cobalt alloy powder designated 2-17, with the aim of bringing it to full scale production. This alloy, which nominally contains 24 per cent samarium, compared with 34.5 per cent in the 1-5 type, has a lower resistance to demagnetization but a higher energy per unit volume (energy product) than the type 1-5. Research on the production of magnets from 2-17 powder has already been successful to the point that this technology has been licensed to a major U.S. magnet producer.

WORLD INDUSTRY

Total world production of rare earth minerals in 1983 was estimated to have been about 41 300 t, of which bastnaesite from the United States accounted for 17 300 t, and monazite 24 000 t. The latter was provided mainly by Australia (15 700 t), India (4 200 t), Brazil (2 000 t) and the United States (1 000 t). The United States and Australia together produced about 80 per cent of total world mineral production, excluding China which has the capacity to produce 9 000 tpy. India's production of monazite increased by 27 per cent from 1980 to 1982.

Estimates of rare earth minerals reserves by the U.S. Bureau of Mines (USBM) indicate about 5 million t of rare earth oxide content in North America, and a world total of about 7 million t. These totals exclude the reported huge reserves of China which amount to 36 million t of rare earth oxides in bastnaesite.

Total world reserves of yttrium, excluding China, were estimated by USBM to amount to 34 000 t of which 18 000 t are in India and 5 000 t in Australia. The potential in Canada is large but so far undefined.

Although the consumptions of rare earths and yttrium are increasing strongly, world reserves are clearly more than sufficient to meet foreseeable demand.

The rare earth elements are principally extracted from two minerals, monazite and bastnaesite. Nearly all of the world's bastnaesite production is from a mine at Mountain Pass, California, which is owned and operated by Molycorp Inc. Very large reserves of bastnaesite also exist in China which also has the processing capability to produce separate and pure rare earth elements. China's proven reserves of bastnaesite amount to 36 million t and occur in a massive iron ore deposit in Inner Mongolia.

Monazite is usually recovered commercially from heavy mineral beach sands with rutile, ilmenite and zircon. It is found in a number of countries but the main producing countries are Australia, the United States, India, Brazil, and Malaysia. Australia produces over one half of the total world output and is the only major mineral-producing country possessing no processing capacity. Both India and Brazil prohibit the export of monazite because of its thorium content, so monazite is processed in these countries to remove thorium and to produce mixed rare earth compounds.

The United States, Japan and France are the leading processors of rare earth minerals, and the two companies Molycorp Inc. in the United States and Rhone-Poulenc Inc. of France dominate this industry. Generally speaking, most of the output from the United States and Japan is from bastnaesite while about one half of the world's output of monazite is processed in western Europe. The United States is also an important processor of monazite. Consumption of rare earth compounds, alloys and individual metals occurs mainly in the United States, Japan and western Europe.

Allied Eneabba Ltd., which accounts for more than half of the total Australian monazite output, has not increased its production capacity since the major expansion of its Geraldton plant in 1980. However, in 1981 Allied acquired property which extended its mine life to 28 years. Because monazite is a coproduct in the mining of heavy mineral sands, any slackness in demand for the associated heavy minerals such as occurred in 1982, can depress monazite output.

India Rare Earths Ltd., (IRE) which processes its own monazite to produce rare earth chlorides, fluorides, and oxides, started a new plant at Orissa in 1984. It also announced plans to diversify into the production of yttrium, gadolinium and samarium concentrates.

China has become a significant producer of rare earth compounds and alloys, with the potential to create a significant impact on the rare earth market.

The China Rare Earth Co. has been collaborating with Japax Inc. of Japan since 1981 to develop new rare earth smelting technology and has been supplying Japax with 2 000 tpy of rare earth concentrates. Mitsui Mining & Smelting Co. Ltd. of Japan was helping the Baotou Iron & Steel Co. to produce rare earth concentrates in Inner Mongolia. Japan is China's leading market for rare earths, importing about 1 000 tpy.

Following the takeover of Molycorp by Union Oil Company of California in 1977, Molycorp REO production statistics have not been made available; only tonnages of rare earth concentrates have since been reported.

In 1981, Molycorp completed a $15 million expansion of its solvent extraction capacity at Mountain Pass. The new circuit has increased by 35 per cent, its capacity to produce samarium and gadolinium oxides, as well as high purity oxides of lanthanum, cerium, praeseodymium, and neodymium. Molycorp also produces high purity oxides and compounds at Washington, Pennsylvania, where a facility was completed in 1980 to produce samarium cobalt alloys. The company has other production facilities for high purity compounds at York, Pennsylvania and at Louviers, Colorado where high purity yttrium oxide is made from xenotime.

Davison Specialty Chemicals Division of W.R. Grace & Co., which processes monazite almost solely for its own catalyst production, started operating a new 23 000 tpy facility at Curtis Bay, Maryland in 1981 to produce zeolite-rare earth catalysts. Catalyst production was also increased at Davison's plants at Lake Charles and South Gate, California.

Ronson Metals Corporation increased its production of mischmetal by 25 per cent in 1980 and a further 20 per cent in 1983. Through its subsidiary Cerium Metals Corporation, which produces alloys of cerium, lanthanum and didymium (a commercial mixture of rare earth metals enriched in neodymium and praeseodymium), cerium free mischmetal, and high-purity samarium metal, Ronson began in 1984 to produce neodymium metal for magnet manufacture.

Reactive Metals and Alloys Corporation installed new arc furnaces at its West Pittsburgh plant during the early 1980s to produce specialty silicon alloys and to triple its capacity for producing rare earth silicide.

Rhône Poulenc Inc. has its main operation at La Rochelle, France where it has an annual production capacity of about 4 600 t of light rare earth elements and 370 t of heavies. Separations of the elements are carried out by solvent extraction in both chloride and nitrate aqueous media employing over 1,000 mixer-settlers. The company completed a 4 000 tpy rare earth separation and finishing plant at Freeport, Texas in 1981; production is from concentrates imported from the parent company.

Th Goldschmidt AG of West Germany which produces by the "Co-reduction Process" a range of alloys based on rare earth elements, ceased production of mischmetal in early-1984. It continued to produce alloys of rare earth elements with cobalt, iron and other transition elements, and in its latest developments, emphasized the production of neodymium, its compounds and alloys.

Treibacher Chemische Werke AG of West Germany is the leading world producer of mischmetal in its various forms, and will probably capture a large part of the mischmetal business relinquished by Th Goldschmidt AG.

Plant closures by British Flint and Cerium Manufacturers Ltd. in 1981, associated with Ronson Metals Corp., and by Steetley Chemicals Ltd. in 1982 after its new plant had operated for only a year, were said to be recession related. Rare Earth Products Ltd., a division of Johnson Matthey Chemicals Ltd. is a relatively small producer of the full range of individual pure rare earth metals and compounds. Its Widnes plant was working to full capacity in 1984, taking advantage of the upturn in the rare earth metals industry. The company was expanding its capacity for rare earth compounds. London and Scandinavian Metallurgical Co. Ltd., a subsidiary of Metallurg Inc., produces cerium oxide polishing powders using bastnaesite as a starting raw material.

There are about a dozen companies in Japan which produce a comprehensive range of rare earth metals, alloys and compounds for all categories of consumption. Japanese companies were active in joint ventures in China, Malaysia and the United States, and in the development of improved extraction and separation technology. The Asahi Chemical Industry Co. Ltd. is developing separation technology based on ion exchange chromatography with the aim of greatly reducing the number of processing steps and equipment units.

CONSUMPTION AND USES

Rare earths have several unique properties which enable them to be employed in distinctly different ways. Because of the similarity of their chemical and some of their physical properties, mixtures of rare earths as alloys or compounds can be applied to certain applications. However, other uses, which depend on individual chemical and nuclear properties, require specific rare earth elements containing no more than about 10 per cent of the other rare earths. Consequently, there is a trend towards the greater use of specific rare earths and of mixtures enriched in individual rare earth components.

Rare earth mixtures are used in catalysts, master alloys, other alloys and glass polishing compounds. Individual rare earth metals or compounds are used in magnetic materials, phosphors, neutron capture applications, glass and ceramics.

Mischmetal, (an alloy containing a mixture of light rare earths generally in the ratio found in the ore), ferrocerium, and cerium silicide are added to some grades of high-strength low-alloy (HSLA) and stainless steels to effect sulphide shape control. This practice has been somewhat in decline in recent years because of a trend towards greater removal of sulphur during steelmaking. However, the need for weldability in these steels appears to be arresting and even reversing the trend towards ever-lower sulphur levels, and the decline in the use of rare metals in steels, accordingly appears to be levelling off.

The development of magnesium alloys with improved hot strength is an example of a trend towards the greater use of specific rare earth metals. Magnesium Elektron Ltd. of the United Kingdom has progressively improved the properties of its range of magnesium alloys, firstly by enriching rare earth additions with yttrium, and later with neodymium and other elements. The rare earth additions combine to form precipitates which strengthen or lock the alloy crystals in place under load.

Another major application for rare earths is as additions to catalysts which are used in the cracking operation of petroleum refining. Rare earths modify the surface activity of other compounds, and are added as natural or cerium-depleted mixtures of rare earth chlorides to zeolite catalysts. They have only recently been available in tonnages large enough to satisfy this large and growing market. Automobile exhaust systems could be another substantial area for the use of natural rare earth mixtures (cerium-rich) because of their ability to stabilize the alumina matrix which carries the active catalysts. Rare earths are also used in many other catalytic applications.

Glass polishing and decolourizing are two distinctly separate applications for rare earths. Natural mixtures of rare earth oxides, high in cerium, are very effective as polishing powders in high-quality optical and plate glass, and mirror applications. The use of plastic lens systems in popular lightweight cameras has reduced demand to some extent, but the increasing demand for video tubes with glass screens could offset this.

Cerium oxide, in small quantities, is an effective glass decolourizer. Owing to its ability to absorb ultraviolet light, cerium oxide is used in the manufacture of transparent bottles to inhibit food spoilage. Neodymium, praseodymium, erbium and holmium are effective in welders' goggles, sunglasses and optical filters because of their absorption characteristics. Lanthanum oxide is used in optical lenses to increase the refractive index of the glass. A recent development involves light polarization and electro-optic switching in pilots' protective goggles for automatically blocking the brightness of a nuclear flash. The switching device is based on ceramic lead-lanthanum zirconate-titanate. For glass colouring, praseodymium imparts a yellow-green colour, neodymium a lilac, europium an orange-red, and erbium a pink colour. Lanthanum is a major component of optical glass, and cerium glass is used for windows in atomic reactors.

Many other important uses for rare earths exist, which so far use only small quantities of the elements. These include laser, nuclear, bubble memory, hydrogen storage, microwave, medicine, jewellery, solar energy, temperature measurement and various other applications. Among the individual rare earths whose sharp line emissions are effective in laser applications, neodymium in an yttrium-bearing host material, is the most commonly used.

Nuclear uses include a role as neutron absorbers in fast breeder reactors. The high captive cross sections for thermal neutrons of europium and its isotopes are utilized when used in reactor control rods. In another nuclear application, gadolinium oxide which has the highest neutron absorption of any element, is used in uranium fuels to reduce uranium consumption and improve the energy output.

Magnetic bubble memory films for data storage and processing, promises to be an important new application, particularly for gadolinium in the form of a gadolinium-gallium garnet. So far the full potential of high efficiency bubble memory systems has not been attained, perhaps because of the comparatively lower cost and greater flexibility of floppy disc systems.

Hydrogen storage is based on the ability of alloys such as lanthanum-nickel to absorb hydrogen at appropriate temperatures and pressures. Lanthanum-nickel can absorb up to 400 times its own volume of hydrogen. However, the need for large scale hydrogen storage has not yet arrived.

Garnets containing yttrium or gadolinium are used in various electronic components to control microwaves in radar, ovens and telecommunications.

There is a potentially large growth application for yttrium oxide in partially stabilized ceramic zirconia. The stabilization, which can also be achieved using magnesia or calcium oxide, toughens the zirconia for wear, heat and corrosion resistance uses. Current applications include extrusion dies, valves and pump parts, and there is active interest in its future use in the wearing surfaces in diesel engines, which would significantly enlarge the market for partially stabilized ceramic zirconia.

Individual rare earth oxides and fluorides are used in significant quantities in carbon-arc lamps where a high intensity white light is desirable. A new type of fluorescent lamp is now on the market that emphasizes three narrow spectral bands around the blue-violet, green and orange-red wavelengths to produce a synthesized white light of great brightness. The new light uses three rare earth phosphors of europium, yttrium, cerium and terbium in combinations with magnesium aluminate.

High-value applications exist in the electronics field where individual rare earth oxides are used as phosphors in colour television tubes, in temperature-compensating capacitors and in associated circuit components. Although the volume of europium and yttrium oxides used in colour television phosphors is comparatively small, the value is disproportionately large because of the high degree of purity required in this application. Europium is used as an activator in an yttrium host material to provide the primary red colour. Other activators are cerium, europium and terbium while other host materials are lanthanum and gadolinium. Phosphors utilizing terbium or thulium as activators in lanthanum and gadolinium hosts are now being used in X-ray intensifying screens to produce sharper images with indirect visible light.

The development of rare earth-cobalt magnets set the trend in broadening the demand for individual rare earths. Samarium-cobalt permanent magnets have four times the power of Alnico magnets and the consequent application of lighter weight magnetic components has played an important role in miniaturization. The volatility of the cobalt market has tended to upset this use, but in the latest development, neodymium-boron-iron magnets promise to supplant many applications for samarium-cobalt. The neodymium-boron-iron alloys have nearly double the magnetic energy of samarium-cobalt alloys, and have advantages in price since neodymium is more abundant and cheaper than samarium, and iron is cheaper and its market more stable than cobalt. However, neodymium-boron-iron alloys tend to lose their magnetism more readily than samarium-cobalt at elevated temperatures, and they are also prone to rusting. Therefore, there will continue to be many special applications for samarium-cobalt magnets which the new alloys cannot fill. Rare earth based magnets are usually fabricated by powder metallurgical methods that facilitate the procedure for inducing a high magnetic flux. Plastic epoxy-bonded samarium-cobalt magnets have also been successfully developed in Japan. High-strength permanent magnets are used in electric motors, generators, meters, speakers, and frictionless bearings.

Since the development of samarium-cobalt magnets, samarium now accounts for nearly two thirds of total consumption of high purity individual rare earth metals. The further growth in demand for rare earth based magnets seems assured. The production of neodymium-cobalt-iron magnets is already under way at Sumitomo in Japan and Crucible Magnetics Div. of Colt Industries in the United States, and production by Magnetic Materials Group in the United Kingdom is expected to begin in mid-1985. General Motors Corp. in the United States is expected to begin in mid-1985. General Motors Corp. in the United States is also developing neodymium-boron-iron magnets for motors to operate windshield wipers, electric windows and car seats, and for generators and larger induction motors.

PRICES

The December 1984 issue of "Industrial Minerals" quotes the following prices for concentrates of rare earth minerals:

Bastnaesite concentrate 70% leached, per lb REO	$US 1.05
Monazite, minimum 55% REO, tonne, fob Australia	$A 510.00-550.00
Yttrium concentrate 60% Y_2O_3 fob Malaysia per kg	$US 46.00

REO Rare earth oxides; fob Free on board.

Prices for rare earth oxide and metal ingots, as quoted in the December 21, 1984 issue of "American Metal Market", were:

Rare earth oxides, U.S. dollars per lb, one-lb lots

	Per cent	$US
Cerium	99.9	8.00
Europium	99.99	650.00
Gadolinium	99.9	55.00
Lanthanum	99.99	7.00
Neodymium	99.9	33.00
Praseodymium	95	16.80
Samarium	96	25.00
Yttrium	99.99	42.00

At the end of 1984, prices of rare earth metal ingots in U.S. dollars per lb for minimum lots of 100 lb fob shipping point were:

	Per cent	$US
Cerium	99	21.00-45.46
Lanthanum	99	29.48-45.00
Samarium		86.37
Mischmetal	98	5.60
Praseodymium		109.10

TABLE 1. RARE EARTH ELEMENTS

Atomic No.	Name	Symbol	Oxide	Abundance in Igneous Rocks (parts per million)	Per Cent Rare Earth Oxide in Total Rare Earth Oxide of Minerals				
					Bastnaesite California	Monazite S. Carolina	Monazite Australia	Xenotime Malaysia	Uranium Residues Elliot Lake, Ont.
(Light rare earths)									
21	Scandium	Sc	Sc_2O_3	5.0	–	–	–	–	–
57	Lanthanum	La	La_2O_3	18.3	32.0	19.5	23.0	0.5	0.8
58	Cerium	Ce	CeO_2	46.0	49.0	44.0	45.5	5.0	3.7
59	Praseodymium	Pr	Pr_6O_{11}	5.5	4.4	5.8	5.0	0.7	1.0
60	Neodymium	Nd	Nd_2O_3	23.8	13.5	19.2	18.0	2.2	4.1
61	Promethium	Pm	Pm_2O_3	(Not measurable)	–	–	–	–	–
62	Samarium	Sm	Sm_2O_3	6.5	0.5	4.0	3.5	1.9	4.5
63	Europium	Eu	Eu_2O_3	1.1	0.1	0.2	0.1	0.2	0.2
64	Gadolinium	Gd	Gd_2O_3	6.3	0.3	2.0	1.8	4.0	8.5
(Heavy rare earths)									
39	Yttrium	Y	Y_2O_3	28.0	0.1	3.0	2.1	60.8	51.4
65	Terbium	Tb	Tb_4O_7	0.9		0.2		1.0	1.2
66	Dysprosium	Dy	Dy_2O_3	4.5		1.3		8.7	11.2
67	Holmium	Ho	Ho_2O_3	1.1		0.1		2.1	2.6
68	Erbium	Er	Er_2O_3	2.5	0.1	0.5	1.0	5.4	5.5
69	Thulium	Tm	Tm_2O_3	0.2		–		0.9	0.9
70	Ytterbium	Yb	Yb_2O_3	2.6		0.2		6.2	4.0
71	Lutetium	Lu	Lu_2O_3	0.7		–		0.4	0.4
	Total			153.0	100.0	100.0	100.0	100.0	100.0

– Nil.

TABLE 2. CANADIAN SHIPMENTS OF
RARE EARTH CONCENTRATES

	Y_2O_3 Concentrates (kilograms)	Values ($)
1978-84	-	-
1977[1]	30 400	..
1976[1]	26 308	..
1975[1]	34 927	..
1974	39 366	..
1973	..	..
1972	-	-
1971	..	..
1970	33 112	657,000
1969	38 756	671,500
1968	51 406	936,067
1967	78 268	1 594,298
1966	9 400	130,223

Source: Statistics Canada.
[1] Annual Reports, Denison Mines Limited.
.. Not available; - Nil.

Rhenium

W. McCUTCHEON

The major end-use for rhenium is as one of the components of a bimetallic catalyst used in the production of low lead and lead-free gasolines. In petroleum refineries, bimetallic rhenium-platinum (Re-Pt) catalysts are usually found to be more economical to use than monometallic platinum catalysts. As most refineries have already switched to Re-Pt catalysts, future growth in this important market is likely to be much less than in the preceding decade.

CANADIAN DEVELOPMENTS

Island Copper Mine is the sole producer of recoverable rhenium in Canada. Island Copper is owned by Utah Mines Ltd., a subsidiary of Utah International Inc. which is a division of The Broken Hill Proprietary Company Limited of Australia.

Island Copper, a copper molybdenum operation, near Port Hardy, Vancouver Island, British Columbia began production in 1972. The ore occurs mainly in altered volcanic rocks and in this respect differs from the porphyry copper deposits which have been the major source of rhenium in the United States and Chile.

Until late 1983, the rhenium contained in the concentrates was treated at the smelters on a toll basis and the recovered rhenium was returned to the company as perrhenic acid for subsequent upgrading and sale. Since 1984, Island Copper has sold the rhenium contained in the molybdenite concentrates.

The production of rhenium from Island Copper is a function of the rhenium content of the molybdenite, the quantity of molybdenite mined and the mill recoveries of molybdenite and rhenium. With present technology, the recovery of rhenium contained in the molybdenite concentrates ranges from about 50 to 60 per cent.

Rhenium has also been identified in the porphyry copper ores of Lornex Mining Corporation Ltd. and Brenda Mines Ltd. in British Columbia. Future recovery of rhenium from these operations is unlikely, given expected future rhenium prices.

Canadian consumption data is not collected. However, the consumption pattern is believed to follow that of the United States, with rhenium-platinum (Re-Pt) catalysts accounting for the vast majority of consumption.

WORLD DEVELOPMENTS

The only known commercial sources for rhenium are molybdenum concentrates recovered from the treatment of low-grade porphyry copper ores and from sedimentary copper deposits in the U.S.S.R. The rhenium content of the copper porphyry ores is relatively low, being only a few parts per million (ppm), whereas the molybdenite concentrates produced from these ores have a rhenium content ranging from 300 to 2 000 ppm. Rhenium has also been identified in certain platinum group metals and ores of manganese, tungsten and uranium, but in concentrations too low to be of economic significance under the present technology and price structure.

As rhenium is produced as a byproduct of copper mining, its production depends upon copper markets, and no operation will increase copper and molybdenum production merely to obtain more rhenium.

Rhenium production statistics are not available for the United States. World production outside the United States is estimated by the United States Bureau of Mines (USBM) to be about 13 550 kg in 1982, 11 250 kg in 1983 and 8 750 kg in 1984.

W. McCutcheon is with the Mineral Policy Sector, Energy, Mines and Resources, Canada. Telephone (613) 995-9466.

In the United States, there are five plants which can recover rhenium. It is believed that plants owned by Kennecott Corporation near Salt Lake City, M&R Refractory Metals, Inc. in New Jersey and Molycorp, Inc. in Pennsylvania did not operate in 1983 or 1984. S.W. Shattuck Chemical Co., Inc. in Denver operated in 1983 and Duval Corporation is believed to have recovered rhenium at its Sierrita property in Arizona in 1984. Some of the rhenium recovered by Shattuck in 1982 and 1983 was material toll refined for Utah from its Island Copper mine. Very low copper and molybdenum prices have resulted in many mine closures in the United States and hence reduced mine production of contained rhenium.

Chile, which has become the world's largest copper producer, was reported by the USBM to be the largest market economy producer of rhenium with a 1983 production of about 3 180 kg. Chilean rhenium mine production from the Corporacion Nacional del Cobre de Chile (Codelco) is estimated by the USBM to be about 2 400 kg in 1984. Prior to 1974, rhenium exported from Chile was contained in molybdenite concentrates shipped for treatment to the United States and elsewhere. In 1974, Chile began to export ammonium perrhenate to the United States. Since 1979, Codelco has had a tolling contract with Molibdenos y Metales S.A. (Molymet) for the recovery of rhenium.

Rhenium is recovered from copper porphyries mined in Iran, Peru, U.S.S.R., Canada, United States and Chile. In the case of the U.S.S.R., some rhenium is recovered from the Dzhezkazgan sedimentary copper deposit in Kazakhstan. Small amounts of rhenium has been produced in the past by Zaire, Bulgaria, Democratic Republic of Germany and Poland.

Besides the United States and Chile, countries which have metallurgical plants for the recovery of rhenium are the U.S.S.R., Sweden, France, the United Kingdom and the Federal Republic of Germany. With the exception of the U.S.S.R., these countries recover rhenium from imported molybdenite concentrates.

Chile is believed to be the largest exporter of rhenium in the form of ammonium perrhenate. The Federal Republic of Germany is believed to be the largest exporter of rhenium in metal form. The United States is the largest importer of rhenium in both forms.

Complete world consumption data are not available, however, the United States is believed to be the largest rhenium consumer accounting for an estimated 2 680 kg in 1982, 4 000 kg in 1983 and 4 100 kg in 1984.

USES

The primary use of rhenium is for Re-Pt calalysts which are used in the production of low lead and lead free high octane gasolines. About 90 per cent of the United States consumption is taken in this use. Other uses for Re-Pt catalytic include production of toluene, benzene and xylene.

The bimetallic Re-Pt catalyst (generally 0.3 per cent Re, 0.3 per cent Pt, 99.4 per cent alumina) generally gives superior performance compared to monometallic platinum catalysts: it can be easily regenerated, is more productive, tolerates greater impurity levels and has a longer life. Hence it has steadily replaced most monometallic catalysts used in petroleum refining since commercial introduction in 1969.

Other applications for rhenium include use in filament alloys, heating elements, electrical contacts, metallic coatings high temperature nickel alloys, and as an alloy with tungsten for use as a catalyst to oxidize SO_2 to SO_3 for the manufacture of sulphuric acid.

TECHNOLOGY

Rhenium is recovered from flue gases emitted from the roasting of byproduct molybdenite concentrates. Under properly controlled temperature, rhenium volatilizes as rhenium heptoxide (Re_2O_7), a compound which is readily soluble in an aqueous solution and can be recovered by subjecting flue gases to wet scrubbing. The rhenium is extracted from this solution as ammonium perrhenate (NH_4ReO_4) by ion-exchange resins or by solvent extraction. Perrhenic acid ($HReO_4$) is also an important commercial product of rhenium. Rhenium metal (99.99 per cent pure) is produced by the reduction of (NH_4ReO_4) with hydrogen to produce rhenium powder. The powdered form is pressed and sintered into bars which are cold-rolled to form different shapes. The cost of producing rhenium metals and salts is high. Research has been directed toward the development of a hydrometallurgical process to recover molybdenum and rhenium from molybdenite concentrates in order to attain a higher rate of recovery and lower costs of production.

PRICES

Rhenium prices rose from a fairly stable base of about $US 600/lb through the 1960s and 1970s to over $US 2,500/lb in May 1980. Prices for the metal averaged $US 410 in 1982, decreased from $US 350 to $US 250/lb in 1983 and are estimated at an average of $US 300/lb for 1984. Prices for the rhenium content in perrhenic acid averaged $US 375 in 1982, decreased from $US 300 to $US 200/lb in 1983 and are estimated at an average of $US 200/lb in 1984.

OUTLOOK

Rhenium has been used as an industrial metal for a short period and has not developed a clearly defined growth pattern. The limited supply of rhenium inhibits development of new uses with widespread applications. The potential supply of the metal is limited to the rhenium contained in byproduct molybdenite concentrates obtained from porphyry copper ores. Under present technology, the overall recovery of molybdenite from the processing of copper ores varies considerably but is relatively low. The recovery of rhenium from the treatment of molybdenite concentrates is about 60 per cent. Any improvement in the recovery rate in either of these areas would increase the supply of available rhenium. Canadian rhenium production is likely to continue until the early 1990s at which time Island Copper's ore will have been depleted.

Not all molybdic oxide producers recover rhenium from the treatment of byproduct molybdenite concentrates due to the high capital costs involved in building a recovery plant. These molybdic oxide operations are potential new sources of rhenium, given a stable price pattern that would justify committing funds for the construction of a recovery plant.

In the short-term, the major use of rhenium will continue to be in bimetallic platinum-rhenium catalysts in the petroleum refining industry. Use in this application could increase as more stringent standards for automotive emissions are introduced reducing the use of tetraethyl lead in gasoline.

Future growth of rhenium demand for petroleum refining is expected to be much less than the rate of the 1970s, due to saturation of the market and technical advances which extend catalyst life and permit high rates of recycling. Future successful technical advances in the development of alternative catalysts could result in substitution away from Re-Pt catalysts. Demand would also be lowered by use of low octane gasoline engines, increased market share of diesel powered automobiles, new octane improving additives.

In the long-term, metallurgical applications for rhenium probably have greater growth potential than do catalytic applications. Iridium, gallium, germanium, indium, selenium, silicon, tin, tungsten and vanadium may all be replaced by rhenium under certain conditions. Rhenium is highly refractory, having a melting point of 3 180°C, second to that of tungsten, and maintains strength and ductility at high temperatures. Its density of 21 grams per cubic centimetre (g/cm^3) is exceeded only by that of the platinum-group metals. Rhenium has good corrosion resistance to halogen acids. Alloyed with tungsten or molybdenum, rhenium improves the ductility and tensile strength of these metals. Stable oxide film on rhenium does not appreciably increase electrical resistance and this property, plus good resistance to wear and arc corrosion, makes the metal ideally suited for electrical contacts.

TABLE 1. CANADIAN RHENIUM STATISTICAL DATA, 1982-1984

	1982	1983	1984[e]
Rhenium content of Island Copper's molybdenite concentrates (ppm)[1]	1075 ppm	1175 ppm	1250 ppm
Shipments of molybdenite concentrates[2]			
United States	757t	–	–
Federal Republic of Germany	1869t	2898t	1701t
Chile	1320t	762t	1696t
United Kingdom	45t	36t	–
Total[3]	3992t	3697t	3397t
Average price[4]	$US550/lb	$US390/lb	$US30-35/lb

Sources: Utah Mines Ltd.; Energy, Mines and Resources.
[1] Rounded to nearest 25 ppm; [2] Dry tonnes; [3] Totals may not add due to independent rounding; [4] Price for ammonium perrhenate in 1982 and 1983; price for rhenium content contained in molybdenite concentrates in 1984.
[e] Estimated; – Nil.

Salt

M. PRUD'HOMME

SUMMARY

Canada produces rock salt from four salt mines and as byproduct from two potash mines. Brine, which accounts for 23 per cent of total production, is produced in 12 plants where it is used to manufacture table salt or chloralkali chemicals.

Production of salt from all sources rose to 8 591 000 t in 1983, and is expected to reach 10 294 000 t in 1984. Increases are mainly due to completion of expansion of the Domtar mine at Goderich, Ontario, and to the operational level of production at Mines Seleine Inc. on Madeleine Islands.

In 1983, the average value of salt in all forms is estimated at $20.28 per t, compared to $19.73 in 1982. Consumption increased mainly in the industrial chemicals sector which has benefitted from an anticipated improvement in the manufacturing sector, namely construction, automobile and the pulp and paper industry. Road de-icing rock salt shipments recorded a slight drop because of high inventories left behind as a result of the mild winter of 1983.

Net exports reached record levels in 1983 with an increase of 98 per cent in current value from 1982. It totalled around $12,988,000. Exports to the United States increased by 11 per cent in tonnage, in 1983, after a 13 per cent rise in 1982. Imports are 57 per cent from the United States, mainly rock salt, and 33 per cent from Mexico, mainly solar-evaporated salt. The latter is classified as wet salt in bulk, with 2 to 6 per cent moisture content.

DOMESTIC PRODUCTION AND DEVELOPMENTS

Atlantic region. Salt deposits occur in isolated sub-basins of a large sedimentary basin that underlies the northern mainland of Nova Scotia and extends westward under the bordering areas of New Brunswick, northeastward under Cape Breton Island, Prince Edward Island, the Madeleine Islands and southwestern Newfoundland. The salt beds occur within the Mississippian Windsor Group and are generally folded and faulted. The deposits appear to be steeply dipping tabular bodies, domes and brecciated structures of rock salt.

Salt production in the Atlantic provinces is from an underground rock salt mine at Pugwash in Nova Scotia, an underground potash and salt mine at Sussex in New Brunswick and a brining operation near Amherst in Nova Scotia.

In New Brunswick, Potash Company of America (PCA) started a $150 million development plan in 1981 to produce potash with salt as byproduct. The mine, located at Plumweseep near Sussex, commenced production in July 1983 and is expected to reach 900 000 tpy of potash. Salt will be extracted at a rate of 400 000 to 500 000 tpy and will be sold mainly to the eastern United States. Reserves are estimated large enough to operate for at least 20 years. Salt is marketed for road de-icing and chemical plants, and in 1983, it was distributed by the International Salt Company, United States, or its Canadian subsidiary, Iroquois Salt Products Ltd.

Denison-Potacan Potash Company produced small amounts of salt from its potash mine now under development at Salt Springs near Sussex. Completion of the project is due in mid-1985. However, the company is not expected to become a regular producer of salt after completion of construction.

In Nova Scotia, the Canadian Salt Company Ltd. operates an underground rock salt mine at Pugwash in Cumberland county, with a rated capacity of nearly 1 000 000 t. Most of the salt is used for snow and ice control, while some high purity salt is dissolved for vacuum pan evaporation and sold for high quality applications including table salt. In 1984, an agreement was signed between the provincial departments

---

M. Prud'homme is with the Mineral Policy Sector, Energy, Mines and Resources, Canada. Telephone (613) 995-9466.

and the Canadian Salt Company Ltd. to upgrade the 14-mile highway route from the mine to the TransCanada Highway. The project will cost $2.7 million and will allow trucks to carry 35 tonne-loads. The project is to be completed by the summer of 1985.

Domtar Chemicals Group has a brining operation at Nappan in Cumberland County. This company has announced, in 1984, a $9.5 million modernization plan, with assistance from the Canadian program of Atlantic Energy Conservation Investment. A $3.4 million contribution will go towards replacement of the present multi-stage evaporators with a single thermocompression evaporator, and the existing top-feed filter dryer by a centrifuge and fluidized bed dryer. Completion of the project is due in January 1986, and is expected to generate substantial energy savings. Meanwhile, Domtar Inc. continues to hold its salt development licence at Kingsville in Inverness County.

In 1984, exploration and development programs have examined salt structures in the Canso Strait area of Cape Breton Island for underground gas storage: Gulf Canada in the McIntyre area of Inverness County and Dow Chemicals at Lower River Inhabitant and at Port Richmond in Richmond County.

Quebec. A salt deposit located on the Archipelago of the Madeleine Islands, in the Gulf of St. Lawrence, is part of the Mississippian Carboniferous Basin. Discovered in 1972, the Rocher-aux-Dauphins deposit is characterized by thick sequences of commercial salt, large sequences of rythmic salt and anhydrite cycles, abundance of low grade potash horizons and some clay. The deposit is a typical piercement salt diapir generated by upward movements of the salt from the underlying anticlinal structure. It contains about 4 billion t of raw salt of which a quarter is above 97 per cent sodium chloride. The salt lies between 30 m and 75 m underneath the surface. The deposit dips about 55 degrees to the southwest. Reserves are 460 million t of which 34.2 per cent are mineable, grading an average of 94.5 per cent NaCl.

Mines Seleine Inc., a subsidiary of Société québécoise d'exploration minière (SOQUEM) mines rock salt commercially since the spring of 1983. This underground operation has a production capacity of 1.23 million tpy, and reserves are sufficient for 20 years.

In 1983, salt production for de-icing purposes has reached 640 443 t. In 1984, production is expected to be 1 million t. Full capacity of production could be achieved by 1986.

In October 1983, the Port of Montreal bought land in Montreal-East for storage and handling of salt from Mines Seleine Inc. The Canadian government purchased the land for $1.4 million and leased it to Mines Seleine Inc. under a long-term contract.

In 1984, a 10-year contract was signed with the Department of Highways of the Quebec Government to supply nearly 90 per cent of their salt requirements. Navigation Sonamar Inc. will provide the service of a 19 000 t capacity ship, M.V. Saunière, under contract until 2002. Underground work has been completed to improve grinding and sizing operations, and to increase the storage capacity to 25 000 t. Current activities involve moisture control to reduce local condensation, and ventilation to provide good environmental working conditions. Mines Seleine Inc. employs 190 persons at the mine site.

The shipping season is 270 days, from April 1 to December 31. All salt produced is for de-icing purposes and is shipped to mainland Quebec, Newfoundland and northeastern United States. In the long-term, investments will be required to control the moisture content in the mine, to improve access to port facilities and to increase surface storage at the mine site.

Ontario. Thick salt beds underlie much of southwestern Ontario, extending from Amherstburg northeastward to London and Kincardine, bordering on what is known geologically as the Michigan Basin. As many as six salt beds, occurring in the Upper Silurian Salina Formation at depths from 275 to 825 m, have been identified and traced from drilling records. Maximum bed thickness is 90 m, with aggregate thickness reaching as much as 215 m. The beds are relatively flat-lying and undisturbed, resulting in low-cost mining.

During 1984, those beds were worked through two rock salt mines, one at Goderich and one at Ojibway, and through brining operations at Goderich, Sarnia, Windsor and Amherstburg.

At Goderich, Domtar Chemicals Group operates an underground rock salt mine.

Since September 1983, an expansion plan has been completed, and the opening of the third shaft is expected to increase production by 55 per cent, from 2 million tpy to 3.15 million tpy. Employment will increase from 275 to 350. The two-year expansion will cost nearly $40 million. Between 60 and 70 per cent of rock salt is used for de-icing, the balance by the chemical industry and for water softening. Domtar Inc. also has a brining operation for production of special salt products. The Canadian Department of Transport has announced in August 1984 that it will proceed with a $17 million harbour development project over the next three years to enhance shipping capabilities of the Port of Goderich.

At Sarnia, Dow Chemicals of Canada Ltd. produces brines from wells for the production of caustic soda and chlorine.

At Windsor, the Canadian Salt Company Ltd. produces both rock salt from an undergound mine near Ogibway and vacuum salt products from brine wells having a total rated capacity of more than 2 million tpy. Installation of new silo facilities to improve salt storage and distribution has been completed in 1984.

In the vicinity of Amherstburg, Allied Chemical Canada Limited operates a brining operation for the manufacture of soda ash and byproduct calcium chloride by the Solvay process.

Prairie Provinces. Salt beds underlie a broad belt of the Prairie Provinces extending from the extreme southwestern corner of Manitoba northwestward across Saskatchewan and into the north-central part of Alberta. Most of the salt deposits occur within the Prairie Evaporite Formation, which constitutes the upper part of the Middle Devonian Elk Point Group, with thinner beds of salt occurring in Upper Devonian rocks. Depths range from 180 m at Fort McMurray, Alberta, to 900 m in eastern Alberta, central Saskatchewan and southwestern Manitoba, and to 1 830 m around Edmonton, Alberta, and in southern Saskatchewan. Cumulative thicknesses reach a maximum of 400 m in east-central Alberta. The beds lie relatively flat and undisturbed. The same rock sequence contains a number of potash beds currently under exploitation in Saskatchewan.

In Saskatchewan, four companies produce salt from the Middle Devonian Prairies formation. International Minerals and Chemical Company (Canada) Limited supplies byproduct rock salt from its potash operation at Esterhazy. Its salt is distributed locally for road de-icing by Kleysen Transport Company. Domtar Inc. operates a brining operation, near Unity, for the production of fine vacuum-pan salt and fusion salt. The Canadian Salt Company Ltd. at Belle Plaine uses byproduct brine from an adjacent potash solution mine owned by PPG Industries Canada, a division of Kalium Chemicals Limited, to extract table salt. Saskatoon Chemicals Company, a division of Prince Albert Pulp Company Ltd., produces brines from wells near Saskatoon for the manufacture of caustic soda and chlorine, mainly used by pulp producers as a bleaching agent.

In Alberta, two producers operate brining operations: at Fort Saskatchewan near Edmonton, Dow Chemical Canada Inc. produces salt brine for the manufacture of chloralkali chemicals; and, at Lindberg, the Canadian Salt Company Ltd. produces fine vacuum pan salt and fusion salt.

British Columbia. There is no production of salt in this province where four companies operate six chloralkali plants: B.C. Chemicals Ltd. in Prince George, ERCO Industries Limited in North Vancouver, FMC of Canada Limited at Squamish and Canadian Occidental Petroleum Limited in North Vancouver, Squamish and Nanaimo. Raw materials are imported from Mexico and the United States.

CANADIAN CONSUMPTION AND TRADE

The salt consumption pattern in Canada is quite different than in the rest of the world. The largest single utilization, for the past five years, has been for snow and ice control. This usage is principally confined to North America and Europe.

Consumption varies from year to year depending on weather conditions and this market is expected to increase marginally over the next decade. For the past eight years, the average proportion for this purpose in Canada was about 45 per cent of total consumption, compared to 20 per cent for the United States and 14 per cent in western Europe. On a world basis, this application accounts for 10 per cent of total world salt consumption. For road de-icing, the American Society for Testing and Materials (ASTM) provides standard specifications for sodium chloride: D632-72 (78). Rates of application are controlled by several factors such as precipitation, temperature, wind effects, traffic density and road conditions. Practices for such usage also include utilization of mixtures with calcium chloride or sand and gravel as abrasive components.

The next largest consumer of salt is the industrial chemicals industry, particularly for the manufacture of chloralkalis, namely caustic soda (sodium hydroxyde), chlorine and soda ash (sodium carbonate). Salt for four caustic soda and chlorine plants in Canada is obtained from on-site brining and natural brines; others use mined rock salt or imported solar-evaporated salt. The chemical industry accounts for over 60 per cent of total world consumption of salt while, in Canada, it has averaged 45 per cent for the past five years. Chlorine is largely used by the plastic industry and as a bleaching agent for the manufacture of bleached pulp and newsprint. The principal uses for caustic soda are in the manufacture of organic and inorganic chemicals, pulp and paper, alumina and textiles. The glass industry is a major user of soda ash. Other industrial chemicals that require significant quantities of salt include sodium chlorate, sodium bicarbonate, sodium chlorite and sodium hypochlorite. The ASTM standard method E-534-75 covers the analytical procedures for chemical analysis of sodium chloride. Strong growth in this market is expected to continue based on domestic demand as well as on export opportunities.

Other consumption areas for salt include the food industry, animal diet, fishery industry and water treatment which all account for less than 10 per cent of total Canadian consumption. Slight growth in these markets should continue in the short-term, although there is some pressure to use less salt for health reasons in the food industry.

The pattern of Canada's salt trade has changed drastically in terms of volume in the past two years. In 1983, Canada exported twice as much as it imported; the United States remained our major trade partner taking 99 per cent of our exports and accounting for 57 per cent of our imports. Mexico accounts for 33 per cent of our imports, especially solar-evaporated salt for chemical plants in British Columbia. Spain and Portugal are our main sources of supply for salt consumed by the fish industry, namely in Newfoundland and Nova Scotia.

Because of its low unit value and availability in most key market areas, salt is seldom hauled over long distances, except in the case of seaborne and intercoastal shipment where longer routes entail little additional cost. Increased capacity in eastern Canada (Ontario, Quebec and New Brunswick) will likely replace salt traditionally imported from Mexico and the northeastern United States. The trade surplus (exports less imports) totalled $12,988,000 in 1983, compared to only $798,000 in 1982.

WORLD PRODUCTION AND SITUATION

In 1983, world production of salt has increased by 0.04 per cent to 165 760 000 t despite a 9 per cent drop in salt production in the United States which accounts for nearly 20 per cent of the world total. The American decline was mainly due to lower demand for de-icing rock salt as a result of large purchases and stocks in 1982.

United States: In 1983, International Salt Co. placed its evaporative salt plant refinery at Avery Island in Louisiana on an indefinite stand-by and its rock salt plant in Detroit, Michigan, was idle most of the year. Diamond Crystal Salt Co. accepted a $32 million settlement for the flooding of its Jefferson's Island rock salt mine, in 1980.

In 1984, Cargill Inc. closed permanently its 1.5 million tpy Belle Isle's rock salt mine at Patterson, Louisiana; it has also announced plans to construct a new solar salt facility by 1985, with a production capacity of about 200 000 tpy, near Freedom, Oklahoma. Morton Thiokol Inc., at Weeks Island in Louisiana, will increase its production of rock salt mine by more than 300 000 tpy, and announced, in August 1984, its intention to close for 3 months its salt plant located in Marysville, Michigan, reducing the output by more than 50 per cent.

Diamond Crystal Salt Co., near St-Clair in Michigan, signed a letter of intent to purchase American Salt Co., a subsidiary of General Host Corp., for $40 million. The purchase is to include a rock salt mine and an evaporated salt facility in Lyons, Kansas, and a solar salt facility near Salt Lake City, Utah. In the summer of 1984, American Salt Co. was found liable in U.S. District Court for damages involving salt pollution of an aquifer in Kansas.

China: Salt deposits have been discovered in the Qamdo prefecture. Reserves are estimated at 90 million t of salt grading 30 per cent of sodium chloride as well as some potassium chloride.

Thailand: ASEAN Soda Ash Co. Ltd. reported plans to build a 400 000 tpy soda ash plant which would require 600 000 tpy of rock salt. Mine development at Bambet Narong in Chaiyaphum Province would take 3 years and investment would be in the range of $32-37 million.

Yugoslavia: A salt deposit with estimated reserves of 400 million t has been discovered near Tulza.

PRICE

Salt is not a standard commodity and its price ranges widely depend on such factors as production methods, purity, scale of operations and transportation costs.

In 1984, Canadian rock salt prices, fob works, for de-icing purposes varied from $21.00 to $41.00 per t.

OUTLOOK

Canada is nearly self-sufficient in salt. Eastern Canadian requirements of rock salt are served locally while imports serve western needs for chloralkalis plants in British Columbia. Current capacity should be sufficient to meet any forecast increase in demand for the next decade. Growth for world production has been estimated at 2 per cent up to 1990.

Consumption of salt in Canada is largely based on two sectors: ice and snow control and industrial chemicals which account together for 90 per cent of the total Canadian consumption.

Road salt de-icing demand should remain stable for the next few years. This usage is highly seasonal but several factors mitigate against any significant increase: slowdowns in road construction, application optimization and environmental considerations. Thus, a slow growth of not more than 1 per cent per annum is forecast for this purpose.

The industrial chemicals industry is definitely the sector which will see some growth in the decade. Chlorine production reached 1 385 000 t in 1983, a 12 per cent increase over 1982. The major share of this rise was in polyvinyl chloride (PVC) production, which rose by 26 per cent from the 1982 level. Chlorine demand is oriented to PVC usage in the construction and automotive sectors, and as a bleaching agent in the pulp and paper industry, the latter accounting for 41 per cent of consumption. Construction of new houses affects demand for PVC pipes and tubing. Up to 1990, a modest growth of 0.2 to 0.6 per cent in residential construction is expected while the non-residential sector should grow at between 4.8 and 6.5 per cent a year. In the automobile industry, the share of plastics in standard American cars is forecast to grow at 9.2 per cent for the period 1981-92; since, automobile demand is expected to grow at a rate of 1.8 to 2.6 per cent in North America for the next five years. The net result should be a greater demand for salt used in these products.

Consumption of chlorine in the pulp and paper industry in 1983 was 480 000 t, an increase of 4 per cent compared to 1982. It should reach about 510 000 t in 1984. However, chlorine producers are reluctant to expand capacity because pulp mills may not increase future purchases as chlorine dioxide bleaching continues its penetration of the chlorine market in paper mills.

Caustic soda markets continued to be saturated in Canada. In 1983, production amounted to 1 528 000 t which is much in excess of domestic needs, estimated at 900 000 t in the pulp and paper industry; exports rose by 33 per cent to reach 228 000 t in 1983. Anticipated recovery in pulp production should result in stronger demand for caustic soda and chlorine for 1984-85. Long-term projections suggest a growth in the range of 2 to 3 per cent for caustic soda and 2 per cent for chlorine.

Sodium chlorate is used in the manufacture of chlorine dioxide. Supply continues to increase as new capacity is projected, especially in Quebec where favorable power rates will permit low-cost production. Canadian production in 1983 was about 294 000 t, of which nearly 196 000 t was consumed in the pulp and paper industry. Canadian exports amounted to 98 000 t in 1983. North American market growth is expected at 2 to 4 per cent per annum over the next decade.

For 1984, the Canadian Pulp and Paper Association has foreseen an 8 per cent rise in overall pulp and paper shipments. Most mill chemicals should show a growth of 5 to

10 per cent in 1984. Corporate investment intentions have been projected to exceed $3 billion over the next few years. Such projects associated with increased production will result in an increase in demand for industrial chemicals, such as caustic soda and sodium chlorate. Chlorine would benefit from this expected growth as well as in the construction and automobile sectors.

Other end-use markets such as animal feeds, human foods and water treatment will likely display slight but steady growth in the short-term, based on population growth.

TARIFFS

Item No.		British Preferential	Most Favoured Nation	General	General Preferential
			(%)		
CANADA					
92501-1	Common salt (including rock salt)	free	free	5¢/100 lb	free
92501-2	Salt for use of the sea or gulf fisheries	free	free	free	free
92501-3	Table salt made by the admixture of other ingredients when containing not less than 90 per cent of pure salt	4.6	4.5	15	3
92501-4	Salt liquors and sea water	free	free	free	free

MFN Reductions under GATT (effective January 1 of year given)	1983	1984	1985	1986	1987
			(%)		
92501-3	4.5	4.4	4.3	4.1	4.0

UNITED STATES, Customs Tariffs (MFN)					
420.92 Salt in brine	4.4	4.2	4.0	3.9	3.7
420.94 Salt in bulk	1.5	1.1	0.8	0.4	free
420.96 Salt, other	Remains free				

Sources: The Customs Tariff, 1983, Revenue Canada, Customs and Excise; Tariff Schedules of the United States Annotated (1983), USITC Publication 1317; U.S. Federal Register Vol. 44, No. 241.

TABLE 1. CANADA, SALT PRODUCTION AND TRADE, 1982-84

	1982		1983P		1984e	
	(tonnes)	($)	(tonnes)	($)	(tonnes)	($)
Production						
By type						
Mined rock salt	5 265 622	..	5 843 000	..		..
Fine vacuum salt	762 850	..	727 000	..		..
Salt content of brines used or shipped	1 944 172	..	2 021 000	..		..
Total	7 972 644	..	8 591 000	..		..
Shipments						
By type						
Mined rock salt	5 223 073	84,901,020	5 842 000	..	..	..
Fine vacuum salt	773 086	65,709,841	727 000	..	..	..
Salt content of brines used or shipped	1 944 172	6,009,493	2 021 000	..	..	..
Total	7 940 331	156,620,354	8 590 000	174,261,000	10 294 000	214,866,000
By province						
Nova Scotia	..	..	..	..	..	..
New Brunswick	..	..	..	..	..	..
Quebec	..	..	..	..	..	..
Ontario	5 461 190	87,505,004	5 059 000	93,180,000	6 502 000	124,400,000
Saskatchewan	434 103	18,675,272	425 000	20,821,000	383 000	21,680,000
Alberta	862 969	15,281,324	1 195 000	15,071,000	1 237 000	18,254,000
Total	7 940 331	156,620,354	8 590 000	174,261,000	10 294 000	214,866,000
Imports					(Jan.-Sept. 1984)	
Salt, wet in bulk						
Mexico	336 470	3,623,000	266 627	2,562,000	184 473	1,921,000
United States	36 066	440,000	11 207	136,000	183 744	2,705,000
Total	372 536	4,062,000	277 834	2,698,000	368 217	4,626,000
Salt, domestic						
United States	8 819	1,033,000	7 956	994,000	7 280	931,000
Switzerland	54	18,000	141	29,000	26	19,000
Netherlands	-	-	128	4,000	1	..
Other countries	4	2,000	52	8,000	44	8,000
Total	8 877	1,052,000	8 277	1,035,000	7 351	958,000
Salt, nes						
United States	948 547	11,976,000	445 304	8,650,000	343 434	6,975,000
Spain	48 894	1,293,000	44 801	716,000	16 401	261,000
Chile	106 873	2,354,000	37 090	307,000	-	-
Other countries	41 152	540,000	944	94,000	39 771	529,000
Total	1 145 466	16,163,000	528 139	9,767,000	399 606	7,765,000
Salt and brine by province of clearance						
Newfoundland	44 563	753,000	25 561	418,000	18 389	300,000
Nova Scotia	20 819	832,000	19 974	337,000	18 755	241,000
New Brunswick	34	1,000	47	7,000	46	24,000
Quebec	414 129	5,529,000	60 500	968,000	96 958	1,668,000
Ontario	543 993	7,123,000	269 531	4,642,000	349 541	6,059,000
Manitoba	785	73,000	2 755	182,000	2 342	152,000
Saskatchewan	1 160	93,000	2 606	206,000	2 296	227,000
Alberta	4 866	326,000	7 693	563,000	5 831	376,000
British Columbia	496 532	6,547,000	425 583	6,177,000	281 016	4,302,000
Total	1 526 879	21,277,000	814 250	13,500,000	775 174	13,349,000
Exports						
Salt and brine						
United States	1 717 973	21,735,000	1 908 385	25,754,000	1 867 074	21,189,000
Guyana	-	-	2 001	309,000	1 001	166,000
Leeward-Windward Islands	1 964	164,000	1 860	178,000	1 126	111,000
Other countries	1 956	176,000	2 383	247,000	1 549	31,000
Total	1 721 893	22,075,000	1 914 629	26,488,000	1 870 750	21,618,000

Sources: Statistics Canada; Energy, Mines and Resources Canada.
P Preliminary; .. Not available; - Nil; nes - Not elsewhere specified; e Estimated.
Note: Totals may not add due to rounding.

TABLE 2. CANADA, SUMMARY OF SALT PRODUCING AND BRINING OPERATIONS

Company	Location	Initial Production	Production[1] 1983P (1982) (000 tonnes)	Employment 1983P (1982)	Remarks
Nova Scotia					
The Canadian Salt Company Limited	Pugwash	1959	667.5 (964.3)	185) (216))	Rock salt mining to a depth of 253 m.
	Pugwash	1962	83.4 (89.6)	))	Dissolving rock salt fines for vacuum pan evaporation.
Domtar Inc.	Amherst	1947	68.6 (72.4)	74 (71)	Brining for vacuum pan evaporation.
New Brunswick					
Denison-Potacan Potash Company	Sussex	1982	29.3 (56.1)	– (-)	Salt from the development of a potash mine.
Potash Company of America	Sussex	1980	377.4 (-)	25[2] (-)	A potash mine in operation since August 1983. Salt shipments resumed in February 1983.
Quebec					
Seleine Mines Inc.	Magdeleine Islands	1982	617.5 (87.8)	190 (150)	Production began in late 1982, mining to depth of up to 275 m.
Ontario					
Allied Chemical Canada, Ltd.	Amherstburg	1919	518.3 (513.1)	8[2] (8)	Brining to produce soda ash.
The Canadian Salt Company Limited	Ojibway	1955	1 784.7 (2 134.3)	221 (256)	Rock salt mining at a depth of 300 m.
	Windsor	1892	123.0 (121.6)	132 (152)	Brining, vacuum-pan evaporation and fusion.
Domtar Inc.	Goderich	1959	2 275.3 (1 906.0)	323 (284)	Rock salt mining at a depth of 536 m.
	Goderich	1880	108.1 (118.2)	70 (63)	Brining for vacuum-pan evaporation.
Dow Chemical Canada Inc.	Sarnia	1950	669.3 (667.6)	5[2] (5)	Brining to produce caustic soda and chlorine.
Prairie Provinces					
International Minerals & Chemical Corporation (Canada) Limited	Esterhazy, Sask.	1962	88.7 (71.5)	3 (3)	Byproduct rock salt from potash mine for use in snow and ice control.
The Canadian Salt Company Limited	Belle Plaine, Sask.	1969	78.7 (77.4)	24 (24)	Producing fine salt from byproduct brine from potash mine.
Domtar Inc.	Unity, Sask.	1949	142.0 (165.5)	85 (87)	Brining, vacuum-pan evaporation and fusion.
Saskatchewan Chemicals Co.	Saskatoon, Sask.	1968	34.0 (34.0)	5 (5)	Brining to produce caustic soda and chlorine.
The Canadian Salt Company Limited	Lindbergh, Alta.	1968	118.3 (133.5)	80 (65)	Brining, vacuum-pan evaporation and fusion.
Dow Chemical Canada Inc.	Fort Sask., Alta.	1968	797.3 (792.5)	3[2] (8)	Brining to produce caustic soda and chlorine.
			8 581.4 (8 005.4)	1 433 (1 398)	

[1] Shipments; [2] Employment part of a chemical complex.
P Preliminary; – Nil.

TABLE 3. CANADA, SALT SHIPMENTS, 1975, 1979-83

	Producers' Shipments					
	Mined Rock	Fine Vacuum	In Brine and Recovered in Chemical Operations	Total	Imports	Exports[1]
			(tonnes)			
1975	3 626 123	578 649	1 291 489	5 496 261	1 183 144	..
1979	4 934 574	735 460	1 645 914	7 315 948	1 276 179	1 822 120
1980	4 507 416	781 428	2 134 010	7 422 854	1 151 203	1 637 601
1981	4 371 314	764 037	2 107 243	7 242 594	1 254 992	1 507 710
1982	5 223 073	773 086	1 944 172	7 940 331	1 526 879	1 721 893
1983P	5 842 000	727 000	2 021 000	8 590 000	814 250	1 914 629

Sources: Statistics Canada; Energy, Mines and Resources Canada.
[1] As of the 1983 review Canadian Exports of salt will be reported in tonnes.
P Preliminary; .. Not available.

TABLE 4. CANADA, AVAILABLE DATA ON SALT CONSUMPTION, 1980-1983

	1980r	1981r	1982P	1983e
		(tonnes)		
Snow and ice control[1]	2 472 849	3 001 260	3 088 315	2 712 088
Industrial chemicals[2]	2 974 520	3 234 020	2 966 218	3 495 000
Fishing industry	65 000	68 000	83 000	64 000
Food processing				
Fruit and vegetable processing	20 619	19 168	18 008	22 300
Bakeries	15 017	14 079	13 746	16 300
Fish products	24 296	33 983	33 582	35 200
Dairy products	13 056	10 740	10 447	12 900
Biscuits	1 892	2 022	2 082	2 600
Miscellaneous food preparation	46 587	24 874	22 680	36 000
Grain mills[3]	77 412	67 036	63 899	79 000
Slaughtering and meat processors	45 611	44 725	37 347	49 000
Pulp and paper mills	28 980	25 344	38 939	35 200
Leather tanneries	7 346	9 964	7 708	9 500
Miscellaneous textiles	2 924	2 664	2 871	3 400
Breweries	294	352	279	300
Other manufacturing industries	8 732	10 492	7 923	10 300
Total	5 805 135	6 568 723	6 397 044	6 583 088

Sources: Statistics Canada; Salt Institute.
[1] Fiscal year ending June 30. [2] Includes rock salt, fine vacuum salt and salt contained in brine. [3] Includes feed and farm stock salt in block and base forms.
e Estimated by Energy, Mines and Resources Canada; P Preliminary; r Revised.

Selenium and Tellurium

W.J. M^cCUTCHEON

Selenium is a nonmetallic element which is chemically similar to sulphur but which has some properties of a metal. Selenium occurs in minerals associated with sulfides of copper, lead and iron. Commercial production is principally from electrolytic copper refinery slimes as well as from flue dusts from copper and lead smelters. A significant amount of selenium is also recovered from secondary sources. In 1984, production and demand were estimated to be nearly in balance in the western world at about 1 540 tonnes (t) and 1 630 t, respectively. Large reductions of primary stocks in 1983 resulted in a rapid price rise in early 1984.

CANADIAN DEVELOPMENTS

Selenium is recovered in Canada as a byproduct of the refining of blister copper and from the retreatment of recycled materials. Annual production (Table 1) is irregular, varying according to operating rates and recoveries at copper refineries and market conditions for selenium. Canadian production in 1983 and 1984 was affected by a ten-month shutdown at Inco Limited which ended in April 1983, by a 25 per cent production cut back at Noranda Inc.'s CCR refinery since mid-1984 and by variations in the content of selenium bearing feed to CCR. Xerographic scrap and other selenium scrap are imported from the United States and other countries to be refined in Canada and reexported. The total amount of selenium refined in Canada from both primary and secondary sources was 352 t in 1983 and is estimated at 450 t in 1984.

Noranda Inc.'s CCR Division copper refinery at Montréal East, Quebec, operates the largest selenium recovery plant in the world. The refinery handles blister copper from the company's Horne and Gaspé smelters in Quebec, from the Flin Flon smelter of Hudson Bay Mining and Smelting Co., Limited in Manitoba, and anode slimes from the copper refinery at KCML Inc. (formerly Kidd Creek Mines Ltd.). The selenium recovery unit produces commercial-grade (99.5 per cent) and high-purity (99.99 per cent) selenium and a variety of selenium compounds. Nominal annual capacity is about 325 t of primary selenium in elemental form and in salts, depending upon the selenium content of the blister copper processed. In addition, production capacity of secondary selenium is nominally 165 tpy, but this too depends upon the selenium content of the feed material. Due to extraordinarily high selenium contents of feed material, selenium recovery increased in 1984 over the 1983 level, although copper refining was reduced by 25 per cent at CCR in mid-1984.

Inco Limited's selenium recovery plant at Copper Cliff, Ontario treats tankhouse slimes from the company's Copper Cliff copper refinery. The capacity of the plant is 67 tpy of minus 200 mesh selenium powder (99.5 per cent Se). The Inco copper refinery reopened in April 1983, following a ten-month shutdown of the company's copper-nickel operations.

Canada consumes only a few per cent of its refined selenium production, mostly in the glass industry. Most selenium exports are to the United States and the United Kingdom with minor amounts to Europe. Shipments to Europe increased sharply in 1984.

WORLD DEVELOPMENTS

Producing countries include the United States, Canada, Japan, the U.S.S.R., Belgium, Sweden, Mexico, Yugoslavia, Finland, Peru, Australia, and Zambia. Noncommunist world reported production of refined selenium was 1 384 t in 1983 and increased to an estimated 1 357 t in 1984 (Table 2). A further 270 t was estimated to have been produced but not reported.

W.J. McCutcheon is with the Mineral Policy Sector, Energy, Mines and Resources, Canada. Telephone (613) 995-9466.

As stocks of semi-processed anode slimes decreased from 227 t at the end of the first quarter 1983 to 116 tpy later, selenium production in the United States increased in 1983 to 354 t from 243 t in 1982. Primary selenium producers in the United States in 1983 were AMAX Copper, Inc., ASARCO Incorporated and Kennecott Corporation. Both domestic and imported material was processed in the United States. In 1984, Phelps Dodge Corporation became the fourth producer in the United States when the company's new selenium recovery facilities were started up at its El Paso, Texas refinery.

With the start-up of the Pasar copper smelter, the Philippines began to recover selenium in 1984. Production figures were unavailable but capacity is thought to be about 70 tpy.

Consumption of selenium in the western world in 1983 was reported at 1 440 t and estimated at 1 630 t for 1984. The United States is the most important consuming nation.

Apparent consumption of selenium in the United States in 1983 was 578 t up from 538 t in 1982. Estimated apparent consumption in 1984 was 570 t. According to the United States Bureau of Mines (USBM), the main end use by industrial categories in 1983 were: electronic and photocopier components, 33 per cent; glass manufacturing, 27 per cent; pigments, 20 per cent; metallurgical applications, 7 per cent; other including animal feed and chemicals, 13 per cent.

The Selenium - Tellurium Development Association, Inc. (STDA) held its Third International Symposium on the industrial uses of selenium and tellurium in Sweden, during October. Papers discussing new applications, health issues and recent developments were presented.

PRICES

Producer prices have not been published since early 1981. Metals Bulletin Inc. prints a "European Free Market" price spread for selenium. The mid-point of the spread averaged $US 4.07/lb in 1983, with a monthly average ranging from $US 3.41 to $US 3.48/lb. In 1984, the European prices increased to an average $US 6.34/lb in the first quarter, to $US 10.74/lb in the second quarter and decreased to $US 9.87/lb in the third quarter. The sharp price increase from $US 4.81/lb in February 1984 to $US 9.80/lb in March was likely a result of tight supplies resulting from the depletion of producer stocks in late 1983. Difficulties by merchants in obtaining supplies could also have contributed to the rapid price rise. The average European Free Market price for 1984 was estimated at $US 9.20/lb.

USES

Selenium is used in the manufacture of glass, steel, electronic components, explosives, batteries, animal and poultry feeds, fungicides and pigments, and in xerography. The 1979 edition of this review contains a more detailed description of selenium uses.

The photoreceptor industry is the major user of selenium. Fully panchromatic organic photoreceptors and amorphorus silicon photoreceptors have the potential to substitute for selenium in new generations of photocopiers. While the final decisions have not been made with respect to which materials are the best photoreceptors for new processes, there is a possibility of a reduction in demand for selenium for its largest end use. Such a substitution, if it were to take place, is thought unlikely to affect selenium demand before the next decade.

Elemental selenium is marketed in two grades: commercial, with a minimum content of 99.5 per cent Se; and high purity, with a minimum content of 99.99 per cent Se. Other forms include ferroselenium, nickel-selenium, selenium dioxide, barium selenite, sodium selenate, sodium selenite and zinc selenite.

OUTLOOK

Selenium is associated with copper minerals and hence its production is dependent upon primary copper production. EMR projects future copper consumption will increase at a rate of between 1.2 per cent and 1.6 per cent annually until the end of the century. The balance between sulfide and porphyry copper production is forecast to shift in favour of porphyry deposits. Porphyrys contain less selenium on average than do sulfide deposits and hence future mine selenium production will increase at a lower rate than the increase in copper mine production forecasted. Recoverable primary selenium production is likely to grow at a rate of about 1 per cent annually.

Given higher prices, production could be increased by improving selenium recovery from the present level of between 50 and 60 per cent. A minor increase in selenium recovery is also likely due to more stringent emission standards at copper and lead smelters.

Scrap supplies are a ready extra source of selenium in case of a significant price rise. Rectifier and xerographic scrap are two components of the estimated 200 to 400 t of scrap stocks which have accumulated.

New large scale uses for selenium in the long term are not predictable. Indeed while existing uses are unlikely to be threatened by substitution at existing prices in the medium term, technological advances such as a new photocopying process or the use of alternative photoreceptors have the potential to seriously reduce consumption. Such technical advances, like new large-scale uses, are difficult to predict.

The introduction of a major new use would be inhibited by the constraint upon supplies, as primary selenium production is a function of copper production. Although selenium recoveries could be improved and significant supplies of inventoried scrap could be processed to meet the increased demand, ultimately supply is constrained. Given significant sustained new demand, prices would rise encouraging the use of substitute materials.

Health related uses are likely to increase. Selenium is now added to vitamins tablets for humans, and to animal and poultry feeds. Selenium has also been studied as dietary cancer preventative agent.

Prices are likely to remain in the range of $US 10/lb in 1984 $US over the medium term. Significant price rises will be inhibited by the stocks of scrap selenium and market strategies of producers. Prices of $US 10/lb are thought to be insufficient to evoke reprocessing of the majority of existing scrap stocks. Much of the existing scrap inventories require a price of $US 15-20/lb before reprocessing is profitable. The long-term interests of major producers and consumers would not be best-served by large-scale price increases which encourage substitution away from selenium.

TARIFFS

CANADA

Item No.		British Preferential	Most Favoured Nation	General	General Preferential	
			(%)			
92804-4	Selenium	5	10	15	5	

MFN Reductions under GATT (effective January 1 of year given)	1983	1984	1985	1986	1987
			(%)		
92804-4	10.0	10.0	10.0	9.9	9.2

UNITED STATES (MFN)

Item No.		1983	1984	1985	1986	1987
				(%)		
420.50	Selenium dioxide	Remains free				
420.52	Selenium salts	Remains free				
420.54	Other selenium compounds	4.4	4.2	4.0	3.9	3.7
632.40	Selenium metal, unwrought, other than alloys, waste and scrap	Remains free				
632.88	Selenium metal alloys, unwrought	7.3	6.8	6.4	5.9	5.5
633.00	Selenium metals, wrought	7.3	6.8	6.4	5.9	5.5

EUROPEAN ECONOMIC COMMUNITY (MFN) 1983

		Base Rate	Concession Rate
28.04 C.2	Selenium free	free	free

Sources: The Customs Tariff 1983, Revenue Canada Customs and Excise; Tariff Schedules of the United States Annotated (1983), USITC Publication 1317; U.S. Federal Register Vol. 44, No. 241; Official Journal of the European Communities, Vol. 25, No. L318, 1982.

Tellurium, like selenium, is recovered in Canada from the tankhouse slimes from electrolytic copper refineries. It is refined by the same two companies who refine selenium: Noranda Inc.'s CCR Division at Montréal East, Quebec, and Inco Limited at Copper Cliff (Sudbury), Ontario. Although more "metallic" than selenium, tellurium resembles selenium and sulphur in chemical properties and, like selenium, is a semiconductor. Tellurium output is related to selenium output because tellurium is a coproduct of selenium recovery.

CANADIAN DEVELOPMENTS

Refined production of tellurium was 23.5 t in 1983 and estimated at 15 t in 1984 (Table 5). The large difference between production (all forms) and refined production in some years is attributable to market conditions. Producers refine according to sales and can stockpile any surplus in less processed forms.

CCR Division has an annual capacity of up to 27.2 t of primary and secondary tellurium in powder, stick, lump and dioxide forms. The Copper Cliff refinery has an annual capacity of up to 8.2 t of tellurium in the form of dioxide (77 per cent Te).

In 1982, Cominco Ltd. built a $3 million plant at Trail, British Columbia to expand its production of mercury-cadmium telluride (MCT) in the form of single crystals. When sliced into thin wafers and polished, this compound is used in a wide range of electronic devices that detect infrared radiation to provide optical images or data. This plant is the only non-captive producer of such crystals and is the largest producer of high-purity detector grade tellurium.

WORLD DEVELOPMENTS

Total refined world production is unavailable: Australia, U.S.S.R. and the Federal Republic of Germany do not report data, production data of the United States' single producer, ASARCO, is withheld for proprietary reasons, and data is insufficient to estimate production from Chile, Zaire and Zambia.

Apparent consumption in the United States increased from 46 t in 1982 to 57 t in 1983 to an estimated 90 t in 1984. Consumption in the United States had been higher (224 t in 1979) until a chemical plant in Texas closed in 1979. This plant had used a large quantity of tellurium as a catalyst for producing ethylene glycol (antifreeze) but experienced problems with its patented tellurium process.

PRICES

Most of the commercial-grade tellurium sold by the primary producers is in the form of slab, stick, lump, tablet or powder. It is also sold in the form of copper-tellurium and iron-tellurium alloys. Normal commercial grades contain a minimum of 99 per cent or 99.5 per cent tellurium. Tellurium dioxide is sold in the form of minus 40 to minus 200-mesh powder containing a minimum of 75 per cent tellurium.

As a result of falling prices, producers suspended publication of tellurium prices on January 5, 1981. Prices in 1983 and 1984 are believed to have ranged between $US 8 and $US 14/lb, depending upon lot size, frequency of purchases and market conditions.

USES

Tellurium supply is related to copper production but the nature of demand justifies only a low rate of recovery. Overexposure to tellurium could be hazardous to health, but fortunately tellurium imparts a disagreeable odor to the breath at low concentrations; this early warning signal has prevented any recorded toxic industrial exposures. Major uses are as additions to ferrous and nonferrous alloys to improve machineability or otherwise improve their metallurgical properties; however, bismuth is increasingly used as a substitute. Tellurium also performs an important role in the manufacture of rubber products, thermoelectric devices, catalysts, electronics, insecticides and germicides, delay blasting caps, glass, ceramics and pigments.

The consumption of tellurium in the United States by end use in 1983 was estimated by the USBM as: iron and steel products, 65 per cent; nonferrous metals, 17 per cent; chemicals including rubber manufacturing, 8 per cent; other, including xerographic and electronic applications, 10 per cent.

OUTLOOK

Supply of tellurium is largely a function of copper output and new copper production is increasingly derived from tellurium-poor ores. In the short to medium term, demand is expected to grow slowly and supply should be adequate to meet requirements. However, as the total available supply of tellurium is even more limited than that of selenium, significant new uses of tellurium, such as in solar collectors, or in the form of MCT in photovoltaic cells could result in the higher prices that would justify a higher recovery from tellurium-bearing copper ores. Military and aerospace applications have the potential to increase MCT demand, presumpresumably even if prices were to rise significantly.

TARIFFS

CANADA

Item No.		British Preferential	Most Favoured Nation	General	General Preferential
			(%)		
92804-5	Tellurium	5	10	15	5

MFN Reductions under GATT (effective January 1 of year given)	1983	1984	1985	1986	1987
			(%)		
92804-5	10.0	10.0	10.0	9.9	9.2

UNITED STATES (MFN)

		1983	1984	1985	1986	1987
				(%)		
427.12	Tellurium salts	4.4	4.2	4.0	3.9	3.7
421.90	Tellurium compounds	4.4	4.2	4.0	3.9	3.7
632.48	Tellurium metals, unwrought other than alloys, and waste and scrap	2.0	1.5	1.0	0.5	free
632.88	Tellurium metal alloys, unwrought	7.3	6.8	6.4	5.9	5.5
633.00	Tellurium metal, wrought	7.3	6.8	6.4	5.9	5.5

EUROPEAN ECONOMIC COMMUNITY

		1983	Base Rate	Concession Rate
28.04 C.3	Tellurium metal	2.3	2.4%	2.1%

Sources: The Customs Tariff 1983, Revenue Canada Customs and Excise; Tariff Schedules of the United States Annotated (1983), USITC Publication 1317; U.S. Federal Register Vol. 44, No. 241; Official Journal of the European Communities, Vol. 25, No. L318, 1982.

TABLE 1. CANADA, SELENIUM PRODUCTION, EXPORTS AND CONSUMPTION, 1982, 1983, 1984e

	1982		1983P		1984e	
	(tonnes)	($000)	(tonnes)	($000)	(tonnes)	($000)
Production						
Refined[1]	234	..	352	..	448	
Exports						
United Kingdom	47	451	111	1,236	99*	1,825*
United States	128	4,055	87	2,321	96*	2,491*
Netherlands	10	128	33	341	92*	1,454*
P.R. China	-	-	-	-	20*	220*
Spain	14	142	14	149	20*	355*
Belgium and Luxembourg	9	79	453	64	13*	343*
Other countries	6	239	9	398	18*	608*
Total	214	5,094	707	4,509	358*	7,296*
Consumption[2]	10.5	..	11.8	..	..	..

Sources: Energy, Mines and Resources Canada; Statistics Canada.

[1] Refinery output from all sources, including imported materials and secondary sources.
[2] Consumption (selenium content), as reported by consumers; other estimates put total consumption in Canada at over 15 t/y for 1982, 1983, 1984.
P Preliminary; .. Not available; - nil; * data for 9 months of exports in 1984, not total exports for the year.

TABLE 2. CANADA, SELENIUM PRODUCTION, EXPORTS AND CONSUMPTION, 1970, 1975, 1980-84

	Production Total Refined[1]	Exports[2]	Consumption[3]
	(tonnes)		
1970	388	311	7.1
1975	342	218	9.9
1980	377	307	10.8
1981	350	299	9.4
1982	234	214	10.5
1983P	352	707	11.8
1984e	448	358*	..

Sources: Energy, Mines and Resources Canada; Statistics Canada.

[1] Refinery output of selenium from all sources, including imported concentrates, blister and scrap and domestic scrap.
[2] Exports of selenium, metal powder, shot, etc.
[3] Consumption (selenium content), as reported by consumers.
P Preliminary; .. Not available; * data for 9 months of exports in 1984, not total exports for the year.

TABLE 3. NON-COMMUNIST WORLD[1,2] REFINERY PRODUCTION OF SELENIUM, 1982-84

	1982	1983	1984e
	(kilograms)		
Japan	410	433	400
Canada	234	352	448
United States	242	354	265
Mexico	29	30	30
Sweden	44	44	44
Belgium and Luxembourge	60	60	60
Other countries[3]	111	111	110
Subtotal[2]	1 130	1 384	1 357

Sources: U.S. Bureau of Mines, Energy Mines and Resources Canada.
Notes:
[1] Includes material from primary plus secondary sources
[2] Australia, Federal Republic of Germany, and USSR refine selenium but do not report outputs. Estimates for these nations' outputs are not included in Table 3.
[3] Peru, Chile, Zambia, Finland and Yugoslavia
e estimates from USBM; other estimates are that Belgian production exceeds 100 t/y.

TABLE 4. CANADA, INDUSTRIAL USE OF SELENIUM, 1980-82

	1982	1983	1984
	(tonnes of contained selenium)		
By end-use[1]			
Glass	7.0	8.5	..
Other[2]	3.4	3.2	..
Total	10.4	11.7	..

[1] Consumption as reported by consumers
[2] Steel, pharmaceuticals.
.. Not available

TABLE 5. CANADA, PRODUCTION AND CONSUMPTION OF TELLURIUM, 1970, 1975, 1978-84

	Production Total Refined[1]	Consumption Refined[2]
	(tonnes)	
1970	29.3	0.4
1975	42.3	..
1978	45.3	..
1979	47.2	..
1980	9.0	..
1981	21.3	..
1982	16.5r	..
1983	23.5r	..
1984e	15.0	..

[1] Refinery output of tellurium from all sources, including imported concentrates, blister, and scrap and domestic scrap.
[2] Consumption (tellurium content), as reported by consumers.
.. Not available, withheld to avoid disclosing company data; e Estimated; r Revised.

TABLE 6. IDENTIFIABLE NON-COMMUNIST PRODUCTION OF TELLURIUM, 1982-84

	1982	1983	1984e
	(tonnes)		
Japan	63	65	65
Peru	21	22	22
Canada	9	23	15
India	0.2	0.2	0.2

Sources: U.S. Bureau of Mines, Energy, Mines and Resources Canada.

[1] Available data. United States withholds its figures to avoid disclosing company data, but accounted for 42 per cent of world output in 1975.
e Estimated.

51.7

Silica

MICHEL A. BOUCHER

SUMMARY

With the exception of sandblasting, flat glass and to some extent fiberglass, most other markets for silica remained weak in 1983 and in 1984 including foundry sand, container glass, artificial abrasives and smelter flux.

This was due to the low level of economic activity in North America in general, and problems in the automotive, steel and base-metal industries.

Imports of cars and automotive engines, a greater use of aluminum diecastings, and the downsizing of cars continued to depress markets for foundry sand, resulting in additional closures of foundries during the past two years. The glass container industry continued to be affected negatively by recycling and by strong competition from plastics and aluminum. New designs for beer bottles, however, helped producers temporarily in late 1983 and early 1984.

A silicon carbide operation was permanently closed due to poor demand for abrasives by the foundry and steel industries.

Demand for smelter flux was also weak as most base-metal smelters continued to operate at low levels of capacity.

Deregulation of transportation through the Staggers Act in the United States lowered freight rates and resulted in stronger competition from U.S. material exported to Canada.

Prices per t of silica shipped from Canadian producers decreased from an average of $18.71 in 1982 to $16.70 in 1983 and $15.95 in 1984 although shipments increased substantially from their very depressed level of 1982.

CANADIAN SCENE

Newfoundland

All silica production from Dunville Mining Company Limited, a subsidiary of ERCO Industries Limited is captive to ERCO, a producer of elemental phosphorous, where silica is used as a flux.

The quartzite quarry at Villa Marie operates from May to December, and production at Dunville increased substantially in recent years due to an expansion at ERCO.

Nova Scotia

Nova Scotia Sand and Gravel Limited produces a high purity silica from sand deposits, for a variety of uses including sandblasting, glass, foundry sand, frac sand etc.

However, due to the closure of The Enterprise Foundry Company Limited in New Brunswick and Fiberglas Canada Inc. also in New Brunswick sales were substantially reduced in 1983 and 1984.

New Brunswick

Chaleur Silica Ltd. produces silica for use as a flux by Brunswick Mining and Smelting Corporation Limited's Belledune lead smelter, for cement plants, and as sandblasting material.

Water cleaning equipment was added to the plant in 1984 and the company also invested $150,000 to increase storage capacity.

Quebec

Indusmin Limited is the largest producer of silica east of Ontario with a reported

---

Michel A. Boucher is with the Mineral Policy Sector, Energy, Mines and Resources, Canada. Telephone (613) 995-9466.

production capacity of some 500 000 tpy. Silica is mined from a quartzite deposit at Saint Donat and from a sandstone deposit at Saint Canut. Silica from Saint Donat is refined at the Saint Canut plant near Montreal.

Most silica produced by Indusmin originates from Saint Canut where the ore is crushed, screened and beneficiated by attrition scrubbing, flotation and magnetic separation. Sales in 1983 and 1984 were reported to be higher than the depressed levels reached in 1982 as more silica was shipped outside of Quebec where Indusmin expanded its market share at the expense of smaller producers.

Baskatong Quartz Inc. continued to operate a high-purity silica deposit from a quartzite deposit north of Saint Urbain. The silica is used mainly by SKW Canada Inc. for the production of ferrosilicon and silicon metal. The small, but very high purity silica operation of Saint Ludger was temporarily closed in 1983. Baskatong continued exploring for new prospects in Quebec and the Maritimes, and in 1984 the company started producing very high purity silica for the photo-voltaic industry.

Melocheville Mining Limited did not produce silica from its Potsdam sandstone deposit of Beauharnois in 1983, but production resumed in 1984. The lump ore is used by Timminco Limited for Fe-Si and Si-Mn production, while the fines portion is sold to cement companies.

The Union Carbide Canada Limited's quartzite sandstone quarry at Beauharnois was not operational in 1983 nor in 1984. The company used the ore that had been mined in previous years for its Si-Mn and Fe-Mn alloy plant at Beauharnois.

Armand Sicotte & Sons Limited mined Potsdam sandstone at Saint Clothilde, south of Montreal. Lump silica is used for ferrosilicon and phosphorus production.

Sable de Silice Crémazie Inc. continued to mine silica sand and gravel from St. Joseph-du-Lac and from Ormstown. The material is used mainly for sandblasting but also for fiberglass, glass and foundries. The company plans to develop a sandstone deposit at Howick where silica would be used by the glass and fiberglass industries. In 1984, Bon Sable purchased a high magnetic separator to lower the iron content of the Howick ore.

Ontario

Indusmin Limited is the largest producer of silica west of Quebec, with a reported capacity of about 500 000 tpy, about the same as its Quebec operation. Lump quartzite from Badgeley Island, north of Georgian Bay is shipped by lake boat to Canadian destinations for the manufacture of Fe-Si. The finer material produced by crushing, is shipped to Midland, south of Georgian Bay where it is further processed to a glass-grade silica sand, and silica flour for the ceramic and other uses.

Sales improved from the very depressed level of 1982 as the glass insulation and sheet glass sectors recovered appreciably. Also, more lumps were shipped to the Fe-Si producers. During 1983, exploration was concentrated on drilling of a new quartzite orebody near Badgeley Island.

Manitoba

Steel Brothers Canada Ltd. continued to operate a high-purity silica sand on Black Island in Lake Winnipeg some 130 km north of Selkirk. The silica is mined from a poorly consolidated white sandstone. The sand is well rounded and suitable for foundry, glass and fiberglass manufacturers. The ore is washed, screened and dewatered at the plant on the island, and is then shipped by barge to a finishing plant at Selkirk on the Red River.

Due to poor sales in farming equipment in recent years some foundries were closed and sales were lost. The company, however, reported increased sales in the glass container industry.

Inco Limited continued to produce a low-grade silica from an impure quartzite from the Manasan quarry for its Thompson smelter.

Production varies from year to year depending on nickel production and silica is used as flux in Inco's converter.

Alberta

Sil Silica a division of Strathcona Resource Industries Ltd. produces silica sand from local sand dunes in the Bruderheim area. Silica is sold mainly as fiberglass and sandblasting material. It is also sold as foundry sand, filtration sand, frac sand and as railway traction sand.

In 1983 the company invested $1 million in a plant to produce silica flour for use as an addition to cement when completing oil wells that require high-temperature steam injection to liberate the oil.

British Columbia

The Mountain Minerals Co. Ltd.'s operation which mines a high-purity sandstone deposit near Golden reported that it was upgrading its processing plant. Silica is sold mainly for the glass and sandblasting industries but sections of the sandstone deposit where silica sand is loosely consolidated may eventually be mined and sold as foundry material. The company is also trying to develop a higher purity silica for high-tech applications.

TRADE

Most silica sand imported into Canada comes from poorly consolidated sandstone or lake sand deposits located near the Great Lakes region of the United States.

The silica sand is used mainly by iron and steel foundries and by the glass industry of Ontario and Quebec.

In recent years, total Canadian imports of silica sand for use in foundries and glass industry were respectively about 400 000 t and 325 000 t. However, these imports have declined steadily since 1979, reflecting lower demand.

The industry claims that recent deregulation of rail freight rates in the United States resulting from the Staggers Act has increased the competitiveness of U.S. companies that ship silica to Canada.

OUTLOOK

Not much improvement is expected in 1985 in Canada's three major markets for silica namely the glass, foundry and fiberglass industries.

In the medium term, competition from U.S. producers of silica for glass and foundry sand will remain strong in Ontario and Quebec because of the proximity of these provinces to the low cost producers of the U.S. Great Lakes region.

Also due to the downsizing of cars, recycling of silica sand at foundries, and other factors, as mentioned before, no growth is expected in the foundry sand industry in Canada.

Competition from substitutes for glass containers such as plastics and aluminum will remain strong across Canada.

Growth in fiberglass production will be restricted by a slow growth expected in construction activity in Canada, and by more dense materials that can be used as substitutes.

In the long-term, there is potential in Canada for the development of a deposit on Iles-de-la-Madeleine in the Gulf of St. Lawrence where silica sand and byproduct feldspar could be sold to foundries, glass and ceramic producers of northeastern United States.

There is also potential for the establishment of a flat glass producing facility in western Canada, where no such plants exist; good quality silica would be readily available as well as cheap natural gas or electricity.

Eventually higher quality silica products could be manufactured in Canada including optical quartz, cultured quartz (based on cheap electricity), solar grade silica, optical fibres and fumed "pyrogenic" silica (produced chemically or through the plasma route).

PRICES

The following tables give the average price of different silica products in 1983.

Silica	U.S./t f.o.b. mine or plant
Metallurgical	7
Glass	9-14
Foundry	10-20
Frac sand	24
Fiberglass	25
Filler	33
Pyrogenic silica	5000-9000
Cultured quartz (as ground bars)	55000-90000

Source: Personal communications with industry.

TARIFFS

Item No.		British Preferential	Most Favoured Nation	General	General Preferential
CANADA					
29500-1	Ganister and sand	free	free	free	free
29700-1	Silex or crystallized quartz, ground or unground	free	free	free	free
UNITED STATES					
513.14	Sand, other		free		
514.91	Quartzite, whether or not manufactured		free		
523.11	Silica, not specially provided for		free		

Item No.		1983	1984	1985	1986	1987
		¢ per long ton				
513.11	Sand containing 95% or more silica, and not more than 0.6% of oxide of iron	12	9	6	3	free

Sources: The Customs Tariff, 1983, Revenue Canada, Customs and Excise; Tariff Schedules of the United States Annotated (1983), USITC Publication 1317; U.S. Federal Register Vol. 44, No. 241.

TABLE 1. CANADA, SILICA PRODUCTION (SHIPMENTS) AND TRADE, 1982-84

	1982		1983		1984P	
	(tonnes)	($'000)	(tonnes)	($'000)	(tonnes)	($'000)
Production (shipments), quartz and silica sand						
By province						
Quebec	661 000	x	709 300	x	758 000	x
Ontario	438 000	8,227	874 548	11,466	1 122 404	11,481
Alberta	x	x	x	x	x	x
Manitoba	x	x	x	x	x	x
Nova Scotia	x	x	x	x	x	x
New Brunswick	x	x	x	x	x	x
Saskatchewan	99 000	1,066	123 062	1,476	120 741	1,510
Newfoundland	x	x	x	x	x	x
British Columbia	x	x	x	x	x	x
Total	1 703 000	31,864	2 303 451	38,467	2 624 002	41,863
By use						
Glass and fiberglass	430 000	11,906	..	..		
Flux	391 000	2,114	..	..		
Ferrosilicon	155 000	925	..	..		
Other uses[1]	727 000	16,919	..	..		
Total	1 703 000	31,864	2 303 451	38,467	2 624 002	41,863
Imports					(Jan.-Sept. 1984)	
Silica sand						
United States	788 468	15,475	982 568	16,864	767 577	13,892
West Germany	-	-	56	17	5	1
Other countries	300	120	38	2	-	-
Total	788 768	15,595	982 662	16,883	767 582	13,893
Silex and crystallized quartz						
United States	230	265	248	237	342	268
Japan	1	1	20	15	9	8
Other countries	10	16	3	3	1	1
Total	241	282	271	255	352	277
Firebrick and similar shapes, silica						
United States	2 584	2,021	1 981	2,983	2 538	2,216
France	219	254	649	454	431	508
West Germany	52	49	360	84	48	48
Other countries	129	82	37	40	-	-
Total	2 984	2,406	3 027	2,671	3 017	2,772
Exports						
Quartzite						
United States	65 314	566	103 944	936	73 547	656
Other countries	19	2	16	2	18	1
Total	65 333	568	103 960	938	73 565	675

Source: Statistics Canada; Energy, Mines and Resources Canada.
[1] Includes foundry use, sand blasting, silica brick, concrete products, chemical manufacture, building products and silicon carbide.
P Preliminary; - Nil; x Confidential.

TABLE 2. IMPORTS BY PROVINCE OF SILICA SAND AND SILEX AND CRYSTALLIZED QUARTZ, 1982 AND 1983

	1982		1983	
	(tonnes)	($)	(tonnes)	($)
Silica sand				
Ontario	583 149	9,202,483	709 149	8,555,711
British Columbia	80 226	2,904,414	144 282	4,769,588
Saskatchewan	16 794	601,072	32 638	1,140,835
Alberta	44 866	1,638,964	31 132	1,134,443
Quebec	61 338	1,142,205	57 279	988,455
Manitoba	1 107	84,305	6 926	270,595
Nova Scotia	435	8,248	658	15,527
New Brunswick	810	12,075	553	7,403
Newfoundland	-	-	45	602
Prince Edward Island	43	1,010	-	-
Total	788 768	15,594,776	982 662	16,883,159
Silex and crystallized quartz				
Ontario	146	167,305	185	156,394
British Columbia	11	17,551	34	49,466
Quebec	6	6,124	48	46,242
Alberta	77	90,567	4	3,181
Saskatchewan	1	665	-	-
Manitoba	..	107	-	-
Total	241	282,319	271	255,283

Source: Statistics Canada.
.. Not available; - Nil.

TABLE 3. CANADA, SILICA PRODUCTION AND TRADE, 1970, 1975, 1979-84

Year	Production Quartz and Silica Sand	Imports Silica Sand	Imports Silex or Crystallized Quartz	Imports Firebrick and Similar Shapes	Exports Quartzite	Consumption Quartz and Silica Sand
			(tonnes)			
1970	2 937 498	1 176 199	186	2 020	58 917	3 979 305
1975	2 491 715	1 044 160	1 550	18 818	39 977	3 510 818
1979	2 368 497	1 651 890	1 259	4 896	60 823	3 611 815
1980	2 252 000	1 200 237	281	4 775	63 166	3 326 956r
1981	2 238 000	1 142 880	251	13 762	119 347	3 079 225r
1982	1 797 000r	788 768	241	2 984	65 333	2 623 263r
1983	2 303 451	982 662	271	3 027	103 960	..
1984P	2 624 002					

Sources: Statistics Canada; Energy, Mines and Resources Canada.
P Preliminary; .. Not available; r Revised.

TABLE 4. CANADA, ESTIMATED CONSUMPTION OF SILICA, BY INDUSTRIES, 1981-82

	1981	1982
	(tonnes)	
Glass manufacture (including glass fibre)	951 442	1 120 565
Smelter flux[1]	724 040[r]	387 482
Foundry sand[1]	467 954[r]	336 119
Refractory brick mixes, cements	359 225[r]	262 541
Metallurgical	143 447	164 987
Concrete products	211 608	149 539
Artificial abrasives	125 706	112 785
Chemicals	39 430	36 877
Fertilizer, stock poultry feed	4 038[r]	3 398
Gypsum products	2 314	2 878
Other[2]	50 021[r]	46 092
Total	3 079 225[r]	2 623 263

[1] Reallocation of data, commencing in 1981, attributable to revision in survey coverage.
[2] Includes asbestos products, ceramic products, frits and enamels, paper and paper products, roofing and other minor uses.
[r] Revised.

Silicon, Ferrosilicon, Silicon Carbide and Fused Alumina

D. PHILLIPS

SUMMARY

The consumption of ferrosilicon is directly related to the consumption of steel and ferrous castings. Canadian ferroalloy companies have remained competitive due to their domestic source of raw material, low energy costs and modern plants. The investment by Elkem A/S in the Canadian ferrosilicon industry, through its aquisition of Union Carbide Canada Ltd. could strengthen the latter organization's international marketing capability and makes accessible the specialized ferroalloy technology necessary to remain competitive.

Silicon is the second most abundant element in the earth's crust and world resources of this metal are almost inexhaustible. Silica deposits (SiO_2) are the main commercial source of silicon. The production of silicon metal, ferrosilicon and silicon carbide from silica ores requires large amounts of electricity and, therefore, production plants are usually situated in areas with an abundant supply of electrical power. In Canada, these silicon products are manufactured in plants in Quebec and southern Ontario.

CANADIAN DEVELOPMENTS

The Canadian ferrosilicon and silicon metal industry is concentrated in Quebec where large supplies of hydroelectric power and raw materials are available. There were three producers of primary ferrosilicon in 1983 and 1984, two of which produced silicon metal. Byproduct ferrosilicon was also produced in the manufacture of fused aluminum oxide abrasives.

Ferrosilicon is offered for sale in several grades, expressed in terms of per cent contained silicon. The more common grades of 50, 75 and 85 per cent are produced for consumption by the steel industry. Byproduct ferrosilicon usually grades below 20 per cent; the most common use for this material is in the flotation circuit of mineral processing operations.

The Canadian ferrosilicon industry operated at approximately 80 per cent of capacity in 1983 and near capacity in 1984, with the exception of the Elkem Metal Canada Inc. Beauharnois, Quebec facility, which closed in May of 1982 and was not reopened by year-end 1984.

SKW Canada Inc.'s production in 1983 and 1984 was estimated to be 25 000 tonnes (t) of 75 per cent grade ferrosilicon and 25 000 t of silicon metal in each of these years. SKW exports most of its production, mainly to the United States, Japan and West Germany.

Elkem Metal Canada Inc. (previously Union Carbide Canada Ltd.), which closed its ferrosilicon and silicon metal facility at Beauharnois in 1982, did not reopened it in 1983 and 1984. Elkem continued to operate its ferrosilicon plant at Chicoutimi, Quebec where an estimated 25 000 t of 75 per cent grade ferrosilicon was produced each year in 1983 and 1984.

Following the 1981 acquisition of the U.S. ferroalloy facilities of Union Carbide Corp. by Elkem A/S, the company exercised its legal option and acquired the assets of Union Carbide Canada Ltd. in 1984. Elkem planned to have its Canadian marketing office based in Toronto. The owners also negotiated a new power contract with Hydro-Québec. This acquisition by Elkem A/S should strengthen the formerly held Union Carbide organization because of Elkem's world-wide operations.

Timminco Canada Limited's (previously Chromasco Limited) total 1984 production of ferrosilicon was estimated at 40 000 t. About one-half of this was in the form of 50 per cent grade and the rest was divided between the 75 and 85 per cent grades. The 1983 production was approximately 20 per cent less than that of 1984.

The abundance of relatively inexpensive electrical energy also enables Canada to produce and export bulk quantities of synthetic

---

D. Phillips is with the Mineral Policy Sector, Energy, Mines and Resources, Canada. Telephone (613) 995-9466.

abrasives, such as silicon carbide (SiC) and fused alumina (Al_2O_3). Producers of these abrasives are located in Quebec and Ontario. The Quebec-based companies, with products shown in brackets, are: Carborundum Canada Inc., Shawinigan (SiC); Norton Company (SiC) and Electro Refractories & Abrasives Canada Ltd., both in Cap-de-la Madeleine (SiC); and Unicorn Abrasives of Canada Limited, Arvida (Al_2O_3). The Ontario-based companies are: Carborundum Canada Inc. (Al_2O_3), Norton Company (Al_2O_3 and SiC) and Usigena (Canada) Limited (Al_2O_3 and SiC), all of Niagara Falls; and The Exolon Company of Canada, Ltd., Thorold (Al_2O_3 and SiC). All Canadian production of synthetic abrasives is exported, principally to the United States where the bulk material is crushed, screened and classified. A small part of the processed material is reimported for the production of bonded abrasives such as abrasive wheels and coated abrasives such as sandpaper.

INTERNATIONAL DEVELOPMENTS

Due to a rising consumption of steel in the Lesser Developed Countries (LDC's), there has been a corresponding increase in demand for ferrosilicon. In order to meet this increased demand, some of the LDC's have increased their ferrosilicon capacities, namely: Brazil, Yugoslavia, the Philippines, Egypt and Venezuela.

In response to predicted shortages of world ferrosilicon production, capacity was increased by an estimated 98 000 t contained silicon in the late 1970s and early 1980s. World utilization of production capacity was estimated at 65 per cent in 1983 and is predicted to be 80 to 85 per cent by 1990, equivalent to 3.7 million t.

Companies in both Norway and the United States have been upgrading their ferrosilicon manufacturing facilities. Canada's ferrosilicon industry, which is relatively new, should remain competitive, especially when its abundance of raw materials and low energy costs are taken into consideration.

In 1984 Elkem A/S transacted the sale of some of its shares in Icelandic Alloys to Sumitomo. The government of Iceland has a controlling interest in the company and Sumitomo has acquired a 15 per cent share interest. It is estimated that Icelandic Alloys will increase its annual shipments of ferrosilicon, to Japan from 8 000 t to 20 000 t. Canada's annual production of ferrosilicon, which is estimated at 100 000 t, remains at approximately one-fifth that of the U.S.S.R. production and approximately one-quarter those of Norway and the United States.

USES

Silicon metal is used mainly as an alloying agent for aluminum. It increases fluidity and corrosion resistance, as well as thermal and electrical conductivity. In addition, silicon metal reduces the specific density and thermal expansion of aluminum alloys. These alloys are used principally to make aluminum castings, and contain on average about 6 per cent silicon. More than one-half of the cast aluminum tonnage is used in the transportation industry. Another important use of silicon metal is in the fabrication of silicones, which are used in oil production and for the manufacture of more than 200 products, including synthetic rubber resins and electric motor insulation. Silicon metal is also used to make silicon bronze, aluminum alloys for coating steel sheets, semiconductor electronic devices and silicon nitride (Si_3N_4).

The iron and steel industry is the largest user of ferrosilicon and other silicon alloys such as silicocalcium, silicochrome and silicomanganese. Ferrosilicon functions primarily as a deoxidizer in molten steel. In addition, it is used as a graphite promoter during the production of carbon steels, as an additive to improve the electrical properties of electric steels and as a reducing agent in the manufacture of nonferrous alloys. Carbon steel contains on average 0.755 kg of silicon per t of steel, and consumes about one-third of Canadian ferrosilicon production. Stainless steels and electric steels, which contain an average of 10 and 20 kg of silicon respectively per t of steel, and other types of steel consume the remaining two-thirds. Ferrosilicon is also used in the silicothermic process for the production of other metals, but only small tonnages are required for this purpose.

OUTLOOK

The outlook for Canadian ferrosilicon and silicon metal production during 1985 is not expected to change from 1984. Three of the four ferrosilicon plants are expected to remain at capacity rates of production except for normal maintenance outages.

Potential areas for expanding silicon metal consumption include the electronics industry where high purity silicon metal is used to produce silicones; the alloy industry, where silicon has scope to substitute for other metals; and the solar energy field, where silicon alloys are widely used in heat exchanger systems.

The future demand for ferrosilicon, which is directly related to the steel and ferrous casting industry (80 per cent of total consumption), is forecast to show moderate increases in growth to the year 1990 of approximately 2.6 per cent per annum. Western world capacity, which presently stands at approximately 1.4 million t, is expected to decrease by approximately 1 per cent. Accordingly, capacity utilization is forecasted to increase approximately 10 per cent by 1990.

PRICES

As published by **METALS WEEK** in December 1982, 1983 and 1984	1982	1983	1984
		(¢US)	
Ferrosilicon, U.S. producer, per pound of silicon content; lump bulk lots, fob shipping point			
High-purity 75% Si	47.00	43.00	47.00
Regular 50% Si	45.00	43.00	45.00

Silicon metal, per pound contained silicon, fob shipping point, lump, bulk and carload lots,

(% max. Fe)	(% max. Ca)			
0.35	0.07	66.80	65.35	68.85
0.50	0.07	64.50	63.05	66.55
1.00	0.07	62.00	60.25	63.75

Prices published by **AMERICAN METAL MARKET** in December 1982, 1983 and 1984	1982	1983	1984
		(¢US)	
SMZ alloy: 60-65% Si, 5-7% Mn, 5-6% Zr, ½ in. x 12 M, per pound of alloy	53.25	55.90	..
Calcium-silicon and calsibar alloy, fob producer, 15-ton lots, per pound	66.00	72.00	72.00
Electric furnace silvery pig iron, fob Keobuck, Iowa			
16% Si, per ton	220.00	220.00	220.00
20% Si, per ton	249.00	249.00	249.00

Prices published by **INDUSTRIAL MINERALS** in December 1982, 1983 and 1984 (tonnes, cif main European port)	1982	1983	1984
		(£)	
Fused alumina, 8-220 mesh, cif			
Brown, min. 94% Al_2O_3	350-420	350-420	350-420
White, min. 99.5% Al_2O_3	410-500	410-500	410-500
Silicon carbide, 8-220 mesh, cif			
Black, about 99% SiC - Grade 1	650-700	700-750	700-750
- Grade 2	580-680	600-700	600-700
Green, over 99.5% SiC	850-850	850-950	850-950

fob Free on board; cif Cost, insurance and freight; .. Not available.

TARIFFS

Item No.		British Preferential	Most Favoured Nation	General	General Preferential
			(cents)		
CANADA					
37502-1	Silicomanganese - silico spiegel and other alloys of manganese and iron containing more than 1%, by weight, of silicon per pound or fraction thereof, on the manganese contained therein	free	0.73	1.75	free
37503-1	Ferrosilicon, being an alloy of iron and silicon containing 8% or more, by weight of silicon and less than 60%, per pound or fraction thereof, on the silicon contained therein	free	free	1.75	free
37504-1	Ferrosilicon, being an alloy of iron and silicon containing 60% or more, by weight, of silicon and less than 90%, per pound or fraction thereof, on the silicon contained therein	free	0.73	2.75	free
37505-1	Ferrosilicon, being an alloy of iron and silicon containing 90% or more, by weight, of silicon per pound or fraction thereof, on the silicon contained therein	free	2.3	5.5	free
92804-1	Silicon metal	10%	12.1%	25%	8%
92815-4	Silicon sulphide	10%	12.1%	25%	8%

MFN Reductions under GATT (effective January 1 of year given)	1983	1984	1985	1986	1987
			(cents)		
37502-1	0.73	0.73	0.72	0.71	0.7
37504-1	0.73	0.73	0.72	0.71	0.7
37505-1	2.3	2.3	2.2	2.1	2.0
			(%)		
92804-1	12.1	11.4	10.7	9.9	9.2
92815-4	12.1	11.4	10.7	9.9	9.2

UNITED STATES (MFN)

519.21	Crude silicon carbide	free
519.37	Silicon carbide in grains, ground, pulverized or refined	0.3¢/lb
606.35	Ferrosilicon, containing 8-60% silicon	free
606.42	Ferrosilicon chromium	10%

TARIFFS (cont'd)

UNITED STATES (cont'd)		1983	1984	1985	1986	1987
		(% unless otherwise specified)				
606.36	Ferrosilicon, containing 60-80% silicon and over 3% calcium	1.1	1.1	1.1	1.1	1.1
606.37	Other ferrosilicon containing 60-80% silicon	1.6	1.6	1.6	1.5	1.5
606.39	Ferrosilicon containing 80-90% silicon	1.9	1.9	1.9	1.9	1.9
606.40	Ferrosilicon containing over 90% silicon	8.6	7.9	7.2	6.5	5.8
606.44	Ferrosilicon manganese	5.0	4.7	4.4	4.2	3.9

EUROPEAN ECONOMIC COMMUNITY (MFN)		1983	Base Rate	Concession Rate
			(%)	
28.13	Silicon dioxide	5.5	6.4	4.6
73.02	Ferrosilicon	8.7	10.0	6.2
	Ferrosilico-manganese	5.5	5.5	5.5
	Ferrosilico-chrome	6.3	7.0	4.9

JAPAN (MFN)				
28.04	Silicon - single crystal	9.0	15.0	7.2
	- other	5.3	7.5	4.9
28.56	Silicon carbide	5.3	7.5	4.9
68.06	Abrasive paper	8.6	15.0	6.5
73.02	Ferrosilicon	3.8	5.0	3.7

Sources: The Customs Tariff 1983, Revenue Canada Customs and Excise; Tariff Schedules of the United States Annotated (1983), USITC Publication 1317; U.S. Federal Register Vol. 44, No. 241; Customs Tariff Schedules of Japan, 1983; Official Journal of the European Communities, Vol. 25, No. L318, 1982.

TABLE 1. CANADA, FERROSILICON, SILICON CARBIDE AND OTHER FERROALLOYS[1], EXPORTS AND IMPORTS, 1982-84

	1982		1983P		1984 (Jan-Sept)	
	(tonnes)	($000)	(tonnes)	($000)	(tonnes)	($000)
Exports						
Ferrosilicon						
United States	14 457	10,473	27 120	17,371	26 050	20,219
Japan	22 340	15,268	14 909	11,536	14 647	13,571
South Korea	2 543	2,321	1 540	1,547	257	259
West Germany	434	551	1 560	1,491	-	-
United Kingdom	601	173	525	351	617	450
Other countries	452	423	61	85	218	200
Total	40 827	29,209	45 715	32,381	41 799	34,699
Silicon carbide, crude and grains						
United States	57 847	30,847	64 707	40,457	52 213	32,144
Other countries	36	45	-	-	-	-
Total	57 883	30,892	64 707	40,457	52 213	32,144
Ferroalloys, nes						
United States	2 603	4,050	2 266	4,520	2 294	5,941
United Kingdom					1 805	173
Netherlands	-	-	73	486	-	-
Mexico	52	86	179	438	-	-
South Korea	21	286	373	201	-	-
Other countries	2 381	703	1 291	379	62	69
Total	5 057	5,125	4 182	6,024	4 161	6,183
Imports						
Ferrosilicon						
United States	9 390	10,462	12 625	12,209	20 671	15,753
South Africa	-	-	-	-	6 077	2,636
West Germany	34	36	171	190	-	-
France	175	244	36	40	198	241
Norway	38	41	19	22	44	53
Other countries	9	9	-	-	-	-
Total	9 860	11,029	13 079	12,729		15,683
Silicomanganese, including silico spiegel						
United States	380	372	453	329	4	251
Brazil	,	-	7	3	2	2
Other countries	2 497	1,348	-	-	-	-
Total	2 877	1,720	460	332		
Ferroalloys, nes						
United States	2 926	5,033	4 094	8,624	4 933	9,794
Brazil	433	3,814	873	4,824	1 021	6,346
France	1 082	2,132	1 172	2,076	754	1,534
Dominican Republic	15	83	600	1,605	1 232	2,688
United Kingdom	112	160	83	300	196	819
West Germany	93	115	87	148	16	18
Other countries	146	821	136	322	307	444
Total	4 806	12,158	7 045	17,899	8 459	21,643

Source: Statistics Canada.
[1] Other important ferroalloys are discussed in the nickel and titanium reviews for 1983-84.
P Preliminary; - Nil; nes Not elsewhere specified.

TABLE 2. CANADA, CONSUMPTION, EXPORTS, IMPORTS AND PRODUCTION OF
FERROSILICON, 1970, 1975, 1979-84

	Consumption[1] (tonnes)	Exports (tonnes)	($ 000)	Imports (tonnes)	($ 000)	Production[2] (tonnes)
1970	50 556	45 345	8,284	9 477	2,386	86 424
1975	54 904	29 029	8,075	26 353	15,665	57 580
1979	61 928	40 732	21,962	19 855	14,041	82 805
1980	63 321	52 164	33,866	18 508	13,869	96 977
1981	62 090	52 410	36,722	18 629	15,605	95 871
1982	46 122	40 827	29,209	9 860	11,029	77 089
1983	50 022	45 715	32,381	13 079	12,729	95 737
1984	60 000P	..	..	..	..	..

Sources: Energy, Mines and Resources Canada; Statistics Canada.
[1] Consumption as reported by consumers. [2] Consumption plus net exports equals derived production.
P Preliminary; .. not available.

TABLE 3. CANADA, MANUFACTURERS'
SHIPMENTS OF CRUDE SILICON CARBIDE
1970, 1975, 1978-82

	(tonnes)	($ 000)
1970	104 113	17,653
1975	89 346	24,597
1978	106 763	38,763
1979	101 265	44,108
1980	86 353	46,897
1981	89 977	50,758
1982	71 518	42,913

Source: Statistics Canada.

TABLE 4. CANADA, EXPORTS OF
SILICON CARBIDE, CRUDE AND GRAINS
1970, 1975, 1979-83

	(tonnes)	($ 000)
1970	96 159	15,976
1975	78 615	17,441
1979	84 436	31,258
1980	72 414	33,244
1981	67 144	34,595
1982	57 884	30,892
1983P	64 707	40,457

Source: Statistics Canada.
P Preliminary.

TABLE 5. CANADA, MANUFACTURERS'
SHIPMENTS OF CRUDE FUSED ALUMINA
1970, 1975, 1978-82

	(tonnes)	($ 000)
1970	131 364	18,088
1975	110 736	26,162
1978	154 303	49,916
1979	152 118	51,206
1980	146 655	56,957
1981	149 840	57,949
1982	114 479	53,816

Source: Statistics Canada.

TABLE 6. CANADA, EXPORTS OF FUSED
ALUMINA, CRUDE AND GRAINS, 1970,
1975, 1979-83

	(tonnes)	($ 000)
1970	152 572	23,234
1975	127 658	26,650
1979	183 124	55,138
1980	166 328	55,867
1981	157 990	67,954
1982	114 551	55,492
1983P	109 864	57,568

Source: Statistics Canada.
P Preliminary.

Silver

D. LAW-WEST

In 1984, silver prices did not retain the gains made during 1983. Higher interest rates coupled with lower inflation reduced investor interest in precious metals and resulted in falling prices.

World mine output of silver was estimated at 12 700 t in 1984, somewhat higher than the 12 100 t in 1983. Canadian silver production remained unchanged from 1983 to 1984, however there was some geographical shifts in production.

CANADIAN DEVELOPMENTS

Canadian primary production of silver in 1984 was estimated at 1 172 000 kg compared to 1 197 000 kg in 1983. A major increase in production occurred in Ontario while several other provinces recorded significant decreases.

Canadian Copper Refiners Limited at Montreal East, Quebec, was Canada's largest producer of refined silver, mainly from the treatment of copper anodes and blister copper and the further refining of lower-grade silver bullion. The silver refinery of Cominco Ltd. at Trail, British Columbia, was the second largest producer, recovering byproduct silver in the processing of its own ores, as well as custom lead, zinc and silver ores and concentrates. Other producers of refined silver were Inco Limited at Copper Cliff, Ontario (from nickel-copper concentrates), and the Royal Canadian Mint at Ottawa, Ontario (from gold bullion). At Cobalt, Ontario, Canadian Smelting & Refining (1974) Limited recovered silver from silver-cobalt ores and concentrates produced in that area. At Belledune, New Brunswick, the Smelting Division of Brunswick Mining and Smelting Corporation Limited recovered byproduct silver from lead concentrates treated in a blast furnace.

At its electronic materials plant at Trail, Cominco also produces a high-purity silver metal with metallic impurities totalling one part per million or less. This specialty metal product is manufactured mainly for applications such as solder preforms, brazing preforms and lead wire in the electronics industry.

Silver production in the Atlantic provinces came almost entirely as a byproduct from the base-metal producing operations of Brunswick Mining and Smelting Corporation Limited in New Brunswick.

The Buchans Mine of ASARCO Inc. in Newfoundland operated between July 1983 and August 1984 when final ore reclamation was completed. The mine has been permanently closed.

In Quebec, silver production is mainly a byproduct of base-metal mining and a minor amount as a byproduct of gold mining. Les Mines Selbaie a subsidiary of Selco Inc. remained the leading silver producer in the province.

In Ontario, Kidd Creek Mines Ltd. (KCML) continued as the province's largest silver producer. Despite weak markets for lead and copper, Kidd Creek operated in 1984 near its normal rate. Agnico Eagle Mines Ltd. increased its silver production in the Cobalt, Ontario area by 38 per cent in 1983 and a further increase of 36 per cent was expected for 1984.

Silver production in Manitoba and Saskatchewan is as a byproduct of base-metal mines operated by Hudson Bay Mining and Smelting Co. Limited in the Flin Flon-Snow Lake area and by Sherritt Gordon Mines Limited from the Fox mine near Lynn Lake and the Ruttan Mine at Leaf Rapids. The Fox mine is expected to close in 1986 and at the Ruttan mine production is planned from the deeper parts of the orebody.

Equity Silver Mining Limited, a subsidiary of Placer Development Ltd. is by far the largest silver producer in British Columbia. The company expects to increase production by some 18 000 kg of silver by the end of 1984, when a precious metals scavanger circuit is fully operational in the mill.

D. Law-West is with the Mineral Policy Sector, Energy, Mines and Resources, Canada. Telephone (613) 995-9466.

Cominco Ltd. continued to recover byproduct silver from the Sullivan lead-zinc mine, which operated normally during 1984. Other important silver producers in the province included Lornex Mining Corporation Ltd. and Westmin Resources Ltd.

Silver production in the Yukon was up from 18 000 kg to 43 900 kg in 1984 mainly as the result of the reopening of United Keno's operations near Elsa. The low silver prices of the last half of 1984 placed the continued operation of United Keno in question, however, the company's exploration program has uncovered some promising discoveries for further evaluation.

During the first nine months of 1984 exports of silver in ores, concentrates and as refined metal totalled 1 118 397 kg compared with 1 165 396 kg in the corresponding period in 1983. Canada's largest market for ores and concentrates was Japan, while the United States remained the largest market for Canadian refined silver.

Canadian imports of refined silver for the first nine months of 1984 amounted to 181 180 kg, mainly from the United States.

Consumption of silver for all uses in 1983 was estimated at about 200 000 kg up slightly from the 180 500 kg consumed during 1982.

INTERNATIONAL DEVELOPMENTS

Estimated world production of silver in 1983 was 12 130 t, slightly above the 1982 figure (Table 5). The Silver Institute has estimated 1984 world mine production at 12 700 t.

Based on preliminary data, Canada dropped to fifth place in 1983 behind Mexico, Peru, U.S.S.R. and the United States. These five countries account for about 64 per cent of the world's total primary silver production.

World consumption of silver during 1983 for both industrial and coinage usage was estimated at 11 400 t by Handy & Harman, compared with 11 250 t in 1982. Most of the difference in consumption is accounted for by the U.S. Olympic coin program.

Peru has emerged as a major world producer. Peruvian production has been growing at an average annual rate of about 3.5 per cent, since 1968. Approximately 10 per cent of Peru's output originates in 28 mines with western Andes, base metal mines account for about 65 per cent with primary silver mine account for the remainder of production. Potential for further growth in silver production depends upon the successful completion of three projects currently under construction, which could add 87 000 kg annually.

During 1983, Mexico reestablished itself as the world's leading silver producer, at the same time setting a new production record. The previous record had been set in 1945. A large part of the production increase was the result of the first full year of production at the 10 000 tpd open-pit mine at the Real de Angeles joint venture. The partners are: Placer Development Ltd. (34 per cent), the Mexican Government (33 per cent), and Frisco SA de CV (33 per cent). Annual silver production from this project will be about 220 000 kg, making it the largest silver producer in Mexico.

Many of the smaller Mexican silver mines have expansions under way or are planning expansions. Mexican authorities have forecast silver production for 1984 at some 3 250 t.

In the United States, Sonora Gold Corp. began preproduction work at its Jamestown mine in California. When completed the operation will produce about 4 300 kg of silver per year.

Hecla Mining Corp. consolidated its position as the largest silver producer in the United States, through a complex trade involving Sunshine Mining Company and Rancher's Exploration and Development. Hecla now produces some 255 000 kg per annum. Sunshine, the country's second largest producer produces about 165 000 kg.

The disposal of silver from the U.S. National Defense Stockpile again became an issue in 1984. Legislation was signed early in U.S. fiscal year 1985 authorizing the disposal of 10 million oz of silver from the stockpile. The disposal however is contingent on an impact report being submitted to Congress. There is some doubt as to whether the approved silver disposal will go ahead.

In Chile, small and medium-scale silver mines will receive support from the government. Enami announced a price support program paid on a sliding scale of around $10 per oz for silver refined at the state owned refinery during the second half of

1984. The mine must repay Enami the difference between their realized sale price and future prices, once the market recovers.

CONSUMPTION AND USES

Western-world silver consumption increased during 1984. This was mainly due to increased demand for the end products associated with silver use, namely photography, electronic equipment and sterlingware and jewelery.

Photography is the largest single use for silver accounting for about 40 per cent of industrial consumption. While silver use by this industry has been rising over the past three years, it remains nearly 10 per cent below consumption in the late 1970s. The high silver prices of 1979 and 1980 led the photographic industry to reduce the amount of silver per exposure while at the same time increasing the silver recovered from spent photographic materials. Colour and black and white photographs account for some 55 per cent of silver used in photography, x-ray photographs account for about 35 per cent, while the remaining 10 per cent is broadly utilized in graphic arts, engineering and other industrial uses. Each of these sectors face different conditions that affects the amount of silver consumed. The colour and black and white print business faces competition from silver-free technology such as electronic or video photo-imaging. However, at the same time the number of amateur photographers are increasing and demand for photographic materials is increasing in developing countries. According to some analysts this sector should experience growth on average.

Since the x-ray sector is mainly centred around hospitals there is fairly complete recycling of used x-ray films, depending upon the length of time which the x-rays are kept on record. The major threat to silver x-ray prints is from digital storage of x-rays using chromium-dioxide video tapes. In addition traditional x-ray machines are facing competition from other machines such as catscans and nuclear magnetic rosand units which could easily be linked with non silver video recording equipment in lieu of the traditional silver-based x-ray prints.

The electrical and electronic industry accounts for about 30 per cent of consumption. Silver's superior electrical conductivity accounts for its extensive use as contacts, conductors, resistors and capacitors in electronic components. Silver is most often required where a high degree of dependability is required. Spacecraft, satellites and aircraft guidance systems are typical examples. Also silver-zinc and silver-cadmium batteries are used in spacecraft and jet aircraft. Consumption in the electronics industry will increase with the increased popularity of home entertainment units such as video casette recorders and home micro computers, however, the move toward miniaturization will largely offset the increased numbers.

Sterlingware and jewellery combined account for about 5 per cent of consumption. Silver usage in these industries has declined substantially since the mid-1970s when in the United States sterlingware producers used 22-30 million ounces, while jewellers required about 13 million ounces. In 1982 U.S. silver consumption by silversmiths and jewellers was slightly over 13 million ounces. Higher metal prices accounted for the substantial drop in consumption by these industries since the silver content often constitutes a large portion of the cost. While silver prices remain relatively high significant increases in silver consumption by these industries is unlikely.

In addition to the photographic, electronic and jewellery and sterlingware the remaining 25 per cent of industrial silver is used mainly in: catalysts for chemical processing, mirrors, brazing alloys and solders, electroplating, dental amalgams, medical equipment bearings, chemicals coins, medallions and commemorative objects.

PRICES

Silver prices increased in 1983, over 1982, but then lost much of this gain during 1984. The 1984 average London fixing dropped to $US 8.14 per oz from $US 11.45 in 1983. In 1982 the average had been $US 7.95. The price spread in 1984 between the high and low prices was $US 3.78. This compares to the much broader range of $US 6.30, which occurred in both 1983 and 1982.

OUTLOOK

Canada's output of primary silver in 1985 is expected to be unchanged from 1984, although there will be some geographical shift in production. In the short- to medium-term Canadian silver production should continue near present levels, as there are no major mine developments on the horizon and little capacity is expected to be closed.

Worldwide silver production in 1985 may increase somewhat from 1984, especially if countries such as Mexico and Peru continue to exploit their silver resources for foreign exchange.

Silver bullion prices will remain depressed as long as high interest rates and low inflation divert investment dollars away from the precious metals market. However, longer-term silver prices are expected to remain in the $5 to $10 per oz ($US 1984).

While actual mine output will continue to fall below fabrication demand, industrial scrap recovery coupled with individual and government dishoarding will keep silver prices under pressure.

Silver usage is expected to either stabilize near current levels or grow by no more than 2.5 per cent per year, depending upon price. The industrial usage of silver accounts for about 55 per cent of its demand, however, in the longer term this portion could grow as silver moves away from its role as a financial investment.

TARIFFS

Item No.		British Preferential	Most Favoured Nation	General	General Preferential
			(%)		
CANADA					
32900-1	Ores of metals, nop	free	free	free	free
35800-1	Anodes of silver	free	free	10	free
35900-1	Silver in ingots, blocks, bars, drops, sheets or plates, unmanufactured; silver sweepings; scrap jewelery	free	free	free	free
35905-1	Scrap silver and metal alloy scrap containing silver	free	free	25	free
36100-1	Silver leaf	12.5	15.7	30	10
36200-1	Articles consisting wholly or in part of sterling or other silverware, nop; manufactures of silver, nop	15.7	16.8	45	11

MFN Reductions under GATT (effective January 1 of year given)	1983	1984	1985	1986	1987
			(%)		
36100-1	15.7	14.6	13.5	12.4	11.3
36200-1	16.8	15.3	13.9	12.4	11.0

UNITED STATES (MFN)

Item No.		
601.39	Precious metal ores, silver content	free
605.20	Silver bullion, silver dore and silver precipitates	free
605.70	Precious metal sweepings and waste and scrap, silver content	free
644.56	Silver leaf	2.5¢ per 100 leaves

Item No.		1983	1984	1985	1986	1987
				(%)		
420.60	Silver compounds	4.4	4.2	4.0	3.9	3.7
605.46	Platinum-plated silver, unwrought or semi-manufactured	11.8	10.7	9.6	8.6	7.5
605.47	Gold-plated silver, unwrought or semi-manufactured	17.5	15.6	13.8	11.9	10.0
605.48	Other unwrought or semi-manufactured silver	8.3	7.7	7.1	6.6	6.0
605.65	Rolled silver, unworked or semi-manufactured	8.3	7.7	7.1	6.6	6.0

TARIFFS (cont'd)

Item No.		1983	Base Rate	Concession Rate
EUROPEAN ECONOMIC COMMUNITY (MFN)			(%)	
28.49	Colloidal silver, amalgams, salts and other compounds of silver			
A.	Colloidal silver	6.7	8.0	5.3
B.	Amalgams of silver	6.7	8.0	5.3
C.	Salts and other compounds, inorganic or organic of silver	7.8	9.6	6.0
71.05	Silver, including silver gilt and platinum-plated silver, unwrought or semi-manufactured			
A.	Unwrought	free	free	free
B.	Bars, rods, wire and sections, plates, sheets, strips	1.9	2.0	1.8
C.	Tubes, pipes and hollow bars	3.2	3.5	2.9
D.	Foil of a thickness, excluding any backing, not exceeding 0.15 mm	5.8	6.5	5.0
E.	Powder, purls, spangles, cuttings and other forms	4.4	5.0	3.8
71.06	Rolled silver, unworked, or semi-manufactured			
A.	Unworked	4.4	5.0	3.8
B.	Semi-manufactured	5.6	6.5	4.6
71.08	Rolled gold on silver, unworked or semi-manufactured	3.2	3.5	2.9
71.10	Rolled platinum or other platinum group metals on silver, unwrought or semi-manufactured	3.2	3.5	2.9
71.11	Silversmiths sweepings, residues and other waste and scrap	free	free	free
71.12	Articles of jewellery and parts thereof, of silver or rolled silver			
A.	Of silver	4.0	4.5	3.5
B.	Of rolled silver	7.4	9.0	5.8
71.13	Articles of silversmiths wares and parts thereof, of silver, other than above			
A.	Of silver	5.3	7.5	3.0
B.	Of rolled silver	4.4	5.0	3.8
71.14	Other articles of silver or rolled silver			
A.	Of silver	6.0	7.5	5.1
B.	Of rolled silver	5.2	6.0	4.4

Sources: The Customs Tariff, 1983, Revenue Canada, Customs and Excise; Tariff Schedules of the United States Annotated (1983), USITC Publication 1317; U.S. Federal Register Vol. 44, No. 241; Official Journal of the European Communities, L318, Vol. 25, 1982.
nop Not otherwise provided for.

TABLE 1. CANADA, SILVER PRODUCTION AND TRADE, 1982 AND 1983 AND CONSUMPTION, 1981 AND 1982

	1982		1983P		Jan.-Sept. 1984	
	(kilograms)	($000)	(kilograms)	($000)	(kilograms)	($000)
Production[1]						
By province and territories						
British Columbia	499 000	157,579	411 000	185,793	355 617	124,315
Ontario	351 000	110,970	347 000	157,026	506 358	177,010
New Brunswick	230 000	72,719	210 000	94,965	136 463	47,704
Northwest Territories	51 000	16,073	51 000	22,993	50 000	17,564
Quebec	55 000	17,507	39 000	17,866	37 789	13,210
Manitoba	25 000	8,014	23 000	10,575	27 974	10,478
Yukon	95 000	29,943	18 000	8,155	43 898	15,346
Saskatchewan	5 000	1,471	5 000	2,222	4 868	1,702
Newfoundland	3 000	928	2 000	846	5 529	1,932
Alberta	--	--	-	-	-	-
Nova Scotia	-	-	-	-	-	-
Total	1 314 000	415,204	1 106 000	500,441	1 168 496	409,262
By source[2]						
Base-metal ores	994 000	314,087	642 000	290,491	..	..
Gold ores	20 000	6,362	77 000	34,841		
Silver ores	299 000	94,563	387 000	175,109		
Placer gold ores	1 000	192	..	..		
Total	1 314 000	415,204	1 106 000	500,441		
Refined silver[3]	790 358	..	1 047 512	..		
Exports						
Silver in ores and concentrates						
Japan	217 259	48,672	185 639	68,119	139 424	38,469
United States	163 120	38,531	89 751	32,831	89 761	24,751
Belgium-Luxembourg	97 487	17,453	93 937	28,639	16 187	3,097
West Germany	21 159	2,797	18 256	3,400	8 527	1,298
Switzerland	-	-	9 919	2,987	17 614	4,868
Taiwan	1 177	192	6 913	2,687	996	217
Sweden	7 503	2,326	5 971	2,327	-	-
Other countries	94 898	26,708	30 523	6,566	49 373	9,541
Total	602 603	136,679	440 909	147,556	321 882	82,231
Refined metal						
United States	1 125 689	367,968	1 041 673	480,533	795 852	227,530
Japan	-	-	1 301	381	-	-
Trinidad-Tobago	538	186	779	378	210	85
Brazil	365	98	768	363	-	-
Other countries	7 755	2,126	1 345	516	453	165
Total	1 134 347	370,378	1 045 866	482,171	795 515	277,780
Imports						
Silver in ores and concentrates						
Peru	19 817	4,438	77 788	26,752	..	..
Chile	26 041	7,915	31 748	11,681		
United States	40 989	10,642	24 887	8,326		
South Korea	22 767	5,535	6 230	2,641		
South Africa	31 399	5,933	5 806	1,860		
Other countries	2 934	912	6 796	2,136		
Total	143 947	35,375	153 255	53,396		

TABLE 1. (cont'd)

	1982		1983P		Jan.-Sept. 1984	
	(kilograms)	($000)	(kilograms)	($000)	(kilograms)	($000)
Refined metal						
United States	256 309	76,606	280 496	125,377	163 194	53,751
Chile	16 000	4,673	33 496	16,106	7 999	3,198
El Salvador	-	-	16 325	7,410	-	-
Japan	-	-	3,225	1,583	-	-
Mexico	3 110	885	1 968	1,016	8 123	4,594
United Kingdom	207 877	66,814	-	-	402	38
Others	944	273	3 929	850	1 462	448
Total	484 240	149,251	339 439	152,342	181 180	62,029
Consumption, by use						
Sterling	32 247	..	12 108	..	..	..
Silver alloys	41 105	..	18 569	..	..	..
Wire rod	3 527	..	3 686	..	..	..
Others[4]	215 251	..	146 096	..	..	..
Total	292 130	..	180 459	..	..	..

Sources: Energy, Mines and Resources Canada; Statistics Canada.
[1] Includes recoverable silver in: ores, concentrates and matte shipped for export; crude gold bullion produced; blister and anode copper produced at Canadian smelters; and base and other bullion produced from domestic ores. [2] Estimated by Energy, Mines and Resources Canada; the base-metal category includes production of some mines normally regarded as silver producers, but which also recover some base-metal. [3] From all sources, domestic and imported materials of both primary and secondary origin. [4] Includes sheet, coinage, fabricated investment bars and miscellaneous uses.
P Preliminary; - Nil; .. Not available; -- Amount too small to be expressed.

TABLE 2. CANADA, SILVER PRODUCTION, TRADE AND CONSUMPTION, 1970, 1975, AND 1979-83

	Production		Exports			Imports, Refined Silver	Consumption[3] Refined Silver
	All Forms[1]	Refined[2] Silver	In Ores and Concentrates	Refined Silver	Total		
	(kilograms)						
1970	1 376 354	955 668	678 676	752 689	1 431 365	134 347	187 679
1975	1 234 642	931 540	471 410	713 566	1 184 976	420 078	642 089
1979	1 146 908	949 778	415 726	911 146	1 326 872	38 308	251 985
1980	1 070 000	985 051	396 690	881 761	1 278 451	339 180	265 938
1981	1 129 394	875 121	546 449	914 800	1 461 249	327 328	292 130
1982	1 314 000	790 358	602 603	1 134 347	1 736 950	484 240	180 459
1983P	1 106 000	1 047 512	440 909	1 045 866	1 486 775	339 439	..

Sources: Energy, Mines and Resources Canada; Statistics Canada.
[1] Includes recoverable silver in: ores, concentrates and matte shipped for export; crude gold bullion produced; blister and anode copper produced at Canadian smelters; and base and other bullion produced from domestic ores. [2] From all sources, domestic and imported materials of both primary and secondary origin. [3] In some years includes only partial consumption for coinage.
P Preliminary; .. Not available.

TABLE 3. WORLD MINE PRODUCTION[1] OF SILVER, 1982 AND 1983

	1982P	1983e
	(kilograms)	
U.S.S.R.e[2]	1 458 000	1 465 000
Mexico	1 550 197	1 832 876
Peru	1 654 161	1 679 588
Canada	1 313 630	1 105 894
United States	1 251 572	1 252 731
Australia	900 290	..
Polande	654 987	653 173
Chile	382 197	..
Japan	306 175	307 173
Republic of South Africa	216 001	172 909
Bolivia	170 188	..
Sweden	167 829	..
Yugoslaviae[2]	103 979	124 012
Spain	117 843	..
Morocco	52 345	..
Zaire	59 239	..
South Korea	44 906	..
Argentina	82 735	..
Philippines	61 689	56 699
People's Republic of China	70 035	..
Greece	49 169	..
Italy	55 701	73 482
France	30 663	21 410
Other countriese	797 924	..
Total	11 551 455	12 130 356

Sources: Energy, Mines and Resources Canada; Nonferrous Metal Data 1983, American Bureau of Metal Statistics Inc.; Mineral Commodity Summaries 1984, U.S. Bureau of Mines.
[1] Recoverable content of ores and concentrates produced unless otherwise noted.
[2] Smelter and refinery production.
P Preliminary; e Estimated; .. Not available.

TABLE 4. ANNUAL AVERAGE SILVER PRICES: CANADA, UNITED STATES AND UNITED KINGDOM, 1973-84

	Canada ($Cdn)	United States Handy & Harman, New York ($US)	United Kingdom London Spot (pence)
	(per troy ounce)		
1973	2.567	2.558[1]	103.783
1974	4.595	4.708	199.819
1975	4.503	4.419	200.118
1976	4.291	4.353	242.423
1977	4.922	4.623	265.512
1978	6.171	5.401	282.203
1979	12.974	11.094	519.607
1980	24.099	20.632	900.778
1981	12.617	10.518	515.303
1982	9.831	7.947	455.331
1983	14.154	11.441	753.644
1984	10.521	8.138	

Sources: Canadian prices as quoted in the Northern Miner (arithmetical average of daily quotations). United States and United Kingdom prices as quoted in Metals Week.
[1] The 60-day general price freeze in effect in the United States from June 13 through August 12, 1973 forced intermittent suspension of Handy & Harman's daily quotation during July and August for a total of 22 days.

TABLE 5. UNITED STATES CONSUMPTION OF SILVER BY END-USE[1], 1982 AND 1983

	1982 (kilograms)	(%)	1983P (kilograms)	(%)
Electroplated ware	101 211	2.7	97 696	2.7
Sterling ware	204 630	5.5	218 409	6.0
Jewellery	194 708	5.3	213 401	5.9
Photographic materials	1 610 196	43.7	1 611 813	44.3
Dental and medical supplies	52 503	1.4	47 557	1.3
Mirrors	30 170	0.8	30 108	0.8
Brazing alloys and solders	229 668	6.2	208 238	5.7
Electrical and electronic products:				
Batteries	129 608	3.5	82 020	2.3
Contacts and conductors	862 499	23.3	817 151	22.4
Bearings	7 092	0.2	5 350	0.1
Catalysts	75 177	2.0	75 053	2.1
Coins, medallions and commemorative objects	56 982	1.5	92 657	2.5
Miscellaneous[2]	141 893	3.9	142 049	3.9
Total net industrial consumption	3 696 337	100.0	3 641 502	100.0
Coinage[3]	57 417		66 188	
Total consumption	3 753 754		3 707 690	

Sources: United States Bureau of Mines, Mineral Industry Surveys, "Gold and Silver in December 1983".
[1] End-use as reported by converters of refined silver. [2] Includes silver-bearing copper, silver-bearing lead anodes, ceramic paints, etc. [3] Includes silver used in minting the George Washington Commemorative Coin and the $1.00 Olympic Coin.
P Preliminary.

TABLE 6. NON-COMMUNIST WORLD CONSUMPTION OF SILVER, 1982 AND 1983

	1982	1983P
	(kilograms)[1]	
Industrial uses		
United States	3 695 093	3 732 417
Japan	1 965 740	2 161 692
West Germany	1 045 077	799 359
India	699 828	678 056
France	578 525	578 525
United Kingdom	622 070	559 863
Belgium	469 662	497 656
Italy	289 262	342 138
Canada	279 931	279 931
Mexico	111 973	59 097
Other countries	1 088 621	1 135 276
Total industrial uses	10 845 782	10 824 010
Coinage		
Austria	124 414	62 207
United States	77 759	348 359
Canada	9 331	12 441
Other countries	186 621	155 518
Total coinage	398 125	578 525
Total consumption	11 243 907	11 402 535

Source: Handy & Harman, The Silver Market, 1983.
[1] One kilogram equals 32.1507 troy ounces.
P Preliminary.

Sodium Sulphate

G.S. BARRY

Sodium sulphate is mainly produced from natural brines and deposits in alkaline lakes in areas with dry climates and restricted drainage, from subsurface deposits and brines, or as a byproduct of chemical processes. Canada's sodium sulphate industry is based on extraction from natural brines and deposits in several alkaline lakes in Saskatchewan and Alberta. Nine plants producing natural sodium sulphate operated in Canada in 1983 and 1984. Byproduct sodium sulphate is recovered at one rayon plant and at three paper mills in Ontario. Production of byproduct sodium sulphate commenced in December 1982 at a new silver mine in British Columbia.

In the United States, natural and byproduct sodium sulphate production is almost evenly split. In Europe, sodium sulphate is produced almost entirely as a byproduct of chemical processes.

PRODUCTION AND DEVELOPMENTS IN CANADA

Markets for sodium sulphate, which were strong for five consecutive years between 1978 and early 1983 started to decline by mid-1983 as the overall North American economy increased and additional secondary supplies became available. The Saskatchewan producers responded by decreasing production by 18 per cent between 1982 and 1983 and by decreasing a further 17 per cent between 1983 and 1984. The unit value of shipments was $86.69 per t in 1982; $93.92 in 1983; and $95.90 in 1984. Exports to the United States declined by 19 per cent for the first nine months of 1984 compared with the same period last year.

Besides natural sodium sulphate, about 90 000 tpy are produced as a byproduct of industrial and chemical processes in central Canada. Between 35 and 40 per cent of the total sodium sulphate produced in Canada is the higher-grade and higher-priced "detergent-grade".

Placer Development Limited brought on-stream the Equity Silver Mines Limited property in British Columbia in December 1982. The capacity of the plant is 5 000 tpy of sodium sulphate but much less was produced in 1983 and 1984.

A new use for sodium sulphate will be the production of potassium sulphate through a reaction of sodium sulphate with potassium chloride (glaserite process). Dry sodium sulphate is actually not needed so Glauber's salt will be used. The Potash Corporation of Saskatchewan announced in April 1983 that it will construct a "demonstration plant" having a capacity of 30 000 tpy at its Cory mine near Saskatoon. The plant is expected to be in production in mid-1985 and capital expenditures will be approximately $6.0 million. As of the end of 1984 construction was on schedule.

Deposits. The sodium sulphate deposits in Saskatchewan and Alberta have formed in shallow, undrained lakes and ponds where in-flow is greater than out-flow. Percolating ground waters carry dissolved salts into the basins from the surrounding soils. High rates of summer evaporation concentrate the brine to near saturation, and cooler fall temperatures cause crystallization and precipitation of sodium sulphate as mirabilite ($Na_2SO_4 \cdot 10H_2O$). The cycle has been repeated year after year and thick deposits of hydrous sodium sulphate, accompanied by other salts and mud, have accumulated.

Deposits in Saskatchewan have been identified that contain, in total, approximately 90 million t of anhydrous sodium sulphate. Of this amount, a total of about 51 million t is in 21 individual deposits, each containing more than 500 000 t of sodium sulphate. Exploitation currently takes place on the following lakes (with reserves, in millions of t, in brackets): Whitehorse Lake (6.5), Horseshoe Lake (3.7), Frederic Lake (2.4), Chaplin Lake (3.0), Ingebrigt Lake (9.0),

---

G.S. Barry is with the Mineral Policy Sector, Energy, Mines and Resources, Canada. Telephone (613) 995-9466.

Alsack Lake (2.6), East Coteau Lake (3.5), and Snakehole Lake (1.7), all in Saskatchewan. Production in Alberta is from Horseshoe Lake (2.0). For Saskatchewan, the reserves calculations were made in 1978 and for Alberta in 1982. Since that time reserves changed only marginally.

Recovery and processing. Because sodium sulphate is recovered by evaporation of concentrated brines or by dredging of the permanent beds of crystals, weather is as important for recovery of sodium sulphate as it is for its deposition. A large supply of fresh water is also essential. One method of sodium sulphate recovery is to pump lake brines that have been concentrated by hot summer weather into evaporating ponds or reservoirs. Continued evaporation produces a saturated or near-saturated solution of mirabilite. Differential crystallization occurs in the fall when the solution cools. Hydrous sodium sulphate crystallizes and precipitates, whereas sodium chloride, magnesium sulphate and other impurities remain in solution. Before freezing weather sets in, the impure solution remaining in the reservoir is drained or pumped back into the source lake. After the crystal bed has become frozen, harvesting is carried out using conventional earth-moving equipment. The harvested crystal is stockpiled adjacent to the plant.

Some operators use floating dredges to mine the permanent crystal bed. The slurry of crystal and brine is transported to a screening house at the plant by pipeline. If sufficiently concentrated, the brine from the screens is collected in an evaporation pond.

One company uses a combination of dredging and solution mining, and pumps a concentrated brine to an air-cooled crystallizer at the plant, where sodium sulphate is separated from other insoluble salts.

Processing of a natural salt consists of dehydration (Glauber's salt contains 55.9 per cent water of crystallization) and drying. Commercial processes used in Saskatchewan include Holland evaporators, gas-fired rotary kilns, submerged combustion and multiple effect evaporators. Salt cake, the product used principally in the pulp and paper industry, contains a minimum of 97 per cent Na_2SO_4. Detergent-grade material analyzes up to 99.7 per cent Na_2SO_4. Uniform grain size and free-flow characteristics are important in material handling and use.

Of the nine plants in the prairies, three are capable of producing detergent-grade sodium sulphate. Each of the three plants has the capacity to produce 80 per cent or more of its output as a high-grade product. The "natural" sodium sulphate industry employs about 300 persons.

Byproduct recovery. Courtaulds (Canada) Inc. produces approximately 21 000 t of detergent-grade sodium sulphate as a by-product of viscose rayon production at its Cornwall, Ontario plant. Ontario Paper Company Limited at Thorold, Ontario produced 47 200 t of salt cake in 1983 and approximately 60 000 t in 1984 as a by-product of paper manufacturing. It is mostly used in the glass industry and 60 per cent is exported. The capacity of the Thorold plant is 77 000 tpy. The Great Lakes Paper Company, Limited at Thunder Bay, produces salt cake for internal consumption (about 10 000 t in 1984).

PRICES

Canadian prices of sodium sulphate were $75 and $95 per t respectively for salt cake and detergent-grade at the beginning of 1983. The prices increased to $81 and $101 per t in April 1983 and remained at this level throughout the rest of 1983 and 1984. Prices for detergent-grade byproduct sodium sulphate in Ontario were in the order of $155 to $170 per t (for bulk) fob plants in the beginning of 1983, but decreased by $20 to $30 per t since that time.

USES

In the chemical pulping of wood the digestion reagents consist of about two-thirds caustic soda and one-third sodium sulphide obtained by using sodium sulphate as makeup. About 33 per cent of sulphur input is retained in the organic chemicals recycled in the process. Lately, technical improvements in the process significantly decreased the consumption of sodium sulphate per t of pulp produced, to about 20 kg/t. More caustic soda and emulsified sulphur is being substituted for salt cake.

Sodium sulphate is used as a builder; or more correctly as a diluent in detergents (supplies "bulk"); it is claimed to improve detergency through its effect on the colloidal properties of the cleaning system. The curtailment in the usage of phosphates on grounds of pollution control in all probability is not going to affect the use of sodium sulphate. The content of sodium sulphate in detergents varies from about 10 to 65 per cent. Roskill Information Services Ltd., suggests that as a very rough estimate sodium sulphate used in detergents of all types would represent some 10 per cent of world consumption.

Some sodium sulphate is used by the glass industry as a source of Na_2O to speed melting. Other end uses of sodium sulphate are in the dyeing industry in the manufacture of viscose sponges, the tanning industry and textiles.

An important new use is linked to pollution abatement measures: sodium sulphate is added to coal as a conditioner, since it improves the efficiency of high-temperature electrostatic precipitators by preventing clogging by fly-ash. Only about 5 kg of sodium sulphate (worth about 50 cents) is used for a tonne of coal. Experiments are being conducted in using sodium sulphate as a heat storage medium in solar energy conservation (heating) projects.

OUTLOOK

Outlook for the Canadian production and sales of natural sodium sulphate in 1985 is not very good. As a result of increased industrial activity, secondary supplies (byproduct sodium sulphate) are plentiful. Furthermore, perhaps as much as 200 000 t of sodium sulphate used by the North American pulp and paper industry was substituted by caustic soda and emulsified sulphur. Caustic prices are very erratic but are currently low; emulsified sulphur may be more expensive but the product is easier to handle.

The longer-term growth in sodium sulphate demand in North America will come mainly from the detergent industry sector (2 per cent to 3 per cent increase per year), although there are indications that the growth in this market may be temporarily levelling off in 1985, and possibly the power industry, where sodium sulphate is increasingly used as a conditioner in coal burning thermal plants. In the United States this new market has the potential to expand substantially perhaps up to 300 000 tpy in the late 1980s, or early 1990s.

U.S. commodity experts, however, still forecast none or little growth in the United States for sodium sulphate consumption in the decade of the 1980s since consumption in many traditional sectors is declining.

TABLE 1. CANADA, SODIUM SULPHATE PRODUCTION AND TRADE, 1982-84

	1982		1983P		1984	
	(tonnes)	($)	(tonnes)	($)	(tonnes)	($)
Production						
Shipments						
Saskatchewan	..	43,777,000	..	38,441,000	..	32,797,000
Alberta	..	3,681,000	..	3,895,000	..	4,133,000
British Columbia	..	4,000	..	300,000	..	146,000
Total	547 000	47,462,000	453 939	42,636,000	386 600	37,076,000
Imports						(Jan.-Sept. 1984)	
Total salt cake and Glauber's salt						
United Kingdom	16 381	1,107,000	21 715	1,497,000	15 749	939,000
United States	912	234,000	713	259,000	424	174,000
West Germany	-	-	51	5,000	56	17,000
Total	17 293	1,341,000	22 479	1,761,000	16 229	1,131,000
Exports						
Crude sodium sulphate						
United States	355 904	34,279,000	265 525	28,718,000	170 799	18,605,000
New Zealand	10 891	1,016,000	-	-	-	-
Other countries	1 129	206,000	227	27,000	-	-
Total	367 924	35,501,000	265 752	28,745,000	170 799	18,605,000

Sources: Energy, Mines and Resources Canada; Statistics Canada.
P Preliminary; .. Not available; - Nil.

TABLE 2. CANADA, NATURAL SODIUM SULPHATE PLANTS, 1982

	Plant Location	Source Lake	Annual Capacity (tonnes)
Alberta			
Agassiz Resources Limited	Metiskow	Horseshoe	75 000
Saskatchewan			
Hudson Bay Mining and Smelting Co., Limited	Grant	Snakehole	63 000
Hudson Bay Mining and Smelting Co., Limited	Hardene	Alsask	42 500
Miller Western Industries Limited	Palo	Whiteshore	109 000
Ormiston Mining and Smelting Co., Ltd.	Ormiston	Horseshoe	90 700
Saskatchewan Minerals	Chaplin	Chaplin	90 000
Saskatchewan Minerals	Bishopric	Frederick	45 000
Saskatchewan Minerals	Fox Valley	Ingebrigt	135 000
Saskatchewan Minerals[1]	Gladmar	East Coteau	45 500
Total			695 600

Source: Company reports.
[1] Since Oct. 1980; formerly Sybouts Sodium Sulphate Co., Ltd.

TABLE 3. CANADA, SODIUM SULPHATE PRODUCTION, TRADE AND CONSUMPTION 1970, 1975, 1979-84

	Production[1]	Imports[2]	Exports	Consumption
	(tonnes)			
1970	445 017	26 449	108 761	291 439
1975	472 196	22 638	178 182	256 385
1979	443 279	23 156	193 268	255 059
1980	496 000	20 211	245 831	232 045
1981	535 000	24 960	284 281	216 298r
1982	547 000	17 293	367 924	195 341
1983	453 939	22 479	265 752	191 618
1984	386 600P	..	..	..

Sources: Energy, Mines and Resources Canada; Statistics Canada.
[1] Producers' shipments of crude sodium sulphate. [2] Includes Glauber's salt and crude salt cake.
P Preliminary; .. Not available;
r Revised.

TABLE 4. CANADA, AVAILABLE DATA ON SODIUM SULPHATE CONSUMPTION, 1981-83

	1981r	1982	1983
	(tonnes)		
Pulp and paper	158 927	142 281	141 172
Soaps	40 855	38 437	40 219
Glass and wool	12 001	11 286	glass 9 551
Other products[1]	4 515	3 337	676
Total	216 298	195 341	191 618

Source: Energy, Mines and Resources Canada.
[1] Colours, pigments, feed supplements and other minor uses.
r Revised.

TABLE 5. CANADA, RAILWAY TRAIN LOADINGS OF SODIUM SULPHATE, 1982 AND 1983

	1982	1983
	(tonnes)	
Eastern Canada[1]	37 483	39 970
Western Canada[2]	515 476	413 463
Total Canada	552 959	453 433

Source: Statistics Canada.
[1] Eastern Canada refers to provinces east of the Ontario-Manitoba border. [2] Final figure has been adjusted to reflect a recalculation of data.

Stone

D.H. STONEHOUSE

SUMMARY

Increased demand for dimension stone to supply the building construction market in the United States has created a renewed interest in the building stone resources in many Canadian provinces. Since 1982 the strong United States dollar has maintained the competitiveness of many imported materials, among them dimension stone finished at more modern European plants. Under such conditions, to establish costly new technology in the United States to produce from domestic sources is not attractive. However, European technology has been set up in Canada to finish both domestic and imported stone for subsequent shipment to the United States. Granicor Inc. brought European sawing and polishing expertise to Quebec in 1982 and has encouraged the production of native stone to the extent that six new granite quarries were opened in Quebec during 1983. The technique of using gang-saws to produce thin panels from large blocks and of applying these panels to steel or concrete construction units to provide aesthetically-pleasing, well-engineered structures, at costs competitive with other cladding, has enabled Granicor and others to avail themselves of a growing United States market for such material. A similar plant is to be established in Cornwall, Ontario by Karnuk Marble Industries Inc., again with the intention of supplying a portion of the demand in the United States. RPS Marbre Ltée of Montreal announced in late-1984 its intention to spend $9.2 million to expand and modernize its stone finishing facilities to enable greater production of marble, travertine and granite panels, mainly for the United States markets. The program will be assisted by a grant of $1.8 million from la Société de développement industriel du Québec.

A number of provinces intend to assess their building stone resources and to determine whether or not markets exist within reach of prospective quarry and plant sites. Studies will be undertaken in some instances under a new round of federal-provincial mineral subagreements as part of the Economic and Regional Development Agreements (ERDA).

Granite, limestone, marble and sandstone are the principal rock types from which building and ornamental stone is fashioned. Over 90 per cent is used in construction-oriented projects, while less than 10 per cent is used as monument stone. Imports of rough blocks, particularly of granite, for sawing and polishing, as well as of finished stones for distribution to retailers, have cut into markets formerly supplied from domestic sources.

CANADIAN DEVELOPMENTS

Production of stone of all types increased by 14 per cent in 1983 to over 67 million t and by a further 5 per cent in 1984 to over 71 million t. These increases are attributed mainly to output from Ontario and Quebec. Stone is produced in direct response to the demands of the construction industry, which utilizes 93 per cent of output principally as crushed stone. Less than 1 per cent of stone production is used as building stone. Since 1979, there has been a growing interest in Canadian stone for building use. Shipments of granite from Quebec, especially black anorthosite, red granite and brownish monzonite, for modular panelling have shown marked increase. The chemical uses are limited to the cement, lime, glass and metal smelting industries and account for about 3 per cent of stone production, mainly limestone. The remaining 3 per cent is consumed in pulverized form as filler and extender materials, and for agricultural purposes.

Most provinces have accumulated data relative to occurrences of stone of all types and in many cases have published this information. The federal government, through the Geological Survey of Canada, has also gathered and published a great

---

D.H. Stonehouse is with the Mineral Policy Sector, Energy, Mines and Resources, Canada. Telephone (613) 995-9466.

number of geological papers pertaining to stone occurrences. Works by W.A. Parks[1] and by M.F. Goudge[2] have become classics in the fields of building stones and limestones, respectively.

Atlantic provinces. Limestone. The many occurrences of limestone in the Atlantic provinces have been systematically catalogued during the past few years[3,4,5]. Deposits of commercial importance are being worked in three of the four provinces.

In Newfoundland limestone is available from small, impure exposures in the eastern portion of the island, from small, high-calcium deposits in the central region, and from large, high-purity, high-calcium occurrences in the west. Other than periodic operation to secure aggregate for highway work, the main exploitation is by North Star Cement Limited at Corner Brook[6]. Large quantities of high-calcium limestone have been outlined in the Port au Port district.

In Nova Scotia limestone occurs in the central and eastern parts of the province and in New Brunswick is quarried at three locations - Brookville, Elm Tree and Havelock - for use as a crushed stone, as an aggregate, for agricultural application, for cement and lime manufacture, and for use as a flux.

Granite. Occurrences of granite in the Atlantic region have been described by Carr[7]. In Nova Scotia, a grey granite is produced from operations near Nictaux and from one quarry at Shelburne for use mainly in the monument industry. A black granite from Shelburne and a diorite from Erinville have been used for monuments and for dimension stone.

Granite is quarried intermittently from a number of deposits in New Brunswick to obtain stone of required colour and texture for specific application. A red, fine- to medium-grained granite is quarried near St. Stephen, and fine-grained, pink, grey and blue-grey granites are available in the Hampstead (Spoon Island) district. In the Bathurst area, a brown-to-grey, coarse-grained granite is quarried upon demand, as is a salmon-coloured, medium-grained granite near Antinouri Lake, and a black, ferromagnesian rock in the Bocabec River area. Red granite is available in the St. George district. Manufacturers of monument stone continue to import dark, crude granite from South Africa.

In Newfoundland, there is a recognized potential for the development of labradorite deposits in the Nain River area of Labrador.

Sandstone. A medium-grained buff sandstone is quarried at Wallace, Nova Scotia, for use as heavy riprap and for dimension stone applications.

In New Brunswick, a red, fine- to medium-grained sandstone has been quarried in Sackville for use in construction. Deposits are exploited from time to time throughout Kent and Westmorland counties for local projects and for highway work.

Quebec. Limestone. Limestone occurs in the St. Lawrence and Ottawa River valleys and in the Eastern Townships. Other major deposits in the province are located in the Gaspé region. The limestones range in age from Precambrian to Carboniferous and vary widely in purity, colour, texture and chemical composition[2]. Limestone blocks and other shapes are produced for the construction trade in the Montreal region and at various locations throughout the province as the need arises. Marble has been produced in the Eastern Townships and the Lac St-Jean areas.

Granite. Quebec, the major Canadian granite producer, accounts for up to 95 per cent of total granite shipments for use as building stone. Since 1979, sales have increased due to improved marketing and advanced processing technology. More than 25 companies quarry granite in Quebec, mainly in the Rivière-à-Pierre, the Lac St-Jean and the Appalachians regions. Six new quarries became operational in 1983 - two at Rivière-à-Pierre, two at Saint Alexis and two in the Rouyn region. Granicor Inc., using advanced technology for cutting and polishing dimension stone, produces brownish monzonite modular block panels from material extracted near the Chamouchouane River in the Lac St-Jean area. Capacity was expanded during 1983. A. Lacroix et fils Granit also expanded cutting and polishing capacity in 1983. In 1983, the North Shore of the St. Lawrence River was investigated by provincial government geologists.[8]

Sandstone. Of six operations producing from sandstone resources in Quebec only one is listed as marketing flagstone and construction blocks, in Hemmingford, Huntingdon County.

Ontario. Limestone. Although limestones in Ontario range from Precambrian through Devonian, the major production comes from Ordovician, Silurian and Devonian deposits[9,10]. Of particular importance are the limestones and dolomite from the following geological sequences: the Black River and Trenton formations, extending from the lower end of Georgian Bay across southern Ontario to Kingston; the Guelph-Lockport Formation, extending from Niagara Falls to the Bruce Peninsula and forming the Niagara Escarpment; and the Middle Devonian limestone extending from Fort Erie through London and Woodstock to Lake Huron. Production of building stone, fluxstone and crushed aggregate from the limestones of these areas normally accounts for about 90 per cent of total stone production in Ontario.

Marble is widely distributed over southeastern Ontario and, according to the Ontario Ministry of Natural Resources reports, underlies as much as 250 square kilometres (km^2)[11].

Steep Rock Calcite, a division of Steep Rock Iron Mines Limited, produces medium- to high-grade calcium carbonate at Tatlock and Perth. The filler markets have become extremely attractive recently, not only to new ventures but also to companies hitherto interested in production of only coarser aggregate materials.

Granite. Granites occur in northern, northwestern and southeastern Ontario[12,13]. Few deposits have been exploited for the production of building stone because the major-consuming centres are in southern and southwestern Ontario where ample, good-quality limestones and sandstones are readily available. The areas most active in granite building stone production have been the Vermilion Bay area near Kenora, the River Valley area near North Bay, and the Lyndhurst-Gananoque area in southeastern Ontario. Rough building blocks were quarried from a gneissic rock near Parry Sound, while at Havelock a massive red-granite rock was quarried. In 1982, Fairmont Granite Limited of Beebe, Quebec, reopened a fine pink granite quarry in Belmont Township for the production of building stone for modular block panels. A study to assess building stone and other industrial minerals in the northwestern region was arranged between the Ontario and the federal governments under the Northern Ontario Rural Development Agreement. A preliminary report of the Building and Ornamental Stone Inventory (open file report 5446) was released in 1984 by the Ontario Geological Survey.

Sandstone. Sandstone quarried near Toronto, Ottawa and Kingston has been used widely in Ontario as building stone[14]. Medina sandstones vary from grey, through buff and brown to red, and some are mottled. They are fine- to medium-grained. The Potsdam stone is medium-grained; the colour ranges from grey-white through salmon-red to purple, and it can also be mottled. Current uses are as rough building stone, mill blocks from which sawn pieces are obtained, ashlar, flagstone and as a source of silica for ferrosilicon and glass.

Western provinces. Limestone. From east to west through the southern half of Manitoba rocks of the following ages are represented: Precambrian, Ordovician, Silurian, Devonian and Cretaceous. Limestones of commercial importance occur in the three middle periods and range from magnesian limestone through dolomite to high-calcium limestones[2,15].

Although building stone does not account for a large percentage of total limestone produced, the best known Manitoba limestone is Tyndall Stone, a mottled dolomitic limestone often referred to as "tapestry" stone. It is widely accepted as an attractive building stone, and is quarried at Garson, Manitoba, about 50 km northeast of Winnipeg. Limestone from Moosehorn, 160 km northwest of Winnipeg and from Mafeking, 40 km east of the Saskatchewan border and 160 km south of The Pas, is transported to Manitoba and Saskatchewan centres for use in the metallurgical, chemical, agricultural and construction industries.

The eastern ranges of the Rocky Mountains contain limestone spanning the geologic ages from Cambrian to Triassic, with major deposits in the Devonian and Carboniferous periods in which a wide variety of types occur[16]. In southwestern Alberta, high-calcium limestone is mined at Exshaw, Kananaskis and Crowsnest, chiefly for the production of cement and lime, for metallurgical and chemical uses and for use as a crushed stone. Similar uses are made of limestone quarried at Cadomin, near Jasper[6].

In British Columbia large volumes of limestone are mined each year for cement and lime manufacture, for use by the pulp and paper industry and for various construction applications[6]. Deposits on Aristazabal Island have been developed for the export market. Other operations at Terrace, Clinton, Westwold, Popkum, Dahl Lake, Doeye River and Cobble Hill produced stone for construction and for filler use[17]. Periodically, interest is revived in the possible use of travertine from a British Columbia source.

Granite. In Manitoba, at Lac du Bonnet northeast of Winnipeg, a durable, red granite is quarried for building and monument use. Grey granite located east of Winnipeg near the Ontario border is a potential source of building stone.

In British Columbia a light-grey, to blue-grey even-grained granodiorite of medium texture is available from Nelson Island. An andesite has been quarried at Haddington Island, off the northeast coast of Vancouver Island, for use as a building stone. Babette Lake Quartzite Products Ltd. produces blocks of massive pink quartzite to make cut and polished facing stone.

Sandstone. Sandstone for building and ornamental uses, quarried near Banff, Alberta is hard, fine-grained, medium-grey and is referred to as "Rundal Stone".

USES

Limestones are widely distributed in Canada and generally are available in sufficient quantity and with such chemical or physical specifications that long transportation hauls are unnecessary. Limestone products are low-priced commodities and only rarely, when a market exists for a high-quality, specialized product such as white portland cement or a high-purity extender, are they beneficiated or moved long distances. Provided the specifications are met, the nearest source is usually considered, regardless of provincial or national boundaries.

Some major uses in the chemical field are: neutralization of acid waste liquors; extraction of aluminium oxide from bauxite; manufacture of soda ash, calcium carbide, calcium nitrate and carbon dioxide; in pharmaceuticals; as a disinfectant; in the manufacture of dyes, rayons, paper, sugar and glass; and in the treatment of water. Dolomitic limestone is used in the production of magnesium chloride and other magnesium compounds.

Agricultural limestone is used to control soil acidity and to add calcium and magnesium to the soil. Limestone and lime are used as soil stabilizers, particularly on highway construction projects.

Dolomite is the source of magnesium metal produced at Haley, Ontario; the company also uses a high-calcium lime from southeastern Ontario in the production of calcium metal. Dead-burned dolomitic limestone for use as a refractory is produced at Dundas, Ontario, by Steetley Industries Limited.

As a dimension stone, granite is processed for interior and exterior floor- and wall-covering, modular block panelling and for monument stone. Uniformity of colour and texture, and durability are the main features sought. Quarrying must take into account geological and structural features as well as topography and accessibility.

OUTLOOK

Dimension stone has been the subject of periodic surges of interest in past years. Currently the industry, especially in Quebec, is in a period of significant growth. Completion of intensive modernization has permitted producers to offer high-quality finished products at competitive prices. Markets for building stone are still under pressure from competitive substitutes such as steel, concrete, glass and ceramics. However, for aesthetic reasons and particular physical characteristics, the demand for granite dimension stone is likely to expand as new markets are developed and producers increase capacity. Efforts have been made on behalf of the industry to illustrate to contractors and architects the availability of a wide range of Canadian building stones and their adaptability in modern building design.

There is justifiable concern for the future development, operation, and rehabilitation of pits and quarries in all locations, especially in and near areas of urban development. Rehabilitation of stone quarries for subsequent land use is generally more difficult and costly than rehabilitation of gravel pits. Although an open-pit mining operation close to residential areas is seldom desirable, nonrenewable mineral resources must be fully and wisely utilized. When urban sprawl has been unexpectedly rapid, conflicts for land use can materialize and potential sources of raw mineral materials for the construction industry can be overrun. Master plans for land use are required to coordinate all phases of development so that mineral exploitation is part of the urban growth pattern.

REFERENCES

1. Parks, W.A., Building and Ornamental Stones of Canada, Canada Department of Mines, Mines Branch, Ottawa, Nos. 100, 203, 279, 388 and 452, Volume 1 (1912) to Volume V (1971) OUT OF PRINT.

2. Goudge, M.F., Limestones of Canada, Canada Department of Mines, Mines Branch, Ottawa, Nos. 733, 742, 755, 781, 811 Part 1. 1934 to Part V, 1946 OUT OF PRINT.

3. DeGrace, John R., Limestone Resources of Newfoundland and Labrador, Department of Mines and Energy, Mineral Development Division, St. John's, Newfoundland, Report 74-2, 1974.

4. Shea, F.S., Murray, D.A., Limestones and Dolomites of Nova Scotia, Department of Mines, Halifax, N.S., Part I, Bulletin No. 2, 1967 and Part II No. 2, 1975.

5. Hamilton, J.B., Limestone in New Brunswick, Department of Natural Resources, Mineral Resources Branch, Fredericton, N.B., Mineral Resources Report No. 2, 1965.

6. Stonehouse, D.H., Cement, Canadian Minerals Yearbook, 1983-1984, Department of Energy, Mines and Resources, Mineral Policy Sector, Ottawa.

7. Carr, G.F., The Granite Industry of Canada, Canada Department of Mines and Technical Surveys, Mines Branch, Ottawa, Ontario, No. 846, 1955.

8. Nantel, S., Dimension Stone of Quebec: Geological Aspects of Commercial Granite Deposits; Ministère de l'Energie et des Ressources du Québec, 1983.

9. Ontario Department of Mines, Toronto, Industrial Mineral Circular No. 5, 1960.

10. Hewitt, D.F., Vos. M.A., The Limestone Industries of Ontario, Ontario Ministry of Natural Resources Division of Mines, Toronto, Industry Mineral Report No. 39, 1972.

11. Hewitt, D.F., Building Stones of Ontario, Part III, Marble, Ontario Department of Mines, Toronto, Industrial Mineral Report No. 16, 1964.

12. Hewitt, D.F., Building Stones of Ontario, Part V, Granite and Gneiss, Ontario Department of Mines, Toronto; Industrial Mineral Report No. 19, 1964.

13. Vos M.A., Smith, B.A., Stevenato, R.J., Industrial Minerals of the Sudbury Area, Ontario Geological Survey, Open File Report No. 5329, 1981, 156p.

14. Hewitt, D.F., Building Stone of Ontario, Part IV, Sandstone, Ontario Department of Mines, Toronto, Industrial Mineral Report No. 17, 1964.

15. Bannatyne, B.B., High-Calcium Limestone deposits of Manitoba, Manitoba Department of Mines, Resources and Environmental Management, Mineral Resources Division, Exploration and Geological Survey Branch, Winnipeg, Publication 75-1, 1975.

16. Holter, M.E., Limestones Resources of Alberta, Transactions, Canadian Institute of Mining and Metallurgy, Bull. V.76, 1971.

17. McCammon, J.W., Sadar, E., Robinson, W.C., Robinson, J.W., Geology Exploration and Mining in British Columbia, 1974, British Columbia Department of Mines and Petroleum Resources.

TARIFFS

Item No.		British Preferential	Most Favoured Nation	General	General Preferential
			(%)		
CANADA					
29635-1	Limestone, not further processed than crushed or screened	free	free	25	free
30500-1	Flagstone, sandstone and all building stone, not hammered, sawn or chiselled	free	free	20	free
30505-1	Marble, rough, not hammered or chiselled	free	free	20	free
30510-1	Granite, rough, not hammered or chiselled	free	free	20	free
30515-1	Marble, sawn or sand rubbed, not polished	free	4.5	35	free
30520-1	Granite, sawn	free	6.5	35	free
30525-1	Paving blocks of stone	free	6.5	35	free
30530-1	Flagstone and building stone, other than marble or granite, sawn on not more than two sides	free	6.5	35	free
30605-1	Building stone, other than marble or granite, sawn on more than two sides but not sawn on more than four sides	5	6.5	10	4.5
30610-1	Building stone, other than marble or granite, planed, turned, cut or further manufactured than sawn on four sides	7.5	10.3	15	6.5
30615-1	Marble, not further manufactured than sawn, when imported by manufacturers of tombstones to be used exclusively in the manufacture of such articles, in their own factories	free	free	20	free
30700-1	Marble, nop	11.1	11.1*	40	7.0
30705-1	Manufactures of marble, nop	11.1	11.1*	40	7.0
30710-1	Granite, nop	14.8	13.9	40	9.0
30715-1	Manufactures of granite, nop	14.8	13.9	40	9.5
30800-1	Manufactures of stone, nop	15.6	15.0	35	10.0
30900-1	Roofing slate, per square of 100 square feet	free	free	75¢	free
30905-1	Granules, whether or not coloured or coated, for use in manufacture of roofing, including shingles and siding	free	free	25	free

MFN Reductions under GATT (effective January 1 of year given)	1983	1984	1985	1986	1987
			(%)		
30515-1	4.5	4.4	4.3	4.1	4.0
30520-1	6.5	6.3	6.0	5.8	5.5
30525-1	6.5	6.3	6.0	5.8	5.5
30530-1	6.5	6.3	6.0	5.8	5.5
30605-1	6.5	6.3	6.0	5.8	5.5
30610-1	10.3	9.7	9.1	8.6	8.0
30700-1	13.3	12.2	11.1	10.1	9.0
30705-1	13.3	12.2	11.1	10.1	9.0
30710-1	13.9	12.9	12.0	11.1	10.2
30715-1	13.9	12.9	12.0	11.1	10.2
30800-1	15.0	14.4	13.8	13.1	12.5

TARIFFS (cont'd)

Item No.		British Preferential	Most Favoured Nation	General	General Preferential
			(%)		
UNITED STATES (MFN)					
513.71	Granite, suitable for use as monumental, paving or building stone: Not pitched, not lined, not pointed, not hewn, not sawed, not dressed, not polished, and not otherwise manufactured		Free		

		1983	1984	1985	1986	1987
				(%)		
513.74	Pitched, lined, pointed, hewn, sawed, dressed, polished, or otherwise manufactured	5.1	4.9	4.7	4.4	4.2
	Limestone, suitable for use as monumental, paving or building stone:					
514.21	Not hewn, not sawed, not dressed, not polished, and not otherwise manufactured, per cubic foot	0.5¢	0.4¢	0.2¢	0.1¢	free
514.24	Hewn, sawed, dressed, polished, or otherwise manufactured	8.3	7.7	7.1	6.6	6.0
514.51	Marble, breccia, in block, rough or squared only, per cubic foot	12.7¢	12.6¢	12.4¢	12.2¢	12.0¢
514.57	Marble, breccia, or onyx, sawed or dressed, over 2 inches thick, per cubic foot	22.5¢	21.9¢	21.2¢	20.6¢	20.0¢
	Stone suitable for use as monumental, paving, or building stone:					
515.51	Not hewn, not sawed, not dressed, not polished, and not otherwise manufactured, per cubic foot	0.5¢	0.4¢	0.2¢	0.1¢	free
515.54	Hewn, sawed, dressed, polished, or otherwise manufactured, per cubic foot	8.3¢	7.7¢	7.1¢	6.6¢	6.0¢

Sources: The Customs Tariff, 1983, Revenue Canada, Customs and Excise; Tariff Schedules of the United States Annotated (1983) USITC Publication 1317; U.S. Federal Register, Vol. 44, No. 241.
* Temporary rate reduction until December 31, 1984.

TABLE 1. CANADA, TOTAL PRODUCTION OF STONE, 1982-84

	1982		1983		1984P	
	(000 t)	($000)	(000 t)	($000)	(000 t)	($000)
By province						
Newfoundland	357	1,763	279	1,431	415	1,608
Nova Scotia	679	4,638	1 296	7,784	1 510	9,400
New Brunswick	2 261	11,556	2 087	11,310	2 005	10,940
Quebec	25 060	106,989	27 303	121,154	28 237	124,581
Ontario	23 582	100,278	27 843	122,272	29 500	131,335
Manitoba	2 345	11,670	1 137	5,452	1 675	9,300
Alberta	264	3,161	286	3,457	300	3,275
British Columbia	4 310	21,926	4 915	27,083	4 885	27,500
Northwest Terrtories	323	1,268	2 409	14,601	2 420	15,750
Canada	59 181	263,249	67 555	314,544	71 047	333,689
By use[1]						
Building stone						
Rough	230	4,828	..	..	..	..
Monumental and ornamental stone	38	4,002	..	..	..	..
Other (flagstone, curbstone, paving blocks, etc.)	26	1,027	..	..	..	..
Chemical and metallurgical						
Cement plants, foreign	598	1,461	..	..	..	..
Lining, open-hearth furnaces	38	141	..	..	..	..
Flux in iron and steel furnaces	742	2,861	..	..	..	..
Flux in nonferrous smelters	114	1,126	..	..	..	..
Glass factories	169	2,271	..	..	..	..
Lime kilns, foreign	512	1,903	..	..	..	..
Pulp and paper mills	295	2,706	..	..	..	..
Sugar refineries	108	586	..	..	..	..
Other chemical uses	137	2,840	..	..	..	..
Pulverized stone						
Whiting (substitute)	71	2,863	..	..	..	..
Asphalt filler	41	238	..	..	..	..
Dusting, coal mines	7	171	..	..	..	..
Agricultural purposes and fertilizer plants	1 037	10,562	..	..	..	..
Other uses	687	2,153	..	..	..	..
Crushed stone for						
Manufacture of artificial stone	7	154	..	..	..	..
Roofing granules	253	16,776	..	..	..	..
Poultry grit	28	721	..	..	..	..
Stucco dash	15	993	..	..	..	..
Terrazzo chips	3	184	..	..	..	..
Rock wool	-	-	..	..	..	..
Rubble and riprap	1 730	6,421	..	..	..	..
Concrete aggregate	4 671	17,571	..	..	..	..
Asphalt aggregate	4 540	17,766	..	..	..	..
Road metal	17 997	62,795	..	..	..	..
Railroad ballast	2 626	12,823	..	..	..	..
Other uses	22 461	80,318	..	..	..	..
Total	59 181	258,261	..	..	..	..

[1] The 1982 value of production includes companies' transportation costs not applicable in the by use category.
P Preliminary; .. Not available; - Nil.

TABLE 2. CANADA, PRODUCTION OF LIMESTONE, 1981 AND 1982

	1981		1982	
	(000 t)	($000)	(000 t)	($000)
By province				
Newfoundland	338	1,223	226	1,098
Nova Scotia	213	2,088	124	1,818
New Brunswick	546	4,565	546	5,178
Quebec	23 155	83,221	19 819	78,663
Ontario	27 889	86,620	21 893	75,284
Manitoba	863	2,817	1 922	7,748
Alberta	271	2,001	262	3,124
British Columbia	2 503	10,611	2 183	10,299
Northwest Territories	..	..	322	1,266
Canada	55 778	193,146	47 297	184,478
By use[1]				
Building stone				
Rough	293	1,428	157	1,360
Monumental and ornamental	1	72	1	51
Other (flagstone, curbstone, paving blocks, etc.)	8	202	10	298
Chemical and metallurgical				
Cement plants, foreign	1 584	2,999	598	1,461
Lining, open-hearth furnaces	20	71	38	141
Flux, iron and steel furnaces	757	2,779	742	2,861
Flux, nonferrous smelters	151	1,337	114	1,124
Glass factories	188	2,370	169	2,272
Lime kilns, foreign	303	1,239	512	1,903
Pulp and paper mills	345	2,886	286	2,590
Sugar refineries	79	378	108	586
Other chemical uses	148	2,277	137	2,840
Pulverized stone				
Whiting substitute	35	1,812	71	2,863
Asphalt filler	34	158	31	202
Dusting, coal mines	8	167	7	171
Agricultural purposes and fertilizer plants	1 020	9,029	1 018	10,293
Other uses	550	466	485	610
Crushed stone for				
Artificial stone	30	123	1	37
Roofing granules	30	312	36	274
Poultry grit	24	726	28	698
Stucco dash	20	1,288	15	993
Rock wool	1	23	–	–
Rubble and riprap	471	1,447	795	2,350
Concrete aggregate	6 038	21,008	4 226	15,162
Asphalt aggregate	3 561	12,795	3 439	12,416
Road metal	18 108	58,906	14 953	50,454
Railroad ballast	999	3,192	1 124	3,759
Other uses	20 972	63,656	18 196	63,399
Total	55 778	193,146	47 297	181,168

[1] The 1982 value of production includes companies' transportation costs not applicable in the by use category.
– Nil; .. Not available.

TABLE 3. CANADA, PRODUCTION OF MARBLE, 1981 AND 1982

	1981		1982	
	(000 t)	($000)	(000 t)	($000)
By province				
Quebec	310	1,881	332	2,189
Ontario	4	135	153	1,028
Canada	314	2,016	485	3,217
By use[1]				
Building stone				
Rough	2	111	-	-
Chemical process stone				
Flux in nonferrous smelters	--	1	--	1
Pulp and paper mills	8	105	8	114
Pulverized stone				
Agricultural purposes and fertilizer plants	11	162	18	269
Other uses	46	507	202	1,543
Crushed stone for				
Artificial stone	7	117	6	117
Roofing granules	2	50	1	32
Poultry grit	--	1	--	1
Stucco dash	--	3	-	-
Terrazzo chips	2	51	4	184
Concrete aggregate	31	184	30	176
Road metal	51	172	125	400
Other uses	153	552	91	363
Total	314	2,016	485	3,200

[1] The 1982 value of production includes companies' transportation costs not applicable in the by use category.
- Nil; -- Amount too small to be expressed.

TABLE 4. CANADA, PRODUCTION OF GRANITE, 1981 AND 1982

	1981		1982	
	(000 t)	($000)	(000 t)	($000)
By province				
Newfoundland	71	369	51	304
Nova Scotia	1	21	--	42
New Brunswick	1 967	5,773	1 536	6,001
Quebec	19 784	62,314	3 815	20,735
Ontario	2 666	29,850	1 480	23,688
Manitoba	982	7,035	423	3,922
Alberta	-	-	1	10
British Columbia	2 541	10,056	2 127	11,607
Northwest Territories	..	..	--	1
Canada	28 012	115,418	9 434	66,310
By use[1]				
Building stone				
Rough	59	3,584	27	2,652
Monumental and ornamental	26	3,131	37	3,952
Other (flagstone, curbstone, paving blocks, etc.)	13	573	6	415
Pulverized stone				
Asphalt filler	7	18	11	37
Crushed stone for				
Roofing granules	234	15,569	215	16,471
Poultry grit	--	18	1	22
Rubble and riprap	10 734	24,151	897	4,001
Concrete aggregate	479	2,183	280	1,453
Asphalt aggregate	844	3,329	898	4,498
Road metal	2 729	8,918	2 485	10,438
Railroad ballast	4 482	26,412	1 486	8,949
Other uses	8 405	27,532	3 091	12,779
Total	28 012	115,418	9 434	65,667

[1] The 1982 value of production includes companies' transportation costs not applicable in the by use category.
- Nil; -- Amount too small to be expressed; .. Not available.

TABLE 5. CANADA, PRODUCTION OF SANDSTONE, 1981 AND 1982

	1981		1982	
	(000 t)	($000)	(000 t)	($000)
By province				
Newfoundland	109	482	80	361
Nova Scotia	612	2,136	554	2,778
New Brunswick	174	326	179	376
Quebec	1 276	6,132	840	4,508
Ontario	4	234	32	259
Alberta	--	16	--	28
British Columbia	-	-	--	20
Canada	2 176	9,326	1 686	8,330
By use[1]				
Building stone				
Rough	22	924	46	816
Monumental and ornamental	--	4	-	-
Other (flagstone, curbstone, paving blocks, etc.)	12	359	10	313
Crushed stone for				
Rubble and riprap	70	164	37	69
Concrete aggregate	190	955	135	780
Asphalt aggregate	145	637	203	852
Road metal	503	1,932	235	964
Railroad ballast	46	341	16	114
Other uses	1 188	4,010	1 004	3,659
Total	2 176	9,326	1 686	7,567

[1] The 1982 value of production includes companies' transportation costs not applicable in the by use category.
- Nil; -- Amount too small to be expressed.

TABLE 6. CANADA, PRODUCTION OF SHALE, 1981 AND 1982

	1981		1982	
	(000 t)	($000)	(000 t)	($000)
By province				
Quebec	436	1,048	254	894
Ontario	144	92	25	19
Northwest Territories	-	-	--	1
Canada	580	1,140	279	914
By use[1]				
Crushed stone for				
Rubble and riprap	-	-	1	1
Road metal	358	893	200	539
Other uses	222	247	78	119
Total	580	1,140	279	659

[1] The 1982 value of production includes companies' transportation costs not applicable in the by use category.
- Nil, -- Amount too small to be expressed.

TABLE 7. CANADA, PRODUCTION OF STONE BY TYPES, 1975, 1980-82

	1975		1980		1981		1982	
	(000 t)	($000)	(000 t)	($000)	(000 t)	($000)	(000 t)	($000)
Granite	11 470	34,913	39 983	140,914	28 012	115,418	9 434	66,310
Limestone	72 284	152,521	58 191	185,085	55 778	193,146	47 297	184,478
Marble	356	1,843	316	1,807	314	2,016	485	3,217
Sandstone	3 753	10,881	3 064	11,540	2 176	9,326	1 686	8,330
Shale	1 551	2,566	1 812	1,810	580	1,140	279	914
Total	89 414	202,724	103 366	341,156	86 860	321,046	59 181	263,249

Sources: Energy, Mines and Resources Canada; Statistics Canada.

TABLE 8. CANADA, STONE EXPORTS AND IMPORTS, 1981-83

	1981		1982		1983P	
	(tonnes)	($000)	(tonnes)	($000)	(tonnes)	($000)
Exports						
Building stone, rough	11 183r	1,222	2 942	576	12 633	1,877
Stone crude, nes	116 782	1,693	16 170	559	45 779	707
Natural stone, basic products	..	10,359	..	19,603	..	22,987
Total	..	13,274	..	20,738	..	25,571
Imports						
Building stone, rough	11 086	1,010r	11 862	890	8 049	1,177
Stone crude, nes	7 233	952	4 180	470	3 263	353
Granite, rough	34 278	4,802	22 033	4,095	24 760	4,447
Marble, rough	7 485	3,053	7 058	3,282	8 251	3,375
Shaped or dressed granite	..	3,880	..	14,831	..	6,952
Shaped or dressed marble	..	2,119	..	1,709	..	2,445
Natural stone basic products	..	3,590	..	3,576	..	4,328
Total	..	19,406r	..	28,853	..	23,077

Source: Statistics Canada.
P Preliminary; nes not elsewhere specified; .. Not available; r Revised.

TABLE 9. CANADA, VALUE OF CONSTRUCTION[1] BY PROVINCE, 1982-84

	1982			1983			1984		
	Building Construction	Engineering Construction	Total	Building Construction	Engineering Construction	Total	Building Construction	Engineering Construction	Total
				($000)					
Newfoundland	414,429	750,073	1,164,502	496,177	920,309	1,416,486	529,042	904,131	1,433,173
Nova Scotia	681,430	884,462	1,565,892	850,097	1,113,145	1,963,242	935,167	1,263,679	2,198,846
New Brunswick	619,611	462,089	1,081,700	749,843	414,249	1,164,092	866,945	503,785	1,370,730
Prince Edward Island	86,981	72,006	158,987	106,406	70,694	177,100	117,272	79,046	196,318
Quebec	5,547,556	4,672,040	10,219,596	6,693,708	4,388,346	11,082,054	7,183,496	4,352,134	11,535,630
Ontario	8,897,137	5,510,574	14,407,711	10,015,802	4,819,861	14,835,663	10,498,275	5,031,231	15,529,506
Manitoba	764,362	657,850	1,422,212	986,418	656,087	1,642,505	1,083,361	698,296	1,781,657
Saskatchewan	1,165,189	1,343,933	2,509,122	1,451,012	1,413,370	2,864,382	1,410,011	1,515,541	2,925,552
Alberta	6,053,165	8,349,406	14,402,571	4,761,621	7,044,529	11,806,150	3,920,440	7,281,719	11,202,159
British Columbia Yukon and Northwest Territories	4,613,640	4,519,456	9,133,096	4,488,816	4,657,297	9,146,113	4,574,138	4,223,386	8,797,524
Canada	28,843,500	27,221,889	56,065,389	30,599,900	25,497,887	56,097,787	31,118,147	25,852,948	56,971,095

Source: Statistics Canada.
[1] Actual expenditures 1982, preliminary actual 1983, intentions 1984.

Sulphur

MICHEL A. BOUCHER

SUMMARY

World consumption of sulphur increased in 1983 and 1984 after a sharp drop in the United States and a smaller decline in western Europe in 1982.

The recovery in sulphur came mainly from the phosphate fertilizer industry which accounts for slightly over half of total consumption, and from improvements in the economy of most industrial nations over the recessionary years of 1981 and 1982.

World production of sulfur continued to decline in 1983 for the third consecutive year but western world production which represents about two thirds of the total, increased substantially in 1984, mainly in North America and the Arabian Gulf.

The rise in consumption was not accompanied by a corresponding rise in production in recent years and as a result stocks of sulphur were reduced in several countries and prices rose considerably in the second half of 1984.

CANADIAN DEVELOPMENTS

Canada's sulphur shipments in 1983 and 1984 were respectively 7.31 million t and 8.57 million t valued at $469 million and $637 million.

Exports in 1983 were 5.67 million t valued at $572 million. Preliminary estimates (based on 9 months figures) for 1984 indicate that exports increased to 6.93 million t valued at $741 million. The increase was mainly due to higher exports to the United States and to some developing countries of South America and Africa.

Western Canada

The Alberta sulphur industry accounts for between 85 to 90 per cent of total Canadian production.

Sulphur production in Alberta peaked in 1973 at 7 million t, and has since been declining, mainly for two reasons: (i) sour gas pools have been generally replaced by sweeter gas pools; (ii) natural gas production and sales have been stagnant since 1973. Production of sulphur in Alberta however, has always exceeded Canadian sales during the 1970s, and as a result inventories have accumulated and reached a peak of 21.3 million t in 1980.

Sales have exceeded production since 1980 and inventories have been drawn down at a rate averaging 2.15 million tpy. At this rate inventories should be depleted in the late 1980s or early 1990s unless gas production in Canada increases substantially or high sulphur gas fields in southwestern Alberta are developed.

There were three additions to sulphur recovery capacity in Alberta during 1983. The Shell Canada Resources Limited sour gas plant at Jumping Pound, which was the first sour gas plant in Canada to recover sulphur when it went into operation in 1952, was expanded to 570 tpd from 510 tpd sulphur; Gulf Canada Resources Inc.'s 1 147 tpd sulphur plant on the Hanlan Robb reservoir came on-stream in March; Chieftain Development Co. Ltd. and Texaco Canada Resources Ltd. brought their 200 tpd sulphur plant at Hythe to the operational stage.

Procor Limited constructed a 500 tpd sulphur granulation facility at the Amoco Canada Petroleum Company Ltd. plant at East Crossfield during the year.

Shell Canada Limited closed its oil refinery and sulphur plant at Oakville, Ontario, during the summer of 1983, while a sulphur recovery unit at the adjacent Petro-Canada refinery was commissioned.

In early 1984, the Alberta Energy Resources Conservation Board gave approval

---

Michel A. Boucher is with the Mineral Policy Sector, Energy, Mines and Resources, Canada. Telephone (613) 995-9466.

to Shell Canada Resources Ltd. for the construction of a pipeline to move sour gas from its Moose Mountain/Whiskey Creek fields to Esso Resources Canada Ltd.'s gas processing facility at Quirk Creek. Initial deliveries of sour gas to Quirk Creek, on completion of the pipeline in 1985, are expected to result in the recovery of 40 000 tpy of sulphur.

Husky Oil Co. announced the planned construction of a Canadian $3.2 billion heavy oil upgrading plant in Saskatchewan with a capacity of around 6 700 m^3/d. Construction is expected to begin in 1986. Plans include a 250 tpd sulphur recovery unit. Canadian Occidental Petroleum Co. received approval from the Alberta Energy Resources Conservation Board (AERCB) and the Ministry of Environment to construct a 1.7 million m^3/d sour gas processing plant which will incorporate a 200 000 tpy sulphur recovery unit to be built at Mazeppa, southeast of Calgary. Production is expected to start in mid-1986.

Canterra Energy Ltd. announced the purchase of a fourteen-section block of land at the north end of the Okotoks gas field. The land is underlain with proven sour gas reserves in the Crossfield and Elkton zones, estimated to be capable of yielding 0.6 billion m^3 of gas and 290 000 t of saleable sulphur. Canterra recently received approval to tie production from this land into its existing Okotoks gas gathering system where a 152 000 tpy sulphur recovery unit is located. Dome Petroleum started production of its 300 tpd sulphur unit on the West Pembina field. The Petro Canada Inc. plant in the Brazeau River field is expected to come on-stream in 1985 with a sulphur production capacity of 78 tpd. The Shell Canada plant in the Progress field is also due to come on-stream in 1985 with a sulphur production capacity of 24 tpd. Western Co-operative Fertilizers Limited's sulphuric acid plant at Medicine Hat, Alberta was closed indefinitely in mid-1983.

As a resumé of these developments, some 1 400 tpd sulphur production capacity was added in 1983, another 300 tpd was added in 1984, and some 215 tpd are planned for 1985.

Other Developments

Several refineries remained idle including: Newfoundland's Come-by-Chance Refinery Co. Ltd.; Gulf Canada Ltd.'s Port Tupper in Nova Scotia; Texaco Canada Resources Ltd. in Calgary; and Shell Canada Resources Ltd. in Shelburn, B.C.

Allied Corporation closed its 140 000 tpy sulphuric acid plant at Valleyfield, Quebec in June 1983 while Canadian Electrolytic Zinc expanded its sulphuric acid production capacity from 210 000 to 480 000 tpy at Valleyfield, Quebec. In 1984 the Federal Government agreed to allow oil and gas companies to negotiate natural gas export prices with buyers instead of adhering to government-posted prices. It is expected that the new measure will result in lower export prices, increased sales of natural gas to the United States and as a consequence an increase in sulphur production. Aberford Resources Ltd. and Sulpak Resources Ltd. joined Cansulex in 1984.

Sales into the offshore market in Canada are handled mainly by Cansulex (an industry marketing agency representing 20 producing companies and handling about 45 per cent of Canadian offshore sales), Shell, Canadian Superior and Amoco.

Cansulex member companies are shown below:

CANSULEX LIMITED (member companies)

Aberford Resources Ltd.
BP Exploration Canada Ltd.
Canadian Occidental Petroleum Ltd.
Canadian Reserve Oil & Gas Ltd.
Canterra Energy Ltd.
    Aquitaine Co. of Canada Ltd.
    CDC Oil and Gas Ltd.
    Texasgulf Inc.
Champlin Canada Ltd.
Chevron Canada Resources Limited
Dome Petroleum Ltd.
Gulf Canada Resources Inc.
Hamilton Brothers Canadian Gas Co. Ltd.
Home Oil Co. Ltd.
Hudson's Bay Oil and Gas Co. Ltd.
Husky Oil Canada, Ltd.
Interedec (USA) Inc.
Mobil Oil Canada, Ltd.
Norcen Energy Resources Ltd.
The Paddon Hughes Development Co. Ltd.
Petrogas Processing Ltd.
Sulfak Resources Ltd.
<u>Union Oil Co. of Canada Ltd.</u>
Source: CANSULEX

WORLD DEVELOPMENTS

In the United States the Frasch mines operating at about 50 per cent of capacity in

1983 and one mine, at Comanche Creek, closed at the end of November. This followed the three Frasch operations that closed in 1982 as a result of rising costs and insufficient prices. In early 1984 the Caillou Island mine owned by Freeport was closed leaving only four Frasch mines in the United States. However in view of strong demand for sulphur Frasch production was reported to have increased substantially in 1984 from their 1983 low level of 3.2 million, but stocks continued to be depleted. Chevron Canada Limited opened a 34 km molten sulphur pipeline to a railhead from its new gas plant in the Overthrust region of Wyoming.

Saudi Arabian sulphur had the greatest impact on world markets in 1983 as most of the 1.4 million t of sulphur stockpiled before 1982 was put on the market. Sales at 1.5 million t were about double the 1982 level.

The U.S.S.R. reportedly increased fertilizer production by about 10 per cent in 1983 and relied on Poland and Canada to supply additional sulphur. The contract for construction of sulphur recovery units at the Astrakhan gas field was awarded to France's Technip. The recovery units will have a combined capacity of about 3.0 million tpy. The plant, to be completed in 1986, will process gas containing an average 24 per cent H_2S.

In Poland, the Oziek mine planned for development in the lat 1980s could add 1.2 million tpy of sulphur.

Polish exports to western countries increased in 1983, mainly as a result of increased imports by Brazil and Tunisia.

In November 1983, a 2 528 000 tpy sulphur recovery plant was brought on-stream at Kirkak for Iraq National Oil Co. Ltd. Production in 1984 was reported to have been only 70 000 t.

In 1984, Azufrera Panamericana SA, Mexico's state-owned sulphur producer, brought on-stream a new Frasch sulphur mine at Petapa, some 30 km west of the port of Coatzacoalas where all sulphur export facilities are located. The mine has a production capacity of 1 000 tpd. By mid-1985, Azufrera plans to open another mine at Otapa with a production capacity of 1 500 tpd. Most of the production at these mines is expected to be consumed internally by Fertimex, the state-owned fertilizer company.

Two 260 tpd sulphur recovery units came on-stream in June 1984 at Mina-al-Ahmadi in Kuwait. In 1987, a 1 000 tpd recovery unit is also expected to come on-stream at the Mina Abdulla refinery.

In Saudi Arabia the Shell/Petromin plant at Inbail on the Persian Gulf is due to come on-stream in early 1985 and produce some 80 000 tpy of sulphur.

Also, the Mobil refinery at Yanbu on the Red Sea is expected to recover 80 000 tpy to 100 000 tpy of sulphur which will be prilled and exported to Jordan. The Petrola refinery at Rabigh is expected to produce 100 000 tpy of sulphur by 1986.

In Iran a gas refinery at the Khangiran gas field near Sarakhs is planned for completion in 1986. The refinery will be able to recover 485 000 tpy sulphur.

PRICES

Contract prices for offshore exports of elemental sulphur fob Vancouver were $Cdn. 118 for the first half of 1983 and down to $104 for the second half of the year.

In 1984 prices increased to $110 for the first half and to $140 for the second half of the year. The major reasons for the sharp increase in the second half of 1984 were as follows: Canadian, Mexican and Polish production were all sold out in early 1984; Canada was the only country in a position to respond to increased demand by remelting sulphur from its stockpile, which commands a premium; remaining inventories of Saudi Arabia were all committed; the war between Iran and Iraq caused disruptions in exports; and increased demand for sulphur particularly in the United States.

Within North America the contract price of sulphur in 1983 varied between $59-62 a t fob Alberta plant. In 1984 price increased steadily and reached $72.00 a t at year-end.

USES

Sulphur, principally in the form of sulphuric acid, is used at some stage in the production of virtually everything we eat, wear or use. As such, its consumption level traditionally has served as an indicator of the state of the economy of an individual nation or of the world. Close to 60 per cent of all sulphur is consumed in the production of phosphate and ammonium sulphate fertilizers.

OUTLOOK

Short-term

The tight supply-demand situation that existed in the sulphur industry in 1984 is expected to continue in 1985 because world consumption should again exceed production as no new major producer is expected to come on-stream. As a result sulphur inventories in Canada should continue to decline.

Prices will likely remain high in 1985. It was reported, however, that in 1984 many fertilizer producers which require approximately half a tonne of sulphur to produce one tonne of ammonium phosphate fertilizer were barely breaking even.

Long-term

On the supply side the major development in the world will be the Astrakhan project in the U.S.S.R. which should add some 6 million tpy of sulphur during the period 1987-1992; most of the new production however is expected to "eventually" be consumed internally.

New projects in the western world such as in Mexico, Saudi Arabia, Kuwait, Iran and Poland will not add significantly to production because they will replace mines that are being depleted in Poland (Grzybow Frasch mine), France (Lacq gas field), and Mexico (Joltipan Frasch mine) among others. In the United States only minor increases in production are forecasted to 1990.

Natural gas price deregulations in Canada are expected to result in increased production and sales of gas to the United States, and consequently higher sulphur production.

Recovery of sulphur from smelters and power plants in North America and western Europe may add 1 to 3 million t to production by the end of the decade.

On the consumption side growth is expected to outpace production for the next three to four years. Strong growth is expected mainly in Asia (China, Israel, Iraq), Africa (Morocco, Tunisia), South America (Brazil), and eastern Europe (U.S.S.R., Yugoslavia).

The imbalance in production and consumption should result in: a reduction of inventories mainly in Canada but also in other producing countries; higher prices; a displacement of shipments to export markets where prices are higher; and increased production from voluntary (Frasch, native and pyrite mines) producers.

PRICES

	1982	1983
	($/tonne)	

Canadian sulphur prices as quoted in Alberta Energy Resources Industries monthly statistics

Sulphur elemental, fob plant
North American deliveries	64.36	52.64
Offshore deliveries	80.44	61.43

Canadian sulphuric acid price as quoted in Corpus Chemical Report

Sulphuric acid, fob plants, East, 66° Be, tanks	98.80	104.0

United States prices, U.S. currency, as quoted in Engineering and Mining Journal

Sulphur elemental
U.S. producers, term contracts fob vessel at Gulf ports, Louisiana and Texas
Bright	137.8	130.4
Dark	137.8	131.4

Export prices, ex terminal Holland
Bright	143.7-150.0	130.9-137.8
Dark	143.7-150.0	130.9-137.8

Mexican export, fob vessel, U.S. currency, from Azufrera Panamericanna S.A.
Bright	108.2-113.1	..
Dark	123.0-132.8	..

fob Free on board; .. Not available.

TARIFFS

Item No.		British Preferential	Most Favoured Nation	General	General Preferential
			(%)		
CANADA					
92503-1	Sulphur of all kinds, other than sublimed sulphur, precipitated sulphur and colloidal sulphur	free	free	free	free
92802-1	Sulphur, sublimed or precipitated; colloidal sulphur	free	free	free	free
92807-1	Sulphur dioxide	free	free	free	free
92808-1	Sulphuric acid, oleum	9.4	7.5	25	5
92813-4	Sulphur trioxide	free	free	free	free

		1983	1984	1985	1986	1987
MFN Reductions under GATT (effective January 1 of year given)				(%)		
92808-1		7.5	5.6	3.8	1.9	free

UNITED STATES

Item No.						
418.90	Pyrites		free			
415.45	Sulphur, elemental		free			
416.35	Sulphuric acid		free			

		1983	1984	1985	1986	1987
				(%)		
422.94	Sulphur dioxide	5.1	4.9	4.7	4.4	4.2

Sources: The Customs Tariff, 1983 Revenue Canada, Customs and Excise; Tariff Schedules of the United States Annotated (1983), USITC Publication 1317; U.S. Federal Register Vol. 44, No. 241.

57.6

TABLE 1. CANADA, SULPHUR SHIPMENTS AND TRADE, 1982-84

	1982		1983		1984P	
	(tonnes)	($000)	(tonnes)	($000)	(tonnes)	($000)
Shipments						
Pyrite and pyrrhotite[1]						
Gross weight	20 000	..	-	-	-	-
Sulphur content	9 000	220	-	-	-	-
Sulphur in smelter gases[2]	627 000	42,027	678 286	42,322	874 906	63,300
Elemental sulphur[3]	6 945 000	569,928	6 631 123	427,358	7 700 455	574,177
Total sulphur content	7 581 000	612,175	7 309 409	469,680	8 575 361	637,477
Imports					Jan.-Sept. 1984	
Sulphur, crude or refined						
United States	2 159	395	2 353	653	2 341	639
Other countries	-	-	12	3	4	1
Total	2 159	395	2 365	656	2 345	640
Sulphuric acid, including oleum						
United States	74 262	4,847	116 567	8,353	22 768	2,122
West Germany	74 405	2,480	7 484	248	3	-
Norway	22 390	913	2 522	172	-	-
Other countries	21 457	840	-	-	10	10
Total	192 514	9,080	126 573	8,952	22 781	2,132
Exports					(Jan.-Sept. 1984)	
Sulphur in ores (pyrite)						
West Germany	-	-	..	63	-	-
United States	..	239	..	14	..	34
Japan	-	429	-	-	-	-
Total	..	668	..	77	..	34
Sulphuric acid, including oleum						
United States	259 716	8,404	250 061	7,744	-	-
Other countries	24	20	11	3	22 984[4]	892
Total	259 740	8,424	250 612	7,747	22 984	892
Sulphur, crude or refined, nes						
United States	1 132 352	85,510	1 112 860	76,797	1 270 171	93,204
Brazil	447 439	60,621	573 145	66,071	323 855	40,797
Morocco	464 889	61,544	358 735	40,153	270 455	33,689
Tunisia	349 830	45,432	310 095	38,261	227 173	30,133
South Africa	453 336	59,748	366 640	38,079	394 812	44,322
Australia	467 268	58,234	364 448	37,878	327 473	37,781
South Korea	194 236	26,058	296 146	33,168	258 830	30,661
People's Republic of China	309 906	41,449	217 027	24,183	189 651	20,841
U.S.S.R.	201 086	27,628	198 075	21,890	136 800	15,596
Finland	96 758	13,303	179 732	21,880	124 572	14,254
India	373 239	43,172	218 170	18,298	150 185	18,783
Israel	152 773	10,802	226 517	18,092	192 462	18,211
Taiwan	125 922	17,638	152 825	16,248	140 741	15,949
Netherlands	53 603	7,223	149 151	15,672	142 787	15,543
France	51 029	7,022	122 754	15,167	71 561	8,577
New Zealand	208 410	25,864	158 021	15,114	163 291	19,500
Other countries[5]	1 029 368	128,581	665 934	75,329	816 384	98,104
Total	6 111 444	719,829	5 670 275	572,280	5 201 203	556,021

Source: Statistics Canada; Energy, Mines and Resources Canada.
[1] Producers' shipments of byproduct pyrite and pyrrhotite from the processing of metallic sulphide ores. [2] Sulphur in liquid SO_2 and H_2SO_4 recovered from the smelting of metallic sulphides and from the roasting of zinc-sulphide concentrates. [3] Producers' shipments of elemental sulphur produced from natural gas; also included are small quantities of sulphur produced in the refining of domestic crude oil and synthetic crude oil. [4] Netherlands and Switzerland.
[5] Mainly Belgium-Luxembourg, Senegal, Indonesia, Argentina, Chile, Cuba, and Mozambique.
P Preliminary; - Nil; .. Not available; nes Not elsewhere specified.

TABLE 2. CANADA, SOUR GAS SULPHUR EXTRACTION PLANTS, 1982 AND 1983

Operating Company	Source Field or Plant Location (Alberta, except where noted)	H_2S in Raw Gas (%)	1982 Daily Sulphur Capacity (tonnes)	1983 Daily Sulphur Capacity (tonnes)
Amerada Hess Corporation	Olds	13	384	389
Amoco Canada Petroleum	Bigstone Creek	19	382	382
Amoco Canada Petroleum	East Crossfield	26	1 757	1 797
Canada-Cities Service, Ltd.	Paddle River	1	19	19
Canadian Superior Oil Ltd.	Harmattan-Elkton	56	490	490
Canadian Superior Oil Ltd.	Lonepine Creek	12	157	157
Sulpetro Limited	Minnehik-Buck Lake	1	45	45
Canterra Energy Ltd.	Brazeau River	2	42	42
Canterra Energy Ltd.	Okotoks	34	459	431
Canterra Energy Ltd.	Rainbow Lake	4	139	139
Canterra Energy Ltd.	Ram River (Ricinus)	19	4 567	4 572
Canterra Energy Ltd.	Windfall	8		
Chevron Standard Limited	Kaybob South	20	3 521	3 537
Chevron Standard Limited	Nevis	7	260	215
Chieftain	Sinclair	5		256
Dome Petroleum Limited	Steelman, Sask.	1	7	7
Esso Resources Canada	Joffre	11	17	17
Esso Resources Canada	Quirk Creek	9	300	293
Esso Resources Canada	Redwater	4	33	33
Gulf Canada Limited	Homeglen-Rimbey	2	333	333
Gulf Canada Limited	Nevis	7	295	297
Gulf Canada Limited	Pincher Creek	5	160	159
Gulf Canada Limited	Strachan	9	943	943
Gulf Canada Limited	Hanlan	9		1 092
Home Oil Company Limited	Carstairs	1	72	65
Hudson's Bay Oil and Gas	Brazeau River	1	110	110
Hudson's Bay Oil and Gas	Caroline	1	22	8
Hudson's Bay Oil and Gas	Edson	2	284.5	284
Hudson's Bay Oil and Gas	Kaybob South (1)	13	1 064	1 086
Hudson's Bay Oil and Gas	Kaybob South (2)	17	1 064	1 085
Hudson's Bay Oil and Gas	Lonepine Creek	10	283	283
Hudson's Bay Oil and Gas	Sturgeon Lake	12	49	98
Hudson's Bay Oil and Gas	Zama	8	74	74
Mobil Oil Canada, Ltd.	Wimborne	14	168	182
Mobil Oil Canada, Ltd.	Teepee	4	29	30
PanCanadian Petroleum Limited	Morley	5	18	18
Petro-Canada	Gold Creek	5	43	43
Petro-Canada	Wildcat Hills	4	177	177
Petrogas Processing Ltd.	Crossfield (Balzac)	14	1 687	1 696
Saratoga Processing Company	Savannah Creek (Coleman)	20	389	389
Shell Canada Limited	Burnt Timber Creek	10	497	489
Shell Canada Limited	Innisfail	23	163	163
Shell Canada Limited	Jumping Pound	6	511	566
Shell Canada Limited	Rosevear	8	153	171
Shell Canada Limited	Simonette River	7	267	95
Shell Canada Limited	Waterton	17	3 066	3 107
Suncor Inc.	Rosevear	8	110	110
Texaco Exploration Company	Bonnie Glen	–	15	12.5
Voyager	Mundare			
Westcoast Transmission	Fort Nelson, B.C.		1 100	1 100
Westcoast Transmission	Taylor Flats, B.C.	3	460	460
Westcoast Transmission	Pine River		1 055	
Western Decalta Petroleum	Turner Valley	1	24	11
Total daily rated capacity December 31, 1982 - 1983.				28 439

Sources: From Alberta Energy Resources Conservation Board publications.
 - Nil

TABLE 3. CANADIAN REFINERY SULPHUR CAPACITIES, 1982 AND 1983

Operating Company	Location	1982	1983
		Daily Capacity (tonnes)	
Gulf Canada Limited	Edmonton, Alberta	103	56
	Port Moody, B.C.	25	25
	Clarkson, Ontario	40	49
	Port Tupper, N.S.	(40)	( )
Husky Oil Ltd.	Prince George, B.C.	5	5
Imperial Oil Ltd.	Edmonton, Alberta	36	40
	Dartmouth, N.S.	40	76
	Sarnia, Ontario	103	100
	Vancouver, B.C.	20	20
Irving Oil Ltd.	Saint John, N.B.	200	200
Sulconam Inc.	Montreal, Quebec	300	300
Petro-Canada	Oakville, Ontario		41
Newfoundland Refining Co. Ltd.	Come-by-Chance, Nfld.	(194)	( )
Shell Canada Res. Ltd.	Shellburn, B.C.	15	( )
	Oakville, Ontario	35	( )
	Sarnia, Ontario	31	31
Suncor Inc.	Sarnia, Ontario	10	10
Texaco Canada Res. Ltd.	Nanticoke, Ontario	8	8
	Calgary, Alta.	10	( )
Canadian Ultramar Limited	Montreal, Quebec	81	81
Total		1 062	942

Sources: Oilweek, April 16, 1984 Chemical Economics Handbook.
( ) Not operational

TABLE 4. CANADA, PRINCIPAL SULPHUR DIOXIDE AND SULPHURIC ACID PRODUCTION CAPACITIES, 1983[1]

Operating Company	Plant Location	Raw Material	Annual Capacity Sulphuric Acid[2] (000 tonnes)	S. equiv. (000 tonnes)
Allied Corporation[3]	Valleyfield, Que.	SO_2 zinc conc.	127	42
Brunswick Mining and Smelting Corporation Limited	Belledune, N.B.	SO_2 lead-zinc	160	52
Canadian Electrolytic Zinc Ltd.[4]	Valleyfield, Que.	SO_2 zinc conc.	440	144
C-I-L Inc.	Beloeil, Que.	Elem. S.	65	21
Inco Metals Company	Copper Cliff, Ont.	SO_2 pyrrhotite	900	294
	Copper Cliff, Ont.	SO_2 copper	Liquified SO_2	45
NL Chem Canada Inc.	Varennes, Que.	Elem. S.	56	18
Falconbridge Limited	Sudbury, Ont.	SO_2 pyrrhotite	355	116
International Minerals & Chemical Corporation (Canada) Limited	Port Maitland, Ont.	Elem. S.	250	82
Gaspé Copper Mines, Limited	Murdochville, Que.	SO_2 copper	160	52
Canada Colors and Chemicals	Elmira, Ont.	Elem. S.	32	11
Kidd Creek Mines Ltd.	Kidd Creek, Ont.	SO_2 zinc conc.	440[6]	144
Subtotal Eastern Canada			2 985	1 021
Border Chemical Company Ltd.	Transcona, Man.	Elem. S.	150	49
Cominco Ltd.	Kimberley, B.C.	SO_2 pyrrhotite	300	98
	Trail, B.C.	SO_2 lead-zinc	430	141
	Trail, B.C.	SO_2 lead-zinc	Liquified SO_2	40
Esso Chemical Canada	Redwater, Alta.	Elem. S.	965	316
Eldorado Resources Limited	Rabbit Lake, Sask.	Elem. S.	45	15
Inland Chemicals Ltd.	Fort Saskatchewan, Alta.	Elem. S.	136	44
	Prince George, B.C.	Elem. S.	35	11
Sherritt Gordon Mines Limited	Fort Saskatchewan, Alta.	Elem. S.	215	70
Western Co-operative Fertilizers Limited	Calgary, Alta.	Elem. S.	417	136
	Medicine Hat, Alta.[5]	Elem. S.	530	173
Subtotal Western Canada			3 223	1 093
TOTAL			6 208	2 114

Source: Company reports.
[1] 100 per cent H_2SO_4. [2] Plant capacities are related to process changes in 1983. [3] Allied Corporation ceased production in June 1983. [4] Canadian Electrolytic Zinc Limited expanded plant capacity in 1983. [5] Western Cooperative Fertilizers Limited closed Medicine Hat plant in April 1983. [6] Expansion to 530 000 tpy planned for mid-1987.

TABLE 5. CANADA, SULPHUR SHIPMENTS AND TRADE, 1966, 1970, 1971, 1975, 1979-84

	Shipments[1]				Imports	Exports	
	Pyrites	In Smelter Gases	Elemental Sulphur	Total	Elemental Sulphur	Pyrites[2]	Elemental Sulphur
	(tonnes)				(tonnes)	($)	(tonnes)
1966	147 226	453 870	1 851 924	2 453 020	131 955	981,000	1 269 157
1970	159 222	640 360	3 218 973	4 018 555	48 494	1,226,000	2 711 069
1971	140 642	561 046	2 856 796	3 558 484	27 923	1,074,000	2 401 975
1975	10 560	694 666	4 078 780	4 784 006	14 335	170,000	3 284 246
1979	13 964	667 265	6 314 244	6 995 473	1 699	281,000	5 154 831
1980	14 328	894 732	7 655 723	8 564 783	1 767	386,000	6 850 143
1981	5 000	783 000	8 018 000	8 806 000	4 633	109,000	7 309 216
1982	9 000	627 000	6 945 000	7 581 000	2 159	668,000	6 111 444
1983	-	678 286	6 631 123	7 309 409	2 365	77,000	5 670 275
1984P	-	874 906	7 700 455	8 575 361			

Sources: Statistics Canada; Energy, Mines and Resources Canada.
[1] See footnotes for Table 1. [2] Quantities of pyrites exported not available.
P Preliminary; - Nil.

TABLE 6. CANADA, SULPHURIC ACID PRODUCTION, TRADE AND APPARENT CONSUMPTION, 1966, 1970, 1971, 1975, 1979-83P

	Production	Imports	Exports	Apparent Consumption
	(tonnes - 100% acid)			
1966	2 267 962	6 303	49 848	2 224 417
1970	2 475 070	9 948	129 327	2 355 691
1971	2 660 773	4 492	91 711	2 573 554
1975	2 723 202	154 020	225 402	2 651 820
1979	3 666 080	170 618	139 425	3 697 273
1980	4 295 366	18 048	323 775	3 989 639
1981	4 116 860	82 495	337 518	3 861 837
1982	3 130 854	192 514	259 740	3 063 628
1983P	3 686 427	126 573	250 612	3 562 388

Sources: Statistics Canada; Energy, Mines and Resources Canada.
P Preliminary.

TABLE 7. WORLD PRODUCTION OF SULPHUR IN ALL FORMS, 1982

	Elemental	Other[1]	Total
	(000 tonnes)		
United States	8 614	2 094	10 708
U.S.S.R.	3 556	5 576	9 132
Canada	5 628	624	6 252
Poland	4 935	159	5 094
Japan	1 062	1 648	2 710
Mexico	1 900	85	1 985
France	1 819	150	1 969
West Germany	1 125	652	1 777
Spain	20	1 014	1 034
Italy	86	400	486
Finland	40	430	470
Sweden	22	312	334
Iraq	200	-	200
Iran	20	-	20
Other countries	2 456	6 282	8 738
Total	31 483	19 426	50 909

Source: British Sulphur Corporation Limited, Sulphur No. 170, January-February 1984.
[1] Sulphur in other forms includes sulphur contained in pyrites and contained sulphur recovered from metallurgical waste gases mostly in the form of sulphuric acid.
- Nil.

TABLE 8. CANADIAN EXPORT MARKETS FOR SULPHUR, 1983P

Country or Area	Exports (million tonnes)	Per cent of Total
United States	1.11	19.6
Europe	.85	15.0
Brazil	.57	10.1
South Africa	.37	6.5
Australia	.36	6.3
Tunisia	.31	5.5
South Korea	.30	5.3
India	.22	3.9
New Zealand	.16	2.8
Taiwan	.15	2.6
Others	1.27	22.4
Total	5.67	100.0

Source: Statistics Canada.
P Preliminary.

TABLE 9. CANADA, SULPHURIC ACID CONSUMPTION BY END USE, 1982 AND 1983

	1982	1983
	(tonnes)	
Uranium mines	339 294	300 236
Miscellaneous metal mines	44 535	12 111
Crude petroleum and natural gas industry	4 449	4 174
Sugar, vegetable oil and miscellaneous food processors	2 253	837
Leather industries	) 2 953	17 530
Textile industries	)	
Pulp and paper mills	257 863	278 126
Iron and steel mills	7 406	2 720
Smelting and refining	219 675	211 649
Electrical products industries	17 150	21 653
Petroleum refineries and coal products	31 201	32 923
Fertilizers and other industrial chemicals	2 353 015	2 404 399
Plastics and synthetic resins	39 299	463
Soap and cleaning compounds	15 323	11 544
Explosives and miscellaneous chemical industries	56 527	38 003
Miscellaneous manufacturing industries	10 861	9 413
Other end uses[1]	33 146	31 927

Source: Reports from producing companies.
[1] Other end uses include miscellaneous non-metal mines; automotive; hydro, municipal utility and water; metal fabricating; and miscellaneous manufacturing industries.

Talc, Soapstone and Pyrophyllite

M. PRUD'HOMME

SUMMARY

Since 1982, Canadian production of talc, soapstone and pyrophyllite has risen continuously due to new installed capacities and aggressive marketing, especially in the United States. In 1984, the value of shipments of talc and pyrophyllite increased by 31 per cent, compared to 57 per cent in 1983. Imports increased slightly to 38 000 t mainly from the United States. Exports of talc to the United States increased by 71 per cent in 1983, and talc and pyrophyllite exports increased by 40 per cent in 1984.

In 1984, Canada Talc Inc. completed the construction of its new 80 000 tpy talc and dolomite plant in Marmora, Ontario. Steetley Talc Limited completed the first phase of an expansion program which will increase its production capacity to 45 000 tpy at Timmins, Ontario. Bakertalc Inc. installed new filtering facilities to improve efficiency at its plant near Highwater in Quebec. BSQ Talc Inc. conducted drilling to assess new reserves near St.-Pierre-de-Broughton in the Eastern Townships, Quebec.

Talc prices have risen by an average of 8 per cent annually since 1982. Over the short-term talc markets have maintained growth because of the versatility of talc and because of competitive prices. Markets related to the construction sector will show steady growth, depending on economic recovery. Appreciable growth is expected in the plastics and the pulp and paper industries for the next few years.

Talc is a hydrous magnesium metasilicate, $Mg_3Si_4O_{10}(OH)_2$, and is usually intimately associated with numerous other minerals such as serpentine, dolomite and quartz. The colour is characteristically a pale green, grey or creamy white. It exhibits a pearly lustre, a low hardness, a greasy feel and an extreme smoothness. Talc is derived from the alteration of magnesian rocks in an intensive metamorphic environment. It occurs as veinlets, tabular bodies or irregular lenses. Talc is valued for its various properties: extreme whiteness, smoothness, high fusion point, low thermal and electrical conductivity and chemical inertness. Talc is produced in various grades which are usually classified by end-use: paint, ceramic, pharmaceutical and cosmetic.

Steatite (soapstone) is an impure, massive, compact form of talc which can be sawn or machined easily. "Steatite grade" is a special block talc suitable for making ceramic insulators. Soapstone is a mixture of talc, serpentine, chlorite, dolomite with, sometimes, small percentages of quartz and calcite. Its durability depends on its chemical inertness and non-absorbency properties. Soapstone has been used since early times in many parts of the world for carving ornaments, pipes, cookware, lamps and other utensils. The art of carving this rock has survived among the Inuit people of Canada up to the present era. Present uses include metalworkers' crayons, refractory bricks, and blocks for sculpturing.

Pyrophyllite is a hydrous aluminum silicate, $Al_2Si_4O_{10}(OH)_2$, formed by hydrothermal alteration of acid igneous rocks, predominantly lavas which are andesitic to rhyolitic in composition. It occurs in low- and medium-grade metamorphic rocks rich in aluminum. Its physical properties are practically identical to those of talc, and, for this reason, pyrophyllite finds industrial uses similar to talc, notably in ceramic bodies and as a filler in paint, rubber and other products.

PRODUCTION AND DEVELOPMENTS IN CANADA

Talc, soapstone. The earliest recorded production of talc or soapstone in Canada was in 1871-72 when 270 t of cut soapstone, valued at $1,800 was shipped from a deposit in Bolton Township, southern Quebec, by Slack and Whitney. In 1896, a deposit in Huntingdon Township, in the Madoc district

---

M. Prud'homme is with the Mineral Policy Sector, Energy, Mines and Resources Canada. Telephone (613) 995-9466.

of Ontario, was opened, and over the next few years numerous deposits were discovered in this area and mined intermittently. Several deposits in southern British Columbia and one in southwestern Alberta were discovered prior to 1920 and some were worked on a small scale.

At present, talc is produced commercially in two provinces, Quebec and Ontario, while pyrophyllite is produced only in Newfoundland.

Bakertalc Inc. produces talc and soapstone from an underground operation at South Bolton, Quebec, 95 km southeast of Montreal. Talc occurs as dykes and sills, associated with serpentine and magnesite, in Cambrian and Lower Ordovician schists. Ore is extracted at the Van Reet mine and is trucked 16 km south to the company's mill facilities at Highwater. It produces around 5 000 tpy of high quality floated material for use principally in the pulp and paper industry, and a similar tonnage of dry-milled talc used as an industrial filler in paints and plastics. Soapstone is also supplied as sculpture blocks. St. Lawrence Chemicals Inc. is the distributor for all Bakertalc's products. In 1983, Bakertalc Inc. assessed the use of talc as a substitute for asbestos in an asbestos manufactured product. Successful tests resulted in increased sales in 1983. In 1984, the company installed new filtering facilities in its mill to increase efficiency and to recover tailings from the production of high-purity talc. Current work with le Centre de Recherches Minérales (CRM) is being done to improve the flotation process and to perfect the determination of talc grades in concentrates.

BSQ Talc Inc., near St.-Pierre-de-Broughton in Quebec, quarries two deposits associated with the Pennington dyke in Leeds and Thetford townships. Occurrences are associated with ultrabasic intrusives, peridotite-serpentinite, in quartz-carbonate-chlorite schists. BSQ Talc Inc. produces ground material containing nearly 70 per cent talc, which is used as a filler in joint cement and auto-body compound and as a dusting agent in asphalt roofing shingles and rubber production. It also supplies soapstone products such as refractory slabs and sculpture blocks through its subsidiary Benmic Inc. Metalworker's crayons are produced by a new process developed by le Centre de Recherche Industrielle du Québec (CRIQ). Rated capacity of the plant is around 35 000 t. In 1984, the company did some development work, such as stripping and drilling, to assess new reserves. Expansion plans are projected to increase the quality of its talc products in the near future.

Canada Talc Inc. operates an underground talc mine and also quarries a newly-discovered talc orebody at Madoc, Ontario. The orebodies occur in crystalline dolomite, where tabular hydrothermal replacements have taken place. The talc is of exceptional whiteness and may contain accessory minerals such as sulphides, mica and prismatic tremolite.

Since 1982, Canada Talc Inc. has conducted mapping, trenching and drilling on its properties in the Queensborough Road area of Elzevir Township and in Cashel Township. In 1983, the W.R. Barnes Company Limited has been custom milling talc ore for Canada Talc at Northbrook to produce materials for the roofing and automotive body filler markets. In 1984, an $825,000 grant, from the Ontario Ministry of Natural Resources under the Small Rural Mineral Development Program, was offered in order to increase talc production at Madoc and to provide assistance for an expansion at Marmora, 17 km west of the mine site. In November 1984, the company completed its 80 000 tpy talc and dolomite processing plant at Marmora. This $3.76 million facility will produce 44-, 20- and 10-micron size talc and dolomite products for the paint, plastics and paper industries. Dolomite terrazzo chips and talcose dolomite are also produced.

Steetley Talc Limited, a division of Steetley Industries Limited, produces talc from an open-pit mine in Penhorwood Township, 70 km southwest of Timmins. Talc occurs in talc-magnesite deposits derived from the alteration of ultrabasic volcanic rocks. The ore is processed by flotation and fine-grinding. The talc is a high purity, platy material and it is used mainly in the pulp industry as a pitch control agent. Other markets are in paints, plastics, paper and cosmetics. Since 1982, R.T. Vanderbilt Co. Inc. is the distributor of their talc products in the American markets. Steetley Talc Limited is engaged in a $3.76 million multi-stage expansion program to increase production and to improve the quality of its talc products. The company received a provincial grant of $940,000 through the Board of Industrial Leadership and Development. Annual capacity of the plant is about 27 000 t with plans to increase to 33 000 t by 1985. Further development work will be carried out to recover magnesite from the magnesium-silicate deposit.

During 1984, several developments undertaken in Canada are worthy of note. Clayburn Co. at Coalmont, British Columbia, produced a small tonnage of low quality talc for use in its refractory plant. International Marble and Stone Company Limited made a trial shipment of low quality talc for application as an industrial filler from the Haas Creek deposit, west of Greenwood, British Columbia. Wabigoon Resources Limited did some bulk sampling for flotation testing to determine the quality of talc from the Wabigoon Lake soapstone-talc property in Zealand Township, Ontario. Ram Petroleum Limited completed a small mill near Clarendon in Palmerston Township, Ontario to process material from a tremolite deposit containing 10 per cent talc. An off-white material will be produced for use in asphalt and concrete applications. In Quebec, geologists of the Energy and Resources department conducted investigations of talc and steatite occurrences in the Eastern Townships; surface mapping and detailed cartography were carried out along the Pennington dyke between Thetford Mines and Black Lake.

Numerous deposits of talc and soapstone occur near the producing areas and in other parts of Canada. A soapstone deposit on Pipestone Lake in Saskatchewan was worked by Indians for the manufacture of pipes and various utensils. Reserves are reported to be considerable. High quality "blue" talc was investigated in the Banff area of Alberta and British Columbia during the 1930s. In the Northwest Territories, a few occurrences of soapstone are known from which Eskimos obtained material for carving. Showings of minor importance occur at several localities in Nova Scotia, Newfoundland and eastern Ontario.

Pyrophyllite. Newfoundland Minerals Limited, a subsidiary of American Olean Tile Company, Inc. (a division of National Gypsum Company), mines pyrophyllite from an open-pit operation near Manuels, 19 km southwest of St. John's, Newfoundland. The deposit appears to be a hydrothermal alteration of sheared rhyolite. Altered zones are associated mainly with extensive fracturing near intrusive granite contacts. Reserves are believed to be sufficient for about 40 years at the present production rate. The mine has operated continuously since 1955. Ore is crushed, sized and hand-cobbed at the mine site prior to being trucked a short distance to tidewater. Annual production varies between 30 000 and 45 000 t. The cut-off grade is 17 per cent aluminum oxide. High-quality crude ore is shipped to the parent company's ceramics plants at Lansdale, Pennsylvania, and Jackson, Mississippi. Some lower grade pyrophyllite has been used in the local manufacture of joint cement, stucco, paints and other products, since 1975.

Other known pyrophyllite deposits in Canada include an extensive area of impure pyrophyllite near Stroud's Pond in the southern part of Burin Peninsula, Newfoundland; occurrences near Senneterre in Abitibi County, Quebec and deposits in British Columbia, near Ashcroft and on Vancouver Island.

TRADE AND MARKETS

Since 1982 the value of shipments of talc and pyrophyllite has risen steadily, by 57 per cent in 1983 and 31 per cent in 1984. The value of pyrophyllite represents 10-13 per cent of the total value of talc minerals. In 1984, the increased value reflects record levels of production and a greater tonnage of high purity talc products.

In 1983, the value of imports increased by 11 per cent while tonnage increased by 3 per cent, compared to 1982. For 1984, imports are expected to increase slightly, near 38 500 t. The average value of imports was $175 per t in 1983, compared to $163 per t in 1982; an increase of 7 per cent in current dollars which is marginal taking into account the inflation effect and the currency exchange rate, as 98 per cent of our imports are from the United States. High quality talc is imported mainly into Ontario (45 per cent), British Columbia (23 per cent), Quebec (18 per cent) and Alberta (10 per cent). Exports to the United States, which account for 99 per cent of total exports, increased by 71 per cent in 1983. For 1984, exports are expected to increase by nearly 40 per cent due to higher shipments of pyrophyllite and the fact that Canadian talc producers captured a greater share of markets in the United States, especially in the northeastern areas.

Since 1982, apparent consumption of talc has increased annually by an average of 10 per cent. In 1983, end-use of ground talc was as follows: roofing products, 19 per cent; paints, 18 per cent; paper, 18 per cent; gypsum products and putty compounds, 17 per cent; rubber and plastics, 9 per cent; ceramics, 9 per cent; chemicals and pharmaceuticals, 5 per cent; and other varied uses, 5 per cent.

USES

Talc is used mostly in a fine-ground state; soapstone in massive or block form. There are many industrial applications for ground talc, but fewer than a dozen countries use ground talc on a major scale.

In pulp and paper manufacture softness, chemical inertness, high reflectance, hydrophobic and organophilic properties and the particle shape of talc, are characteristics that permit its use as a pitch-adsorbing agent, as a paper filler and as a coating pigment. For filler usage, maximum particle size must be below 20 microns; however, 40 micron grades are also used. For coating applications, particle size must be below 10 microns and close to 1 micron for pitch control.

The ceramic industry utilizes very finely ground talc to increase the translucence and toughness of the finished product and to aid in promoting crack-free glazing. Talc must be low in iron, manganese and other impurities which would discolour the fired product. Average particle size for most ceramics must range between 6 and 14 microns, with 90-98 per cent of material passing through 325 mesh.

In plastics, talc improves dimensional stability, chemical and heat resistances, impact and tensile strengths, electrical and insulation properties. It is used in thermoplastics and in thermosets, mainly in polypropylene, nylon and polyester. Chemical coupling agents are used to enhance the bond between the talc filler and the resin matrix in plastic materials. Talc must be free of iron impurities and grits, and must be superfine with an average particle size below 8 microns.

High-quality talc is used as an extender pigment in paints. Specifications for a talc pigment, as established in American Society for Testing and Materials (ASTM) designation D605-69 (1976), relate to its chemical composition, colour, particle size, oil absorbency and fineness of dispersion. A low carbonate content, a nearly white colour, a fine particle size with controlled particle size distribution and a specific oil-absorption are important. However, because of the variety of paints, precise specifications for talc pigments are generally based on agreement between consumer and supplier. Paint characteristics influenced by the use of talc as extender are gloss, adhesion, flow, hardness and hiding power.

Pharmaceutical industries are well-known users of high-purity talc for pharmaceutical preparations and cosmetics, relying on its softness, hydrophobic property and chemical inertness. Finely ground, it is used as a filler in tablets and as an additive in medical pastes, creams and soaps.

Lower-grade talc is used as a dusting agent for asphalt roofing and rubber products, as a filler in drywall sealing compounds, floor tiles, asphalt pipeline enamels, auto-body patching compounds, and as a carrier for insecticides. Other applications for talc include use in cleaning compounds, polishes, electric cable coating, foundry facings, adhesives, linoleum, textiles and in the food industry.

Soapstone has now only very limited use as a refractory brick or block, but, because of its softness and resistance to heat, it is still used by metalworkers as marking crayons. The ease with which it can be carved makes it an excellent artistic medium.

Pyrophyllite can be ground and used in much the same way as talc. In ceramics, it imparts a very low coefficient of thermal expansion to tiles. It must be graded minus 325 mesh and contain a minimum of quartz and sericite which are common impurities. It may also be used in refractories as its expansion on heating tends to counteract the shrinkage of the plastic fraction. Massive pyrophyllite, the compact and homogenous variety, is chiefly used in the manufacture of refractories, although small amounts of the crystalline or radiating variety find similar use. Foliated or micaceous pyrophyllite find use as a filler and ceramic raw material.

WORLD PRODUCTION AND REVIEW

Talc is widely distributed throughout the world and many countries have been developing deposits. These widespread occurrences enjoy limited international trade except for high-grade materials, where small shipments compete with substitutes. The majority of international trade takes place within Europe; in the Far East between Japan, the People's Republic of China and Korea; and in North America between Canada and the United States.

In 1983, world production of talc and pyrophyllite remained unchanged at about 6.85 million t. Japan is the largest supplier of pyrophyllite with an estimated production of 1.35 million t, and is also the world's largest importer of talc for use in the paper industry as a filler and coating material. The United States is the major producer of talc accounting for 20 per cent of the world total, followed by China with 18 per cent.

In 1983 United States production decreased by 6 per cent although apparent consumption increased by 20 per cent to 857 000 t. For 1984 production of talc minerals is estimated at 1.18 million t. Vermont Talc Co., a division of Omya Inc., is planning to double its capacity, adding a Raymond roller mill and a froth flotation plant to its facilities located near Chester, Vermont. In 1983, Engelhard Corporation closed its 80 000 tpy subsidiary, Eastern Magnesia Talc Company, at Johnson, Vermont.

In France, la Société anonyme des talcs de Luzenac (SATL) has been operating, since 1982, an optical sorting pilot plant to increase productivity and to obtain higher quality products from its mines at Trimouns.

China reported exports of 523 000 t of talc in 1982, mainly to Japan, southeast Asian countries and England. Domestic consumption accounts for 68 per cent of the total output; paper making is the major consumer with 60 per cent followed by asphalt products with 14 per cent.

PRICE

Prices of talc vary according to quality, method of processing, specifications and transportation cost. Since 1982, prices have remained steady with a small average annual increase of about 8 per cent. Canadian prices vary between $35-70 per t for medium-grade talc, $95-160 per t for high-grade talc, $180-250 per t for highly-beneficiated talc, and around $1,000 per t for steatite blocks. However, list prices and actual prices differ as negotiations occur between producers and consumers.

OUTLOOK

In 1984 Canada, as a supplier of fine talc, benefit from the gap in the market created by the closure of Engelhard's facilities, in Vermont. Aggressive marketing from Canadian producers in the pulp and paper industry also increased sales in the United States. Producers should maintain their market share due to transportation deregulation in the United States, currency exchange rate and economies of scale from their newly installed capacities. In the short run, the demand for talc will maintain a steady growth because of its versatility and because of competitive pricing.

Consumption trends look promising in the pulp and paper, the plastics and the ceramics industries. Markets should hold steady in the manufacturing sectors subjected to cyclical effects such as in the construction area: paints, roofing and gypsum products will show a slight growth relative to economic recovery. The Canadian Construction Association anticipates, for the period 1984-1990, a growth around 0.2-0.6 per cent for investment in the residential sector, and around 4.8-6.5 per cent in the non-residential sector. Appreciable growth is expected in the pulp and paper industry due to the versatility of the talc as a multiple-effects functional filler and pitch-adsorbing agent. Talc will benefit from increased usage as a filler in plastics for the automotive sector, which is also expected to grow in the next few years. Talc consumption will be steady in automotive body patching compounds.

From a 1979 base, the U.S. Bureau of Mines has estimated that the demand for talc and related minerals is expected to increase at an annual rate of about 2 per cent through 1990.

Substitutes for talc are numerous in its major markets: nepheline syenite, kaolin and calcium carbonate in paints; pyrophyllite and feldspar in ceramics; mica and calcium carbonate in plastics; kaolin and calcium carbonate in paper. However, talc is still the primary pitch control agent in the pulp industries.

PRICES OF TALC

Talc; free on board mine, carload lots, containers included unless otherwise specified: U.S. $ per short ton.

New Jersey	
mineral pulp, ground;	
(bags extra)	18.50-20.50
Vermont	
98% through 325 mesh, bulk	70
99.99% through 325 mesh, dry processed, bags	147
99.99% through 325 mesh, water beneficiated, bags	213-228
New York	
96% through 200 mesh	62-70
98-99.25% through 325 mesh (fluid energy ground)	160
California	
Standard	69.50
Fractionated	37-71
Micronized	62-104
Cosmetic/Steatite	33-65
Georgia	
98% through 200 mesh	50
99% through 325 mesh	60
100% through 325 mesh (fluid energy ground)	100

Source: Engineering and Mining Journal, December 1984.

TARIFFS

Item No.		British Preferential	Most Favoured Nation	General	General Preferential
			(%)		
CANADA					
71100-3	Talc or soapstone	10	12.1	25	8.0
29646-1	Talc for use in manufacture of pottery or ceramic tile (expires June 30, 1984)	free	free	25	free
29647-1	Micronized talc, not exceeding 20 microns	free	4.5	25	free
29655-1	Pyrophyllite	free	free	25	free

MFN Reductions under GATT (effective January 1 of year given)	1983	1984	1985	1986	1987
			(%)		
71100-3	12.1	11.4	10.7	9.9	9.2
29647-1	4.5	4.4	4.3	4.1	4.0

UNITED STATES

523.31	Talc and soapstone, crude and not ground	0.02¢ per lb.				

		1983	1984	1985	1986	1987
523.33	Talc and soapstone, ground, washed, powered or pulverized	4.2%	3.8%	3.3%	2.9%	2.4%
523.35	Talc and soapstone, cut or sawed, or in blanks, crayons, cubes, disks or other forms, per lb.	.1¢	free	free	free	free
523.37	All other, not provided for	4.8%	4.8%	4.8%	4.8%	4.8%

Sources: The Customs Tariff, 1983, Revenue Canada, Customs and Excise; Tariff Schedules of the United States Annotated 1983, USITC Publication 1317; U.S. Federal Register Vol. 44, No. 241.

TABLE 1. TALC, SOAPSTONE AND PYROPHYLLITE PRODUCTION, TRADE 1982-84 AND CONSUMPTION 1981-83

	1982		1983		1984P	
	(tonnes)	($)	(tonnes)	($)	(tonnes)	($)
Production (shipments)						
Talc and soapstone						
Quebec[1]	..	1,708,516	..	1,960,844	..	2,430,000
Ontario[2]	..	2,753,746	..	4,894,441	..	6,698,000
Total	..	4,462,262	..	6,958,685	..	9,128,000
Pyrophyllite						
Newfoundland	..	603,446	..	1,140,659	..	1,120,323
Total production	70 523	5,065,708	97 030	7,995,914	125 650	10,530,300
					(Jan.-Sept. 1984)	
Imports		($000)		($000)		($000)
Talc, incl. micronized						
United States	33 604	5,508	34 718	6,123	27 544	5,241
United Kingdom	65	8	46	19	75	11
Italy	75	9	42	14	–	–
Japan	151	18	2	3	24	3
France	–	–	–	–	36	11
Sub-total, talc	33 895	5,543	34 808	6,159	27 679	5,266
Soapstone, exc. slabs						
United States	139	31	32	4	26	9
Taiwan	18	2	–	–	–	–
Sub-total, soapstone	157	33	32	4	26	9
Pyrophyllite						
United States	470	40	548	41	483	35
Sub-total, pyrophyllite	470	40	548	41	483	35
Total talc, soapstone and pyrophyllite	34 522	5,616	35 390	6,207	28 188	5,410

	1981	1982	1983
		(tonnes)	
Consumption[3] (ground talc available data)			
Paints and varnish	9 724	8 612	7 959
Gypsum products	5 233	2 735	3 133
Pulp and paper products	5 762	6 660	9 660
Roofing products	5 565	6 631	5 671
Ceramic products	5 300	5 546	3 376
Toilet preparations	1 671	1 513	1 722
Chemicals	2 479	2 734	2 577
Rubber products	1 559	2 470	3 400
Other products[4]	1 691	1 732	1 999
Total	38 984	38 633	39 497

Sources: Statistics Canada; Energy, Mines and Resources Canada.
[1] Ground talc, soapstone, blocks and crayons. [2] Ground talc. [3] Breakdown by Energy, Mines and Resources, Canada. [4] Adhesives, floor covering, insecticides and other miscellaneous uses.
P Preliminary; .. Not available; – Nil.

TABLE 2. CANADA, TALC AND
PYROPHYLLITE PRODUCTION AND
IMPORTS, 1970, 1975, 1979-84

	Production[1]	Imports
	(tonnes)	
1970	65 367	29 999
1975	66 029	30 428
1979	90 330	50 322
1980	91 848	50 774
1981	82 715	30 322
1982	70 523	34 522
1983	97 030	35 390
1984P	125 650	..

Sources: Statistics Canada; Energy, Mines and Resources Canada.
[1] Producers' shipments.
P Preliminary; .. Not available.

TABLE 3. WORLD PRODUCTION OF TALC, SOAPSTONE AND PYROPHYLLITE, 1980-83

	1980	1981	1982P	1983e
	(000 tonnes)			
Japan	1 749	1 545	1 492	1 464
United States	1 125	1 218	1 030	967
People's Republic of China[e]	916	898	898	898
Republic of Korea	719	620	591	632
U.S.S.R.[e]	490	500	510	510
Brazil	413	503	384	500
India	369	367	336	322
Finland	317	307	325	300
France	302	309	277	277
North Korea	168	168	168	168
Italy	166	163	164	163
Austria	117	116	117	116
Canada	92	83	70	97
Australia	171	91	93	93
Norway	88	33	32	36
Other countries	332	303	351	308
Total	7 534	7 224	6 838	6 851

Sources: U.S. Bureau of Mines Preprints 1983; Energy, Mines and Resources Canada.
P Preliminary; e Estimated.

Tantalum

D.G. FONG

CANADIAN DEVELOPMENTS

Tantalum mining operations at Tantalum Mining Corporation of Canada Limited's (TANCO) Bernic Lake, Manitoba property remained suspended throughout 1983 and 1984. TANCO, Canada's only tantalum producer, was closed at year-end 1982 and has been maintained on a standby basis pending an improvement in the tantalum market. No sales of tantalum pentoxide were made during the last two years because of low prices and high inventories. Prior to the closure, the company had normally shipped 60 per cent of its output to KBI Division of Cabot Corporation at Reading, Pennsylvania under a long-term contract.

TANCO is jointly owned by KBI (37.5 per cent), Hudson Bay Mining and Smelting Co., Limited (HBM&S) (37.5 per cent) and the Manitoba government (25 per cent), and it is managed by HBM&S under a management contract. In 1984, the company was awarded a grant from the Federal Department of Regional Industrial Expansion to assist in the temporary conversion of the tantalum mill into a pilot mill to produce ceramic grade lithium concentrate. Test samples were shipped to end users for trial production during the later part of 1984.

Highwood Resources Ltd. resumed exploration drilling in the winter of 1983 at the Thor Lake deposit in the Northwest Territories. Thor Lake is located 110 km southeast of Yellowknife on the north shore of Great Slave Lake. The property contains several multi-mineral deposits, each with slightly different mineralization. Exploration work between 1977 and 1981 outlined 70 million t of mineralization with a grade of 0.30 per cent tantalum (Ta), 0.40 per cent columbium (Cb), 3.5 per cent zirconium (Zr) and 1.7 per cent combined rare earths. Additional drilling in the 1983-84 seasons identified a beryllium zone containing 1.789 million t of 0.86 per cent beryllium (BeO).

During 1984, Highwood Resources conducted metallurgical test work on the recovery of tantalum and columbium from the Thor Lake zones. A high recovery of ultrafine tantalo-columbite was achieved using a novel fine particle recovery process. Metallurgical testing on the recovery of beryllium was being carried out at an Ontario laboratory. In view of the attractive grade, Highwood Resources began a pre-feasibility study to assess the economics and viability of an open-pit mine and concentrator at the beryllium zone. Tantalum, columbium, rare earths and other associated elements in this deposit could be produced as byproducts or coproducts.

WORLD DEVELOPMENTS

The world's major tantalum producing countries are Canada, Australia, Brazil, Thailand, Nigeria, Zaire and Mozambique. Canada was, until 1982 the world's largest tantalite producer, supplying about 15 per cent of the world's tantalum requirements, while Thailand was the world's largest supplier of tantalum-bearing tin slag, a byproduct of tin mining. During 1983, when Canada temporarily ceased production of tantalum, Brazil continued to produce at an annual rate of 68 t. Australian production in the year to June 1983 was 29.64 t, about 35 per cent of plant capacity.

Australian tantalum production to June 1984 increased 61 per cent to 47.74 t. The higher output in Australia stemmed from an increased emphasis on the production of tantalite rather than tin. Australian tantalum is produced with tin which, during the last few years, has faced export quotas imposed by the International Tin Council (ITC).

Greenbushes Tin N.L. of Perth, Australia's largest tantalum producer, was proceeding with its mine expansion project. At the end of 1983, the company borrowed $20 million from two banks to finance a new processing plant and underground mine

---

D.G. Fong is with the Mineral Policy Sector, Energy, Mines and Resources Canada. Telephone (613) 995-9466.

developments. The mine expansion is expected to raise Greenbushes' annual production capacity to 204 t of tantalum pentoxide (Ta_2O_5) by mid-1985. Production could be further increased to 272 tpy whenever markets warrant.

Greenbushes also commenced sales of tantalite to the U.S. General Services Administration (GSA) during the last quarter of 1983. The company was awarded a contract to supply 54.4 t of tantalite concentrates, with the stipulation that shipments be completed before October 14, 1984.

Thailand Tantalum Industry Corp. Ltd. (TTIC) signed an agreement in late 1983 to borrow $US 53.5 million from International Finance Corporation (IFC) to build a tantalum recovery plant in Phuket, Thailand. The new plant will include a smelter and a chemical treatment plant with an annual capacity to produce 300 t of Ta_2O_5 and 300 t of columbium pentoxide (Cb_2O_5). Feedstocks will consist of high-grade tin slags, tantalo-columbite, and struverite concentrates. TTIC also planned to further process up to 7 000 tpy low- to medium-grade tin slags to recover an additional 100 to 150 tpy of Ta_2O_5. The IFC, an affiliate of the World Bank, has agreed to buy a 12 per cent interest in the project for $US 3.9 million.

Production at the Phuket plant had originally been scheduled to commence in 1984 but this was later postponed to 1986. When the project was proposed in 1980, the Thai Board of Investment gave TTIC exclusive rights for Ta_2O_5 production for eight years, and agreed to place a ban on the export of tin slags one year before the start-up of the Phuket plant.

Bureau des Recherches Geologiques et Minières (BRGM) completed an initial feasibility study in 1983 of Echassieres, a new tin-tantalum deposit near Vichy, in southern France. The Echassieres deposit contains 55 million t grading 0.023 per cent Ta_2O_5, 0.13 per cent tin (Sn), 0.022 per cent Cb_2O_5 and 0.71 per cent lithium oxide (Li_2O). The initial plan was to have a 170 000 tpy mining operation together with a processing plant to recover 150 to 160 tpy of tin and 35 tpy of Ta_2O_5.

USES

Tantalum is a refractory metal with unique physical, electrical and chemical properties which make it useful in a number of industrial applications including electronic, metal making, chemical equipment and high temperature alloy industries.

Capacitors are by far the largest users of tantalum, accounting for about 55 per cent of the world demand in 1984. Tantalum has proved its superiority as a metal for electronic capacitor anodes because of its inertness and the stability of its electrolytic oxide film. Tantalum capacitors are used in all forms of electronic systems such as computers, communication systems and military applications where compactness and reliability are key factors.

Tantalum cemented carbides are used mainly in mixtures with other carbides such as tungsten, titanium, columbium, chromium, vanadium, molybdenum and hafnium. The addition of tantalum carbide to other metallic carbides imparts a greater cratering resistance, as well as the ability to be machined at much higher cutting edge temperatures. Cemented carbides are used in cutting, turning, and boring tools, and wear resistant parts and dies.

At low to moderate temperatures, tantalum's high resistance to corrosion by most acids and the complete inertness to many chemicals has made it the preferred material in severe environments. The metal is used mainly as thin sheet linings in chemical equipment. Explosive cladding techniques are used to clad thin tantalum sheets on container walls, which enable it to be competitive with substituting materials.

Tantalum is an important addition in special nickel- and cobalt-base superalloys that are used in high-temperature applications such as jet engines and gas turbine parts. The addition of tantalum increases the strength of these superalloys and improves the high temperature performance in terms of fuel efficiency and durability. Pratt and Whitney of United Technology Inc. has developed a high tantalum content single crystal superalloy, PW1480, for use in jet engines. The single crystal technology has been licensed to the U.S. government for applications in the space shuttle program.

PRICES

TANCO's published tantalite price remained unchanged at $US 45 a lb contained Ta_2O_5 during 1983 and 1984. The spot market price increased to $US 28-31 in October, 1983 from $US 20-25 a lb at the beginning of the year. It increased again to $US 31-33 in August 1984, as consumption continued on a recovery course. The published producer price for tantalum carbide, in 1984, quoted at $US 51 a lb, was the first published list price since 1979, an evidence of growing stability in the carbide market.

OUTLOOK

The tantalum market is expected to continue its recovery in 1985. Western world production is forecast to increase by 10 per cent in 1985 to about 2.7 million lbs. A modest improvement of 2 to 3 per cent is indicated for 1986. Inventories can be expected to decline to a normal level in relation to demand. However, prices are unlikely to surpass the $US 40 a lb mark because consumption growth will likely be matched by increases in output.

Among the end uses, capacitors will remain the largest consuming segment. However, the growth in consumption of tantalum powder for capacitors will continue to be hampered by technology advances, the continuing improvement of capacitance value of the tantalum powder and miniaturization, each of which will contribute to a lower powder requirement per unit capacitor. Also, because of a previous history of widely fluctuating prices and limited supplies, consumption growth has been affected by substituting materials. In this regard, tantalum capacitors are not expected to recapture much of their lost markets because end use applications such as personal computers are relatively inflexible to design changes. Once tantalum is designed out of a particular application, it is unlikely that the production process will be changed just because of a lower price. Nevertheless, the fast growing electronics industry is expected to provide new opportunities for expanding the tantalum capacitor market. The increase in consumption of tantalum units for new applications could more than offset the miniaturization and substitution effect.

Tantalum cemented carbide is used mainly in cutting tools in the metal working industries. The demand for tantalum from this segment has been greatly hampered by the recessionary effect of the early 1980s and the increased usage of titanium coating and other substitution materials. Despite a recent strong recovery in the automobile industry, consumption in this segment has remained low. The consumption of tantalum in cutting tools is expected to grow at a slower rate than the averaged 3.25 per cent a year in other end-use sectors. However, the carbide cutting tool technology is continuously advancing and the development of new alloys of tantalum and columbium could change the demand pattern in the future.

Markets for tantalum mill products are highly sensitive to the business cycle because the use of these products is closely linked with capital investments, mainly in chemical process equipment. Capital investments have been depressed for several years and, even with the current economic recovery, the chemical industry continues to buy equipment at a low rate. In the longer term, the demand for tantalum from this sector is expected to outperform the other end-uses. The consumption of tantalum mill products is expected to grow at 4 to 5 per cent a year.

Tantalum-bearing superalloys are used mainly in jet engines. In recent years, tantalum consumption in this sector has expanded rapidly, despite a poor performance in the aircraft manufacturing industry. The use of substantial quantities of tantalum in jet engines for the Boeing and Airbus aircrafts was the main factor. In 1984, tantalum usage in this sector accounted for about 10 per cent of the total consumption, as compared with 3 per cent five years ago. A significant growth rate is forecast in this application, in view of the high temperature efficiency achieved from tantalum.

The main source of western world tantalum supply, accounting for about three-quarters, is the tin mining and smelting industry. Tantalum is produced in association with tin operations in countries such as Thailand, Australia, Malaysia, Brazil and Africa. Because of this relationship, future supplies of tantalum will, to a great extent, continue to hinge on the well-being of the tin industry. Also, the continued imposition of tin export quotas by the International Tin Council could be a major barrier to any significant increase in tantalum supply.

The majority of tantalum processing is carried out in the highly industrialized nations: the United States, West Germany and Japan. Processors in these countries normally carry a large stock of tantalite concentrates and tin slags, often equivalent to one to two years of supply. The processors thus can continue to have a significant influence over the tantalum market.

The tantalum processing industry could soon undergo an important structural change. Thailand has been the largest single source of tantalum raw materials for the western world processors. However, with the start-up of the TTIC upgrading plant in Thailand, tantalum-bearing tin slags will cease to come from that country. This will likely result in a shortage of feed materials for the processing industry in other parts of the world. Although the shortage could be partly offset by increased output from Australian sources, some processors may have to make adjustments by either switching to intermediate products or by phasing down their operations.

Canadian tantalum production could increase during the next decade, particularly if Highwood Resources proceeds with its beryllium and rare earth mine development. Highland Resources could produce tantalum as a byproduct or coproduct of beryllium.

TABLE 1. CANADA, TANTALUM PRODUCTION, TRADE AND CONSUMPTION, 1970, 1975 AND 1979-84

	Production[1] Ta_2O_5 Content	Imports[2] Primary forms and fabricated metals		Consumption Ferrocolumbium and Ferrotantalumcolumbium, Cb and Ta-Cb Content
		Tantalum	Tantalum Alloys	
		(kilograms)		
1970	143 800	..	..	132 449
1975	178 304	..	..	215 910
1979	158 845	6 901	2 503	360 152
1980	115 261	21 280	12 112	486 251
1981	103 949	2 769	5 043[r]	455 500[r]
1982	59 276	1 759	1 146	356 000
1983	-	1 742	332	352 000
1984P	-	4 428	1 389	..

Sources: Energy, Mines and Resources Canada; Statistics Canada; U.S. Department of Commerce.
[1] Producers' shipments of tantalum ores and concentrates and primary products, Ta_2O_5 content. [2] 1984 imports based on nine month statistics.
P Preliminary; - Nil; .. Not available; r Revised.

PRICES

Prices as quoted in Metals Week in December 1983 and 1984, U.S. currency.

	1983	1984
	($)	
Tantalum ore		
Tantalite, per kg of pentoxide, Tanco price	99.21	99.21
Spot tantalite ore	61.73-68.34	68.34-72.75

TARIFFS

Item No.		British Preferential	Most Favoured Nation	General	General Preferential
			(%)		
CANADA					
32900-1	Columbium and tantalum ores and concentrates	free	free	free	free
35120-1	Columbium (niobium) and tantalum metal and alloys in powder, pellets, scrap, ingots, sheets, plates, strips, bars, rods, tubing or wire for use in Canadian manufactures (expires June 30, 1984)	free	free	25	free
37506-1	Ferrocolumbium, ferrotantalum, ferro-tantalum-columbium	free	4.7	5	free

MFN Reductions under GATT (effective January 1 of year given)	1983	1984	1985	1986	1987
			(%)		
37506-1	4.7	4.5	4.3	4.2	4.0

Item No.						
UNITED STATES						
601.42	Tantalum ore	free				

		1983	1984	1985	1986	1987
				(%)		
629.05	Tantalum metal, unwrought and waste and scrap (duty on waste and scrap suspended to June 30, 1982)	4.4	4.2	4.0	3.9	3.7
629.07	Tantalum, unwrought alloys	6.2	5.9	5.6	5.2	4.9
629.10	Tantalum metal, wrought	7.3	6.8	6.4	5.9	5.5

Sources: The Customs Tariff, 1983, Revenue Canada, Customs and Excise; Tariff Schedules of the United States Annotated 1983, USITC Publication 1317; U.S. Federal Register, Vol. 44, No. 241.

Tin

A. BOURASSA

World tin consumption in 1983 marked the end of the downward trend which began in 1973. Preliminary data even suggests a slight recovery in world consumption in 1984. The domestic market followed this world trend. Since April 1982, the tin producing members of the Sixth International Tin Agreement (ITA) have been subject to export controls, in order to re-establish the balance between the supply and demand of tin on the markets. These controls caused a marked decline in production in 1983 and an additional drop in 1984. The operations of the buffer stock manager have helped keep tin prices relatively stable during the past two years. However, these operations have been complicated by the major fluctuations in a number of currencies. Development work is continuing on a large tin deposit in Nova Scotia. Commercial production is expected to begin in October 1985.

CANADA

Canada still produces relatively little tin, but ranks among the dozen largest, non-communist consuming countries. Production of tin in concentrates and tin-lead alloy increased in 1983 and 1984.

Canada relies on imports for its tin requirements, except for small amounts recovered from recycled solders and detinning, and in primary tin-lead alloy production. Consumption was falling for several years, but this trend slowed down in 1983 and could be reversed in 1984.

Tin concentrates are recovered as byproducts of base-metal mining by Cominco Ltd. at Kimberley, British Columbia. Cominco also recovers a tin-lead alloy containing about 8 per cent tin at its Trail, British Columbia smelter and produces small quantities of special high purity tin from imported commercial-grade metal. Some Yukon placer gold deposits contain tin and tungsten and small quantities of these metals have been recovered in placer mining operations.

Tin mineralization is known in various parts of Canada, and higher prices in recent years encouraged exploration. The most promising reported discovery is the East Kemptville tin deposit near Yarmouth, Nova Scotia, discovered by Shell Canada Resources Limited in 1978. The deposit is estimated to contain some 56 million t grading about 0.16 per cent tin, recoverable by open-pit mining. In October 1982, the property was purchased by Rio Algom Ltd., Toronto. Work is in progress to begin production by October 1985.

The mine will produce concentrates that will be exported for smelting. The mill's capacity is set at 9 000 tpd of ore. The concentrates produced will be the equivalent of over 4 000 tpy of tin, or close to the current domestic tin consumption level. The life of the mine is expected to be 17 years. Concentrate production will peak early in the operations, then fall off with the passing years.

The total cost of the project is assessed at $150 million. Although there are other known tin deposits and exploration work continues across the country, no other plans were announced for their development.

THE INTERNATIONAL TIN AGREEMENT

Tin is the only metal for which there is an intergovernmental agreement involving producing and consuming countries that contains economic provisions for market stabilization. Successive five-year pacts have been in force since 1956. The Sixth International Tin Agreement entered provisionally into force on July 1, 1982, to replace the Fifth Agreement that had been extended by one year to allow more time to negotiate its successor. Provision is made in the agreements for market stabilization measures, including purchases and sales under a buffer stock arrangement, and the implementation of export controls on producing members if buffer stock operations are insufficient to protect the floor price.

A. Bourassa is with the Mineral Policy Sector, Energy, Mines and Resources Canada. Telephone (613) 995-9466.

Upon its entry into force, countries that had either signed or ratified the Sixth Agreement comprised six producers (Australia, Indonesia, Malaysia, Nigeria, Thailand and Zaire), which together accounted for 70 per cent of reported 1982 world tin mine production, and 18 consuming members, including Canada, which together accounted for 51 per cent of 1982 world tin consumption. Leading members of the Fifth Agreement that did not join the Sixth included the United States, U.S.S.R. and Bolivia.

The Sixth Agreement, as negotiated, provided for a buffer stock of up to 50 000 t of tin: 30 000 t financed by mandatory cash contributions from both producing and consuming members and 20 000 t by borrowing, with member government guarantees if necessary. Mandatory contributions from consuming members were introduced for the first time in the Sixth Agreement, replacing voluntary consumer contributions in the Fifth Agreement. The Sixth Agreement also provides for the imposition of export controls on producer members by a two-thirds majority vote when the buffer stock contains at least 35 000 t of tin, or by a simple majority vote when the buffer stock level reaches 40 000 t. Export controls are reviewed at each quarterly Tin Council meeting but may be eased automatically as the price improves.

Implementation of the Sixth Agreement required that countries accounting for at least 65 per cent of both production and consumption ratify the agreement by April 30, 1982. Although this level was not achieved on the consumption side, countries that had signed the agreement agreed to implement it provisionally on July 1, 1982. The member-financed buffer stock was reduced from 30 000 t to 19 666 t but the loan-financed portion remained at 20 000 t. Stockholdings necessary to permit implementation of export controls were reduced proportionately. Price levels established under the new agreement remained unchanged, with a floor of 29.15 Malaysian ringgets ($M) per kg and a ceiling of $M 37.89. The buffer stock must be a net buyer in the lower range ($M 29.15 - $M 32.06) and a net seller in the upper range ($M 34.98 - $M 37.89). These ranges were last changed in October 1981. Under the export control scheme, producers may stockpile excess tin in concentrates up to a maximum of about 25 per cent of their base annual production, to be held for smelting and sale upon removal of the controls.

ASSOCIATION OF TIN PRODUCING COUNTRIES

The Governments of Malaysia, Indonesia and Thailand were reported to have reached agreement in principle in mid-1982 to establish a tin producers association. This initiative was generally perceived as an indication of concern about the implementation of the Sixth ITA and about its ability to effectively enforce tin prices, if it were implemented. Thus, from the outset, Malaysia pressed for market stabilization measures, should the international agreement prove ineffective.

The Association was officially formed on August 13, 1983, after lengthy negotiations. It had five members: Bolivia, Malaysia, Indonesia, Thailand and Zaire. Nigeria joined on August 31 and Australia in November of the same year. These seven participating members produce 75 per cent of the tin in non-communist countries. The Association's head office is in Kuala Lumpur, Malaysia.

The main objectives proposed by the Association are the promotion of tin use through research and technological development, support of the Sixth ITA's market stabilization activities and an increase in the economic fallout from tin production in the economies of member countries.

Members agree on objectives and the nature of the problems affecting tin markets, but do not all agree on what measures to take if the Sixth ITA is unable to ensure market stability.

The Association works closely with the International Tin Research Council in London, England and the South East Asia Tin Research and Development Centre, in Malaysia. Both of these organizations are already financed by these same tin producers.

TIN MARKETS AND PRICES

Trends in reported annual world tin production, consumption and prices since 1970 are shown in Table 3. Since 1973, world tin consumption has trended downwards because of substitution away from tin in some end uses and because of technological developments that have decreased the quantities of tin for other uses. Although this decline halted momentarily in 1983 and 1984, it is still too early to speak of a recovery. The structural factors that caused the decline in previous years are apparently still at work. However, one can probably speak of consumption stabilization around the 1984 level, due to the growth potential of certain tin uses.

The production of tin in concentrates peaked in 1982, and primary production reached a high in 1979. Both recorded a marked decline in 1983 and 1984. These drops are the result of export restrictions imposed on the producing members of the Sixth ITA. Although Bolivia did not ratify this agreement, it had agreed to apply the same constraints to its own production, if necessary. Brazil practically doubled its production since 1982. The reader will note the growing difference in Table 3 between tin production and consumption from 1978 to 1982.

Statistics in the accompanying tables do not include information on most centrally planned countries. Leading producers among these countries include the U.S.S.R. and the People's Republic of China, for which the United States Bureau of Mines estimates production in 1983 at 37 000 t and 15 000 t, respectively. The Democratic Republic of Germany is estimated to have produced 1 700 t in 1983. This tin is generally consumed domestically, although China is a net exporter to the west (estimated at 2 500 t in 1983) while the U.S.S.R. and East Germany had estimated combined net imports from the west of over 17 000 t in 1983.

A substantial tonnage of tin production, which is included in the accompanying tables, is of unspecified origin. This was estimated at 16 000 t in 1983 and 10 000 t in 1984. These tin concentrates are generally channelled through the free port of Singapore where they are smelted or re-exported to smelters elsewhere. Their origin is believed to be largely southeast Asian countries, particularly Thailand, having been smuggled out of these countries to avoid payment of royalties and export duties and, since 1982, to bypass export controls.

In 1983 and 1984, tin prices were subject to the fluctuations of various foreign currencies. Two currencies are especially relevant to the tin market: the Malaysian dollar and the British pound. The latter is important because most tin market financial transactions occur on the London Metal Exchange. London is therefore an influential centre for determining the international price of tin.

The Malaysian market is an important physical market; in addition, the floor and ceiling prices used by the buffer stock manager are set in Malaysian dollars. This is why watching the price of tin on the Malaysian market is a priority for the manager. This price, in turn, influences the London market, where the manager will also intervene to attempt to maintain parity between both markets. In the past two years, the Malaysian and American dollars have evolved in a similar manner on the exchange market, but the British pound recorded major falls compared to American currency. This is why, depending on the market under consideration, tin prices have been relatively stable or on the rise.

Since October 17, 1981, there has been no change in the floor and ceiling prices set under the aegis of International Tin Agreements. The floor price has remained at \$M 29.15/kg and the ceiling at \$M 29.15/kg. In 1983, the price of tin on the Malaysian market stayed mainly in the margin of \$M 30.00 to \$M 31.00/kg. The 1984 prices, however, remained almost constantly at the \$M 29.15 floor price. However, the behaviour of tin on the London market is entirely different. On this market, tin rose from its January 1983 price of £7,669/t to almost £10,000/t by the end of 1984: an historic record for this metal. Despite this rise, the London price was frequently below the floor price equivalent in Malaysian dollars. The buffer stock manager's interventions on the London market to maintain a certain parity between both markets, also contributed to this rise in the price of tin. The price of tin for the majority of world tin consumers therefore rose considerably in the past two years, despite a very abundant supply. Accordingly, the competitive position of tin is at a disadvantage. One encouraging sign on the market is that tin stocks are on the decline. In 1984, they dropped by one-half on the London market, or by approximately 20 000 t. This reduction is the result of export restrictions imposed on producing members of the Sixth ITA. Export cutbacks have been raised from 36 per cent in early 1983, to 39.6 per cent in July, 1983. Unfortunately, the stocks held by the buffer stock manager did not experience these declines and are still at over 55 000 t, of which approximately 40 per cent are held under the terms of the Fifth ITA; he has been unable to bring them back on the market to date.

The United States General Services Administration (GSA) continued offerings of tin from the strategic stockpile. Sales in 1983 totalled 2 860 t, and should not exceed 2 400 t in 1984 (all GSA figures are in long tons). Total sales by the end of 1984 were about 15 382 t, under a program begun in 1980. The goal for the United States strategic stockpile is 42 700 t, whereas the stockpile actually contained over 188 000 t at the end of 1984. By October 1, 1984, sales of 20 000 t were authorized. Although GSA sales represent only a small percentage of the total market, producers believe that they place great downward pressure on prices, especially under the current unfavourable circumstances. Pressure has been exerted on the GSA to suspend sales until the market has recovered.

The Penang market in Malaysia was a physical market in tin. Prices were set by the two large tin smelters in the region, using a relatively closed process. Since 1980, palm oil and rubber have been traded on the exchange in Kuala Lumpur. In October 1984, this exchange replaced the Penang market and also deals in tin. It operates like the London exchange and members of the Sixth ITA recognized this market for the operations of the buffer stock manager. This new market, like Penang's, still allows only physical transactions.

WORLD DEVELOPMENTS

For two years, tin concentrate production in Australia has fallen by 30 per cent, to 8 600 t in 1984, while metal production fell by 10 per cent to 2 800 t. This reduction was caused by export controls. Tasmania is still the main tin mining centre, with the Renison Ltd. mine, which yields over 40 per cent of the total production. This mine produced 3 800 t in 1984. Because of tin production restrictions, Renison Goldfields Consolidatd Ltd., the parent company, is increasingly orienting its exploration toward base metals and gold. Aberfoyle Ltd., controlled by Cominco, which has 47 per cent of the shares, has two tin mines: in Cleveland and Ardlethan. Despite temporary closures and a reduction in operations, production exceeded the allowable quota. Therefore, by the end of 1983, tin inventories had been assessed at close to $A 10 million. The company, with 50 per cent of its revenues still coming from tin, is turning toward diversification in order to expand. The Tin Creek Mining alluvial mining project at Granite Creek began production in early-1984. This combined gold and tin extraction project received a tin production quota of 76 tpy. Greenbushes Tin Mining obtained a $15 million loan for its tin-tantalum mine at Greenbushes. The company also concluded an agreement with Barbara Mining Corp. for 50 per cent participation in the tin-tantalum project in Bynoe, Northern Territory. In June 1984, Endeavour Resources Ltd. started up its alluvial gold-tin-tantalum mine in Moolyella, Western Australia. Great Northern Mining Corp. stopped its copper production in February 1984, but continues to mine tin according to the allowable quota of 40 t quarterly. In August 1984, Metals Exploration Ltd. announced the beginning of production at the tin alluvial project in Gibsonvale, New South Wales. In June 1984, Shell Company of Australia Ltd. purchased from North Broken Hill Holdings a 50 per cent share in the Collingwood tin property in Northern Queensland. A three-year exploration program is being prepared.

Burma's production continues to rise. During the fiscal year ending March 31, 1984, tin in concentrates production rose to 2 240 t, compared to 647 t the previous year. The Rangoon smelter, in operation since 1982 and built with assistance from North Korea, produced 500 t of metal last year. In 1983, Burma exported 717 t of concentrates and 320 t of metal.

Bolivia was a member of the first five International Tin Agreements, but refused to ratify the sixth. However, it maintains close ties with the producing members of the Sixth ITA, as it belongs to the Association of Tin Producing Countries. It has also agreed to make its tin production policy conform with that of the Sixth ITA members. However, the reductions in tin concentrate and metal production over the past two years are probably as much the result of export controls as of Bolivia's serious economic problems. It can no longer meet the deadlines for its enormous foreign debt of $3,700 million, and the tin industry is suffering the effects of this difficult situation.

The mining and metallurgical industry is important for Bolivia. It accounts for 80,000 jobs, 25 per cent of government revenue and one-half of foreign exchange earnings. The tin industry represents 60 per cent of the country's mineral industry. Most of Bolivia's mining industry belongs to the state-owned Corporacion Minera de Bolivia (COMIBOL).

However, there are many small mining developers and cooperatives.

COMIBOL produces approximately 75 per cent of tin concentrates in Bolivia, but production has been declining steadily for a number of years. This fall has been caused by decreasing ore grades and reserves, more difficult mining conditions, production costs that exceed prices, and labour problems. COMIBOL was unable to obtain funds to modernize its operations and purchase spare parts for equipment. After many strikes, COMIBOL was restructured; workers are now in the majority on the board of directors.

Tin concentrate exporting by private producers is prohibited. All concentrates must be sold at the smelter in Vinto or the Banco Minera, both state-owned and the only authorized exporters of excess concentrate.

An ambitious $750 million five-year development plan was proposed for the country's mining and metallurgical industry. However, it is doubtful whether aid organizations will grant a large portion of the aid requested. Bolivia recently obtained a $24 million loan from the Inter-American Development Bank to develop the country's small- and medium-sized mining operations.

Tin production in Brazil increased considerably over the past two years. Tin concentrate production doubled to 17 700 t in 1984, while metal production increased by 80 per cent to 16 900 t. This increase occurred at the same time as large surpluses appeared on the market and coincides with the implementation of export restrictions by producing members of the Sixth ITA. Brazil did not ratify this Agreement.

The government encourages growth of this industry, as it acquires currencies that help repay the country's massive foreign debt. Rising mining production comes mainly from the new mine in Pitingo, in the Amazon, 200 km north of Manaus. This mine belongs to Paranapanema, Brazil's largest producer; in 1984, its production was estimated at 10 000 t. Paranapanema has its own smelter, with a capacity of 12 000 tpy in Sao Paulo, where it treats all the concentrates coming from the alluvial mines in Randonia, Mato Grosso and Para.

The second largest producer is Brascan Recursos Naturais, whose interests are held jointly by British Petroleum and Brascan of Canada. In 1984-85, Brascan will invest $66 million in the Rondonia tin operations. Brascan production in 1984 was estimated at 3 500 t. Brumadinho is the third producer in Brazil. Its production was estimated at 2 400 t in 1984. The country has a number of smelters with a total capacity estimated at 24 000 tpy. Exporting of tin concentrates is prohibited in order to maximize the added value of exports. The country consumes approximately 4 000 tpy of tin; the remainder is exported. It is generally believed that Brazil will continue to increase its production in the future, although at a slower rate. Production is expected to reach 20 000 tpy by the end of the decade. The country's mine operating costs are unknown, but the remoteness of the deposits probably generates high costs. The main Brazilian producers are continuing important exploration programs in the country's alluvial regions. Tin is still being sought, although the emphasis is now more on gold.

India has been a tin producer since 1984. A small mine was opened in Koraput, Orissa. Production will be only 30 tpy. India consumes over 2 000 tpy of tin.

Indonesia's concentrate production fell by one-third between 1982 and 1984, due to the export controls. Metal production was reduced by one-quarter. However, Indonesia is the largest tin producer after Malaysia. In order to increase tin producers' profitability, the government lifted the 10 per cent tax on all tin exports in July 1984. Government-owned P.T. Tambang Timah accounts for over 75 per cent of Indonesian production. Like the other producers, it suffered 60 per cent increases in the cost of its petroleum products in early-1984. These products account for about 50 per cent of production costs. At the end of 1984, it received delivery of a new dredge, Singkep 1, which will operate in the Kundur Laut region near Singkep, where a washing plant has already been built. In cooperation with P.T. Krakatau Steel and P.T. Nusantara Ampera Bakti, it is continuing construction at Cilegon, Java, of a tinplate mill with a capacity of 130 000 t, which will ensure the country's self-sufficiency in tinplate. P.T. Koba Tin, the second largest tin producer, has produced approximately 5 000 t of tin at its mine on Banyka Island. This company is jointly-owned by P.T. Tambang Timah and Kajaura Mining Corp. Ltd. of Australia. Broken Hill Proprietary Indonesia was placed on sale by its Australian parent company. It operates the Kilapa Kampit mine on Belitaung Island. There are enough attractive tin deposits identified in Indonesia to allow for a substantial increase in production when the market permits.

Malaysia is still the world's largest producer of tin. In absolute terms, it is therefore the producer that suffers most from export restrictions. Since 1981, the last year of operation without controls, 10,000 jobs were lost, almost 200 mines were closed and 25 dredges became inactive. Annual tin in concentrate production fell by 20 000 t. In 1984, there were fewer than 25,000 jobs, 530 mines, 34 dredges and 39 000 t of production. Financial losses were widespread among the small producers, who had to stop operations; the profitability rate of the largest companies fell considerably. Kuala Langat Mining, formed to mine the largest tin deposit in the state of Selangor, continues construction of its facilities. However, the absence of production quotas is causing uncertainty for this project, which is to begin in 1986. Known reserves in Malaysia are falling, but a number of apparently promising territories still remain; exploration of these areas will begin once government permission is issued.

Nigeria's tin industry continues to face serious difficulties. The alluvial industry has disappeared, and the only remaining mines are located in the central plateau region. The allowable quotas are too low to generate sufficient revenues for the return on the investments that would be needed to modernize or increase the production of increasingly deep mines. The government is studying the possibility of merging the five major producers into a single state-owned company, Nigerian Tin Mining Co., in order to rationalize operations. The government already holds majority shares in the five companies. The Makeri Smelting Co. smelter, the only one in Nigeria, now receives so little tin that it operates for only 10 per cent of the year.

The production of concentrates at the San Rafael mine south of Peru has now exceeded 2 200 tpy since the introduction of a new flotation circuit in 1982. This mine, owned by Minsur S.A., is located at an altitude of 4 800 m, 180 km from Juliaca. Originally a copper mine, its tin grades have been improving as the depth of the mine increases. This region is the continuation of Bolivia's tin belt and there appears to be potential for new discoveries.

In the United Kingdom, tin production sagged by a few tonnes in 1983, but rose again to 4 400 t in 1984. This is the highest level since World War I. Although the United Kingdom is a member of the ITA, it is classed as a consumer and therefore its production is not constrained by export controls. Each of the three major tin producing companies in the country implemented expansion programs. The Geevor mine has just completed expansion of its mill, thus increasing its capacity by 25 per cent. With this additional capacity, it will be possible to treat surface tailings and increase underground production if the exploration program in the Batallack region uncovers sufficient reserves. Geevor has also requested permission to mine the alluvial deposits in the Hayle River. The recent £9 million modernization program at the South Crofty mine increased production to 1 800 t in 1984. Wheal Pendarves, a subsidiary of South Crofty, suspended its operations at the end of 1984, as its reserves were depleted. An exploration program currently in progress could, however, result in the mine's reopening in one or two years. Carnon Consolidated Tin mines anticipates the Wheal Jane/Mt. Wellington complex production to increase to 2 500 t in 1985; this is a rise of 850 t over current capacity. The company is also considering development of the Bissoe Valley project, which involves the treatment of old tailings and estuarine muds. This production could reach 650 tpy. The Wheal Concord mine, in the hands of a receiver for 15 months, was repurchased by CTS Mining in June 1984. The latter wishes to reopen the mine and plans an ambitious expansion program over two or three years. Marine Mining Consortium Ltd. continues construction of its mill in Gwithian; the company will dredge the ocean floor, where residues from mines in the Cornwall region were deposited for centuries. It would grade one kg of tin for two t of residue. The amount of tin produced in the country's smelters has fallen drastically for a number of years. However, the increased local production of concentrates and the treatment of concentrates from Canada will cause a rise by 1985.

Rwanda has requested financial assistance from the European Economic Community's Sysmin program, in order to modernize its tin industry. It has also approached the European Development Bank and the International Finance Corp. Small mining operations produce 60 per cent of the tin concentrate, estimated at 1 275 t.

Thailand's production, like that of other producers in southeast Asia, fell in the past two years. After a 23 per cent drop to 19 942 t, tin in concentrate production rose slightly in 1984 to 22 000 t. The decline was therefore smaller than in neighbouring countries, but this could contribute to the increase in stocks. Metal production fell by 7 000 t to 18 467 t and should increase by only 400 t in 1984. In September 1983, the government cut royalties on tin by the equivalent of $US 0.35 to $US 1.15 per pound. This measure came in response to pressures from producers who were caught between high taxes and costs, and falling prices on the London market. In March 1984, the government also cancelled the new tin mining permits in order to facilitate the decrease in production. Despite a difficult situation, a number of companies, among them, Metals Exploration and Development Co., a subsidiary of Billiton Int'l, Sea Minerals Ltd., a subsidiary of Aakom Thai, Tongkak Harbour and IFC Ltd., maintained important exploration programs, particularly along the Andaman coast. In 1983, Sea Minerals announced an interesting discovery 25 km from Phuket. In June 1984, it announced a 36 million baht increase in its capitalization, in order to continue its exploration programs. In 1983, the government signed a $25 million agreement for geophysical surveys of the entire territory. The project was granted to Kenting Earth Sciences Ltd. of Canada. The Thaisarco smelter in Phuket operated at only one-half of its 38 000 t capacity. The Thai Pioneer smelter was closed because of financial problems. The Cha-am Pineapple Tinplate Co.'s tinplate production project has been delayed and will probably not be completed before 1986. The only current tinplate producer, Thai Tinplate Mfg. Co., has installed a new production line that will increase its capacity to 150 000 tpy.

In 1983, Zimbabwe produced 1 235 t of tin, or 38 t more than in 1982. This slight increase is due to the operation of a residual tin recovery unit at the Kamativi mine.

OCCURRENCE AND RECOVERY

About 80 per cent of the world's tin output is derived from alluvial deposits. The principal production methods are bucket-line dredging and gravel pump operations. Suction dredges are also used, but they tend to be less efficient than the bucket-line method. Other methods are hydraulicking and dulang washing. Tin is recovered as cassiterite (SnO_2) and is often associated with other heavy minerals such as ilmenite, zircon, wolframite (tungsten), tantalite and others.

Economic grades of placer deposits generally range from 0.15 to 0.40 kg of tin per cubic metre of sand, or from 0.008 to 0.02 per cent tin. Leaders in placer tin production are Malaysia, Indonesia, Thailand and, more recently, Brazil.

Lode mining, though less important than alluvial mining, still accounts for most of the tin output of Bolivia, Australia, Britain and South Africa. Countries of the communist bloc, notably The People's Republic of China and the U.S.S.R., are also important producers of tin from lode as well as alluvial deposits. Viable lode deposits normally range in tin grade from 0.4 per cent or less in open-pit mines to 0.9-1.0 per cent or more in underground mines. Silver, tungsten, bismuth and lead are common byproducts of lode mines. Cassiterite is the predominant tin-bearing mineral of lode deposits but stannite, a copper-tin-iron-bearing sulphide, is of some importance.

Average grades in both placer and lode mining tended to decline during the 1970s and early-1980s, and this trend is expected to continue. Productivity improvements have offset part but not always all of this decline and real tin production costs have also risen, exacerbated by high royalty and tax rates levied by some producing countries.

Concentrating processes for alluvial and most lode tin are based on relatively simple gravity separation methods that produce concentrates ranging from 50 to 76 per cent tin. Magnetic and electrostatic separation are also used. However, mill recoveries of tin from lode deposits often are quite low by base-metal standards and some companies have installed flotation cells in their beneficiating plants to complement gravity separation and improve the recovery of tin and other metals. Fuming processes, which can recover tin as tin oxide from slags, residues, low-grade concentrates and even directly from ores, are being used increasingly to improve overall tin recovery. The impure oxide is converted to metal in conventional smelters.

USES

The major use of tin is in tinplate and tinning, which account for over 40 per cent of the world's consumption. The manufacture of solders is the second-largest use of tin, accounting for just over one-quarter of the world's total. Tin is also used in the manufacture of babbit, bronze and brass alloys, pewter, and a wide range of tin chemicals.

Tin use in tinplate generally remained flat over the past few years in most industrial countries or has declined. In the United States, tinplate consumption has fallen as aluminum almost totally replaced tinplate in the large beverage can market, but there has so far been relatively little penetration of aluminum into the food can market. For the remaining market, tin is being used more efficiently as tin used in tinplating declined from about $5\frac{1}{2}$ t per thousand t of steel in the mid-1960s to about 4 t in 1982. Tinplate consumption has remained relatively stable in western Europe and Japan, where there has been only limited penetration by aluminum into can markets. Moreover, tin used per thousand t of steel in tinplate is higher in both these markets, at about 5 t and over 6 t, respectively. Both regions are also significant exporters of tinplate, but the growing production of tinplate in developing countries is likely to curtail this trade in the future.

The solder and bronze/brass markets, other important uses for tin, are both relatively mature. In solder uses, the strong growth in electronics applications is partially tempered by increasing miniaturization, which reduces the amount of solder used per unit. However, the electronics component market is growing rapidly, and tin consumption should also follow this trend. The use of solders in automobile production is declining as alternative materials and fabricating techniques are introduced. Any large-scale substitution of aluminum for copper radiators would significantly reduce solder and therefore tin use. Bronze, brass and other tin-containing alloys are widely used in construction, machinery and equipment and consumer durables. Growth in these applications has tended to be relatively slow and some are vulnerable to substitution, for example by plastics in plumbing and aluminum in refrigeration and air conditioning.

Tin consumption prospects are more promising in a wide range of chemical, including agricultural, applications. While no particular developments stand out as individually larger tin users, future market growth in this area is likely to be stronger than in the traditional tinplate and other alloy applications. Organotin chemicals in particular have a wide range of applications in wood preservatives, anti-fouling paints, plastics and crop protection.

The International Tin Research Council, which is the only major organization conducting research and development on tin applications and promoting the use of tin, celebrated its 50th anniversary in 1982. Its headquarters and research facilities are on the outskirts of London, England, and Tin Information Centres are located in Australia, Belgium, West Germany, Japan and United States, with part-time representatives in Brazil, Italy and the Netherlands. Funding is contributed by the governments of Indonesia, Malaysia, Nigeria, Thailand and Zaire. Bolivia was a member until 1981, and Australian tin producing companies contribute voluntarily.

PRICING MECHANISMS AND TARIFFS

The principal tin markets are centred on the Kuala Lumpur market in Malaysia and the London Metal Exchange, which trades in both cash (spot) metal and three months future contracts. In October 1984, the traditional market in Penang was replaced by the Kuala Lumpur market, which still trades only in cash (spot) metal. It is less restrictive than the Penang market, where prices were fixed jointly by the two main smelters. Its workings are based on those of the London Exchange.

The tariffs of Canada and the United States are listed in the Tariff table. Neither has tariffs on tin ores, concentrates or wrought tin, and both agreed during the Tokyo Round of GATT to reduce MFN rates on tin-containing manufactures over an eight-year period beginning in 1980. Tariffs levied by the European Economic Community and Japan are broadly similar to those of the United States, being free for ore, concentrates and unwrought metal from all sources and mostly between 4 and 8 per cent on tin products (MFN), but free from developing countries.

OUTLOOK

World tin consumption, as defined in Table 4, peaked in 1973 at 214 900 t, a level that has never since been matched. Rising tin prices during the 1970s encouraged efforts to find substitutes and it is only in the developing countries, where the use of tinplate for containers is rising, that growth has been at all consistent. Rising tin prices were attributed to cost pressures (energy, equipment, etc.) and generally falling ore grades. Higher prices have encouraged greater interest in tin exploration, but this was initially slow to impact on tin supply. Unfortunately, it occurred at a time when world consumption would accelerate its fall. From 1979, world production exceeded consumption, and this gap widened until 1982, when export controls were implemented. Excess inventories should now decline slowly, but export controls are likely to be necessary for several more years.

Since 1982 and 1983, when it bottomed out, tin consumption should be able to increase very slowly during the remainder of the 1980s. Growth in tinplate production and consumption in developing countries will perhaps be able to compensate for the progressive decline that will continue to be felt in industrialized countries. On the solder market, the heavy growth of electronics applications should more than compensate for the miniaturization, and by 1987 will show an 8 000 t net increase over the 1983 figure for tin consumption in this sector. Tin consumption for the tinning of wires and electronics components and for chemicals should also increase from 5 000 to 6 000 t by 1987. Tin consumption is therefore expected to reach 167 000 t by 1987.

In the same period, prices could show a slight nominal increase, but the large remaining inventories will probably force the real price downward each year. A real increase in prices would probably harm the competitive position of tin at a time of growing substitution away from tin for its major use on traditional markets. By the end of 1985, discussions should begin on a Seventh International Tin Agreement. Current difficulties on the markets and the fact that the Sixth ITA will probably never be fully implemented appear to indicate that negotiations may be complicated and difficult.

TARIFFS

Item No.		British Preferential	Most Favoured Nation	General	General Preferential
			(%)		
CANADA					
32900-1	Tin in ores and concentrates	free	free	free	free
33507-1	Tin oxides	free	13.8	25	free
33910-1	Collapsible tubes of tin or lead coated with tin	10	13.9	30	free
34200-1	Phosphor tin	5	6.5	10	4.0
34300-1	Tin in blocks, pigs, bars or granual form	free	free	free	free
34400-1	Tin strip waste and tin foil	free	free	free	free
38203-1	Sheet or strip, iron or steel, corrugated or not, coated with tin	10	11.0	25	7.0
43220-1	Manufacturers of tin plate	14.8	13.9	30	9.0

MFN: Reductions under GATT (effective January 1 or year given)	1983	1984	1985	1986	1987
			(%)		
33507-1	13.8	13.4	13.1	12.8	12.5
33910-1	13.9	12.9	12.0	11.1	10.2
34200-1	6.5	6.3	6.0	5.8	5.5
38203-1	11.0	10.3	9.5	8.8	8.0
43220-1	13.9	12.9	12.0	11.1	10.2

UNITED STATES (MFN)

Item No.					
601.48	Tin ore and black oxide in tin			free	
622.02	Unwrought tin other than alloys of tin			free	
622.04	Unwrought tin, alloys of tin			free	
622.06	Unwrought tin, other			free	
622.10	Tin waste and scrap			free	

		1983	1984	1985	1986	1987
				(%)		
622.15	Tin plates, sheets and strips, not clad	4.2	3.8	3.3	2.9	2.4
622.17	Tin plates, sheets and strips, clad	8.4	7.5	6.6	5.7	4.8
622.20	Tin wire, not coated or plated with metal	2.4	2.4	2.4	2.4	2.4
622.22	Tin wire, coated or plated with metal	5.1	4.9	4.7	4.4	4.2
622.25	Tin bars, rods, angles shapes and sections	5.1	4.9	4.7	4.4	4.2
622.35	Tin powder and flakes	5.1	4.9	4.7	4.4	4.2
622.40	Tin pipes, tubes and blanks	4.2	3.8	3.3	2.9	2.4
644.15	Tin foil	12.3	10.9	9.6	8.3	7.0

Sources: The Customs Tariff and Commodities Index, 1982, Revenue Canada; Tariff Schedules of the United States Annotated 1982, USITC Publication 1200; U.S. Federal Register, Vol.44, No. 241.

TABLE 1. CANADA, TIN PRODUCTION, IMPORTS AND CONSUMPTION, 1982-84

	1982		1983P		1984	
	(tonnes)	($)	(tonnes)	($)	(tonnes)	($)
Production						
Tin content of tin concentrates and lead-tin alloys	135	1,915,000	140	2,013,010	217e	2,997,972e
Imports					(Jan.-Sept.)	
Blocks, pigs, bars						
United States	1 920	33,200,000	1 393	21,324,000	1 348	21,216,000
Brazil	602	9,939,000	980	15,933,000	508	8,371,000
Bolivia	451	6,993,000	798	13,028,000	334	5,468,000
Malaysia	-	-	240	4,068,000	360	5,966,000
Belgium-Luxembourg	210	3,522,000	160	2,611,000	65	1,069,000
Other countries	52	901,000	178	2,270,000	564	8,352,000
Total	3 235	54,555,000	3 749	59,234,000	3 179	50,442,000
Tinplate						
United States	2 049	2,002,000	1 899	1,906,000	2 010	1,818,000
West Germany	2 295	1,882,000	298	239,000	-	-
United Kingdom	43	75,000	3	3,000	-	-
Total	4 387	3,959,000	2 200	2,148,000	2 010	1,818,000
Tin, fabricated materials, nes						
United States	294	1,137,000	320	1,432,000	196	1,012,000
West Germany	2	11,000	9	58,000	3	17,000
United Kingdom	7	42,000	7	49,000	14	70,000
Other countries	4	21,000	13	48,000	7	32,000
Total	307	1,211,000	349	1,587,000	220	1,131,000
Exports						
Tin in ores, concentrates and scrap[1]						
United Kingdom	16	5,000	272	1,647,000	231	1,211,000
United States	386	959,000	49	262,000	19	108,000
Spain	68	452,000	52	225,000	-	-
U.S.S.R.	46	672,000	-	-	-	-
Other countries	85	602,000	-	-	-	-
Total	601	2,690,000	373	2,134,000	250	1,319,000
Tinplate scrap						
United States	2 145	222,000	4 984	226,000	3 204	137,000
Indonesia	-	-	305	125,000	-	-
Italy	-	-	94	38,000	-	-
Taiwan	-	-	34	13,000	36	9,000
Other countries	105	22,000	-	-	-	-
Total	2 250	244,000	5 417	402,000	3 240	146,000
Consumption						
Tinplate and tinning	2 034	..	2 049	..		
Solder	1 212	..	1 059	..		
Babbit	131	..	174	..		
Bronze	37	..	60	..		
Other uses (including collapsible containers, foil, etc.)	114	..	73	..		
Total	3 528	..	3 415	..		

Sources: Energy, Mines and Resources Canada; Statistics Canada.
[1] Tin content of ores and concentrates plus gross weight of tin scrap.
P Preliminary; .. Not available; - Nil; e Estimate.

TABLE 2. CANADA, TIN PRODUCTION, EXPORTS, IMPORTS AND CONSUMPTION, 1970, 1975 AND 1979-84

	Production[1]	Exports[2]	Imports[3]	Consumption[4]
		(tonnes)		
1970	120	268	5 111	4 565
1975	319	1 052	4 487	4 315
1979	337	712	4 689	4 675
1980	243	883r	4 527	4 517
1981	239	513	3 791	3 766
1982	135	601	3 235	3 528
1983P	140	373	3 749	3 415
1984P	217	250[5]	3 179[5]	..

Sources: Energy, Mines and Resources Canada; Statistics Canada.
[1] Tin content of tin concentrates shipped plus tin content of lead-tin alloys produced. [2] Tin in ores and concentrates and tin scrap, and re-exported primary tin. [3] Tin metal.
[4] Current coverage exceeds 90 per cent, whereas until 1972, coverage was in the order of 80 to 85 per cent. [5] Jan.-Sept. only.
P Preliminary; r Revised.

TABLE 3. WORLD[1] TIN PRODUCTION, CONSUMPTION AND PRICES, 1970 TO 1984

	Production		Consumption	Prices	
	Tin in Conc.	Primary Metal		Malaysia[2]	NY Dealer[3]
		(000 t)			
1970	185	185	185	10.99	1.74
1971	188	187	189	10.44	1.67
1972	196	191	192	10.36	1.77
1973	189	188	215	11.35	2.27
1974	184	182	200	18.79	3.96
1975	181	179	173	15.94	3.40
1976	180	183	194	18.96	3.75
1977	188	180	185	26.26	5.33
1978	197	194	185	28.82	5.89
1979	200	201	186	32.42	7.07
1980	201	198	174	35.72	7.86
1981	205	197	163	32.34	6.80
1982	190	180	157	30.09	6.20
1983	172	159	155	30.19	6.17
1984e	164	151	161	29.16	5.90

Source: International Tin Council.
[1] Coverage is the same as in Tables 4 and 5. [2] Cash price ex-smelter for Grade A tin, shipment within 60 days, in Malaysian ringgits per kg, the ringgit being the unit used to define price levels under successive International Tin Agreements. [3] Median of prices for Grade A tin, in U.S. dollars per pound, ex-dock New York, submitted by participating dealers for delivery within seven business days.
e Estimate.

TABLE 4. WORLD[1] CONSUMPTION OF PRIMARY[2] TIN, 1970, 1982, 1983 AND 1984

	1970	1982	1983	1984e
		(tonnes)		
EEC, total[3]	58 246	39 936	38 214	38 500
West Germany	14 062	13 163	13 792	15 100
France	10 500	8 187	7 564	7 800
United Kingdom	16 951	6 979	6 123	4 600
Netherlands	5 467	5 142	4 400	5 000
Italy	7 200	4 200	4 200	4 400
Belgium/Luxembourg	3 000	1 889	1 804	1 600
United States	53 807	33 000	34 300	34 900
Japan	24 710	28 707	30 504	32 400
Spain	3 040	4 400	4 400	4 400
Poland	..	4 575	4 351	4 300
Brazil	2 139	4 953	3 942	..
Canada	4 640	3 400	3 776	4 000
Czechoslovakia	3 420	3 500	3 550	3 500
Republic of Korea	394	2 093	2 628	..
Australia	3 837	2 700	2 500	2 500
Total, incl. Others	184 800	153 500	154 700	160 800

Source: International Tin Council.
[1] Excludes countries with centrally planned economies, except Bulgaria, Czechoslovakia, Hungary, Poland, Romania and Yugoslavia. [2] May include secondary tin in some countries. [3] Includes all 1982 members in all years except Greece in 1970.
.. Not available; e Estimate.

TABLE 5. WORLD[1] PRODUCTION OF TIN-IN-CONCENTRATES, 1970, 1982, 1983 AND 1984

	1970	1982	1983	1984e
		(tonnes)		
Malaysia	73 794	52 342	41 367	39 000
Indonesia	19 092	33 800	26 554	23 300
Bolivia	30 100	26 773	24 736	20 000
Thailand	21 779	26 207	19 942	22 100
Brazil	3 610	8 218	13 083	17 700
Australia	8 828	12 615	9 578	8 600
United Kingdom	1 722	4 175	4 067	4 900
South Africa	1 986	3 035	2 668	..
Peru	20	1 700	2 200	..
Zaire	6 458	2 174	2 004	2 300
Total, incl. Others	184 900	190 500	171 700	164 300

Source: International Tin Council.
[1] Excludes countries with centrally planned economies, except Czechoslovakia, Poland and Hungary.
.. Not available; e Estimate.

TABLE 6. WORLD[1] PRODUCTION OF PRIMARY TIN METAL, 1970, 1982, 1983 AND 1984

	1970	1982	1983	1984e
	(tonnes)			
Malaysia	91 945	62 836	53 338	44 800
Indonesia	5 190	29 755	28 390	22 800
Thailand	22 040	25 479	18 467	18 900
Bolivia	300	18 980	14 293	12 000
Brazil	3 100	9 297	12 560	16 900
United Kingdom	22 035	8 164	6 498	6 600
Netherlands	5 937	2 757	3 650	7 100
Australia	5 211	3 105	2 878	2 800
Spain	3 908	2 750	2 783	3 000
United States	4 540	3 500	2 500	3 500
South Africa	1 491	2 197	2 200	..
Singapore	..	4 000	1 800	..
Nigeria	8 069	1 691	1 400	1 300
Total, incl. Others	184 900	180 000	158 800	151 000

Sources: International Tin Council.
[1] Excludes countries with centrally planned economies, except Czechoslavakia, Poland and Hungary.
.. Not available; e estimate.

TABLE 7. MONTHLY AVERAGE TIN PRICES[1], 1983 AND 1984

	Canada Cdn ¢/lb		Dealer NY US ¢/lb		London Metals Exch. US Equiv. ¢/lb		Malaysia US Equiv. ¢/lb	
	1983	1984	1983	1984	1983	1984	1983	1984
January	772.78	791.01	553.43	569.05	547.86	549.50	579.80	565.12
February	801.37	794.44	600.53	574.81	591.09	557.64	585.52	566.54
March	818.87	807.69	619.57	581.68	609.19	562.57	598.60	576.83
April	847.93	817.15	633.86	584.19	625.37	564.34	620.41	577.66
May	835.86	839.03	620.43	586.00	613.23	570.55	612.95	573.92
June	824.76	842.28	613.91	588.33	605.56	574.20	600.67	574.01
July	812.92	845.83	609.85	574.81	598.31	564.00	593.89	565.42
August	799.97	828.78	593.74	565.94	581.14	557.86	583.84	566.87
September	797.48	823.50	593.62	555.68	578.43	548.28	580.00	562.60
October	796.99	807.87	596.00	540.05	578.20	532.53	580.94	551.02
November	814.46	815.73	602.32	553.00	579.19	543.79	583.85	551.87
December	799.77	803.15	578.14	540.00	559.81	531.61	569.00	547.49
Yearly Average	810.26	818.04	601.28	567.80	589.19	556.55	590.79	564.95

Sources: Metals Week; US General Services Administration; Northern Miner.
[1] Prices are for Grade A (in the U.S.) or High Grade - 99.85 per cent tin or more - except the LME price which is for Standard Grade - 99.75 per cent tin or more.

Titanium and Titanium Dioxide

D.E.C. KING

CANADA

Canadian industries based on titanium, include ilmenite mining and smelting, titanium oxide and pigment production, titanium metal fabrication to finished parts, coating of welding rods, and the manufacture of titanium carbide and nitride coated parts. Also, titanium-bearing master alloys are incorporated into special steel and aluminum alloys. The mining, smelting and pigment operations are carried out exclusively in Quebec, whereas the downstream activities are located in several provinces. Canada does not have any capacity for producing primary titanium (in the form of sponge or granules), ferrotitanium, or for vacuum melting of primary titanium to produce billets, although Atlas Steels division of Rio Algom Limited has facilities at Welland, Ontario, to custom forge and roll billets.

QIT - Fer et Titane Inc. (QIT) is the only company that mines titanium ore in Canada. Ilmenite, a mineral containing somewhat more iron than titanium, is mined at Havre St. Pierre, Quebec. The raw ore is shipped to Tracy, Quebec, where it is beneficiated, and the concentrate smelted to produce high quality pig iron and titania (TiO_2) slag (Sorelslag). QIT completed a $9.4 million expansion to its beneficiation plant during 1983, enabling a further removal of gangue material by dry magnetic separation. By smelting the improved concentrate (96 per cent ilmenite content compared with the former 92-93 per cent content) to a greater degree of reduction, the company has been able to produce Sorelslag containing about 80 per cent TiO_2 compared with the 70-72 per cent TiO_2 slag previously marketed. This operating mode has reduced furnace operating capacity to some degree, but except for this limitation QIT has been able to operate at full capacity in response to the recovery in the market for slag, particularly in the second half of 1984.

QIT in late-1984 announced plans for a $150 million capital investment at its Tracy operation. The electric furnaces will be up-graded and basic oxygen furnaces added, so that liquid iron from the electric furnaces can be further blown to produce steel. Of the 400 000 t of steel capacity in the new mode of operation, expected to start in 1986, a part will be used to produce steel powder. QIT already produces and markets Atomet iron powder at the plant. The rated capacity of the modified plant will be 850 000 tpy of 80 per cent TiO_2 Sorelslag.

Most of QIT's output of Sorelslag is exported to the United States and Europe, while approximately 10 to 15 per cent is sold in Canada to two pigment producers, NL Chem Canada Inc. and Tioxide Canada Inc. Both pigment producers employ the sulphate process.

In 1984, both Canadian pigment producers were operating at their full capacity of about 36 000 tpy of TiO_2 pigment each. Tioxide lost production for a period of several weeks in the second half of 1984 when the plant was shut down because of a strike during negotiations for a collective agreement. The labour contract at NL Chem expires in June of 1985. Pigment exports to the United States amounted to over 25 per cent of Canadian production, which helped the Canadian producers while the demand in Canada lagged behind the buoyant U.S. pigment market. However, Canadian imports of TiO_2 pigment from the United States and Europe offset approximately one half of Canadian exports. Both companies purchase sulphuric acid at prices which reflect the present escalating price of elemental sulphur. NL Chem buys elemental sulphur to produce about 60 per cent of its acid requirement.

Tioxide was negotiating with Federal and Provincial departments for approval to construct a new sulphate pigment plant adjacent to its existing plant.

A small number of Canadian companies which make finished products from titanium forgings, castings, bar, pipe, tube, plate

---

D.E.C. King is with the Mineral Policy Sector, Energy, Mines and Resources Canada. Telephone (613) 995-9466.

and sheet. Walbar of Canada Inc. of Toronto, Ontario and Pratt and Whitney Aircraft of Canada Ltd. of Longueuil, Quebec, machine forgings, investment castings, and bar stock to produce parts for turbine engines. The shop scrap is sold to U.S. producers offerrotitanium and briquettes which are made from Titanium scrap and offgrade sponge. The total amount of titanium forgings, castings and bar stock consumed by these two companies in 1984 was of the order of 240 t.

Titanium Ltée of St. Laurent, Quebec and Ellett Copper and Brass Co. Ltd. of Port Coquitlam, British Columbia custom produce titanium tanks, pressure vessels, heat exchangers, fans, and other equipment for pulp, chemical, petrochemical and metallurgical industries.

The aircraft companies de Havilland Aircraft of Canada Ltd., Downsview, Ontario, Canadair Ltd, Montreal, Quebec, and McDonnell Douglas Canada Ltd., Malton, Ontario, produce airframe parts, such as firewalls, motor mounts, nacelles and wings. The quantities of titanium used in making chemical equipment and airframes in Canada vary widely, but appear to be of the order of 50 to 125 tpy for chemical equipment and 10 to 30 tpy for airframes.

The quantities of titanium added as ferrotitanium and composite master alloys to specific grades of steels are small compared with other alloying elements. They nevertheless accounted for an estimated 150 to 190 t of contained titanium in 1984. Used as an alloying agent, titanium is beneficial in controlling nitrogen and acts as a grain refiner in high-strength, low-alloy steel plate; it is also used as a carbide stabilizer in type 409 stainless steel. By comparison, the quantities of titanium added to aluminum alloys are of a much smaller order, possibly about 10 tpy of titanium in 5-10 per cent titanium-aluminum master alloys.

The quantities of titanium used by Canadian companies producing wear resistant parts for the mining and other industries are very small and are not separately reported in statistics. Titanium is used in mixed carbides with tungsten, and in titanium nitride coatings. Canadian companies producing carbides and nitrides include Kennametal Ltd. Canadian General Electric Ltd., and Valenite Modco Ltd.

WORLD DEVELOPMENTS

Titanium Minerals

Ilmenite is the source for 90 per cent of the world supply of titanium dioxide pigment production. The more expensive rutile (TiO_2) is favoured by producers of primary titanium metal. Anatase (TiO_2) deposits of Brazil are also potentially important. Two other raw materials, titaniferous slag and synthetic rutile produced from ilmenite, are used extensively. The relative scarcity and high cost of rutile supplies are creating an increasing demand for beneficiated ilmenite, although in the past rutile producers have tended to curb competition from beneficiates by cutting the price of rutile. There was by the end of 1984 an incipient world shortage of titanium raw materials although ore reserves were generally abundant. Accordingly, the following increases in capacity were in progress:

Australia: Consolidated Rutile will double its rutile and zircon production capacity to 80 000 tpy for each mineral, by installing a new dredge. By early-1985, it will also add a further 15 000 tpy capacity by purchasing the dredge now operated by Associated Minerals Consolidated (AMC). AMC has reported that it may proceed with plans to install a synthetic rutile plant to produceabout 60 000 tpy, and Westralian Sands Ltd. has similar plans. Allied Aneabba's 1983 and 1984 outputs were lower than 1982 due to mining problems, but are expected to approach 1982 levels again in 1985.

Ilmenite production by Allied Aneabba, Westralian, AMC and Cable has been running close to capacity.

Sierra Leone: Sierra Rutile Ltd. produced nearly 72 000 t of rutile in 1983, the largest output of any single mine. The company resumed production in January 1983 after Nord Resources Corp. of Dayton, Ohio acquired an 85 per cent interest in it. Assuming continued strong demand, production Sierra expects to build up to nearly its full capacity of 100 000 t in 1985.

Brazil: The Companhia Vale do Rio Doce (CVRD) intends to go ahead with a 180 000 tpy plant at Tapira at a cost of $200 million to produce a 90 per cent TiO_2 concentrate from anatase ore. CVRD had built a pilot concentrator and has been testing the circuit since late-1983. Engineering for the new plant is expected to be finalized in two years time.

Norway: Supported by the Norwegian government Kronos Titan A/S, a subsidiary of NL Industries Inc., is installing an ilmenite smelting process which is expected to go on-stream at Tyssedal in 1986 with a capacity of 200 000 tpy of 75 per cent TiO_2 slag.

India: India Rare Earths Ltd. is expected to produce some synthetic rutile in 1985 from a plant being constructed at Orissa. No further information was available on an announcement by Kerala Minerals Ltd. to build a new rutile plant.

United States: Kerr McGee has scheduled the start-up of a new synthetic rutile plant at Mobile in 1985.

Titanium Dioxide Pigment

The consumption of pigment has recovered from the recession of 1982 and is expected to remain strong throughout 1985. While the long-term growth in demand by the mature markets of North America and western Europe is estimated to be in the range of 1 to 2 per cent per year, worldwide demand should grow at a rate of 2.5 to 3.0 per cent per year, being influenced by greater growth rates in the less developed regions of South America and Asia. The approximate worldwide distribution of titanium pigment usage (60 per cent in paint, 13 per cent in paper, and 15 per cent in plastics with the remainder spread more or less evenly between, rubber, ink, textiles and ceramics), is somewhat distorted by North America's higher usage in paper (20 per cent).

There has recently been a considerable corporate re-structuring of the pigment industry involving acquisitions of existing plants. However, virtually no overall new capacity has been added. Because of this and the fact that existing facilities are operating at or near full capacity, a strong upward pressure on prices may be anticipated for several years, if demand remains firm. Total world titanium pigment capacity in 1983 was about 2.5 million t of which the United States had about 32 per cent, other western countries about 47 per cent, Japan 9 per cent and communist countries 12 per cent.

In recent years, environmental regulations have been a major consideration in the closure of several sulphate plants and the lost capacity has been replaced by chloride plants. Both processes produce volumes of effluent which are indirectly proportional to the grade of feedstock. Hence, rutile, high-grade slag, and beneficiated ilmenite produce less effluent and, except for price, would normally be preferred to ilmenite. In the chloride process the chlorine reagent is recycled and the economy so achieved enables the use of higher-grade feedstock than in the sulphate process.

Operating costs in the chloride process are roughly in the proportions 40 per cent fixed costs, 40 per cent variable costs, 10 per cent laboratory and 10 per cent plant materials. The largest variable cost is in raw materials, but the biggest unit operating cost is in "finishing", which involve re-slurrying and re-drying in order to coat the pigment particles with compounds which reduce the absorption of ultraviolet light, which would otherwise break down the organic paint base.

Apart from raw materials, a large component of sulphate process operating cost is the cost of sulphuric acid. Fossil fuel energy is the next largest cost component. While fuel costs are currently fairly stable, the price of acid is escalating partly because of the weakness in base-metals production.

Europe still has some excess pigment capacity, and has become very competitive in North America because of the relative weakness of European currencies against the dollar.

Titanium Metal

World consumption increased in the late-1970s and reached an all-time peak of 51 412 tpy in 1981. This rapid growth stimulated increases in production capacity, which in 1983 totalled about 68 000 tpy of primary titanium in the market economies, including about 33 400 tpy in the United States, 36 600 tpy in Japan where the greatest expansions took place, and 5 000 tpy in the United Kingdom. However, the USSR has the world's largest production capacity, estimated at 42 000 tpy by the U.S. Bureau of Mines and 57 000 to 60 000 tpy by Wogen Resources Ltd. China's capacity was estimated at 3 500 tpy.

The western world capacity for melting, semi-fabricating, and casting amounted to a total of 65 000 tpy in 1983, including 45 000 tpy in the United States, 13 000 tpy in Japan, 5 000 tpy in the United Kingdom, 2 000 tpy in West Germany and 1 000 tpy in France.

Western world consumption slumped after 1981 under the influence of the recession, reaching a low in the United States of 17 600 t in 1983 but recovering to an estimated 20 000 t in 1984. About 45 per cent of titanium consumed in the United States in 1983 (excluding alloying) went to military applications, 34 per cent to commercial aircraft and 21 per cent to industrial applications. Japan's consumption depends far less on the volatile military market; less than 10 per cent is used in aerospace applications and more than 90 per cent in industrial applications. In western Europe, industrial applications account for 40 to 50 per cent of consumption.

The main loss of demand in 1982-83 was caused by a fall in demand in the civilian and military aircraft industries. A recovery in the demand for civil aircraft and industrial products was the main element in market strength in 1984, which is forecast by several sources to continue into 1985 with a gain in consumption of about 15 per cent over 1984. Average growth of about 10 per cent per annum is forcast for the next 3 years during which time the military market is expected to play a reduced role.

The United States General Service Administration (GSA) let contracts for the purchase of 4 500 t of titanium sponge in 1983-84. Dumping penalties were charged against Japanese sponge suppliers in connection with their participation in this purchase. This action has lead to complaints, by non-integrated producers which depended on Japanese sponge and a claim that 'U.S. integrated producers can now control domestic supply.

A new process for producing titanium sponge and powder will be pilot tested by Albany Titanium Inc. in two pilot plants now under construction, one a 2,000 lbs per month titanium powder plant already under test and the other a 2,000 lbs per month titanium sponge plant due to start in February or March 1985. Both units employ a patented technology for the continuous reduction of ilmenite by zinc and aluminum, following a pretreatment in which the ilmenite is briquetted with other compounds. Albany claims that the process is low cost and the end product of high purity. Plans for a 5 000 tpy plant are well advanced and start-up is forecast for the summer of 1986.

A joint venture between Martin Marietta Corp. and Nippon Kokan to produce fabricated titanium products, was announced during 1984. The new corporation International Light Metals Inc. plans to establish its operations near the MMC plant at Torrance, California.

The Titanium Development Association of Dayton, Ohio, was formed in 1983 to improve communications between North American producers and consumers and held its first annual meeting in November 1984. A monthly newsletter has been started and a 1985 Buyers' Guide completed. Periodic distribution of titanium industry statistics is planned.

PRICES

Prices of mill products fell in 1982-83 and, expressed in constant dollars, were in 1984 about 30 per cent less than 1981 prices. However, titanium sponge prices did not fall to the same extent. An overall lowering of titanium prices would open the way to dramatically increased usage. In this regard, progress has been made, notably in Japan, towards more efficient and less costly processing. However, even at present prices, titanium is cheaper than special nickel alloys and its light-weight and high-corrosion resistance can often permit design economies in the thickness and weight of components.

The current excess of sponge capacity over consumption should eliminate erratic price movements for the next 5 years, with prices reacting normally to market demand. An increase in sponge price of about 15 to 20 per cent in 1985 appears likely.

USES

Titanium metal usage is based on its relative abundance, unique physical properties and corrosion resistance. Initially, uses were found in military aircraft where cost was not the main factor, and its high-strength, lightness and high melting point, could be utilized for engine and airframe applications. Greater availability and lower prices have led to expanding usage in commercial and private aircraft. Specifications for aircraft quality are high, and since titanium has a strong tendency to combine with oxygen and nitrogen, melting has to be carried out in vacuum, sometimes twice or three times before an ingot is produced for fabrication.

Commercial titanium produced to less demanding specifications, is used in industrial applications. Titanium's high corrosion resistance lends itself to a wide range of uses in the chemical, metallurgical and paper industries, power plants and desalination plants. In these applications about 50 per cent of the total quantity of titanium consumed is used in heat transfer and seawater cooling applications, about 25 per cent in chemical process equipment, and about 20 per cent as electrodes in electrolytic plants. However, a vast number of minor applications are developing, such as spectacle eye frames, camera parts, yacht rigging, and medical uses such as hip joints.

OUTLOOK

Environmental regulations will probably continue to favour the chloride process in new pigment plants. Although fluid-bed roasting is a standard way of oxidizing titanium chloride in the chloride process, plasma oxidation such as is used by Tioxide UK might be favoured where electric power costs are low. Effluent treatment and disposal will require further development for both the chloride and sulphate processes.

In the production of titanium sponge, developments toward the use of less power is a trend exemplified by the new plant of Showa Titanium, Japan where power consumption is reportedly 15 000 to 18 000 kWh per ton instead of the 25 600 to 30 000 kWh per ton in most plants. Since existing sponge plants are based on batch operations, there are also considerable potential cost savings to be gained from the development of a continuous process.

The development of semi-fabrication to near-net-shape would eliminate much subsequent machining, particularly in the case of turbine blades which currently require the removal of up to 85 per cent of the initial forging weight.

Titanium turnings and cuttings are difficult to recycle for use in rotating jet engine parts. One of the main problems is the removal of tiny tungsten carbide particles which contaminate the scrap. Suisman Titanium Corp. in the United States has patented a process to remove high density contaminants, which would greatly enhance the value of the recycled titanium.

British Aerospace Plc. plans to increase the use of superplastic forming and diffusion bonding of titanium alloys for aircraft parts. While slow, the process reduces necking and local failures, and leaves the material with a fine uniform grain size and limited texturing.

A dramatic increase in the demand for titanium metal has been anticipated for some years. Such a development would likely occur if there was a major price breakthrough, which could happen if there was a significant lowering of production costs.

TARIFFS

Item No.		British Preferential	Most Favoured Nation	General	General Preferential
			(%)		
CANADA					
32900-1	Titanium ore	free	free	free	free
34715-1	Sponge and sponge briquettes, ingots, blooms, slabs, billets, and castings in the rough, of titanium or titanium alloys for use in Canadian manufactures (expires June 30, 1984)	free	free	25	free
34735-1	Tubing of titanium or titanium alloys for use in Canadian manufactures (expires June 30, 1984)	free	free	25	free
34736-1	Sheet, strip or plate of titanium or titanium alloys, cold-rolled, not more than 0.2015 inch in thickness, for use in the manufacture of tubes (expires June 30, 1984)	free	free	25	free
34745-1	Bars, rods, plate, sheet, strip, foil, wire, coated or not; forgings and mesh of titanium or titanium alloys, for use in Canadian manufactures (expires June 30, 1984)	7.5	7.5	25	5
37506-1	Ferrotitanium	free	4.7	5	free
92825-1	Titanium oxides	free	11.3	25	free
93207-6	Titanium whites, not including pure titanium dioxide	free	11.3	25	free

MFN Reductions under GATT (effective January 1 of year given)	1983	1984	1985	1986	1987
			(%)		
37506-1	4.7	4.5	4.3	4.2	4.0
92825-1	11.3	10.9	10.6	10.3	10.0
93207-6	11.3	10.9	10.6	10.3	10.0

UNITED STATES (MFN)

Item No.		1983	1984	1985	1986	1987
422.30	Titanium compounds	6.2	5.9	5.6	5.2	4.9
473.70	Titanium dioxide	6.8	6.6	6.4	6.2	6.0
601.51	Titanium ore	Remains free				
606.46	Ferrotitanium and ferro-silicon titanium	4.6	4.4	4.1	3.9	3.7
629.12	Titanium metal, waste and scrap	12.6	11.3	9.9	8.6	7.2
629.14	Titanium metal, unwrought	17.0	16.5	16.0	15.5	15.0
629.20	Titanium metal, wrought	17.0	16.5	16.0	15.5	15.0

Sources: The Customs Tariff, 1983, Revenue Canada, Customs and Excise; Tariff Schedules of the United States Annotated (1983), USITC Publication 1317; U.S. Federal Register Vol. 44, No. 241.

TABLE 1. CANADA, TITANIUM PRODUCTION AND TRADE, 1982-84

	1982P		1983P		1984	
	(tonnes)	($000)	(tonnes)	($000)	(tonnes)	($000)
					(Jan.-Oct. 1984)	
Production (shipments)						
Titanium dioxide, slag	x	x	x	x		
Imports						
Titanium in ores and concentrates						
United States					1 347	825
Australia					2 058	854
South Africa					36	2
Total					3 441	1,681
Titanium dioxide, pure						
United States	3 344	6,470	7 101	12,641	6 956	12,444
West Germany	1 351	1,794	2 797	3,990	2 402	3,319
Australia	17	30	592	1,181	582	1,178
France	74	104	790	1,155	545	744
Belgium-Luxembourg	297	418	584	798	475	663
United Kingdom	182	281	321	458	581	803
Spain	438	753	278	296	240	299
Other countries	34	42	505	666	1 495	2,141
Total	5 737	9,892	12 968	21,185	13 276	21,591
Titanium dioxide, extended						
West Germany	-	-	2 599	2,955	4 733	5,862
United States	135	340	611	1,186	865	1,820
Belgium-Luxembourg	-	-	481	832	94	140
Spain	163	284	454	646	249	402
Other countries	71	120	1 410	2,122	1 698	2,352
Total	369	744	5 555	7,741	7 639	10,576
Titanium metal						
United States	389	15,881	227	8,903	197	6,017
Belgium-Luxembourg	3	321	5	624	5	398
United Kingdom	18	334	20	500	17	385
Japan	91	1,708	18	203	5	460
Other countries	3	188	5	413	59	1,307
Total	504	18,432	275	10,643	284	8,567
Ferrotitanium[1]						
United States	-	-	14	39	10	32
Belgium-Luxembourg	-	-	5	28	28	126
United Kingdom	-	-	298	1,045	188	721
Total	110	-	317	1,112	226	879
Exports[2] to the United States						
Titanium metal, unwrought including waste and scrap	211	1,364	415	2,342	5	62
Titanium metal, wrought	432	7,616	287	5,180	150	2,908
Titanium dioxide	19 880	25,135	23 190	27,396	21 427	26,467

Sources: Energy, Mines and Resources Canada; Statistics Canada.
1 Total alloy weight. 2 U.S. Department of Commerce, U.S. General Imports, Report F.T. 135. Canadian export statistics do not provide separate categories.
P Preliminary; - Nil; x Confidential.

TABLE 2. CANADIAN TITANIUM PRODUCTION AND IMPORTS 1970, 1975, 1979-84

	Production		Imports		Total
	Ilmenite[1]	Titanium Dioxide Slag[2]	Titanium Dioxide Pure	Titanium Dioxide Extended[3]	Titanium Dioxide Pigments
			(tonnes)		
1970	1 892 290	766 300	2 523	7 415	9 938
1975	1 543 480	749 840	2 467	241	2 708
1979	1 004 260	477 030	9 815	1 515	11 330
1980	1 853 270	874 710	6 135	148	6 283
1981	2 008 117	759 191	6 986	314	7 300
1982	1 735 000	669 000	5 737	369	6 106
1983	x	x	12 968	5 555	18 523
1984[4]	x	x	13 276	7 639	20 915

Sources: Energy, Mines and Resources Canada; Statistics Canada; Company reports.
[1] Ore treated at Sorel; from company reports. [2] Slag with 70 to 72 per cent TiO_2; from company reports. [3] About 35 per cent TiO_2. [4] Jan.-Oct. 1984.
x Confidential.

TABLE 3. PRODUCTION OF ILMENITE CONCENTRATE BY COUNTRIES, 1981-1983

	1981	1982P	1983e
		(000 tonnes)	
Australia	1 337	1 179	1 134
Canada[1]	759	680	612
Norway	658	552	544
U.S.S.R.e	426	431	431
Republic of South Africa	370	381	363
United States	462	239	W
Indiae	189	190	181
Finland	159	160	154
China	136	136	136
Malaysia	145	121	109
Sri Lanka	80	80	73
Other countries	18	15	18
Total	4 739	4 164	3 755

Sources: U.S. Bureau of Mines, Minerals Yearbook Preprint, 1983; U.S. Bureau of Mines, Mineral Commodity Summaries, 1984.
[1] Titanium slag containing 70-71 per cent TiO_2.
P Preliminary; e Estimated; W withheld to avoid disclosing company proprietary data.

TABLE 4. PRODUCTION OF RUTILE BY COUNTRIES, 1981-1983

	1981	1982P	1983e
		(000 tonnes)	
Australia	229	220	218
Sierra Leone	51	48	72
Republic of South Africa	50	47	54
United States	W	W	W
Sri Lanka	13	13	14
U.S.S.R.e	9	9	9
Indiae	9	8	9
Brazil	--	--	--
Total	361	345	376

Sources: U.S. Bureau of Mines, Minerals Yearbook Preprint, 1982; U.S. Bureau of Mines, Mineral Commodity Summaries, 1984.
P Preliminary; e Estimated; -- Amount too small to be expressed; W withheld to avoid disclosing company proprietary data.

TABLE 5. PRICES OF SELECTED TITANIUM COMMODITIES, 1982-84

	1982	1983	1984
		($US)	
Titanium ore, fob cars Atlantic and Great Lake ports			
Rutile, 96%, per short ton, delivered within 12 months	450.00-475.00	400.00-430.00	460.00-490.00
Ilmenite, 54%, per long ton, shiploads	70.00-75.00	70.00-75.00	70.00-75.00
Slag, 70%, per long ton, fob Quebec	150.00	150.00	155.00
Titanium sponge, per lb	5.56	5.56	5.55-5.85
Mill products, per lb delivered			
Billet (Ti - 6AL-4V)	15.00	15.00	8.01
Bar (Ti - 6AL-4V)	18.00	18.00	10.06
Titanium dioxide, anatase[1],			
Bags, 20-ton lots, freight allowed, per lb	0.69-0.70	0.69-0.70	0.69-0.70
Titanium dioxide, rutile, regular grades, per lb	0.75	0.75	0.75

Source: Metals Week, December.
[1] Chemical Marketing Report, December.
fob - Free on board.

Tungsten

D.R. PHILLIPS

SUMMARY

The rebound of tungsten prices in 1984 from their historic low in 1983 led to a strengthening of the market and increased production by Canadian mines.

Canada's two producing tungsten mines, Canada Tungsten Mining Corporation Ltd. and Mount Pleasant Tungsten Mine reached near-capacity rates of operation in the first half of 1984. The opening of the latter in 1983 marks the beginning of a major new supplier to world markets of high-grade wolframite, in addition to the high-grade scheelite produced by Canada Tungsten.

CANADIAN DEVELOPMENTS

Canada maintained its rank in fifth place of world shipments of tungsten ores and concentrates. However, 1983 producer shipments, estimated at 1 537 tonnes (t) tungsten trioxide (WO_3) content, were approximately 50 per cent less than 1982. Producer shipments for 1984 were estimated to exceed 1982 shipments of 3 029 t.

Canadian production of tungsten trioxide in 1984 was estimated at 465 300 tonne units (tu) compared to 41 399 tu in 1983 and to 358 300 tu in 1982. The decrease in 1983 is a result of the long closure at Canada Tungsten (Cantung) from January 22 to November 30, 1983 because of weak markets and continued low prices throughout the year. Cantung restarted its operation in December 1983 at half capacity and returned to full capacity in August of 1984. During the 1983 closure, programs were initiated to improve efficiencies, including high productivity mining methods, which resulted in a 20 per cent reduction in the work force on reopening.

Amax, through its wholly-owned subsidiary Amax North West Mining Company Ltd., continued to evaluate the Mactung scheelite deposit on the Yukon-Northwest Territories boundary. Development and production plans in 1983 had been temporarily delayed because of weakness of the market. The project is rescheduled to be completed in 1987.

Construction at the Mount Pleasant Tungsten mine/mill complex was completed in early 1983 and fine tuning of the 2 000 tpd mill operation was under way in late 1983. Mount Pleasant Tungsten Mine is a joint venture between Sullivan Mines Inc. (through its 89 per cent ownership of Brunswick Mines Ltd.) and Billiton Canada Ltd. The latter manages the mine and will market an estimated 1 000 t of WO_3 and 600 t of molybdenite (MoS_2) annually. Initially, the company will direct its attention to the production of tungsten, which will be marketed in the United States and Europe. Due to low prices and some technical problems, Mount Pleasant Tungsten reduced mining, beginning in October 1984, to 50 per cent of capacity. During the period of this reduction, the company will continue work on metallurgical improvements to increase recoveries.

Dimac Resources Corp., located in British Columbia, closed its 100 tpd mine/mill complex in 1982 because of poor market conditions and operational problems.

The resumption of production at Cantung and the opening of Mount Pleasant Tungsten in 1983 coincided with a major cyclical contraction in the consumption of tungsten. If the present economic recovery faulters, reductions in operations at Cantung and Mount Pleasant Tungsten could occur.

INTERNATIONAL DEVELOPMENTS

World production of tungsten ores and concentrates in 1983 declined 19 per cent, to 38 320 t of contained tungsten, compared to 44 328 t in 1982. World production of ores and concentrates in 1984 was estimated to exceed 1982 production levels. Western world production declined in 1983 by approximately 17 per cent to 19 000 t of contained tungsten. Production in 1984 was estimated to be 25 000 t.

D.R. Phillips is with the Mineral Policy Sector, Energy, Mines and Resources, Canada. Telephone (613) 995-9466.

At the end of 1983, all major North American mines, except Mount Pleasant Tungsten and Cantung, were closed or had plans to close. A modest recovery in tungsten prices in 1984 resulted in a return to higher production throughout the world. In the United States the Strawberry mine, operated by Teledyne Wah Chang, resumed production on a regular basis and produced close to record amounts in 1984. Union Carbide Corporation kept its Emerson mine closed throughout 1984 and operated its Pine Creek Bishop mine at approximately 50 per cent capacity on an intermittent basis. The Springer Mine, owned by the General Electric Company, remained closed throughout 1984.

The People's Republic of China (PRC) continued to be the world's largest producer, followed closely by the U.S.S.R. Tungsten production of the PRC decreased by 20 per cent in 1983 to an estimated 10 000 t, compared to a slight increase for the U.S.S.R. of 0.2 per cent. All other producing countries shared a decrease in production except Bolivia, which showed an increase of 18 per cent to approximately 3 000 t contained tungsten. Bolivia's plans to start-up its Anschutz mine in Chicote Grande in 1983 was delayed due to the weakening of the market following the 1981/82 recession. The 1 000 tpd mine, with an estimated investment of $25 million, plans to export tungsten to Europe and the United States.

Both major producers in Australia, Peko-Wallsend Ltd. at King Island and Queensland Wolfram at Mt. Carbine, reduced production by about one third during 1983. All but one of the other six smaller Australian operations were closed and will remain idle until the market situation improves.

Australian production of tungsten in 1983 was estimated at 2 000 t tungsten content, 17 per cent less than its 1982 production of 2 588 t. Approximately 45 per cent of Australia's total production came from King Island scheelite mine of the Peko-Wallsend Group and 42 per cent from the Mount Caroline wolframite mine of Wolfram Pty. Ltd.

Peko-Wallsend reduced production twice in 1983 at its King Island mine. However, Australia remained the western world's leading producer. Australian production of tungsten in 1984 is estimated to be approximately 3 000 t contained tungsten.

Tungsten producers in Thailand, Bolivia and South Korea all announced production cutbacks of about 30 per cent during 1982 and continued to produce at reduced capacity in 1983. It is estimated that production in these countries increased in 1984.

MARKET STABILIZATION

International discussions on stabilizing the tungten market were held at the 15th Session of the United Nations Committee on Tungsten (COT) in Geneva on December 12-16, 1983. The Committee concluded that a Sessional Working Group should be established to assist the COT, and instructed the secretariat to prepare a discussion paper outlining the composition and organization of such a group. The Sessional Working Group is intended to provide a forum for more open discussion, including the active participation of industry advisors, and to prepare each year a reasonably detailed analysis of the market situation for the coming year.

Activity has declined in regards to a proposal for a producer-only group, especially since the Chinese voiced their belief that membership in the United Nations COT was more comprehensive and there was no need for a separate producer group. The Chinese are presently submitting statistics to the COT.

The 16th Session of the COT was held in Geneva December 10-14, 1984. The Committee agreed to establish a Sessional Working Group. A consensus was also reached by the Committee on the terms of reference for the Sessional Working Group, which includes the study of statistics and all matters relating to the tungsten market.

PRICES

Tungsten prices, which started to decline in late-1981 continued to fall in the first half of 1983. There was a slight recovery in mid-1983, suggesting a strengthening of the market. However, a renewed weakness occurred in the last quarter of 1983 before prices resumed their upward trend, which continued throughout most of 1984. Nevertheless, prices at the end of 1984 were still well below those that existed before the 1981/82 recession.

The slow recovery of prices in the United States may have been influenced by imports from China and the auctioning of General Services Administration (GSA) stocks at a time when consumption of tungsten continued to decline.

PRICES

	December 31, 1982	December 31, 1983
	($US)	
Tungsten ore, 65% minimum WO_3		
G.S.A. domestic, duty excluded, per short ton unit of WO_3	99.600	64.480
G.S.A. export, per short ton unit of WO_3	95.090	74.690
L.M.B. ore quoted by **London Metal Bulletin**, cif Europe, per metric tonne unit of WO_3	76.00-84.00	68.250-72.875
Ferrotungsten, per pound W, fob Niagara Falls, low-molybdenum	list price suspended	list price suspended
Tungsten metal, per pound, fob shipping point Hydrogen reduced: 99.5%, depending on Fisher No. range	13.100-13.720	13.100-13.720

Source: Metals Week.
cif Cost, insurance and freight; fob Free on board.

In May of 1984, the London Metal Bulletin (LMB) began to report the price of high-quality scheelite, in addition to the price of wolframite which has been available for several years.

Prices reported by the LMB and the International Tungsten Indicator (ITI) for the months February, July and November 1984 are summarized as follows.

1984	LMB		ITI
	Wolframite $US/tu WO_3	Scheelite $US/tu WO_3	$US/tu WO_3
Feb.	74-79	-	75.28
July	82-87	99-102	85.25 - 85.19
Nov.	81-84	85-87	83.43

- One tonne unit (tu) of WO_3 contains 7.93 kilograms of tungsten;
- Not published prior to May 1984

USES

Approximately 80 per cent of the western world tungsten consumption in 1983 and 1984 was accounted for in the manufacture of cemented carbide and tool steel products, the former amounting to approximately 50 per cent of total consumption. Tungsten metal, superalloys and miscellaneous end uses accounted for the remaining 20 per cent.

Tungsten materials can be divided into several major classes, depending upon the product form and its use. The main product classes include tungsten carbide, tungsten-bearing steels, superalloys, mill products made essentially from pure metal, and chemicals.

Tungsten carbide (WC) is one of the hardest materials known and accordingly, has widespread applications where intense wear and abrasion are encountered. This product is the preferred metalworking material for the cutting edges of machine tools and as a metal surface in forming and shaping dies. It is produced by the chemical combination of tungsten metal powder and finely divided carbon. Tungsten carbide is compacted to the desired form, using cobalt as a binder, and sintered to produce cemented tungsten carbide. Cutting tools of cemented tungsten carbide are used for machining steel, cast iron and nonferrous metals, and for shaping in the woodworking and plastics industries. Cemented tungsten carbide is also used to make dies for wire and tube drawing, punches and dies for metal forming, and bits and tools for drilling equipment and wear-resistant parts. With the addition of tantalum, titanium and columbium carbides, the coefficient of friction of cemented tungsten carbides is lowered, thereby producing grades better suited to the machining of specific items, particularly steel products. Other uses of tungsten carbide are in tire studs, spikes for golf shoes, armour-piercing projectiles and welding electrodes.

As an alloy constituent, tungsten is used primarily in the production of high-speed steels, and tool and die steels. Tungsten is added to steels either as ferrotungsten (80 per cent tungsten), melting base (30-35 per cent tungsten), scheelite ($CaWO_4$) or as tungsten-bearing scrap. Tungsten-bearing steels are used for the same applications as carbides, especially where lower operating temperatures are encountered. Tungsten is also used in some stainless steels for application in high-temperature environments.

Tungsten is an important constitutent in a wide variety of superalloys and nonferrous alloys. Tungsten-containing superalloys are being used increasingly in high-temperature applications and in highly corrosive environments because of their high-temperature strength and oxidation resistance. In making the alloys, tungsten is usually added in the form of metal powder, although scrap can be used to satisfy part of the requirements. Superalloys can be classified into three principal types: nickel base, iron base and cobalt base or "Stellite" superalloys. While only small amounts of tungsten are currently used in the nickel and iron base superalloys, several companies are developing new superalloys containing larger amounts of tungsten, a factor which could significantly expand the market for tungsten.

Mill products made from pure or nearly pure tungsten metal powder are used in significant quantities by the electrical industries. The relevant important properties of tungsten for electrical applications include its high-melting point, low-vapour pressure, hardness, good electrical conductivity and low coefficient of thermal expansion. Tungsten mill products such as rods, wire and flat products are made by compressing tungsten metal powder into the desired shape and then sintering.

Discs cut from tungsten rods are used as electrical contacts to improve resistance to heat deformation resulting from sparking and associated high temperatures. Pure tungsten contacts are used principally in ignition circuits of automobiles and aircraft. However, the trend to electronic ignition systems without tungsten contacts has resulted in a decline in its use for this application. Tungsten discs are also used as heat sinks in semiconductor applications and, in combination with other elements, as electrical contacts and breakers for industrial use.

Tungsten wire is used for filaments in incandescent lamps, and heating elements in both fluorescent lamps and vacuum tubes. The overall demand for tungsten wire is increasing in response to the upward trend in the manufacture of lamps and new uses such as de-icing and defogging elements in automobile windshields.

Flat products are used for various parts of electron tubes and radiation shields as well as for very high-temperature applications in reducing or inert atmospheres.

Tungsten is used for counterweights and balances, especially by the aircraft industry, but it is being replaced by depleted uranium which has about the same density.

Minor amounts of tungsten are used to make chemicals and compounds for nonmetallurgical applications. Some of the end-uses include dyes, toners, phosphors, chemical reagents, corrosion inhibitors and catalysts.

OUTLOOK

Currently, the world tungsten industry is characterized by a large amount of unutilized capacity (approximately 50 per cent utilization with considerable differences between countries). Although production is severely constrained, supply generally exceeds demand. Mine production capacity and utilization for the western world and selected countries for 1983, and projected capacity for 1988 are summarized in Table 4.

The demand for tungsten, like most other minerals and metals, is a derived demand. It derives from the demand for goods (end-use products) in which tungsten in one of its many forms is a component. Hence, growth in tungsten uses will be dictated by the level of economic activity in countries that are the principal users of this mineral commodity.

Although there has been an economic revival in North America, Japan and, to a lesser degree, western Europe the market price of tungsten is still at depressed levels despite improvements in the first half of 1984.

Due to the unused capacity in the world tungsten industry and given the fact that a significant portion of world supply is produced in a number of countries, the future market will remain, as it has been in the past decade, highly competitive.

The future growth rate of the Canadian industry will depend on Canada's ability to remain cost competitive in international markets. Canada's share of the world market, which stood at 20 per cent in 1982, could increase to 25 per cent by 1990 due to the high efficiency of its mining operations, large reserves of high-grade ore and aggressive marketing. These factors will contribute to Canada's ability to compete on world markets in the foreseeable future and probably beyond the year 2000.

A moderate growth rate of 2.4 per cent per annum to the year 2000 is forecast for world tungsten consumption, largely because world economic growth in general is unlikely to exceed 3.5 per cent in real terms. A relevant factor in this modest growth rate is the impact of reduced consumption due to the use in cutting tool applications of ceramics and coatings on tungsten carbide inserts. There is also substitution between tungsten products themselves, for example tungsten carbide for tungsten-alloyed high-speed tools in metal machining operations.

Further in the future, there are numerous other technological developments aimed at substituting other materials for tungsten, all in an early stage of development.

It is difficult to project the long-term effect of substitution. However, it should be recognized that new applications which emerge from ongoing tungsten reasearch and development could expand the demand for tungsten more than enough to offset the substitution effect.

Developments in the recycling of scrap could have a major impact on the future consumption of tungsten ores and concentrates. Figures for scrap recycling are sparse and incomplete. However, it is estimated that of the total tungsten consumed in Canada in 1983, 20 per cent was recycled and that approximately 30 per cent of the total U.S. consumption was accounted for by recycled material. Secondary tungsten and its compounds are currently recovered from items with a high content of tungsten, such as tungsten carbide.

The degree to which partial substitution for tungsten occurs in traditional uses and the extent to which different metals are substituted for tungsten carbide, either directly or as a coating, could make the recovery of tungsten from these items both technically difficult and costly. Accordingly, a smaller proportion of tungsten is likely to be sourced from scrap in the future and a correspondingly larger proportion from tungsten ores.

TARIFFS

Item No.		British Preferential	Most Favoured Nation	General	General Preferential
			(%)		
CANADA					
32900-1	Tungsten ores and concentrates	free	free	free	free
34700-1	Tungsten metal in lumps, powder, ingots, blocks or bars and scrap of tungsten alloy metal, for alloying purposes	free	free	free	free
34710-1	Tungsten rod and tungsten wire	free	free	25	free
35120-1	Tungsten and alloys in powder, pellets, scrap, ingots, sheets, strips, plates, bars, rods, tubing, wire, for use in Canadian manufactures (expires June 30, 1984)	free	free	25	free
37506-1	Ferrotungsten	free	4.7	5	free
37520-1	Tungsten oxide in powder, lumps and briquettes, for use in the manufacture of iron and steel	free	free	5	free
82900-1	Tungsten carbide in metal tubes for use in Canadian manufactures	free	free	free	free

MFN Reductions under GATT (effective January 1 of year given)	1983	1984	1985	1986	1987
			(%)		
37506-1	4.7	4.5	4.3	4.2	4.0

UNITED STATES (MFN)

601.54	Tungsten ore, per pound tungsten content	17¢				

Item No.		1983	1984	1985	1986	1987
		(% unless otherwise specified)				
422.40	Tungsten carbide, on tungsten content	12.5	12.0	11.5	11.0	10.5
422.42	Other tungsten compounds	11.0	10.7	10.5	10.2	10.0
606.48	Ferrotungsten and ferro-silicon tungsten, on tungsten content	8.2	7.5	6.9	6.2	5.6
629.25	Tungsten metal waste and scrap, not over 50% tungsten	6.3	5.9	5.6	5.2	4.9
629.26	Tungsten metal waste and scrap, over 50% tungsten	4.2	4.2	4.2	4.2	4.2
629.28	Tungsten metal, unwrought, other than alloys: lumps grains and powders, on tungsten content	9¢/lb + 12.5	3¢/lb + 12.5	12.1	11.3	10.5
629.29	Tungsten metal, unwrought, other than alloys: ingots and shot	9.0	8.3	7.5	6.8	6.0
629.30	Other unwrought tungsten metal	10.5	9.6	8.6	7.6	6.6
629.32	Unwrought tungsten alloys, not over 50% tungsten	5.9	5.6	5.3	5.0	4.7
629.33	Unwrought tungsten alloys, over 50% tungsten	10.5	9.6	8.6	7.6	6.6
629.35	Wrought tungsten metal	9.5	8.8	8.0	7.3	6.5

Sources: The Customs Tariff, 1983, Revenue Canada and Excise Canada; Tariff Schedules of the United States Annotated 1983, USITC Publication 1317; U.S. Federal Register Vol. 44, No. 241.

TABLE 1. CANADA, TUNGSTEN PRODUCTION, IMPORTS, 1982-84 AND CONSUMPTION 1981-83

	1982		1983P		1984	
	(kilograms)	($)	(kilograms)	($)	(kilograms)	($)
Production[1] (WO_3)	3 029 730	..	1 537 880	..		
Imports					(Jan.-Sept. 1984)	
Tungsten in ores and concentrates						
United States	7 620	104,000	9 000	121,000	7 000	108,000
Peoples Republic of China	-	-	3 000	15,000	-	-
Total	7 620	104,000	12 000	136,000	7 000	108,000
Ferrotungsten[2]						
United States	4 536	160,000	3 000	78,000	5 000	124,000
West Germany	--	5,000	-	-	-	-
Total	4 536	165,000	3 000	78,000	5 000	124,000
Tungsten carbide powder						
United States	249 000	4,973,000	197 000	5,170,000	218 222	5,472,000
Other countries	25 000	857,000	23 000	618,000	22 544	550,000
Total	274 000	5,830,000	220 000	5,788,000	240 766	6,022,000
	(number)	($)	(number)	($)	(number)	($)
Tungsten carbide rotary rock drill bits						
United States	6 829	32,327,000	9 187	46,127,000	6 291	26,295,000
Other countries	3 395	3,616,000	560	1,825,000	794	3,636,000
Total	10 224	35,943,000	9 747	47,925,000	7 085	29,931,000
Tungsten carbide percussion rock drill bits						
Ireland	68 744	1,277,000	139 654	2,587,000	87 579	1,486,000
United States	19 043	1,452,000	42 114	1,467,000	38 942	1,405,000
Other countries	1 738	109,000	3 589	107,000	7 762	183,000
Total	89 525	2,838,000	185 357	4,161,000	134 283	3,056,000
Tungsten carbide tools for metal work						
United States	..	5,835,000	..	6,152,000	..	8,211,000
Other countries	..	1,595,000	..	2,722,000	..	2,785,000
Total	..	7,430,000	..	8,874,000	..	10,996,000
	1981		1982		1983	
	(kilograms)	($)	(kilograms)	($)	(kilograms)	($)
Consumption (W content)						
Tungsten metal and metal powder	377 815r	..	466 672	..	487 463	..
Other tungsten products[3]	23 632r	..	18 934r	..	16 188	..
Total	401 447r	..	485 606	..	503 651	..

Sources: Energy, Mines and Resources Canada; Statistics Canada.
[1] Producers' shipments. [2] Gross weight. [3] Includes tungsten ore, tungsten carbide and tungsten wire.
P Preliminary; r Revised; - Nil; .. Not available; -- Small amount not quantified.

TABLE 2. CANADA, TUNGSTEN PRODUCTION, TRADE AND CONSUMPTION, 1970, 1975, 1979-83

	Production[1]	Imports Tungsten Ore[2]	Ferro-tungsten[3]	Consumption[2]
		(kilograms)		
1970	1 690 448	82 645	90 718	446 687
1975	1 477 731	1 000	45 359	451 336
1979	3 254 000	11 000	28 000	380 229
1980	4 007 000	6 000	7 000	290 479
1981	2 515 000	14 000	6 000	401 447r
1982	3 029 730	7 620	4 536	507 606
1983P	1 537 880	12 000	3 000	..

Sources: Energy, Mines and Resources Canada; Statistics Canada.

[1] Producers' shipments of scheelite (WO_3 content); [2] W content; [3] Gross weight.
P Preliminary; r Revised; .. Not available.

TABLE 3. WORLD TUNGSTEN PRODUCTION IN ORES AND CONCENTRATES, 1981-83

	1981	1982P	1983e
	(tonnes of contained tungsten: W content)		
People's Republic of China	13 517	12 519	10 000
U.S.S.R.	8 845	8 981	9 000
Bolivia	2 779	2 534	3 000
Australia	3 517	2 588	2 000
Republic of Korea	2 642	2 233	2 000
Canada	1 995	2 403	1 220
Austria	1 450	1 406	1 200
Portugal	1 395	1 361	1 200
United States	3 605	1 521	1 100
Brazil	1 249	1 089	1 000
Thailand	1 210	856	700
Burma	825	844	500
Turkey	153	150	200
Mexico	263	99	100
Other central economy countries	2 279	2 279	2 300
Other market economy countries	3 484	3 465	2 800
World total	49 208	44 328	38 320

Sources: United States Bureau of Mines Minerals Yearbook Preprint 1982; USBM Mineral Commodity Summaries, 1984; Energy Mines and Resources Canada.
P Preliminary; e Estimated.

TABLE 4. WESTERN WORLD AND SELECTED COUNTRIES, 1983 MINE CAPACITY AND PER CENT UTILIZATION, AND 1988 FORECASTED CAPACITY

	1983		1988
	Capacitye	% Utilization[1]	Capacitye
	(tonnes W content)		
Canada	5 200	4	6 440
United States	4 575	24	4 575
Bolivia	3 500	86	3 550
Brazil	1 280	78	1 280
Austria	1 600	75	1 600
France	840	89	840
Portugal	1 570	76	1 570
Spain	460	..	460
Sweden	400	..	400
United Kingdom	75	67	75
South Africa	420	..	1 130
Japan	700	..	2 800
South Korea	2 800	71	2 800
Thailand	1 750	40	1 750
Turkey	1 000	20	1 000
Australia	3 400	59	3 400

Sources: Chase Econometrics World Ferroalloy Report, January 1984 Update, Tungsten; USBM Mineral Commodity Summaries, 1984; Energy, Mines and Resources Canada.
[1] Per cent utilization calculated.
e Estimated; .. Not available.

Uranium

R.T. WHILLANS

The outlook for the uranium market is expected to remain unsettled throughout the 1980s with little sign of improvement in the short term. Further cancellations and delays in planned nuclear power programs and the continued accumulation of uranium inventories, will prolong the uncertainty within the uranium industry worldwide.

Despite a decline in annual uranium production from the peak levels of 1980-81, output is expected to exceed requirements to the end of the decade and swell stockpiles now estimated to be equivalent to 4 or 5 years forward supply.

Uranium requirements are forecast to grow at a rate of 6 per cent per annum to 1990 as the planned doubling in installed nuclear capacity is realized. However, market opportunities in the next few years will be limited, as much of this growth can be supplied from inventory. In Canada, this growth continues to sustain uranium exploration efforts which, though diminished in comparison with the 1979-80 peak, remain significant.

In Ontario, Rio Algom Limited and Denison Mines Limited will continue their efforts beyond 1985 to reduce costs and improve overall productivity; underground leaching is expected to contribute significantly toward achieving these objectives. In Saskatchewan, Eldorado Resources Limited is developing its Collins Bay "B" orebody for production in 1986, Cluff Mining is proceeding with Phase II of the Cluff Lake operation, Key Lake Mining Corporation (KLMC) could commence development of the Deilmann orebody before the end of the decade, and COGEMA Canada Limited is continuing the evaluation of the very promising Cigar Lake deposit, the discovery of which was announced in early 1983.

PRODUCTION AND DEVELOPMENT

During 1984, Canada's five primary uranium producers, Denison Mines Limited, Rio Algom Limited, Eldorado Resources Limited, Cluff Mining and Key Lake Mining Corporation produced concentrates containing an estimated 11 170 tU.* With the phasing in of committed development plans in Saskatchewan and recently completed projects in Ontario, annual production capability is expected to stabilize at about 12 000 tU through the mid to late 1980s. As domestic requirements are small, some 15 per cent of current output, most of Canada's production will be exported. (See Table 1 for a comparison of primary uranium production for the years 1982 and 1983, and Table 2 for a summary of the 1983 operating characteristics of Canada's uranium production centres).

Of Canada's total uranium shipments in 1984, some 55 per cent was attributable to the three Saskatchewan operations with the balance coming from the two Ontario producers at Elliot Lake (see Table 3).

A milling rate of some 10 000 tpd of ore has been deemed adequate at the Elliot Lake, Ontario, operations of Denison Mines Limited to meet production and delivery schedules required under long-term sales contracts. These approach 60 000 tU and extend to 2012. Such commitments suggest that for the next several years, Denison will likely maintain its current production rate of approximately 2 300 tU per year.

The underground, in-place leaching tests being conducted by the company continue to show favourable results. Together with better training, improved machinery and high-grade-pillar recoveries, the leaching program will become an increasingly important part of Denison's efforts to reduce costs and enhance productivity. It is expected to ave contributed up to 10 per cent of the company's 1984 production and may account for up to 20 per cent of future production as more working areas become available. All major works for the expansion of the Denison

---

* tU = One metric ton (tonne) of elemental uranium (U), written as tU, is equivalent in terms of uranium content to 1.2999 short tons of uranium oxide (U_3O_8).

R.T. Whillans is with the Energy Sector, Energy, Mines and Resources Canada. Telephone (613) 995-1118.

main property were completed by year-end 1983. At that time the adjoining Stanrock/Can-Met mine was placed on a care-and-maintenance basis, since production from this area will not be required to meet commitments to Ontario Hydro until a future date.

At the Elliot Lake operations of Rio Algom Limited, overall production during 1984 was expected to reach 2 700 tU. As part of a plan implemented in late 1982 to match production more closely with deliveries made under existing contracts, combined ore output from the three production centres (Quirke, Panel and Stanleigh) has been established at about 12 000 tpd. This rate is adequate to meet deliveries under long-term sales contracts which total some 44 000 tU through to the year 2020.

Rio Algom's intensive and ongoing efforts to reduce costs and increase productivity have proven effective. Costs are being trimmed through mechanization, improved training and by driving development headings in ore. The company's underground in-place leaching program is also of major importance. Production from leaching could reach 20 per cent of total output in the near future.

Completed below budget, the rehabilitation of the Stanleigh property marked the end of Rio Algom's multi-million dollar expansion program started in 1975. The Stanleigh mill commenced production on schedule in July 1983 and achieved targetted throughput by year-end 1984. During the 40-year life of the operation, some 28 000 tU will be delivered to Ontario Hydro which has financed the project and will purchase its entire output.

At Bancroft, Ontario, Madawaska Mines Limited submitted its decommissioning and close-out proposal to the Atomic Energy Control Board (AECB) in June 1983. The following November, the AECB announced that decommissioning approvals for the Madawaska mining facility had been authorized, subject to certain conditions. During 1984, work agreed upon under the approved decommissioning licence was substantially completed. After a period of monitoring the property can revert to the Crown.

The salvage leaching operation of Agnew Lake Mines Limited, 90 km east of Elliot commencement of operations in 1977 was 752 tU. The company reported that its uranium loan from Eldorado was repaid and that the AECB had approved the company's plan to decommission and close out the property. Most of the surface rehabilitation and tailings area stabilization was completed by early 1984. If compliance with environmental standards is satisfactory after the prescribed monitoring period, the property can be returned to the Crown.

Production in 1984 from the Rabbit Lake, Saskatchewan, operation of Eldorado Resources Limited was expected to match previous year output levels. The Rabbit Lake orebody was mined out in August 1984 and the company is completing the preparation of the depleted pit as a disposal facility for the tailings from the Collins Bay "B" deposit.

After signing a surface lease agreement with the Government of Saskatchewan in March 1983 (see Government Affairs), Eldorado commenced work on the development of its Collins Bay "B" orebody, some 9 km northeast of the Rabbit Lake mill, spending in excess of $26 million during 1983. Retaining dykes are in place at the site, damming off the small bay which overlies the deposit. The bay has been drained and stripping of the overburden is well underway. The project includes modifications to the 1 500 tpd capacity Rabbit Lake mill to facilitate the processing of a wider range of complex uranium-bearing ores. Hydrogen peroxide will be used instead of ammonia in the precipitation of mine concentrates in order to meet stringent environmental criteria.

The completion of the Collins Bay "B" expansion will return Eldorado's annual production capability to 2 000 tU. Ore stockpiled from the Rabbit Lake open pit will provide feed until 1986, when first production from the "B" deposit is expected. The reported cost of the project is $100 million of which approximately $60 million had been committed by the end of 1984.

In northwestern Saskatchewan, Cluff Mining - owned 80 per cent by Amok Ltd. and 20 per cent by Saskatchewan Mining Development Corporation (SMDC) - completed construction of its Phase II facilities in August 1984, two months ahead of schedule. Cluff Mining's Environmental Impact Assessment and surface lease arrangement were accepted in early 1983, by the Government of Saskatchewan (see Government Affairs). The company was issued an Underground Exploration Permit in July and application for a Mining Facility

Operating Licence (MFOL) was before the AECB at year-end 1983. A two-year MFOL for Phase II was granted effective August 1, 1984.

Stripping of the Claude deposit commenced in early 1983 and dewatering of the "O-P" workings started in the second quarter. Work on the decline from surface of the new "O-P" mine was completed in November 1983 in preparation for the underground development of the deposit, and to permit the eventual advancement to the much larger "Dominique-Peter" orebody nearby. Phase II production began in April 1984 from the exploitation of the "O-P" underground and "Claude" open-pit mines and will shift to the "Dominique-Peter" underground mine around 1985, as the "O-P" deposit is depleted. The "N40" underground and "N" open-pit mines could be brought into production in the 1990s. Phase II facilities, designed to process ore of a more conventional grade than the very high-grade "D" orebody exploited during Phase I of the operation, will have the capability of mining and milling some 230 000 tpy of ore to produce between 850 and 1 270 tU annually.

During January and February of 1983, some 700 t of preconcentrated ore from the Phase I gravimetric separation circuit was processed; this material graded over 310 kgU/t. From March 1983 to year-end, a small solvent extraction circuit, designed to reprocess Phase I residues, treated some 30 000 t of gravimetric separation rejects, grading over 15 kgU/t.

Key Lake Mining Corporation (KLMC) commenced ore processing at its Key Lake operation in northern Saskatchewan in early October 1983. By month-end, the first uranium concentrates had been packaged and the facility was operating at 40 per cent capacity. Commissioning continued through November toward commercial production and mill recovery rates of 95 per cent were attained by December 1983. Operated by KLMC - jointly owned by Saskatchewan Mining Development Corporation (one-half), Uranerz Exploration and Mining Limited (one-third), and Eldor Resources Limited, wholly-owned by Eldorado Nuclear Limited, (one-sixth) - the $500 million project came on-stream close to schedule and 11 per cent under budget.

The ceremonial opening of the Key Lake operation took place on June 1-2, 1984. In July, production surpassed the nominal monthly output capacity of 385 tU, prompting estimates that output in 1984 could total 4 200 tU. Upon achieving its annual design output level of some 4 600 tU, the Key Lake operation will become the world's largest uranium production centre.

To overcome problems caused by the high clay content of the ore, installation of a semi-autogenous grinding (SAG) mill was undertaken in May 1984. Completed late in the year at a cost of some $7 million, the independent SAG circuit should ensure continuity of production at design throughput levels.

Production resumed in mid-1983 at the byproduct uranium recovery plant operated by Earth Sciences Extraction Company at Calgary, Alberta. The facility, placed on standby in late October 1981, underwent major equipment modifications to improve recoveries of uranium from phosphoric acid produced at an adjacent plant operated by Western Co-operative Fertilizers Limited. Testing of the modified circuits commenced in June 1983 and by September the plant was operating at 90 per cent of its 40-60 tU/yr* design capacity. The facility is owned by a partnership of ESI Resources Limited, a wholly-owned subsidiary of Earth Sciences Inc. of Golden, Colorado, and Urangesellschaft Canada Limited, a subsidiary of Urangesellschaft mbH of Frankfurt, Federal Republic of Germany.

As shown in Table 4, the work force at Canada's producing uranium operations as of January 1984 totalled some 5,850 employees. Of this total, over 2,500 worked in the mines, both open-pit and underground, and some 780 in the mills, with the balance described as general employees. Head-office and construction-related employment is not included. The 5,850 figure, showing close similarity with the 1981 total of some 6,000 employees, reveals that many of the 1,200 jobs lost during 1982/83, primarily as a result of the cessation of production at Madawaska, the closure of Beaverlodge in northern Saskatchewan, and the phasing out of operations at Agnew Lake, have been regained as a result of the start-up of new and expanded uranium production facilities.

---

* Output from ESI is not included in Canadian production totals because the uranium is recovered from phosphate rock imported from the United States. This uranium is contracted to American utilities.

EXPLORATION

The Uranium Resource Appraisal Group (URAG) of Energy, Mines and Resources, Canada, (EMR) completed its tenth (1983) resource assessment and exploration survey during 1984. EMR reported* that overall uranium exploration activity in Canada declined in 1983 for the third consecutive year, in terms of both drilling effort and level of total expenditures. Responses to the 1983 URAG questionnaire indicated the exploration activities of 71 companies or joint ventures representing essentially all the major participants active in uranium exploration in Canada. The survey showed that 52 companies or joint ventures were actively involved in uranium exploration in 1983, of which 29 acted as project operators. Total expenditures reached $41 million, distributed amongst some 90 active projects.

The 10 operators** with the largest exploration budgets in 1983 - accounting in aggregate for some 94 per cent of the $41 million total, were, in alphabetical order, AGIP Canada Ltd., Amok Ltd., Anaconda Canada Exploration Ltd., Eldor Resources Limited, PNC Exploration (Canada) Co. Ltd., Saarberg Interplan Canada Ltd., Saskatchewan Mining Development Corporation (SMDC), SERU Nuclear (Canada) Limited (now COGEMA Canada Limited), Uranerz Exploration and Mining Limited, and Urangesellschaft Canada Limited. Six of those companies were amongst the top 10 from 1979 to 1983, inclusive.

Eight of the above 10 operators are companies whose majority interests are held outside of Canada; seven of those eight are supported, directly or indirectly, by their national governments in their uranium exploration efforts. Thus, in spite of the falling uranium market price, Canada continues to attract foreign investment for uranium exploration.

Not all companies that responded to the 1983 URAG questionnaire had planned their exploration expenditures for 1984 at the time of the survey. However, from the preliminary estimate of $35 million, it appears that the decline in exploration expenditures in 1984 has been less severe than in the past few years.

Indeed, the preliminary estimates for 1984 reveal that drilling activity will be in the order of 165 000 metres, representing an increase of some 8 per cent over the effort in 1983.

As in the previous three years, virtually all the surface development drilling in 1983 was associated with major programs in the Athabasca Basin of Saskatchewan. Most other exploration drilling was done in Quebec and the Northwest Territories.

Despite the downturn in the level of uranium exploration that began in 1980, and the prohibition of such activity in favourable areas because of provincial moratoria (British Columbia and Nova Scotia) or their consideration as park/wildlife areas (Yukon and Northwest Territories), annual exploration expenditures in the order of $40 million continue to represent a significant level of activity compared with other uranium producing areas of the world.

Figure 1 illustrates the responsiveness of uranium exploration expenditures in Canada to uranium price movements since the early 1970s. The decline in uranium exploration expenditures, in an apparent response to softening in the uranium market, has become increasingly evident with each successive survey since 1980.

Of particular significance, as a boost to the uranium exploration industry, was the announcement in March 1983 by SERU Nuclear (Canada) Limited, operator of the Waterbury Lake joint venture in the eastern Athabasca Basin of Saskatchewan, that significant uranium mineralization had been discovered at Cigar Lake. In May 1984, project majority-owner SMDC announced that continued drilling at Cigar Lake had demonstrated the existence of resources estimated to contain 88 500 tU grading an average 8.5 per cent U. Some 70 per cent of these resources were identified as drill-indicated, the balance as inferred. Drilling has extended the strike length of the deposit to 1 850 m, over widths from 25 to 115 m, and thicknesses from 1 to 19 m. The orebody is located at a depth of 410 to 440 m. High grade material averaging up to 32 per cent U across 19 m has been intersected.

---

* Energy, Mines and Resources Reports Uranium Supply Assessment Results, Communiqué 84/98, October 11, 1984.
** An operator may incur expenditures on a project either alone or in a joint venture. In the latter case, the combined expenditure of all participants is attributed to the project operator; as such, contributions by other parties not responding directly to the URAG survey are accounted for in the total.

The preliminary engineering and mining studies that were initiated in 1983 and continued in 1984 will confirm the dimensions and grades in the richest part of the deposit. Pre-development studies are underway, including an environmental baseline program. The joint venture partners are SMDC (50.75 per cent), COGEMA (37.375 per cent), and Idemitsu Uranium Exploration Canada Ltd. (11.875 per cent). COGEMA acquired the minority interest share in Cigar Lake held by Reserve Oil and Minerals Corporation (3.75 per cent) in July 1984.

To the northeast of Waterbury Lake, equal partners Canadian Occidental Petroleum Ltd. and Inco Limited followed up the conceptual study, which Inco completed in December 1982, for the underground mining of their McClean deposits. These occur at depths varying from 152 to 175 m and reportedly contain in excess of 5 000 tU. It was planned to test the proposed mining method over a two-year pilot program commencing in 1984. Although approval of an Underground Exploration Permit from the AECB was granted by the end of 1983, the program did not get underway during 1984.

URANIUM RESOURCES

The results of EMR's interim-year URAG uranium resource assessment, released in October 1984, are summarized in Table 6. For comparison, the results of the 1982 assessment* are also presented. Uranium resource estimates are divided by URAG into separate resource categories which reflect different levels of confidence in the quantities reported. For the 1983 assessment, these categories were subdivided into three levels of economic exploitability related to the market price of uranium. The low price range (A) was limited at the upper bound by the uranium market price estimated in Canadian dollars at $100/kg U in December 1983, when most of the data for the assessment were gathered (see Markets and Prices for derivation). Tonnages assigned to the A price category, therefore, provide a measure of those Canadian uranium resources that were of economic significance, as of the date of the assessment. The second price range (B) and third price range (C) spanned the $100-$150/kg U, and the $150-$300/kg U intervals,

---
* Uranium in Canada: 1982 Assessment of Supply and Requirements, Report EP 83-3, Energy, Mines and Resources Canada, September 1983.

respectively. All quantities are reported in tonnes of elemental uranium, consistent with international practice. Prices are given in Canadian dollars/kg U*.

In comparing the 1983 estimates of Canada's mineable uranium resources with those of the 1982 resource assessment, the most significant change to note is the 22 per cent increase in Indicated resources in contrast to the 10 and 7 per cent decreases in the Measured and Inferred resource categories, respectively; the increase primarily reflects the results of continued exploration and development work in northern Saskatchewan. The sum of Measured, Indicated, and Inferred resources, within all three price categories, was 591 000 tU. This is slightly higher than the comparable 1980 figure despite production of some 23 000 tU in the intervening three-year period!

To provide an illustration of uranium availability in the short term, a projection of Canadian production capability to 1996 was made, as illustrated in Figure 2. This scenario is an illustration of firm production capability. It is based on existing and committed production centres only, and assumes levels of production that can be practically and realistically achieved, under favourable circumstances. Only resources in the Measured, Indicated and Inferred categories, in the A plus B price range (i.e., mineable at uranium prices of $150/kg U or less), were incorporated into the projection. The lives of these production centres could be extended, in certain cases, by the exploitation of associated higher-priced resources, or through additions of resources in the A plus B price range resulting from continued exploration and development work.

This production capability scenario is not intended to represent a projection of actual production. Rather, it is intended to illustrate a level of production that could be supported by known deposits, given favourable developments in the uranium market. Actual levels of production from these centres will depend on a number of operational variables, and could be different from the capabilities projected.

GOVERNMENT AFFAIRS

On March 28, 1983, Northern Saskatchewan Minister George McLeod announced the signing of a surface lease agreement by the

---
* $1/lb U_3O_8 = $2.6/kg U

provincial government and Eldorado Resources Limited. It gave Eldorado the go-ahead to develop the Collins Bay "B" orebody and thereby expand mining operations at its Rabbit Lake uranium mine/mill complex near Wollaston Lake, northern Saskatchewan. Granting of the surface lease also signified the agreement between the government and Eldorado concerning environmental protection, health and safety of workers, and development of economic and employment opportunities.

Cluff Mining received environmental approval on June 7, 1983 from the Saskatchewan Minister of the Environment to proceed with Phase II of its Cluff Lake operation; on December 5 the surface lease agreement was finalized with the provincial government. Phase II will involve the exploitation of two open-pit and three underground mines adjacent to the present facility, and permit production to continue into the 1990s.

In September 1983, the Government of Canada made known the results of its recent review of Canadian uranium export policy*. While deciding to retain the basic principles of the existing policy, the intent of which was to ensure that Canada remains a reliable supplier of uranium to world markets, Ministers agreed to make several modifications to its implementation. The review covered three elements of the policy: security of domestic supply, further processing and commercial terms and conditions.

As the security of domestic supply seems to be less urgent than in 1974, when the last review of uranium export policy took place, Ministers decided that the supply situation should be monitored on a national rather than on an individual producer basis. As long as there is no urgent security of supply problem, individual producers would be free to sell uranium for export. Contracts can now be approved for up to 15 years, and in the case of uranium exports associated with new CANDU sales, approval can be granted for contracts which provide a uranium supply for up to 30 years for each CANDU exported.

---
* See "**Canada's Role as a Uranium Supplier**", an address by Dr. R.W. Morrison, Director-General, Uranium & Nuclear Energy Branch, EMR, to the Fourth Pacific Basin Conference, Vancouver, September 13, 1983.

In the case of further processing, Ministers reaffirmed the policy that uranium should be upgraded to the greatest extent possible in Canada before export. In practice, this means conversion to UF_6. The policy provides for the regulating agencies to consider requests for exemptions from the further processing provision if the Canadian facilities do not have the capacity or if they are not generally competitive in the world scene. Exemptions for other reasons would only be granted in the most exceptional circumstances. The further processing objective is a long-standing one and is not limited to uranium.

With respect to commercial terms and conditions, Ministers decided to reaffirm the requirement for a floor price or similar mechanism that would protect investment and employment in uranium production facilities. At the same time, the terms of sale should equitably balance benefits and risks and should generally be in accord with those being obtained by Canadian and international producers for uranium under contracts of similar duration.

It was also agreed that uranium export contracts will continue to be reviewed by the Uranium Exports Review Panel to ensure that they are consistent with Canada's policy. Acceptance by Ministers of the contract terms and conditions will remain a necessary step prior to the consideration by the regulating agencies of applications for export permits. The Panel will make every effort to ensure that the contract review is as expeditious as possible.

In the United States, an amendment to the Nuclear Regulatory Commission (NRC) Authorization Act for fiscal years 1982 and 1983 was passed in January 1983 that requires the Department of Energy (DOE) to monitor the uranium industry and make, for the years 1983 to 1992, an annual determination of its viability. A structure for the government's monitoring of that viability was established. The criteria to be used for the determination were published in the **Federal Register** in October 1983. A mechanism provides for the initiation of investigations by the U.S. International Trade Commission and the Secretary of Commerce, respectively, if it is determined that the level of uranium imports will be a substantive cause of serious injury to the U.S. uranium industry, or if imports exceed 37.5 per cent of domestic uranium requirements for two consecutive years, or if the level of imports threatens or impairs national security. The actual determination of viability rests with the Secretary of Energy.

Pursuant to the requirements set forth in the NRC Authorization Act, a comprehensive one-time review on the status of the U.S. uranium mining and milling industry was prepared and submitted to Congress on behalf of the President in May 1984. In addition to presenting projections of industry behaviour under current policy, the report presents projections under alternative policy scenarios, in the event that foreign uranium import restrictions were enacted by Congress.

The release date of the Secretary of Energy's first annual determination of industry viability, based on the criteria of resource capability, supply response capability, financial capability, and import commitment dependency, may be delayed in light of a December 1984 lawsuit, filed against DOE by certain U.S. uranium producers, that challenges the criteria of industry viability.

In December 1983, federal Justice Minister Mark MacGuigan indicated that the federal government was dropping charges against Denison Mines Limited, Rio Algom Limited, Gulf Minerals Canada Limited and Uranerz Canada Limited, following an earlier decision by the Supreme Court of Canada that the alleged co-conspirators, Eldorado Nuclear Limited and Uranium Canada, Limited, as agents of the Crown, were not subject to prosecution under Section 32 of the Combines Investigation Act.

The six companies, charged in July 1981 under the Combines Investigation Act, were alleged to have unlawfully conspired, combined, agreed or arranged to prevent or lessen, unduly, competition in the production, manufacture, sale or supply in Canada of uranium oxide and other uranium substances between September 1, 1970 and April 1, 1978.

MARKETS AND PRICES

Canadian producers continue to play an active role in the uranium market. During 1983 and 1984, a significant number of new export contracts were reviewed and accepted by the federal government. As shown in Table 7, net additions resulting from new and revised contracts in 1983 and 1984, brought to some 104 000 tU, the total amount of uranium under export contracts reviewed since September 5, 1974. The year-end 1984 total reflects scheduled deliveries under more than 110 contracts, over one third of which remain active. As of December 1984, forward commitments under all active contracts were estimated at 63 000 tU. Forward domestic commitments exceed 75 000 tU.

Actual exports in 1983 exceeded 3 800 tU, and were primarily to the United States, Japan and the United Kingdom (see Table 8). Japan has been Canada's most important single customer, receiving about 31 per cent of Canada's total exports over the last decade. Most of the remaining exports have gone to the European Economic Community (33 per cent), other countries in western Europe (18 per cent), and the United States (17 per cent).

In January 1983, SMDC and the New Brunswick Electric Power Commission (NBEPC) announced the signing of a long-term contract whereby SMDC will supply uranium concentrates required for NBEPC's Point Lepreau CANDU reactor. Some 80 tU will be required annually during normal operation of the station.

In Canada the average price for 1983 under all export contracts made by Canadian producers for deliveries in 1983 was $Cdn 98.30/kg U ($US 30.40/lb U_3O_8). In the United States data gathered from essentially all of the principal companies involved in domestic uranium marketing activities indicate that the reported price of U.S. uranium for 1983 delivery averaged $US 37.81/lb U_3O_8 ($Cdn 122/kg U). Similar to the above-mentioned Canadian and U.S. average prices, the weighted average price for 1983 deliveries to consumers in the European Economic Community, under medium- to long-term contracts, was reported at $US 31/lb U_3O8 ($Cdn 100/kg U).

By comparison, uranium spot market prices were significantly lower, as reflected by the Nuclear Exchange Corporation* (Nuexco) monthly exchange value** (EV). The EV rose from $US 17/lb U_3O_8 in September 1982 to a high of $US 24 in July, before slipping to $US 22/lb U_3O_8 by year-end 1983. It hit a 10-year low of $US 15.25 lb U_3O_8 by the end of 1984.

---
* A California-based uranium brokerage firm.
** Nuexco's judgement of the price at which transactions for significant quantities of natural uranium concentrates could be concluded as of the last day of the month.

REFINING

Eldorado Resources Limited, operates Canada's only uranium refining/conversion facilities. To keep pace with increasing domestic uranium concentrate production capability, expansions of the company's refining and conversion capacity have been under way during the past few years. These commitments were made in anticipation of greatly increased world requirements for uranium and for associated further processing services. Canada is now in a position to refine and convert the full output of its expanded uranium production industry.

At a new Blind River, Ontario, plant, uranium concentrates from mines in Canada and other countries are refined to high purity uranium trioxide (UO_3)*. The UO_3 is then converted at Port Hope, Ontario, into either uranium hexafluoride (UF_6)** for foreign utilities that operate light water reactors, or ceramic-grade uranium dioxide (UO_2) for CANDU-type heavy water reactors.

Construction of the 18 000 tU/year Blind River facility was completed, within budget, in July 1983 at a total cost of approximately $145 million. Production commenced in October after a short commissioning phase to test the design capacity and the product quality, and reached commercial levels in early 1984. Start-up of this new facility permitted the closure of Eldorado's old refinery at Port Hope.

The refined UO_3 is transported from Blind River to Eldorado's conversion plants in Port Hope. There the new $120 million UF_6 facility, with an annual output capacity of 9 000 tU as UF_6, was essentially completed, within budget, by the end of 1983. Commissioning of the facility began in early 1984, first production was achieved in June, and commercial production levels were anticipated in January 1985.

The new refinery at Blind River also gives Eldorado the capability to utilize fully its enlarged UO_2 conversion facility at Port Hope, the capacity of which was doubled to 2 800 tU as UO_2 during the upgrading/expansion program of 1980.

In total, Eldorado processed mine concentrates containing some 5 564 tU during 1983, a 12 per cent decrease from 1982; the decline was largely the result of lower demand in the world market for UF_6.

NUCLEAR POWER DEVELOPMENTS

Despite the very slow rate of new reactor orders in the world, the number of operating reactors will continue to grow steadily into the 1990s. The International Atomic Energy Agency (IAEA) reported that at the end of 1983, 317 nuclear power reactors, with a combined generating capacity of some 191 electrical gigawatts (GWe)*, were on-line in national grids in 25 countries. Although nuclear programs in some countries are still small, in others the nuclear share in electricity production at times of low demand often exceeds 50 per cent (Table 11).

At year-end 1983, a further 209 reactors with a combined capacity of 194 GWe were under construction. The IAEA estimates that the total world installed nuclear capacity will amount to 275 GWe by 1985, and will grow to between 370 and 400 GWe by 1990 and to between 580 and 850 GWe by the turn of the century.

In Canada, 14 CANDU reactors with an aggregate net output capacity of 8,097 electrical megawatts (MWe) were in service (i.e. commercially operable) at year-end 1984 and a further 9 reactors with an additional capacity of some 6,806 MWe were either in the pre start-up phase or under construction (see Table 12). Nuclear generation in Canada exceeded 46 TWh** in 1983, an increase of some 28 per cent from 1982; it accounted for 12 per cent of total Canadian electricity generation. Canada's commitment to nuclear power remains firm.

Ontario Hydro's nuclear reactors maintained their standing among the world's best performers. To the end of 1983, 7 of Hydro's 9 in-service CANDU's were in the top 10 in terms of lifetime capacity factor┬ out of some 168 commercial power reactors, rated at 500 MWe or greater, in service around the world.

---

* Uranium trioxide is the initial refined product from which UO_2 or UF_6 is produced.
** Uranium hexafluoride is the required feed material for the uranium enrichment process.

---

* GWe = 10^9 watts.
** Terawatts-hours = 10^{12} watt-hours.
┬ Lifetime capacity factor is the ratio of electricity produced, from the in-service date of the reactor, relative to that which could have been produced had the reactor operated at 100 per cent power output for 100 per cent of the time.

Some 35 per cent of the total electrical energy generated by Ontario Hydro during 1983 came from nuclear-electric units; 33 per cent was derived from hydro-electric sources and 32 per cent came from coal-fired plants.

At Ontario Hydro's four-reactor Pickering "B" Nuclear Generating Station (NGS) east of Toronto, unit 5 was declared in service on May 10, 1983. The commissioning of unit 6 proceeded favourably through 1983; the reactor reached 95 per cent of full power by year-end and was declared in service on February 1, 1984. The most probable in-service dates for units 7 and 8 are December 1984 and July 1985, respectively.

On August 1, 1983, a pressure tube rupture in unit 2 of the adjacent Pickering "A" NGS resulted in the manual shut down of the reactor; unit 1 was shut down on November 14 so that the extent and severity of the problem could be assessed. The cause of the rupture has been identified and is limited to these two early units. The fuel channel replacement program, started in 1984, will see Pickering Units 1 and 2 back on line by November 1, 1986 and February 1, 1987, respectively.

Ontario Hydro received approval in principle in November 1983 from the AECB for an operating licence for full electric power at unit 6 of the Bruce "B" NGS near Kincardine. The reactor, the first completed of the four-unit station, is capable of generating 750 MWe. It produced first electricity in June 1984 and was declared in service on September 14, 1984. Unit 5, the next reactor to be commissioned, achieved criticality on November 15, 1984 and is scheduled to go into service by April 1985. The expected in-service dates for Units 7 and 8 are April 1, 1986 and January 1, 1987, respectively.

On May 5, 1984, the Douglas Point CANDU prototype, located in the Bruce Nuclear Power Development, was taken out of service. In addition to electricity generation, the Douglas Point facility, owned and operated by Atomic Energy of Canada Limited (AECL), supplied process heat to Ontario Hydro's heavy water production plants and served as a training and R&D centre.

At Ontario Hydro's Darlington NGS near Bowmanville, construction remained on schedule; the most probable in-service dates for units 1 to 4 are February 1989, May 1988, November 1991 and August 1992, respectively.

In New Brunswick, Canada's first 600 MWe series CANDU reactor to go into commercial operation, Point Lepreau I, was declared in service on February 1, 1983. The AECB authorized the New Brunswick Electric Power Commission on March 22, 1983 to operate the reactor at 100 per cent of full power; the facility reached full power on March 27. The unit achieved essentially 100 per cent operating capacity throughout most of 1984, except for required service inspections in April and May.

Under unit-participation power export contracts, United States utilities in New England purchase 230 MWe or 36 per cent of Point Lepreau I's total output; the remaining 400 MWe of low-cost power from Point Lepreau is used within the Province of New Brunswick.

On September 28, 1983, the federal and provincial Ministers of the Environment announced that a proposal for a second 600 MWe CANDU at Point Lepreau would undergo an environmental assessment review. The AECB had approved the Point Lepreau site, some 40 km southwest of Saint John, for two CANDU's of the 600 MWe design in 1974. It is envisaged that Point Lepreau II would produce electricity for the export market on a fully commercial basis.

In late 1982, Hydro-Québec's Gentilly 2 Nuclear Power Station near Bécancour went on-line. Difficulties with the station turbine were overcome, permitting the rescheduling of the in-service date of Canada's second 600 MWe CANDU to October 1, 1983.

CANDU's of the same size and design as those installed in New Brunswick and Quebec began operation during 1983 in the Republic of South Korea and Argentina; the in-service dates of the Wolsung and Cordoba reactors are December 28, 1983 and January 20, 1984, respectively.

On August 10, 1983, AECL announced that manufacturing orders would commence on the two-unit CANDU station in Romania. A substantial portion of the exterior construction for the two CANDU's has been completed by Romania and site preparation is in progress for a third unit.

OUTLOOK

In the years ahead, the requirements for Canadian uranium will depend upon the growth of nuclear power capacity in Canada and its trading partners. Nuclear energy is already making a significant contribution to electricity requirements in many countries, and this installed nuclear capacity will require supplies of uranium well into the next century.

From data released by the IAEA it can be determined that total installed nuclear power capacity in the world increased by 10 per cent during 1983. Nuclear plants accounted for around 8 per cent of the world's total electrical generating capacity but, because they are generally used for base-load operation, nuclear units produced about 12 per cent of the world's electricity. This percentage breakdown held true for Canada in 1983.

By the year 2000, Canada's total installed nuclear capacity is expected to have grown to between 15 and 20 GWe, requiring approximately 2 500 tU per year. By that time, nuclear energy will be providing about 18 per cent of Canada's electricity supply; the level in Ontario alone will be over 60 per cent.

Since 1959, Canada has ranked second behind the United States in terms of world* uranium production. In 1983, Canada accounted for about 20 per cent of the world* total, which was estimated at some 37 000 tU; from the 1984 data available, it appears that Canada will replace the United States as the world's* leading producer. Although challenged by South Africa and Australia, Canada's position as the world's* leading exporter of uranium is expected to be maintained at least in the short term.

Canada's annual uranium production capability will expand to some 12 000 tU within the next year or two, reflecting the recent completion of projects in Ontario and the phasing in of committed development plans in Saskatchewan. Because of the bleak outlook for the uranium market over the next decade, it is unlikely that further new domestic production projects will be brought on-stream before the early 1990s. Actual production over the intervening period will be dependent on a number of factors, the most important of which will be delivery commitment levels and corporate decisions by individual companies concerning inventory accumulation. It is quite possible that production over the next few years will not match Canada's full 12 000 tU/year capability.

Whatever the magnitude of future uranium requirements turns out to be, Canada has the capability to provide for its own needs while maintaining its place among the leading suppliers of uranium to world markets.

---

* World, as used here, excludes the U.S.S.R., Eastern Europe and the People's Republic of China.

* World, as used here, excludes the U.S.S.R., Eastern Europe and the People's Republic of China.

TABLE 1. URANIUM PRODUCTION IN CANADA, BY COMPANY, 1982 AND 1983

Company	Location	Production 1982 (tonnes U[1])	Production 1983 (tonnes U[1])
Agnew Lake Mines Limited	Agnew Lake, Ont.	65	15
Cluff Mining (Amok Ltd/SMDC)	Cluff Lake, Sask.	1 469	682
Denison Mines Limited	Elliot Lake, Ont.	2 359	2 298
Eldorado Resources Limited	Eldorado, Sask.[2]	282	1[3]
	Rabbit Lake, Sask.	1 210[4]	1 244
Key Lake Mining Corporation	Key Lake, Sask.	-	423[5]
Madawaska Mines Limited	Bancroft, Ont.	153	-
Rio Algom Limited - Quirke	Elliot Lake, Ont.	1 672[6]	1 446
- Panel		865	831
- Stanleigh		-	203[7]
Total Canada[8]		8 075	7 143

Source: Company annual reports.
[1] One metric ton (tonne) of elemental uranium (U), written as tU, is equivalent in terms of uranium content to 1.2999 short tons of uranium oxide (U_3O_8). [2] Beaverlodge operation only. [3] Final clean-up of product from precipitation circuit. [4] Joint operation - Gulf Minerals Canada Limited/Uranerz Canada Limited - acquired by Eldorado in October 1982. [5] Milling commenced in October 1983. [6] Does not include uranium recovered from Panel ore processed at Quirke, or from acid raffinate from Eldorado. [7] Milling commenced in July 1983. [8] Primary uranium production only; does not include uranium recovered from raffinates and sludges by Rio Algom and Denison, which during 1983 totalled some 20 tU.

TABLE 2. OPERATIONAL CHARACTERISTICS OF CANADIAN URANIUM PRODUCTION CENTRES IN 1983

Company Name/ Production Centre	Nominal Mill Capacity/Actual Throughput (tonnes/day)	Total Ore Processed (tonnes)	Grade of Ore Processed (kgU/t)	Overall Mill Recovery (%)
Amok Ltd.-SMDC/ Cluff Lake	100 in Phase I (800 in Phase II)	see text	see text	91
Denison Mines Limited/ Elliot Lake	13 610 / 9 612	3 191 079	0.77	93
Eldorado Resources Limited/Rabbit Lake	1 500 / 2 400	620 239	2.15	94
Key Lake Mining Corporation/Key Lake*	500-700/ NA	NA	NA	95
Rio Algom Limited/ Elliot Lake				
- Quirke	6 350 / 5 138	1 705 752	0.85	95
- Panel	2 990 / 3 010	987 215	0.89	95
- Stanleigh*	4 540 / 3 028	467 032	0.45	83

Sources: Corporate Annual Reports and the Atomic Energy Control Board (AECB).
* Mill tune-up and pre-operational testing during 1983. NA Not available.

TABLE 3. VALUE OF URANIUM SHIPMENTS[1] IN CANADA BY PROVINCE, 1982-84

	1982		1983		1984P	
	(t)	($000)	(t)	($000)	(t)	($000)
Ontario	5 092	589,057	4 767	546,306	4 381	538,733
Saskatchewan	2 551	248,411	2 056	121,366	5 312	377,561
Total	7 643	837,468	6 823	667,672	9 693	916,294

[1] Shipments of uranium (U) in concentrate from ore processing plants.
P Preliminary.

TABLE 4. WORK FORCE SUMMARY - CANADIAN URANIUM PRODUCING OPERATIONS

Company Name (Mine Name)	Total Number of Employees (Mine, Mill, General)		
	1/1/81	1/1/83	1/1/84
Agnew Lake Mines Limited (Agnew Lake)	79	53	-
Cluff Mining (Cluff Lake)	241	304	272
Denison Mines Limited (Denison)	2,027	2,027	2,199
Eldorado Resources Limited (Beaverlodge)	845	120	-
Key Lake Mining Corporation (Key Lake)	-	-	489
Gulf Minerals Canada Limited/ Uranerz Canada Limited (Rabbit Lake)*	320	330	337
Madawaska Mines Limited (Faraday)	381	9	-
Rio Algom Limited (Quirke)	1,404	1,271	1,079
(Panel)	771	713	687
(Stanleigh)	-	-	789
Total all producers	6,068	4,827	5,845

* Acquired by Eldorado Resources Limited in October 1982.

TABLE 5. PRODUCTION OF URANIUM IN CONCENTRATES BY MAJOR PRODUCING COUNTRIES, 1975-83

	United States	Canada	South Africa	Namibia	France	Niger	Gabon	Australia	Other[1]	Total[2]
					(tonnes U)					
1975	8 900	3 560	2 490	-	1 730	1 310	800	-	330	19 120
1976	9 800	4 850	2 760	650	1 870	1 460	..	360	340	22 090
1977	11 500	5 790	3 360	2 340	2 100	1 610	910	355	385	28 350
1978	14 200	6 800	3 960	2 700	2 180	2 060	1 020	515	455	33 890
1979	14 400	6 820	4 800	3 840	2 360	3 620	1 100	705	465	38 110
1980	16 800	7 150	6 150	4 040	2 630	4 100	1 030	1 560	510	43 970
1981	14 800	7 720	6 130	3 970	2 560	4 360	1 020	2 920	670[3]	44 150
1982	10 330	8 080	5 820	3 780	2 860	4 260	970	4 420	970[4]	41 490
1983	8 140	7 140	6 060	3 800	3 240	3 420	1 040	3 220	1 020[4]	37 110

Sources: Data derived principally from "Uranium: Resources, Production and Demand, December 1983, a biennial report jointly produced by the Nuclear Energy Agency of the Organization for Economic Co-operation and Development, and the International Atomic Energy Agency, with supplements from the annual "MINEMET" report of IMETAL SA, for 1981, and from miscellaneous sources, for 1982 and 1983. From 1980, country totals are rounded to the nearest 10 tU.

[1] Includes Argentina, Federal Republic of Germany, Japan, Portugal, Spain, and Sweden (1975 only). [2] Totals (rounded) are of listed figures only. [3] Includes Belgium, Brazil, India and Israel. [4] Includes Belgium and Brazil plus estimates for India and Israel.
- Nil; .. Not available.

TABLE 6. ESTIMATES[a] OF CANADA'S URANIUM RESOURCES CONTAINED IN MINEABLE ORE[b], 1982, 1983

Price ranges within which mineable ore is assessed[c]	Reasonably Assured Resources				Estimated Additional Resources - Category I (Inferred)	
	(Measured)		(Indicated)			
(Canadian dollars)	1983[d]	1982[e]	1983[d]	1982[e]	1983[d]	1982[e]
			('000 tonnes U)			
A	30	35	162	150	141	191
B	-	1	41	9	78	50
A + B	30	36	203	159	219	241
C	27	28	48	46	64	63
A + B + C	57	64	251	205	283	304

[a] Interim revisions for 1983; comprehensive assessments for selected properties only.
[b] Milling losses have not been deducted; uranium recoverable from such ore will be less.
[c] The price figures reflect the price of a quantity of uranium concentrate containing 1 kg of elemental uranium. The prices were used in determining the cut-off grade at each deposit assessed, taking into account the mining method used in the milling losses expected.
[d] For the 1983 assessment, the price ranges were (A) $100/kg U or less, (B) between $100 and $150/kg U, and (C) between $150 and $300/kg U.
[e] For the 1982 assessment, the price ranges were (A) $115/kg U or less, (B) between $115 and $170/kg U, and (C) between $170 and $340/kg U.
[f] Only the resources mineable within the lower two price ranges (A and B) are used for the purpose of ensuring that sufficient uranium is reserved to meet domestic requirements.
- Nil.

TABLE 7. URANIUM UNDER EXPORT CONTRACTS REVIEWED[1] SINCE SEPTEMBER 5, 1974

Country of buyer	1983	1984
	(tonnes U)	
Belgium	3 030	3 030
Finland	2 200	3 510
France	9 390	9 390
Italy	1 120	1 120
Japan	22 680	22 740
South Korea	5 140	5 140
Spain	4 040	3 940
Sweden	4 960	5 310
Switzerland	150	150
United Kingdom	7 700	7 700
United States	28 670	30 360
West Germany	11 480	11 580
Total	100 560	103 970

[1] Reviewed and accepted under Canadian uranium export policy. Totals adjusted to reflect new and amended contracts as of December of given year.

TABLE 8. EXPORTS OF URANIUM OF CANADIAN ORIGIN

Country of Final Destination	Tonnes of contained uranium*	
	1982	1983
Belgium	85	–
Finland	97	179
France	–	435
Italy	143	–
Japan	718	663
South Korea	74	94
Spain	110	–
Sweden	889	612
United Kingdom	379	674
United States	4 852**	672
West Germany	471	490
Total	7 818	3 819

Source: The Atomic Energy Control Board.

* Some of this uranium was first exported to intermediate countries, namely France, USA and USSR, for enrichment and then forwarded to the country of final destination.
** The bulk of this material represents concentrates exchanged by Eldorado in the purchase of the Rabbit Lake operation.

TABLE 9. EXPORTS[1] OF RADIOACTIVE ORES AND CONCENTRATES[2] FROM CANADA, 1976-83

	United States[3]	U.S.S.R.	United Kingdom	Italy	France	Japan	Norway	South Korea	Total
					($000)				
1976	46,850	–	20,541	–	–	–	–	–	67,392
1977	72,848	–	2,590	–	–	–	–	–	75,438
1978	163,911	–	39,106	3,348	–	791	–	–	207,156
1979	347,388	–	18,851	12,613	–	9	–	–	378,862
1980	218,013	–	10,319	–	1	–	–	–	230,662
1981	152,473	3,182	18,845	–	–	–	–	2,329	179,384
1982	346,891	–	11,690	–	–	–	2,862	2,022	358,581
1983	25,400	–	37,175	–	–	–	–	–	62,575

Source: Statistics Canada.
1 Material that cleared Canadian customs with destination as indicated. 2 Includes uranium in concentrates. 3 For 1976, uranium almost entirely destined for transshipment, primarily to western Europe and Japan, following conversion and enrichment; for subsequent years, figures represent a mixture of sales to U.S. and others, primarily in western Europe and Japan.
– Nil.

TABLE 10. EXPORTS[1] OF RADIOACTIVE ELEMENTS[2] AND ISOTOPES FROM CANADA, 1976-83

	United States[3]	U.S.S.R.[4]	UK	West Germany	France	Belgium Luxembourg	Netherlands	Finland	Argentina	Japan	South Korea	Other	Total
						($000)							
1976	151,427	24,471	3,786	288	375	–	–	–	84	1,068	–	4,198	185,697
1977	151,869	6,133	356	384	685	75	–	10	287	288	–	1,078	161,165
1978	269,903	101,619	38,602	6,918	19,046	23	–	10	12,177	1,017	–	1,668	450,983
1979	293,577	170,500	5,147	26,159	1,762	221	629	5,493	94,038	1,101	87	3,363	602,077
1980	199,001	77,235	2,104	20,406	144,013	4,847	374	6,408	27,766	1,911	137,002	4,312	625,379
1981	382,418	20,192	2,081	40,092	213,051	339	7,506	–	248	1,577	67	2,915	670,486
1982	299,246	34,854	796	37,250	36,213	291	–	199	214	19,617	123	5,230	434,033
1983	261,168	8,148	2,303	32,208	39,037	232	1,517	11	315	12,371	3,057	7,248	367,615

Source: Statistics Canada.
1 Material that cleared Canadian customs with destination as indicated. 2 Includes uranium hexafluoride (UF_6) and radio-isotopes for medical and industrial purposes. 3 For year 1976, UF_6 component destined for transshipment, primarily to western Europe and Japan, following enrichment; for subsequent years, figures would also include UF_6 sales to the U.S. market. 4 UF_6 component destined entirely for transshipment to western Europe, following enrichment.
– Nil.

TABLE 11. NUCLEAR SHARE OF ELECTRICITY PRODUCTION IN SELECTED COUNTRIES IN 1983

	%		%
France	48.3	Japan	18*
Belgium	45.9	Federal Republic	
Finland	41.5	of Germany	17.8
Sweden	36.9	United Kingdom	17.0
Taiwan	37*	Canada	12.9
Bulgaria	32.3	United States	12.6
Switzerland	29.3	German Democratic	
Republic of		Republic	12
Korea	18.4	Hungary	10*

Source: International Atomic Energy Agency.
* Estimate as no official data provided.

TABLE 12. NUCLEAR POWER PLANTS IN CANADA, DECEMBER 1984

Reactors	Owner	Net Output (MWe)	In-Service Dates (Expected)
Nuclear Power Demonstration	Atomic Energy of Canada Limited	22	1962
Pickering 1 to 4	Ontario Hydro	2 060	1971-73
Bruce 1 to 4	Ontario Hydro	2 960	1977-79
Point Lepreau	New Brunswick Electric Power Commission	635	1983
Gentilly 2	Hydro-Québec	638	1983
Pickering 5 and 6	Ontario Hydro	1 032	1983-84
Pickering 7 and 8	Ontario Hydro	1 032	(1984-85)
Bruce 6	Ontario Hydro	750	1984
Bruce 5, 7 and 8	Ontario Hydro	2 250	(1985-87)
Darlington 1 to 4	Ontario Hydro	3 524	(1988-92)
Total net output expected by 1993		14 903 MWe	

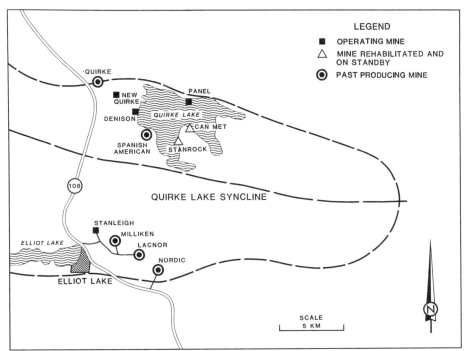

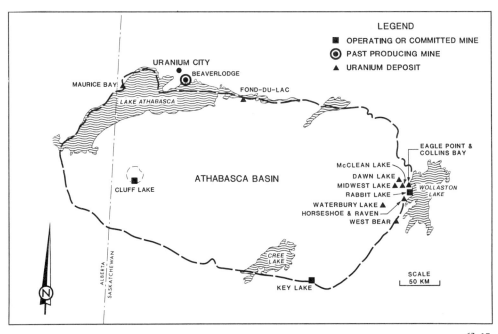

DISTRIBUTION OF URANIUM RESOURCES AMONG PRINCIPAL PRODUCING COUNTRIES

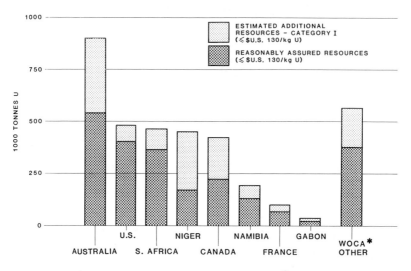

SOURCE: NEA/IAEA DECEMBER 1983 "REDBOOK" PLUS 1984 REVISIONS FOR AUSTRALIA, U.S., NIGER AND CANADA

* WORLD OUTSIDE CENTRALLY PLANNED-ECONOMY AREAS

PROJECTED CANADIAN PRODUCTION CAPABILITY COMPARED WITH ESTIMATED ANNUAL URANIUM REQUIREMENTS

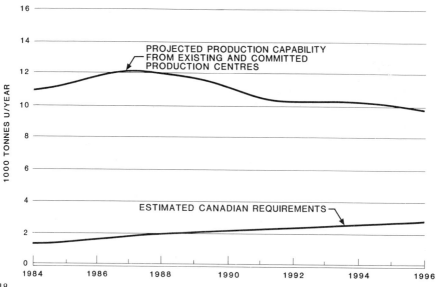

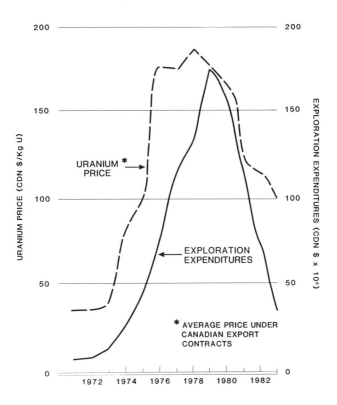

Zinc

M.J. GAUVIN

With the recovery in the world economy from the deep recession of 1981-82, improved market conditions for zinc began during the second half of 1983 and continued during the first half of 1984. Demand and prices then declined and producers were forced to continue evaluating the effects of their production decisions on the marketplace.

CANADIAN DEVELOPMENTS

Mining

Canadian zinc mine production increased sharply during 1984 to an estimated 1 213 000 t from the 1 069 709 t produced in 1983, an increase of 13.4 per cent.

Production at the Buchans, Newfoundland copper-lead-zinc mine of Abitibi-Price Inc. (51 per cent) and ASARCO Incorporated (49 per cent) ceased permanently in September, because of exhaustion of ore reserves. The fifty-six year old mine had discontinued production in December 1981 but reopened in July 1983 to salvage the remaining reserves in its deepest levels. Normal production was maintained throughout the year at the Daniel's Harbour mine of Newfoundland Zinc Mines Limited. Diamond drilling in the mine area discovered two new ore zones which will be further evaluated by underground exploration from the main L-Zone workings.

At Brunswick Mining and Smelting Corporation Limited's No. 12 mine near Bathurst, production was at slightly higher rates than in previous years. Heath Steele Mines Limited and ASARCO Incorporated have not operated their Little River Joint Venture mine located near Newcastle since April 1983, but during 1984 some production was obtained from a new open-pit on the property.

In October Noranda Mines and Les Mines reopened the Gallen zinc-precious metal mine near Noranda. Production at the Gallen had been suspended since July 1982.

Corporation Falconbridge Copper started a shaft-sinking and exploration program on the Ansil deposit which is located close to the plant of its Lake Dufault operations near Noranda, Quebec. The company is also proceeding with the sinking of a 510 m shaft at its Winston Lake deposit near Schreiber, Ontario. The property, under option from Zenmac Explorations Limited, has ore reserves estimated at 2.7 million t averaging 17.8 per cent zinc, 0.95 per cent copper, 0.74 oz silver and 0.025 oz gold per t. Underground development and diamond drilling will follow the initial shaft-sinking.

In early 1984 Sherritt Gordon Mines Limited announced the closure of the Ruttan copper-zinc mine in northern Manitoba, but reconsidered and authorized continued operations after it experienced improved productivity and received a $10 million project loan from the Manitoba government, for a development program at depth. Sherritt placed increased emphasis on zinc production at both its Ruttan and Fox mines. The Fox is expected to close late in 1985 because of depletion of ore reserves.

In British Columbia, Noranda Mines suspended operations in April, 1984 at its Goldstream property near Revelstoke. This new mine started production in the first half of 1983. Westmin Resources resumed production in April after a 4½ month strike at its Myra Falls operations, and continued to work towards bringing its nearby H-W mine into production. A new 2 700 tpd mill will process ore from the company's mines and production of zinc in concentrates is expected to increase by 30 000 tpd early in 1985.

In the Yukon, production is still suspended at the Faro lead-zinc mine of Cyprus Anvil Mining Corporation. A two-year overburden stripping program

---

M.J. Gauvin is with the Mineral Policy Sector, Energy, Mines and Resources, Canada. Telephone (613) 995-9466.

begun in June 1983, with financial assistance from the Yukon territorial and federal governments, is continuing. In the far north, the Polaris zinc-lead mine of Cominco Ltd. on Little Cornwallis Island was closed for one month starting in mid-December 1984, because of low zinc prices. The Polaris mine has an annual capacity of 100 000 t of zinc and 30 000 t of lead in concentrates.

Smelting and Refining

Refined zinc production in Canada estimated at 683 000 t in 1984, is an all-time high, and up considerably from the 617 033 t produced in 1983. Canadian Electrolytic Zinc Limited (CEZ) started operating its new zinc roaster at Valleyfield, Quebec in September 1983. The facility allows CEZ to roast all its zinc concentrates at the plant. It also raised refined zinc capacity by 9 000 t to 227 000 t. Kidd Creek Mines during 1983 completed a 19 000 t expansion of its zinc facility at Timmins, Ontario raising annual capacity to 127 000 t. The expansion included a pressure leaching plant similar to that installed by Cominco Ltd. at Trail, British Columbia. Kidd Creek is now in the process of adding an additional 7 000 t of capacity which is expected to be in operation in 1986.

Consumption

Canadian consumption of refined zinc, as measured by domestic shipments, was estimated at 141 600 t in 1984 compared with 134 424 t in 1983. However, in addition to metal received from Canadian producers, consumers imported almost 10 000 t of zinc from foreign producers during 1983.

WORLD DEVELOPMENTS

Mining

Non-socialist world zinc mine production in 1984 is estimated at 5.0 million t, a 4.6 per cent increase from the 4.79 million t produced in 1983. Large increases in production in Canada and Spain more than offset small decreases in Australia and the United States. During 1984, seven zinc mines were opened or expanded which increased capacity by 152 000 t. Most of the increase was accounted for by mines in Thailand and Mexico. Three mine closures or suspensions of production were reported, two of these in Canada. During 1983, seventeen zinc mines were opened or expanded which increased total annual capacity by about 280 000 t of zinc. Half of this increase was in Australia. Closures in Canada and the United States, mainly through suspensions, totalled about 225 000 t of annual capacity in 1983.

United States mine production in 1984 was estimated to have fallen slightly as a result of strikes at the lead-zinc mines in Missouri. The permanent closure of the Friedenville mine in Pennsylvania in late 1983 was largely offset by the reopening of Inspiration's Beaver Creek Mine in Tennessee and the opening of St. Joe's Viburnum No. 35 in Missouri. ASARCO resumed development of its West Fork lead-zinc mine in Missouri, which is expected to start producing in 1985 with a capacity of 7 000 tpy of zinc in concentrates. Cominco American Incorporated completed feasibility studies for a joint venture with Nana Regional Corp., an Alaskan native company, to develop the Red Dog zinc-lead-silver deposit in northwest Alaska. Located some 100 km from the Chukchi Sea, ore reserves are estimated at 85 million t with grades of 17.1 per cent zinc, 5.0 per cent lead and 74 g/t of silver. This large project is expected to come on-stream in the late 1980s. It will be operated and financed by Cominco and after the investment is recovered, the partners will eventually share proceeds on a 50:50 basis.

Mexican zinc production was at a new high of 300 000 t in 1984. Industrial Minera Mexico S.A. completed the expansion of its San Martin mine which added 45 000 tpy to its capacity. In 1983, the company doubled the capacity of its Santa Barbara zinc-lead-silver underground mine to 48 000 tpy of zinc in concentrates. An additional 68 000 t of zinc mine capacity is expected to come on-stream in Mexico in the next four years. In Peru, three mine expansions will bring an additional 30 000 t of mine capacity on-stream in 1985.

Bula Mines Ltd. announced plans to bring its Navan, Ireland zinc-lead deposit into production in 1987. The mine will have a capacity of 55 000 tpy of zinc in concentrates. In Italy, SAMIM S.p.A. of Italy is in the process of expanding its Montiponi lead-zinc mine, which will increase the mine's zinc capacity by 42 000 t in 1986. Minas De Almagrera has increased the capacity of the Sotiel mine in Spain by 6 000 tpy and Austuriana de Zinc is proceeding with an expansion of its Reocin lead-zinc mine which will add 20 000 t to its capacity in 1985. Two small expansions in Yugoslavia added 7 000 t to the country's zinc mine capacity.

The main development in Australia in 1983 was the opening, by EZ Industries Ltd., of the large Elura mine in New South Wales. This $A 200 million underground mining project has a capacity to produce 70 000 t of zinc and 40 000 t of lead in concentrates. MIM Holdings Limited completed the expansion of its Mt. Isa, Queensland mine. Capacity increased by 65 000 tpy to 175 000 tpy of zinc in concentrates.

Strikes in Australia reduced 1984 zinc mine production to a little below the 1983 level. MIM Holdings is continuing development of the Hilton zinc-lead-silver mine with trial stoping scheduled for 1985. This project is required to maintain production levels at the Mt. Isa complex. Exploration continued at the Woodcutters zinc-lead-silver deposit in the Northern Territory, which is scheduled to start producing in 1985 with a capacity of 23 000 tpy of zinc. The Mae Sot zinc mine of Padaeng Industry in Thailand, which came into production in March in preparation for the opening of the new zinc smelter, has a capacity of 60 000 tpy of zinc. Hindustan Zinc Inc. is considering bringing the large Rampura-Agucha zinc deposit into production in 1989 at a rate of 70 000 tpy of zinc in concentrates. With this additional tonnage, India will become almost self-sufficient in zinc.

Smelting and Refining

Non-socialist world zinc metal production was estimated at 4.8 million t in 1984, or 3.8 per cent above the 1983 level. All principal producing countries except Australia recorded substantial increases over 1983. Total output in 1984 was at a record level, well above the previous peak of 4.71 million t in 1979.

ASARCO Inc. reopened its Corpus Christi, Texas plant in May. Production at the 104 000 tpy plant had been suspended since October 1982. It was operating at 50 per cent of capacity prior to the closure. Gulf and Western Industries sold its 60 per cent interest in Jersey Minière Zinc to its Belgium partner, a subsidiary of Union Minière. Jersey Minière operates a zinc mining and refining complex at Clarksville, Gulf and Western Industries sold its 60 per cent interest in Jersey Minière Zinc to its Belgium partner, a subsidiary of Union Minière. Jersey Minière operates a zinc mining and refining complex at Clarksville, Tennessee. The acquisition by St. Joe Minerals Corp. of National Zinc Co. resulted in the creation of the largest Unites States zinc producing company with a capacity of 142 000 tpy. National operates a 51 000 t electrolytic refinery in Bartlesville, Oklahoma while St. Joe has long operated an electrothermic plant at Monaco, Pennsylvania where a 14 000 t expansion has just been completed, raising its capacity to 91 000 tpy.

Electrolytic Zinc Company of Australasia Ltd. completed a small 4 000 t expansion of its Risdon refinery, bringing its capacity up to 214 000 t. In Thailand, Padaeng Industry started production in late 1984 at its new 60 000 t electrolytic refinery.

European zinc producers in 1983 received approval from the EC Commission for a capacity shutdown agreement, whereby companies would be paid for each tonne of zinc smelting capacity permanently closed. As no nominations for closure were received by the Commission, the agreement expired. It had been anticipated that up to 200 000 tpy capacity would be nominated for closure and the industry hoped that this concerted action would reverse the series of financial losses most companies had been experiencing. Rather than reducing capacity, expansions and a new plant will add 163 000 tpy capacity in Europe in the next two years. SAMIM S.p.A. is building a new 83 000 t electrolytic plant at Porto Vesme, Sardinia. Expansions of 20 000 t are under way by Pertusola at Crotone, Italy, and by Norzink at Odda, Norway, while in Yugoslavia, Trepca is adding 40 000 t capacity to its Kosovska Mitrovica refinery.

In India, Cominco-Binani is expanding its electrolytic plant by 6 000 t in 1987. Iran has announced the start of construction in the northern part of the country of a metallurgical plant expected to produce about 60 000 t of lead and 40 000 t of zinc towards the end of this decade. The Zinc Corporation of South Africa Ltd. completed in 1983 the 20 000 t expansion of its zinc electrolytic plant at Springs. This brought its total capacity to 105 000 t.

Consumption

World zinc metal consumption in 1984 was estimated at 4.7 million t, up 96 000 t from 1983. The peak level of consumption in the non-socialist world was 4.88 million t in 1973.

PRICES

At the beginning of 1983, the price of high grade zinc was 40¢ (U.S.) in the United States while the European producer price (EPP) was split at $US 800-850/t. Prices reached their low point in March when the U.S. price averaged 37.9¢ and the EPP was $750. Periodic price increases, the last in late December brought the producer price for high grade zinc up to 49¢/lb in the United States, 60¢ in Canada and $980/t outside North America. A similar price pattern prevailed on the LME where at the end of the year, zinc traded at £617.5 a t.

Early in January, 1984, the U.S. producer price was raised to 51¢ for high grade and the EPP moved up to $1,010 and then to $1,050 at the end of the month. A further increase in March brought the U.S. price of high grade up to 53¢ and in Canada it was quoted at 67¢/lb, their highest prices for the year. During April and part of June most producers quoted their EPP at $1,090 a t but a split price prevailed as Preussag and Metallgesellschaft maintained their $1,050 quote. At the end of June the EPP was lowered to $1,040 a t and the U.S. producer price of high grade dropped to 51 and then 50¢/lb. A further reduction of the EPP to $990 in mid-July was followed in early August by a lowering of the United States and Canadian producer prices of high grade to 48¢ and 62¾¢, respectively. In early September the EPP was lowered to $940 a t and later in the month the U.S. producer price for high grade dropped to 45¢/lb. The price in Canada was 59¼¢/lb. During the first week of October the EPP was lowered further, to $900 (U.S.) a t, which was maintained until the end of the year.

On September 3, the LME introduced a new zinc contract based on metal with a minimum purity of 99.95 per cent (high grade). This new contract will run concurrently with the old contract until the end of November 1985 when the old g.o.b. contract will be terminated.

OTHER DEVELOPMENTS

Zinc usage in zinc diecast parts in the average North American built automobile has been declining over the years due to the increased use of thin wall zinc castings and competition from plastic parts, both in response to the demand for lighter-weight cars. The weight of zinc diecastings per car has dropped from almost 22.7 kilos in 1970 to 11.4 kilos in 1980 but appears to have levelled out at about 10.5 kilos in currently built vehicles. The strong demand for a better rust-resistant automobile has resulted in increased usage of galvanized steel and zinc consumption in galvanized iron for automobiles is expected to almost double in 1990 from the 55 000 t used in 1983.

The first copper-plated zinc penny was struck by the U.S. Mint in November 1981 and placed in circulation in January 1982. The penny blanks are made of an alloy containing 99.2 per cent SHG zinc and 0.8 per cent copper; the total penny including plating is 97.6 per cent zinc and 2.4 per cent copper. This new usage for zinc now consumes about 40 000 tpy of zinc and other countries are considering converting to this form of coinage. The regular tenders conducted by the Mint are looked upon by many merchants and producers as a barometer of the zinc market.

Galfan, a new and improved galvanizing alloy developed by the International Lead Zinc Research Organization, Inc. (ILZRO) was first used commercially in 1983 in Japan. The alloy contains about 95 per cent zinc, five per cent aluminum and a small but significant amount of rare earth metals. The new alloy outperforms conventional galvanizing and Galvalume in corrosion resistance and several other characteristics. Another advantage is that only minor modifications are necessary to adapt existing galvanizing lines compared with the major cost of converting a line for Galvalume. Galvalume, a 55 per cent aluminum, 43.4 per cent zinc and 1.6 per cent silicon alloy, developed by Bethlehem Steel Corporation was introduced to the U.S. market in 1976 and is being used in specialized applications.

These alloys are complementary to galvanizing and increase the potential market for zinc.

OUTLOOK

Overcapacity in all sectors of the industry is expected to persist until at least the end of the decade. At the same time the industry's need to remain competitive will require a close balance between supply and demand. This implies a further reduction of costs and a low level of inventory. Pressure on prices will continue and prices are expected to change little in constant dollar terms at least until overcapacity in all segments of the industry has been worked off. Any substantial upturn in the world economy, particularly in the key construction and automobile industries would result in improved demand and would increase real prices for a wide range of zinc mineral products.

TARIFFS

Item No.		British Preferential	Most Favoured Nation	General	General Preferential
		(% unless otherwise specified)			
CANADA					
32900-1	Zinc in ores and concentrates	free	free	free	free
34500-1	Zinc dross and zinc scrap for remelting, or for processing into zinc dust	free	free	10	free
34505-1	Zinc spelter, zinc and zinc alloys containing not more than 10% by weight of other metal or metals, in the form of pigs, slabs, blocks, dust or granules	free	free	2¢/lb	free
35800-1	Zinc anodes	free	free	10	free

UNITED STATES (MFN)

626.04	Zinc, unwrought, alloyed			19.0%		

Item No.		1983	1984	1985	1986	1987
		(% unless otherwise specified)				
602.20	Zinc in ores and concentrates	0.48¢/lb	0.44¢/lb	0.39¢/lb	0.35¢/lb	0.30¢/lb
626.02	Zinc, unwrought, unalloyed	1.8	1.7	1.6	1.6	1.5
626.10	Zinc, waste and scrap (suspended until June 30, 1984)	3.7	3.3	2.9	2.5	2.1

EUROPEAN ECONOMIC COMMUNITY (MFN)

Item No.		Base Rate	Concession Rate
		(% unless otherwise specified)	
26.01	Zinc, ores and concentrates	free	free
79.01	Zinc, unwrought	3.5	3.5
	Zinc, waste and scrap	free	free

JAPAN (MFN)

Item No.		1983	Base Rate	Concession Rate
		(% unless otherwise specified)		
26.01	Zinc, ores and concentrates	free	free	free
79.01	Zinc, unwrought, unalloyed	2.3	2.5	2.1
	Zinc, unwrought, alloyed	7.5 yen/kg	10 yen/kg	7 yen/kg
	Zinc, waste and scrap	1.9	2.5	1.9

Sources: The Customs Tariff, 1983, Revenue Canada, Customs and Excise; Tariff Schedules of the United States Annotated 1983, USITC Publication 1317; U.S. Federal Register Vol. 44, No. 241; Official Journal of the European Communities, Vol. 25, No. L 318, 1982; Customs Tariff Schedules of Japan, 1983.

TABLE 1. CANADA, ZINC PRODUCTION AND TRADE, 1982-84 AND CONSUMPTION 1981 and 1982

	1982		1983		1984P	
	(tonnes)	($000)	(tonnes)	($000)	(tonnes)	($000)
Production						
All forms[1]						
Ontario	260 544	279,563	288 528	331,605	294 237	413,991
New Brunswick	230 299	247,110	225 054	258,655	246 424	346,719
Northwest Territories	213 523	229,110	234 883	269,951	245 357	345,217
British Columbia	75 183	80,671	95 289	109,516	92 619	130,315
Manitoba	31 435	33,730	49 007	56,324	48 564	68,330
Quebec	67 002	71,893	53 688	61,703	48 900	68,802
Newfoundland	28 139	30,194	35 358	40,637	39 910	56,153
Saskatchewan	4 945	5,306	5 879	6,757	6 043	8,503
Yukon	54 537	58,519	27	30	-	-
Total	965 607	1,036,096	987 713	1,135,178	1 022 054	1,438,030
Mine output[2]	1 189 129	..	1 069 709	..	1 213 000	..
Refined[3]	511 870	..	617 033	..	683 000	..
					(Jan.-Sept. 1984)	
Exports						
Zinc blocks, pigs and slabs						
United States	263 588	266,028	309 490	328,684	251 086	343,177
People's Republic of China	21 495	20,289	54 244	45,255	14 688	17,135
United Kingdom	44 692	42,708	25 697	23,454	29 755	36,562
Taiwan	9 855	9,140	16 231	13,791	9 703	11,289
Philippines	10 877	9,876	9 397	7,760	5 140	5,710
Hong Kong	2 778	2,714	8 486	7,404	6 645	8,218
West Germany	12 021	11,588	6 197	5,751	7 380	8,844
Indonesia	11 192	10,129	6 422	5,393	4 502	5,390
Italy	5 926	5,210	5 059	4,301	5 682	5,886
Singapore	7 572	7,049	6 092	5,096	3 566	3,936
New Zealand	6 905	7,306	5 560	4,960	6 300	7,519
Other countries	73 489	67,623	47 568	41,709	35 673	41,310
Total	470 390	459,660	500 443	493,558	380 120	494,976
Zinc contained in ores and concentrates						
Belgium-Luxembourg	214 056	98,249	344 672	139,853	239 159	116,001
West Germany	30 563	12,975	113 698	30 287	25 213	10,908
Japan	83 748	32,352	47 817	21,403	51 917	28,738
Italy	9 859	5,336	23 355	15,190	4 110	2,170
France	18 086	11,133	30 191	14,934	16 702	11,465
Yugoslavia	-	-	22 826	8,707	..	..
United Kingdom	34 602	20,271	17 660	8,497	11 644	6,275
United States	4 953	2,553	8 939	6,639	19 462	14,402
Finland	12 305	4,230	10 478	5,492	..	..
Other countries	49 579	30,109	41 156	21 216	15 999	11,059
Total	457 751	217,208	660 792	272,218	384 206	201,018
Zinc alloy scrap, dross and ash[4]						
United States	10 155	4,714	12 541	6,003	7 021	5,555
United Kingdom	7 992	4,699	2 549	858	1 209	446
West Germany	7 048	2,889	1 610	310	6 097	2,805
Japan	34	16	392	204	136	88
Belgium-Luxembourg	22 996	13,831	194	85	266	200
Other countries	25 364	15,313	643	273	1 320	1,083
Total	73 589	41,462	17 929	7,733	16 049	10,177

TABLE 1. (cont'd.)

	1982		1983		1984P	
	(tonnes)	($000)	(tonnes)	($000)	(tonnes)	($000)
Zinc dust and granules						
United States	2 296	3,061	4 090	4,602	2 379	3,714
Colombia	14	22	169	272	28	41
United Kingdom	-	-	128	90	..	..
Other countries	1	2	28	55	112	76
Total	2 311	3,085	4 415	5,019	2 519	3,831
Zinc fabricated material, nes						
United States	1 020	2,925	1 762	3,734	669	2,167
Chile	-	-	80	74	..	..
France	18	60	24	69	..	..
Hong Kong	9	59	74	67	..	2
Taiwan	-	-	35	31	-	-
Other countries	49	83	16	62	104	561
Total	1 096	3,127	1 991	4,037	773	2,730
Imports					(Jan.-Sept. 1984)	
In ores, concentrates and scrap	29 492	15,038	78 100	37,155	34 778	21,851
Dust and granules	616	875	445	669	522	966
Slabs, blocks, pigs and anodes	689	753	9 964	10,845	4 646	6,293
Bars, rods, plates, strip and sheet	298	786	575	1,226	317	857
Slugs, discs and shells	211	128	58	48	13	14
Zinc oxide	1 367	1,604	1 257	1,313	926	1,020
Zinc sulphate	1 966	827	1 688	771	1 624	876
Zinc fabricated materials, nes	841	2,142	859	2,139	511	1,576
Total	35 480	22,153	92 946	54,166	43 337	33,453

	1981			1982		
	Primary	Secondary	Total	Primary	Secondary	Total
			(tonnes)			
Consumption[5]						
Zinc used for, or in the manufacture of:						
Copper alloys (brass, bronze, etc.)	9 019)			6 141)		
Galvanizing: electro	1 565)	1 336	74 966	2 078)	219	66 571
hot dip	63 046)			58 133)		
Zinc die-cast alloy	12 553	X	X	22 041	X	X
Other products (including rolled and ribbon zinc, zinc oxide)	21 286	X	X	7 518	X	X
Total	107 469	5 592	113 061	95 911	4 321	100 232
Consumer stocks, year-end	12 117r	370	12 487r	5 483	376	5 859

Sources: Energy, Mines and Resources Canada; Statistics Canada.
[1] New refined zinc produced from domestic primary materials (concentrates, slags, residues etc.) plus estimated recoverable zinc in ores and concentrates shipped for export. [2] Zinc content of ores and concentrates produced. [3] Refined zinc produced from domestic and imported ores. [4] Gross weight. [5] Consumer survey does not represent 100 per cent of Canadian consumption and is therefore consistently less than apparent consumption.
P Preliminary; r Revised; - Nil; .. Not available; nes Not elsewhere specified; X Confidential.

TABLE 2. CANADA, ZINC MINE OUTPUT, 1982-84

	1982	1983	1984P
	(tonnes)		(Jan.-Sept.)
Newfoundland	33 157	40 905	36 981
New Brunswick	304 619	258 731	198 507
Quebec	76 050	52 061	41 929
Ontario	286 691	317 438	252 776
Manitoba-Saskatchewan	46 390	58 816	43 415
British Columbia	88 577	83 730	77 441
Yukon Territory	60 210	-	-
Northwest Territories	293 435	258 028	256 296
Total	1 189 129	1 069 709	907 345

P Preliminary.

TABLE 3. CANADA, ZINC PRODUCTION, EXPORTS AND DOMESTIC SHIPMENTS, 1970, 1975, 1979-84

	Production		Exports			Producers'
	All Forms[1]	Refined[2]	In Ores and Concentrates	Refined	Total	Domestic Shipments
			(tonnes)			
1970	1 135 714	417 906	809 248	318 834	1 128 082	106 405
1975	1 055 151	426 902	705 088	247 474	952 562	149 214
1979	1 099 926	580 449	598 279	429 353	1 027 632	153 744
1980	883 697	591 565	434 178	471 949	906 127	132 543
1981	911 178	618 650	516 210	453 526	969 736	131 859
1982	965 607	511 870	457 751	470 390	928 141	119 714
1983	970 803	617 033	660 792	500 443	1 161 235	134 424
1984P	1 022 054	683 000				141 600

Sources: Energy, Mines and Resources Canada; Statistics Canada.

[1] New refined zinc produced from domestic primary materials (concentrates, slags, residues, etc.) plus estimated recoverable zinc in ores and concentrates shipped for export. [2] Refined zinc produced from domestic and imported ores.
P Preliminary.

TABLE 4. CANADA, PRODUCERS' DOMESTIC SHIPMENTS OF REFINED ZINC, 1981-84

Quarter	1981	1982	1983	1984P
	(tonnes)			
1st	35 044	39 767	33 717	41 405
2nd	39 151	30 429	38 444	38 730
3rd	27 910	21 580	29 188	30 100
4th	29 754	27 938	33 075	31 365
Total	131 859	119 714	134 424	141 600

P Preliminary.

TABLE 5. WESTERN WORLD PRIMARY ZINC STATISTICS, 1981-84

	1981	1982	1983	1984e
	(000 tonnes)			
Mine production (Zn content)	4 459	4 801	4 791	5 000
Metal production	4 542	4 318	4 643	4 900
Metal consumption	4 428	4 250	4 601	4 750

Source: International Lead and Zinc Study Group.
e Estimated by Energy, Mines and Resources Canada.

TABLE 6. CANADA, ZINC-BEARING DEPOSITS CONSIDERED MOST PROMISING FOR FUTURE PRODUCTION

Company and Province	Deposit Name	Indicated Tonnage (000 tonnes)	Per Cent Zinc	Zinc Content (000 tonnes)
New Brunswick				
Billiton Canada Ltd. and Gowganda Resources Inc.	Restigouche	2 900	6.00	174.0
Caribou-Chaleur Bay Mines Ltd.	Caribou	44 800	4.48	2 007.0
Cominco Ltd.	Stratmat 61	2 040	6.29	128.3
Key Anacon Mines Limited	Middle Landing	1 690	7.43	125.6
Kidd Creek Mines Ltd. and Bay Copper Mines Limited	Halfmile Lake	10 160	7.51	763.0
		61 590	5.19	3 197.9
Quebec				
Les Mines d'Argent Abcourt Inc. and Antiquois Mining Corporation	Barraute	3 270	2.50	81.8
Noranda Mines Limited	Magusi	2 130	3.55	75.6
Les Mines Selbaie	A-2 Zone	5 000	1.33	66.5
		10 400	2.15	223.9
Ontario				
Corporation Falconbridge Copper	Winston Lake	2 950	17.8	525.1
British Columbia				
Cyprus Anvil Mining Corporation	Cirque	39 920	7.80	3 113.8
Yukon Territory				
Cyprus Anvil Mining Corporation	DY zone	14 700	7.10	1 043.7
	Swim Lake	4 540	5.50	249.7
Hudson Bay Mining and Smelting Co., Limited	Tom	7 840	8.40	658.6
Aberford Resources Ltd. and Ogilvie Joint Venture	Jason	11 790	7.00[e]	825.3
Placer Development Limited and United States Steel Corporation	Howard's Pass	272 160[e]	6.40[e]	17 418.2
Sulpetro Minerals Limited and Sovereign Metals Corporation	MEL	4 780	5.10	243.8
		315 760	6.51	20 439.3
Northwest Territories				
Cominco Ltd. and Bathurst Norsemines Ltd.	Seven deposits	19 050	4.98	948.7
Kidd Creek Mines Ltd.	Izok Lake	11 020	13.77	1 517.5
Westmin Resources Limited,	X-25	3 450	9.10	314.0
Du Pont Canada Inc. and Philipp Brothers (Canada) Ltd.	R-190	1 270	11.90	151.1
		34 790	8.43	2 931.3
Canada		465 560	6.54	30 431.3

Source: MR 191 Canadian Reserves of Copper, Nickel, Lead, Zinc, Molybdenum, Silver and Gold, as of January 1, 1981; Energy, Mines and Resources Canada, 1981 and company reports.
[e] Estimated.

TABLE 7. WESTERN WORLD ZINC INDUSTRY, PRODUCTION AND CONSUMPTION, 1983

	Mine Production	Metal Production	Metal Consumption
	(000 tonnes)		
Europe			
Austria	19	24	27
Belgium	-	263	166
Denmark[1]	73	-	9
Finland	56	155	26
France	34	249	271
Germany F.R.	114	356	405
Greece	21	-	13
Ireland	186	-	2
Italy	43	156	208
Netherlands	-	187	54
Norway	32	91	22
Portugal	-	4	9
Spain	176	198	106
Sweden	205	-	34
Switzerland	-	-	20
United Kingdom	9	88	181
Yugoslavia	73	93	93
Total	1 041	1 864	1 647
Africa			
Algeria	14	31	22
Egypt	-	-	17
Morocco	8	-	4
South Africa[2]	137	82	83
Tunisia	8	-	1
Zaire	75	63	-
Zambia	41	38	1
Others	-	-	35
Total	283	214	163
Americas			
Argentina	37	34	34
Bolivia	47	-	-
Brazil	73	100	102
Canada	1 070	617	144
Colombia	-	-	12
Honduras	38	-	-
Mexico	275	175	89
Peru	553	154	16
United States	293	305	933
Venezuela	-	-	15
Others	6	-	24
Total	2 392	1 385	1 369
Asia			
Hong Kong	-	-	30
India	36	54	125
Indonesia	-	-	60
Iran	25	-	16
Japan	256	701	771
Korea, Rep.	57	108	113
Philippines	2	-	27
Taiwan	-	-	60
Thailand	3	-	41
Turkey	31	14	22
Others	5	-	59
Total	415	877	1 324
Oceania			
Australia	660	303	79
New Zealand	-	-	19
Total	660	303	98
Total Non-Socialist World	4 791	4 643	4 601

Source: International Lead and Zinc Study Group.

[1] Includes Greenland. [2] Includes Namibia.
- Nil.

TABLE 8. CANADA, PRIMARY ZINC METAL CAPACITY, 1984

Company and Location	Annual Rated Capacity
	(tonnes of slab zinc)
Canadian Electrolytic Zinc Limited (CEZ) Valleyfield, Quebec	227 000
Kidd Creek Mines Ltd. Hoyle, Ontario	127 000
Hudson Bay Mining and Smelting Co., Limited Flin Flon, Manitoba	73 000
Cominco Ltd. Trail, British Columbia	272 000
Canada total	699 000

TABLE 9. MONTHLY AVERAGE ZINC PRICES

	Overseas[1] Producer ($US/tonne)	U.S.[1] Producer (¢US/lb)	Canadian[1] Producer (¢Cdn./lb)	LME[2] Settlement (£/tonne)
1983				
January	$800.0	38.6	49.0	443.7
February	760.0	38.6	48.5	444.5
March	750.0	37.9	46.5	455.3
April	750.0	38.0	46.8	454.0
May	759.0	38.1	49.0	467.9
June	780.0	39.5	49.0	464.1
July	787.6	40.0	50.4	484.4
August	830.4	40.6	53.7	538.3
September	880.0	43.0	56.0	559.1
October	892.9	46.1	57.0	573.8
November	937.5	47.5	60.0	582.2
December	958.6	48.7	60.0	597.7
Year Average	823.8	41.4	52.2	506.2
1984				
January	990.9	49.2	62.9	680.4
February	1 048.1	50.6	63.5	692.6
March	1 051.8	51.1	66.0	714.8
April	1 090.0	51.9	67.0	706.6
May	1 090.0	52.8	67.0	720.6
June	1 080.5	52.5	66.6	683.7
July	1 010.5	49.5	65.5	646.7
August	990.0	47.8	63.4	635.1
September	957.5	46.4	61.5	611.4
October	907.0	44.2	59.3	623.3
November	900.0	43.6	59.3	635.6
December	900.0	43.6	59.3	669.2
Year Average	1 001.4	48.6	63.4	668.3

Source: Metals Week, ILZSG, Northern Miner.
[1] High grade zinc. [2] G.O.B. zinc.

Principal Producers
(numbers refer to numbers on map above)

1. ASARCO Incorporated (Buchans Unit)
2. Newfoundland Zinc Mines Limited
3. Brunswick Mining and Smelting Corporation Limited
   Heath Steele Mines Limited
4. Corporation Falconbridge Copper Lake Dufault Division
   Noranda Inc. and Les Mines Gallen Limitée (Gallen Mine)
5. Mattagami Lake Mines Limited
   Noranda Mines Limited (Orchan mine)
6. Kidd Creek Mines Ltd.
7. Noranda Inc. (Geco Division)
8. Mattabi Mines Limited
   Noranda Inc. (Lyon Lake)
9. Hudson Bay Mining and Smelting Co., Limited (Chisel Lake, Osborne Lake, Stall Lake, Ghost Lake, Anderson Lake, Westarm, Flin Flon, White Lake, Centennial, Trout Lake, Spruce Point)
10. Sherritt Gordon Mines Limited (Fox Lake mine and Ruttan mine)
11. Cominco Ltd. (Sullivan mine)
    Teck Corporation (Beaverdell mine)
    Dickenson Mines Limited (Silmonac mine)
12. Westmin Resources Limited
13. Pine Point Mines Limited
14. United Keno Hill Mines Limited
15. Northair Mines Ltd.
16. Nanisivik Mines Ltd.
17. Cominco Ltd. (Polaris mine)

Metallurgical Plants

6. Kidd Creek Mines Ltd., Hoyle
9. Hudson Bay Mining and Smelting Co., Limited, Flin Flon
18. Canadian Electrolytic Zinc Limited, Valleyfield
19. Cominco Ltd., Trail

PRINCIPAL CANADIAN NONFERROUS AND PRECIOUS METAL PRODUCERS IN 1983, WITH HIGHLIGHTS FOR 1984

Company and Mine Location	Capacity (tonnes per day)	Grades of Ore Mined					Ore Milled (tonnes)	Metal Contained in Concentrates Produced						Highlights
		Cu	Ni	Pb	Zn	Ag	Au		Copper	Nickel	Zinc	Lead	Gold	Silver	
		(per cent)				(g/tonne)			(tonnes)				(kilograms)		
NEWFOUNDLAND															
ASARCO Incorporated Buchans	1 089	0.79	-	4.79	-	62.40	0.55	38 102	96	-	1 953	1 315	12	1 705	Mine closed permanently, September 1984.
Newfoundland Zinc Mines Limited Daniel's Harbour	1 497	-	-	-	8.18	-	-	486 646	-	-	38 947	-	-	-	
NEW BRUNSWICK															
Anaconda Canada Exploration Ltd. Caribou property Restigouche	-	-	-	-	-	48.00	5.14	60 346	-	-	-	-	104	2 036	Heap leach operation.
Brunswick Mining and Smelting Corporation Limited, Nos. 6 and 12 mines Bathurst	10 500	0.30	-	3.51	8.88	98.16	-	3 438 859	4 795	-	245 690	80 473	-	233 701	Ore extraction at No. 6 mine completed in 1983.
Heath Steele Mines Limited Newcastle	3 629	0.97	-	1.50	3.71	58.63	0.69	449 615	2 752	-	12 637	3 097	77	18 121	Underground production suspended May 1983.
QUEBEC															
Agnico-Eagle Mines Limited Joutel	998	-	-	-	-	0.86	5.97	291 528	-	-	-	-	1 536	208	
Bachelor Lake Gold Mines Inc. Desmaraisville	454	-	-	-	-	0.69	5.69	151 404	-	-	-	-	798	88	
Belmoral Mines Ltd. Ferderber and Dumont mines Val d'Or	816	-	-	-	-	0.72	6.27	219 866	-	-	-	-	1 299	140	
Camflo Mines Limited Malartic	1 134	-	-	-	-	0.34	5.73	423 101	-	-	-	-	2 405	110	

TABLE (cont'd)

Company and Mine Location	Capacity (tonnes per day)	Grades of Ore Mined						Ore Milled (tonnes)	Metal Contained in Concentrates Produced						Highlights
		Cu	Ni	Pb	Zn	Ag	Au		Copper	Nickel	Zinc	Lead	Gold	Silver	
		(per cent)				(g/tonne)			(tonnes)				(kilograms)		
QUEBEC (cont'd)															
Campbell Resources Inc. Chibougamau	3 402	0.78	-	-	-	6.45	2.92	273 561	2 015	-	-	-	729	976	
Corporation Falconbridge Copper Lake Dufault Division Millenbach and Corbet mines Noranda	1 451	3.32	-	-	1.19	18.36	0.93	379 118	12 213	-	3 141	-	278	4 682	Production resumed April 1983 after a seven-month shutdown.
Opemiska Division Perry, Springer & Cooke mines Chapais	2 585	1.44	-	-	-	9.94	1.65	922 654	12 874	-	-	-	1 374	7 324	
Gaspé Copper Mines Limited Copper Mountain and Needle Mountain mines Murdochville	227	-	-	-	-	-	-	-	-	-	-	-	-	-	Both mines remained shut through 1983. Needle Mountain (underground) reopened at a reduced level in September 1984. Copper Mountain remained shut through 1984.
Muscocho Explorations Limited Mont-auban mine		-	-	-	-	13.61	4.39	57 024	-	-	-	-	196	250	
Noranda Inc. Horne Division	3 447	3.13	-	-	-	0.88	1.54	630 560	18 660	-	-	-	848	297	Smelter slag milled and then resmelted.
Mattagami Division	4 082	0.90	-	0.04	4.85	17.55	0.41	1 106 031	8 208	-	48 443	-	193	10 102	
Northgate Patino Mines Inc. Copper Rand and Portage mines	2 903	1.65	-	-	-	9.67	3.22	681 083	10 948	-	-	-	1 871	4 176	
Lemoine mine Chibougamau	327	2.30	-	-	4.35	44.57	2.57	19 061	418	-	640	-	43	740	Closed, March 1983.
Selco Division - BP Resources Canada Ltd. Joutel	1 497	3.43	-	-	0.59	37.20	1.71	535 300	17 477	-	-	-	792	16 552	
Sigma Mines (Quebec) Limited Val d'Or	1 270	-	-	-	-	0.86	4.56	436 103	-	-	-	-	1 914	340	

Company / Division														Notes
Société Minière Louvem Inc. (Soquem)	544	-	-	-	-	-	1.71	4.01	151 250	-	532	202		
Lac Minerals Ltd. Doyon Division	1 089	-	-	-	-	-	0.41	5.28	413 580	-	1 981	150		
Est-Malartic Division	1 633	-	-	-	-	-	0.37	4.97	617 091	-	2 833	209		
Terrains Aurifères Division	1 814	-	-	-	-	-	1.54	6.21	549 392	-	3 200	718		
Teck Corporation Lamaque Division Val d'Or	1 905	-	-	-	-	-	1.12	6.06	445 782	-	2 548	489		

ONTARIO

Company / Division														Notes
Agnico-Eagle Mines Limited Silver Division Cobalt	363	-	-	-	-	-	672.69	-	38 111	-	-	24 171		
Campbell Red Lake Mines Limited Red Lake	975	-	-	-	-	-	2.06	20.50	353 330	-	6 818	688		
Campbell Red Lake Mines Limited Detour Joint Venture James Bay	2 500	-	-	-	-	-	0.38	2.37	265 693	-	536	85		
Consolidated Louanna Gold Mines Limited Nakina	181	-	-	-	-	-	3.09	7.44	18 606	-	127	47		
Cullaton Lake Gold Mines Limited Sudbury	454	-	-	-	-	-	1.99	5.01	166 423	-	766	233		
Dome Mines Limited South Porcupine	1 996	-	-	-	-	-	0.93	4.73	691 638	-	4 293	600		
Dickenson-Sullivan Joint Venture Red Lake	635	-	-	-	-	-	1.03	10.49	186 670	-	1 613	154		
Falconbridge Limited (6 mines) Sudbury district	10 342	1.13	1.28	-	-	-	6.86	0.14	2 668 351	28 555	28 224	-	189	9 424	Falconbridge mine closed permanently after rockburst in June. Operations resumed Jan. 1985 after a 6-month shutdown.
Goldlund Mines Limited Sioux Lookout	318	-	-	-	-	-	-	5.73	56 406	-	248	-		

TABLE (cont'd)

Company and Mine Location	Capacity (tonnes per day)	Grades of Ore Mined						Ore Milled (tonnes)	Metal Contained in Concentrates Produced					Highlights
		Cu	Ni	Pb	Zn	Ag	Au		Copper	Nickel	Zinc	Lead	Gold	Silver	
		(per cent)				(g/tonne)			(tonnes)				(kilograms)		
ONTARIO (cont'd)															
Inco Limited 10 mines, Sudbury area and Shebandowan	49 442	1.29	1.30	-	-	5.14	0.17	6 352 956	63 942	63 731	-	-	665	25 486	Copper Cliff North mine opened in January. Operations reopened April 1983 after a 10-month shutdown.
Kerr Addison Mines Limited Virginiatown	1 225	-	-	-	-	0.38	4.80	362 575	-	-	-	-	1 735	117	
Lac Minerals Ltd. Macassa Division Kirkland Lake	386	-	-	-	-	2.37	17.11	124 144	-	-	-	-	2 043	250	
Kidd Creek Mines Ltd. Limited Timmins	13 500	2.57	-	0.25	5.90	81.22	0.06	4 215 441	100 008	-	204 657	3 461	-	277 196	Expansion of zinc plant continuing; expansion of copper smelter and refinery to 90 000 tpy Cu scheduled for completion in 1988.
Mattabi Mines Limited Lyon Lake mine "F" Group mine Sturgeon Lake	2 722	0.90	-	0.94	8.55	127.89	0.41	934 451	5 467	-	72 532	4 438	249	100 085	
Noranda Inc. Geco Division Manitouwadge	3 856	1.60	-	0.11	3.48	46.63	0.10	1 246 993	18 318	-	37 022	494	88	46 144	
Pamour Porcupine Mines Limited Pamour Division Timmins	2 585	-	-	-	-	0.71	3.09	751 331	-	-	-	-	2 055	373	
Pamour Porcupine Mines, Limited Schumacher Division Timmins	2 722	0.04	-	-	-	5.62	3.50	641 910	131	-	-	-	1 843	1 903	
Sulpetro Minerals Limited	454	-	-	-	-	105.94	-	35 393	-	-	-	-	-	2 521	
Teck Corporation Silverfields Division Cobalt	222	0.60	0.25	-	-	257.14	-	18 144	-	-	-	-	-	4 385	
MANITOBA															
Brinco Limited San Antonio mine Bissett	408	-	-	-	-	0.89	-	28 263	-	-	-	-	129	22	

udson Bay Mining and melting Co. Limited 9 mines), Flin Flon concentrators & Snow Lake concentrators	10 523	2.20	-	0.10	2.86	16.32	1 918 688	38 566	41 442	705	1 618	22 026	
nco Limited Thompson and Pipe mines Thompson district	12 701	0.14	1.90	-	-	5.14	1 874 537	1 292	-	-	117	7 520	The Pipe open-pit mine closed, but stockpiled are sufficient to last through 1985. Production from Thompson open-pit expected in early 1986.
erritt Gordon Mines mited ox mine Lynn Lake district	2 722	2.10	-	-	1.94	14.06	554 122	10 801	7 996	-	210	6 140	
uttan mine Leaf Rapids	6 804	1.62	-	-	-	0.73	8.57	1 295 570	19 819	6 601	-	311	8 412	
TISH COLUMBIA													
ton Operating rporation Afton mine Kamloops	7 711	0.53	-	-	-	2.09	1 075 853	4 626	-	-	261	1 330	
nda Mines Ltd. Peachland	27 216	0.14	-	-	-	1.20	8 185 403	9 927	-	-	100	5 776	
thlehem Copper rporation ona and Jersey mines Highland Valley	17 690	0.38	-	-	-	2.40	3 112 829						
minco Ltd. pper Division Valley Copper mine	19 051	0.52	-	-	-	4.63	7 171 985	33 336	-	-	108	13 240	
minco Ltd. ullivan mine Kimberley	10 886	-	-	4.39	3.50	51.09	2 017 383	-	60 846	79 330	-	97 123	Mechanization continuing.
arolin Mines Ltd. Hope	1 361	-	-	-	-	0.10	359 017	2.71	-	-	552	7	
ickenson Mines Ltd. ilvana Division ilmonac mine New Denver	109	-	-	3.02	3.23	311.31	28 234	-	560	749	-	7 795	
ankoe Mines Ltd. Keremeos	408	0.42	-	0.05	0.13	182.40	5 583	5.62	-	-	27	956	

TABLE (cont'd)

Company and Mine Location	Capacity (tonnes per day)	Grades of Ore Mined (per cent)					Ore Milled (tonnes)	Metal Contained in Concentrates Produced					Highlights
		Cu	Ni	Pb	Zn	Ag (g/tonne)	Au (g/tonne)		Copper (tonnes)	Nickel (tonnes)	Zinc (tonnes)	Lead (tonnes)	Gold (kilograms)	Silver (kilograms)	

Company and Mine Location	Capacity	Cu	Ni	Pb	Zn	Ag	Au	Ore Milled	Copper	Nickel	Zinc	Lead	Gold	Silver	Highlights
BRITISH COLUMBIA (cont'd)															
Du Pont of Canada Inc. Exploration Limited Baker Mine	91	-	-	-	-	290.06	14.06	29 022	-	-	-	-	374	7 048	
Erickson Gold Mining Limited Cassiar	181	-	-	-	-	15.09	17.49	63 367	-	-	-	-	1 061	838	Closed April 6, 1984.
Esso Resources Canada Limited Esso Minerals Canada Division Granduc mine	3 629	1.48	-	-	-	11.31	0.20	1 031 805	14 976	-	-	-	161	9 606	
Falconbridge Limited Wesfrob Mining Division Tasu mine Tasu	5 443	0.29	-	-	-	2.98	0.07	906 563	2 262	-	-	-	46	2 034	Closed October 1983.
Gibraltar Mines Limited McLeese Lake Cariboo district	37 195	0.27	-	-	-	0.69	-	13 437 210	27 905	-	-	-	-	3 630	
Highmont Operating Corporation (Teck Corp.) Highland Valley	22 680	0.21	-	-	-	-	-	9 799 692	16 027	-	-	-	-	-	On standby October 19, 1984 due to low metal prices.
Lornex Mining Corporation Ltd. Lornex mine Highland Valley	72 575	0.34	-	-	-	2.56	-	28 766 769	85 196	-	-	-	-	22 083	
The Mosquito Creek Gold Mining Company Limited Cariboo	91	-	-	-	-	4.29	14.06	17 188	-	-	-	-	206	62	
Noranda Inc. Bell Copper mine Granisle mine Babine Lake																Closed through 1983 and 1984.
Noranda Inc. Goldstream Division	1 361	3.14	-	-	1.70	17.14	-	293 928	8 203	-	187	-	-	2 642	Production suspended April 1984.
Newmont Mines Limited Similkameen Division Princeton	19 958	0.36	-	-	-	1.23	0.32	6 809 286	21 570	-	-	-	587	4 907	

Company / Mine / Location															Remarks
Placer Development Limited/Equity Silver mines	29 937	0.48	-	-	91.99	1.30	2 180 000	8 080	-	-	-	880	154 206	-	
Scottie Gold Mines Ltd. Summit Lake	181	-	-	-	6.86	18.79	61 158	-	-	-	-	1 060	352	-	
Taurus Resources Ltd. Cassiar	113	-	-	-	3.70	5.01	42 256	-	-	-	-	174	87	-	
Teck Corporation Beaverdell mine Beaverdell	109	-	0.28	0.51	299.66	-	32 946	-	-	-	81	63	9 549	-	
Utah Mines Ltd. Island Copper mine Coal Harbour	38 102	0.44	-	-	1.37	0.21	16 330 081	60 026	-	-	-	1 713	14 718	-	
Westmin Resources Limited Lynx and Myra mines Buttle Lake	907	1.10	1.08	7.45	121.03	2.57	215 718	2 173	-	-	15 462	2 081	580	27 110	Development to bring H-W orebody into production in 1985 continuing.
YUKON TERRITORY															
Cyprus Anvil Mining Corporation Faro	-	-	-	-	-	-	-	-	-	-	-	-	-	-	Production suspended June 1982. Waste-stripping program in progress.
United Keno Hill Mines Limited Elsa, Husky, No Cash, Keno mines Elsa	450	-	-	0.65	3.70	-	778.29	-	-	-	124	953	-	36 958	
NORTHWEST TERRITORIES															
Cominco Ltd. Con and Rycon mines Yellowknife	658	-	-	-	3.43	12.45	191 721	-	-	-	-	2 193	604	-	
Pine Point mine Pine Point	9 979	-	-	2.68	8.16	-	-	893 577	-	-	67 104	21 424	-	-	
Polaris mine Little Cornwallis Island	2 041	-	-	5.17	16.81	-	-	827 582	-	-	131 866	38 975	-	-	New mine, production started early 1982, production suspended for 1 month commencing mid-December 1984.
Cullaton Lake Gold Mines Ltd. Cullaton Lake	299	-	-	-	0.69	12.72	98 553	-	-	-	-	1 194	62	-	

TABLE (cont'd)

Company and Mine Location	Capacity (tonnes per day)	Grades of Ore Mined (per cent.)					Ore Milled (tonnes)	Metal Contained in Concentrates Produced						Highlights
		Cu	Ni	Pb	Zn	Ag (g/tonne)	Au (g/tonne)		Copper	Nickel	Zinc (tonnes)	Lead (tonnes)	Gold (kilograms)	Silver (kilograms)	
NORTHWEST TERRITORIES (cont'd)															
Echo Bay Mines Ltd. Lupin mine Contwoyto Lake	1 034	-	-	-	-	1.20	12.03	323 034	-	-	-	-	3 666	366	
Giant Yellowknife Mines Limited Giant mine Yellowknife	1 134	-	-	-	-	2.33	7.70	296 989	-	-	-	-	1 943	658	
Salmita mine	145	-	-	-	-	3.43	15.63	11 926	-	-	-	-	159	35	
Nanisivik Mines Ltd. Nanisivik	1 996	-	-	1.08	10.18	50.78	-	619 300	-	-	61 042	6 147	-	26 299	
Terra Mines Ltd. Smallwood and Norex mines Camsell River	363	0.19	-	0.34	0.34	661.71	-	72 316	-	-	-	-	-	46 610	

Statistical Report

The statistical material contained in this summary was principally derived from surveys conducted by the Information Systems Division of the Mineral Policy Sector, Energy, Mines and Resources Canada.

The statistical survey program of Energy, Mines and Resources Canada is conducted jointly with the provincial governments and Statistics Canada. This joint program is intended to minimize the reporting burden on the mineral companies. The cooperation of the companies that provide information is greatly appreciated. Without this cooperation, a statistical report of this nature would not be possible. International mineral statistics contained in this summary are derived from the U.S. Bureau of Mines, The American Bureau of Metal Statistics, The World Bureau of Metal Statistics, **Metals Week, Engineering and Mining Journal,** The United Nations and the Organization for Economic Co-operation and Development (OECD).

This statistical summary of the mineral industry in Canada for the year 1983-84 was prepared by J.T. Brennan and staff, Statistics Section, Mineral Policy Sector, Energy, Mines and Resources Canada, Ottawa. Telephone (613) 995-9466.

STATISTICAL TABLES

Table No.
Canada, general economic indicators, 1969-83.

SECTION 1: PRODUCTION

1 Mineral production of Canada, 1982 and 1983, and average 1979-83.
2 Canada, value of mineral production, per capita value of mineral production, and population, 1954-83.
3 Canada, value of mineral production by provinces, territories and mineral classes, 1983.
4 Canada, production of leading minerals by provinces and territories, 1983.
5 Canada, percentage contribution of leading minerals to total value of mineral production, 1977-83.
6 Canada, value of mineral production by provinces and territories, 1977-83.
7 Canada, percentage contribution of provinces and territories to total value of mineral production, 1977-83.
8 Canada's world role as a producer of certain important minerals, 1982.
9 Canada, gross domestic product by industry in constant 1971 dollars, 1977-83.
10 Canada, census value added, total activity, mining and mineral manufacturing industries, 1976-82.
11 Canada, indexes of gross domestic product of industrial production, mining and mineral manufacturing, 1969-83.
12 Canada, indexes of gross domestic product by industries, 1969-83.
13 Canada, gross domestic product for selected industries by province, 1981.
14 Canada, gross domestic product for mining by province, 1975-81.

Table No.
SECTION 2: TRADE

15	Canada, value of exports of crude minerals and fabricated mineral products, 1977-83.
16	Canada, value of imports of crude minerals and fabricated mineral products, 1977-83.
17	Canada, value of exports of crude minerals and fabricated mineral products in relation to total export trade, 1973, 1978 and 1983.
18	Canada, value of imports of crude minerals and fabricated mineral products in relation to total import trade, 1973, 1978 and 1983.
19	Canada, value of exports of crude minerals and fabricated mineral products, by main groups and destination, 1983.
20	Canada, value of imports of crude minerals and fabricated mineral products, by main groups and origin, 1983.
21	Canada, value of exports of crude minerals and fabricated mineral products, by commodity and destination, 1983.
22	Canada, physical volume of import trade for selected commodities, 1977-83.
23	Canada, physical volume of export trade for selected commodities, 1977-83.

SECTION 3: CONSUMPTION

24	Canada, apparent consumption of some minerals and relation to production, 1981-83.
25	Canada, reported consumption of minerals and relation to production, 1980-82.
26	Canada, domestic consumption of principal refined metals in relation to refinery production, 1976-82.

SECTION 4: PRICES

27A	Average annual prices of selected minerals, 1977-83.
27B	Canadian average annual prices of selected minerals, 1977-83.
28	Canada, mineral products industries, selling price indexes, 1977-83.

SECTION 5: PRINCIPAL STATISTICS

29	Canada, principal statistics of the mining industry, 1982.
30	Canada, principal statistics of the mineral manufacturing industries, 1982.
31	Canada, principal statistics of the mineral industry by region, 1982.
32	Canada, principal statistics of the mineral manufacturing industry by region, 1982.
33	Canada, principal statistics of the mining industry, 1976-82.
34	Canada, principal statistics of the mineral manufacturing industries, 1976-82.
35	Canada, consumption of fuel and electricity in the mining industry, 1982.
36	Canada, consumption of fuel and electricity in the mineral manufacturing industries, 1982.
37	Canada, cost of fuel and electricity used in the mining industry, 1976-82.
38	Canada, cost of fuel and electricity used in the mineral manufacturing industries, 1976-82.

SECTION 6: EMPLOYMENT, SALARIES AND WAGES

39	Canada, employment, salaries and wages in the mining industry, 1976-82.
40	Canada, employment, salaries and wages in the mineral manufacturing industries, 1976-82.
41	Canada, number of wage earners employed in the mining industry (surface, underground and mill), 1976-82.
42	Canada, mine and mill workers by sex, 1982
43	Canada, labour costs in relation to tonnes mined, metal mines, 1980-82.
44	Canada, man-hours paid, production and related workers, tonnes of ore mined and rock quarried, metal mines and nonmetallic mineral operations, 1976-82.
45	Canada, average weekly wages and hours worked, hourly-rated employees in mining, manufacturing and construction industries, 1977-83.

Table No.

46	Canada, average weekly wages of hourly-rated employees in the mining industry, in current and 1971 dollars, 1977-83.
47	Canada, industrial fatalities per thousand workers, by industry groups, 1981-83.
48	Canada, industrial fatalities per thousand workers, by industry groups, 1977-83.
49	Canada, industrial fatalities by occupational injuries and illnesses, 1981-83.
50	Canada, number of strikes and lockouts by industries, 1981-83.
51	Canada, number of strikes and lockouts by mining and mineral manufacturing industries 1981-83.

SECTION 7: MINING, EXPLORATION AND DRILLING

52	Canada, source of ores hoisted or removed from selected types of mines, 1980-82.
53	Canada, source of material hoisted or removed from metal mines, 1982.
54	Canada, ore mined and rock quarried, in the mining industry, 1976-82.
55	Canada, exploration and capital expenditures in the mining industry, by provinces and territories, 1980-82.
56	Canada, exploration and capital expenditures in the mining industry, by type of mining, 1980-82.
57	Canada, diamond drilling in the mining industry, by mining companies with own equipment and by drilling contractors, 1980-82.
58	Canada, ore mined and rock quarried, in the mining industry, 1953-82.
59	Canada, total diamond drilling, metal deposits, 1953-82.
60	Canada, exploration diamond drilling, metal deposits, 1953-82.
61	Canada, diamond drilling other than for exploration, metal deposits, 1953-82.

SECTION 8: TRANSPORTATION

62	Canada, crude minerals transported by Canadian railways, 1980-82.
63	Canada, fabricated mineral products transported by Canadian railways, 1980-82.
64	Canada, crude and fabricated minerals transported by Canadian railways, 1953-82.
65	Canada, crude and fabricated minerals transported through the St. Lawrence Seaway, 1981-83.
66	Canada, crude and fabricated minerals transported through the St. Lawrence Seaway, 1954-1983.
67	Canada, crude minerals loaded and unloaded in coastwise shipping, 1982.
68	Canada, fabricated minerals loaded and unloaded in coastwise shipping, 1982.
69	Canada, crude and fabricated minerals loaded at Canadian ports in coastwise shipping, 1953-82.
70	Canada, crude minerals loaded and unloaded at Canadian ports in international shipping trade, 1980-82.
71	Canada, fabricated mineral products loaded and unloaded at Canadian ports in international shipping trade, 1980-82.
72	Canada, crude and fabricated minerals loaded at Canadian ports in international shipping trade, 1953-82.

SECTION 9: INVESTMENT AND FINANCE

73	Canada, financial statistics of corporations in the mining industry, by degree of non-resident ownership, 1981.
74	Canada, financial statistics of corporations in the mineral manufacturing industries, by degree of non-resident ownership, 1981.
75	Canada, financial statistics of corporations in non-financial industries, by major industry group and by control, 1980 and 1981.
76	Canada, capital and repair expenditures by selected industrial sector; 1982-84.
77	Canada, capital and repair expenditures in mining by geographical region; 1982-84.
78	Canada, capital and repair expenditures in mining and mineral manufacturing industries, 1982-84.
79	Canada, capital and repair expenditures in the mining industry, 1978-84.
80	Canada, capital and repair expenditures in the mineral manufacturing industries, 1978-84.
81	Canada, capital expenditures in the petroleum, natural gas and allied industries, 1978-84.

CANADA, GENERAL ECONOMIC

		1969	1970	1971	1972	1973
Gross national product, current dollars	$ million	79,815	85,685	94,450	105,234	123,560
Gross national product, constant dollars (1971 = 100)	"	86,225	88,390	94,450	100,248	107,812
Value of manufacturing industry shipments	"	45,930	46,381	50,276	56,191	66,674
Value of mineral production	"	4,734	5,722	5,963	6,408	8,370
Merchandise exports	"	14,498	16,401	17,397	19,671	24,838
Merchandise imports	"	14,130	13,952	15,618	18,669	23,325
Balance of payments, current account	"	-917	+1,106	+431	-386	+108
Corporation profits before taxes	"	8,294	7,699	8,681	10,799	15,417
Capital investment current dollars[r]	"	17,232	18,015	20,800	23,051	27,848
Capital investment, constant dollars[r] (1971 = 100)	"	18,850	18,904	20,800	21,955	24,384
Population	000's	21,001	21,297	21,568	21,802	22,043
Labour force	"	8,194	8,395	8,639	8,897	9,276
Employed	"	7,832	7,919	8,104	8,344	8,761
Unemployed	"	362	476	535	553	515
Unemployment rate	per cent	4.4	5.7	6.2	6.2	5.5
Labour income	$ million	43,065	46,706	51,528	57,570	135.9
Index industrial production	1971=100	93.6	94.9	100.0	107.6	119.0
Index manufacturing production	"	95.8	94.5	100.0	107.7	119.1
Index mining production	"	86.9	98.7	100.0	104.4	117.8
Index gross domestic product	1971=100	92.2	94.4	100.0	105.2	114.1
Consumer price index	1981=100	39.7	41.0	42.2	44.2	47.6

[p] Preliminary; [r] Revised.

INDICATORS, 1969-83

1974	1975	1976r	1977r	1978r	1979r	1980r	1981r	1982	1983p
147,528	165,343	191,857	210,189	232,211	264,279	297,556	339,797	358,302	209,340
111,678	113,005	119,612	121,988	126,347	130,362	131,765	136,108	130,065	134,353
82,455	88,427	98,076	109,747	129,019	152,133	165,985	188,212	183,432	199,560
11,754	13,347	15,693	18,473	20,319	26,135	31,926	32,420	33,837	35,976
31,739	32,587	37,651	43,685	52,259	64,317	74,446	81,203	84,540	90,825
31,722	34,716	37,494	42,363	50,108	62,871	69,274	79,129	66,726	73,120
-1,460	-4,757	-3,842	-4,301	-4,935	-4,962	-1,096	-5,346	2,665	1,686
20,062	19,663	19,985	21,090	25,360	34,884	36,456	32,638	21,110	32,684
34,260	40,044	44,927	48,376	52,482	60,921	69,196	82,058	79,330	77,647
25,694	26,661	27,731	27,606	27,585	29,448	30,461	32,401	29,265	27,844
22,364	22,697	22,993	23,258	23,476	23,671	23,936	24,342	24,634	24,890
9,639	9,974	10,203	10,500	10,895	11,231	11,573	11,904	11,958	12,183
9,125	9,284	9,477	9,651	9,987	10,395	10,708	11,006	10,644	10,734
514	690	726	849	908	836	865	898	1,314	1,448
5.3	6.9	7.1	8.1	8.3	7.4	7.5	7.5	11.0	11.9
79,846	93,299	109,053	120,508	131,702	148,256	167,936	193,874	208,180	219,793
122.8	115.5	122.6	125.7	129.9	137.9	135.9	136.5	123.0	129.7
123.4	116.2	123.1	125.5	131.9	138.1	133.7	137.8	121.3	128.6
114.0	100.9	103.1	106.1	97.8	107.1	109.6	104.6	92.7	96.2
119.3	120.4	126.4	130.1	134.4	139.3	139.8	145.5	139.2	142.9
52.8	58.5	62.9	67.9	73.9	80.7	88.9	100.0	110.8	117.2

TABLE 1. MINERAL PRODUCTION[1] OF CANADA, 1982 AND 1983, AND AVERAGE 1979-83

	Unit of Measure	1982		1983P		Average 1979-83	
		(Quantity)	($000)	(Quantity)	($000)	(Quantity)	($000)
Metals							
Antimony	t	..	2,455	..	2,350	1 477	4,667
Bismuth	t	189	1,057	202	1,031	169	1,040
Cadmium	t	836	2,684	1 107	3,388	1 014	5,276
Cobalt	t	1 274	38,741	1 584	53,760	1 739	88,995
Columbium (Cb_2O_5)	t	3 086	20,832	1 600	11,200	2 480	16,248
Copper	000 t	612	1,195,083	625	1,307,307	656	1,480,600
Gold	kg	64 735	968,012	70 746	1,186,411	57 856	966,655
Iron ore	000 t	33 198	1,201,256	32 382	1,143,380	44 763	1,520,212
Iron remelt	000 t		103,614		94,000	407	98,159
Lead	000 t	272	197,335	259	152,883	272	259,618
Molybdenum	t	13 961	159,142	10 523	103,651	12 080	236,523
Nickel	000 t	89	600,936	122	766,351	136	986,294
Platinum group	kg	7 105	82,253	5 195	67,885	8 627	100,349
Selenium	t	222	2,294	190	1,916	233	5,712
Silver	t	1 134	415,204	1 106	500,441	1 153	536,197
Tantalum (Ta_2O_5)	t	59	7,243	-	-	87	13,348
Tellurium	t	18	577	12	319	24	982
Tin	t	135	1,915	141	2,519	219	3,771
Tungsten (WO_3)	t	3 030	..	1 538	..	2 868	..
Uranium (U)	t	7 643	837,468	7 035	722,727	7 032	734,523
Zinc	000 t	966	1,036,096	971	1,116,423	966	1,032,080
Total metals			6,874,197		7,237,942		8,091,250
Nonmetals							
Asbestos	000 t	834	364,795	829	402,280	1 120	508,287
Barite	000 t	..	2,966	..	2,970	59	3,617
Gemstone	t	..	405	..	363	241	785
Gypsum	000 t	5 987	46,608	7 481	56,790	7 186	46,183
Magnesitic dolomite and brucite	000 t	..	8,216	..	8,108	53	9,438
Nepheline syenite	000 t	550	17,324	528	15,590	574	16,160
Peat	000 t	487	54,261	544	52,503	488	48,808
Potash (K_2O)	000 t	5 309	630,562	6 203	620,912	6 472	799,819
Pyrite, pyrrhotite	000 t	20	220	-	-	9	190
Quartz	000 t	1 784	32,424	1 988	38,793	2 126	32,335
Salt	000 t	7 940	156,620	8 590	174,261	7 825	139,620

Soapstone, talc & pyrophyllite	000 t	71	5,511	97	7,785	87	5,014
Sodium sulphate	000 t	547	47,462	447	43,363	491	37,182
Sulphur in smelter gas	000 t	627	42,027	803	50,691	755	36,708
Sulphur, elemental	000 t	6 945	569,928	6 327	428,119	7 053	450,004
Total nonmetals	000 t		1,979,329		1,902,528		2,134,150

Fuels

Coal	000 t	42 811	1,294,476	44 250	1,300,000	39 407	1,091,795
Natural gas	000 m³	75 977 000	7,262,446	69 266 000	6,623,158	80 120 000	6,262,181
Natural gas by-products	000 m³	18 466	2,302,000	17 408	2,568,635	18 714	2,058,635
Petroleum, crude	000 m³	73 790	12,179,454	76 874	14,470,793	79 121	10,518,823
Total fuels			23,038,376		24,962,586		19,931,434

Structural materials

Clay products	000 $		95,993		127,357		114,489
Cement	000 t	8 426	673,653	7 828	650,833	9 688	645,134
Lime	000 t	2 197	142,081	2 126	139,638	2 258	129,520
Sand and gravel	000 t	216 274	554,608	199 293	514,609	247 380	510,341
Stone	000 t	59 181	263,249	62 359	269,394	84 297	305,111
Total structural materials			1,729,584		1,701,831		1,704,595
Other minerals²			215,536		171,590		169,925
Total, all minerals			33,837,022		35,976,477		32,031,354

Notes: 1 Production statistics for the following are not available for publication: helium, nitrogen and yttrium.
2 Other minerals include arseneous trioxide, bentonite, calcium, cesium, diatomite, graphite, ilmenite, indium, magnesium, mica, pumice, rhenium, rhubidium, serpentine, strontium, titanium dioxide for which the value of production is confidential.
p Preliminary; .. Not available; - Nil.

TABLE 2. CANADA, VALUE OF MINERAL PRODUCTION, PER CAPITA VALUE OF MINERAL PRODUCTION, AND POPULATION, 1954-83

	Metallics	Industrial Minerals ($ million)	Fuels	Other Minerals[1]	Total	Per Capita Value of Mineral Production ($)	Population of Canada (000)
1954	802	333	353		1,488	97.36	15,287
1955	1,008	373	414		1,795	114.37	15,698
1956	1,146	420	519		2,085	129.65	16,081
1957	1,159	466	565		2,190	131.87	16,610
1958	1,130	460	511		2,101	122.99	17,080
1959	1,371	503	535		2,409	137.79	17,483
1960	1,407	520	566		2,493	139.48	17,870
1961	1,387	542	674		2,603	142.72	18,238
1962	1,496	574	811		2,881	155.05	18,583
1963	1,510	632	885		3,027	159.91	18,931
1964	1,702	690	973		3,365	174.45	19,291
1965	1,908	761	1,046		3,715	189.11	19,644
1966	1,985	844	1,152		3,981	198.88	20,015
1967	2,285	861	1,235		4,381	214.99	20,378
1968	2,493	886	1,343		4,722	228.10	20,701
1969	2,378	891	1,465		4,734	225.42	21,001
1970	3,073	931	1,718		5,722	268.68	21,297
1971	2,940	1,008	2,015		5,963	276.46	21,568
1972	2,956	1,085	2,367		6,408	293.92	21,802
1973	3,850	1,293	3,227		8,370	379.69	22,043
1974	4,821	1,731	5,202		11,754	525.55	22,364
1975	4,796	1,898	6,653		13,347	588.05	22,697
1976	5,315	2,269	8,109		15,693	682.51	22,993
1977	5,988	2,612	9,873		18,473	794.26	23,258
1978	5,682	2,986	11,578	73	20,319	865.51	23,476
1979	7,924	3,514	14,617	81	26,135	1,104.11	23,671
1980	9,666	4,201	17,944	115	31,926	1,333.79	23,936
1981r	8,753	4,486	19,012	136	32,420	1,331.85	24,342
1982	6,874	3,709	23,038	215	33,837	1,373.59	24,634
1983P	7,238	3,604	24,963	172	35,976	1,445.40	24,890

[1] Other minerals include arseneous trioxide, bentonite, calcium, cesium, diatomite, graphite, ilmenite, indium, magnesium, mica, pumice, rhenium, rhubidium, serpentine, strontium, titanium dioxide for which the value of production is confidential.
P Preliminary; r - Revised.

TABLE 3. CANADA, VALUE OF MINERAL PRODUCTION BY PROVINCES, TERRITORIES AND MINERAL CLASSES, 1983P

	Metals ($000)	(% of total)	Industrial Minerals ($000)	(% of total)	Fuels ($000)	(% of total)	Other Minerals[1] ($000)	(% of total)	Total ($000)	(% of total)
Alberta	352	x	721,611	20.0	21,496,250	86.1	-	-	22,218,213	61.8
Ontario	2,671,512	36.9	801,046	22.2	65,108	0.3	29,030	16.9	3,566,696	9.9
British Columbia	1,225,964	16.9	308,856	8.6	1,291,189	5.2	721	0.4	2,826,730	7.9
Saskatchewan	229,739	3.2	730,276	20.3	1,774,469	7.1	1,652	1.0	2,736,136	7.6
Quebec	1,086,987	15.0	715,027	19.8	-	-	114,620	66.8	1,916,634	5.3
Newfoundland	645,514	8.9	44,852	1.2	-	-	-	-	690,366	1.9
Manitoba	414,309	5.7	86,688	2.4	131,066	0.5	5,566	3.2	637,629	1.7
Northwest Territories	473,573	6.5	41,025	1.1	28,856	0.1	17,583	10.2	561,037	1.6
New Brunswick	431,155	6.0	51,734	1.4	28,948	0.1	1,836	1.1	513,673	1.4
Nova Scotia	-	-	101,219	2.8	146,700	0.6	582	0.3	248,501	0.7
Yukon	58,837	0.8	525	x	-	-	-	-	59,362	0.2
Prince Edward Island	-	-	1,500	x	-	-	-	-	1,500	x
Total, Canada	7,237,942	100.0	3,604,359	100.0	24,962,586	100.0	171,590	100.0	35,976,477	100.0

1 Other minerals include arseneous trioxide, bentonite, calcium, cesium, diatomite, graphite, ilmenite, indium, magnesium, mica, pumice, rhenium, rhubidium, serpentine, strontium, titanium dioxide for which the value of production is confidential.
P Preliminary; - Nil; x - Amount too small to be expressed.

TABLE 4. CANADA, PRODUCTION OF LEADING MINERALS,

	Unit of Measure	Nfld.	P.E.I.	Nova Scotia	New Brunswick	Quebec	Ontario
Petroleum, crude	000 m^3	-	-	-	x	-	85
	$000	-	-	-	13	-	16,552
Natural gas	million m^3	-	-	-	2	-	448
	$000	-	-	-	35	-	48,556
Natural gas byproducts	000 m^3	-	-	-	-	-	-
	$000	-	-	-	-	-	-
Copper	000 t	x	-	-	9	63	197
	$000	384	-	-	19,037	132,352	412,140
Coal	000 t	-	-	3 000	550	-	-
	$000	-	-	146,700	28,900	-	-
Gold	000 kg	x	-	-	x	26	22
	$000	162	-	-	2,383	435,735	362,574
Iron ore	000 t	18 123	-	-	-	9 980	3 664
	$000	601,078	-	-	-	346,191	180,519
Zinc	000 t	37	-	-	234	42	283
	$000	42,146	-	-	269,302	48,645	325,249
Nickel	000 t	-	-	-	-	-	95
	$000	-	-	-	-	-	595,165
Uranium (U)	000 t	-	-	-	-	-	5
	$000	-	-	-	-	-	518,364
Cement	000 t	..	-	..	..	2 138	2 845
	$000	10,575	-	8,800	11,200	124,108	229,850
Potash (K$_2$O)	000 t	-	-	-	..	-	-
	$000	-	-	-	..	-	-
Sand and gravel	000 t	2 900	935	6 200	5 400	29 838	59 250
	$000	9,320	1,500	15,300	7,650	47,599	142,500
Silver	t	2	-	-	210	39	347
	$000	846	-	-	94,965	17,866	157,026
Sulphur, elemental	000 t	-	-	..	-	-	25
	$000	-	-	..	-	-	2,216
Asbestos	000 t	32	-	-	-	717	-
	$000	16,845	-	-	-	315,696	-
Stone	000 t	280	-	650	2 140	26 514	25 000
	$000	1,300	-	3,463	9,750	111,358	102,475
Salt	000 t	-	-	..	..	..	5 059
	$000	-	-	..	..	..	93,180
Lead	000 t	2	-	-	74	-	7
	$000	898	-	-	43,617	-	4,007
Lime	000 t	-	-	-	..	297	1 418
	$000	-	-	-	3,540	21,673	89,546
Clay products	$000	1,385	-	2,225	1,260	24,482	75,430
Molybdenum	000 t	-	-	-	-	-	-
	$000	-	-	-	-	-	-
Total leading minerals	$000	684,939	1,500	176,488[1]	491,652[1]	1,625,705[1]	3,355,349
Total all minerals	$000	690,366	1,500	248,501	513,673	1,916,634	3,566,696
Leading minerals as % of all minerals		99.2	100.0	71.0	95.7	84.8	94.1

[1] Value of salt production excluded.
P Preliminary; - Nil; .. Not available; x Less than 1 unit.

BY PROVINCES AND TERRITORIES, 1983P

Manitoba	Saskat-chewan	Alberta	British Columbia	Yukon	N.W.T.	Total Canada
714	9 319	64 464	2 119	-	173	76 874
131,066	1,612,399	12,282,894	408,504	-	19,365	14,479,793
-	1 329	61 632	5 729	-	126	69 266
-	60,415	6,227,597	277,064	-	9,491	6,623,158
-	-	74	17 097	237	-	-	17 408
-	10,055	2,523,359	35,221	-	-	2,568,635
63	7	-	286	-	-	625
132,008	13,676	-	597,710	-	-	1,307,307
-	7 500	21 400	11 800	-	-	44 250
-	91,600	462,400	570,400	-	-	1,300,000
2	x	x	8	3	9	71
37,903	2,573	352	141,211	50,451	153,067	1,186,411
-	-	-	615	-	-	32 382
-	-	-	15,586	-	-	1,143,380
46	6	-	99	-	224	971
53,077	6,782	-	114,153	-	257,069	1,116,423
27	-	-	-	-	-	122
171,186	-	-	-	-	-	766,351
-	2	-	-	-	-	7
-	204,363	-	-	-	-	722,727
300	200	1 260	720	-	-	7 828
28,100	24,000	138,600	75,600	-	-	650,833
-	6 203	-	-	-	-	6 203
-	620,912	-	-	-	-	620,912
9 850	7 700	42 500	28 300	420	6 000	199 293
26,575	19,000	116,000	88,860	525	39,780	514,609
23	5	-	411	18	51	1 106
10,575	2,222	-	185,793	8,155	22,993	500,441
-	-	6 015	287	-	-	6 327
-	-	413,393	12,510	-	-	428,119
-	-	-	80	-	-	829
-	-	-	69,739	-	-	402,280
2 400	-	125	4 950	-	300	62 359
11,800	-	1,600	26,403	-	1,245	269,394
-	425	1 195	-	-	-	8 590
-	20,821	15,071	-	-	-	174,261
x	-	-	108	x	68	259
183	-	-	63,503	231	40,444	152,883
..	-	159	129	-	-	2 126
5,436	-	10,717	8,726	-	-	139,638
1,900	3,610	12,565	4,500	-	-	127,357
-	-	-	11	-	-	11
-	-	-	103,651	-	-	103,651
609,809	2,692,428	22,204,548	2,799,134	59,362	543,454	35,289,563
637,629	2,736,136	22,218,213	2,826,730	59,362	561,037	35,976,477
95.6	98.4	99.9	99.0	100.0	96.9	98.1

TABLE 5. CANADA, PERCENTAGE CONTRIBUTION OF LEADING MINERALS TO TOTAL VALUE OF MINERAL PRODUCTION, 1977-83

	1977	1978	1979	1980	1981	1982	1983P
Oil, crude	26.4	28.7	28.6	28.4	29.2	36.0	40.2
Natural gas	18.5	19.4	18.6	19.3	19.8	21.5	18.4
Natural gas byproduct	5.3	5.3	5.5	5.7	6.5	6.8	7.1
Copper	6.3	5.4	5.8	5.8	4.7	3.5	3.6
Coal	3.3	3.8	3.3	2.9	3.3	3.8	3.6
Gold	1.5	1.9	2.3	3.7	2.8	2.9	3.3
Iron ore	7.5	6.0	6.9	5.3	5.4	3.6	3.2
Zinc	4.5	4.0	4.1	2.7	3.4	3.1	3.1
Nickel	6.6	3.1	3.2	4.7	3.8	1.8	2.1
Uranium (U)	1.9	3.1	2.4	2.2	2.5	2.5	2.0
Cement	2.3	2.8	2.5	1.8	2.1	2.0	1.8
Potash (K_2O)	2.2	2.5	2.8	3.2	3.1	1.9	1.7
Sand and gravel	2.0	2.1	1.8	1.6	1.6	1.6	1.4
Silver	1.1	1.2	1.8	2.6	1.4	1.2	1.4
Sulphur, elemental	0.4	0.5	0.6	1.4	2.0	1.7	1.2
Asbestos	3.1	2.6	2.3	1.9	1.7	1.1	1.1
Stone	1.6	1.6	1.3	1.1	1.0	0.8	0.7
Salt	0.5	0.5	0.4	0.4	0.4	0.5	0.5
Lead	1.1	1.3	1.6	0.9	0.8	0.6	0.4
Lime	0.4	0.4	0.3	0.4	0.5	0.4	0.4
Clay products	0.6	0.5	0.5	0.3	0.4	0.3	0.4
Molybdenum	0.8	0.9	1.3	0.9	0.9	0.5	0.3
Other minerals	2.1	2.4	2.1	2.8	2.7	1.9	1.9
Total	100.0	100.0	100.0	100.0	100.0	100.0	100.0

P Preliminary.

TABLE 6. CANADA, VALUE OF MINERAL PRODUCTION BY PROVINCES AND TERRITORIES, 1977-83

	1977	1978	1979	1980	1981	1982	1983P
				($ million)			
Alberta	8,576	10,087	12,899	16,379	17,559	20,913	22,218
Ontario	2,980	2,698	3,265	4,640	4,160	3,148	3,567
British Columbia	1,687	1,883	2,677	2,795	2,822	2,769	2,827
Saskatchewan	1,208	1,582	1,874	2,315	2,293	2,313	2,736
Quebec	1,675	1,796	2,165	2,467	2,420	2,065	1,917
Newfoundland	867	675	1,125	1,036	1,030	647	690
Manitoba	564	459	653	803	642	530	638
Northwest Territories	256	310	435	425	447	503	561
New Brunswick	289	339	480	373	531	498	514
Nova Scotia	159	211	210	247	269	281	249
Yukon	210	219	299	361	236	169	59
Prince Edward Island	2	2	2	2	2	2	2
Total	18,473	20,261	26,084	31,842	32,410	33,837	35,976

P Preliminary.

TABLE 7. CANADA, PERCENTAGE CONTRIBUTION OF PROVINCES AND TERRITORIES TO TOTAL VALUE OF MINERAL PRODUCTION, 1977-83

	1977	1978	1979	1980	1981	1982	1983P
Alberta	46.4	49.8	49.5	51.4	54.2	61.8	61.8
Ontario	16.1	13.3	12.5	14.6	12.8	9.3	9.9
British Columbia	9.1	9.3	10.3	8.8	8.7	8.2	7.9
Saskatchewan	6.5	7.8	7.2	7.2	7.0	6.8	7.6
Quebec	9.1	8.9	8.3	7.7	7.5	6.1	5.3
Newfoundland	4.7	3.3	4.3	3.3	3.2	1.9	1.9
Manitoba	3.1	2.3	2.5	2.5	2.0	1.6	1.7
Northwest Territories	1.4	1.5	1.7	1.3	1.4	1.5	1.6
New Brunswick	1.6	1.7	1.8	1.2	1.6	1.5	1.4
Nova Scotia	0.9	1.0	0.8	0.8	0.8	0.8	0.7
Yukon	1.1	1.1	1.1	1.1	0.7	0.5	0.2
Prince Edward Island	0.01	0.01	0.01	0.01	0.01	0.01	x
Total	100.0	100.0	100.0	100.0	100.0	100.0	100.0

P Preliminary. x - Too small to be expressed.

TABLE 8. CANADA'S WORLD ROLE AS A PRODUCER OF

		World
Zinc (mine production)	000 t	6 503
	% of world total	
Potash (K$_2$O equivalent)	000 t	25 950
	% of world total	
Uranium (U concentrates)	t	48 662
	% of world total	
Asbestos	000 t	4 668
	% of world total	
Sulphur, elemental	000 t	45 904
	% of world total	
Titanium concentrates (ilmenite)	000 t	4 577
	% of world total	
Nickel (mine production)	000 t	625
	% of world total	
Molybdenum (Mo content)	000 t	96
	% of world total	
Aluminum (primary metal)	000 t	13 989
	% of world total	
Gypsum	000 t	73 119
	% of world total	
Gold (mine production)	t	1 278
	% of world total	
Platinum group metals (mine production)	kg	199 466
	% of world total	
Silver (mine production)	t	11 725.7
	% of world total	
Lead (mine production)	000 t	3 581
	% of world total	
Cadmium (smelter production)	t	16 455
	% of world total	
Copper (mine production)	000 t	8 220
	% of world total	
Iron ore	000 t	803 287
	% of world total	

P Preliminary;   e Estimated.

CERTAIN IMPORTANT MINERALS, 1982P

Rank of Six Leading Countries					
1	2	3	4	5	6
Canada	U.S.S.R.	Australia	Peru	U.S.A.	Japan
1 189	1 020e	665	541	330	251
18.3	15.7	10.2	8.3	5.1	3.9
U.S.S.R.	**Canada**	East Germany	West Germany	France	U.S.A.
9 000	5 196	3 500	2 500	1 823	1 784
34.7	20.0	13.5	9.5	7.0	6.9
U.S.A.	**Canada**	South Africa	Australia	Nigeria	Namibia
12 156.3	9 625.4	6 858.3	5 250.8	5 023.1	4 453.4
25.0	19.8	14.1	10.8	10.3	9.2
U.S.S.R.	**Canada**	South Africa	Zimbabwe	Brazil	China
2 800	834.2	211.9	194.4	144.8	140.0
60.0	17.9	4.5	4.2	3.1	3.0
U.S.A.	**Canada**	Poland	U.S.S.R.	Japan	France
8 878.6	5 715.3	4 935	3 556	2 268.0	1 905.1
19.3	12.5	10.8	7.7	5.4	4.2
Australia	**Canada**	Norway	U.S.S.R.	South Africa	U.S.A.
1 300	750	608	475	420	263
28.4	16.4	13.3	10.4	9.2	5.7
U.S.S.R.	**Canada**	Australia	New Caledonia	Indonesia	Cuba
170.0	88.7	88.6	60.1	48.5	37.6
27.2	14.2	14.2	9.6	7.8	6.0
U.S.A.	Chile	**Canada**	U.S.S.R.	Mexico	Peru
37.7	20.0	16.5	11.0	5.2	2.6
39.4	20.9	17.3	11.5	5.4	2.7
U.S.A.	U.S.S.R.	**Canada**	West Germany	Norway	France
3 274.0	2 400.1	1 118.1	722.8	645.1	390.4
23.4	17.2	8.0	5.2	4.6	2.8
U.S.A.	France	**Canada**	U.S.S.R.	Spain	Iran
9 559.9	6 168.9	5 726.2	5 443.1	5 261.7	4 989.5
13.1	8.4	7.8	7.4	7.2	6.8
South Africa	U.S.S.R.	**Canada**	China	U.S.A.	Brazil
664.2	265.9	64.7	56.0	45.0	45.0
52.0	20.8	5.1	4.4	3.5	3.5
U.S.S.R.	South Africa	**Canada**	Japan	Australia	Colombia
108 862.2	80 869.0	7 105.0	1 345.6	436.4	373.2
54.6	40.5	3.6	0.7	0.2	0.2
Peru	U.S.S.R.	Mexico	**Canada**	U.S.A.	Australia
1 691.0	1 595.0	1 550.2	1 313.6	1 251.6	906.9
14.4	13.6	13.2	11.2	10.7	7.7
U.S.S.R.	U.S.A.	Australia	**Canada**	Peru	Mexico
575.0	522.9	455.3	341.2	201.4	167.9
16.1	14.6	12.7	9.5	5.6	4.7
U.S.S.R.	Japan	U.S.A.	**Canada**	West Germany	Australia
2 800	2 021.2	1 351.8	1 058.2	1 030.1	1 010.2
16.7	12.1	8.1	6.6	6.1	6.0
Chile	U.S.S.R.	U.S.A.	**Canada**	Zambia	Zaire
1 240.7	1 180.0	1 139.6	612.4	529.6	502.8
15.1	14.4	13.9	7.4	6.4	6.1
U.S.S.R.	Brazil	Australia	China	**Canada**	India
243 952.9	110 037.9	88 294.5	70 005.6	41 861.1	40 946.7
30.4	13.7	11.0	8.7	5.2	5.1

TABLE 9. CANADA, GROSS DOMESTIC PRODUCT BY INDUSTRY IN CONSTANT 1971 DOLLARS, 1977-83

	1977	1978	1979r	1980r	1981r	1982	1983p
	($ million)						
Goods producing industries							
Agriculture	3,069.7	2,996.5	2,702.8	2,921.3	3,158.5	3,250.2	3,311.0
Forestry	741.9	794.9	800.8	826.7	759.0	621.8	761.5
Fishing and trapping	162.3	179.5	182.8	172.5	188.1	183.2	181.9
Mining[1]	3,337.3	3,015.1	3,347.9	3,465.4	3,290.8	2,889.1	3,062.8
Manufacturing	23,901.6	25,139.9	26,587.7	25,830.9	26,235.8	23,066.7	24,496.2
Construction	6,856.2	6,706.0	7,108.6	7,042.0	7,477.5	6,640.6	6,457.4
Electrical power, gas and water utilities	3,311.3	3,521.6	3,792.6	3,832.4	3,900.5	3,906.3	4,051.3
Total	41,380.3	42,353.6	44,523.2	44,091.2	45,010.2	40,557.9	42,322.1
Service producing industries							
Transportation, storage and communication	10,972.8	11,462.3	15,905.1	16,419.9	16,882.1	16,377.1	16,733.4
Trade	13,710.4	14,206.5	14,998.2	15,011.8	15,136.4	14,121.7	14,543.3
Finance, insurance and real estate	13,444.8	14,119.9	14,768.5	15,331.7	16,019.4	16,108.4	16,324.9
Community, business and personnel services	21,096.3	21,888.1	22,007.6	22,744.4	23,876.1	23,866.3	24,171.5
Public administration and defense	7,736.2	7,927.5	7,886.7	7,980.0	8,137.0	8,404.9	8,517.0
Total	66,960.5	69,604.3	75,566.1	77,487.8	80,051.0	78,878.4	80,290.1
Grand total	108,340.8	111,957.9	120,089.3	121,579.0	125,061.2	119,436.3	122,612.2

[1] Cement, lime, clay and clay products (domestic clays) industries are included under "manufacturing".
p Preliminary; r Revised.

TABLE 10. CANADA, CENSUS VALUE ADDED, TOTAL ACTIVITY, MINING AND
MINERAL MANUFACTURING INDUSTRIES, 1976-82

	1976	1977	1978	1979	1980	1981	1982
	($ million)						
Mining							
Metallic minerals							
Gold-quartz	113.7	152.0	207.6	322.8	588.8	519.0	566.2
Silver-lead-zinc	233.7	279.8	372.7	671.9	513.6	380.3	351.1
Nickel-copper-gold-silver	1,488.8	1,244.3	1,288.5	2,469.7	2,992.2	2,007.9	1,144.9
Iron	732.1	807.3	717.0	1,022.2	1,005.0	1,036.0	761.4
Uranium	195.8	300.1	501.7	525.4	559.3	865.8	600.1
Miscellaneous metal mines	74.2	118.0	138.6	179.7	243.3	150.2	73.7
Total	2,838.4	2,901.4	3,226.1	5,191.6	5,902.2	4,959.3	3,497.4
Industrial minerals							
Asbestos	373.2	474.8	401.6	456.8	473.4	431.5	267.3
Gypsum	15.8	21.0	25.9	27.5	26.9	31.3	26.6
Peat	23.7	27.4	33.7	38.8	42.7	47.8	41.1
Potash	262.1	301.4	360.2	613.5	900.4	889.7	488.5
Sand and gravel	99.0	91.3	85.8	91.5	92.0	98.3	75.6
Stone	111.0	106.1	110.2	121.7	123.4	122.5	109.4
Miscellaneous nonmetals	113.1	116.5	122.6	140.1	152.8	171.0	183.5
Total	997.8	1,138.4	1,139.9	1,489.8	1,811.5	1,791.9	1,192.0
Fuels							
Coal	474.3	508.5	566.8	658.6	621.6	671.1	838.0
Petroleum and natural gas	7,052.0	8,698.3	10,078.6	12,554.1	14,917.3	15,924.6	18,915.5
Total	7,526.3	9,206.9	10,645.4	13,212.7	15,538.9	16,595.7	19,753.5
Total mining industry	11,362.5	13,246.7	15,011.4	19,894.1	23,252.6	23,347.0	24,442.9
Mineral manufacturing							
Primary metal industries							
Iron and steel mills	1,498.8	1,677.6	1,924.9	2,424.3	2,537.9	2,750.9	2,149.9
Steel pipe & tube mills	148.8	160.3	225.1	280.4	297.6	378.3	320.3
Iron foundries	241.9	257.7	273.8	298.2	266.9	266.0	279.9
Smelting and refining	812.7	1,176.1	1,387.2	1,401.0	1,976.9	1,808.9	1,493.0
Aluminum rolling, casting and extruding	149.4	193.7	154.3	249.0	273.5	292.8	289.9
Copper and alloy rolling, casting and extruding	71.4	78.5	93.1	131.5	103.7	129.3	101.6
Metal rolling, casting and extruding, nes	113.3	110.2	136.2	198.9	203.6	210.4	169.2
Total	3,036.3	3,654.0	4,194.7	4,983.3	5,660.1	5,836.6	4,803.8

(continued on following page)

TABLE 10. (cont'd)

	1976	1977	1978	1979	1980	1981	1982
				($ million)			
Mineral manufacturing (cont'd)							
Nonmetallic mineral products industries							
Cement manufacturers	249.1	275.0	319.9	388.8	357.3	422.2	387.4
Lime manufacturers	30.0	36.6	44.6	49.3	59.5	62.8	60.1
Concrete products manufacturers	282.1	273.5	309.3	328.7	324.6	378.5	349.7
Ready-mix concrete manufacturers	282.6	292.8	317.3	341.6	352.4	430.1	388.6
Clay products (domestic clay)	65.9	69.6	73.6	87.5	84.6	82.0	57.1
Clay products (imported clay)	39.1	39.8	43.1	44.9	51.6	50.9	37.9
Refractories manufacturers	44.4	32.5	45.3	66.6	73.6	54.5	61.8
Stone products manufacturers	16.3	19.6	22.4	28.2	33.2	40.9	39.5
Glass manufacturers	205.1	199.2	266.8	294.9	308.1	364.6	339.6
Glass products manufacturers	87.4	96.6	122.9	141.0	143.6	141.0	144.9
Abrasive manufacturers	55.1	64.1	70.6	79.4	92.1	95.9	80.4
Other nonmetallic mineral products industries	270.2	253.6	341.0	375.2	370.7	388.0	325.4
Total	1,627.3	1,652.9	1,976.8	2,226.2	2,251.3	2,510.5	2,272.4
Petroleum and coal products industries							
Petroleum refining	945.8	1,206.7	1,180.4	1,390.9	1,750.1	2,641.5	2,108.4
Manufacturers of lubricating oil and greases	32.6	36.8	36.9	38.3	26.7	35.0	31.7
Other petroleum and coal products industries	45.7	44.4	33.1	30.5	36.0	39.3	39.9
Total	1,024.2	1,287.9	1,250.4	1,459.8	1,812.8	2,715.8	2,180.1
Total mineral manufacturing	5,687.8	6,594.8	7,421.9	8,669.2	9,724.2	11,062.9	9,256.2
Total mining and mineral manufacturing	17,050.3	19,841.5	22,433.3	28,563.3	32,976.9	34,409.9	33,699.2

nes Not elsewhere specified.

TABLE 11. CANADA, INDEXES OF GROSS DOMESTIC PRODUCT OF INDUSTRIAL PRODUCTION, MINING AND MINERAL MANUFACTURING, 1969-83
(1971=100)

	1969	1970	1971	1972	1973	1974	1975	1976	1977	1978	1979r	1980r	1981r	1982r	1983p
Total industrial production	93.6	94.9	100.0	107.6	119.0	122.8	115.5	122.2	125.3	129.9	137.9	135.9	137.1	122.5	129.7
Total mining	86.9	98.7	100.0	104.4	117.8	114.0	100.9	103.1	106.1	95.8	106.4	110.2	104.6	91.8	97.2
Metals															
All metals	88.4	105.4	100.0	94.3	105.7	101.8	91.2	96.7	99.5	73.8	79.1	85.9	83.9	63.7	72.1
Placer gold and gold quartz mines	118.2	105.3	100.0	90.1	80.0	68.4	67.4	69.1	68.2	65.5	59.8	59.0	60.2	81.7	105.1
Iron mines	91.9	116.1	100.0	78.7	97.4	80.4	71.4	104.6	94.7	41.5	82.2	77.4	78.8	53.6	51.4
Other metal mines	85.3	103.0	100.0	98.6r	109.3r	109.3r	97.7r	96.0r	102.4r	82.8	81.9	92.6	89.4	67.9	78.4
Fuels															
All fuels	80.8	92.6	100.0	114.7	130.1	124.7	112.4	107.5	108.6	109.5	123.0	121.3	113.5	113.1	116.8
Coal	68.4	87.5	100.0	105.4	115.5	116.8	137.5	128.5	125.2	138.9	167.8	184.5	193.7	204.4	208.8
Crude oil and natural gas	81.7	93.0	100.0	115.4	131.2	125.3	110.5	105.9	107.3	107.3	119.6	116.5	107.5	106.2	109.9
Nonmetals															
All metals	92.8	95.0	100.0	99.7	107.8	119.7	88.9	103.6	109.4	103.2	116.6	113.3	106.5	82.7	88.6
Asbestos	89.8	95.2	100.0	101.0	102.1	102.0	63.7	85.5	85.5	64.6	69.9	63.4	53.8	37.7	36.5
Mineral manufacturing															
Primary metals	94.9	100.9	100.0	101.3	112.2	118.7	107.0	105.6	113.2	119.5	121.6	121.2	121.3	97.4	106.8
Nonmetallic mineral products	90.5	86.6	100.0	109.1	119.5	125.2	117.7	120.5	119.4	127.3	134.6	122.7	119.9	95.3	104.8
Petroleum and coal products	92.1	94.4	100.0	115.3	136.1	136.8	130.9	120.0	112.1	110.8	97.7	97.6	102.9	89.6	86.8

p Preliminary; r Revised.

TABLE 12. CANADA, INDEXES OF GROSS DOMESTIC PRODUCT BY INDUSTRIES, 1969-83 (1971=100)

	1969	1970	1971	1972	1973	1974	1975	1976	1977	1978	1979r	1980r	1981r	1982r	1983p
Gross domestic product, all industries	92.2	94.4	100.0	105.9	114.1	119.3	120.4	126.4	130.1	134.5	144.2	146.0	150.2	143.4	147.3
Agriculture	90.6	89.0	100.0	88.7	96.9	89.5	103.0	109.3	113.9	111.2	100.3	108.3	117.2	120.5	122.8
Forestry	102.4	103.3	100.0	105.7	113.7	112.1	97.8	105.4	110.8	118.7	119.6	123.5	113.4	92.9	113.8
Fishing and trapping	102.6	105.4	100.0	95.7	101.6	90.2	85.8	98.0	110.1	121.8	123.8	116.8	127.9	124.1	123.5
Mines (incl. milling), quarries and oil wells	86.9	98.7	100.0	104.4	117.8	114.0	100.9	103.1	106.1	95.8	106.4	110.2	104.6	91.8	97.4
Electric power, gas and water utilities	85.4	93.3	100.0	111.1	120.3	130.1	130.5	142.0	150.9	160.5	172.8	174.7	177.8	178.0	184.6
Manufacturing	95.8	94.5	100.0	107.7	119.1	123.4	116.2	123.1	125.5	132.0	140.1	136.1	138.3	121.6	129.1
Construction	92.5	90.9	100.0	103.0	106.1	110.3	116.0	119.6	117.3	114.7	121.6	120.5	127.9	113.6	110.5
Transportation, storage and communications	89.0	94.2	100.0	108.5	117.9	125.0	126.5	134.2	141.6	148.6	204.9	211.5	217.4	210.9	215.5
Trade	91.7	93.2	100.0	109.9	119.8	129.5	132.5	138.0	139.8	144.9	153.0	153.1	154.4	144.0	148.3
Community, business and personal service	91.6	95.5	100.0	104.8	109.5	115.8	121.1	127.3	131.2	136.1	136.9	141.5	148.5	148.4	150.3
Finance, insurance and real estate	92.4	94.6	100.0	105.3	114.0	120.9	125.9	132.3	140.2	147.3	154.1	159.9	167.1	168.0	170.3
Public administration and defence	91.6	95.2	100.0	104.2	109.7	113.9	119.4	123.0	125.7	128.9	128.2	129.8	132.3	136.7	138.5

p Preliminary; r Revised.

TABLE 13. CANADA, GROSS DOMESTIC PRODUCT FOR SELECTED INDUSTRIES BY PROVINCE, 1981

	New-found-land	Prince Edward Island	Nova Scotia	New Brunswick	Quebec	Ontario	Manitoba	Saskat-chewan	Alberta	British Columbia	Yukon and Terri-tories	Northwest Canada
					($ million)							
Agriculture	17.5	90.2	112.9	86.4	1,315.8	2,554.1	902.2	2,468.1	2,128.5	506.0	..	10,181.7
Forestry	54.0	0.1	30.0	126.6	389.7	355.1	16.5	34.4	53.8	970.0	-	2,030.2
Fishing, Hunting and Trapping	117.6	22.5	185.7	39.4	40.4	35.0	17.5	6.8	7.9	168.0	4.8	645.6
Mining[1]	444.6	-	126.8	169.8	1,059.9	2,317.7	397.5	1,329.2	9,782.4	1,484.5	220.4	17,288.7
Manufacturing	463.6	63.0	1,074.5	884.6	17,208.6	31,329.6	1,760.2	679.1	3,379.2	5,699.1	6.0	62,548.2
Construction	280.1	46.4	421.7	339.4	3,317.8	4,762.4	474.5	820.9	4,530.1	2,967.1	279.2	18,239.6
Electric power, gas and water utilities	204.2	14.5	194.8	249.3	2,869.9	3,293.9	431.8	256.8	833.6	1,003.8	45.4	9,398.0
Goods-producing industries	1,581.6	236.7	2,146.4	1,895.5	26,202.1	44,647.8	4,000.2	5,595.3	20,715.5	12,798.5	555.8	120,332.0

1 Cement, lime, clay and clay products (domestic clays) industries are included under "manufacturing".
x Confidential; .. Not available; - Nil.

TABLE 14. CANADA, GROSS DOMESTIC PRODUCT FOR MINING BY PROVINCE, 1975-81

	New-foundland	Prince Edward Island	Nova Scotia	New Brunswick	Quebec	Ontario	Manitoba	Saskatchewan	Alberta	British Columbia	Yukon and Northwest Territories	Canada
						($ million)						
1975	212.3	-	63.5	66.9	503.2	1,128.9	170.3	445.3	3,474.6	613.2	104.8	6,771.7
1976	309.6	-	80.5	59.2	677.7	1,261.1	207.5	504.5	3,860.6	849.1	68.0	7,865.9
1977	346.6	-	113.4	65.6	737.1	1,203.1	125.4	660.5	4,804.2	866.9	155.2	9,064.6
1978	230.7	-	103.9	113.2	708.3	1,217.0	184.9	861.4	5,245.9	924.5	215.2	9,794.3
1979	459.2	-	111.1	206.4	1,175.2	1,519.9	426.4	1,045.3	7,120.6	1,507.3	262.2	13,921.7
1980	410.3	-	120.0	88.6	1,123.6	2,806.1	522.6	1,333.0	9,641.6	1,464.3	368.0	17,851.2
1981	444.6	-	126.8	169.8	1,059.9	2,317.7	397.5	1,329.2	9,782.4	1,484.5	220.4	17,288.7

- Nil.

TABLE 15. CANADA, VALUE OF EXPORTS OF CRUDE MINERALS AND FABRICATED MINERAL PRODUCTS, 1977-83

	1977	1978	1979	1980	1981	1982	1983P
			($ million)				
Ferrous							
Crude material	1,114.9	854.5	1,469.5	1,342.9	1,540.0	1,098.3	1,054.3
Fabricated material	1,242.9	1,696.0	1,947.6	2,358.0	2,664.9	2,299.2	2,011.6
Total	2,357.9	2,550.6	3,417.1	3,701.1	4,205.0	3,397.5	3,065.9
Nonferrous							
Crude material	1,614.9	1,549.2	2,425.1	2,866.6	2,544.0	2,088.8	1,845.9
Fabricated material	2,578.4	3,360.9	3,807.1	6,273.8	5,615.6	4,977.9	5,624.5
Total	4,193.4	4,910.1	6,232.1	9,140.4	8,159.6	7,066.7	7,470.5
Nonmetals							
Crude material	1,276.1	1,369.7	1,715.3	2,305.0	2,618.7	2,171.1	2,103.5
Fabricated material	253.6	377.2	455.9	412.5	439.7	409.3	424.8
Total	1,529.6	1,746.8	2,171.2	2,717.5	3,058.3	2,580.5	2,528.3
Mineral fuels							
Crude material	4,428.9	4,514.9	6,128.9	7,816.8	8,022.0	8,752.4	8,727.9
Fabricated material	649.1	1,022.7	1,885.3	2,324.2	2,642.0	2,534.9	2,815.6
Total	5,078.0	5,537.6	8,014.2	10,141.0	10,664.0	11,287.3	11,534.5
Total minerals and products							
Crude material	8,434.9	8,288.2	11,738.8	14,331.4	14,724.6	14,110.6	13,731.6
Fabricated material	4,724.1	6,456.8	8,095.8	11,368.7	11,362.3	9,685.2	10,876.6
Total	13,158.9	14,745.0	19,834.7	25,700.1	26,086.9	24,332.0	24,608.3

P Preliminary.

TABLE 16. CANADA, VALUE OF IMPORTS OF CRUDE MINERALS AND FABRICATED MINERAL PRODUCTS, 1977-83

	1977	1978	1979	1980	1981	1982	1983P
			($ million)				
Ferrous							
Crude material	106.0	223.8	322.1	354.2	373.2	227.5	285.2
Fabricated material	1,501.0	1,838.3	2,533.9	2,329.0	3,303.2	2,115.1	2,004.6
Total	1,607.0	2,062.1	2,856.0	2,683.2	3,676.4	2,342.5	2,289.8
Nonferrous							
Crude material	409.0	480.9	808.1	1,778.3	1,509.4	1,263.1	1,365.9
Fabricated material	662.1	949.1	2,122.7	2,784.6	2,433.4	1,862.4	2,358.7
Total	1,071.1	1,430.0	2,930.8	4,562.9	3,942.8	3,125.5	3,724.6
Nonmetals							
Crude material	170.6	231.0	284.5	329.3	339.3	282.2	271.9
Fabricated material	472.0	526.8	644.7	724.2	805.3	671.9	746.3
Total	642.6	757.8	929.2	1,053.5	1,144.6	954.1	1,018.2
Mineral fuels							
Crude material	3,876.4	4,092.8	5,364.3	7,732.3	8,696.9	5,906.3	4,115.8
Fabricated material	299.7	344.8	394.0	687.7	881.3	863.6	1,046.4
Total	4,176.1	4,437.6	5,758.3	8,420.0	9,578.2	6,769.9	5,162.3
Total minerals and products							
Crude material	4,562.0	5,028.6	6,779.0	10,194.1	10,918.7	7,679.0	6,038.8
Fabricated material	2,934.8	3,659.0	5,695.3	6,525.4	7,423.3	5,513.1	6,156.0
Total	7,496.8	8,687.6	12,474.3	16,719.5	18,342.0	13,192.1	12,194.8

P Preliminary.

TABLE 17. CANADA, VALUE OF EXPORTS OF CRUDE MINERALS AND FABRICATED MINERAL PRODUCTS IN RELATION TO TOTAL EXPORT TRADE, 1973, 1978 AND 1983

	1973		1978		1983P	
	($ million)	(%)	($ million)	(%)	($ million)	(%)
Crude material	4,593.4	18.4	8,288.2	11.1	13,731.6	15.5
Fabricated material	2,974.3	11.9	6,456.8	8.7	10,876.6	12.3
Total	7,567.7	30.4	14,745.0	19.8	24,608.3	27.8
Total exports, all products	24,837.9	100.0	74,259.3	100.0	88,506.2	100.0

P Preliminary.

TABLE 18. CANADA, VALUE OF IMPORTS OF CRUDE MINERALS AND FABRICATED MINERAL PRODUCTS IN RELATION TO TOTAL IMPORT TRADE, 1973, 1978 AND 1983

	1973		1978		1983P	
	($ million)	(%)	($ million)	(%)	($ million)	(%)
Crude material	1,535.5	6.6	5,028.6	10.0	6,038.8	8.0
Fabricated material	1,953.7	8.4	3,659.0	7.3	6,156.0	8.1
Total	3,489.2	15.0	8,687.6	17.3	12,194.8	16.1
Total imports, all products	23,325.3	100.0	50,107.7	100.0	75,586.6	100.0

P Preliminary.

TABLE 19. CANADA, VALUE OF EXPORTS OF CRUDE MINERALS AND FABRICATED MINERAL PRODUCTS, BY MAIN GROUPS AND DESTINATION, 1983P

	United States	United Kingdom	EFTA[1]	EEC[2]	Japan	Other Countries	Total
	($ million)						
Ferrous materials and products	2,225.4	239.6	12.5	273.1	105.4	210.0	3,065.9
Nonferrous materials and products	4,496.2	462.5	232.9	708.1	738.6	832.2	7,470.5
Nonmetallic mineral materials and products	1,147.0	38.9	42.6	229.6	142.0	928.2	2,528.3
Mineral fuels, materials and products	10,170.9	0.9	15.2	94.6	949.5	312.4	11,534.5
Total	18,039.5	741.8	303.3	1,305.4	1,935.5	2,282.8	24,608.3
Percentage of total mineral exports	73.3	3.0	1.2	5.3	7.9	9.3	100.0

[1] European Free Trade Association includes Austria, Norway, Portugal, Sweden, Switzerland, Finland and Iceland. [2] European Economic Community includes Belgium-Luxembourg, France, Italy, Netherlands, West Germany, Greece, Denmark and Ireland.
P Preliminary.

TABLE 20. CANADA, VALUE OF IMPORTS OF CRUDE MINERALS AND FABRICATED MINERAL PRODUCTS, BY MAIN GROUPS AND ORIGIN, 1983P

	United States	United Kingdom	EFTA[1]	EEC[2]	Japan	Other Countries	Total
	($ million)						
Ferrous materials and products	1,591.3	87.9	58.8	214.9	181.1	155.8	2,289.8
Nonferrous materials and products	2,552.4	45.9	138.8	168.4	77.3	741.6	3,724.6
Nonmetallic mineral materials and products	710.5	19.5	13.8	155.2	30.4	88.9	1,018.2
Mineral fuels, materials and products	1,930.0	209.7	53.8	52.9	0.2	2,915.7	5,162.3
Total	6,784.3	363.1	265.1	591.4	289.0	3,901.9	12,194.8
Percentage of total mineral imports	55.6	3.0	2.1	4.9	2.4	32.0	100.0

[1] European Free Trade Association includes Austria, Norway, Portugal, Sweden, Switzerland, Finland and Iceland. [2] European Economic Community includes Belgium-Luxembourg, France, Italy, Netherlands, West Germany, Greece, Denmark and Ireland.
P Preliminary.

TABLE 21. CANADA, VALUE OF EXPORTS OF CRUDE MINERALS AND FABRICATED MINERAL PRODUCTS, BY COMMODITY AND DESTINATION, 1983P

	United States	United Kingdom	EFTA[1]	EEC[2]	Japan	Other Countries	Total
				($000)			
Aluminum	1,278,351	6,043	7,742	31,934	229,323	306,831	1,860,224
Asbestos	87,642	25,590	17,949	93,177	42,008	217,088	483,454
Copper	360,957	96,722	43,946	171,857	277,865	234,052	1,185,399
Fuels	10,170,877	860	15,238	94,591	949,525	312,435	11,543,526
Iron Ore	479,970	226,770	8,780	212,577	83,124	43,109	1,054,330
Lead	46,212	13,827	2,473	28,032	1,752	10,771	103,067
Molybdenum	3,584	24,374	3,098	58,243	14,940	3,538	107,777
Nickel	364,010	124,499	146,899	113,416	37,600	46,733	833,157
Primary ferrous metals	219,208	540	20	41,267	19,462	16,248	296,745
Uranium	25,400	37,175	–	–	–	–	62,575
Zinc	349,662	32,917	9,999	216,378	24,023	149,586	782,565
All other minerals	4,653,647	152,482	47,130	243,924	255,836	942,416	6,295,435
Total	18,039,520	741,799	303,274	1,305,396	1,935,458	2,282,807	24,608,254

1 European Free Trade Association includes Austria, Norway, Portugal, Sweden, Switzerland, Finland and Iceland. 2 European Economic Community includes Belgium-Luxembourg, France, Italy, Netherlands, West Germany, Greece, Denmark and Ireland.
P Preliminary; – Nil.

TABLE 22. CANADA, PHYSICAL VOLUME OF IMPORT TRADE FOR SELECTED COMMODITIES, 1977-83

	Units of Weight	1977	1978	1979	1980	1981	1982	1983p
Crude materials								
Metals								
Alumina	t	821 596	1 056 190	952 584	983 972	1 020 550	939 267	1 063 181
Bauxite ore	t	2 764 286	2 434 435	2 149 636	3 504 368	2 734 665	2 574 718	2 329 910
Iron ore	t	2 505 203	4 685 868	5 912 581	5 875 292	5 794 634	3 359 303	4 013 109
Manganese ore	t	57 644	136 446	45 150	95 161	119 746	71 656	42 260
Nonmetals								
Bentonite	t	358 724	353 790	638 307	471 684	311 250	238 069	187 221
Clay, ground & unground	t	334 431	381 486	445 231	403 282	413 040	345 382	369 019
Fluorspar	t	124 494	170 237	167 904	223 940	173 599	126 954	141 928
Limestone, crushed	t	2 922 684	2 873 601	3 215 717	2 418 330	2 526 876	1 485 428	1 799 861
Phosphate rock	t	2 439 021	3 043 899	3 341 039	3 816 514	3 245 446	2 511 723	2 662 725
Salt & brine	t	1 126 225	1 330 474	1 275 627	1 151 203	1 254 992	1 526 881	814 254
Sand & gravel	t	1 645 663	1 810 989	1 201 915	1 209 582	1 446 872	1 179 285	878 614
Silica sand	t	1 101 186	1 242 444	1 651 890	1 200 237	1 142 880	788 768	982 662
Fuels								
Coal	t	15 026 358	13 000 320	17 381 794	15 719 025	14 687 279	15 488 113	14 509 685
Petroleum, crude	m³	38 042 718	36 754 037	35 330 535	32 710 030	30 752 166	19 671 109	14 412 728
Fabricated materials								
Metals								
Aluminum & aluminum alloy	t	118 216	119 154	168 125	128 150	139 385	131 322	152 591
Ferroalloys	t	93 672	101 160	167 232	118 516	117 907	64 662	71 577
Steel:								
bars & rods	t	301 502	318 336	300 069	189 853	340 772	219 638	278 151
castings & forgings	t	113 365	116 473	139 095	129 363	118 473	70 150	92 432
pipes & tubes	t	203 238	317 031	285 144	322 121	364 865	249 581	217 425
sheets & strips	t	552 606	704 502	1 039 054	582 233	1 717 134	540 390	535 546
structural shapes	t	225 869	151 502	273 111	207 657	362 891	120 360	162 231
Nonmetals								
Cement	t	263 528	256 721	248 422	223 247	721 205	231 829	253 015
Fire bricks	t	242 720	156 002	227 156	236 205	187 016	132 600	154 795
Phosphate fertilizers	t	200 445	286 744	381 887	248 328	307 215	249 828	360 304
Fuels								
Coke	t	1 267 895	1 527 342	1 366 182	1 311 535	1 436 074	1 064 536	1 345 806
Fuel oil	000 l	1 260 034	1 277 077	871 425	1 617 606	1 256 790	1 571 003	1 468 464

p Preliminary.

TABLE 23. CANADA, PHYSICAL VOLUME OF EXPORT TRADE FOR SELECTED COMMODITIES, 1977-83

	Unit of Weight	1977	1978	1979	1980	1981	1982	1983p
Crude material								
Metals								
Copper, ores & concentrates	t	279 582	282 159	315 211	286 076	276 810	257 930	313 796
Iron, ores	t	45 060 391	31 929 094	48 849 270	39 020 922	41 452 044	27 281 396	25 527 960
Lead, ores & concentrates	t	137 820	142 693	151 485	147 008	146 304	106 744	85 460
Zinc, ores & concentrates	t	598 451	688 186	598 279	435 831	516 210	457 751	660 790
Nonmetals								
Asbestos, crude & fibers	t	1 415 482	1 398 081	1 461 042	1 217 737	1 062 189	880 696	753 912
Crude refractory materials	t	747 938	1 081 684	1 023 734	803 892	629 770	40 839	241 131
Gypsum	t	4 994 323	5 178 631	5 474 764	4 960 240	5 094 873	4 775 780	5 187 032
Limestone, crushed	t	1 502 492	1 710 348	2 296 295	2 214 489	1 758 299	1 517 499	1 390 795
Nepheline syenite	t	443 763	420 961	471 056	448 468	476 281	414 787	398 299
Salt and brine	t	1 163 163	1 608 582	1 822 120	1 655 768	1 507 710	1 721 892	1 914 629
Sand and gravel	t	273 745	269 216	323 639	383 533	318 635	168 692	95 632
Sulphur, crude	t	4 291 032	4 984 545	5 154 831	6 850 143	7 309 216	6 111 444	5 670 275
Fuels								
Coal	t	12 068 905	13 657 514	13 852 848	14 310 782	16 285 102	15 528 541	16 978 451
Natural gas	000 m³	28 141 415	24 992 242	28 047 648	22 963 134	21 687 359	22 072 136	20 023 253
Fabricated materials								
Metals								
Aluminum, pig ingots	t	655 353	863 320	551 957	784 720	725 441	896 365	925 403
Copper, refinery shapes	t	294 490	247 727	191 211	335 200	263 046	232 623	298 528
Iron, pig ingots	t	505 277	544 716	255 523	562 351	466 360	485 620	348 278
Lead, pig ingots	t	130 819	131 950	117 992	126 538	119 815	146 126	147 265
Zinc, pig ingots	t	295 358	439 260	429 352	472 148	453 526	470 390	500 443
Nonmetals								
Cement	t	1 274 652	1 634 582	2 288 822	1 550 562	1 578 659	1 752 129	1 561 080
Lime, quick & hydrated	t	359 540	478 551	490 863	403 166	432 845	281 247	215 942
Peat	t	303 414	312 903	358 267	390 457	326 826	326 826	396 884
Fuels								
Butane gas, liquified	000 l	2 432 188	2 208 682	2 926 459	2 563 406	3 137 545	3 572 545	3 011 710
Coke	t	355 919	352 358	354 016	470 496	392 664	234 690	110 929
Fuel oil	000 l	1 456 991	4 232 409	4 654 162	4 273 510	3 846 907	2 718 769	3 825 520
Gasoline	000 l	388 080	972 282	913 271	706 539	600 969	536 268	1 240 028
Propane gas, liquified	000 l	5 019 524	3 543 782	4 858 175	3 879 915	3 867 950	4 513 307	3 534 562

p Preliminary.

TABLE 24. CANADA, APPARENT CONSUMPTION[1] OF SOME MINERALS, AND RELATION TO PRODUCTION[2], 1981-83

	Unit of Measure	1981			1982			1983P		
		Apparent Consumption	Production	Consumption as % of production	Apparent Consumption	Production	Consumption as % of production	Apparent Consumption	Production	Consumption as % of production
Asbestos	t	60 590	1 121 845	5.4	-	834 249	-	87 462	840 277	10.4
Cement	t	9 294 745	10 152 199	91.6	6 636 084	8 156 391	81.4	6 470 832	7 778 897	83.2
Gypsum	t	2 074 045	7 025 418	29.5	1 674 259	5 987 396	28.1	3 123 950	7 693 759	40.6
Iron ore	t	13 893 389	49 550 799	28.0	8 666 497	33 197 561	26.1	12 020 643	32 869 627	36.6
Lime	t	2 145 087	2 554 788	84.0	1 932 124	2 197 298	87.9	2 032 615	2 225 713	91.3
Quartz silica	t	3 262 119	2 238 333	145.7	2 447 231	1 703 059	143.7	2 900 722	1 988 086	145.9
Salt	t	6 986 743	7 239 461	96.5	7 749 081	7 940 331	97.6	7 441 894	8 542 269	87.1

1 "Apparent consumption" is production, plus imports, less exports. 2 "Production" refers to producers' shipments.
P Preliminary; - Nil.

TABLE 25. CANADA, REPORTED CONSUMPTION OF MINERALS AND RELATION TO PRODUCTION, 1980-82

	Unit of Measure	1980			1981			1982p		
		Consumption	Production	Consumption as % of production	Consumption	Production	Consumption as % of production	Consumption	Production	Consumption as % of production
Metals										
Aluminum	t	329 400	1 068 197	30.8	336 989	1 115 691	30.2	237 534	1 064 795	25.7
Antimony	kg	336 105	..	..	209 829	..	..	161 034	..	..
Bismuth	kg	10 271	149 566	6.9	10 094	167 885	6.0	10 074	189 132	5.3
Cadmium	kg	61 011	1 033 097	5.9	34 092	833 788	4.1	33 818	886 055	3.8
Chromium (chromite)	t	27 900	–	–	24 771	–	–	15 330	–	–
Cobalt	kg	105 225	2 118 154	5.0	101 334	2 080 395	4.9	86 399	1 274 484	6.8
Copper[1]	t	195 124	716 363	27.2	216 799	691 327	31.4	130 559	612 455	21.3
Lead[2]	t	130 988	251 627	52.1	137 245	268 556	51.1	116 432	272 187	42.8
Magnesium	t	5 412	9 252	58.5	6 387	–	..	5 005	–	..
Manganese ore	t	157 680	–	..	288 908	–	..	130 826	–	..
Mercury	kg	36 326	–	..	35 635	–	..	38 746	–	..
Molybdenum (Mo content)	t	1 055	11 889	8.9	1 315	12 850	10.2	681	13 961	4.9
Nickel	t	9 676	184 802	5.2	8 603	160 247	5.4	6 657	88 581	7.5
Selenium	kg	10 795	279 626	3.9	9 414	255 369	3.7	10 469	222 323	4.7
Silver	kg	265 998	1 069 635	24.9	292 130	1 129 394	25.9	180 459	1 313 630	13.7
Tellurium	kg	..	15 011	..	..	31 145	..	..	18 423	..
Tin	t	4 517	243	1 858.8	3 766	239	1 575.7	3 528	135	2 613.3
Tungsten (W content)	kg	290 479	4 006 647	7.2	401 447	2 515 165	16.0	507 606	3 029 780	16.8
Zinc	t	116 618	883 697	13.2	113 061	911 178	12.4	100 232	965 607	10.4
Nonmetals										
Barite	t	138 829	94 317	147.2	94 027	78 154	120.3	24 359	23 552	103.4
Feldspar	t	4 051	–	..	4 606	–	..	2 790	–	..
Fluorspar	t	65 492	–	..	135 091	–	..	252 859	–	..
Mica	kg	2 576	–	..	2 259	–	..	1 745	–	..
Nepheline syenite	t	84 873	999 699	14.2	97 734	587 565	16.6	102 609	550 480	18.6
Phosphate rock	t	3 546 636	–	..	3 264 779	–	..	2 581 671	–	..
Potash (K$_2$O)	t	223 222	7 201 217	..	216 298	6 548 701	..	195 061	5 308 532	..
Sodium sulphate	t	223 222	480 666	46.4	216 298	535 214	40.4	195 061	547 208	35.6
Sulphur	t	808 618	7 655 723	10.6	847 230	8 017 885	10.6	1 082 248	6 945 183	15.6
Talc, etc.	t	42 217	91 848	46.0	38 984	82 715	47.1	38 633	70 523	54.8
Fuels										
Coal	000t	57 333	36 688	101.8	38 367	40 088	95.7	41 500	42 906	96.7
Natural gas[3]	million m^3	43 255	87 108	49.7	42 886	73 824	58.1	46 143	69 288	66.6
Crude oil[4]	000 m^3	109 802	83 477	131.5	100 777	74 553	135.2	86 528	79 255	109.2

Note: Unless otherwise stated, consumption refers to reported consumption of refined metals or nonmetallic minerals by consumers. Production of metals, in most cases, refers to production in all forms, and includes the recoverable content of ores, concentrates, matte, etc., and metal content of primary products recoverable at domestic smelters and refineries. Production of nonmetals refers to producers' shipments. For fuels, production is equivalent to actual output less waste.
[1] Consumption defined as producers domestic shipments of refined metal. [2] Consumption includes primary and secondary refined metal. [3] Consumption defined as domestic sales. [4] Consumption defined as refinery receipts.

TABLE 26. CANADA, DOMESTIC CONSUMPTION OF PRINCIPAL REFINED METALS IN RELATION TO REFINERY PRODUCTION[1], 1976-82

	Unit of Measure	1976	1977	1978	1979	1980	1981	1982p
Copper								
Domestic consumption[2]	t	206 205	200 372	228 694	210 689	195 124	216 759	130 559
Production	t	510 469	508 767	446 278	397 263	505 238	476 655	298 290
Consumption of production	%	40.4	39.4	51.2	53.0	38.6	45.5	43.8
Zinc								
Domestic consumption[3]	t	98 897	105 412	121 375	131 317	116 618	113 061	100 232
Production	t	472 316	494 938	495 243	580 449	591 565	618 650	511 870
Consumption of production	%	20.9	21.3	24.5	22.6	19.7	18.3	19.6
Lead								
Domestic consumption[3]	t	107 654	106 962	100 762	126 464	130 988	137 245	116 432
Production	t	175 720	187 457	194 054	183 769	162 463	168 450	174 310
Consumption of production	%	61.3	57.1	51.9	68.8	80.6	81.5	66.8
Aluminum								
Domestic consumption[4]	t	332 206	322 393	380 291	398 834	329 400	336 989	237 534
Production	t	628 049	973 524	1 048 469	860 287	1 068 197	1 115 691	1 064 795
Consumption of production	%	51.3	34.1	36.3	46.4	30.8	30.2	25.7

[1] Production of refined metal from all sources, including metal derived from secondary materials at primary refineries. [2] Producers' domestic shipments of refined metal. [3] Consumption of primary and secondary refined metal, reported by consumers. [4] Consumption of primary refined metal, reported by consumers. p Preliminary.

TABLE 27A. AVERAGE ANNUAL PRICES[1] OF SELECTED MINERALS, 1977-83[2]

	Unit of Measure	1977	1978	1979	1980	1981	1982	1983
Aluminum, major U.S. producer[3]	cents/lb	51.339	53.075	59.395	69.566	57.274	44.966	65.342
Antimony, New York dealer	$/lb	1.237	1.145	1.407	1.508	1.355	1.072	0.913
Asbestos, No. 4 cement fibre	Cdn $/st	551.000	642.000	687.000	769.000	850.000	876.000	1,003.000
Bismuth, U.S. producer	$/lb	6.010	3.378	3.011	2.637	2.044	2.300	2.300
Cadmium, U.S. producer	$/lb	2.962	2.450	2.760	2.843	1.927	1.113	1.129
Calcium, metal crowns	$/lb	1.482	1.680	1.868	2.502	2.831	3.050	3.050
Chrome, U.S. metal, 9% carbon	$/lb	2.900	3.080	3.375	4.017	4.450	4.450	4.450
Cobalt metal, shot/cathode/250 kg	$/lb	5.633	12.246	24.583	25.000	21.4297	12.500	12.500
Columbium, pyrochlore	$/lb	n	2.550	2.550	2.550	3.250	3.250	3.250
Copper, electrolytic cathode	Cdn $/lb	0.695	0.746	1.076	1.178	1.004	0.885	0.769
Gold, London[4]	Cdn $/troy oz	157.089	220.407	359.289	716.087	551.178	465.102	520.792
Iridium, major producer	$/troy oz	300.000	300.000	258.333	505.833	600.000	600.000	600.000
Iron ore, taconite pellets	cents/ltu	55.300	57.108	63.966	69.562	80.073	80.500	80.500
Lead, producer	Cdn cents/lb	31.420	36.820	59.920	49.350	44.520	32.887	26.770
Manganese, U.S. metal, regular	cents/lb	58.000	58.000	58.333	65.267	70.000	86.274	67.583
Magnesium, U.S. primary ingot	cents/lb	97.487	100.500	105.758	116.667	130.250	134.000	136.508
Mercury, New York	$/flask (76 lb)	135.710	153.322	281.096	389.447	413.885	370.934	322.505
Molybdenum, climax concentrate	$/lb	3.730	4.644	7.762	9.768	8.493	9.740	..
Nickel, major producer cathode	$/lb	2.360	2.091	2.707	3.415	3.429	3.200	3.200
Osmium, major producer	$/troy oz	170.000	150.000	150.000	150.000	150.000	139.167	110.000
Palladium, major producer	$/troy oz	59.702	70.873	113.143	213.975	129.500	110.000	130.000
Platinum, major producer	$/troy oz	162.544	237.250	351.649	439.425	475.000	475.000	475.000
Potash, K₂O, coarse major producer	cents/stu	76.000	80.583	100.417	112.667	120.750	119.615	116.000
Rhodium, major producer	$/troy oz	441.667	516.667	737.500	764.583	639.583	600.000	600.000
Ruthenium, major producer	$/troy oz	60.000	60.000	45.750	45.000	45.000	45.000	45.000
Selenium, major producer commercial	$/lb	17.000	15.000	12.250	9.654	..	..	..
Silver, Handy & Harman, Toronto	Cdn $/troy oz	4.920	6.171	12.974	24.099	12.617	9.831	14.154
Sulphur, elemental, major producer[5]	Cdn $/lt	15.678	17.913	25.665	30.740	59.323	66.923	58.663
Tantalum, Tanco	$/lb	17.750	26.479	60.014	97.604	100.830	48.958	45.000
Tellurium, major producer, slab	$/lb	17.416	20.000	20.000	19.500	..	..	..
Tin	Cdn $/lb	5.779	7.265	8.898	10.008	8.893	8.144	8.103
Titanium, ilmenite ore	$/lt	55.000	53.229	51.083	55.000	68.021	70.000	70.000
Tungsten, U.S. hydrogen red	$/lb	14.065	13.900	13.900	13.900	13.900	13.350	13.100
Uranium, U₃O₈[6]	Cdn $/lb	42.311	48.081	50.004	51.927	42.311	44.234	38.500
Vanadium, pentoxide metallurgical	$/lb	2.750	2.900	3.050	3.050	3.250	3.350	3.350
Zinc, special high grade	Cdn cents/lb	35.530	34.757	43.717	44.050	54.240	49.167	52.632

[1] Prices except for noted, are in United States currency. [2] Sources: Alberta Energy Resource Industries Monthly Statistics, Asbestos, Engineering and Mining Journal, Metals Week and Northern Miner. [3] Starting 1981, London Metal Exchange. [4] Average afternoon fixings of London bullion dealers, converted to Canadian dollar. [5] Starting 1980, North American deliveries. [6] From Energy Mines and Resources Canada publications on assessment of Canada's uranium supply and demand series EP 77-3 to EP 83-3. [7] Seven month average.
.. Not available; n Nominal

TABLE 27B. CANADIAN AVERAGE ANNUAL PRICES OF SELECTED MINERALS, 1977-83[1]

	Unit of Measure	1977	1978	1979	1980	1981	1982	1983
Aluminum, major U.S. producer[2]	$/kg	1.204	1.334	1.534	1.793	1.514	1.223	1.775
Antimony, New York dealer	$/kg	2.900	2.879	3.634	3.887	3.582	2.917	2.481
Asbestos, No. 4 cement fibre	$/mt	607.373	707.684	757.288	847.677	936.964	965.625	1,105.618
Bismuth, U.S. producer	$/kg	14.091	8.495	7.777	6.796	5.403	6.258	6.249
Cadmium, U.S. producer	$/kg	6.945	6.161	7.128	7.327	5.094	3.028	3.067
Calcium, metal crowns	$/kg	3.475	4.225	4.825	6.448	7.483	8.298	8.287
Chrome, U.S. metal, 9% carbon	$/kg	6.799	7.745	8.717	10.353	11.763	12.107	12.090
Cobalt metal, shot/cathode/250 kg	$/kg	13.207	30.795	63.492	64.430	56.6106	34.009	33.961
Columbium, pyrochlore	$/kg	n	6.413	6.586	6.572	8.591	8.842	8.830
Copper, electrolytic cathode	$/kg	1.532	1.645	2.372	2.597	2.213	1.951	1.695
Gold, London[3]	$/gm	5.051	7.086	11.551	23.023	17.721	14.953	16.744
Iridium, major producer	$/gm	10.258	11.002	9.730	19.011	23.129	23.806	23.773
Iron Ore, taconite pellets	$/kg	57.883	64.086	73.754	80.034	94.490	97.776	97.638
Lead, producer	¢/kg	69.269	81.174	132.101	108.798	98.150	72.503	59.018
Manganese, U.S. metal, regular	$/kg	1.360	1.459	1.507	1.682	1.850	2.347	1.836
Magnesium, U.S. primary ingot	$/kg	2.286	2.527	2.731	3.007	3.443	3.646	3.709
Mercury, New York	$/kg	3.316	5.073	9.553	13.206	14.395	13.279	11.529
Molybdenum, climax concentrate	$/kg	8.745	11.678	20.047	25.174	22.450	26.500	..
Nickel, major producer cathode	$/kg	5.533	5.258	6.992	8.801	9.064	8.706	8.694
Osmium, major producer	$/gm	5.813	5.501	5.650	5.638	5.782	5.522	4.358
Palladium, major producer	$/gm	2.041	2.599	4.262	8.042	4.992	4.364	5.151
Platinum, major producer	$/gm	5.558	8.701	13.245	16.515	18.310	18.847	18.820
Potash, K₂O, coarse major producer	$/mt	53.454	60.793	87.445	87.110	95.754	97.632	94.547
Rhodium, major producer	$/gm	15.102	18.948	27.778	28.736	24.655	23.806	23.773
Ruthenium, major producer	$/gm	2.052	2.200	1.723	1.691	1.735	1.785	1.783
Selenium, major producer commercial	$/kg	39.858	37.721	31.639	24.880	..	..	..
Silver, Handy & Harman, Toronto	$/kg	158.182	198.402	417.124	774.801	405.646	316.074	455.062
Sulphur, elemental, major producer[4]	$/mt	15.430	17.630	25.260	30.255	58.386	65.866	59.604
Tantalum, Tanco	$/kg	41.617	66.587	155.002	251.545	266.524	133.201	122.259
Tellurium, major producer, slab	$/kg	40.834	50.294	51.655	50.255	..	..	..
Tin	$/kg	12.740	16.017	19.617	22.064	19.606	17.954	17.864
Titanium, ilmenite ore	$/mt	57.566	61.791	58.900	63.280	80.268	85.022	84.902
Uranium, U[5]	$/kg	110.000	125.000	130.000	135.000	110.000	115.000	100.000
Vanadium, pentoxide metallurgical	$/kg	6.448	7.293	7.877	7.861	8.591	9.114	9.102
Zinc, special high grade	$/kg	0.783	0.766	0.964	0.971	1.196	1.084	1.160

1 Sources: Alberta Energy Resource Industries Monthly Statistics, Asbestos, Engineering and Mining Journal, Metals Week and Northern Miner. 2 Starting 1981, London Metal Exchange. 3 Average afternoon fixings of London bullion dealers, converted to Canadian dollar. 4 Starting 1980, North American deliveries. 5 From Energy Mines and Resources Canada publications on assessment of Canada's uranium supply and demand series EP 77-3 to EP 83-3. 6 Seven month average.
.. Not available; n Nominal.

TABLE 28. CANADA, MINERAL PRODUCTS INDUSTRIES, SELLING PRICE INDEXES, 1977-83 (1971 = 100)

	1977	1978	1979	1980	1981	1982	1983p
Iron and steel products industries							
Agricultural implements industry	177.6	188.7	206.0	224.9	260.2	293.1	310.9
Hardware, tool and cutlery manufacturers	162.6	179.1	207.3	238.4	268.2	296.0	308.2
Heating equipment manufacturers	156.5	169.8	188.0	213.2	236.5	267.7	280.4
Primary metal industries	190.5	207.7	258.8	308.3	312.6	310.7	320.5
Iron and steel mills	187.9	203.9	233.7	261.7	290.3	314.2	319.2
Steel pipe and tube mills	197.8	218.0	248.1	276.9	322.1	362.6	359.7
Iron foundries	189.6	200.1	223.3	243.2	261.8	268.9	272.3
Wire and wire products manufacturers	175.4	185.8	206.4	226.9	242.4	249.6	252.7
Nonferrous metal products industries							
Aluminum rolling, casting and extruding	173.6	191.5	234.0	271.0	292.6	290.9	291.7
Copper and alloy, rolling, casting and extruding	144.5	153.0	201.8	219.7	205.8	193.0	206.4
Jewellery and silverware manufacturers	277.8	337.6	507.3	871.3	676.1	609.5	698.4
Metal rolling, casting and extruding, nes	216.3	239.8	310.4	327.3	325.7	314.0	324.3
Nonmetallic mineral products industries							
Abrasives manufacturers	194.7	223.6	255.3	290.6	325.1	361.8	371.0
Cement manufacturers	186.7	207.5	233.2	265.7	308.0	359.7	374.1
Clay products and manufacturers from imported clay	164.7	173.7	190.1	215.2	251.9	278.0	290.6
Glass and glass products manufacturers	150.4	162.1	173.4	197.0	223.2	250.2	259.3
Lime manufacturers	228.7	252.9	292.7	338.3	396.1	453.2	514.4
Concrete products manufacturers	173.7	187.7	200.1	222.5	259.4	296.7	310.5
Clay products from domestic clay	182.8	196.4	214.3	226.9	243.0	269.9	286.5
Petroleum and coal products industries	244.5	275.4	321.3	404.6	551.7	634.4	675.3
Petroleum refineries	246.7	278.7	325.8	410.6	559.8	643.7	685.2
Mixed fertilizers	180.2	191.0	229.0	280.3	289.5	294.6	284.1

nes Not elsewhere specified; p Preliminary.

TABLE 29. CANADA, PRINCIPAL STATISTICS OF THE MINING INDUSTRY[1], 1982

		Mining Activity								Total Activity[2]		
		Production and Related Workers			Costs							
	Establish-ments	Employees	Man-hours paid	Wages	Fuel and Electri-city	Materials and Supplies	Value of Production	Value Added	Employees	Salaries and Wages	Value Added
	(number)	(number)	(000)	($000)	($000)	($000)	($000)	($000)	(number)	($000)	($000)
Metals											
Gold quartz	38	5,809	11,992	163,619	40,132	178,743	781,306	564,798	7,350	213,191	566,201
Silver-lead-zinc	22	4,812	10,260	153,782	62,993	538,067	963,324	362,264	6,837	226,671	351,126
Nickel-copper-gold-silver	40	21,365	35,134	495,302	186,426	1,287,897	2,599,607	1,125,284	28,851	742,653	1,144,859
Iron	12	6,578	11,843	181,650	154,237	432,317	1,377,020	790,466	10,676	320,149	761,429
Uranium	8	4,401	8,775	149,445	48,364	179,178	822,971	595,429	6,035	208,706	600,120
Misc. metal mines	7	1,296	2,367	36,687	15,789	61,968	146,622	68,864	1,754	54,365	73,705
Total	128	44,261	80,370	1,180,485	507,942	2,678,170	6,693,217	3,507,105	61,503	1,765,734	3,497,440
Nonmetals											
Asbestos	8	3,938	7,677	100,077	48,213	57,898	367,465	261,354	4,973	132,613	267,342
Gypsum	10	508	1,071	8,944	3,773	13,967	44,458	26,718	614	11,392	26,609
Peat	54	1,073	2,135	15,814	3,594	13,619	58,445	41,232	1,323	21,170	41,099
Potash	9	2,924	5,810	80,737	77,935	79,170	645,638	488,532	4,076	125,705	488,513
Sand and gravel	98	981	2,173	21,867	12,588	22,760	105,712	70,362	1,463	31,462	75,639
Stone	115	1,541	3,402	35,198	18,181	45,311	169,031	105,540	2,028	46,985	109,358
Misc. nonmetals	40	1,883	4,080	47,099	36,675	43,474	263,893	183,743	2,694	69,525	183,516
Total	334	12,848	26,348	309,736	200,959	276,199	1,654,642	1,177,481	17,171	438,852	1,192,076
Fuels											
Coal	30	10,281	20,268	292,976	94,129	296,456	1,232,968	842,383	13,113	393,582	838,012
Oil, crude and natural gas	756	7,568	15,639	242,697	153,266	517,946	19,520,791	18,849,579	31,699	1,049,836	18,915,469
Total	786	17,849	35,907	535,673	247,395	814,402	20,753,759	19,691,962	44,812	1,443,418	19,753,481
Total mining industry	1,248	74,958	142,625	2,025,894	956,296	3,768,771	29,101,618	24,376,548	123,486	3,648,004	24,442,997

[1] Cement manufacturing, lime manufacturers, clay and clay products (domestic clays) are included in the mineral manufacturing industry. Industry coverage is the same as in Tables 31, 33, 35 and 37. [2] Total activity includes sales and head offices.

TABLE 30. CANADA, PRINCIPAL STATISTICS OF THE MINERAL MANUFACTURING INDUSTRIES[1], 1982

		Mineral Manufacturing Activity								Total Activity[2]		
		Production and Related Workers			Costs							
	Establish-ments (number)	Employees (number)	Man-hours paid (000)	Wages ($000)	Fuel and Electri-city ($000)	Materials and Supplies ($000)	Value of Production ($000)	Value Added ($000)	Employees (number)	Salaries and Wages ($000)	Value Added ($000)
Primary metal industries											
Iron and steel mills	53	38,692	79,264	1,060,835	420,627	3,166,203	5,714,870	2,145,189	52,330	1,512,490	2,149,877
Steel pipe and tube mills	41	4,829	10,266	125,390	27,148	668,996	1,015,833	316,557	6,017	158,723	320,270
Iron foundries	114	6,587	13,030	137,059	35,185	224,899	518,849	280,198	8,163	181,159	279,944
Smelting and refining	33	21,986	45,298	620,008	329,218	1,512,489	3,369,389	1,389,160	33,215	1,003,852	1,492,967
Aluminum rolling, cast-ing and extruding	73	4,435	9,480	100,224	30,442	581,770	899,339	290,484	6,255	154,649	289,900
Copper and alloy roll-ing, casting and extruding	38	2,036	3,992	44,223	11,235	278,421	399,500	104,211	2,541	57,955	101,632
Metal rolling, casting and extruding, nes	94	3,621	7,437	69,447	17,832	291,407	484,670	164,489	4,694	99,205	169,162
Total	446	82,186	168,768	2,157,186	871,687	6,724,186	12,402,450	4,690,917	113,215	3,168,033	4,803,751
Nonmetallic mineral products industries											
Cement manufacturers	25	2,623	5,612	78,074	141,560	122,846	640,176	379,811	4,317	130,038	387,358
Lime manufacturers	15	653	1,408	17,290	44,328	18,421	122,352	59,770	895	24,622	60,126
Concrete products manufacturers	447	6,123	12,524	131,809	23,404	233,445	609,077	344,428	8,245	188,175	349,738
Ready-mix concrete manufacturers	530	6,061	12,609	147,927	45,074	575,697	991,888	368,475	8,034	199,972	388,623
Clay products manu-facturers (domestic)	67	1,200	2,569	22,375	19,690	17,448	94,386	56,740	1,630	35,220	57,078
Clay products manu-facturers (imported)	53	1,091	2,104	17,984	4,583	19,999	63,374	36,888	1,374	25,380	37,894
Refractories manufacturers	21	730	1,463	15,059	5,040	54,735	113,136	53,319	1,367	31,844	61,823
Stone products manufacturers	124	1,012	2,082	17,563	2,383	30,107	71,853	39,733	1,217	21,986	39,461
Glass manufacturers	15	5,790	11,995	125,290	64,480	161,329	567,065	341,932	7,756	180,400	339,628
Glass products manufacturers	108	2,668	5,156	51,540	9,990	134,946	289,990	143,316	3,260	66,592	144,873
Abrasive manufacturers	29	1,572	3,300	31,454	25,741	83,131	183,386	78,403	2,170	48,381	80,359
Other nonmetallic mineral products industries	101	4,475	9,230	95,949	58,537	273,175	638,586	301,106	7,684	182,710	325,444
Total	1,535	33,997	70,052	751,915	444,809	1,725,280	4,385,269	2,203,922	47,949	1,135,320	2,272,405
Petroleum and coal products industries											
Petroleum refining industry	41	7,453	16,316	247,274	213,349	19,079,135	21,369,959	2,119,257	20,155	734,016	2,108,423
Manufacture of lubri-cating oils & greases	22	454	998	10,460	2,188	169,848	194,815	26,821	775	19,697	31,679
Other petroleum & coal products industries	62	368	766	8,288	5,215	103,038	144,381	37,336	571	13,695	39,949
Total	125	8,121	18,080	266,022	220,751	19,352,020	21,709,154	2,183,414	21,501	767,407	2,180,051
Total, mineral manu-facturing industries	2,106	124,304	256,900	3,175,123	1,537,247	27,801,486	38,496,873	9,078,253	182,665	5,070,760	9,256,207

[1] Industry coverage is the same as in Tables 32, 34, 36 and 38. [2] Includes sales and head offices. nes Not elsewhere specified.

TABLE 31. CANADA, PRINCIPAL STATISTICS OF THE MINERAL INDUSTRY[1] BY REGION, 1982

		Mines, Quarries and Oil Well Activity							Total Activity[2]		
		Production and Related Workers			Costs						
	Establish- ments (number)	Employees (number)	Man- hours paid (000)	Wages ($000)	Fuel and Electri- city ($000)	Materials and Supplies ($000)	Value of Production ($000)	Value Added ($000)	Employees (number)	Salaries and Wages ($000)	Value Added ($000)
Atlantic[3]	99	10,431	20,205	250,977	107,513	512,284	1,337,576	717,776	12,444	316,117	720,984
Quebec	182	13,025	25,876	331,138	160,019	509,859	1,683,770	1,013,892	18,777	513,260	1,007,576
Ontario	159	21,372	35,838	504,606	146,530	899,110	2,613,502	1,567,864	29,448	766,007	1,597,661
Prairies	555	16,912	33,714	480,988	313,484	874,567	20,546,439	19,358,387	43,585	1,369,696	19,405,619
British Columbia[4]	192	10,546	20,904	336,496	181,255	711,450	2,392,784	1,500,079	15,184	515,537	1,498,407
Yukon and Northwest Territories[5]	61	2,672	6,089	121,686	47,495	261,501	527,547	218,550	4,048	167,387	212,750
Canada	1,248	74,958	142,625	2,025,894	956,296	3,768,771	29,101,618	24,376,548	123,486	3,648,004	24,442,997

[1] Cement manufacturing, lime manufacturing, clay and clay products are included in the mineral manufacturing industry. Industry coverage is the same as in Tables 29, 33, 35 and 37. [2] Total activity includes sales and head offices. [3] Includes eastern Canada offshore. [4] Includes western Canada offshore. [5] Includes Arctic Islands and offshore.

TABLE 32. CANADA STATISTICS OF THE MINERAL MANUFACTURING INDUSTRY[1] BY REGION, 1982

		Mineral Manufacturing Activity							Total Activity[2]		
		Production and Related Workers			Costs					Salaries	
	Establish-ments (number)	Employees (number)	Man-hours paid (000)	Wages ($000)	Fuel and Electri-city ($000)	Materials and Supplies ($000)	Value of Production ($000)	Value Added ($000)	Employees (number)	and Wages ($000)	Value Added ($000)
Atlantic Provinces	129	(3)	(3)	(3)	(3)	(3)	(3)	(3)	(3)	(3)	(3)
Quebec	527	28,490	58,831	748,101	476,139	7,164,363	9,905,848	2,216,934	43,120	1,203,081	2,285,976
Ontario	847	68,716	141,202	1,714,080	704,654	11,515,881	16,781,029	4,581,394	101,401	2,808,298	4,588,152
Prairie Provinces	352	(3)	(3)	(3)	(3)	(3)	(3)	(3)	(3)	(3)	(3)
British Columbia	249	9,685	20,390	276,862	87,070	2,478,265	3,511,214	916,048	13,913	415,564	987,459
Yukon and Northwest Territories	2	(3)	(3)	(3)	(3)	(3)	(3)	(3)	(3)	(3)	(3)
Canada	2,106	124,304	256,900	3,175,123	1,537,247	27,801,486	38,496,873	9,078,253	182,665	5,070,760	9,256,207

[1] Industry coverage same as in Tables 30, 34, 36 and 38. [2] Includes sales and head offices. [3] Confidential, included in Canadian total.

TABLE 33. CANADA, PRINCIPAL STATISTICS OF THE MINING INDUSTRY[1], 1976-82

		Mineral Manufacturing Activity							Total Activity[2]		
		Production and Related Workers			Costs						
	Establish-ments (number)	Employees (number)	Man-hours paid (000)	Wages ($000)	Fuel and Electri-city ($000)	Materials and Supplies ($000)	Value of Production ($000)	Value Added ($000)	Employees (number)	Salaries and Wages ($000)	Value Added ($000)
1976	1,244	78,989	163,426	1,185,184	401,899	2,438,672	14,178,010	11,337,439	117,694	1,902,682	11,360,511
1977	1,232	79,902	167,884	1,342,508	473,202	2,715,468	16,400,460	13,211,792	119,061	2,137,523	13,246,689
1978	1,179	70,306	150,291	1,275,008	501,335	2,766,072	18,201,459	14,934,052	109,948	2,118,342	15,016,214
1979	1,150	72,580	152,560	1,493,773	600,448	3,252,991	23,546,742	19,693,303	115,245	2,492,715	19,899,635
1980	1,323	80,066	166,427	1,779,389	706,405	3,802,062	27,661,246	23,152,778	126,422	2,979,470	23,347,682
1981	1,361	81,136	167,308	2,053,761	888,554	4,266,634	28,460,030	23,304,775	129,251	3,439,945	23,346,991
1982	1,248	74,958	142,625	2,025,894	956,296	3,768,771	29,101,618	24,376,548	123,486	3,648,004	24,442,997

1 Cement manufacturing, lime manufacturers, clay and clay products (domestic clays) are included in the mineral manufacturing industries. Industry coverage is the same as in Tables 29, 31, 35 and 37. 2 Includes sales and head offices.

TABLE 34. CANADA, PRINCIPAL STATISTICS OF THE MINERAL MANUFACTURING INDUSTRIES[1], 1976-82

		Mineral Manufacturing Activity							Total Activity[2]		
		Production and Related Workers			Costs						
	Establish-ments (number)	Employees (number)	Man-hours paid (000)	Wages ($000)	Fuel and Electri-city ($000)	Materials and Supplies ($000)	Value of Production ($000)	Value Added ($000)	Employees (number)	Salaries and Wages ($000)	Value Added ($000)
1976	1,662	137,310	284,392	1,898,753	655,828	10,798,653	16,793,147	5,548,868	188,751	2,820,873	5,687,750
1977	1,616	138,700	288,409	2,110,400	798,486	12,743,217	19,725,082	6,489,111	189,576	3,114,744	6,594,794
1978	2,022	143,917	297,554	2,365,782	981,506	15,700,614	24,036,539	7,272,298	198,085	3,494,336	7,421,897
1979	2,115	145,929	308,770	2,614,816	1,118,146	19,116,369	28,318,690	8,522,128	202,695	3,910,454	8,669,240
1980	2,143	146,606	308,312	2,927,363	1,272,902	22,045,572	32,177,335	9,417,966	204,872	4,386,065	9,599,868
1981	2,124	140,914	293,781	3,187,784	1,560,453	28,125,138	39,495,229	10,862,006	203,051	4,932,893	11,062,937
1982	2,106	124,304	256,900	3,175,123	1,537,247	27,801,486	38,496,873	9,078,253	182,665	5,070,760	9,256,207

[1] Industry coverage is the same as in Tables 30, 32, 36 and 38. [2] Includes sales and head offices.

TABLE 35. CANADA, CONSUMPTION OF FUEL AND ELECTRICITY IN THE MINING INDUSTRY[1], 1982

	Unit	Metals	Nonmetals	Fuels	Total
Coal and coke	000 t	109	16	-	125
	$000	4,778	408	-	5,186
Gasoline	000 litres	23 445	15 562	14 851	53 868
	$000	9,352	6,172	5,215	20,739
Fuel oil, kerosene, diesel oil	000 litres	935 507	224 392	155 464	1 315 363
	$000	221,129	63,044	48,729	332,902
Liquefied petroleum gas	000 litres	102 395	6 000	7 910	116 205
	$000	19,310	1,520	1,541	22,371
Natural gas	000 m^3	144 660	675 784	175 000	995 444
	$000	19,503	70,531	14,200	104,234
Other fuels[2]	$000	1,733	1,718	800	4,251
Total value of fuels	$000	275,805	143,393	70,484	489,683
Electricity purchased	million kWh	9 891	1 782	5 780	17 453
	$000	232,137	57,567	176,911	466,614
Total value of fuels and electricity purchased, all reporting companies	$000	507,942	200,960	247,395	956,297

Note: Totals may not add due to rounding.
[1] Cement and lime manufacturing and manufacturers of clay products (domestic clays) are included under mineral manufacturing, Tables 36 and 38. Industry coverage is the same as in Tables 29, 31, 33 and 37. [2] Includes wood, manufactured gas, steam purchased and other miscellaneous fuels.
- Nil.

TABLE 36.   CANADA, CONSUMPTION OF FUEL AND ELECTRICITY IN THE MINERAL MANUFACTURING INDUSTRIES[1], 1982

	Unit	Primary Metal Industries	Nonmetallic Mineral Products Industries	Petroleum and Coal Products Industries	Total
Coal and coke	000 t	294	602	–	896
	$000	36,019	36,127	–	72,146
Gasoline	000 litres	15 530	26 761	4 416	46 707
	$000	5,805	10,253	2,016	18,074
Fuel oil, kerosene, diesel oil	000 litres	938 364	384 636	13 375	1 336 375
	$000	175,912	86,077	3,656	265,645
Liquefied petroleum gas	000 litres	48 180	19 430	3 184	70 794
	$000	8,899	3,823	310	13,032
Natural gas	000 m^3	2 266 900	1 405 800	1 033 200	4 705 900
	$000	292,061	173,877	122,797	588,735
Other fuels	$000	7,377	18,409	5,524	31,310
Total value of fuels	$000	526,073	328,566	134,303	988,942
Electricity purchased	million kWh	16 848	3 973	3 476	24 297
	$000	345,614	116,243	86,448	548,305
Total value of fuels and electricity purchased, all reporting companies	$000	871,687	444,809	220,751	1,537,247

[1] Industry coverage is the same as in Tables 30, 32, 34 and 38.
- Nil.

TABLE 37. CANADA, COST OF FUEL AND ELECTRICITY USED IN THE MINING INDUSTRY[1], 1976-82

	Unit	1976	1977	1978	1979	1980	1981	1982
Metals								
Fuel	$000	128,637	148,578	153,608	193,828	220,052	293,979	275,205
Electricity purchased	million kWh	11 326	11 713	10 739	11 459	11 024	10 494	9 891
	$000	107,318	135,014	132,100	153,905	174,837	209,316	232,137
Total cost of fuel and electricity	$000	235,955	283,591	285,708	347,733	394,889	503,295	507,942
Nonmetals[2]								
Fuel	$000	62,453	72,946	79,090	92,499	112,672	142,169	143,393
Electricity purchased	million kWh	1 959	2 457	2 082	2 244	2 269	2 100	1 782
	$000	23,401	29,510	35,141	42,982	48,336	56,297	57,567
Total cost of fuel and electricity	$000	85,854	102,456	114,231	135,481	161,008	198,466	200,960
Fuels								
Fuels	$000	12,015	15,117	19,774	23,988	32,582	46,991	70,484
Electricity purchased	million kWh	2 770	2 791	2 699	3 238	3 504	3 740	5 780
	$000	68,075	72,035	81,624	98,783	117,927	139,802	176,911
Total cost of fuel and electricity	$000	80,090	87,152	101,398	122,771	150,509	186,793	247,395
Total mining industry								
Fuel	$000	203,105	236,642	252,470	310,315	365,306	483,139	489,683
Electricity purchased	million kWh	16 055	16 961	15 520	16 941	16 797	16 334	17 453
	$000	198,794	236,559	248,865	295,670	341,100	405,415	466,614
Total cost of fuel and electricity	$000	401,899	473,201	501,335	605,985	706,406	888,554	956,297

[1] Cement and lime manufacturing and manufacture of clay products (domestic clays) are included in mineral manufacturing, Tables 36 and 38. Industry coverage is the same as in Tables 29, 31, 33 and 35. [2] Includes structural materials.

TABLE 38. CANADA, COST OF FUEL AND ELECTRICITY USED IN THE MINERAL MANUFACTURING INDUSTRIES[1], 1976-82

	Unit	1976	1977	1978	1979	1980	1981	1982
Primary metals								
Fuel	$000	224,928	279,172	336,684	357,775	421,426	538,175	526,073
Electricity purchased	million kWh	16,497	15,352	17,257	18,451	20,535	20,429	16,848
	$000	151,011	183,574	226,313	260,317	316,884	357,186	345,614
Total cost of fuel and electricity	$000	375,939	462,746	562,997	618,092	738,317	895,361	871,687
Nonmetallic mineral products								
Fuel	$000	162,312	181,952	221,855	280,846	271,481	333,061	328,566
Electricity purchased	million kWh	4,137	4,190	4,782	5,163	4,633	4,573	3,973
	$000	52,113	65,553	79,606	98,296	102,765	114,062	116,243
Total cost of fuel and electricity	$000	214,425	247,507	301,461	379,142	374,248	447,123	444,809
Petroleum and coal products								
Fuel	$000	30,474	42,184	61,891	74,968	88,311	137,463	134,303
Electricity purchased	million kWh	3,010	3,205	3,505	3,555	3,705	3,669	3,476
	$000	34,988	46,050	55,303	63,395	72,186	80,517	86,448
Total cost of fuel and electricity	$000	65,462	88,233	117,194	138,363	160,498	217,980	220,751
Total mineral manufacturing industries								
Fuel	$000	417,714	503,308	620,430	713,589	781,218	1,008,699	988,942
Electricity purchased	million kWh	23,644	22,747	25,544	27,169	28,873	28,671	24,297
	$000	238,112	295,177	361,222	422,008	491,834	551,765	548,305
Total cost of fuel and electricity	$000	655,826	798,486	981,652	1,135,597	1,273,063	1,560,464	1,537,247

[1] Industry coverage is the same as in Tables 30, 32, 34 and 36.

TABLE 39. CANADA, EMPLOYMENT, SALARIES AND WAGES IN THE MINING INDUSTRY[1], 1976-82

	Unit	1976	1977	1978	1979	1980	1981	1982
Metals								
Production and related workers	Number	49,834	49,414	39,977	41,541	47,592	49,586	44,261
Salaries and wages	$000	759,499	849,345	757,258	879,383	1,091,848	1,265,547	1,180,485
Annual average salary and wage	$	15,241	17,188	18,942	21,169	22,942	25,522	26,671
Administrative and office workers	Number	18,435	17,831	16,470	17,419	18,526	19,126	17,242
Salaries and wages	$000	352,847	377,714	358,680	428,639	504,316	585,120	585,249
Annual average salary and wage	$	19,140	21,183	21,778	24,608	27,222	30,593	33,943
Total metals								
Employees	Number	68,269	67,245	56,447	58,960	66,118	68,712	61,503
Salaries and wages	$000	1,112,346	1,227,059	1,115,938	1,308,022	1,596,165	1,850,667	1,765,734
Annual average salary and wage	$	16,294	18,248	19,770	22,185	24,141	26,933	28,710
Nonmetals								
Production and related workers	Number	16,447	16,812	16,133	16,633	16,645	15,666	12,848
Salaries and wages	$000	237,982	266,294	274,037	321,303	343,004	352,302	309,736
Annual average salary and wage	$	14,470	15,840	16,986	19,317	20,607	22,488	24,108
Administrative and office workers	Number	4,887	4,986	4,749	4,829	4,795	4,908	4,323
Salaries and wages	$000	82,861	89,757	95,659	106,776	116,932	128,852	129,116
Annual average salary and wage	$	16,955	18,002	20,143	22,114	24,386	26,253	29,867
Total nonmetals								
Employees	Number	21,334	21,798	20,882	21,462	21,440	20,574	17,171
Salaries and wages	$000	320,843	356,051	369,696	428,079	459,936	481,154	438,852
Annual average salary and wage	$	15,039	16,334	17,704	19,946	21,452	23,387	25,558
Fuels								
Production and related workers	Number	12,708	13,679	14,196	14,406	15,829	15,884	17,849
Salaries and wages	$000	187,704	226,869	243,713	293,087	344,537	435,911	535,673
Annual average salary and wage	$	14,771	16,585	17,168	20,345	21,766	27,443	30,011
Administrative and office workers	Number	15,383	16,342	18,423	20,417	23,035	24,081	26,963
Salaries and wages	$000	281,789	327,544	388,995	463,527	578,832	672,213	907,745
Annual average salary and wage	$	18,318	20,043	21,115	22,703	25,128	27,915	33,666
Total fuels								
Employees	Number	28,091	30,021	32,619	34,823	38,864	39,965	44,812
Salaries and wages	$000	469,493	554,413	632,708	756,614	923,369	1,108,124	1,443,418
Average annual salary and wage	$	16,713	18,468	19,397	21,727	23,759	27,727	32,211
Total mining								
Production and related workers	Number	78,989	79,905	70,306	72,580	80,066	81,136.	74,958
Salaries and wages	$000	1,185,184	1,342,508	1,275,008	1,493,773	1,779,389	2,053,761	2,025,894
Average annual salary and wage	$	15,004	16,801	18,135	20,581	22,224	25,313	27,027
Administrative and office workers	Number	38,705	39,159	39,642	42,665	46,356	48,115	48,528
Salaries and wages	$000	717,498	795,015	843,335	998,942	1,200,081	1,386,184	1,622,110
Annual average salary and wage	$	18,538	20,302	21,274	23,414	25,888	28,810	33,426
Total mining								
Employees	Number	117,694	119,064	109,948	115,245	126,422	129,251	123,486
Salaries and wages	$000	1,902,682	2,137,523	2,118,343	2,492,715	2,979,470	3,439,945	3,648,004
Annual average salary and wage	$	16,166	17,954	19,267	21,630	23,568	26,614	29,542

[1] Does not include cement and lime manufacturing and clay products (domestic clays) manufacturing. These industries are included in the following table under "Nonmetallic mineral products industries".

TABLE 40. CANADA, EMPLOYMENT, SALARIES AND WAGES IN THE MINERAL MANUFACTURING INDUSTRIES, 1976-82

	Unit	1976	1977	1978	1979	1980	1981	1982
Primary metal industries								
Production and related workers	Number	88,939	91,683	93,798	95,942	97,530	92,337	82,186
Salaries and wages	$000	1,241,893	1,399,390	1,544,412	1,725,904	1,980,423	2,120,019	2,157,186
Annual average salary and wage	$	13,963	15,263	16,465	17,989	20,306	22,960	26,248
Administrative and office workers	Number	28,102	27,536	28,198	30,812	28,920	32,831	31,029
Salaries and wages	$000	511,236	545,957	597,544	713,279	787,022	938,790	1,010,847
Annual average salary and wage	$	18,192	19,827	21,191	23,149	27,214	28,595	32,577
Total primary metal industries								
Employees	Number	117,041	119,219	121,996	126,754	126,450	125,168	113,215
Salaries and wages	$000	1,753,128	1,945,347	2,140,956	2,432,183	2,767,445	3,058,809	3,168,033
Annual average salary and wage	$	14,979	16,317	17,549	19,188	21,886	24,438	27,982
Nonmetallic mineral products industries								
Production and related workers	Number	41,272	39,321	41,297	41,813	40,799	40,145	33,997
Salaries and wages	$000	529,264	564,444	638,152	710,622	743,254	818,566	751,915
Annual average salary and wage	$	12,824	14,355	15,452	16,995	18,217	20,390	22,117
Administrative and office workers	Number	13,749	13,187	14,439	14,935	15,287	15,124	13,952
Salaries and wages	$000	218,164	229,855	264,166	297,211	333,815	369,899	383,405
Annual average salary and wage	$	15,868	17,430	18,295	19,900	21,837	24,458	27,480
Total nonmetallic mineral products								
Employees	Number	55,021	52,508	55,736	56,748	56,086	55,269	47,949
Salaries and wages	$000	747,428	794,299	902,318	1,007,833	1,077,069	1,188,455	1,135,320
Annual average salary and wage	$	13,584	15,127	16,189	17,760	19,203	21,503	23,678
Petroleum and coal products industries								
Production and related workers	Number	7,099	7,696	8,822	8,174	8,277	8,432	8,121
Salaries and wages	$000	127,594	146,566	183,218	185,290	203,686	249,199	266,022
Annual average salary and wage	$	17,974	19,044	20,768	22,668	24,609	29,554	32,757
Administrative and office workers	Number	9,590	10,153	11,531	11,019	11,769	14,182	13,380
Salaries and wages	$000	192,722	228,532	267,844	285,148	337,865	436,430	501,385
Annual average salary and wage	$	20,096	22,509	23,228	25,887	28,708	30,773	37,473
Total petroleum and coal products								
Employees	Number	16,689	17,849	20,353	19,193	20,046	22,614	21,501
Salaries and wages	$000	320,316	375,098	451,062	470,438	541,551	685,629	767,407
Annual average salary and wage	$	19,193	21,015	22,162	24,511	27,015	30,319	35,692
Total mineral manufacturing								
Production and related workers	Number	137,310	138,700	143,917	145,929	146,606	140,914	124,304
Salaries and wages	$000	1,898,751	2,110,400	2,365,782	2,621,816	2,927,363	3,187,784	3,175,123
Annual average salary and wage	$	13,828	15,216	16,439	17,966	19,968	22,622	25,543
Administrative and office workers	Number	51,441	50,876	54,168	56,766	55,976	62,137	58,359
Salaries and wages	$000	922,122	1,004,344	1,129,554	1,295,638	1,458,702	1,745,109	1,895,637
Annual average salary and wage	$	17,926	19,741	20,853	22,824	26,059	28,085	32,482
Total mineral manufacturing industries								
Employees	Number	188,751	189,576	198,085	202,695	202,582	203,051	182,665
Salaries and wages	$000	2,820,872	3,114,744	3,494,336	3,910,454	4,386,065	4,932,893	5,070,760
Annual average salary and wage	$	14,945	16,430	17,641	19,292	21,651	24,294	27,760

TABLE 41. CANADA, NUMBER OF WAGE EARNERS EMPLOYED IN THE MINING INDUSTRY, (SURFACE, UNDERGROUND AND MILL), 1976-82

	1976	1977	1978	1979	1980	1981	1982
Metals							
Surface	16,143	16,115	12,901	12,664	14,347	14,043	12,133
Underground	20,043	19,482	15,682	15,906	19,308	19,784	18,673
Mill	13,648	13,817	11,394	12,971	13,937	15,759	13,455
Total	49,834	49,414	39,977	41,541	47,592	49,586	44,261
Nonmetals							
Surface	7,264	7,166	6,660	6,877	6,510	6,015	4,833
Underground	2,180	2,245	2,275	2,370	2,550	2,606	2,055
Mill	7,003	7,401	7,198	7,386	7,585	7,045	5,960
Total	16,447	16,812	16,133	16,633	16,645	15,666	12,848
Fuels							
Surface	9,705	10,510	11,045	11,535	12,929	12,958	14,623
Underground	3,003	3,169	3,151	2,871	2,900	2,926	3,226
Total	12,708	13,679	14,196	14,406	15,829	15,884	17,849
Total mining industry							
Surface	33,112	33,791	30,606	31,076	33,786	33,016	31,589
Underground	25,226	24,896	21,108	21,147	24,758	25,316	23,954
Mill	20,651	21,218	18,592	20,357	21,522	22,804	19,415
Total	78,989	79,905	70,306	72,580	80,066	81,136	74,958

TABLE 42. CANADA, MINE AND MILL WORKERS BY SEX, 1982

	Mine workers				Mill workers		Total	
	Underground		Surface					
	Male	Female	Male	Female	Male	Female	Male	Female
Metallic minerals								
Gold-quartz	3,297	4	1,101	38	1,334	35	5,732	77
Silver-lead-zinc	2,173	1	1,114	32	1,454	38	4,741	71
Nickel-copper-gold-silver	10,575	17	5,588	127	4,816	242	20,979	386
Iron ore	192	2	2,029	49	4,133	173	6,354	224
Uranium	2,189	5	1,356	46	748	57	4,293	108
Miscellaneous metal mines	218	-	602	51	391	34	1,211	85
Total	18,644	29	11,790	343	12,876	579	43,310	951
Industrial minerals								
Asbestos	282	-	1,184	7	2,374	91	3,840	98
Gypsum	81	-	368	1	58	-	507	1
Peat	-	-	620	15	431	7	1,051	22
Potash	1,313	14	68	1	1,495	33	2,876	48
Sand and gravel	-	-	954	6	20	1	974	7
Stone	6	-	1,273	7	253	2	1,532	9
Miscellaneous nonmetals	359	-	326	3	1,172	23	1,857	26
Total	2,041	14	4,793	40	5,803	157	12,637	211
Mining Total	20,685	43	16,583	383	18,679	736	55,947	1,162

- Nil.

TABLE 43. CANADA, LABOUR COSTS IN RELATION TO TONNES MINED, METAL MINES, 1980-82

Type of Metal Mine	Number of Wage Earners	Total Wages ($000)	Average Annual Wage ($)	Tonnage of Ore Mined (kilotonnes)	Average Annual Tonnes Mined per Wage Earner	Wage Cost per Tonne Mined ($)
1980						
Gold-quartz	3,946	85,102	21,567	6 346	1 608	13.41
Nickel-copper-gold-silver	18,377	398,677	21,694	121 399	6 606	3.28
Silver-lead-zinc	3,862	91,265	23,632	16 219	4 200	5.63
Iron ore	3,081	80,637	26,172	123 107	39 957	0.66
Uranium	3,577	87,594	24,488	7 152	2 000	12.25
Miscellaneous metals	812	20,604	25,374	15 871	19 546	1.30
Total	33,655	763,879	22,697	290 095	8 620	2.63
1981						
Gold-quartz	4,349	105,802	24,328	6 810	1 566	15.54
Nickel-copper-gold-silver	18,398	433,026	23,537	137 710	7 485	3.14
Silver-lead-zinc	3,832	105,381	27,500	15 964	4 166	6.60
Iron ore	2,755	86,303	31,326	118 579	43 041	0.73
Uranium	3,796	107,707	28,374	7 454	1 964	14.45
Miscellaneous metals	697	17,586	25,231	15 014	21 541	1.17
Total	33,827	855,805	25,299	301 530	8 914	2.84
1982						
Gold-quartz	4,440	125,178	28,193	8 368	1 885	14.96
Nickel-copper-gold-silver	16,307	365,743	22,429	117 833	7 226	3.10
Silver-lead-zinc	3,320	106,834	32,179	14 113	4 251	7.57
Iron ore	2,272	66,205	29,139	81 963	36 075	0.81
Uranium	3,596	124,024	34,489	7 609	2 116	16.30
Miscellaneous metals	871	25,987	29,836	8 477	9 732	3.07
Total	30,806	813,971	26,422	238 362	7 738	3.41

TABLE 44. CANADA, PERSON-HOURS PAID, PRODUCTION AND RELATED WORKERS, TONNES OF ORE MINED AND ROCK QUARRIED, METAL MINES AND NONMETALLIC MINERAL OPERATIONS, 1976-82

	Unit	1976	1977	1978	1979	1980	1981	1982
Metal mines[1]								
Ore mined	million tonnes	296.5	299.5	248.1	274.8	290.1	301.5	238.4
Person-hours paid[2]	million	100.6	101.2	84.9	85.1	97.5	100.6	80.4
Person-hours paid per tonne mine	number	0.34	0.34	0.34	0.31	0.34	0.33	0.34
Tonnes mined per person-hour paid	tonnes	2.95	2.96	2.92	3.23	2.98	3.00	2.97
Nonmetallic mineral operations[3]								
Ore mined and rock quarried	million tonnes	162.0	200.2	200.4	192.1	185.0	164.8	113.4
Person-hours paid[2]	million	26.9	27.7	26.3	27.8	26.5	23.5	18.0
Person-hours paid per tonne mine	number	0.17	0.14	0.13	0.14	0.14	0.14	0.16
Tonnes mined per person-hour paid	tonnes	6.02	7.23	7.62	6.91	6.98	7.01	6.30

[1] Excludes placer mining. [2] Man-hours paid for production and related workers only. [3] Includes asbestos, potash, gypsum and stone.

TABLE 45. CANADA, AVERAGE WEEKLY WAGES AND HOURS WORKED, HOURLY-RATED EMPLOYEES IN MINING, MANUFACTURING AND CONSTRUCTION INDUSTRIES, 1977-83

	1977	1978	1979	1980	1981	1982	1983[1]
Mining							
Average hours per week	40.6	40.5	41.1	40.8	40.4	39.6	38.8
Average weekly wage ($)	329.45	354.51	396.58	440.61	494.62	551.68	552.79
Metals							
Average hours per week	39.8	39.4	40.4	40.1	40.2	39.0	38.3
Average weekly wage ($)	325.75	344.94	387.14	425.08	485.03	535.92	565.60
Mineral fuels							
Average hours per week	41.3	41.0	40.8	41.2	41.3	42.1	39.7
Average weekly wage ($)	333.51	367.34	410.38	476.30	553.71	631.91	626.12
Nonmetals							
Average hours per week	40.3	40.5	40.3	39.5	38.7	37.2	37.5
Average weekly wage ($)	301.92	326.23	366.03	402.98	445.02	479.44	468.05
Manufacturing							
Average hours per week	38.7	38.8	38.8	38.5	38.5	37.7	38.4
Average weekly wage ($)	246.63	265.06	287.82	314.80	352.08	384.79	406.76
Construction							
Average hours per week	38.7	39.0	39.4	39.0	38.9	38.1	36.9
Average weekly wage ($)	378.50	400.58	433.51	470.45	531.54	564.33	512.26

Note: Wages reflect seasonally adjusted figures.
[1] Ten-month average; new time series.

TABLE 46. CANADA, AVERAGE WEEKLY WAGES OF HOURLY-RATED EMPLOYEES IN THE MINING INDUSTRY, IN CURRENT AND 1971 DOLLARS, 1977-83

	1977	1978	1979	1980	1981	1982	1983[1]
Current dollars							
All mining	329.45	354.51	396.58	440.61	494.62	551.68	552.79
Metals	325.75	344.94	387.14	425.08	485.03	535.92	565.60
Mineral fuels	333.51	367.34	414.96	476.30	553.11	631.91	626.12
Coal	303.53	323.49	362.20	430.16	485.03	562.12	564.18
Nonmetals except fuel	301.92	326.23	330.47	402.98	445.02	479.44	468.05
1971 dollars							
All mining	204.88	202.35	207.42	209.22	208.79	210.16	199.04
Metals	202.58	196.88	202.48	226.16	244.74	204.16	203.65
Mineral fuels	207.41	209.67	217.03	220.82	233.48	240.73	225.44
Coal	188.76	184.64	189.44	204.25	204.74	214.14	203.14
Industrial minerals	187.76	186.20	172.84	191.35	187.85	182.64	168.53

Note: Wages reflect seasonally adjusted figures.
[1] Ten month average; new time series.

TABLE 47. CANADA, INDUSTRIAL FATALITIES PER THOUSAND WORKERS, BY INDUSTRY GROUPS 1981-83[1]

	Fatalities (number)			Number of Workers (000)			Rate per 1,000 Workers[2]		
	1981	1982	1983p	1981	1982	1983p	1981	1982	1983p
Agriculture	17	19	17	151.0	149.0	156.7	0.11	0.13	0.11
Forestry	60	65	59	65.6	54.3	55.3	0.91	1.20	1.07
Fishing	20	18	15	13.8	11.4	14.7	1.45	1.58	1.02
Mining	126	146	90	178.0	155.5	146.7	0.71	0.94	0.61
Manufacturing	146	170	128	1,883.9	1,709.2	1,712.6	0.08	0.10	0.07
Construction	174	140	100	475.1	409.7	351.4	0.37	0.34	0.28
Transportation	198	172	121	849.6	826.4	79.0	0.23	0.21	0.15
Trade	60	66	54	1,629.0	1,575.9	1,491.2	0.04	0.04	0.04
Finance	9	5	2	533.1	534.7	520.0	0.02	0.01	0.04
Service	83	79	56	2,932.4	2,965.9	2,844.0	0.03	0.03	0.02
Public administration	62	53	48	628.3	646.6	655.8	0.10	0.08	0.07
Unknown	5	7	2	..	..	..	..	..	..
Total	960	940	694	9,339.8	9,038.6	8,738.4	0.10	0.10	0.08

Note: See footnotes on next table.
[1] Includes fatalities resulting from occupational chest diseases such as silicosis, lung cancer, etc. [2] The rates may be understated because only 80 per cent of workers in the Statistics Canada employment estimates are covered by workers' compensation.
p Preliminary; .. Not available.

TABLE 48. CANADA, INDUSTRIAL FATALITIES PER THOUSAND WORKERS, BY INDUSTRY GROUPS, 1977-83

	1977	1978	1979	1980	1981	1982	1983P
Agriculture	0.11	0.05	0.10	0.05	0.11	0.13	0.11
Forestry	0.92	1.28	1.51	1.11	0.91	1.20	1.07
Fishing[1]	2.37	1.44	1.25	1.47	1.45	1.58	1.02
Mining[2]	0.92	0.82	0.93	0.99	0.71	0.94	0.61
Manufacturing	0.10	0.10	0.09	0.08	0.08	0.10	0.07
Construction	0.37	0.38	0.38	0.40	0.37	0.34	0.28
Transportation[3]	0.22	0.25	0.26	0.26	0.23	0.21	0.15
Trade	0.05	0.04	0.04	0.05	0.04	0.04	0.04
Finance[4]	0.02	0.01	0.01	0.02	0.02	0.01	--
Service[5]	0.03	0.02	0.03	0.03	0.03	0.03	0.02
Public administration	0.08	0.12	0.10	0.07	0.10	0.08	0.07
Total	0.11	0.12	0.12	0.12	0.10	0.10	0.08

[1] Includes trapping and hunting. [2] Includes quarrying and oil wells. [3] Includes storage, communication, electric power and water utilities and highway maintenance. [4] Includes insurance and real estate. [5] Includes community, business and personal service.
P Preliminary; -- Too small to be expressed.

TABLE 49. CANADA, INDUSTRIAL FATALITIES BY OCCUPATIONAL INJURIES AND ILLNESSES[1], 1981-83

	Occupational Injuries			Occupational Illnesses			Total		
	1981	1982	1983P	1981	1982	1983P	1981	1982	1983P
Agriculture	12	13	12	0	0	0	12	13	12
Forestry	49	54	54	0	0	0	49	54	54
Fishing	20	17	15	0	0	0	20	17	15
Mining	70	96	39	52	48	48	122	144	87
Manufacturing	83	99	82	40	49	32	123	148	114
Construction	149	107	79	6	13	14	155	120	93
Transportation	176	155	107	1	6	3	177	161	110
Trade	47	57	42	1	0	1	48	57	43
Finance	6	4	2	0	0	0	6	4	2
Service	64	57	49	3	3	2	67	60	51
Public administration	48	42	39	2	1	1	50	43	40
Unknown	1	0	0	0	0	0	1	0	0
Total	725	701	520	105	120	101	830	821	621

[1] Excludes the Province of Quebec for which data is unavailable.
P Preliminary.

TABLE 50. CANADA, NUMBER OF STRIKES AND LOCKOUTS BY INDUSTRIES, 1981-83

	1981			1982			1983		
	Strikes and Lockouts	Workers Involved	Duration in Person-days	Strikes and Lockouts	Workers Involved	Duration in Person-days	Strikes and Lockouts	Workers Involved	Duration in Person-days
Agriculture	3	65	7,750	3	64	7,320	2	26	770
Forestry	14	3,292	349,400	3	215	7,840	5	1,326	13,890
Fishing and trapping	1	400	330	0	0	0	1	3,000	3,000
Mines	42	24,359	580,720	8	12,686	257,140	12	11,889	178,390
Manufacturing	423	152,207	4,638,290	292	63,959	1,690,560	311	64,206	1,385,290
Construction	44	5,780	43,280	63	94,228	2,199,610	24	9,394	243,680
Transportation and utilities	101	58,135	1,513,970	67	24,005	565,740	63	15,257	275,000
Trade	90	4,886	149,170	72	4,465	171,180	74	14,831	251,690
Finance, insurance and real estate	18	3,480	294,760	15	746	49,620	17	606	9,600
Service	221	57,248	577,400	110	27,846	415,380	104	168,376	1,770,710
Public administration	90	17,696	717,420	43	36,088	251,030	32	40,398	311,940
Various industries	1	6,000	6,000	1	180,000	180,000	0	0	0
All industries	1,048	338,548	8,878,490	677	444,302	5,795,420	645	329,309	4,443,960

TABLE 51. CANADA, NUMBER OF STRIKES AND LOCKOUTS BY MINING AND MINERAL MANUFACTURING INDUSTRIES, 1981-83

	1981			1982			1983		
	Strikes and Lockouts	Workers Involved	Duration in Person-days	Strikes and Lockouts	Workers Involved	Duration in Person-days	Strikes and Lock-outs	Workers Involved	Duration in Person-days
Mines	42	24,359	580,720	8	12,686	257,140	12	11,889	178,390
Metal	25	11,457	248,930	2	10,211	248,300	6	6,046	91,500
Mineral fuels	9	11,159	306,690	2	2,400	4,670	3	4,991	80,950
Nonmetals	5	1,674	16,130	0	0	0	2	847	5,540
Quarries	3	69	8,970	4	75	4,170	1	5	400
Mineral manufacturing	62	30,770	1,553,000	29	6,839	291,600	32	4,334	118,540
Primary metals	29	27,169	1,429,150	11	4,259	199,900	15	2,609	88,070
Nonmetallic mineral products	33	3,601	123,850	17	2,576	91,600	17	1,725	30,470
Petroleum and coal products	0	0	0	1	4	100	0	0	0

TABLE 52. CANADA, SOURCE OF ORES HOISTED OR REMOVED FROM SELECTED TYPES OF MINES, 1980-82

Mines	1980			1981			1982		
	Under-ground	Open-pit	Total	Under-ground	Open-pit	Total	Under-ground	Open-pit	Total
	(kilotonnes)								
Asbestos	1 997	26 106	28 103	1 789	23 874	25 664	1 308	16 184	17 492
Gold-quartz	5 193	1 153	6 346	5 835	975	6 810	6 710	1 657	8 367
Gypsum	1 062	6 549	7 611	685	5 535	6 220	475	5 355	5 830
Iron ore	3 222	119 886	123 107	3 269	115 309	118 579	2 448	79 515	81 963
Nickel-copper-gold-silver	30 840	90 559	121 399	31 193	106 516	137 710	21 431	96 402	117 833
Silver-lead-zinc	9 822	6 397	16 219	9 943	6 021	15 964	9 950	4 163	14 113
Uranium	5 981	1 171	7 152	6 664	790	7 454	6 900	709	7 609
Miscellaneous metals	1 491	14 381	15 871	1 518	13 496	15 014	1 517	6 959	8 476
Total	59 608	266 201	325 809	60 896	272 516	333 415	50 739	210 944	261 683
Percentage	18.3	81.7	100.0	18.3	81.7	100.0	19.4	80.6	100.0

TABLE 53. CANADA, SOURCE OF MATERIAL HOISTED OR REMOVED FROM METAL MINES, 1982

	Underground		Open-pit		
	Ore	Waste	Ore	Waste	Overburden
	(kilotonnes)				
Gold-quartz	6 710	767	1 657	1 228	10 558
Nickel-copper-gold-silver	21 431	2 152	96 402	98 990	52 046
Silver-lead-zinc	9 950	1 089	4 163	22 152	16 537
Iron	2 448	71	79 515	30 223	12 568
Uranium	6 900	300	709	1 181	-
Miscellaneous metals	1 517	101	6 959	13 645	10
Total	48 956	4 480	189 405	167 419	91 719

- Nil.

TABLE 54. CANADA, ORE MINED AND ROCK QUARRIED IN THE MINING INDUSTRY, 1976-82

	1976	1977	1978	1979	1980	1981	1982
				(kilotonnes)			
Metals							
Gold-quartz	5 921	5 768	5 914	5 478	6 346	6 810	8 368
Silver-lead-zinc	14 309	16 730	15 859	15 078	16 219	15 964	14 113
Nickel-copper-gold-silver	125 062	129 361	109 613	109 437	121 399	137 709	117 833
Iron	133 073	127 057	96 323	130 799	123 107	118 579	81 963
Uranium	3 663	5 014	6 126	6 141	7 152	7 454	7 608
Miscellaneous metals	14 499	15 599	14 221	7 822	15 871	15 014	8 477
Total	296 527	299 528	248 056	274 755	290 095	301 530	238 362
Nonmetals							
Asbestos	31 055	31 912	28 788	31 522	28 103	25 664	17 493
Potash	20 277	24 813	24 856	25 511	26 988	30 344	16 946
Gypsum	5 978	7 216	8 393	8 310	7 611	6 220	5 830
Rock salt	5 080	4 974	5 050	5 639	5 321	4 927	5 723
Total	62 390	68 915	67 087	70 982	68 023	67 155	45 992
Structural materials							
Stone, all kinds quarried[1]	87 876	120 163	122 144	109 719	103 366	86 860	59 181
Stone used to make cement	13 350	12 614	13 051	13 982	14 138	14 047	10 593
Stone used to make lime	3 442	3 534	3 178	3 028	4 751	1 626	3 411
Total	104 668	136 310	138 373	126 729	122 255	102 533	73 085
Total ore mined and rock quarried	463 585	504 753	453 516	472 466	480 373	471 218	357 439

[1] Excludes stone used to manufacture cement and lime.

TABLE 55. CANADA, EXPLORATION AND CAPITAL EXPENDITURES IN THE MINING INDUSTRY[1], BY PROVINCES AND TERRITORIES, 1980-82

| | | Capital | | | | | | | Repair | | | | | |
| | | Construction | | | | | | | | | | | | |
		On-Property Exploration	On-Property Development	Structures	Total	Machinery and Equipment	Total Capital Construction	Machinery and Equipment	Total Repair	Total Capital and Repair	Outside or General Exploration	Land and Mining Rights	Total, all Expenditures
							($ million)						
Atlantic Provinces	1980	2.7	60.3	22.4	85.4	60.0	145.4	14.8	168.2	183.0	328.4	35.5	0.2	364.1
	1981	6.3	63.5	80.7	150.5	115.4	265.9	11.0	185.2	196.2	462.1	43.7	1.5	507.3
	1982P	9.4	78.5	103.7	191.6	105.1	296.7	16.3	174.1	190.4	487.1	20.0	0.7	507.8
Quebec	1980	15.6	151.6	81.3	248.5	98.8	347.3	45.4	281.8	327.2	674.5	58.5	9.2	742.2
	1981	28.0	156.1	106.5	290.6	135.9	426.5	49.3	261.7	311.0	737.5	81.7	2.1	821.3
	1982P	32.5	135.5	54.6	222.6	81.7	304.3	43.5	197.0	240.5	544.8	61.5	0.4	606.7
Ontario	1980	12.1	179.3	124.5	315.9	120.2	436.1	66.2	235.9	302.1	738.2	58.5	3.4	800.1
	1981	17.9	206.2	148.8	372.9	177.2	550.1	70.6	281.7	352.3	902.4	79.5	6.4	988.3
	1982P	21.6	206.0	153.8	381.4	115.3	496.7	30.9	268.4	299.3	796.0	66.4	1.7	864.1
Manitoba	1980	(2)	(2)	(2)	39.2	11.3	50.5	6.6	44.2	50.8	101.3	21.2	0.3	122.8
	1981	8.3	27.3	13.5	49.1	34.0	83.1	5.1	44.2	49.3	132.4	20.6	0.3	153.3
	1982P	(2)	(2)	(2)	47.3	17.0	64.3	4.1	29.6	33.7	98.0	13.9	(2)	(2)
Saskatchewan	1980	7.0	40.4	62.1	109.5	87.1	196.6	9.1	90.3	99.4	296.0	56.4	4.7	357.1
	1981	20.2	39.0	101.6	160.8	175.7	336.5	11.5	120.5	132.0	468.5	45.4	8.1	522.0
	1982P	16.2	42.2	163.0	221.4	189.6	411.0	9.6	117.4	127.0	538.0	44.4	1.4	583.8
Alberta	1980	(2)	(2)	(2)	34.5	41.8	76.3	1.2	57.5	58.7	135.0	14.2	(2)	(2)
	1981	2.6	20.1	52.6	75.3	52.2	127.5	0.9	59.0	59.9	187.4	23.9	(2)	(2)
	1982P	(2)	(2)	(2)	65.5	141.5	207.0	3.6	76.3	79.9	286.3	21.9	(2)	(2)
British Columbia	1980	31.1	154.1	302.6	487.8	233.3	721.1	21.8	232.5	254.3	975.4	91.0	3.7	1,070.1
	1981	34.9	139.7	490.3	664.9	197.2	862.1	24.1	338.9	363.0	1,225.1	111.7	1.5	1,338.3
	1982P	19.5	186.1	474.8	680.4	203.3	883.7	25.4	317.9	343.3	1,227.0	61.0	1.8	1,289.8
Yukon and Northwest Territories	1980	8.6	26.9	99.2	134.7	82.3	217.0	4.7	50.4	55.1	272.1	68.3	(2)	(2)
	1981	16.3	43.4	155.3	215.0	106.5	321.5	5.4	57.4	63.8	384.3	78.2	(2)	(2)
	1982P	7.7	35.0	36.7	79.4	80.4	159.8	7.6	56.2	63.8	223.6	73.3	(2)	(2)
Canada	1980	85.4	646.8	723.3	1,455.5	734.8	2,190.3	169.8	1,160.8	1,330.6	3,520.9	403.6	43.6	3,968.1
	1981	134.5	695.3	1,149.3	1,979.1	994.1	2,973.2	177.9	1,348.6	1,526.5	4,499.7	484.7	29.8	5,014.2
	1982P	115.6	724.5	1,048.9	1,889.0	933.9	2,822.9	141.0	1,236.9	1,377.9	4,200.8	362.4	32.5	4,595.7

[1] Excludes the crude oil and natural gas industries as well as overhead expenditures; (2) Confidential, included in total.
P Preliminary.

TABLE 56. CANADA, EXPLORATION AND CAPITAL EXPENDITURES[1] IN THE MINING INDUSTRY, BY TYPE OF MINING, 1980-82

		Capital						Repair					
		On-Property Exploration	On-Property Development	Structures	Total	Construction Machinery and Equipment	Total Capital	Construction	Machinery and Equipment	Total Repair	Total Capital and Repair	Outside or General Exploration	Land and Mining Rights	Total, all Expenditures
		($ million)											
Metal Mining														
Gold	1980	22.6	63.4	36.7	122.7	38.2	160.9	6.8	27.9	34.7	195.6	20.0	(2)	215.6
	1981	21.7	111.8	179.7	313.2	96.3	409.5	13.9	44.7	58.6	468.1	40.1	2.7	510.9
	1982P	27.8	118.0	135.4	281.2	98.2	379.4	11.5	47.2	58.7	438.1	10.8	(2)	448.9
Copper-gold-silver	1980	24.1	93.3	187.6	305.0	185.9	490.9	24.2	211.0	235.2	726.1	8.4	(2)	734.5
	1981	28.2	91.2	157.1	276.5	161.6	438.1	29.7	292.2	321.9	760.0	13.5	0.6	774.1
	1982P	28.9	82.0	42.9	153.8	52.2	206.0	22.3	241.9	264.2	470.2	12.3	(2)	482.5
Silver-lead-zinc	1980	9.4	35.5	97.1	141.9	86.1	228.0	7.0	61.4	68.4	296.4	10.7	(2)	307.1
	1981	21.5	55.2	95.4	172.1	104.7	276.8	6.8	75.4	82.2	359.0	15.4	0.6	375.0
	1982P	11.3	48.8	27.0	87.1	57.2	144.3	13.6	106.0	119.6	263.9	6.2	(2)	270.1
Iron	1980	(2)	(2)	26.3	123.9	44.1	168.0	39.2	298.0	337.2	505.2	(2)	(2)	505.2
	1981	(2)	(2)	19.9	127.9	60.4	188.3	35.6	302.8	338.4	526.7	(2)	(2)	526.7
	1982P	(2)	(2)	23.3	98.1	40.0	138.1	37.7	232.7	270.4	408.5	(2)	(2)	408.5
Other metal mining	1980	14.7	178.8	214.3	407.8	109.3	517.1	60.1	169.1	229.2	746.3	(2)	(2)	746.3
	1981	37.3	198.6	204.0	439.9	149.1	589.0	65.8	184.8	250.6	839.6	(2)	(2)	839.6
	1982P	21.8	194.3	172.3	388.4	118.7	507.1	27.3	176.9	204.2	711.3	(2)	(2)	711.3
Total metal mining	1980	(2)	(2)	562.0	1,101.3	463.6	1,564.9	137.3	767.4	904.7	2,469.6	54.4	24.9	2,548.9
	1981	(2)	(2)	656.1	1,329.6	572.1	1,901.7	151.8	899.9	1,051.7	2,953.4	97.0	4.4	3,054.8
	1982P	(2)	(2)	400.9	1,008.6	366.3	1,374.9	112.4	804.7	917.1	2,292.0	35.2	1.2	2,328.4
Nonmetal mining														
Asbestos	1980	0.7	56.4	8.0	65.1	23.1	88.2	7.4	106.0	113.4	201.6	(2)	(2)	201.6
	1981	(2)	(2)	5.5	53.7	15.3	69.0	4.0	79.5	83.5	152.5	(2)	(2)	152.5
	1982P	(2)	(2)	3.2	36.6	8.9	45.5	3.7	55.7	59.4	104.9	(2)	(2)	104.9
Other non-metal mining	1980	9.6	120.8	150.9	281.3	244.5	525.8	25.1	287.1	312.2	838.0	(2)	(2)	838.0
	1981	21.3	85.4	487.9	594.1	402.4	996.5	22.0	368.3	390.3	1,386.8	(2)	(2)	1,399.4
	1982P	19.6	174.4	644.2	838.2	554.4	1,392.6	24.9	376.1	401.0	1,793.6	(2)	(2)	1,822.2
Total non-metal mining	1980	10.3	177.2	158.9	346.4	267.6	614.0	32.5	393.1	425.6	1,039.6	18.4	9.4	1,067.4
	1981	(2)	(2)	492.9	647.8	417.7	1,065.5	26.0	447.8	473.8	1,539.3	38.5	12.6	1,590.4
	1982P	(2)	(2)	647.4	874.8	563.3	1,438.1	28.6	431.8	460.4	1,898.5	31.3	28.6	1,998.4
Metal and nonmetal exploration	1980	(2)	(2)	2.4	7.8	3.6	11.4	–	0.3	0.3	11.7	330.8	9.3	351.8
	1981	(2)	(2)	0.3	1.7	4.3	6.0	0.1	0.9	1.0	7.0	349.2	12.8	369.0
	1982P	(2)	(2)	0.6	5.6	4.3	9.9	–	0.4	0.4	10.3	295.9	2.7	308.9
Total mining	1980	85.4	646.8	723.3	1,455.5	734.8	2,190.3	169.8	1,160.8	1,330.6	3,520.9	403.6	43.6	3,968.1
	1981	134.5	695.3	1,149.3	1,979.1	994.1	2,773.2	177.9	1,348.6	1,526.5	4,499.7	484.7	29.8	5,014.2
	1982P	115.6	724.5	1,048.9	1,889.0	933.9	2,822.9	141.0	1,236.9	1,377.9	4,200.8	362.4	32.5	4,595.7

[1] Excludes expenditures in the petroleum and natural gas industries as well as overhead expenditures. (2) Confidential, included in total.
P Preliminary; – Nil.

TABLE 57. CANADA, DIAMOND DRILLING IN THE MINING INDUSTRY, BY MINING COMPANIES WITH OWN EQUIPMENT AND BY DRILLING CONTRACTORS, 1980-82

		1980			1981			1982		
		Exploration	Other	Total	Exploration	Other	Total	Exploration	Other	Total
					(metres)					
Metal mining										
Gold-quartz	Own equipment	27 775	1 000	28 775	45 162	1 524	46 686	57 957	3 262	61 219
	Contractors	154 812	4 048	158 860	234 432	25 079	259 511	227 202	-	227 202
	Total	182 587	5 048	187 635	279 594	26 603	306 197	285 159	3 262	288 421
Nickel-copper-	Own equipment	239 469	-	239 469	318 530	223	318 753	111 189	13 423	124 612
gold-silver	Contractors	286 536	40 605	327 141	355 586	1 373	356 959	203 357	58 971	262 328
	Total	526 005	40 605	566 610	674 116	1 596	674 712	314 546	72 394	386 940
Silver-lead-zinc and	Own equipment	42 161	19 545	61 706	68 716	199 151	267 867	79 110	171 989	251 099
silver-cobalt	Contractors	198 171	-	198 171	207 126	3 761	210 887	173 119	-	173 119
	Total	240 332	19 545	259 877	275 842	202 912	478 754	252 229	171 989	424 218
Iron mines	Own equipment	38 424	-	38 424	15 817	-	15 817	22 067	-	22 067
	Contractors	30 007	27 474	57 481	-	-	-	-	-	-
	Total	68 431	27 474	95 905	15 817	-	15 817	22 067	-	22 067
Uranium	Own equipment	-	-	-	28 279	-	28 279	41 645	-	41 645
	Contractors	10 884	-	10 884	59 232	21 668	80 900	45 714	13 362	59 076
	Total	10 884	-	10 884	87 511	21 668	109 179	87 359	13 362	100 721
Miscellaneous metal	Own equipment	-	-	-	-	-	-	-	-	-
mining	Contractors	67 156	-	67 156	45 373	-	45 373	41 954	-	41 954
	Total	67 156	-	67 156	45 373	-	45 373	41 954	-	41 954
Total metal mining	Own equipment	347 829	20 545	368 374	460 687	200 898	661 585	289 901	188 674	478 575
	Contractors	747 566	72 127	819 093	917 566	51 881	969 447	713 413	72 333	785 746
	Total	1 095 395	92 672	1 188 067	1 378 253	252 779	1 631 032	1 003 314	261 007	1 264 321
Nonmetal mining										
Asbestos	Own equipment	28 790	-	28 790	10 814	-	10 814	8 400	-	8 400
	Contractors	-	-	-	-	-	-	-	-	-
	Total	28 790	-	28 790	10 814	-	10 814	8 400	-	8 400
Gypsum	Own equipment	1 314	-	1 314	-	-	-	-	-	-
	Contractors	4 463	-	4 463	1 841	-	1 841	-	-	-
	Total	5 777	-	5 777	1 841	-	1 841	-	-	-
Salt	Own equipment	-	-	-	1 552	-	1 552	-	-	-
	Contractors	-	-	-	-	-	-	-	-	-
	Total	-	-	-	1 552	-	1 552	-	-	-
Miscellaneous	Own equipment	2 844	-	2 844	404	-	404	1 073	-	1 073
nonmetal mining	Contractors	798	-	798	1 128	-	1 128	3 596	-	3 596
	Total	3 642	-	3 642	1 532	-	1 532	4 669	-	4 669
Total nonmetal	Own equipment	4 158	-	4 158	1 956	-	1 956	1 073	-	1 073
mining	Contractors	34 051	-	34 051	13 783	-	13 783	11 996	-	11 996
	Total	38 209	-	38 209	15 739	-	15 739	13 069	-	13 069
Total mining	Own equipment	351 987	20 545	372 532	462 648	200 898	663 541	290 974	188 674	479 648
industry	Contractors	781 617	72 127	853 744	931 349	51 881	983 230	725 409	72 333	797 742
	Total	1 133 604	92 672	1 226 276	1 393 992	252 779	1 641 771	1 016 383	261 007	1 277 390

- Nil.

TABLE 58. CANADA, ORE MINED AND ROCK QUARRIED IN THE MINING INDUSTRY, 1953-82

	Metals	Nonmetal[1]	Total
		(million tonnes)	
1953	49.3	42.8	92.1
1954	53.5	55.7	109.2
1955	62.7	57.6	120.3
1956	70.2	66.2	136.4
1957	76.4	74.5	150.9
1958	71.4	71.2	142.6
1959	89.9	82.2	172.1
1960	92.1	88.7	180.8
1961	90.1	96.7	186.8
1962	103.6	103.8	207.4
1963	112.7	120.4	233.1
1964	128.0	134.1	262.1
1965	151.0	146.5	297.5
1966	147.6	171.8	319.4
1967	169.1	177.5	346.6
1968	186.9	172.7	359.6
1969	172.0	178.8	350.8
1970	213.0	179.1	392.1
1971	211.5	185.8	397.3
1972	206.0	189.7	395.7
1973	274.8	162.6	437.3
1974	278.7	178.8	457.6
1975	264.2	158.7	422.9
1976	296.5	167.1	463.6
1977	299.5	205.2	504.8
1978	248.1	205.5	453.5
1979	274.8	197.7	472.5
1980	290.1	190.3	480.4
1981	301.5	169.7	471.2
1982	238.4	119.1	357.4

[1] Includes nonmetallic mineral mining and all stone quarried, including stone used to make cement and lime. From 1973 onwards, coverage is the same as in Table 54.

TABLE 59. CANADA, TOTAL DIAMOND DRILLING, METAL DEPOSITS, 1953-82

	Gold-quartz Deposits	Copper-gold-silver and nickel-copper Deposits	Silver-lead-zinc and silver-cobalt Deposits	Other Metal Bearing Deposits[1]	Total Metal Deposits
			(metres)		
1953	675 598	976 514	367 864	65 279	2 085 255
1954	737 266	826 288	271 873	199 097	2 034 524
1955	717 674	875 942	341 857	537 612	2 473 085
1956	682 600	1 490 298	399 679	383 431	2 956 008
1957	706 273	1 098 490	323 704	287 364	2 415 831
1958	546 861	923 026	297 792	286 970	2 054 649
1959	558 160	1 110 664	282 088	383 471	2 334 383
1960	628 016	1 267 792	226 027	315 067	2 436 902
1961	503 741	1 128 091	255 101	221 079	2 199 452
1962	902 288	1 025 048	350 180	358 679	2 636 195
1963	529 958	977 257	288 204	148 703	1 944 122
1964	458 933	709 588	401 099	104 738	1 674 358
1965	440 020	779 536	331 294	275 917	1 826 727
1966	442 447	729 148	292 223	164 253	1 628 071
1967	391 347	947 955	230 182	120 350	1 689 834
1968	375 263	935 716	198 038	56 780	1 565 797
1969	274 410	923 452	197 670	109 592	1 505 124
1970	214 717	1 132 915	375 019	99 373	1 822 024
1971	193 291	1 089 103	308 798	83 851	1 675 043
1972	229 771	967 640	240 195	50 225	1 487 831
1973	243 708	713 134	185 946	57 730	1 200 518
1974	250 248	798 564	197 322	83 484	1 329 618
1975	216 158	532 991	184 203	97 971	1 031 323
1976	156 030	507 620	166 366	97 735	927 751
1977	175 643	515 780	213 279	124 329	1 029 031
1978	209 335	227 065	490 489	135 197	1 181 743
1979	198 955	437 562	131 032	150 018	917 567
1980	187 635	566 610	259 877	173 945	1 188 067
1981	306 197	675 712	478 754	170 369	1 631 032
1982	288 421	386 940	424 218	164 742	1 264 321

[1] Includes iron, titanium, uranium, molybdenum and other metal deposits.

TABLE 60. CANADA, EXPLORATION DIAMOND DRILLING, METAL DEPOSITS, 1953-82

	Mining Companies With Own Personnel and Equipment	Diamond Drill Contractors	Total
		(metres)	
1953	318 970	872 668	1 191 638
1954	295 613	1 109 844	1 405 457
1955	464 118	1 546 025	2 010 143
1956	474 562	1 644 735	2 119 297
1957	358 300	1 233 323	1 591 623
1958	237 133	1 200 625	1 437 758
1959	239 786	1 367 061	1 606 847
1960	268 381	1 409 416	1 677 797
1961	302 696	1 337 173	1 639 869
1962	167 214	1 748 023	1 915 237
1963	361 180	1 169 292	1 530 472
1964	143 013	1 072 985	1 215 998
1965	209 002	1 176 996	1 385 998
1966	163 379	1 044 860	1 208 239
1967	93 164	1 123 137	1 216 301
1968	159 341	990 690	1 150 031
1969	135 311	1 072 328	1 207 639
1970	62 147	1 228 061	1 290 208
1971	86 838	1 053 330	1 140 168
1972	251 651	839 753	1 091 404
1973	321 333	742 899	1 064 232
1974	357 823	892 557	1 250 380
1975	346 770	618 161	964 931
1976	335 919	532 036	867 955
1977	327 241	638 327	965 568
1978	237 250	534 557	771 807
1979	311 221	571 721	882 942
1980	347 829	747 566	1 095 395
1981	460 687	917 566	1 378 253
1982	289 901	713 413	1 003 314

TABLE 61. CANADA, DIAMOND DRILLING, OTHER THAN FOR EXPLORATION, METAL DEPOSITS, 1953-82

	Mining Companies With Own Personnel and Equipment	Diamond Drill Contractors	Total
	(metres)		
1953	..	..	893 617
1954	..	..	629 067
1955	410 925	52 017	462 942
1956	790 522	46 188	836 710
1957	524 724	156 060	680 784
1958	444 376	172 516	616 892
1959	488 783	238 753	727 536
1960	450 246	308 860	759 105
1961	384 432	175 149	559 581
1962	528 700	192 259	720 959
1963	388 228	25 422	413 650
1964	385 765	72 594	458 359
1965	393 947	46 822	440 769
1966	227 968	191 863	419 831
1967	186 463	287 071	473 534
1968	122 851	292 914	415 765
1969	87 552	209 933	297 485
1970	290 363	241 453	531 816
1971	295 966	238 910	534 876
1972	304 523	91 903	396 426
1973	77 162	59 124	136 286
1974	54 353	24 885	79 238
1975	31 917	34 475	66 392
1976	31 413	28 383	59 796
1977	24 303	39 160	63 463
1978	351 344	58 592	409 936
1979	4 090	30 535	34 625
1980	20 545	72 127	92 672
1981	200 898	51 881	252 779
1982	188 674	72 333	261 007

Nonproducing companies excluded since 1964.
.. Not available.

TABLE 62. CANADA, CRUDE MINERALS TRANSPORTED BY CANADIAN RAILWAYS, 1980-82

	1980	1981	1982
	(000 tonnes)		
Metallic minerals			
Alumina and bauxite	2 752	3 133	2 793
Copper ores and concentrates	1 546	1 624	1 507
Iron ores and concentrates	54 167	49 788	35 101
Iron pyrite	46	30	295
Lead ores and concentrates	515	511	545
Lead-zinc ores and concentrates	353	3	1
Manganese ores	7	8	5
Nickel-copper ores and concentrates	4 983	4 457	1 890
Nickel ores and concentrates	628	612	228
Tungsten ores and concentrates	2	2	4
Zinc ores an concentrates	1 442	1 630	1 638
Metallic ores and concentrates, nes	32	29	40
Total metallic minerals	66 473	61 827	44 047
Nonmetallic minerals			
Abrasives, natural	70	61	37
Asbestos	400	332	190
Barite	133	72	21
Clay	621	606	485
Gravel	13	7	4
Gypsum	4 652	4 767	3 591
Limestone, agricultural	72	61	42
Limestone, industrial	331	299	177
Limestone, nes	3 801	4 139	3 049
Nepheline syenite	340	340	274
Phosphate rock	2 912	2 572	1 665
Potash (KCl)	10 652	9 703	7 681
Refractory materials, nes	4	4	3
Salt, rock	1 015	909	1 078
Salt, nes	120	102	83
Sand, industrial	1 105	986	743
Sand, nes	13	11	10
Silica	33	16	12
Sodium carbonate	581	552	481
Sodium sulphate	547	600	623
Stone, building, rough	62	9	6
Stone, nes	236	185	87
Sulphur, liquid	1 750	1 905	1 518
Sulphur, nes	5 728	5 931	4 855
Nonmetallic minerals, nes	178	221	145
Total nonmetallic minerals	35 369	34 390	26 860
Mineral fuels			
Coal, anthracite	125	69	56
Coal, bituminous	22 177	23 054	23 293
Coal, lignite	486	1 148	1 312
Coal, nes	18	21	12
Natural gas and other crude bituminous substances	4	4	7
Oil, crude	172	163	91
Total mineral fuels	22 982	24 459	24 771
Total crude minerals	124 824	120 676	95 678
Total revenue freight moved by Canadian railways	254 447	246 643	212 542
Per cent crude minerals of total revenue freight	49.1	48.9	45.0

nes Not elsewhere specified.

TABLE 63. CANADA, FABRICATED MINERAL PRODUCTS TRANSPORTED BY CANADIAN RAILWAYS, 1980-82

	1980	1981	1982
	(000 tonnes)		
Metallic mineral products			
Ferrous mineral products			
Ferroalloys	75	102	47
Pig iron	80	134	42
Ingots, blooms, billets, slabs of iron and steel	425	933	630
Other primary iron and steel	64	210	21
Castings and forgings, iron and steel	198	179	114
Bars and rods, steel	728	825	521
Plates, steel	553	590	314
Sheet and strip, steel	992	1 016	666
Structural shapes and sheet piling, iron and steel	445	467	216
Rails and railway track material	101	131	94
Pipes and tubes, iron and steel	546	767	448
Wire, iron or steel	39	29	21
Iron and steel scrap	2 087	1 806	1 162
Slag, dross, etc.	128	162	52
Total ferrous mineral products	6 461	7 351	4 348
Nonferrous mineral products			
Aluminum paste, powder, pigs, ingots, shot	128	115	291
Aluminum and aluminum alloy fabricated material, nes	230	229	234
Copper matte and precipitates	3	1	351
Copper and alloys, in primary form	389	379	327
Copper and alloys, nes	58	44	23
Lead and alloys	128	126	119
Nickel and nickel-copper matte	96	94	46
Nickel and alloys	30	35	15
Zinc and alloys	447	453	406
Other nonferrous base metals and alloys	29	19	13
Nonferrous metal scrap	103	60	48
Total nonferrous mineral products	1 641	1 555	1 873
Total metallic mineral products	8 102	8 906	6 221
Nonmetallic mineral products			
Natural stone basic products, chiefly structural	227	196	160
Bricks and tiles, clay	45	46	20
Fire brick and similar shapes	111	86	47
Dolomite and magnesite, calcined	85	71	39
Refractories, nes	36	33	16
Glass basic products	102	91	84
Asbestos and asbestos-cement basic products	33	36	23
Portland cement, standard	1 763	1 804	1 349
Concrete pipe	20	10	4
Cement and concrete basic products, nes	324	333	169
Plaster	21	18	13
Gypsum wallboard and sheathing	22	25	14
Gypsum basic products, nes	3	7	7
Lime, hydrated and quick	303	219	186
Nonmetallic mineral basic products, nes	458	424	299
Fertilizers and fertilizer materials, nes	2 092	1 937	1 581
Total nonmetallic mineral products	5 645	5 336	4 011

(continued on following page)

TABLE 63. (cont'd)

	1980	1981	1982
	(000 tonnes)		
Mineral fuel products			
Gasoline	1 455	1 511	1 376
Aviation turbine fuel	54	63	32
Diesel fuel	2 898	2 778	2 223
Kerosene	1	1	2
Fuel oil, nes	1 000	1 080	890
Lubricating oils and greases	389	342	296
Petroleum coke	626	463	537
Coke, nes	708	701	567
Refined and manufactured gases, fuel type	2 737	3 010	2 991
Asphalts and road oils	187	214	256
Bituminous pressed or molded fabricated material	1	1	1
Other petroleum and coal products	747	766	641
Total mineral fuel products	10 803	10 930	9 812
Total fabricated mineral products	24 550	25 172	20 044
Total revenue freight moved by Canadian railways	254 447	246 643	212 542
Fabricated mineral products as a percentage of total revenue freight	9.6	10.2	9.4

nes Not elsewhere specified.

TABLE 64. CANADA, CRUDE AND FABRICATED MINERALS TRANSPORTED BY CANADIAN RAILWAYS, 1953-82

	Total Revenue Freight	Total Crude Minerals	Total Fabricated Minerals	Total Crude and Fabricated Minerals	Crude and Fabricated Minerals as % of Revenue Freight
			(million tonnes)		
1953	141.7	44.7	16.4	61.1	43.1
1954	129.8	45.0	16.8	61.8	47.6
1955	152.2	61.2	19.0	80.2	52.7
1956	172.0	68.7	21.8	90.5	52.6
1957	157.9	64.2	17.1	81.3	51.5
1958	139.2	52.4	15.2	67.6	48.6
1959	150.6	62.8	15.3	78.1	52.9
1960	142.8	57.1	14.5	71.6	50.1
1961	138.9	54.1	13.6	67.7	48.7
1962	146.0	60.3	13.8	74.1	50.8
1963	154.6	62.9	15.5	78.3	50.6
1964	180.0	74.6	15.9	90.5	50.3
1965	186.2	80.9	17.3	98.2	52.7
1966	194.5	80.6	17.8	98.4	50.6
1967	190.0	81.2	17.7	98.9	52.1
1968	195.4	86.7	18.8	105.5	54.0
1969	189.0	81.9	27.6	109.5	57.9
1970	211.6	97.5	28.4	127.9	60.4
1971	214.5	95.6	27.4	123.0	57.3
1972	215.8	89.4	27.6	117.0	54.2
1973	241.2	113.1	29.1	142.2	59.0
1974	246.3	115.3	30.9	146.2	59.4
1975	226.0	110.6	26.6	137.2	60.7
1976	238.5	116.6	25.5	142.1	59.6
1977	247.2	121.1	25.7	146.8	59.4
1978	238.8	107.7	26.2	133.9	45.1
1979	257.9	127.2	26.6	153.8	59.6
1980	254.4	124.8	24.6	149.4	58.8
1981	246.6	120.7	25.2	145.9	59.2
1982	212.5	95.7	20.0	115.7	54.4

TABLE 65. CANADA, CRUDE AND FABRICATED MINERALS TRANSPORTED THROUGH THE ST. LAWRENCE SEAWAY, 1981-83

	Montreal-Lake Ontario Section			Welland Canal Section		
	1981	1982	1983	1981	1982	1983
	(tonnes)					
Crude minerals						
Coal	1 519 188	1 046 580	350 170	5 935 727	6 478 426	5 494 597
Iron ore	11 727 044	6 740 758	10 280 210	12 468 808	6 364 815	9 229 290
Aluminum ores and concentrates	149 932	96 024	115 345	144 525	96 024	115 345
Clay and bentonite	180 280	129 267	76 849	180 280	129 266	76 849
Gravel and sand	36 651	33	7 975	203 970	118 341	203 063
Stone, ground or crushed	23 036	30 839	47 462	952 603	102 695	401 719
Stone, rough	122	2 025	292	122	2 026	289
Salt	1 029 608	648 547	878 535	1 599 337	1 287 540	1 455 070
Phosphate rock	27 432	–	35 156	–	–	16 326
Sulphur	25 615	2 733	–	25 613	2 733	–
Other crude minerals	706 831	449 397	651 140	620 819	475 377	419 199
Total crude minerals	15 452 739	9 146 203	12 443 134	22 131 804	15 057 243	17 411 747
Fabricated mineral products						
Coke	773 992	617 617	638 042	880 911	686 590	683 081
Gasoline	112 348	144 035	249 993	136 566	157 842	218 092
Fuel oil	1 667 865	909 030	936 121	1 652 474	972 930	835 488
Lubricating oils and greases	64 677	44 330	13 070	51 026	34 414	12 889
Other petroleum products	151 924	157 202	110 029	111 501	139 305	116 155
Tar, pitch and creosote	39 613	38 236	25 154	37 482	45 328	43 015
Pig iron	183 752	138 048	161 017	173 884	128 814	150 896
Iron and steel: bars, rods, slabs	314 656	103 714	286 838	299 479	99 304	361 841
Iron and steel: nails, wire	7 364	15 005	4 184	6 949	10 705	3 305
Iron and steel: manufactured	2 313 521	2 412 338	2 605 115	1 861 767	1 459 619	2 416 949
Scrap iron and steel	79 254	414 788	390 006	57 564	382 445	366 974
Cement	2 512	3 129	2 522	259 002	215 523	409 794
Total fabricated minerals	5 711 478	4 997 472	5 422 091	5 528 605	4 332 819	5 618 479
Total crude and fabricated minerals	21 137 217	14 143 675	17 865 225	27 660 409	19 390 062	23 030 226
Total all products	45 875 658	38 841 399	45 060 981	53 388 616	44 473 919	50 145 086
Crude and fabricated minerals as a per cent of total	46.1	36.4	39.6	51.8	43.6	45.9

– Nil.

TABLE 66. CANADA, CRUDE AND FABRICATED MINERALS TRANSPORTED THROUGH THE ST. LAWRENCE SEAWAY, 1954-83

	Montreal – Lake Ontario Section				Welland Canal Section			
	Total All Products	Total Crude Minerals	Total Fabricated Minerals	Crude and Fabricated Minerals as % of All Products	Total All Products	Total Crude Minerals	Total Fabricated Minerals	Crude and Fabricated Minerals as % of All Products
	(kilotonnes)				(kilotonnes)			
1954	8 742	1 920	1 077	34.3	15 888	6 996	2 308	58.6
1955	10 384	3 859	1 244	49.1	18 954	10 257	2 097	65.2
1956	12 247	4 807	1 314	50.0	20 925	11 405	2 169	64.8
1957	11 059	4 439	1 392	52.7	20 296	11 305	2 421	67.6
1958	10 670	3 064	1 020	38.3	19 300	8 994	2 107	57.5
1959	19 252	7 725	2 197	51.5	24 953	12 117	2 246	57.6
1960	18 460	5 760	2 904	46.9	26 563	12 679	2 606	57.5
1961	21 212	6 706	2 358	42.7	28 490	12 599	2 378	52.7
1962	23 271	7 531	2 522	43.2	32 215	15 625	2 342	55.8
1963	28 198	9 507	2 804	43.7	37 490	18 094	2 524	55.0
1964	35 701	13 127	3 558	46.7	46 644	23 489	3 095	57.0
1965	39 352	13 788	6 024	50.3	48 477	23 555	4 933	58.8
1966	44 538	16 376	6 340	51.0	53 648	25 712	5 329	57.8
1967	39 918	17 800	6 430	60.7	47 945	26 010	5 459	65.6
1968	43 496	19 312	8 425	63.8	52 712	29 075	7 587	69.6
1969	37 256	12 682	8 263	56.2	48 601	25 090	6 715	65.4
1970	46 445	15 554	8 932	52.7	57 121	27 233	7 156	60.2
1971	48 069	14 204	9 263	48.8	57 205	23 903	7 914	55.6
1972	48 607	13 425	9 837	47.9	58 146	24 808	7 701	55.9
1973	52 285	17 111	9 639	51.1	60 958	26 907	7 718	56.8
1974	40 049	16 137	7 018	57.8	47 500	23 952	5 437	61.9
1975	43 554	15 698	6 071	50.0	53 387	26 100	5 129	58.5
1976	49 348	20 884	7 181	56.9	58 368	29 914	6 323	62.1
1977	57 456	23 008	9 918	57.3	65 079	30 459	8 933	60.5
1978	51 658	15 057	8 558	45.7	59 576	22 700	7 759	51.1
1979	50 187	16 408	8 104	48.8	60 023	24 851	7 940	54.6
1980	42 142	12 248	6 009	43.3	54 074	20 487	5 405	47.9
1981	45 876	15 453	5 711	46.1	53 389	22 132	5 529	51.8
1982	38 841	9 146	4 997	36.4	44 474	15 057	4 333	45.9
1983	45 061	12 443	5 422	39.6	50 145	17 412	5 618	45.9

TABLE 67. CANADA, CRUDE MINERALS LOADED AND UNLOADED IN COASTWISE SHIPPING, 1982

	Loaded				Unloaded			
	Atlantic	Great Lakes	Pacific	Total	Atlantic	Great Lakes	Pacific	Total
	(tonnes)							
Metallic minerals								
Copper ore and concentrates	17 786	-	-	17 786	17 786	-	-	17 786
Iron ore and concentrates	3 036 890	812 385	-	3 849 275	772 198	3 077 077	-	3 849 275
Titanium ore	1 497 413	-	-	1 497 413	1 497 413	-	-	1 497 413
Zinc ore and concentrates	-	-	20 838	20 838	-	-	20 838	20 838
Total metals	4 552 089	812 385	20 838	5 385 312	2 287 397	3 077 077	20 838	5 385 312
Nonmetallic minerals								
Dolomite	-	31 433	-	31 433	31 433	-	-	31 433
Gypsum	345 175	-	21 562	366 737	281 395	63 780	21 562	366 737
Limestone	4 325	1 451 063	803 220	2 258 608	19 321	1 436 067	803 220	2 258 608
Potash	59	203 932	-	203 991	59	203 932	-	203 991
Salt	476 852	940 630	41 769	1 459 251	1 019 669	397 813	41 769	1 459 251
Sand and gravel	104 289	-	2 705 725	2 810 014	104 289	-	2 705 725	2 810 014
Stone, crude, nes	23 263	336 143	119 874	479 280	509	358 897	119 874	479 280
Sulphur in ores	8 019	-	4 659	12 678	8 019	-	4 659	12 678
Crude nonmetallic minerals, nes	5 036	-	316	5 352	5 036	-	316	5 352
Total nonmetals	967 018	2 963 201	3 697 125	7 627 344	1 469 730	2 460 489	3 697 125	7 627 344
Mineral fuels								
Coal, bituminous	58 085	2 099 383	-	2 157 468	58 085	2 099 383	-	2 157 468
Petroleum, crude	1 303 155	-	-	1 303 155	1 303 155	-	-	1 303 155
Total mineral fuels	1 361 240	2 099 383	-	3 460 623	1 361 240	2 099 383	-	3 460 623
Total crude minerals	6 880 347	5 874 969	3 717 963	16 473 279	5 118 367	7 636 949	3 717 963	16 473 279
Total all commodities	16 860 211	25 600 810	23 420 561	65 881 582	30 081 985	12 425 200	23 374 397	65 881 582
Crude minerals as a per cent of all commodities	40.8	22.9	15.9	25.0	17.0	61.5	15.9	25.0

- Nil; nes Not elsewhere specified.

TABLE 68. CANADA, FABRICATED MINERALS LOADED AND UNLOADED IN COASTWISE SHIPPING, 1982

	Loaded				Unloaded			
	Atlantic	Great Lakes	Pacific	Total	Atlantic	Great Lakes	Pacific	Total
	(tonnes)							

Metallic mineral products

Ferrous mineral products								
Primary iron, steel	6 915	3 831	-	10 746	36	10 710	-	10 746
Castings and forgings, steel	3 161	900	4 069	8 130	4 061	-	4 069	8 130
Bars and rods, steel	3 491	6 616	-	10 107	3 491	6 616	-	10 107
Plates and sheets, steel	6 912	16 371	-	23 283	6 912	16 371	-	23 283
Structural shapes, iron and steel	18 037	14 018	2 096	34 151	18 037	14 018	2 096	34 151
Rails and railway track material	2 513	-	-	2 513	2 513	-	-	2 513
Pipes and tubes, iron and steel	4 512	-	771	5 283	4 512	-	771	5 283
Wire, iron and steel	649	-	36	685	649	-	36	685
Iron and steel scrap	129	-	1 226	1 355	129	-	1 226	1 355
Total ferrous mineral products	46 319	41 736	8 198	96 253	40 340	47 715	8 198	96 253
Nonferrous mineral products								
Aluminum and aluminum products	64 550	-	-	65 550	64 550	-	-	64 550
Copper and alloys	5	-	-	5	5	-	-	5
Nickel and alloys	38	-	-	38	38	-	-	38
Other nonferrous metals and alloys	6 049	-	-	6 049	5 691	358	-	6 049
Nonferrous metal scrap	1 845	-	-	1 845	1 845	-	-	1 845
Total nonferrous mineral products	72 487	-	-	72 487	72 129	358	-	72 487
Total metallic mineral products	118 806	41 736	8 198	168 740	112 469	48 073	8 198	168 740

Nonmetallic mineral products

Asbestos basic products	62	-	-	62	62	-	-	62
Bricks, tiles and pipes, clay	4 401	-	-	4 401	4 401	-	-	4 401
Cement	8 598	407 184	173 055	588 837	8 598	407 184	173 055	588 837
Cement basic products	2 535	-	1 362	3 897	2 535	-	1 362	3 897
Fertilizers and fertilizer material nes	15 205	-	5 038	20 243	7 789	7 416	5 038	20 243
Glass basic products	181	-	-	181	181	-	-	181
Lime, quick and hydrated	3 218	-	7 510	10 728	3 218	-	7 510	10 728
Sulphur acid	42 926	-	18 136	61 062	6 360	36 566	18 136	61 062
Other nonmetallic mineral products	4 472	-	-	4 472	4 472	-	-	4 472
Total nonmetallic mineral products	81 598	407 184	205 101	693 883	37 616	451 166	205 101	693 883

Mineral fuel products

Asphalts and road oils	48 883	-	26 160	75 043	37 802	11 081	26 160	75 043
Fuel oil	5 463 014	1 815 151	1 154 014	8 432 179	6 242 017	1 064 740	1 125 422	8 432 179
Gasoline	2 440 035	639 622	630 542	3 710 199	2 532 660	546 997	630 542	3 710 199
Lubricating oils and greases	27 929	9 498	-	37 427	12 773	24 654	-	37 427
Petroleum coke	16 690	14 384	-	31 074	31 074	-	-	31 074
Other petroleum and coal products	13 338	52 319	-	65 657	37 296	28 361	-	65 657
Total mineral fuel products	8 009 889	2 530 974	1 810 716	12 351 579	8 893 622	1 675 833	1 782 124	12 351 579
Total fabricated mineral products	8 210 293	2 979 894	2 024 015	13 214 202	9 043 707	2 175 072	1 995 423	13 214 202
Total all commodities	16 860 211	25 600 810	23 420 561	65 881 582	30 081 985	12 425 200	23 374 997	65 881 582
Fabricated mineral products as a per cent of all commodities	48.7	11.6	8.6	20.1	30.0	17.5	8.5	20.1

- Nil; nes Not elsewhere specified.

TABLE 69. CANADA, CRUDE AND FABRICATED MINERALS LOADED AT CANADIAN PORTS IN COASTWISE SHIPPING; 1953-82

	Total All Commodities	Total Crude Minerals	Total Fabricated Minerals	Crude and Fabricated Minerals as % of All Products
	(kilotonnes)			
1953	25 922	4 271	5 449	37.5
1954	23 402	4 101	5 552	41 2
1955	25 050	4 371	6 229	42.3
1956	31 303	6 750	7 275	44.8
1957	34 354	8 696	7 832	48.1
1958	34 808	7 673	7 258	42.9
1959	36 494	9 984	7 819	48.8
1960	37 058	8 786	8 229	45.9
1961	41 861	9 527	8 857	43.9
1962	39 763	8 361	9 768	45.6
1963	40 328	7 998	9 942	44.5
1964	47 171	8 522	11 194	41.8
1965	48 200	9 183	11 766	43.5
1966	55 122	10 155	12 653	41.4
1967	49 799	11 509	12 207	47.6
1968	50 921	13 698	13 245	52.9
1969	51 890	12 746	14 181	51.9
1970	57 301	14 415	14 818	51.0
1971	55 128	14 783	15 374	54.7
1972	55 326	14 197	15 290	53.3
1973	55 314	16 573	15 615	58.2
1974	53 633	11 723	16 575	52.8
1975	54 373	15 687	17 510	61.1
1976	53 882	15 924	16 208	59.6
1977	58 309	18 131	17 435	61.0
1978	60 668	18 318	16 619	57.6
1979	79 950	22 130	17 486	50.2
1980	82 761	22 947	17 134	48.4
1981	71 271	17 849	16 669	48.4
1982	65 881	16 473	13 214	45.1

TRADE, 1980-82

	1980 Loaded	1980 Unloaded	1981 Loaded	1981 Unloaded	1982 Loaded	1982 Unloaded
			(tonnes)			
Metallic minerals						
Alumina, bauxite ore	15 945	3 934 926	6 595	3 886 501	7 336	3 367 797
Copper ores and concentrates	587 352	26 223	1 034 946	78 240	1 097 233	108 646
Iron ore and concentrates	35 594 404	5 209 050	41 909 908	7 713 979	27 770 684	3 322 648
Lead ore and concentrates	74 749	5 092	124 939	3 833	206 261	6 119
Manganese ore	19 800	129 682	25 959	168 395	-	165 332
Nickel ore and concentrates	71 262	1 463	85 603	2 620	39 089	3 531
Titanium ore	130 913	-	855 586	14 936	845 861	5 518
Zinc ore and concentrates	292 799	524	728 140	-	940 419	-
Other nonferrous ores, concentrates and metal scrap, nes	681 518	611 841	119 493	107 307	29 311	31 211
Total metals	37 468 742	9 918 801	44 891 169	11 975 811	30 936 194	7 010 802
Nonmetallic minerals						
Asbestos	891 831	10 682	706 622	25 286	605 982	25 564
Barite	-	36	-	8 158	25	14 573
Bentonite	14 317	151 649	-	176 559	18	96 908
China clay	93	19 059	4	34 693	-	6 409
Clay materials, nes	15 258	78 405	1 334	5 533	1 756	50 242
Dolomite	907 715	38 413	948 552	-	117 788	10 724
Fluorspar	-	145 838	-	190 592	-	125 789
Gypsum	4 733 725	175 759	5 062 237	134 252	4 475 409	80 864
Limestone	1 842 439	1 365 421	1 711 487	2 261 324	1 443 482	1 266 945
Phosphate rock	-	1 368 116	-	1 197 106	-	1 353 595
Potash (KCl)	3 843 013	32 723	4 253 511	18	4 103 313	-
Salt	1 879 269	991 855	1 431 460	1 327 244	1 664 815	1 164 624
Sand and gravel	78 678	804 079	151 833	1 322 115	98 179	935 763
Stone, crude, nes	235 805	548 113	95 377	27 290	17 037	50 911
Stone, crushed	100 974	330 230	13 442	62 766	-	5 315
Sulphur	5 011 131	43 550	5 726 661	3	4 869 230	-
Crude, nonmetallic minerals, nes	60 891	120 844	145 860	26 201	97 002	10 151
Total nonmetals	15 772 126	6 192 049	20 248 380	6 799 140	17 494 036	5 198 377
Mineral fuels						
Coal, bituminous	13 735 346	15 137 034	17 458 453	16 066 286	17 162 442	15 142 357
Coal, nes	1 093	13	194	3	101	1
Oil, crude	920 578	15 198 039	408 408	14 070 091	891	8 246 236
Total fuels	14 657 017	30 335 086	17 867 055	30 136 380	17 163 434	23 388 594
Total crude minerals	67 897 885	46 445 936	83 006 604	48 911 331	65 593 664	35 597 773
Total all commodities	138 161 219	67 834 656	145 445 080	68 187 889	125 281 616	48 729 336
Crude minerals as a per cent of all commodities	49.1	68.5	57.1	71.7	52.4	73.1

- Nil; nes Not elsewhere specified.

TABLE 71. CANADA, FABRICATED MINERAL PRODUCTS LOADED AND UNLOADED AT CANADIAN PORTS IN INTERNATIONAL SHIPPING TRADE, 1980-82

	1980		1981		1982	
	Loaded	Unloaded	Loaded	Unloaded	Loaded	Unloaded
	(tonnes)					
Metallic products						
Aluminum	398 230	174 109	272 585	47 503	557 593	42 200
Copper and alloys	480 212	25 843	224 600	44 540	157 620	36 606
Ferroalloys	18 426	28 958	24 858	50 890	19 764	19 805
Iron and steel, primary	28 884	53 666	2 737	29 898	1 002	7 916
Iron, pig	468 308	20	458 534	7 717	431 916	–
Iron and steel, other bars and rods	343 034	103 467	79 921	199 244	131 415	127 193
castings and forgings	225 155	62 617	120 633	64 419	109 329	52 690
pipes and tubes	58 664	191 210	62 462	278 956	27 845	173 819
plates and sheet	1 438 646	442 783	191 667	1 282 572	1 013 763	351 119
rails and track material	99 726	7 028	97 644	12 433	42 095	16 105
structural shapes	97 094	69 109	24 030	240 887	38 170	41 690
wire	35 685	70 625	15 910	132 814	31 558	106 943
Lead and alloys	103 421	21 173	53 320	3 781	57 668	1 479
Nickel and alloys	52 520	12 385	40 847	7 661	44 979	5 489
Zinc and alloys	388 341	3 707	140 043	19 277	133 918	7 065
Nonferrous metals, nes	115 726	144 951	68 487	155 811	23 887	11 443
Metal fabricated basic products	470 038	607 827	56 351	170 980	72 131	121 232
Total metals	4 822 110	2 019 478	1 934 629	2 749 383	2 894 653	1 122 794
Nonmetallic products						
Asbestos basic products	5 349	1 345	5 606	1 907	1 878	1 194
Building blocks, nes	38 490	25 126	31 527	36 057	18 681	45 736
Cement	1 704 324	75 130	1 719 170	130 990	187 272	7 599
Cement basic products	42 639	4 289	850	681	22 724	129
Glass basic products	32 801	15 773	35 226	15 631	30 271	13 131
Nonmetallic mineral basic products	45 406	201 882	54 739	73 732	61 800	204 060
Fertilizers, nes	148 320	57 843	138 603	125 364	71 921	92 572
Total nonmetals	1 869 009	323 545	1 985 721	384 362	1 394 547	364 421
Mineral fuel products						
Asphalts, road oils	16 366	14 001	44 512	36 388	9 650	12 109
Coal tar, pitch	9 819	42 693	17 028	83 515	3 625	52 687
Coke	1 059 856	1 319 773	666 609	1 110 170	403 347	781 671
Fuel oil	2 101 989	2 352 355	3 380 547	1 888 349	1 612 410	1 721 714
Gasoline	1 250 230	221 458	615 796	63 450	487 160	41 047
Lubricating oils and greases	355 314	457 521	14 801	9 051	12 609	34 193
Petroleum and coal products, nes	285 609	242 793	266 081	47 448	275 031	106 462
Total fuels	5 079 183	4 650 594	5 005 374	3 238 371	2 803 832	2 749 885
Total fabricated mineral products	11 770 302	6 993 617	8 925 724	6 372 116	7 093 032	4 237 100
Total all commodities	138 161 219	67 834 656	145 445 080	68 187 889	125 281 616	48 729 336
Fabricated mineral products as a per cent of all commodities	8.5	10.3	6.1	9.3	5.7	8.7

– Nil; nes Not elsewhere specified.

TABLE 72. CANADA, CRUDE AND FABRICATED MINERALS LOADED AT CANADIAN PORTS IN INTERNATIONAL SHIPPING TRADE, 1953-82

	Total All Commodities	Total Crude Minerals	Total Fabricated Minerals	Crude and Fabricated Minerals as % of All Products
	(kilotonnes)			
1953	29 213	8 251	1 024	31.7
1954	27 878	9 316	1 108	37.4
1955	35 836	17 126	1 684	52.5
1956	44 791	23 284	1 904	56.2
1957	44 539	24 210	2 588	60.2
1958	36 559	16 602	1 642	49.9
1959	45 772	25 789	1 619	59.9
1960	45 872	24 671	2 039	58.2
1961	48 771	23 241	2 133	52.0
1962	54 676	30 446	2 296	59.9
1963	62 031	32 214	2 503	56.0
1964	75 760	42 087	2 602	59.0
1965	74 521	41 338	2 746	59.2
1966	76 192	41 374	3 350	58.7
1967	72 598	42 704	3 701	63.9
1968	78 663	48 680	2 960	65.6
1969	70 432	42 442	3 456	65.1
1970	95 807	55 849	4 965	63.5
1971	95 887	53 245	5 022	60.7
1972	98 988	51 912	9 091	61.6
1973	112 434	64 195	10 103	66.1
1974	106 110	64 093	9 041	68.9
1975	102 444	61 970	7 495	67.8
1976	114 815	71 527	6 108	67.6
1977	119 770	70 257	5 979	63.7
1978	116 522	62 291	.7 556	59.9
1979	134 639	79 685	8 901	65.8
1980	138 161	67 898	11 770	57.7
1981	145 445	83 007	8 926	63.2
1982	125 282	65 594	7 093	58.1

TABLE 73. CANADA, FINANCIAL STATISTICS OF CORPORATIONS IN THE MINING INDUSTRY[1], BY DEGREE OF NON-RESIDENT OWNERSHIP, 1981

	Corporations (number)	(%)	Assets ($ million)	(%)	Equity ($ million)	(%)	Sales ($ million)	(%)	Profits ($ million)	(%)	Taxable Income ($ million)	(%)
Metal mines												
Reporting corporations												
Canadian	68	51.5	17,555	71.8	9,390	72.8	6,208	63.5	1,398	78.0	240	60.0
Foreign	33	25.0	6,885	28.2	3,505	27.2	3,572	36.5	394	22.0	159	40.0
Unclassified	31	23.5	2	--	--	--	1	--	--	--	--	--
Total all corporations	132	100.0	24,443	100.0	12,895	100.0	9,782	100.0	1,792	100.0	400	100.0
Mineral fuels												
Reporting corporations												
Canadian	680	61.4	29,159	56.8	12,452	53.3	8,818	41.5	1,943	39.4	564	18.5
Foreign	157	14.2	22,151	43.1	10,904	46.7	12,390	58.4	2,980	60.5	2,485	81.4
Unclassified	271	24.4	25	.1	10	--	15	.1	2	.1	4	.1
Total all corporations	1,108	100.0	51,335	100.0	23,367	100.0	21,222	100.0	4,926	100.0	3,053	100.0
Other mining (including mining services)												
Reporting corporations												
Canadian	2,209	47.2	7,421	64.4	3,036	58.4	3,583	59.0	403	45.9	409	68.0
Foreign	237	5.1	3,917	34.0	2,129	40.9	2,306	38.0	474	54.0	173	28.6
Unclassified	2,236	47.7	182	1.6	37	.7	185	3.0	1	.1	20	3.4
Total all corporations	4,682	100.0	11,520	100.0	5,202	100.0	6,074	100.0	878	100.0	602	100.0
Total mining												
Reporting corporations												
Canadian	2,957	49.9	54,135	62.0	24,878	60.0	18,609	50.2	3,744	49.3	1,213	29.9
Foreign	427	7.2	32,953	37.7	16,538	39.9	18,268	49.3	3,848	50.7	2,817	69.5
Unclassified	2,538	42.9	209	.3	47	.1	201	.5	3	--	24	.6
Total all corporations	5,922	100.0	87,296	100.0	41,463	100.0	37,078	100.0	7,595	100.0	4,055	100.0

Note: Footnotes for Table 74 apply to this table. Figures may not add to totals due to rounding.
[1] Classification of the industry is the same as in Table 29.
-- Amount too small to be expressed.

TABLE 74. CANADA, FINANCIAL STATISTICS OF CORPORATIONS IN THE MINERAL MANUFACTURING INDUSTRIES[1], BY DEGREE OF NON-RESIDENT OWNERSHIP, 1981

	Corporations[2] (number)	(%)	Assets[4] ($ million)	(%)	Equity[5] ($ million)	(%)	Sales[6] ($ million)	(%)	Profits[7] ($ million)	(%)	Taxable income[8] ($ million)	(%)
Primary metal products												
Reporting corporations[2]												
Canadian	240	67.6	12,689	86.9	5,480	85.4	11,184	85.2	1,160	89.7	298	80.1
Foreign	41	11.5	1,904	13.0	938	14.6	1,925	14.8	132	10.3	73	19.7
Unclassified[3]	74	20.9	7	.1	2	–	12	–	1	–	1	.2
Total all corporations	355	100.0	14,600	100.0	6,420	100.0	13,121	100.0	1,293	100.0	372	100.0
Nonmetallic mineral products												
Reporting corporations[2]												
Canadian	798	52.0	1,964	30.6	615	23.5	2,195	40.1	101	24.8	69	40.1
Foreign	82	5.3	4,390	68.5	1,987	76.1	3,186	58.2	305	74.9	99	57.6
Unclassified[3]	655	42.7	58	.9	10	.4	92	1.7	1	.3	4	2.3
Total all corporations	1,535	100.0	6,412	100.0	2,612	100.0	5,473	100.0	407	100.0	172	100.0
Petroleum and coal products												
Reporting corporations[2]												
Canadian	39	63.9	11,079	39.5	4,131	30.6	6,409	22.0	801	27.0	325	17.8
Foreign	14	23.0	16,986	60.5	9,381	69.4	22,680	78.0	2,164	73.0	1,498	82.2
Unclassified[3]	8	13.1	1	–	–	–	1	–	–	–	–	–
Total all corporations	61	100.0	28,066	100.0	13,512	100.0	29,090	100.0	2,965	100.0	1,823	100.0
Total mineral manufacturing industries												
Reporting corporations[2]												
Canadian	1,077	55.2	25,732	52.4	10,226	45.4	19,788	41.5	2,062	44.2	692	29.2
Foreign	137	7.0	23,280	47.4	12,306	54.6	27,791	58.3	2,601	55.8	1,670	70.6
Unclassified[3]	737	37.8	66	.2	12	–	105	.2	2	–	5	.2
Total all corporations	1,951	100.0	49,078	100.0	22,544	100.0	47,684	100.0	4,665	100.0	2,367	100.0

[1] Classification of industries is the same as in Table 30. [2] Corporations reporting under the Corporations and Labour Unions Returns Act. A corporation is considered to be foreign controlled if 50 per cent or more of its voting rights are known to be held outside Canada, and/or by one or more Canadian corporations which are, in turn, foreign controlled. Each corporation is classified according to the percentage of its voting rights which are owned by non-residents, either directly or through other Canadian corporations, and the whole of the corporation is assigned to this particular degree of foreign ownership. [3] Corporations exempt from reporting under the Corporations and Labour Unions Returns Act. These include corporations reporting under other acts, small companies and corporations and non-profit organizations. [4] Included are cash, marketable securities, accounts receivable, inventories, fixed assets, investments in affiliated corporations and other assets. The amounts tabulated are those shown on the balance sheets of corporations after deducting allowances for doubtful accounts, amortization, depletion and depreciation. [5] Equity represents the shareholders' interest in the net assets of the corporation and includes the total amount of all issued and paid-up share capital, earnings retained in the business and other surplus accounts such as contributed and capital surplus. [6] For non-financial corporations, sales are gross revenues from non-financial operations. For financial corporations sales include income from financial as well as non-financial sources. [7] The net earnings from operations, investment income and net capital gains. Profits are tabulated after deducting allowances for amortization, depletion and depreciation, but before income tax provisions or declaration of dividends. [8] Taxable income figures are as reported by corporations prior to assessment by the Department of National Revenue. They include earnings in the reference year after the deduction of applicable losses of other years.
-- Amount too small to be expressed; - Nil.

TABLE 75. CANADA, FINANCIAL STATISTICS OF CORPORATIONS IN NON-FINANCIAL INDUSTRIES, BY MAJOR INDUSTRY GROUP AND BY CONTROL, 1980 AND 1981

	Agriculture, Forestry, Fishing and Trapping		Mines Quarries & Oil Wells		Manufacturing		Construction		Transportation, Communication and Other Utilities		Trade		Services		Total	
	1980	1981P	1980	1981P	1980	1981P	1980	1981P	1980	1981P	1980	1981P	1980	1981P	1980	1981P
	(number)															
Number of corporations																
Canadian control	8,285	9,572	2,489	2,957	16,243	17,172	15,705	16,964	5,119	5,616	44,345	47,938	5,844	6,341	98,030	106,560
Foreign control	101	95	481	427	2,096	1,922	216	191	180	161	1,951	1,721	303	269	5,328	4,786
Other corporations	10,526	10,446	2,411	2,538	18,138	17,882	39,830	40,561	13,054	13,823	73,859	72,468	13,342	14,092	171,160	171,810
Total corporations	18,912	20,103	5,381	5,922	36,477	36,976	55,751	57,716	18,353	19,600	120,155	122,127	19,489	20,702	274,518	283,146
	($ million)															
Assets																
Canadian control	6,390	8,475	38,260	54,135	73,561	90,073	16,409	18,715	5,032	5,289	55,434	65,400	107,766	120,429	302,852	362,516
Foreign control	334	401	30,705	32,953	69,093	73,997	2,217	2,504	106,022	117,893	14,576	16,120	5,129	5,377	228,076	249,245
Other corporations	973	989	193	209	1,407	1,402	2,450	2,522	924	982	5,324	5,286	947	1,004	12,218	12,394
Total corporations	7,697	9,865	69,158	87,297	144,061	165,472	21,076	23,741	111,978	124,164	75,334	86,806	113,842	126,810	543,146	624,155
Equity																
Canadian control	1,965	2,671	17,765	24,878	28,450	31,629	3,743	4,109	1,738	1,801	16,349	19,481	29,938	32,166	99,948	116,735
Foreign control	111	139	16,297	16,538	35,017	35,891	710	812	29,434	31,559	5,013	5,531	1,754	1,818	88,336	92,288
Other corporations	206	195	51	47	263	249	533	501	168	161	1,167	1,029	171	164	2,559	2,346
Total corporations	2,282	3,005	34,113	41,463	63,730	67,768	4,986	5,422	31,340	33,521	22,529	26,042	31,863	34,148	190,843	211,369
Sales																
Canadian control	5,663	6,759	14,519	18,609	94,194	109,336	23,693	27,476	3,212	3,656	140,023	160,088	45,453	52,026	326,757	377,950
Foreign control	290	310	20,451	18,268	101,405	108,125	2,769	3,728	43,442	49,749	39,479	41,982	3,265	3,717	211,101	225,879
Other corporations	1,053	1,088	183	201	2,602	2,641	5,082	5,310	1,453	1,572	11,040	11,169	1,485	1,601	22,898	23,582
Total corporations	7,006	8,157	35,153	37,078	198,201	220,102	31,544	36,514	48,107	54,977	190,542	213,239	50,203	57,344	560,756	627,411
Profits																
Canadian control	512	357	4,786	3,744	6,872	6,891	987	1,095	450	435	5,159	5,240	4,557	5,123	23,323	22,885
Foreign control	24	22	5,568	3,848	8,137	7,468	154	160	4,411	4,918	1,242	1,213	455	440	19,991	18,069
Other corporations	84	72	17	3	100	92	194	189	62	52	360	293	64	54	881	755
Total corporations	620	451	10,371	7,595	15,109	14,451	1,335	1,444	4,923	5,405	6,761	6,746	5,076	5,617	44,195	41,709

Note: Figures may not add to totals due to rounding.
P Preliminary.

TABLE 76. CANADA, CAPITAL AND REPAIR EXPENDITURES BY SELECTED INDUSTRIAL SECTOR; 1982-84

		Capital Expenditures			Repair Expenditures			Capital and Repair Expenditures		
		Construction	Machinery and Equipment	Total	Construction	Machinery and Equipment	Total	Construction	Machinery and Equipment	Total
					($ million)					
Agriculture	1982	1,314.5	3,027.1	4,341.6	347.6	1,065.0	1,412.6	1,662.1	4,092.1	5,754.2
	1983P	1,389.3	3,152.4	4,541.7	369.1	1,070.3	1,439.4	1,758.4	4,222.7	5,981.1
	1984f	1,421.3	3,383.7	4,805.0	386.5	1,098.1	1,484.6	1,807.8	4,481.8	6,289.6
Forestry	1982	93.5	54.5	148.0	65.9	189.0	254.9	159.4	243.5	402.9
	1983P	91.6	51.7	144.3	74.7	230.9	305.6	166.3	283.6	449.9
	1984f	119.7	105.5	225.2	82.2	238.3	320.5	201.9	343.8	545.7
Mining[1]	1982	8,007.2	2,354.4	10,361.6	625.4	1,935.2	2,560.6	8,632.6	4,289.6	12,922.2
	1983P	7,777.3	1,830.0	9,607.3	606.5	1,899.9	2,506.4	8,383.8	3,729.9	12,113.7
	1984f	8,391.6	1,629.9	10,021.5	669.3	2,129.1	2,798.4	9,060.9	3,759.0	12,819.9
Construction	1982	206.9	1,086.5	1,293.4	28.5	830.0	858.5	235.4	1,916.5	2,151.9
	1983P	207.1	1,088.9	1,296.0	28.6	831.7	860.3	235.7	1,920.6	2,156.3
	1984f	207.3	1,089.6	1,296.9	28.7	832.3	861.0	236.0	1,921.9	2,157.9
Housing	1982	10,148.8	-	10,148.8	3,432.8	-	3,432.8	13,581.6	-	13,581.6
	1983P	12,871.8	-	12,871.8	3,811.2	-	3,811.2	16,683.0	-	16,683.0
	1984f	13,163.7	-	13,163.7	4,076.2	-	4,076.3	17,239.9	-	17,239.9
Manufacturing	1982	2,908.7	8,583.8	11,492.5	797.6	4,056.7	4,854.3	3,706.3	12,640.5	16,346.8
	1983P	1,869.1	6,932.6	8,801.7	768.1	4,129.4	4,897.5	2,637.2	11,062.0	13,699.2
	1984f	1,912.8	6,671.6	8,584.4	820.2	4,264.5	5,084.7	2,733.0	10,936.1	13,669.1
Utilities	1982	9,609.7	8,243.4	17,853.1	1,462.5	3,979.5	5,442.0	11,072.2	12,222.9	23,295.1
	1983P	8,078.3	7,955.4	16,033.7	1,560.3	4,185.2	5,745.5	9,638.6	12,140.6	21,779.2
	1984f	7,566.1	7,692.6	15,258.7	1,731.6	4,538.7	6,271.3	9,297.7	12,232.2	21,530.0
Trade	1982	629.1	1,307.4	1,936.5	241.2	315.9	557.1	870.3	1,623.3	2,493.6
	1983P	607.8	1,340.7	1,948.5	226.1	308.2	534.3	833.9	1,648.9	2,482.8
	1984f	488.6	1,479.4	1,968.0	239.0	315.8	554.8	727.6	1,795.2	2,522.8
Other[2]	1982	13,572.1	5,586.4	19,158.2	2,545.8	1,143.1	3,688.9	16,144.9	6,729.5	22,874.4
	1983P	13,211.8	5,968.4	19,180.2	2,547.1	1,091.2	3,638.3	15,758.9	7,059.6	22,818.5
	1984f	13,046.9	6,690.6	19,737.5	2,619.4	1,182.2	3,801.6	15,666.3	7,872.8	23,539.1
Total	1982	46,517.5	30,243.5	76,761.0	9,547.3	13,514.4	23,061.7	56,064.8	43,757.9	99,822.7
	1983P	46,104.1	28,321.1	74,425.2	9,991.7	13,746.5	23,738.5	56,095.8	42,067.9	98,163.7
	1984f	46,318.0	28,742.9	75,060.9	10,653.1	14,600.0	25,253.1	56,971.1	43,342.9	100,314.0
Mining as a percentage of total	1982	17.2	7.8	13.5	6.6	14.3	11.1	15.4	9.8	12.9
	1983P	16.9	6.5	12.9	6.1	13.8	10.6	14.9	8.9	12.3
	1984f	18.1	5.7	13.4	6.3	14.6	11.9	15.9	8.7	12.8

[1] Includes mines, quarries and oil wells. [2] Includes finance, real estate, insurance, commercial services, institutions and government departments.
P Preliminary; f Forecast; - Nil.

TABLE 77. CANADA, CAPITAL AND REPAIR EXPENDITURES IN MINING[1] BY GEOGRAPHICAL REGION; 1982-84

		Capital Expenditures			Repair Expenditures			Capital and Repair Expenditures		
		Construction	Machinery and Equipment	Total	Construction	Machinery and Equipment	Total	Construction	Machinery and Equipment	Total
		($ million)								
Atlantic Region	1982	878.6	516.0	1,394.1	16.3	176.5	192.8	894.9	692.5	1,587.4
	1983p	1,289.4	379.6	1,669.0	13.3	165.8	179.1	1,302.7	545.4	1,848.1
	1984f	1,524.3	173.2	1,697.5	15.1	182.0	197.1	1,539.4	355.2	1,894.6
Quebec	1982	245.3	81.7	327.0	43.5	197.1	240.6	288.8	278.8	567.6
	1983p	167.8	53.1	220.9	41.3	209.3	250.6	209.1	262.4	471.5
	1984f	210.0	75.5	285.5	40.9	222.5	263.4	250.9	298.0	548.9
Ontario	1982	447.6	116.9	564.5	35.6	270.6	306.2	483.2	387.5	870.7
	1983p	389.9	128.1	518.0	38.0	286.6	324.6	427.9	414.7	842.6
	1984f	425.0	194.2	619.2	47.1	328.0	375.1	472.1	522.2	994.3
Prairie Region	1982	4,415.6	855.7	5,271.3	458.3	891.4	1,349.7	4,873.9	1,747.1	6,621.0
	1983p	3,988.2	802.0	4,790.2	458.2	840.7	1,298.9	4,446.4	1,642.7	6,089.1
	1984f	4,668.5	875.1	5,543.6	505.9	927.1	1,433.0	5,174.4	1,802.2	6,976.6
British Columbia	1982	889.5	211.2	1,100.7	61.4	340.8	402.2	950.9	552.0	1,502.9
	1983p	872.5	196.3	1,068.8	47.3	347.3	394.6	919.8	543.6	1,463.4
	1984f	348.2	214.9	563.1	48.3	411.9	460.2	396.5	626.8	1,023.3
Yukon and Northwest Territories	1982	1,130.6	572.9	1,703.5	10.3	58.8	69.1	1,140.9	631.7	1,772.6
	1983p	1,069.5	270.9	1,340.4	8.4	50.2	58.6	1,077.9	321.1	1,399.0
	1984f	1,215.6	97.0	1,312.6	12.0	57.6	69.6	1,227.6	154.6	1,382.2
Canada, total	1982	8,007.2	2,354.4	10,361.6	625.4	1,935.2	2,560.6	8,632.6	4,289.6	12,922.2
	1983p	7,777.3	1,830.0	9,607.3	606.5	1,899.9	2,506.4	8,383.8	3,729.9	12,113.7
	1984f	8,391.6	1,629.9	10,021.5	669.3	2,129.1	2,798.4	9,060.9	3,759.0	12,819.9

[1] Includes mines, quarries and oil wells.
p Preliminary; f Forecast.

TABLE 78. CANADA, CAPITAL AND REPAIR EXPENDITURES IN MINING[1] AND MINERAL MANUFACTURING INDUSTRIES, 1982-84

	1982			1983p			1984f		
	Capital	Repair	Total	Capital	Repair	Total	Capital	Repair	Total
				($ million)					
Mining industry									
Metal mines									
Gold	397.4	58.7	456.1	326.0	69.7	395.7	503.8	79.0	582.8
Silver-lead-zinc	145.1	119.6	264.7	64.7	105.3	170.0	94.8	101.9	196.7
Copper-gold-silver	212.0	264.2	476.2	152.5	257.5	410.0	288.6	297.7	586.3
Iron	141.5	270.4	411.9	81.1	244.0	325.1	91.2	244.2	335.4
Other metal mines	574.0	204.6	778.6	462.8	214.6	677.4	421.0	280.2	701.2
Total metal mines	1,470.0	917.5	2,387.5	1,087.1	891.1	1,978.2	1,399.4	1,003.0	2,402.4
Nonmetal mines									
Asbestos	52.2	59.4	111.6	36.9	76.5	113.4	35.7	81.3	117.0
Other nonmetal mines[2]	1,399.7	401.0	1,800.7	1,391.8	414.0	1,805.8	788.9	488.6	1,277.5
Total nonmetal mines	1,451.9	460.4	1,912.3	1,428.7	490.5	1,919.2	824.6	569.9	1,394.5
Mineral fuels									
Oil, crude and gas[3]	7,439.7	1,182.7	8,622.4	7,091.5	1,124.8	8,216.3	7,797.5	1,225.5	9,023.0
Total mining industries	10,361.6	2,560.6	12,922.2	9,607.3	2,506.4	12,113.7	10,021.5	2,798.4	12,819.9
Mineral manufacturing									
Primary metal industries									
Iron and steel mills	416.3	685.8	1,102.1	198.3	555.1	753.4	226.5	577.6	804.1
Steel pipe and tube mills	191.5	66.8	258.3	70.2	40.4	110.6	56.4	40.5	96.9
Iron foundries	13.3	39.0	52.3	9.7	38.3	48.0	24.1	39.1	63.2
Smelting and refining	525.5	281.8	807.3	377.7	377.1	754.8	772.1	386.1	1,158.2
Aluminum rolling, casting and extruding	20.1	31.5	51.6	22.2	44.5	66.7	42.3	46.5	88.8
Copper and copper alloy rolling, casting and extruding	22.9	4.8	27.7	6.1	6.3	12.4	13.5	7.3	20.8
Metal rolling, casting and extruding	16.2	11.1	27.3	9.0	7.1	16.1	8.5	7.4	15.9
Total primary metal industries	1,205.8	1,120.8	2,326.6	693.2	1,068.8	1,762.0	1,143.4	1,104.5	2,247.9
Nonmetallic mineral products									
Cement	47.1	64.5	111.6	36.3	65.6	101.9	29.5	66.9	96.4
Stone products	3.7	1.2	4.9	0.6	0.6	1.2	1.6	0.7	2.3
Concrete products	9.8	26.2	36.0	11.4	23.1	34.5	12.3	25.7	38.0
Ready-mix concrete	20.0	54.2	74.2	12.3	45.1	57.4	11.8	43.5	55.3
Clay products	9.3	6.7	16.0	5.9	6.5	12.4	5.3	7.0	12.3
Glass and glass products	29.1	19.9	49.0	37.8	23.4	61.2	53.4	32.3	85.7
Abrasives	12.2	13.1	25.3	6.3	11.2	17.5	7.9	11.4	19.3
Lime	7.9	3.3	11.2	2.2	2.8	5.0	3.6	2.5	6.1
Other nonmetallic mineral products	27.3	42.7	70.0	20.1	42.0	62.1	34.6	41.9	76.5
Total nonmetallic mineral products	166.4	231.8	398.2	132.9	220.3	353.2	160.0	231.9	391.9
Petroleum and coal products	1,224.5	319.7	1,544.2	734.1	272.6	1,006.7	519.9	271.7	791.6
Total mineral manufacturing industries	2,596.7	1,672.3	4,269.0	1,560.2	1,561.7	3,121.9	1,823.3	1,608.1	3,431.4
Total mining and mineral manufacturing industries	12,958.3	4,232.9	17,191.2	11,167.5	4,068.1	15,235.6	11,844.8	4,406.5	16,251.3

[1] Does not include cement, lime and clay products (domestic clay) manufacturing, smelting and refining. [2] Includes coal mines, gypsum, salt, potash and miscellaneous nonmetal mines and quarrying. [3] The total of capital expenditures shown under "petroleum and gas" is equal to the total capital expenditure under the column entitled "petroleum and natural gas extraction" and under the column "natural gas processing plants" of Table 80.
p Preliminary; f Forecast.

TABLE 79. CANADA, CAPITAL AND REPAIR EXPENDITURES IN THE MINING INDUSTRY[1], 1978-84

	1978	1979	1980	1981	1982	1983P	1984f
				($ million)			
Metal mines							
Capital							
Construction	407.3	606.4	1,109.1	1,331.3	1,099.4	806.4	912.4
Machinery	169.3	281.6	467.2	576.4	370.6	280.4	487.0
Total	576.6	888.0	1,576.3	1,907.7	1,470.0	1,087.1	1,399.4
Repair							
Construction	53.7	70.2	137.3	151.9	112.4	104.6	119.4
Machinery	487.6	632.1	767.7	900.8	805.1	786.5	883.6
Total	541.3	702.3	905.0	1,052.7	917.5	891.1	1,003.0
Total capital and repair	1,117.9	1,590.3	2,481.3	2,960.4	2,387.5	1,978.2	2,402.4
Nonmetal mines[2]							
Capital							
Construction	187.5	248.8	346.4	647.8	888.6	948.8	373.8
Machinery	236.4	202.6	267.6	417.7	563.3	479.9	450.8
Total	423.9	451.4	614.0	1,065.5	1,451.9	1,428.7	824.6
Repair							
Construction	18.2	14.6	32.5	26.0	28.6	21.0	22.6
Material	289.1	332.5	393.1	447.8	431.8	469.5	547.3
Total	307.3	347.1	425.6	473.8	460.4	490.5	569.9
Total capital and repair	731.2	798.5	1,039.6	1,539.3	1,912.3	1,919.2	1,394.5
Mineral fuels							
Capital							
Construction	2,520.9	3,820.3	5,453.1	5,825.1	6,019.2	6,021.8	7,105.4
Machinery	382.0	494.9	800.3	1,206.3	1,420.5	1,069.7	692.1
Total	2,902.9	4,315.2	6,253.4	7,031.4	7,439.7	7,091.5	7,797.5
Repair							
Construction	389.6	444.1	627.6	514.4	484.4	480.9	527.3
Machinery	100.2	242.1	313.6	639.0	698.3	643.9	698.2
Total	489.8	686.2	941.2	1,153.4	1,182.7	1,124.8	1,225.5
Total capital and repair	3,392.7	5,001.4	7,194.6	8,184.8	8,622.4	8,216.3	9,023.0
Total mining							
Capital							
Construction	3,115.7	4,675.5	6,908.6	7,804.2	8,007.2	7,777.3	8,391.6
Machinery	787.7	979.1	1,535.1	2,200.4	2,354.4	1,830.0	1,629.9
Total	3,903.4	5,654.6	8,443.7	10,004.6	10,361.6	9,607.3	10,021.5
Repair							
Construction	461.5	528.9	797.4	692.5	625.4	606.5	669.3
Machinery	876.9	1,206.7	1,474.4	1,987.6	1,935.2	1,899.9	2,129.1
Total	1,338.4	1,735.6	2,271.8	2,680.1	2,560.6	2,506.4	2,798.4
Total capital and repair	5,241.8	7,390.2	10,715.5	12,684.7	12,922.2	12,113.7	12,819.9

[1] Does not include cement, lime and clay products (domestic clays) manufacturing, smelting and refining.   [2] Includes coal mines, asbestos, gypsum, salt, potash, miscellaneous nonmetals, quarrying and sand pits.
P Preliminary;   f Forecast.

TABLE 80. CANADA, CAPITAL AND REPAIR EXPENDITURES IN THE MINERAL MANUFACTURING INDUSTRIES[1], 1978-84

	1978	1979	1980	1981	1982	1983P	1984f
				($ million)			
Primary metal industries[2]							
Capital							
Construction	130.7	153.4	328.2	330.1	278.3	207.4	606.7
Machinery	475.4	621.1	960.9	1,289.6	927.5	485.8	536.7
Total	606.1	774.5	1,289.1	1,619.7	1,205.8	693.2	1,143.4
Repair							
Construction	80.8	87.6	122.1	139.0	99.2	110.2	118.6
Machinery	780.1	887.7	998.5	1,053.3	1,021.6	958.6	985.9
Total	860.9	975.3	1,120.6	1,192.3	1,120.8	1,068.8	1,104.5
Total capital and repair	1,467.0	1,749.8	2,409.7	2,812.0	2,326.6	1,762.0	2,247.9
Nonmetallic mineral products[3]							
Capital							
Construction	62.0	102.0	70.0	93.4	32.0	15.2	13.6
Machinery	217.9	293.5	249.7	254.0	134.4	117.7	146.4
Total	279.9	395.5	319.7	347.4	166.4	132.9	160.0
Repair							
Construction	17.5	20.2	16.7	23.7	20.7	20.5	20.2
Machinery	190.3	206.1	213.8	227.5	211.1	199.8	211.7
Total	207.8	226.3	230.5	251.2	231.8	220.3	231.9
Total capital and repair	487.7	621.8	550.2	598.6	398.2	353.2	391.9
Petroleum and coal products							
Capital							
Construction	215.6	180.0	215.6	629.9	890.8	542.9	378.3
Machinery	99.5	94.0	109.1	215.0	333.7	191.2	141.6
Total	315.1	274.0	324.7	844.9	1,224.5	734.1	519.9
Repair							
Construction	117.5	158.1	190.5	212.9	218.5	198.1	197.2
Machinery	57.4	61.3	76.2	89.1	101.2	74.5	74.5
Total	174.9	219.4	266.7	302.0	319.7	272.6	271.7
Total capital and repair	490.0	493.4	591.4	1,146.9	1,544.2	1,006.7	791.6
Total mineral manufacturing industries							
Capital							
Construction	408.3	435.4	613.8	1,053.4	1,201.1	765.5	998.6
Machinery	792.8	1,008.6	1,319.7	1,758.6	1,395.6	794.7	824.7
Total	1,201.1	1,444.0	1,933.5	2,812.0	2,596.7	1,560.2	1,823.3
Repair							
Construction	215.8	256.9	329.3	375.6	338.4	328.8	336.0
Machinery	1,027.8	1,155.1	1,288.5	1,369.9	1,333.9	1,232.9	1,272.1
Total	1,243.6	1,412.0	1,617.8	1,745.5	1,672.3	1,561.7	1,608.1
Total capital and repair	2,444.7	2,856.0	3,551.3	4,557.5	4,269.0	3,121.9	3,431.4

[1] Industry groups are the same as in Table 28. [2] Includes smelting and refining. [3] Includes cement, lime and clay products manufacturing.
P Preliminary; f Forecast.

TABLE 81. CANADA, CAPITAL EXPENDITURES IN THE PETROLEUM, NATURAL GAS AND ALLIED INDUSTRIES[1], 1978-84

	Petroleum and Natural Gas Extraction[2]	Transportation Including Rail, Water and Pipelines	Marketing (Chiefly Outlets of Oil Companies)	Natural Gas Distribution	Petroleum and Coal Products Industries	Natural Gas Processing Plants	Total Capital Expenditures
				($ million)			
1978	2,684.1	312.4	145.6	246.6	315.1	218.8	3,922.6
1979	4,013.4	229.3	134.3	262.5	274.0	301.8	5,215.3
1980	5,744.2	602.1	205.2	386.4	324.7	311.5	7,574.1
1981	6,444.9	1,745.7	264.1	408.7	844.9	311.6	10,046.9
1982	6,743.4	1,994.3	320.5	517.6	1,224.5	522.8	11,323.1
1983p	6,624.7	630.2	384.4	525.5	734.1	277.8	9,176.7
1984f	7,464.0	728.1	380.2	555.4	519.9	307.0	9,954.6

[1] The petroleum and natural gas industries in this table include all companies engaged in whole or in part in oil and gas activities. [2] Does not include expenditures for geological and geophysical operations. See also Footnote 3 to Table 75.
p Preliminary; f Forecast.